Advances in

Pharmacology

Volume 27

Conjugation-Dependent Carcinogenicity and Toxicity of Foreign Compounds

Advances in

Pharmacology

Volume 27

Conjugation-Dependent Carcinogenicity and Toxicity of Foreign Compounds

Edited by

M. W. Anders
Department of Pharmacology
University of Rochester
Rochester, New York

Wolfgang Dekant
Institut für Toxikologie
Universität Würzburg
Würzburg, Germany

Academic Press
A Division of Harcourt Brace & Company

San Diego New York Boston London Sydney Tokyo Toronto

Front cover illustration: Geometry-optimized (HyperChem, AM1) structure of *S*-(E-1,2-dichlorovinyl)-L-cysteine.

This book is printed on acid-free paper. ∞

Academic Press, Inc.
A Division of Harcourt Brace & Company
525 B Street, Suite 1900, San Diego, California 92101-4495

United Kingdom Edition published by
Academic Press Limited
24–28 Oval Road, London NW1 7DX

International Standard Serial Number: 0065-3144

International Standard Book Number: 0-12-032927-1

PRINTED IN THE UNITED STATES OF AMERICA
94 95 96 97 98 99 QW 9 8 7 6 5 4 3 2 1

Contents

Enzymology of Cytosolic Glutathione *S*-Transferases
Brian Ketterer and Lucia G. Christodoulides

Enzymology of Cysteine *S*-Conjugate β-Lyases
Arthur J. L. Cooper

Formation and Fate of Nephrotoxic and Cytotoxic Glutathione S-Conjugates: Cysteine Conjugate β-Lyase Pathway

Wolfgang Dekant, Spyridon Vamvakas, and M. W. Anders

Reversibility in Glutathione-Conjugate Formation

Thomas A. Baillie and Kelem Kassahun

Glutathione Conjugation as a Mechanism for the Transport of Reactive Metabolites

Terrence J. Monks and Serrine S. Lau

Metabolism and Genotoxicity of Dihaloalkanes

F. Peter Guengerich

Bioactivation of Thiols by One-Electron Oxidation

Rex Munday

Glutathione Mercaptides as Transport Forms of Metals
N. Ballatori

Part II
Sulfate Conjugate-Dependent Toxicity

Biochemistry of Cytosolic Sulfotransferases Involved in Bioactivation
Charles N. Falany and Teresa W. Wilborn

Carcinogen Activation by Sulfate Conjugate Formation
Christopher J. Michejda and Marilyn B. Kroeger Koepke

Part III
Glucuronide Conjugate-Dependent Toxicity

UDP-Glucuronosyltransferases and Their Role in Metabolism and Disposition of Carcinogens
Karl Walter Bock

Bioactivation by Glucuronide-Conjugate Formation

Parnian Zia-Amirhosseini, Hildegard Spahn-Langguth, and Leslie Z. Benet

Part IV
Bioactivation and Bioconversion

N-Acetyltransferases, *O*-Acetyltransferases, and *N,O*-Acetyltransferases: Enzymology and Bioactivation

Patrick E. Hanna

Aminoacylases

M. W. Anders and Wolfgang Dekant

Contributors

Numbers in parentheses indicate the pages on which the authors' contributions begin.

M. W. Anders (115, 431, 479), Department of Pharmacology, University of Rochester, Rochester, New York 14642

Claes Andersson (19), Institute of Environmental Medicine, Department of Toxicology, Karolinska Institutet, 17177 Stockholm, Sweden

Thomas A. Baillie (163), Department of Medicinal Chemistry, School of Pharmacy, University of Washington, Seattle, Washington 98195

N. Ballatori (271), Department of Environmental Medicine, University of Rochester School of Medicine, Rochester, New York 14642

Leslie Z. Benet (385), Department of Pharmacy, University of California, San Francisco, San Francisco, California 94143

Karl Walter Bock (367), Institute of Toxicology, University of Tübingen, 72074 Tübingen, Germany

Lucia G. Christodoulides (37), Department of Biochemistry and Molecular Biology, University College London, London W1P 6DB, United Kingdom

Arthur J. L. Cooper (71), Departments of Biochemistry and Neurology, Cornell University Medical College, New York, New York 10021

Wolfgang Dekant (115, 431), Institut für Toxikologie, Universität Würzburg, 97078 Würzburg, Germany

Charles N. Falany (301), Department of Pharmacology, University of Alabama at Birmingham, Birmingham, Alabama 35294

F. Peter Guengerich (211), Department of Biochemistry and Center in Molecular Toxicology, Vanderbilt University School of Medicine, Nashville, Tennessee 37232

Patrick E. Hanna (401), Departments of Medicinal Chemistry and Pharmacology, University of Minnesota, Minneapolis, Minnesota 55455

Jerald L. Hoffman (449), Department of Biochemistry, School of Medicine, University of Louisville, Louisville, Kentucky 40292

Kelem Kassahun (163), Department of Medicinal Chemistry, School of Pharmacy, University of Washington, Seattle, Washington 98195

Brian Ketterer (37), Department of Biochemistry and Molecular Biology, University College London, London W1P 6DB, United Kingdom

Marilyn B. Kroeger Koepke (331), Molecular Aspects of Drug Design Section, Macromolecular Structure Laboratory, ABL-Basic Research Program, NCI–Frederick Cancer Research and Development Center, Frederick, Maryland 21702

Serrine S. Lau (183), Division of Pharmacology and Toxicology, College of Pharmacy, University of Texas at Austin, Austin, Texas 78712

Christopher J. Michejda (331), Molecular Aspects of Drug Design Section, Macromolecular Structure Laboratory, ABL-Basic Research Program, NCI–Frederick Cancer Research and Development Center, Frederick, Maryland 21702

James A. Miller (1), McArdle Laboratory for Cancer Research, University of Wisconsin Medical School, Madison, Wisconsin 53706

Terrence J. Monks (183), Division of Pharmacology and Toxicology, College of Pharmacy, University of Texas at Austin, Austin, Texas 78712

Ralf Morgenstern (19), Institute of Environmental Medicine, Department of Toxicology, Karolinska Institutet, 17177 Stockholm, Sweden

Erifili Mosialou (19), Institute of Environmental Medicine, Department of Toxicology, Karolinska Institutet, 17177 Stockholm, Sweden

Rex Munday (237), Ruakura Agricultural Research Centre, Hamilton, New Zealand

Hildegard Spahn-Langguth (385), Department of Pharmacology, University of Frankfurt, D-6000 Frankfurt, Germany

Young-Joon Surh (1), Department of Epidemiology and Public Health, Yale University School of Medicine, New Haven, Connecticut 06510

Spyridon Vamvakas (115, 479), Institut für Toxikologie, Universität Würzburg, 97078 Würzburg, Germany

Rolf Weinander (19), Institute of Environmental Medicine, Department of Toxicology, Karolinska Institutet, 17177 Stockholm, Sweden

Teresa W. Wilborn (301), Department of Pharmacology, University of Alabama at Birmingham, Birmingham, Alabama 35294

Parnian Zia-Amirhosseini (385), Department of Pharmacy, University of California, San Francisco, San Francisco, California 94143

Historical Perspectives on Conjugation-Dependent Bioactivation of Foreign Compounds

James A. Miller* and Young-Joon Surh†

**McArdle Laboratory for Cancer Research*
University of Wisconsin Medical School
Madison, Wisconsin 53706

†*Department of Epidemiology and Public Health*
Yale University School of Medicine
New Haven, Connecticut 06510

I. Metabolic Research Leading to Conjugation-Dependent Bioactivation of 4-Dimethylaminoazobenzene and 2-Acetylaminofluorene

A. Introduction

The conjugation-dependent bioactivation of foreign compounds for carcinogenesis and toxicity had its origin in research conducted by the senior author with his late co-worker, Professor Elizabeth C. Miller, and our associates on the mechanisms of action of chemical carcinogens. These studies started in the early 1940s in graduate school in the Department of Biochemistry in the School of Agriculture at the University of Wisconsin in Madison and continued for many years in the nearby McArdle Laboratory for Cancer Research in the Medical School. For about 25 years, these investigations concerned primarily the metabolism and carcinogenicity of two compounds: 4-dimethylaminoazobenzene (DAB) and 2-acetylaminofluorene (AAF) (Fig. 1). The carcinogenic activity of DAB in the

Advances in Pharmacology, Volume 27

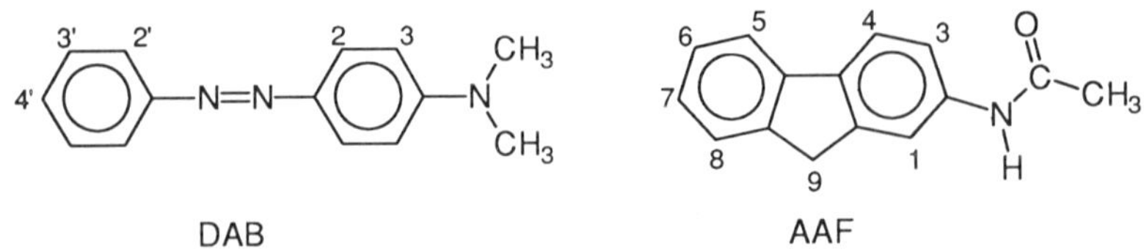

Fig. 1 The structures of 4-dimethylaminoazobenzene (DAB) and 2-acetylaminofluorene (AAF).

rat was first reported by Kinosita (1937), and the corresponding activity of AAF was discovered by Wilson *et al.* (1941). The long-term objective of our studies was the elucidation of the biochemical mechanisms of action of these carcinogens in the induction of tumors, especially in the liver, in rats after prolonged feeding of these compounds. Surprisingly, these studies led to the first observations of (*a*) the covalent binding of carcinogens to the cellular macromolecules of tissues undergoing carcinogenesis, (*b*) the hepatic microsomal oxidases that metabolize many carcinogens and drugs and the induction of the synthesis of these enzymes *in vivo* following the administration of many different chemicals, and (*c*) the activation of many carcinogens by hepatic oxidases and conjugases to form electrophilic ultimate carcinogens that covalently bind to informational macromolecules in cells and initiate the development of carcinogenic, mutagenic, and other toxic lesions. These findings are summarized in brief detail below.

B. Covalent Binding of Carcinogens to Tissue Macromolecules

The detection of the covalent binding of aminoazo dyes to liver protein in rats fed DAB and similar dyes (Miller and Miller, 1947) depended on the intense red color of these dyes in acid as a marker for the whole *p*-aminoazobenzene structure (Cilento *et al.*, 1956). The protein-bound dye from DAB was the first identification of an adduct of a metabolite of a carcinogen to a macromolecule in a tissue undergoing carcinogenesis. The amount of these adducts correlated strongly with the carcinogenicities of DAB and its ring-methyl derivatives in the rat liver (Miller *et al.*, 1949). Surprisingly, no protein-bound dye could be detected in the liver tumors caused by DAB, even during its administration. This led to a deletion theory of the mechanism of tumor formation by this carcinogen (E. C. Miller and Miller, 1947; J. A. Miller and Miller, 1953). Thus, if dye bound to proteins that controlled growth led to a loss of these proteins in certain cells, then these cells would become tumor cells. This research was done in the 1940s when the structures and functions of DNA and RNA were poorly understood. Attempts were made to find nucleic acid-bound dye but none was found (Miller and Miller, 1952). Only much later was it found that the structure of the principal nucleic acid-bound dye from DAB did not show a red color in acid (Lin *et al.*, 1975b). In the 1950s, ^{14}C-labeled carcinogens became available and it was shown that various chemical carcinogens became covalently bound to proteins, RNAs, and DNA *in vivo* (Miller and Miller, 1985). The seminal paper by Watson and Crick (1953) on the structure and genetic function of DNA soon directed studies on covalent binding of carcinogens to DNA, and the protein-binding of these agents was less studied. Attention also turned to the metabolism and possible reactive forms of chemical carcinogens.

C. Hepatic Microsomal Metabolism of Carcinogens and Its Induction by Unrelated Chemicals

As noted by Trager (1980), the carcinogen DAB was the first xenobiotic to be studied *in vitro* for its metabolism by liver homogenates or microsomes. These studies by Mueller and Miller (1948) showed the *N*-demethylation, reduction of the azo linkage, and aromatic ring-hydroxylation occurred on aerobic incubation of DAB with liver homogenates fortified with pyridine nucleotides and hexose diphosphate. Enzymatic reduction of the azo linkage was found predominantly in the microsome fraction (then called the small granule fraction) and was dependent on both NADPH and a flavin adenine dinucleotide-containing protein (Mueller and Miller, 1949, 1950). It was further shown that the enzymatic demethylation of *N*-monomethylaminoazo dyes depended on oxygen and NADPH with the formation of stoichiometric amounts of formaldehyde (Mueller and Miller, 1953). One of the substrates used in these studies was 3-methyl-4-monomethylaminoazobenzene (3-methyl-MAB). It proved to be very resistant to reduction of the azo linkage and formed equimolar amounts of 3-methyl-4-aminoazobenzene and formaldehyde. Further study of this oxidative demethylation enzyme system led unexpectedly to the finding that the liver homogenates of mice fed certain crude diets had activities two to three times greater than the liver homogenates of mice fed a purified diet (Brown *et al.*, 1954). Several pure compounds, including pinane hydroperoxide and 3-methylcholanthrene, were active at low levels in the purified diet, but none were active in homogenates. With rat liver homogenates and microsomes, even larger increases up to 5-fold were observed with the inclusion of these pure compounds in the purified diet (Conney *et al.*, 1956, 1957a). Several lines of evidence, including the strong inhibition by ethionine in the diet, suggested that these increases in demethylase activity were increases in the amount of the enzyme. Still larger increases of up to 10-fold were found in benzpyrene hydroxylase activity after ip injections of benzpyrene (Conney *et al.*, 1957b). Subsequent studies with protein synthesis inhibitors demonstrated that these increases in enzyme activity were increases in the amount of enzyme (Conney *et al.*, 1956; von der Decken and Hultin, 1960; Gelboin and Sokoloff, 1961; Gelboin and Blackburn, 1963; Conney and Gilman, 1963; Gelboin, 1964).

Today, a variety of microsomal oxidases dependent on NADPH and molecular oxygen are known (the large family of cytochromes P450 and flavoprotein enzymes) that metabolize a range of endogenous and exogenous substrates, including many carcinogens and other xenobiotics. Many of these enzymes are inducible by a range of chemicals.

D. Conjugation-Dependent Bioactivation of DAB and AAF

The early findings of protein- and nucleic acid-bound derivatives of chemical carcinogens and the lack of demonstrable reactivity of most of these carcinogens in the absence of *in vivo* conditions suggested that reactive metabolites of chemical carcinogens were formed *in vivo*. The approximate correlations between the carcinogenic activities and the levels of their macromolecular-bound derivatives further suggested that reactive metabolites and their precursors would be ultimate or proximate carcinogens. Early in our studies on the protein-bound aminoazo dyes, we hoped to characterize these dyes and get clues about the nature of reactive metabolites that formed these bound derivatives. However, this approach proved to be very difficult since analytical methods for the characterization of small amounts of the bound dyes were not available. A more fruitful lead came instead from our studies on the urinary metabolites of AAF in rats. AAF was known to form a series of ring-hydroxylated urinary metabolites in the rat after the administration of single doses of this carcinogen (Weisburger and Weisburger, 1958). It was also known that low dietary levels of 3-methylcholanthrene strongly inhibited hepatocarcinogenesis by aminoazo dyes and AAF (Richardson *et al.*, 1952; Miller *et al.*, 1958). In an attempt (Cramer *et al.*, 1959, 1960; Miller *et al.*, 1960) to associate hepatocarcinogenesis with some of the ring-hydroxylated metabolites of AAF, 0.03% of AAF was fed with and without 0.003% 3-methylcholanthrene and analyses of the hydroxylated metabolites in the urine were made during continuous feeding of the two diets until liver tumors appeared. As it turned out, 3-methylcholanthrene raised the levels of some of the ring-hydroxylated metabolites, but after feeding of AAF alone a new spot appeared on the paper chromatogram of the glucuronidase-treated urine extract. It tested positive with a molybdate-based test for phenols, but was negative in a diazo-coupling test for phenols. Most importantly, it increased greatly in amount as the feeding of AAF alone proceeded and as damage to the liver increased. The new metabolite was shown to be a metabolite of AAF since it contained ^{14}C when ^{14}C-labeled AAF was fed and it amounted to over 10% of the AAF fed. The low excretion of this metabolite in the urine of rats protected against the effects of AAF by 3-methylcholanthrene in the diet suggested that the new metabolite might be important in carcinogenesis by AAF. Isolation of the metabolite from large amounts of urine by column chromatography yielded milligram quantities of a crystalline compound that was different from all the possible mono ring-hydroxylated derivatives of AAF, yet its elementary analysis showed that it contained one oxygen atom more than found in AAF. After some consideration, it was realized that it was *N*-hydroxy-AAF, a hydroxamic acid that was easily oxidized by the molybdate reagent but did not respond to the diazo-coupling test for phenols. Synthesis of *N*-hydroxy-AAF was achieved by the catalytic hydrogenation of 2-nitrofluorene in excess acetic anhydride to form *N*-acetoxy-AAF. Removal of the *O*-acetyl group with base yielded *N*-hydroxy-AAF. It proved to be more toxic than AAF and to be more carcinogenic than AAF, not only in the usual tissues, but also in tissues that do not respond to AAF. It was also carcinogenic in the guinea pig in which AAF is inactive (Miller *et al.,* 1961, 1964). However, although it formed more covalently bound adducts with proteins and nucleic acids in the liver *in vivo* than did AAF, it was not reactive with these macromolecules *in vitro*. Thus it became the first example of proximate carcinogenic metabolite of a carcinogen. Subsequent studies with a variety of aromatic amines and amides similarly indicate that N-hydroxylation is the first step in the bioactivation for carcinogenic activity (Beland and Kadlubar, 1990).

Several years passed before the nature of the second step in the bioactivation of AAF became clear. The key observation came from further studies on the aminoazo dyes. Our first efforts to prepare *N*-hydroxy-MAB proved to be very difficult, and as a substitute we prepared the ester *N*-benzoyloxy-MAB (Poirier *et al.,* 1967) hoping that it would be hydrolyzed *in vivo* to *N*-hydroxy-MAB. Instead, this ester was found to be directly reactive with methionine, proteins, and nucleic acids to form covalently bound adducts. Then it was realized that this ester was a strong electrophile and contained a strong benzoyloxy leaving group forming an electrophilic residue at the nitrogen atom that combined covalently with nucleophilic sites in methionine, protein, and nucleic acids. Analogous studies showed that *N*-acetoxy-AAF has similar and even stronger electrophilic activity. Similarly, enzymatic degradation of the DNA and RNA from the livers of rats given [9-^{14}C]AAF yielded products chromatographically indistinguishable from the *N*-(guan-8-yl)AAF derivatives obtained on reaction of *N*-acetoxy-AAF with deoxyguanosine and guanosine (Kriek *et al.*, 1967; Kriek, 1969).

These results, as well as the carcinogenicities of *N*-benzoyloxy-MAB and *N*-acetoxy-AAF in the subcutaneous tissue of the rat where MAB and AAF are not active (Poirier *et al.*, 1967; Miller and Miller, 1969), stimulated searches for the enzymatic formation of reactive esters of *N*-hydroxy-AAF by tissue preparations. Studies from our laboratory (DeBaun *et al.*, 1968, 1970a,b) and that of King and Phillips (1968) soon provided evidence for the formation of an electrophilic *N*-sulfate from *N*-hydroxy-AAF by liver cytosols supplemented with 3′-phosphoadenosine-5′-phosphosulfate (PAPS). Studies *in vivo* showed that the hepatic toxicity of *N*-hydroxy-AAF was reduced by the simultaneous feeding of *p*-hydroxyacetanilide that is excreted as a stable *O*-sulfate and leads to a depletion of PAPS (DeBaun *et al.*, 1970a). Likewise, the hepatocarcinogenicity of *N*-hydroxy-AAF was inhibited by the administration of acetanilide, a precursor of *p*-hydroxyacetanilide (Weisburger *et al.*, 1972). These studies suggested that the sulfuric acid ester of *N*-hydroxy-AAF has important biological activity.

Later studies have shown that *N*-hydroxy-AAF forms multiple electrophilic metabolites in rat and mouse livers. Today it is known that six metabolic pathways give rise to five strong electrophilic metabolites of *N*-hydroxy-AAF:

1. Cytosolic sulfotransferase activity for *N*-hydroxy-AAF gives rise to a sulfuric acid ester that reacts with DNA to form two adducts of the guanine ring, *N*-(deoxyguanosin-8-yl)-AAF and 3-(deoxyguanosin-N^2-yl)-AAF (Kriek *et al.*, 1967; Kriek, 1969).
2. A microsomal one-electron peroxidatic reaction forms the nitroxide free radical of *N*-hydroxy-AAF that dismutates to form 2-nitrosofluorene and *N*-acetoxy-AAF (Bartsch and Hecker, 1971). The latter ester reacts with DNA to form the guanine–DNA adducts just described.
3. Cytosolic *N,O*-transacetylase activity for *N*-hydroxy-AAF transfers the *N*-acetyl group to the *N*-hydroxyl oxygen to form *N*-acetoxy-2-aminofluorene (Bartsch *et al.*, 1972) that reacts with DNA to form N-(deoxyguanosin-8-yl)-2-aminofluorene, the major DNA adduct found after the administration of AAF or *N*-hydroxy-AAF.
4. A microsomal deacetylase activity for *N*-hydroxy-AAF (Irving, 1966) forms *N*-hydroxy-2-aminofluorene (*N*-hydroxy-AF), which can be activated to form *N*-(deoxyguanosin-8-yl)-AF adducts in DNA by three cytosolic pathways: (*a*) acetyl-CoA-transacetylase activity to form *N*-acetoxy-AF (Flammang and Kadlubar, 1986), (*b*) sulfotransferase activity to form

the sulfuric acid ester of *N*-hydroxy-AF (Lai *et al.*, 1985), and (*c*) protonation at slightly acid pHs with loss of water to form AF-nitrenium ion (Kriek, 1960).

These complex pathways are well reviewed by Beland and Kadlubar (1990) and are outlined in Figure 2.

In addition, *N*-hydroxy-AAF is found in the urine of rats given AAF or *N*-hydroxy-AAF as a *N,O*-glucuronide that has weak electrophilic activity. It reacts at neutral pH with nucleic acids *in vitro* and with partial loss of the *N*-acetyl groups to form *N*-(guanosin-8-yl)-AAF and -AF adducts (Miller *et al.*, 1968; Irving *et al.*, 1969). This was the first glucuronide found with electrophilic activity for cellular constituents.

Although less extensive, similar findings were made with DAB following the characterization of the polar aminoazo dyes derived from the alkaline hydrolysis of the hepatic protein-bound dye in the livers of rats fed DAB or MAB (Lin *et al.*, 1968, 1969). The synthesis of *N*-hydroxy-MAB was achieved by Kadlubar *et al.* (1976a). This metabolite was shown to form an electrophilic *N*-sulfate with hepatic sulfotransferase activity plus PAPS;

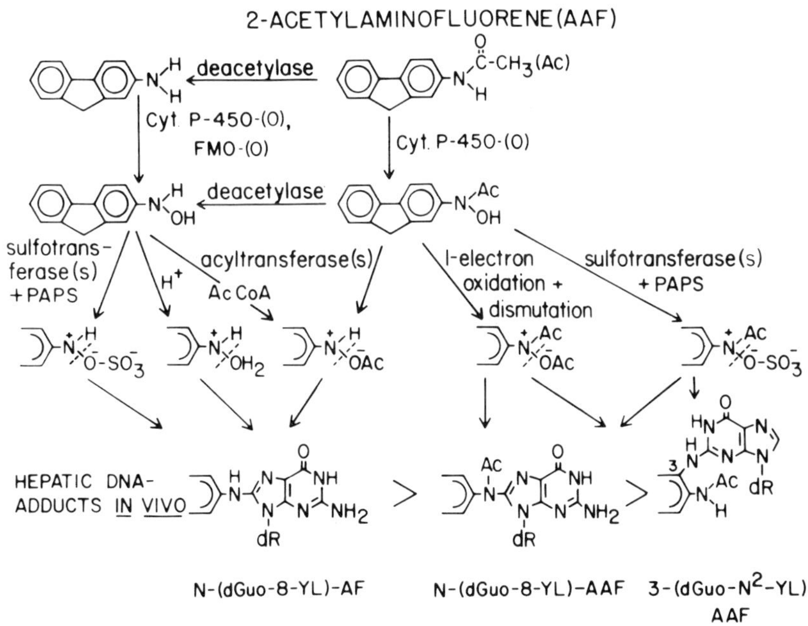

Fig. 2 Multiple pathways in the metabolism of AAF to strong electrophilic reactants in rat and mouse liver.

this ester reacted with guanosine to form the *N*-(guanosin-8-yl)-MAB adduct (Kadlubar *et al.,* 1976b). The availability of *N*-benzoyloxy-MAB facilitated the characterization of the nucleic acid-bound dyes formed from MAB *in vitro* and *in vivo* (Lin *et al.,* 1975a,b; Beland *et al.,* 1980; Tullis *et al.,* 1981).

E. Conclusion

By 1968–1972, the studies described above provided the first examples of the conjugation-dependent bioactivation by enzymatic sulfonation, acetylation, and glucuronidation of xenobiotics. The central role of electrophilic metabolites in carcinogenesis, mutagenesis, and toxicity of xenobiotics has been discussed in a previous historical perspective (Miller and Miller, 1985) and the prominent role of sulfonation in chemical carcinogenesis has recently been reviewed (Miller and Surh, 1993).

II. Important Observations Leading to Bioactivation of Foreign Compounds via Gluthathione Conjugation, Methylation, and Epoxide Formation and Hydrolysis

A. Introduction

As described in the above section, three major conjugation reactions, sulfonation, acetylation, and glucuronidation, have been shown to be implicated in the toxication by certain xenobiotics such as carcinogenic arylamines. These reactions give rise to the formation of electrophilic esters that generate reactive species, such as nitrenium and benzylic or allylic carbonium ions. However, other phase-II reactions may also play roles in the bioactivation of other types of toxicants. Some important findings and current knowledge about glutathione conjugation-, methylation-, and epoxide formation and hydration-dependent activation of foreign compounds will be briefly presented here.

B. Glutathione Conjugation

The formation of glutathione conjugates, in general, has been regarded an important cellular defense mechanism against a variety of toxic xenobiotics or their reactive metabolites (Ketterer, 1986). More recently, however, it has become apparent that glutathione conjugation of certain types of chemicals leads to the transformation of the parent compounds into more reactive and toxic derivatives (Anders *et al.,* 1992). Investigations on the glutathione-dependent toxification of xenobiotics started with an earlier finding of the hazardous nature of *S*-(1,2-dichlorovinyl)-L-cysteine (DCVC). This cysteine conjugate of trichloroethylene produced hematopoietic toxicity in calves (McKinney *et al.*, 1957; Schultze *et al.*, 1959). DCVC was also found to be a potent nephrotoxic agent, which produced lesions in the proximal kidney tubules in rodents (Teracini and Parker, 1965). In addition to DCVC, cysteine *S* conjugates of certain polyhalogenated olefinic hydrocarbons such as trichloroethene, chlorotrifluoroethene, tetrafluoroethene, and hexachloro-1,3-butadiene have also been found to be nephrotoxic, indicating that conjugation with glutathione may be involved in the selective renal toxicity of these halogenated alkenes (reviewed by Dekant *et al.*, 1989; Lock, 1988). The target-organ toxicity of these hydrocarbons appears to be associated with selective uptake and subsequent processing of glutathione conjugates by the kidney. Hence the glutathione conjugates formed in the liver by glutathione *S*-transferase activity are further metabolized by γ-glutamyl transpeptidase and cysteinyl glycine dipeptidase to the corresponding cysteine *S* conjugates, which then undergo bioactivation by cysteine conjugate β-lyase (β-lyase) predominantly localized in proximal tubular epithelium (reviewed by Anders *et al.*, 1987; van Bladeren, 1988). β-Lyase cleaves DCVC and other nephrotoxic cysteine conjugates to produce putative reactive thiols, which can react covalently with nucleophiles of target cells. Another early example of the involvement of glutathione in the formation of reactive species is the increased DNA methylation by *N*-methyl-*N'*-nitro-*N*-nitroguanidine (MNNG) in the presence of glutathione (Lawley and Thatcher, 1970). Glutathione conjugation also plays an important role in the toxic action of vicinal halodialkanes such as 1,2-dibromoethane and 1,2-dichloroethane. This concept was based on the initial observation by Rannug *et al.* (1978) of enhanced mutagenicity of 1,2-dichloroethane in the presence of rat liver cytosol containing glutathione *S*-transferase activity. The same result occurred when the analog 1,2-dibromoethane was activated by glutathione conjugation (van Bladeren *et al.*, 1980). In both cases, the resulting *S*-(2-haloethyl)glutathione conjugates are considered to yield highly electrophilic episulfonium (thiiranium) ion intermediates that alkylate cellular nucleophiles, such as DNA (Hill *et al.*, 1978; Peterson *et al.*, 1988; Kim and Guengerich, 1990). In this respect, glutathione-dependent formation of sulfur half-mustards appears to be responsible for genotoxicity and possibly carcinogenicity of above vicinal dihaloalkanes and other related hydrocarbons. There are other examples in which adverse effects of toxicants are mediated through glutathione conjugation (Anders *et al.*, 1992; Baillie and Slatter, 1991; Koob and Dekant, 1991; Monks *et al.*, 1990). More detailed considerations of these matters will be covered elsewhere in this volume.

C. Methylation

Methyl conjugation is common in the biotransformation of endogenous compounds, such as catecholamine neurotransmitters, but some xenobiotics are also methylated by nonspecific or physiological methyltransferases under certain conditions. Enzymatic methylation, in general, involves the transfer of an activated methyl group from the coenzyme *S*-adenosylmethionine to hydroxyl, amino, and thiol functions of acceptor molecules. Methylation differs from other conjugation reactions in that the reaction products formed, in the majority of cases, are less polar or more lipophilic than the parent molecules, and thus their excretion may not be facilitated. The products formed by methylation reactions may in some cases have as great or greater pharmacological activity than the parent compounds. An example is the conversion of norepinephrine to epinephrine. Methylated metabolites may interfere with the actions of their parent compounds. Thus, methylation of the β-agonist isoprenalin, which is used as a bronchodilator in the treatment of asthma, produces 3-*O*-methylisoprenalin with β-receptor blocking properties (β-blocker). In this case, the methylated metabolite exerts a pharmacological effect opposed to that of the parent drug, thereby aggravating the symptoms if given to patients at high doses (Dollery and Davies, 1971). In addition, methylation may result in the formation of toxic metabolites. For example, amine-*N*-methyltransferase in the brain tissues of humans and animals catalyzes the methylation of 4-phenyl-1,2,3,6-tetrahydropyridine and 4-phenylpyridine to 1-methyl-4-phenyltetrahydropyridine (MPTP) and 1-methyl-4-phenylpyridinium ion (MPP^+), respectively (Ansher *et al.*, 1986). The metabolite in each instance is neurotoxic and regarded as a causative factor in a Parkinson-like syndrome. Methylation also appears to play a role in carcinogenesis by some aromatic amines. The importance of methylation in chemical carcinogenesis by certain aromatic amines was emphasized in the early 1950s by the Millers, who found that *N*-methyl-4-aminoazobenzene was carcinogenic in rats, whereas the parent compound 4-aminoazobenzene was very weakly active (Miller and Miller, 1953). Later studies by other investigators have demonstrated that methylation and subsequent oxygenation at nitrogen are necessary to achieve carcinogenicity of a certain type of arylamines. For instance, N-methylation of 4-aminoazobenzene renders it oxidizable by microsomal flavin-containing monooxygenase (Ziegler *et al.*, 1988). Esterification of the resulting *N*-hydroxylamine then would convert it to an ultimate electrophilic and carcinogenic metabolite. In the case of thiols, methyl transfer mainly represents a detoxification step. In addition to O-, N-, and S-methylation reactions, direct incorporation of the methyl group to an aromatic ring system (C-methylation) has been reported. Earlier studies on the structure–activity relationships clearly demonstrated dramatic increases in the carcinogenicity of several polynuclear aromatic hydrocarbons following substitution (e.g., methylation) in the mesoposition (Fieser, 1938). On the basis of these early observations, Flesher and co-workers have proposed a unified hypothesis that the chemical or enzymatic introduction of a methyl group, most favorably in the meso-anthracenic carbon center with high chemical reactivity, is important in the activation of a polycyclic aromatic hydrocarbon to an ultimate carcinogen (Flesher *et al.*, 1988). Further studies by Flesher's group have demonstrated the methylation of hydrocarbons *in vivo* as well as in reactions *in vitro* with cytosols fortified with *S*-adenosylmethionine as a methyl group donor (recently reviewed by Flesher and Myers, 1991). The oxidation at the methyl group with subsequent esterification of the resulting hydroxymethyl derivative would produce a highly reactive benzylic carbocation capable of interacting covalently with critical cellular macromolecules such as DNA, RNA, and protein (Flesher and Myers, 1991). However, few quantitative data have been provided on the extent of aforementioned bioalkylation reactions, and it remains unclear whether such reactions play a role in carcinogenesis by unsubstituted polycyclic aromatic hydrocarbons that generally have no or weak carcinogenic activities.

D. Epoxide Formation and Hydrolysis

Epoxides formed by oxidation of aromatic rings or alkenes are, in general, chemically reactive and play roles as ultimate electrophiles in various toxification reactions, including necrosis, mutagenesis, carcinogenesis, and teratogenesis (Daly *et al.*, 1972). These active oxiranes are subject to further transformation in the body. Conjugation with a water molecule, that is, hydration, converts epoxides to dihydrodiols. The hydration of an epoxide may proceed nonenzymatically, but the reaction is facilitated by the action of the enzyme epoxide hydrolase (EC 3.3.2.3), formerly known as epoxide hydratase or epoxide hydrase. Two distinct epoxide hydrolases are known to exist in the cell; one in the microsomal fractions and the other in the cytosol. Epoxide hydrolases have dual roles in the epoxide metabolism. In general, the dihydrodiols that result from the oxirane cleavage are less reactive and less toxic than the parent epoxides, and the process hence represents detoxification. Under certain conditions, however, dihydrodiols may undergo further oxidation to form intermediates far more reactive than the parent oxiranes. Thus, benzo[*a*]pyrene 7,8-dihydrodiol derived from benzo[*a*]pyrene 7,8-oxide is a precursor of

7,8-dihydroxy-9,10-oxo-7,8,9,10-tetrahydrobenzo[*a*]pyrene, an ultimate mutagen and carcinogenic metabolite of benzo[*a*]pyrene (Wood *et al.*, 1976).

References

Anders, M. W., Elfarra, A. A., and Lash, L. H. (1987). Cellular effects of reactive intermediates: Nephrotoxicity of S-conjugates of amino acids. *Arch. Toxicol.* **60,** 103–108.

Anders, M. W., Dekant, W., and Vamvakas, S. (1992). Glutathione-dependent toxicity. *Xenobiotica* **22,** 1135–1145.

Ansher, S. S., Cadet, J. L., Jakoby, W. B., and Baker, J. K. (1986). Role of N-methyltransferases in the neurotoxicity associated with the metabolism of 1-methyl-4-phenyl-1,2,3,6-tetrahydropyridine (MPTP) and other 4-substituted pyridines present in the environment. *Biochem. Pharmacol.* **35,** 3359–3363.

Baillie, T. A., and Slatter, J. G. (1991). Glutathione: A vehicle for the transport of chemically reactive metabolites in vivo. *Acc. Chem. Res.* **24,** 264–270.

Bartsch, H., and Hecker, E. (1971). On the metabolic activation of the carcinogen N-hydroxy-N-2-acetylaminofluorene. III. Oxidation with horseradish peroxidase to yield 2-nitrosofluorene and N-acetoxy-N-2-acetylaminofluorene. *Biochim. Biophys. Acta* **237,** 567–578.

Bartsch, H., Dworkin, M., Miller, J. A., and Miller, E. C. (1972). Electrophilic N-acetoxy-aminoarenes derived from carcinogenic N-hydroxy-N-acetylaminoarenes by enzymatic deacetylation and transacetylation in liver. *Biochim. Biophys. Acta* **286,** 272–298.

Beland, F. A., and Kadlubar, F. F. (1990). Metabolic activation and DNA adducts of aromatic amines and nitroaromatic hydrocarbons. *In* "Chemical Carcinogenesis and Mutagenesis" (C. S. Cooper and P. L. Grover, eds.), Vol. I, pp. 267–325. Springer-Verlag, Berlin.

Beland, F. A., Tullis, D. L., Kadlubar, F. F., Straub, K. M., and Evans, F. E. (1980). Characterization of DNA adducts of the carcinogen N-methyl-4-aminoazobenzene *in vitro* and *in vivo*. *Chem.-Biol. Interact.* **31,** 1–17.

Brown, R. R., Miller, J. A., and Miller, E. C. (1954). The metabolism of methylated aminoazo dyes. IV. Dietary factors enhancing demethylation *in vitro*. *J. Biol. Chem.* **209,** 211–222.

Cilento, G., Miller, E. C., and Miller, J. A. (1956). On the addition of protons to derivatives of 4-aminoazobenzene. *J. Am. Chem. Soc.* **78,** 1718–1722.

Conney, A. H., and Gilman, A. E. (1963). Puromycin inhibition of enzyme induction by 3-methycholanthrene and phenobarbital. *J. Biol. Chem.* **238,** 3682–3685.

Conney, A. H., Miller, E. C., and Miller, J. A. (1956). The metabolism of methylated aminoazo dyes. V. Evidence for induction of enzyme synthesis in the rat by 3-methylcholanthrene. *Cancer Res.* **16,** 450–459.

Conney, A. H., Brown, R. R., Miller, J. A., and Miller, E. C. (1957a). The metabolism of methylated aminoazo dyes. VI. Intracellular distribution and properties of the demethylase system. *Cancer Res.* **17,** 628–633.

Conney, A. H., Miller, E. C., and Miller, J. A. (1957b). Substrate-induced synthesis and other properties of benzpyrene hydroxylase in rat liver. *J. Biol. Chem.* **228,** 753–766.

Cramer, J. W., Miller, J. A., and Miller, E. C. (1959). The effect of a low dietary level of 3-methylcholanthrene on the urinary derivatives of 2-acetylamino-fluorene in the rat. *Proc. Am. Assoc. Cancer Res.* **3,** 13–14.

Cramer, J. W., Miller, J. A., and Miller, E. C. (1960). N-Hydroxylation: A new metabolic reaction observed in the rat with the carcinogen 2-acetylamino-fluorene. *J. Biol. Chem.* **235,** 885–888.

Daly, J. W., Jerina, D. M., and Witkop, B. (1972). Arene oxides and NIH shift: The metabolism, toxicity and carcinogenicity of aromatic hydrocarbons. *Experientia* **28,** 1129–1149.

DeBaun, J. R., Rowley, J. Y., Miller, E. C., and Miller, J. A. (1968). Sulfotransferase activation of N-hydroxy-2-acetylaminofluorene in rodent livers susceptible and resistant to this carcinogen. *Proc. Soc. Exp. Biol. Med.* **129,** 268–273.

DeBaun, J. R., Miller, E. C., and Miller, J. A. (1970a). N-Hydroxy-2-acetylaminofluorene sulfotransferase: Its probable role in carcinogenesis and protein-(methion-S-yl) binding in rat liver. *Cancer Res.* **30,** 577–595.

DeBaun, J. R., Smith, J. Y. R., Miller, E. C., and Miller, J. A. (1970b). Reactivity *in vivo* of the carcinogen N-hydroxy-2-acetylaminofluorene: Increase by sulfate ion. *Science* **167,** 184–186.

Dekant, W., Vamvakas, S., and Anders, M. W. (1989). Bioactivation of nephrotoxic haloalkenes by glutathione conjugation: Formation of toxic and mutagenic intermediates by cysteine conjugate β-lyase. *Drug Metab. Rev.* **20,** 43–83.

Dollery, C. T., and Davies, D. S. (1971). Differences in the metabolism of drugs depending upon their routes of administration. *Ann. N. Y. Acad. Sci.* **179,** 108–114.

Fieser, L. F. (1938). Carcinogenic activity, structure and chemical reactivity of polycyclic aromatic hydrocarbons. *Am. J. Cancer* **34,** 37–124.

Flammang, T. J., and Kadlubar, F. F. (1986). Acetyl coenzyme A-dependent metabolic activation of N-hydroxy-3,2′-dimethyl-4-aminobiphenyl and several carcinogenic N-hydroxy arylamines in relation to tissue and species differences, other acyl donors, and aryl hydroxamic acid-dependent acyltransferases. *Carcinogenesis* (*London*) **7,** 919–926.

Flesher, J. W., and Myers, S. R. (1991). Rules of molecular geometry for predicting carcinogenic activity of unsubstituted polynuclear aromatic hydrocarbons. *Teratog. Carcinog. Mutagen.* **11,** 41–54.

Flesher, J. W., Myers, S. R., and Blake, J. W. (1988). Bioalkylation of polynuclear aromatic hydrocarbons in vivo: A predictor of carcinogenic activity. *In* "Polynuclear Aromatic Hydrocarbons: A Decade of Progress" (M. Cooke and A. J. Dennis, eds.), pp. 261–276. Battelle Press, Columbus, OH.

Gelboin, H. V. (1964). Studies on the mechanism of methylcholanthrene induction of enzyme activities. II. Stimulation of microsomal and ribosomal amino acid incorporation: The effects of polyuridylic acid and actinomycin D. *Biochim. Biophys. Acta* **91,** 130–144.

Gelboin, H. V., and Blackburn, N. R. (1963). The stimulatory effect of 3-methylcholanthrene on microsomal amino acid incorporation and benzpyrene hydroxylase activity and its inhibition by actinomycin D. *Biochim. Biophys. Acta* **91,** 130–144.

Gelboin, H. V., and Sokoloff, L. (1961). Effects of 3-methylcholanthrene and phenobarbital on amino acid incorporation in protein. *Science* **134,** 611–612.

Hill, D. L., Shih, T.-W., Johnston, T. P., and Struck, R. F. (1978). Macromolecular binding and metabolism of the carcinogen 1,2-dibromoethane. *Cancer Res.* **38,** 2438–2442.

Irving, C. C. (1966). Enzymatic deacetylation of N-hydroxy-2-acetylaminofluorene by liver microsomes. *Cancer Res.* **26,** 1390–1396.

Irving, C. C., Veazey, R. A., and Hill, J. T. (1969). Reaction of the glucuronide of the carcinogen N-hydroxy-2-acetylaminofluorene with nucleic acids. *Biochim. Biophys. Acta* **179,** 189–198.

Kadlubar, F. F., Miller, J. A., and Miller, E. C. (1976a). Microsomal N-oxidation of the hepatocarcinogen N-methyl-4-aminoazobenzene and the reactivity of N-hydroxy-N-methyl-4-aminoazobenzene. *Cancer Res.* **36,** 1196–1206.

Kadlubar, F. F., Miller, J. A., and Miller, E. C. (1976b). Hepatic metabolism of N-hydroxy-4-aminoazobenzene and other N-hydroxy arylamines to reactive sulfuric acid esters. *Cancer Res.* **36,** 2350–2359.

Ketterer, B. (1986). Detoxication reactions of glutathione and glutathione transferases. *Xenobiotica* **16,** 957–973.

Kim, D.-H., and Guengerich, F. P. (1990). Formation of DNA adduct S-(2-(N^7-guanyl)ethyl)-glutathione from ethylene dibromide: Effects of modulation of glutathione and glutathione S-transferase levels and the lack of a role for sulfation. *Carcinogenesis (London)* **11,** 419–424.

King, C. M., and Phillips, B. (1968). Enzyme-catalyzed reactions of the carcingen N-hydroxy-2-fluorenylacetamide with nucleic acid. *Science* **159,** 1351–1353.

Kinosita, R. (1937). Studies on the cancerogenic chemical substances. *Trans. Jpn. Pathol. Soc.* **27,** 665–727.

Koob, M., and Dekant, W. (1991). Bioactivation of xenobiotics by formation of toxic glutathione conjugates. *Chem.-Biol. Interact.* **77,** 107–136.

Kriek, E. (1960). On the interaction of N-fluorenylhydroxylamine with nucleic acids *in vitro. Biochem. Biophys. Res. Commun.* **20,** 793–799.

Kriek, E. (1969). On the mechanism of action of carcinogenic aromatic amines. I. Binding of 2-acetylaminofluorene and N-hydroxy-2-acetylaminofluorene to rat liver nucleic acids *in vivo. Chem.-Biol. Interact.* **1,** 3–17.

Kriek, E., Miller, J. A., Juhl, U., and Miller, E. C. (1967). 8-(N-Fluorenylacetamido) guanosine, an arylamidation product of guanosine and the carcinogen *N*-acetoxy-N-fluorenylacetamide in neutral solution. *Biochemistry* **6,** 177–182.

Lai, C.-C., Miller, J. A., Miller, E. C., and Liem, A. (1985). Sulfooxy-2-aminofluorene is the major ultimate electrophilic and carcinogenic metabolite of N-hydroxy-2-acetyl-aminofluorene in the livers of infant male C57BL/6J × C3H/HeJ F_1 (B6C3F_1) mice. *Carcinogenesis (London)* **6,** 1037–1045.

Lawley, P. D., and Thatcher, C. J. (1970). Methylation of deoxyribonucleic acid in cultured mammalian cells by N-methyl-N′-nitro-N-nitrosoguanidine. *Biochem. J.* **116,** 693–707.

Lin, J.-K., Miller, J. A., and Miller, E. C. (1968). Studies on structures of polar dyes derived from liver proteins of rats fed N-methyl-4-aminoazobenzene. II. Identity of synthetic 3-(homocytein-S-yl)-N-methyl-4-aminoazobenzene with the major polar dye P2b. *Biochemistry* **7,** 1889–1895.

Lin, J.-K., Miller, J. A., and Miller, E. C. (1969). Studies on structures of polar dyes derived from liver proteins of rats fed N-methyl-4-aminoazobenzene. III. Tyrosine and homocysteine sulfoxide polar dyes. *Biochemistry* **8,** 1573–1582.

Lin, J.-K., Schmall, B., Sharpe, I. D., Miura, I., Miller, J. A., and Miller, E. C. (1975a). N-Substitution of carbon 8 in guanosine and deoxyguanosine by the carcinogen N-benzoyloxy-N-methyl-4-aminoazobenzene *in vitro. Cancer Res.* **35,** 832–843.

Lin, J.-K., Miller, J. A., and Miller, E. C. (1975b). Structures of hepatic nucleic acid-bound dyes in rats given the carcinogen N-methyl-4-aminoazobenzene. *Cancer Res.* **35,** 844–850.

Lock, E. A. (1988). Studies on the mechanism of nephrotoxicity and nephrocarcinogenicity of halogenated alkenes. *CRC Crit. Rev. Toxicol.* **19,** 23–42.

McKinney, L. L., Weakley, F. B., Eldridge, A. C., Campbell, R. E., Cowan, J. C., Pickens, J. C., Jr., and Biester, H. E. (1957). S-(Dichlorovinyl)-L-cysteine: An agent causing fatal aplastic anemia in calves. *J. Am. Chem. Soc.* **79,** 3932–3933.

Miller, E. C., and Miller, J. A. (1947). The presence and significance of bound aminoazo dyes in the livers of rats fed p-dimethylaminoazobenzene. *Cancer Res.* **7,** 468–480.

Miller, E. C., and Miller, J. A. (1952). *In vivo* combinations between carcinogens and tissue constituents and their possible role in carcinogenesis. *Cancer Res.* **12,** 547–556.

Miller, E. C., and Miller, J. A. (1985). Some historical perspectives on the metabolism of xenobiotic chemicals to reactive electrophiles. *In* "Bioactivation of Foreign Compounds" (M. W. Anders, ed.), pp. 3–28. Academic Press, Orlando, FL.

Miller, E. C., Miller, J. A., Sapp, R. W., and Weber, G. M. (1949). Studies on the protein-bound aminoazo dyes formed *in vivo* from 4-dimethylaminoazobenzene and its C-monomethyl derivatives. *Cancer Res.* **9,** 336–343.

Miller, E. C., Miller, J. A., Brown, R. R., and MacDonald, J. C. (1958). On the protective action of certain polycyclic aromatic hydrocarbons against carcinogenesis by aminoazo dyes and 2-acetylaminofluorene. *Cancer Res.* **18,** 469–477.

Miller, E. C., Miller, J. A., and Hartmann, H. A. (1961). N-Hydroxy-2-acetylaminofluorene: A metabolite of 2-acetylaminofluorene with increased carcinogenic activity in the rat. *Cancer Res.* **21,** 815–824.

Miller, E. C., Miller, J. A., and Enomoto, M. (1964). The comparative carcinogenicities of 2-acetylaminofluorene and its N-hydroxy metabolite in mice, hamsters, and guinea pigs. *Cancer Res.* **24,** 2018–2032.

Miller, E. C., Lotlikar, P. D., Miller, J. A., Butler, B. W., Irving, C. C., and Hill, J. T. (1968). Reactions *in vitro* of some tissue nucleophiles with the glucuronide of the carcinogen N-hydroxy-2-acetylaminofluorene. *Mol. Pharmacol.* **4,** 147–154.

Miller, J. A., and Miller, E. C. (1953). The carcinogenic amino-azo dyes. *Adv. Cancer Res.* **1,** 339–396.

Miller, J. A., and Miller, E. C. (1969). The metabolic activation of carcinogenic aromatic amines and amides. *Prog. Exp. Tumor Res.* **11,** 273–301.

Miller, J. A., and Surh, Y.-J. (1993). Sulfonation in chemical carcinogenesis. *In* "Handbook of Experimental Pharmacology" (F. C. Kauffman, ed.). Springer-Verlag, Berlin.

Miller, J. A., Cramer, J. W., and Miller, E. C. (1960). The N- and ring-hydroxylation of 2-acetylaminofluorene during carcinogenesis in the rat. *Cancer Res.* **20,** 950–962.

Monks, T. J., Anders, M. W., Dekant, W., Stevens, J. L., Lau, S. S., and van Bladeren, P. J. (1990). Glutathione conjugate mediated toxicities. *Toxicol. Appl. Pharmacol.* **106,** 1–19.

Mueller, G. C., and Miller, J. A. (1948). The metabolism of 4-dimethylaminoazobenzene by rat liver homogenates. *J. Biol. Chem.* **176,** 535–544.

Mueller, G. C., and Miller, J. A. (1949). The reductive cleavage of 4-dimethylaminoazobenzene by rat liver: The intracellular distribution of the enzyme system and its requirement for triphosphopyridine nucleotide. *J. Biol. Chem.* **180,** 1125–1136.

Mueller, G. C., and Miller, J. A. (1950). The reductive cleavage of 4-dimethylaminoazobenzene by rat liver: Reactivation of carbon dioxide treated homogenates by riboflavin adenine dinucleotide. *J. Biol. Chem.* **185,** 145–154.

Mueller, G. C., and Miller, J. A. (1953). The metabolism of methylated aminoazo dyes. II. Oxidative demethylation by rat liver homogenates. *J. Biol. Chem.* **202,** 579–587.

Peterson, L. A., Harris, T. M., and Guengerich, F. P. (1988). Evidence for an episulfonium ion intermediate in the formation of S-[2-(N^7-guanyl)ethyl]-glutathione in DNA. *J. Am. Chem. Soc.* **110,** 3284–3291.

Poirier, L. A., Miller, J. A., Miller E. C., and Sato, K. (1967). N-Benzoyloxy-N-methyl-4-aminoazobenzene: Its carcinogenicity in the rat and its reactions with proteins and nucleic acids and their constituents *in vitro*. *Cancer Res.* **27,** 1600–1613.

Rannug, U., Sundvall, A., and Ramel, C. (1978). The mutagenic effect of 1,2-dichloroethane on *Salmonella typhimurium*. I. Activation through conjugation with glutathione *in vitro*. *Chem.-Biol. Interact.* **20,** 1–16.

Richardson, H. L., Stier, A. R., and Borsos-Nachnebel, E. (1952). Liver tumor inhibition and adrenal histologic responses in rats to which 3′-methyl-4-dimethylaminoazobenzene and 20-methylcholanthrene were simultaneously administered. *Cancer Res.* **12,** 356–361.

Schultze, M. O., Klubes, P., Perman, V., Mizuno, N. S., Bates, F. W., and Sautter, J. H. (1959). Blood dyscrasia in calves induced by S-(dichlorovinyl)-L-cysteine. *Blood* **14,** 1015–1025.

Teracini, B., and Parker, V. H. (1965). A pathological study on the toxicity of dichlorovinyl-L-cysteine. *Food Cosmet. Toxicol.* **3,** 67–74.

Trager, W. F. (1980). Oxidative functionalization reactions. *Drugs Pharm. Sci.* **10,** Part A, 177–210.

Tullis, D. L., Straub, K. M., and Kadlubar, F. F. (1981). A comparison of the carcinogen-DNA adducts formed in rat liver *in vivo* after administration of single or multiple doses of N-methyl-4-aminoazobenzene. *Chem.-Biol. Interact.* **38,** 15–27.

van Bladeren, P. J. (1988). Formation of toxic metabolites from drugs and other xenobiotics by glutathione conjugation. *Trends Pharmacol. Sci.* **9,** 295–299.

van Bladeren, P. J., Breimer, D. D., Rotteveel-Smijs, G. M. T., de Jong, R. A. W., van der Gen, A., and Mohn, G. R. (1980). The role of glutathione conjugation in the mutagenicity of 1,3-dibromoethane. *Biochem. Pharmacol.* **29,** 2975–2982.

von der Decken, A., and Hultin, T. (1960). Inductive effects of 3-methylcholanthrene on enzyme activities and amino acid incorporation capacity of rat liver microsomes. *Arch. Biochem. Biophys.* **90,** 201–207.

Watson, J. D., and Crick, F. H. C. (1953). Molecular structure of nucleic acids: A structure for deoxyribose nucleic acid. *Nature* (*London*) **171,** 737–738.

Weisburger, E. K., and Weisburger, J. H. (1958). Chemistry, carcinogenicity, and metabolism of 2-fluorenamine and related compounds. *Adv. Cancer Res.* **5,** 331–431.

Weisburger, J. H., Yamamoto, R. S., Williams, G. M., Grantham, P. H., Matsushima, T., and Weisburger, E. K. (1972). On the sulfate ester of N-hydroxy-N-2-fluorenylacetamide as a key ultimate hepatocarcinogen in the rat. *Cancer Res.* **32,** 491–500.

Wilson, R. H., DeEds, F., and Cox, A. J., Jr. (1941). The toxicity and carcinogenic activity of 2-acetylaminofluorene. *Cancer Res.* **1,** 595–608.

Wood, A. W., Levin, W., Lu, A. Y. H., Yagi, H., Hernandez, O., Jerina, D. M., and Conney, A. H. (1976). Metabolism of benzo[*a*]pyrene and benzo[*a*]pyrene derivatives to mutagenic products by highly purified hepatic microsomal enzymes. *J. Biol. Chem.* **251,** 4882–4890.

Ziegler, D. M., Ansher, S. S., Nagata, T., Kadlubar, F. F., and Jakoby, W. B. (1988). N-Methylation: Potential mechanism for metabolic activation of carcinogenic primary arylamines. *Proc. Natl. Acad. Sci. U.S.A.* **85,** 2514–2517.

Part I

Glutathione-Dependent Toxicity

Enzymology of Microsomal Glutathione S-Transferase

Claes Andersson, Erifili Mosialou, Rolf Weinander, and Ralf Morgenstern
Institute of Environmental Medicine
Division of Toxicology
Karolinska Institutet
17177 Stockholm, Sweden

I. Introduction

Glutathione transferases can be regarded the most important and versatile enzymes that protect organisms from electrophilic compounds. The versatility is manifested in many isoenzymes with different subcellular locations displaying broad and overlapping substrate specificities (for reviews, see Armstrong, 1991; Chasseaud, 1979; Mannervik, 1985; Mannervik and Danielson, 1988; Morgenstern and DePierre, 1985, 1988). Microsomal glutathione transferase was first purified from rat liver microsomes (Morgenstern *et al.*, 1982). The key factor that made this possible was the observation that the glutathione transferase activity toward 1-chloro-2,4-dinitrobenzene (CDNB)[1] could be activated severalfold by sulfhdryl reagents in microsomes but not in the cytosol (Morgenstern *et al.*, 1979). Thus microsomes that had been treated with *N*-ethylmaleimide (NEM) were used as the starting material for purification. This resulted not only in an increased activity but also permitted the selective purification of the enzyme without contamination by cytosolic glutathione transferases. The

[1] Abbreviations used: CDNB, 1-chloro-2,4-dinitrobenzene; NEM, *N*-ethylmaleimide; HCBD, hexachlorobutadiene.

Advances in Pharmacology, Volume 27

cytosolic transferases are bound to the endoplasmic reticulum in amounts much higher than can be expected from simple contamination (Morgenstern *et al.*, 1983). Not surprisingly, there are reports describing the purification of *cytosolic* glutathione transferases from microsomes (Friedberg *et al.*, 1980; Lee and McKinney, 1982). Once the purification of the activated enzyme had been achieved, the identical procedure was used to purify the microsomal glutathione transferase in its unactivated form (Morgenstern and DePierre, 1983). The unactivated purified enzyme could be activated by sulfhydryl reagents to the same extent as the enzyme activity in microsomes (allowing for the contribution of bound cytosolic glutathione transferases). It also became clear that NEM treatment of the enzyme does not increase its activity to all electrophilic substrates and that substrate reactivity seemed to be an important factor. Thus, in general, but certainly not always, reactive substrates are "activatable."

A. Molecular Properties

The subunit molecular mass of the purified protein is 17.3 kDa, as determined by amino acid and cDNA sequence analysis (Morgenstern *et al.*, 1985; DeJong *et al.*, 1988). After SDS–PAGE, the mass varies between 14 and 17 kDa, depending on the gel electrophoretic system used and choice of molecular mass markers. The isoelectric point is 10.1, which facilitates the purification of the enzyme by cation exchange chromatography. The molecular properties together with the lack of immunological cross-reactivity (Morgenstern *et al.*, 1982) clearly distinguish the enzyme from cytosolic glutathione transferases to which there is no obvious relationship at the amino acid or cDNA sequence level (Morgenstern *et al.*, 1985; DeJong *et al.*, 1988).

The quaternary structure of the microsomal glutathione transferase is trimeric. Hydrodynamic, radiation inactivation, and cross-linking studies as well as two-dimensional crystal structure determination all support this conclusion (Morgenstern *et al.*, 1982; Boyer *et al.*, 1986; Lundqvist *et al.*, 1992; Andersson *et al.*, 1993). The trimeric structure is present in the purified detergent complex, in proteoliposomes, and in the endoplasmic reticulum.

B. Subcellular, Extrahepatic, and Phylogenetic Distribution

Microsomal glutathione transferase is present in very high amounts in liver microsomes (3.1% of the protein) as well as in the outer mitochondrial membrane (4.8%) (Morgenstern *et al.*, 1984). The concentration of the enzyme in the endoplasmic reticulum has been estimated at 1 m*M* (Morgenstern and DePierre, 1985). These levels are not increased by common inducers of drug-metabolizing enzymes (Morgenstern *et al.*, 1980). A recent report shows hepatocyte plasma membrane location with differing amounts in canalicular and sinusoidal membrane fractions (2% and 0.5–1%, respectively) (Horbach *et al.*, 1993). The claim that microsomal glutathione transferase is present in peroxisomes is difficult to ascertain, since high levels of microsomal contamination have to be taken into account (Nishino and Ito, 1990). In general, the amount of microsomal glutathione transferase in other intracellular membranes is not higher than that calculated from microsomal contamination (Morgenstern *et al.*, 1984). Nevertheless, improvements in obtaining extremely pure subcellular fractions could make quantification in other locations possible, as was the case with plasma membrane fractions.

The extrahepatic distribution of the microsomal glutathione transferase has been probed by activity measurements and by Western and Northern blot analysis (DeJong *et al.*, 1988; Morgenstern *et al.*, 1984). The activity measurements revealed, somewhat surprisingly, that no NEM activation could be detected in microsomes from any of the organs tested, although Western blots showed the presence of significant amounts of protein [10- to 100-fold lower than in liver (Table I)]. The reason for the lack of activation is not clear, although the presence of endogenous inhibitors, which have been suggested to be present in liver microsomes (Boyer *et*

Table I

Extrahepatic Distribution of Rat Microsomal Glutathione Transferase as Measured by Western- and Northern-Blot Analysis

	Microsomal glutathione transferase	
Organ	Protein in microsomes (μg/mg)	mRNA expression (% of liver)
Liver	31	100
Kidney	≤0.3	20
Lung	0.9	1–8[a]
Intestine	4.0	
Adrenal	3.2	
Testes	2.5	20
Spleen	0.5	1–8[b]
Brain	≤0.3	1–8[c]
Heart	<0.3	
Thymus	1.1	

[a–c] Relative amount $a > b > c$.

al., 1982), could explain the phenomena. The activated enzyme is, in general, more sensitive to inhibition than the unactivated enzyme (Mosialou and Morgenstern, 1990). Northern blot analysis indicated that mRNAs were expressed at 20% of the liver level in kidney and testis with lower levels being present in other organs (Table I). The high amount in testis agrees with the Western blot analysis, whereas the kidney contained very little protein. The reason for the discrepancies between Western and Northern blot analysis is not known. Different subcellular localization of the protein in different organs (only the microsomal fraction was analyzed) or different turnover rates, or both, are possible explanations. Studies on the half-life and extrahepatic cellular and subcellular localization of the microsomal glutathione transferase are warranted, especially regarding the possible contribution of the enzyme to the production of glutathione conjugates that are carcinogenic and toxic.

The phylogenetic distribution, as probed by immunological cross-reaction and activity measurements, shows that all mammalian species examined contain a related enzyme of similar size (Morgenstern *et al.*, 1984). The human microsomal glutathione transferase is, of course, of special interest and has been purified and characterized with respect to substrate specificity, activation characteristics, size, isoelectric point, and cDNA sequence (McLellan *et al.*, 1989; Mosialou *et al.*, 1993a; DeJong *et al.*, 1988). The human enzyme is similar to the rat enzyme with a 95% conservation at the amino acid sequence level. The gene for human microsomal glutathione transferase is located on chromosome 12 (DeJong *et al.*, 1990).

II. Catalysis

The substrates and, consequently, the assays for glutathione transferases are numerous. For routine work, the substrate of choice is CDNB, because it displays high activity with most glutathione transferases. In principle, assays for cytosolic and microsomal glutathione transferase differ only by the inclusion of detergent (0.1% Triton X-100) when the purified microsomal enzyme is used (representative substrates are given in Table II). Microsomal fractions are normally assayed in the absence of detergent, although the inclusion of detergent can be used to decrease turbidity. Any inhibitory or stimulatory effect has to be checked because, with human liver microsomes, the inclusion of detergent increases the activity toward CDNB up to threefold. Immobilized microsomal (and cytosolic) fractions have also been used for studying conjugate formation and offer advantages in the ease of product separation (Pallante *et al.*, 1986). When the microsomal fraction is assayed, it must always be borne in mind that cytosolic

Table II
Specific Activity of Purified Rat Liver Microsomal Glutathione Transferase with Selected Substrates

Substrate	Specific activity (μmol/min mg protein)		Reference
	Unactivated	NEM-treated	
CDNB	2.0	30	Morgenstern and DePierre (1983)
(+*N*-acetyl-L-cysteine)	5.6	2.0	[a]
2,5-Dichloronitrobenzene	1.3	1.3	Morgenstern *et al.* (1988)
(+*N*-acetyl-L-cysteine)	0.11	1.6	[a]
p-Nitrophenylacetate	0.2	0.9	Morgenstern *et al.* (1988)
Dilinoleoylphosphatidyl-choline hydroperoxide	0.3	0.8	Mosialou and Morgenstern (1989)
γ-Linoleoyl, β-palmitoyl-phosphatidylcholine hydroperoxide	0.2	ND[b]	[a]
Cholesteryl linoleate hydroperoxide	0.2	ND	[a]
4-Hydroxy-2-nonenal	0.5	ND	Mosialou and Morgenstern (1989)
Methyl iodide	0.61	0.61	Morgenstern and DePierre (1983)
Hexachlorobutadiene	0.005	0.005	Morgenstern and DePierre (1983)
Hexachlorobutadiene[c]	0.44	0.44	Wallin *et al.* (1988)

[a] Weinander *et al.*, 1993.
[b] Not determined.
[c] Improved assay.

glutathione transferases are present in low amounts (Morgenstern *et al.*, 1983). When the presence of detergent is not compatible with the experimental design, the purified microsomal glutathione transferase can be reconstituted into proteoliposomes by cholate dialysis in the presence of phosphatidyl choline (Lundqvist *et al.*, 1992). Other detergents might also be used to exchange Triton X-100, which absorbs strongly around 275 nm. Reduced Triton X-100, lauryldiaminoxide, and zwittergent 3-14 have been used successfully for this purpose.

Characterization of the catalytic activity of the microsomal glutathione transferase has shown that it follows a random sequential mechanism, both in the activated and unactivated state (R. Morgenstern, unpublished observation). The ability of the enzyme to lower the pK_a of the thiol in glutathione to 6.4–6.8 seems to be responsible for part of the catalytic enhancement in both the activated and the unactivated enzymes (unpublished observation). The finding that only the activity with more reactive substrates is increased upon activation indicates that activation augments the rate of product release (Morgenstern *et al.*, 1988). It has been noted that the NEM-activated enzyme is much more susceptible to trypsin cleavage than the unactivated enzyme, indicating that activation is accompanied by a conformational change in the enzyme (Morgenstern *et al.*, 1989).

Substrates for glutathione transferases are lipophilic and electrophilic and are important because of their potential carcinogenic, tumorigenic, toxic, and pharmacological properties. In comparison with cytosolic glutathione transferases, the microsomal enzyme shares their broad substrate specificity, and the catalytic efficiency ranges from lower to higher in certain cases. Substrates that display selectivity for the microsomal glutathione transferase include many polyhalogenated, unsaturated hydrocarbons as well as lipid hydroperoxides (Tables II and III). Other substrates include aromatic hydrocarbons with activated ring structures and halogen-leaving groups (i.e., CDNB), α,β-unsatured carbonyls, activated esters (thiolysis of *p*-nitrophenylacetate), and simple halocarbons (Table II).

Table III

Relative Contribution of Rat Liver Microsomal and Cytosolic Fractions to the Glutathione Conjugate Formation from Various Polyhalogenated Hydrocarbons

	Specific activity (nmol/min mg protein)		
Substrate	Microsomes	Cytosol	Reference
Trichloroethene	0.002	—	Dekant *et al.* (1989)
Tetrachloroethene	0.23	0.13	Dekant *et al.* (1987)
Tetrafluoroethene	3	0.7	Odum and Green (1984)
Chlorotrifluoroethene	170	90	Dohn *et al.* (1985)
Perfluoropropene	276	136	Koob and Dekant (1990)
Trichlorotrifluoropropene	530	120	Vamvakas *et al.* (1989)
Dichloroacetylene	3160	660	Kanhai *et al.* (1989)
	3030(kidney)	127(kidney)	
Hexachlorobutadiene	1.12	0.04	Dekant *et al.* (1988)
	38[a]	0.09[a]	Wallin *et al.* (1988)
	1.17[b]	0.031[b]	Oesch and Wolf (1989)

[a] Improved assay.
[b] Human.

Epoxides are, in general, poor substrates (Morgenstern *et al.*, 1988). Although the glutathione transferase reaction generally represents a detoxification reaction, a growing number of glutathione conjugates with toxic properties have been discovered (Koob and Dekant, 1991). Many of these harmful substances are halocarbons (Dekant *et al.*, 1990a,b). Around the same time that microsomal glutathione transferase was purified, it was reported that some halogenated hydrocarbons were conjugated to glutathione preferentially in the microsomal fraction of liver (Table III) (Dohn and Anders, 1982; Odum and Green, 1984; Wolf *et al.*, 1984). The question arose whether the enzyme responsible for this metabolism was the microsomal glutathione transferase and whether the higher reaction rates observed were inherent to the enzyme or a consequence of its location in the membrane, where lipophilic substances may be expected to accumulate. The following examples illustrate the importance of the microsomal glutathione transferase and its catalytic efficiency as well as membrane location. Hexachlorobutadiene (HCBD), which is nephrotoxic, was found to be metabolized more efficiently in microsomes than in cytosol (Wolf *et al.*, 1984). In experiments designed to pinpoint the role of the microsomal glutathione transferase, an observed lack of mutual inhibition of CDNB and HCBD conjugation was interpreted as an indication of additional membrane bound enzymes (Oesch and Wolf, 1989). However, experiments where the activity was followed during purification of the human enzyme clearly established that the microsomal glutathione transferase was responsible for the HCBD conjugation (McLellan *et al.*, 1989). In addition, experiments utilizing an improved assay procedure showed a very high conjugation rate by the pure enzyme, which accounts for the activity observed in microsomes (allowing for an inhibitory effect of Triton X-100) (Tables II and III) (Wallin *et al.*, 1988). It is important to note that the activity toward HCBD is not influenced when the microsomal glutathione transferase is treated with NEM. In general, lack of NEM activation can never be taken as evidence for the lack of involvement of the microsomal glutathione transferase, whereas activation is a good indication that the enzyme catalyzes the activity under study. In the case of chlorotrifluoroethene, where the specific activities in the cytosol and microsomal fractions are comparable (Table III), experiments examining diastereomeric product distributions have shown that the membrane location of the microsomal glutathione transferase is very favorable (Hargus *et al.*, 1991).

Perhaps one reason for the membrane location of microsomal glutathione transferase is its activity toward lipid hydroperoxides and, unique to glutathione transferases, toward phospholipid hydroperoxides (Table II). The ability of the enzyme to reduce hydroperoxides formed in the membrane as well as its capability to conjugate hydroxyalkenals may serve to protect cells against one consequence of oxidative stress, namely lipid peroxidation (Mosialou and Morgenstern, 1989; Mosialou *et al.*, 1993b). It is also interesting to note that activation of the enzyme increases its peroxidase activity toward phospholipid hydroperoxides (Mosialou and Morgenstern, 1989) and that the enzyme is activated by hydrogen peroxide in the purified form as well as by hydrogen peroxide and by ischemia-reperfusion in the liver (Aniya and Anders, 1992; Aniya and Naito, 1993).

III. Activation

There are now numerous examples of activation of the microsomal glutathione transferase *in vitro* as well as *in vivo* (Table IV). The purified enzyme can be activated by sulfhydryl reagents, including various disulfides, proteolysis, and bromosulfophtalein, which is also an inhibitor. In microsomes, the enzyme can be activated by removal of an endogenous inhibitor, by heat treatment and by reactive metabolites formed during the metabolism of phenol and α-methyldopa. In hepatocytes, the GSH depletor phorone, but not diethyl maleate, activates the enzyme efficiently, but in an unknown manner. Activation by the substrate CDNB as well as a slight activation by *tert*-butyl hydroperoxide was also observed in this system. As mentioned above, hydrogen peroxide and ischemia-reperfusion activate the enzyme in the isolated perfused liver. In whole animals, activation has been observed with compounds that form reactive intermediates capable of binding sulfhydryl groups as well as with phorone and diethyl maleate (Table IV). In general, the observed increases in activity in more complex systems are small, but taking into account that only 30% of the control CDNB activity of liver microsomes stems from the microsomal glutathione transferase, an increase of the overall activity by 30% is the result of a doubling of the specific activity of the microsomal glutathione transferase (Morgenstern *et al.*, 1983).

The principle ways to activate the enzyme include modification of the single thiol group of cysteine-49, proteolysis, or modulation of non-covalently acting inhibitory/activating compounds (Fig. 1). It has been shown that reactive metabolites activate the enzyme and, when they are substrates (CDNB), thereby increase their own metabolism (Lundqvist and Morgenstern, 1992a). In addition, oxidative stress activates the enzyme, which increases its ability to reduce lipid hydroperoxides. The mechanism by which oxidative stress activates the enzyme may involve dimer formation (by intersubunit disulfide formation) or thiol-disulfide interchange (Aniya and Anders, 1992; Aniya and Naito, 1993). It is signifi-

Table IV

Activation of Microsomal Glutathione Transferase in Various Systems

System	Treatment	Mechanism/hypothesis	Reference
Purified enzyme	*N*-ethylmaleimide (other sulfhydryl reagents)	Thioether to Cys-49	Morgenstern and DePierre (1983)
	Cystamine	Mixed disulfide to Cys-49	Aniya and Anders (1989)
	Diamide ± GSH	Mixed disulfide to Cys-49 Intersubunit dimer (P-SS-P)	Aniya and Anders (1989)
	Hydrogen peroxide	Intersubunit dimer (P-SS-P)	Aniya and Anders (1992)
	Reactive metabolites of phenol and allyl alcohol	Covalent binding (Cys-49)	Haenen *et al.* (1988); Wallin and Morgenstern (1990)
	Trypsin	Proteolysis at Lys-41	Morgenstern *et al.* (1989); Mosialou *et al.* (1993a)
	Bromosulfophthalein	Binding/conformational change	Andersson *et al.* (1988); Mosialou and Morgenstern (1990)
Liver microsomes	All the above except bromosulfophthalein		
	GSSG + other disulfides	Mixed disulfide to Cys-49	Morgenstern and DePierre (1985); Nishino and Ito (1989); Aniya and Anders (1989)
	GSSG + 37–45°C	Mixed disulfide to Cys-49 enzyme catalyzed	Aniya (1989)
	Radiation	Radiation-induced reactive oxygen species	Boyer *et al.* (1986)
	Phenol metabolites (P450)	Covalent binding (Cys-49)	Wallin and Morgenstern (1990)
	α-Methyldopa metabolites (P450)	Covalent binding (Cys-49)	Haenen *et al.* (1991)
	Autoxidation products of noradrenaline	Covalent binding (Cys-49)	Lundqvist and Morgenstern (1992b)
	Small unilammellar phospholipid vesicles	Removal of endogenous inhibitors	Boyer *et al.* (1982)
	Heat	Unknown/not disulfide formation	Aniya (1989)
Hepatocytes	Phorone	Unknown	Lundqvist and Morgenstern (1992a)
	CDNB, 1,2-dibromoethane	Covalent binding (Cys-49)	Lundqvist and Morgenstern (1992a)
	tert-Butyl hydroperoxide	Unknown	Lundqvist and Morgenstern (1992a)

(*continued*)

Table IV *Continued*

System	Treatment	Mechanism/hypothesis	Reference
Perfused liver	Hydrogen peroxide	Mixed disulfide to Cys-49 Intersubunit dimer (P-SS-P)	Aniya and Naito (1993)
	Ischemia–reperfusion	Mixed disulfide to Cys-49 Intersubunit dimer (P-SS-P)	Aniya and Naito (1993)
Whole animals	Phorone, diethylmaleate	Mixed disulfide to Cys-49 by increased GSSG	Masukawa and Iwata (1986)
	Carbon tetrachloride	Covalent binding (Cys-49) of reactive metabolite	Botti *et al.* (1982)
	1,2-Dibromoethane	Covalent binding (Cys-49)	Botti *et al.*, (1982)
	Allyl alcohol	Covalent binding (Cys-49) of reactive metabolite (acrolein)	Haenen *et al.* (1988)

cant that this type of activation could be reversed by dithiothreitol, indicating the involvement of a disulfide bond (Aniya and Anders, 1992; Aniya and Naito, 1993). The putative dimer observed on treatment of the purified enzyme with hydrogen peroxide and in the ischemia-reperfusion experiments migrates, as determined by SDS–PAGE, and Western blot analysis, close to the position of the chemically cross-linked trimer (Lundqvist *et al.*, 1992). Furthermore, it is difficult to generate significant amounts of a dimer from the purified native protein by use of diamide oxidation, as compared with the SDS-denatured protein (Lundqvist *et al.*, 1992). This defined dimer migrates below the "dimer" observed in the oxidative stress experiments. However, dimer formation resulting from oxidative stress cannot be disregarded because of possible differences in the migration of cross-linked enzyme subunits in different electrophoretic systems. Until this point is clarified, however, thiol-disulfide interchange with a sulfhydryl-containing compound remains a more likely mechanism of activation. Such thiol-disulfide interchange could be reversible and catalyzed by enzymes, for which there exists some evidence (Aniya and Anders, 1989), to allow for control of the activity of microsomal glutathione transferase.

In some studies, proteolytic modification was investigated by Western

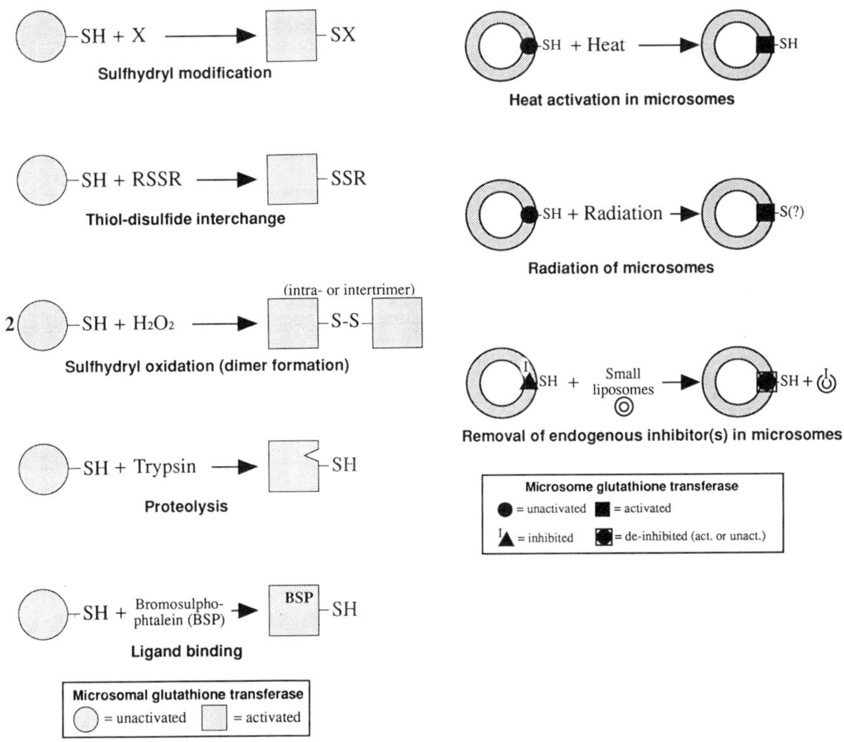

Fig. 1 Summary of conditions and proposed mechanisms for activation of the microsomal glutathione transferase. For references, see Table IV.

blot analysis and was not observed (Aniya and Anders, 1992; Lundqvist and Morgenstern, 1992b; Wallin and Morgenstern, 1990). Therefore, proteolysis appears unlikely as a general mechanism of activation.

The possibility of modulating the enzyme activity by small molecules that bind non-covalently is difficult to ascertain *in vivo*, and it would be interesting to find ways to measure the activity selectively in whole cells with a substrate that is activable.

There is no reason to postulate that regulation of the microsomal glutathione transferase activity should involve only one mechanism. Different mechanisms could function in response to various forms of toxic insult or unknown physiological demands and could even overlap. The elucidation of these mechanisms, considering their potential generality in enzyme regulation, is an attractive area for future research.

IV. The Activated Enzyme

What are the functional consequences of activation under different conditions of toxic insult? The conclusions that can be drawn from steady-state kinetic studies involving a series of increasingly reactive electrophilic substrate analogues and glutathione at low (R. Morgenstern, unpublished observation) or physiological concentrations (Morgenstern *et al.*, 1988) are as follows. At physiological glutathione concentrations and at high concentrations of reactive electrophiles, activation increases conjugation dramatically. On the other hand, if the reactive electrophile is present in low concentrations, activation does not increase the enzymatic rate appreciably. Of course, activation by covalent modification is less likely in this case. The activity toward less reactive electrophilic substrates does not change regardless of the activation state of the enzyme when the glutathione concentration is high. When the glutathione concentration is low ($\approx$0.1 m*M*), reactive electrophiles are conjugated at slower rates than at high glutathione concentrations but are still acted on more efficiently by the activated enzyme, a situation that can be envisioned during severe toxic insult. In this case a peculiar phenomenon is seen in that the activated enzyme also becomes more efficient at conjugating less reactive electrophiles. The observation that activation increases the activity to less reactive substrates at low glutathione concentrations has been made with three substrates so far. Whether this is a general phenomenon that is beneficial under toxic insult awaits further studies.

In steady-state kinetics, the behavior of the enzyme is examined with excess substrate concentrations. In the liver, however, the concentration of the enzyme is 1 m*M* in the endoplasmic reticulum. This concentration is probably higher than normal concentrations of electrophilic substrates. Thus, depending on the affinity of the enzyme toward different substrates, a considerable and rapid sequestration of harmful lipophilic compounds can occur. Such sequestration could protect other critical cellular targets from modification by electrophiles. The finding that the microsomal glutathione transferase is preferentially alkylated by certain reactive metabolites speaks in favor of this reasoning (Lundqvist and Morgenstern, 1992b; Wallin and Morgenstern, 1990).

V. Thiol Substrates

Because the toxicity of many glutathione conjugates of halogenated hydrocarbons depends on further processing of the products to the L-cysteine conjugates it is of interest that the microsomal glutathione transferase can utilize γ-L-glutamyl-L-cysteine and *N*-acetyl-L-cysteine as substrates (Weinander *et al.*, 1993). L-Cysteine can also function as a substrate, but at an extremely low rate. There is no reason to believe that these substrates function physiologically, but they could be used for the enzymatic synthesis of compounds of toxicological interest. Furthermore we have observed that *N*-acetyl-L-cysteine is not utilized by cytosolic glutathione transferase (assayed in the cytosol fraction), as was demonstrated previously with purified enzymes (Habig *et al.*, 1974). The activity of the microsomal glutathione transferase toward *N*-acetyl-L-cysteine is higher than that with glutathione when CDNB is used as the second substrate, but lower when less reactive analogues of CDNB are assayed (i.e., 2,5-dichloronitrobenzene, Table II). Thus, *N*-acetyl-L-cysteine conjugation offers the possibility to assay selectively the microsomal glutathione transferase, a property that could be useful, for instance, in the determination of extrahepatic subcellular activity and intraorgan distribution. The finding that dichloroacetylene is metabolized at the same rate in kidney and liver microsomes (Table III) is intriguing in view of the small amount of cross-reactive microsomal glutathione transferase that was quantified in the kidney (Morgenstern *et al.*, 1984). Either different preparation procedures or the existence of a new isozyme could explain the results. The latter is an interesting possibility that deserves investigation.

VI. Concluding Remarks

The study of the microsomal glutathione transferase regarding molecular properties, substrate specificity, kinetic behavior, and activation mechanisms has advanced considerably over the past few years. Nevertheless, polyhalogenated hydrocarbons that form toxic and carcinogenic glutathione conjugates need to be characterized with the purified enzyme with respect to the molecular properties that determine catalysis. These and additional studies on the extrahepatic distribution of microsomal glutathione transferase will likely yield information on the sites of formation of conjugates and on the enzymes involved. Are there additional enzyme forms, as indicated by the high rate of dichloroacetylene conjugation in kidney microsomes? The glutathione peroxidase activity of the microsomal glutathione transferase may protect the organism from reactive hydroperoxides formed during oxidative stress. The significance of the extrahepatic enzyme needs to be studied in this respect. Finally, studies on the interplay and functional significance of different activation mechanisms are interesting regarding enzyme regulation in general and protection against toxic insult in particular.

Acknowledgments

Studies from the authors' laboratories were supported by the Swedish Cancer Society, the Swedish Medical Research Council, the Swedish National Board of Laboratory Animals, and funds from Karolinska Institutet and Magnus Bergvalls Foundation.

References

Andersson, C., Söderström, M., and Mannervik, B. (1988). Activation and inhibition of microsomal glutathione transferase from mouse liver. *Biochem. J.* **249,** 819–823.

Andersson, C., Mosialou, E., Weinander, R., Hebert, H., and Morgenstern, R. (1993). Rat liver microsomal glutathione transferase: Studies on structure and function. *In* "Structure and function of glutathione transferases" (K. D. Tew, C. B. Picket, T. J. Mantle, B. Mannervik, and J. D. Hayes, eds.), pp. 109–116. CRC Press, Boca Raton.

Aniya, Y. (1989). Activation of liver microsomal glutathione *S*-transferase by heating. *J. Pharmacobio-Dyn.* **12,** 235–240.

Aniya, Y., and Anders, M. W. (1989). Regulation of rat liver microsomal glutathione *S*-transferase activity by thiol/disulfide exchange. *Arch. Biochem. Biophys.* **270,** 330–334.

Aniya, Y., and Anders, M. W. (1992). Activation of rat liver microsomal glutathione *S*-transferase by hydrogen peroxide: Role for protein dimer formation. *Arch. Biochem. Biophys.* **296,** 611–616.

Aniya, Y., and Naito, A. (1993). Oxidative stress-induced activation of microsomal glutathione *S*-transferase in isolated rat liver. *Biochem. Pharmacol.* **45,** 37–42.

Armstrong, R. N. (1991). Glutathione *S*-transferases—reaction mechanism, structure, and function. *Chem. Res. Toxicol.* **4,** 131–140.

Botti, B., Moslen, M. T., Trieff, N. M., and Reynolds, E. S. (1982). Transient decrease of liver cytosolic glutathione *S*-transferase activities in rats given 1,2-dibromoethane or CCl_4. *Chem.-Biol. Interact.* **42,** 259–270.

Boyer, T. D., Zakim, D., and Vessey, D. A. (1982). Studies of endogenous inhibitors of microsomal glutathione *S*-transferase. *Biochem. J.* **207,** 57–64.

Boyer, T. D., Vessey, D. A., and Kempner, E. (1986). Radiation inactivation of microsomal glutathione transferase. *J. Biol. Chem.* **261,** 16963–16968.

Chasseaud, L. F. (1979). The role of glutathione and glutathione *S*-transferases in the metabolism of chemical carcinogens and other electrophilic agents. *Adv. Cancer Res.* **29,** 175–274.

DeJong, J. L., Morgenstern, R., Jörnvall, H., DePierre, J. W., and Tu, C.-P. D. (1988). Gene expression of rat and human microsomal glutathione S-transferases. *J. Biol. Chem.* **263,** 8430–8436.

DeJong, J. L., Mohandas, T., and Tu, C.-P. D. (1990). The gene for the microsomal glutathione *S*-transferase is on human chromosome-12. *Genomics* **6,** 379–382.

Dekant, W., Martens, G., Vamvakas, S., Metzler, M., and Henschler, D. (1987). Bioactivation of tetrechloroethylene- Role of glutathione *S*-transferase catalyzed conjugation versus cytochrome P-450 dependent phospholipid alkylation. *Drug Metab. Dispos.* **15,** 702–709.

Dekant, W., Vamvakas, S., Henschler, D., and Anders, M. W. (1988). Enzymatic conjugation of hexachloro-1,3-butadiene with glutathione: Formation of 1-(glutathion-*S*-yl)-1,2,3,4,4-pentachlorobuta-1,3-diene and 1,4-bis-glutathion-*S*-yl)-1,2,3,4,-tetrachlorobuta-1,3-diene. *Drug Metab. Dispos.* **16,** 701–706.

Dekant, W., Koob, M., and Henschler, D. (1989). Metabolism of trichloroethene- *in vivo* and *in vitro* evidence for activation by glutathione conjugation. *Chem.-Biol. Interact.* **73,** 89–101.

Dekant, W., Vamvakas, S., and Anders, M. W. (1990a). Biosynthesis, bioactivation, and mutagenicity of *S*-conjugates. *Toxicol. Lett.* **53,** 53–58.

Dekant, W., Vamvakas, S., Koob, M., Kochling, A., Kanhai, W., Muller, D., and Henschler, D. (1990b). A mechanism of haloalkene-induced renal carcinogenesis. *Environ. Health Perspect.* **88,** 107–110.

Dohn, D. R., and Anders, M. W. (1982). The enzymatic reaction of chlorotrifluoroethylene with glutathione. *Biochem. Biophys. Res. Commun.* **109,** 1339–1345.

Dohn, D. R., Quebbemann, A. J., Borch, R. F., and Anders, M. W. (1985). Enzymatic reaction of chlorotrifluoroethene with glutathione ^{19}F NMR evidence for stereochemical control of the reaction. *Biochemistry* **24,** 5137–5143.

Friedberg, T., Bentley, P., Stasiecki, P., Glatt, H. R., Raphael, D., and Oesch, F. (1980). Identification, solubilization, and characterization of microsome associated glutathione *S*-transferases. *J. Biol. Chem.* **254,** 12028–12033.

Habig, W. H., Pabst, M. J., and Jakoby, W. B. (1974). Glutathione *S*-transferases. The first enzymatic step in mercapturic acid formation. *J. Biol. Chem.* **249,** 7130–7139.

Haenen, G. R. M. M., Vermeulen, N. P. E., Jacintha, N. L., Tsoi, T. T., Ragetli, H. M. N., Timmerman, H., and Bast, A. (1988). Activation of the microsomal glutathione *S*-transferase and reduction of the glutathione dependent protection against lipid peroxidation by acrolein. *Biochem. Pharmacol.* **37,** 1933–1938.

Haenen, G. R. M. M., Jansen, F. P., Vermeulen, N. P. E., and Bast, A. (1991). Activation of the microsomal glutathione *S*-transferase by metabolites of alpha-methyldopa. *Arch. Biochem. Biophys.* **287,** 48–52.

Hargus, S. J., Fitzsimmons, M. E., Aniya, Y., and Anders, M. W. (1991). Stereochemistry of the microsomal glutathione *S*-transferase-catalyzed addition of glutathione to chlorotrifluoroethene. *Biochemistry* **30,** 717–721.

Horbach, M. E., Sies, H., and Akerboom, T. P. M. (1993). Identification of a hepatic plasma membrane glutathione *S*-transferase activated by *N*-ethylmaleimide. *Biochim. Biophys. Acta,* **1148,** 61–66.

Kanhai, W., Dekant, W., and Henschler, D. (1989). Metabolism of the nephrotoxin dichloroacetylene by glutathione conjugation. *Chem. Res. Toxicol.* **2,** 51–56.

Koob, M., and Dekant, W. (1990). Metabolism of hexafluoropropene. Evidence for bioactivation by glutathione conjugate formation in the kidney. *Drug Metab. Dispos.* **18,** 911–916.

Koob, M., and Dekant, W. (1991). Bioactivation of xenobiotics by formation of toxic glutathione conjugates. *Chem.-Biol. Interact.* **77,** 107–136.

Lee, C. Y., and McKinney, J. D. (1982). Identity of microsomal glutathione *S*-transferases. *Mol. Cell. Biochem.* **48,** 91–96.

Lundqvist, G., and Morgenstern, R. (1992a). Studies on the activation of rat liver microsomal glutathione transferase in isolated hepatocytes. *Biochem. Pharmacol.* **43,** 131–135.

Lundqvist, G., and Morgenstern, R. (1992b). Mechanism of activation of rat liver microsomal glutathione transferase by noradrenaline and xanthine oxidase. *Biochem. Pharmacol.* **43,** 1725–1728.

Lundqvist, G., Yucel-Lindberg, T., and Morgenstern, R. (1992). The oligomeric structure of rat liver microsomal glutathione transferase studied by chemical cross-linking. *Biochim. Biophys. Acta* **1159,** 103–108.

Mannervik, B. (1985). The isoenzymes of glutathione transferase. *Adv. Enzymol. Relat. Areas Mol. Biol.* **57,** 357–417.

Mannervik, B., and Danielson, U. H. (1988). Glutathione transferases—Structure and catalytic activity. *CRC Crit. Rev. Biochem.* **23,** 283–337.

Masukawa, T., and Iwata, H. (1986). Possible regulation mechanism of microsomal glutathione *S*-transferase activity in rat liver. *Biochem. Pharmacol.* **35,** 435–438.

McLellan, L. I., Wolf, C. R., and Hayes, J. D. (1989). Human microsomal glutathione *S*-transferase, Its involvement in the conjugation of hexachloro-1.3-butadiene with glutathione. *Biochem. J.* 258, 87–93.

Morgenstern, R., and DePierre, J. W. (1983). Microsomal glutathione transferase, Purification in unactivated form and further characterization of the activation process, substrate specificity and amino acid composition. *Eur. J. Biochem.* **134,** 591–597.

Morgenstern, R., and DePierre, J. W. (1985). Microsomal glutathione transferases. *In* "Reviews in Biochemical Toxicology" (E. Hodgson, J. R. Bend, and R. M. Philpot, eds.), pp. 67–84. Elsevier, New York.

Morgenstern, R., and DePierre, J. W. (1988). Membrane-bound glutathione transferases. *In* "Glutathione Conjugation: Its Mechanism and Biological Significance" (B. Ketterer and H. Sies, eds.), pp. 157–174. Academic Press, London.

Morgenstern, R., DePierre, J. W., and Ernster, L. (1979). Activation of microsomal glutathione transferase activity by sulfhydryl reagents. *Biochem. Biophys. Res. Commun.* **87,** 657–663.

Morgenstern, R., Meijer, J., DePierre, J. W., and Ernster, L. (1980). Characterization of rat liver microsomal glutathione transferase activity. *Eur. J. Biochem.* **104,** 167–174.

Morgenstern, R., Guthenberg, C., and DePierre, J. W. (1982). Microsomal glutathione transferase. Purification, initial characterization and demonstration that it is not identical to the cytosolic glutathione transferases A, B and C. *Eur. J. Biochem.* **128,** 243–248.

Morgenstern, R., Guthenberg, C., Mannervik, B., and DePierre, J. W. (1983). The amount and nature of glutathione transferases in rat liver microsomes determined by immunochemical methods. *FEBS Lett.* **160,** 264–268.

Morgenstern, R., Lundqvist, G., Andersson, G., Balk, L., and DePierre, J. W. (1984). The distribution of microsomal glutathione transferase among different organelles, different organs and different organisms. *Biochem. Pharmacol.* **33,** 3609–3614.

Morgenstern, R., DePierre, J. W., and Jörnvall, H. (1985). Microsomal glutathione transferase, primary structure. *J. Biol. Chem.* **260,** 13976–13983.

Morgenstern, R., Lundqvist, G., Hancock, V., and DePierre, J. W. (1988). Studies on the activity and activation of rat liver microsomal glutathione transferase, in particular with a substrate analogue series. *J. Biol. Chem.* **263,** 6671–6675.

Morgenstern, R., Lundqvist, G., Jörnvall, H., and DePierre, J. W. (1989). Activation of rat liver microsomal glutathione transferase by limited proteolysis. *Biochem. J.* **260,** 577–582.

Mosialou, E., and Morgenstern, R. (1989). Activity of rat liver microsomal glutathione transferase towards products of lipid peroxidation and studies on the effect of inhibitors on glutathione-dependent protection against lipid peroxidation. *Arch. Biochem. Biophys.* **275,** 289–294.

Mosialou, E., and Morgenstern, R. (1990). Inhibition studies on rat liver microsomal glutathione transferase. *Chem.-Biol. Interact.* **74,** 275–280.

Mosialou, E., Andersson, C., Lundqvist, G., Andersson, G., Bergman, T., Jörnvall, H., and Morgenstern, R. (1993a). Human liver microsomal glutathione transferase - Substrate specificity and important protein sites. *FEBS Lett.* **315,** 77–80.

Mosialou, E., Ekström, G., Adang, A. E. P., and Morgenstern, R. (1993b). Evidence that rat liver microsomal glutathione transferase is responsible for glutathione dependent protection against lipid peroxidation. *Biochem. Pharmacol.* **45,** 1645–1651.

Nishino, H., and Ito, A. (1989). Increase in glutathione disulfide level regulates the activity of microsomal glutathione *S*-transferase in rat liver. *Biochem. Int.* **19,** 731–735.

Nishino, H., and Ito, A. (1990). Subcellular distribution of *N*-ethylmaleimide-stimulatable glutathione *S*-transferase activity in rat liver—Evidence of localization of glutathione *S*-transferase in peroxisomal membrane. *Biochem. Int.* **21,** 331–340.

Odum, J., and Green, T. (1984). The metabolism and nephrotoxicity of tetrafluoroethylene in the rat. *Toxicol. Appl. Pharmacol.* **76,** 306–318.

Oesch, F., and Wolf, C. R. (1989). Properties of the microsomal and cytosolic glutathione transferases involved in hexachloro-1 : 3-butadiene conjugation. *Biochem. Pharmacol.* **38,** 353–359.

Pallante, S. L., Lisek, C. A., Dulik, D. M., and Fenselau, C. (1986). Immobilized enzyme synthesis and characterization by fast atom bombardment mass spectrometry. *Drug Metab. Dispos.* **14,** 313–318.

Vamvakas, S., Kremling, E., and Dekant, W. (1989). Metabolic activation of the nephrotoxic haloalkene 1,1,2-trichloro-3,3,3-trifluoro-1-propene by glutathione conjugation. *Biochem. Pharmacol.* **38,** 2297–2304.

Wallin, A., Gerdes, R. G., Morgenstern, R., Jones, T. W., and Ormstad, K. (1988). Features of microsomal and cytosolic glutathione conjugation of hexachlorobutadiene in rat liver. *Chem.-Biol. Interact.* **68,** 1-12.

Wallin, H., and Morgenstern, R. (1990). Activation of microsomal glutathione transferase activity by reactive intermediates formed during the metabolism of phenol. *Chem.-Biol. Interact.* **75,** 185–199.

Weinander, R., Andersson, C., and Morgenstern, R. (1993). Identification of *N*-Acetylcysteine as a new substrate for rat liver microsomal glutathione transferase. A study of thiol ligands. *J. Biol. Chem.* (in press).

Wolf, C. R., Berry, P. N., Nash, J. A., Green, T., and Lock, E. A. (1984). Role of microsomal and cytosolic glutathione *S*-transferases in the conjugation of hexachloro-1 : 3-butadiene and its possible relevance to toxicity. *J. Pharmacol. Exp. Ther.* **228,** 202–208.

Enzymology of Cytosolic Glutathione S-Transferases

Brian Ketterer and Lucia G. Christodoulides
Department of Biochemistry and Molecular Biology
University College London
London W1P 6DB, United Kingdom

I. Introduction

The glutathione *S*-transferases (GSTs) catalyze reactions between glutathione and a number of electrophiles, mostly of xenobiotic origin. The substrates have in common a degree of hydrophobicity and possess electrophilic centers that undergo nucleophilic substitution, nucleophilic addition to α,β-unsaturated ketones or epoxides, or, in the case of hydroperoxides, nucleophilic attack on electrophilic oxygen resulting in reduction. In addition, they catalyze certain glutathione-dependent isomerizations and behave as binding proteins for a range of nonsubstrate ligands. There is a remarkable variety of structures among substrates and ligands (Fig. 1). A number of enzymes have evolved to deal with them. The enzymes tend to have broad substrate specificities, but are often particularly active toward specific substrates (Ketterer *et al.*, 1988). (This is illustrated in Tables I–IV.)

The attack of electrophiles on cellular nucleophiles is frequently associated with a toxic outcome. The more polarized the charge in the electrophile (i.e., the "harder" the electrophile), the more likely it is to attack either nucleophilic oxygen or nitrogen in DNA, with the potential for causing genotoxicity. With hard electrophiles, the spontaneous rate of their reaction with glutathione (GSH), even at normal cellular concentrations of GSH, is low, with the result that GSTs are especially valuable

Advances in Pharmacology, Volume 27

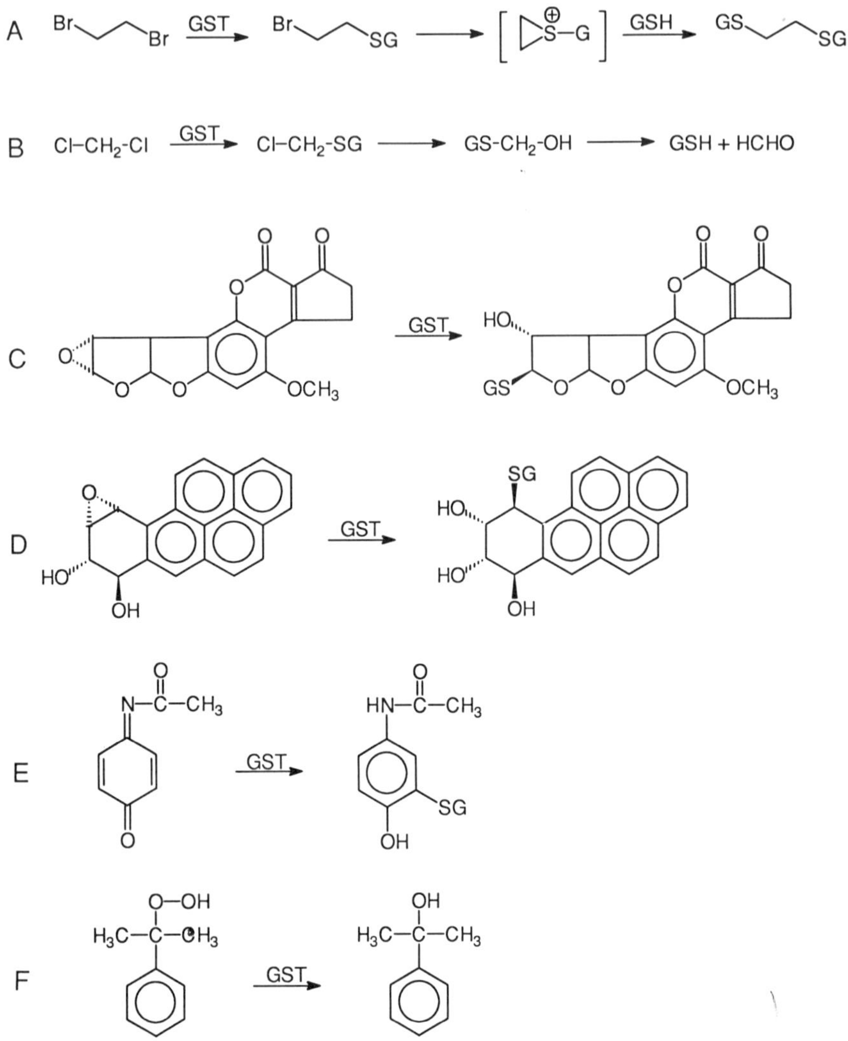

Fig. 1 Examples of GST-catalyzed reactions. The following substrates are illustrated (a) dibromethane; (b) dichloromethane; (c) *exo*-aflatoxin B_1-8,9-oxide; (d) (+)-*anti*-benzo[*a*]pyrene-7,8-diol-9,10-oxide; (e) *N*-acetylbenzoquinone imine; (f) cumene hydroperoxide.

in conferring protection. For this reason, GSTs are regarded as being particularly important in anticarcinogenesis (Coles, 1984; Coles and Ketterer, 1990). Hard electrophiles may also react with nucleophilic nitrogen or thioether sulfur in proteins. This reaction, although frequently highly

Table I

Rat Glutathione Transferases: Nucleophilic Displacement

Substrate	GSH transferase isoenzyme										
	1-1	2-2	3-3	4-4	5-5	6-6 (μmol/min/mg)	7-7	8-8	10-10	12-12	13-13
CDNB	40.0	38.0	50.0	20.0	<0.5		20.0	10.0		<0.5	82
NQO[a]	nil	nil	4.8	5.1		9.9		2.6			
DCP[b]	0.004	0.008	0.017	0.001	nil		0.001	0.002		0.008	
EDB[c]	0.011	0.117	0.070	0.021	+ + +		0.003				
DCM[d]	nil	nil	nil	nil	11.0						
5-SMC[e]										0.08	
7SM-12MB[e]										0.15	
7SM-12OHMB[e]										0.05	

Note. Abbreviations used: CDNB, 1-chloro-2,4-dinitrobenzene; NQO, nitroquinoline *N*-oxide; DCP, 1,2-dibromo-3-chloropropane; EDB, ethylene dibromide; DCM, methylene dichloride; 5-SMC, 5-sulfonyloxymethyl chrysene; 7SM-12MB, 7-sulfonyloxymethyl, 12-methylbenzanthracene; 7SM-12OHMB, 7-sulfonyloxymethyl, 12-hydroxymethylbenzanthracene.

[a] Aceto *et al*., 1990.

[b] E. Soderland and D. J. Meyer, unpublished data.

[c] Cmarik *et al*., 1990.

[d] Meyer *et al*., 1991a.

[e] Hiratsuka *et al*., 1990.

Table II

Rat Glutathione Transferases: Conjugation of Epoxides

Substrate	GSH transferase isoenzyme (μmol/min/mg)										
	1-1	2-2	3-3	4-4	5-5	6-6	7-7	8-8	10-10	12-12	13-13
exo-AFBO[a]	0.002	0.0005	0.0013	0.0015			nil	nil	0.10		
endo-AFBO[b]	nil	nil	nil	0.014			nil	nil	nil		
BP-4,5-O[c]	0.011	0.004	0.087		0.069						
(+)*anti*-BPDE[d]		0.006	0.012	0.7			4.5				
(−)*anti*-BPDE[d]		0.0012	0.004	0.007			nil				
1-NP-4,5-O[e]	0.01	0.03	0.30	0.30			0.02				
1-NP-9,10-O[e]	0.06	0.01	0.40	0.20			0.08				
Ch-5,6-O[f]	0.0001		nil	nil							

Note. Abbreviations used: *exo*-AFBO, aflatoxin B_1-8,9-*exo*-oxide; *endo*-AFBO, aflatoxin B_1-8,9-*endo*-oxide; BP-4,5-0, benzo[a]pyrene-4,5-oxide; (+)*anti*-BPDE, (+)*anti* benzo[a]pyrene-7,8-diol-9,10-oxide; (−)*anti*-BPDE, (−)*anti*-benzo[a]pyrene-7,8-diol-9,10-oxide; 1-NP-4,5-0, 1-nitropyrene-4,5-oxide; 1-NP-9,10-0, 1-nitropyrene-9,10-oxide; Ch-5,6-O, cholesterol-5,6-oxide.

[a,b] Raney *et al.*, 1992.

[c] Nemoto *et al.*, 1975.

[d] Jernström *et al.*, 1985; Robertson *et al.*, 1986.

[e] Djuric *et al.*, 1987.

[f] Meyer and Ketterer, 1982.

Table III

Rat Glutathione Transferases: Michael Addition

	GSH transferase isoenzyme										
Substrate	1-1	2-2	3-3	4-4	5-5	6-6	7-7 (μmol/min/mg)	8-8	10-10	12-12	13-13
NABQI[a]	108	48.0	6.0	3.0			190				
ETA[b]	0.3	2.1	0.4	1.0			4.0	7.0			26[c]
HNE[b]	2.6	0.7	2.7	6.9				170.0			nil[c]

Note. Abbreviations used: NABQI, *N*-acetyl-*p*-benzoquinone imine; ETA, ethacrynic acid; HNE, 4-hydroxynon-2-enal.

[a] Coles *et al.*, 1988.

[b] Coles and Ketterer, 1990.

[c] Harris *et al.*, 1991.

specific, may not always have striking effects on biological activity (Coles *et al.*, 1987).

The more polarizable the electrons in the electrophile (*i.e.*, the "softer" the electrophile), the more likely it is to attack protein thiols causing cytotoxicity, due to the frequent involvement of thiols in the biological activity of proteins. Glutathione reacts so readily with soft electrophiles (e.g., *N*-acetylbenzoquinone imine, the toxic metabolite of paracetamol; Coles *et al.*, 1988) that at normal cellular concentrations of glutathione (2–10 m*M*) the spontaneous reaction is sufficient to provide significant

Table IV

Rat Glutathione Transferases: Products of Peroxidation

	GSH transferase isoenzyme										
Substrate	1-1	2-2	3-3	4-4	5-5	6-6	7-7 (μmol/min/mg)	8-8	10-10	12-12	13-13
LAOOH[a]	3.0	1.6	0.2	0.2	5.3		1.5				nil
DNAOOH[a]	nil	nil	0.02	0.03	0.03		0.01				nil
CuOOH[a]	1.4	3.0	0.1	0.4			<0.02	1.1			nil
PGH_2-PGE_2[b]	1.4	0.42	0.057	0.092			0.063	0.033			

Note. Abbreviations used: LAOOH, linoleic acid hydroperoxide; DNAOOH, DNA hydroperoxide; CuOOH, cumene hydroperoxide; PGH_2-PGE_2, prostaglandin H2 → E2.

[a] Ketterer and Meyer, 1989.

[b] Ujihara *et al.*, 1988.

protection from toxicity; however, should GSH become depleted, the GSTs are much more important in detoxication.

II. The GST Supergene Family

In mammals, cytoprotection provided by GSTs is made available by at least four multigene families referred to as alpha, mu, pi, and theta (Coles and Ketterer, 1990; Armstrong, 1991; Meyer *et al.*, 1991a). Our knowledge of these gene families depends largely on work with the rat and humans. As with other gene superfamilies, there have been problems in arriving at a common nomenclature. Several nomenclatures exist for the rat, but the one most favored gives each GST subunit (the GSTs are dimers) a number in the chronological order of its structural characterization (Jakoby *et al.*, 1984; see Table V). However, this nomenclature is insufficiently informative in that it does not indicate the gene family to which the subunit belongs. This is not the case with the new nomenclature for human GSTs, which gives each gene or gene product a number (largely determined by the chronology of its characterization by cloning) preceded by the designation of the gene family to which it belongs (see Table VI; Mannervik *et al.*, 1992).

The identity in amino acid sequence within the GST gene families is considerable (at least 70%) and between gene families is sufficient (20–30%) to confirm their common origin. The alpha, mu, and pi families seem

Table V
Rat Glutathione Transferases Nomenclature by Subunit[a]

Family			
alpha	mu	pi	theta
1a (Ya_1)[b]	3 (Yb_1)	7 (Yf, Yp)	5
1b (Ya_2)	4 (Yb_2)		12 (Y_{rs})
2 (Yc)	6 (Yn_1)		13[c]
8 (Yk)	9 (Yn_2)		
10 (Yc_2)	11 (Yo)		

[a] This nomenclature was devised by Jakoby *et al.* (1984).
[b] The Ya, etc., is an alternative nomenclature originated by Bass *et al.* (1977).
[c] Specific location in mitochondrial matrix.

Table VI
Human Glutathione Transferase Nomenclature by Subunit

Family			
alpha	mu	pi	theta
GSTA1[a] (B_1, ε, α_x)[b]	GSTM1a (μ, GST1)	GSTP1 (π)	GSTT1 (θ)
GSTA2 (B_2, δ, α_y)	GSTM1b (ψ)		GSTT2
	GSTM1 null (μ null, GST 1 0)		
	GSTM2		
	GSTM3		

[a] This nomenclature was devised by Mannervik *et al.* (1992).
[b] This nomenclature is from a range of sources.

to be restricted to fungi and animals, but the theta family is much more widespread, having been detected in both the plant and the animal kingdoms and also in bacteria. It appears to be a GST prototype (Pemble and Taylor, 1992). Recent analysis of the primary structure of a GST from the bacterium *Proteus mirabilis* shows that it represents yet another related gene family; however, the phylogenetic distribution of this new family is unknown (Mignogna *et al.*, 1993).

III. Structure of GSTs Including Their Active Sites

That GSTs are dimers has been known for some time (Ketterer and Christodoulides, 1969) and also that within each gene family homo- and heterodimers may form (Mannervik and Danielson, 1988; Coles and Ketterer, 1990). The first GST to have its three-dimensional structure solved by X-ray crystallography was a porcine pi GST gene (Reinemer *et al.*, 1991). Subsequently structures of rat GST3-3 (Ji *et al.*, 1992), human GSTP1-1 (Reinemer *et al.*, 1992), and human GSTA1-1 (Sinning *et al.*, 1993) have also been published.

Despite considerable differences in amino acid sequence, the conformations of these three GSTs are very similar (Fig. 2). All subunits have two domains, with the active site lying between them. The smaller *N*-terminal domain (domain 1) contains the glutathione binding site and part of the electrophile binding site. The larger, domain 2, contributes further to electrophilic substrate binding, particularly in the region of the C-terminus. Domain 1 includes residues 1–82 forming four β-strands and 3 α-helices in the folding pattern βαβαββα. Domain 2, which extends to

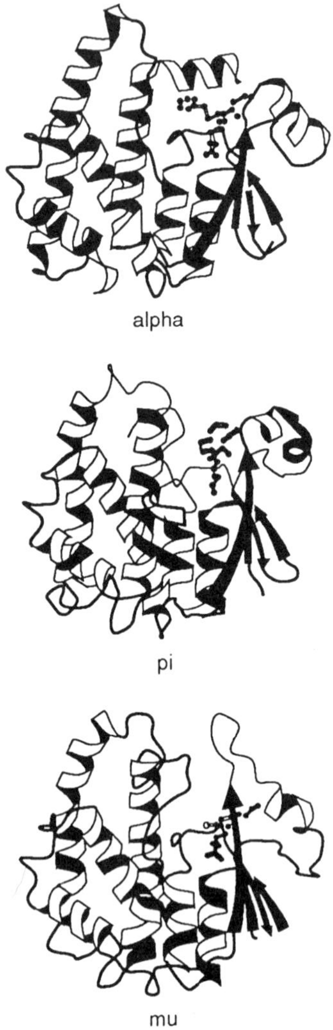

Fig. 2 The conformation of GST subunits of the alpha, mu, and pi classes showing bound conjugate at the active sites (Sinning *et al.*, 1993).

residues 207, 217, and 222 in GSTP1-1, GST3-3, and GSTA1-1 respectively, is completely helical.

A high affinity for glutathione is a characteristic of enzymes in the alpha, mu, and pi families (Meyer, 1993). The amino and carboxylate groups of the γ-glutamyl residue, the carboxylate group of the glycyl residue, the cysteinyl sulfur and the peptide backbone are all involved in binding. For example, in GSTP1-1 the results of X-ray crystallographic analysis show that the γ-glutamyl carboxylate group interacts with the side chains of Arg_{13} and Ser_{63}, the γ-glutamyl amino group with the side chains of Gln_{64} and Asp_{98B}, (*i.e.*, Asp_{98} in the other subunit), the cysteinyl sulfur with Tyr_7, and the carboxylate of glycine with Trp_{38}, Lys_{44}, and Gln_{51}. In addition, a short antiparallel β-pleated sheet results from the interaction of amide nitrogen and the carboxyl oxygen of Leu_{52} with the amide nitrogen and carboxyl oxygen of the cysteinyl group (Reinemer *et al.,* 1992). The essential nature of interactions with Tyr_7, Arg_{13}, Glu_{62}, and Asp_{98} have been confirmed by site-directed mutagenesis (Manoharan *et al.*, 1992) and in the case of Tyr_7, Arg_{13}, Lys_{44}, and Asp_{98} by chemical modification (Xia *et al.*, 1993; Meyer *et al.*, 1993a).

Most interesting is Tyr_7. The hydroxyl moiety of this tyrosine interacts with the thiol of glutathione and is important for its activation. Site-directed mutagenesis substituting phenylalanine for this tyrosine causes almost complete loss of activity (Manoharan *et al.*, 1992; Kolm *et al.*, 1992), as does loss of the the homologous Tyr residues in GSTA1, and rat GST3 (Stenberg *et al.*, 1991; Liu *et al.*, 1992). Diethylpyrocarbonate (DEPC), a reagent normally used to modify histidine residues, also reacts with Tyr_7 in the native protein (the same residue in the isolated N-terminal undecapeptide is unreactive). The reaction of DEPC with the enzyme causes complete loss of activity together with the loss of a second-derivative spectrum associated with tyrosinate anion, but interestingly, glutathione, at the active site, causes little inhibition of this reaction (Meyer *et al.*, 1993a).

Most important in the enzyme mechanism of GSTs is the reduction of the pK_a of glutathione thiol from approximately 9, the value in free solution, to a value in the region of neutrality at the active site, with the result that glutathione bound to the enzyme possesses the much more nucleophilic thiolate anion. Tyr_7 appears to be largely responsible for this essential reduction in pK_a. Ji *et al.* (1992), who determined the three-dimensional structure of rat GST3-3, proposed three possible mechanisms whereby Tyr could assist the ionization of glutathione: (*a*) the hydrogen bond Tyr_6-OH-H · · $^-$SG assists stabilization of the thiolate anion at the active site; (*b*) Tyr_6-O^- · · H-S-G occurs with a very low barrier to proton transfer; and (*c*) the proton resides in a single-well potential Tyr_6–O^- · · H^+ · · $^-$SG. In the case of GSTP1-1, option (*b*) is favored.

When the amino acid residues essential to the active site in rat GST3-3 and GSTA1-1 are compared with those in GSTP1-1, little is found to be conserved (Ji *et al.*, 1992; Sinning *et al.*, 1993). All the charged groups of glutathione are involved in salt links, but only those equivalent to Asp_{98B}

and Gln_{64} are identical. As with GSTP1-1, glutathione in GSTA1-1 also forms two major chain hydrogen bonds with the enzyme protein, whereas in GST3-3, glutathione has a somewhat different conformation such that its cysteinyl carbonyl hydrogen bond arises from a Trp side chain. However, in all enzymes the tyrosine equivalent to Tyr_7 in GSTP1-1 interacts with the cysteinyl sulfur. A similar mechanism of activation of glutathione appears to occur in each of these three classes and probably also in the theta class where tyrosine residues occur in position 5 in both rat GSTs 5 and 12 (Meyer, 1993) (see Fig. 3).

Because of the wide ranging substrate specificity of the electrophile binding site, its structure is presumed to be less defined. Contacts are not as intimate or specific as for glutathione (Reinemer *et al.*, 1992). In GST3-3, the active-site cleft is deeper, and as with GSTP1-1, both the immediate surroundings of Tyr_6 and the C-terminal domain interact with the electrophilic substrate (Ji *et al.*, 1992). GSTA1-1 differs somewhat in that its longer polypeptide chain results in a C-terminal α-helix containing two conserved Phe residues, which packs onto and helps form a hydrophobic binding site less available to solvent than of those of the other two enzymes (Sinning *et al.*, 1993).

The interface between subunits in the dimer is relatively extensive and has a twofold axis. Interactions occur between domain 1 in one monomer and domain 2 in the other and are more numerous in the periphery. The interacting structures tend to be conserved within each of the classes, thus enabling the formation of both hetero- and homodimers (Sinning *et al.*, 1993).

IV. Enzyme Kinetics

A characteristic of reactions catalyzed by GSTs of the alpha, mu, and pi family is product inhibition. This appears to be associated with low K_m values (i.e., high binding affinity) for both the glutathione and electrophilic substrates. However, in theta enzymes, where the K_m values for GSH appear to be much higher, product inhibition may be absent (Meyer, 1993).

Table VII shows the K_m values for GSH of a number of GSTs. With 1-chloro-2,4-dinitrobenzene as a substrate, it is seen that with the alpha, mu, and pi class enzymes, the K_m values range from 81 to 160 μM, but with rat GST13-13, a theta class enzyme, the K_m is 1900 μM, an order of magnitude higher. The K_m value for GSH may vary with the nature of the electrophilic substrate; thus GSTM1a-1a has a K_m value of 95 μM with 1-chloro-2,4-dinitrobenzene but of a K_m value an order of magnitude greater with *p*-nitrobenzyl chloride. Rat GST5-5 does not utilize 1-chloro-

GST subunit	C-SH / E-COO⁻	G-COO⁻	-HN-C-CO-	E-COO⁻	E-NH$_3^+$
hP1 (π)	V V Y F P V - - - R G R C [14]	T W Q E G S L K A S	C - L Y G - Q L P [53]	S N A I L R [70]	Q E A A L V D M V N D G V E D L R [100]
r2 (α)	L H Y F D G - - - R G R M [15]	R D D L A R L R N D	G S L M F Q Q V P [55]	T R A I L N [72]	K E R A L I D M Y A E G V A D L D [102]
r3 (μ)	L G Y W N V - - - R G L A [13]	R S Q W L N E K F K	L G L D F P N L P [60]	S N A I M R [77]	E E R I R A D I V E N Q V M D N R [107]
r5 (θ)	E L Y L D L L S Q P C R A [15]	K G E H L S D A F A	Q V N P M K K V P [54]	S V A I L L [71]	Q A R A R V D E Y L A W Q H T T L [105]
r12 (θ)	E L Y L D L L S Q PC/SR A [15]	K G Q H L S E Q F S	Q V N C L K K V P [54]	S T A I L I [71]	Q A R A Q V H E Y L G W H A D N I [105]

Fig. 3 Interactions between glutathione and several GSTs at the active site. C — SH is the cysteinly thiol of glutathione; E — COO^- is the α-glutamyl carboxylate; G — COO^- is the glycyl carboxylate; —HN — CO^- is the peptide chain contributed by the cysteinyl residue; E — NH_3^+ is the glutamyl α-ammonium ion.

Table VII

The K_m Values for Glutathione [a]

GST	K_m (GSH) (μM)	Electrophile
P1-1	110	1-Chloro-2,4-dinitrobenzene
A1-1	81	1-Chloro-2,4-dinitrobenzene
A2-2	160	1-Chloro-2,4-dinitrobenzene
M1a-1a	95	1-Chloro-2,4-dinitrobenzene
Rat GST 13-13[b]	1900	1-Chloro-2,4-dinitrobenzene
M1a-1a	820	*p*-Nitrobenzyl chloride
Rat 5-5	4200	*p*-Nitrobenzyl chloride
Rat 5-5	5900	*p*-Nitrophenethyl bromide

[a] Meyer, 1993.
[b] Harris *et al.*, 1991.

2,4-dinitrobenzene as a substrate, but the high K_m values for GSH associated with the theta family are seen with *p*-nitrobenzyl chloride and *p*-nitrophenethyl bromide. Figure 3 gives a rationale for the low affinity of rat GST5-5 for GSH, namely the lack of Lys (or Arg) associated with binding of the glycyl carboxylate residue of glutathione in the other GSTs.

A high affinity for GSH is useful in laboratory practice since it enables GSH affinity matrices to isolate alpha, mu, and pi class GSTs from complex protein mixtures. Furthermore, because their affinities differ to some extent, they can be differentially eluted by a glutathione concentration gradient (Hayes, 1988). This advantage in purification does not pertain to theta enzymes.

The K_m values for electrophilic substrates range from 0.7 to 600 μM (Table VIII). Some substrates important in toxicology, such as *N*-acetylbenzoquinone imine, 1-nitropyrene-4,5-oxide, 4-hydroxynonenal, and ($\pm$)*anti*-benzo[*a*]pyrene-7,8-diol-9,10-oxide, have very low K_m values, which should maintain low intracellular levels of the free electrophiles and therefore be advantageous in detoxication.

In GSTs, low K_m values for GSH and/or electrophilic substrates, lead to product inhibition and to the accumulation of GST-bound GSH conjugates *in vivo*. The GSTs appear to behave as reservoirs for both GSH conjugates and nonsubstrate ligands, presumably releasing them slowly for export in the case of the conjugates or, in the case of nonsubstrate ligands, such as heme and bilirubin, for further metabolism in the cell. It is noteworthy that the first GST to be purified, rat GST1-1, was isolated from livers given the carcinogen 3′-methyl-*N*,*N*-dimethyl-4-aminoazobenzene. It was originally believed to be a protein–carcinogen adduct; how-

Table VIII

Kinetic Constants for Some Electrophilic Substrates

Substrate	Isoenzyme	k_{cat} (s^{-1})	K_m ($M \times 10^{-6}$)	k_{cat}/K_m ($M^{-1} \cdot s^{-1}$)
1-Chloro-2,4-dinitrobenzene[1]	3-3	50	60.0	8.3×10^5
N-Acetyl-benzoquinone imine[2]	2-2	35	0.7	5×10^7
4-Hydroxynon-2-enal[3]	8-8	—	—	4.2×10^6
1-Nitropyrene-4,5-oxide[4]	4-4	0.3	1.0	3×10^5
Linoleate hydroperoxide[5]	1-1	3.2	20.0	1.6×10^5
α-Bromoisovaleryl urea[6]	2-2	0.1	600	167
(±)*anti*-benzo(*a*)pyrene-7,8-diol-9,10-oxide[7]	7-7	2.45	14.0	1.8×10^5

Note. Data are from [1] Habig *et al.*, 1974; [2] Coles *et al.*, 1988; [3] Danielson *et al.*, 1987; [4] Djuric *et al.*, 1987; [5] D. J. Meyer, unpublished information; [6] te Koppele *et al.*, 1988; [7] Robertson *et al.*, 1986.

ever, it transpires that the dye was not covalently bound, but that a GSH conjugate formed by spontaneous reaction of the electrophilic metabolite of the arylamine and GSH was bound noncovalently with high affinity (Ketterer *et al.*, 1971). *N*-Sulfonyloxy-*N*-methyl-4-aminoazobenzene, the putative carcinogenic electrophile, does not appear to be a substrate for GSTs.

Intrinsically, the GSTs have random sequential kinetics (Jakobson *et al.*, 1979) but *in vivo* at physiologically concentrations of glutathione, the glutathione binding site is always occupied.

V. Gene Structure and Gene Regulation

cDNA clones have been obtained for the GSTs shown in Tables V and VI and information available regarding genomic structure is beginning to accumulate. Genes for rat GST subunits 1b (Telakowski-Hopkins *et al.*, 1986), 3 (Morton *et al.*, 1990), 4 (Lai *et al.*, 1988), and 7 (Okuda *et al.*, 1987) and for human GST subunits A1 (Klöne *et al.*, 1992; Rozen *et al.*, 1992), A2 (Röhrdanz *et al.*, 1992), M1 (Zhong *et al.*, 1993), M2 (Taylor *et al.*, 1991), M4 (Taylor *et al.*, 1991; Zhong *et al.*, 1993), "M5" (Pemble *et al.*, 1993), and P1 (Cowell *et al.*, 1988; Morrow *et al.*, 1989) are known. "M5" is enclosed in quotation marks because the enzymes it codes for have yet to be isolated. As yet, no information regarding gene structure is available for the theta family.

The regulation of these genes is of considerable interest for several reasons: (*a*) There is considerable variation in the distribution of GSTs from one tissue to another (Ketterer *et al.*, 1988) (Fig. 4). (*b*) In any tissue, the distribution of GSTs varies during development; e.g., the fetal rat liver expresses GSTs 7 and 10 but these are of very low occurrence in the normal male rat liver (Tee *et al.*, 1992) (Fig. 5). (*c*) GST distribution

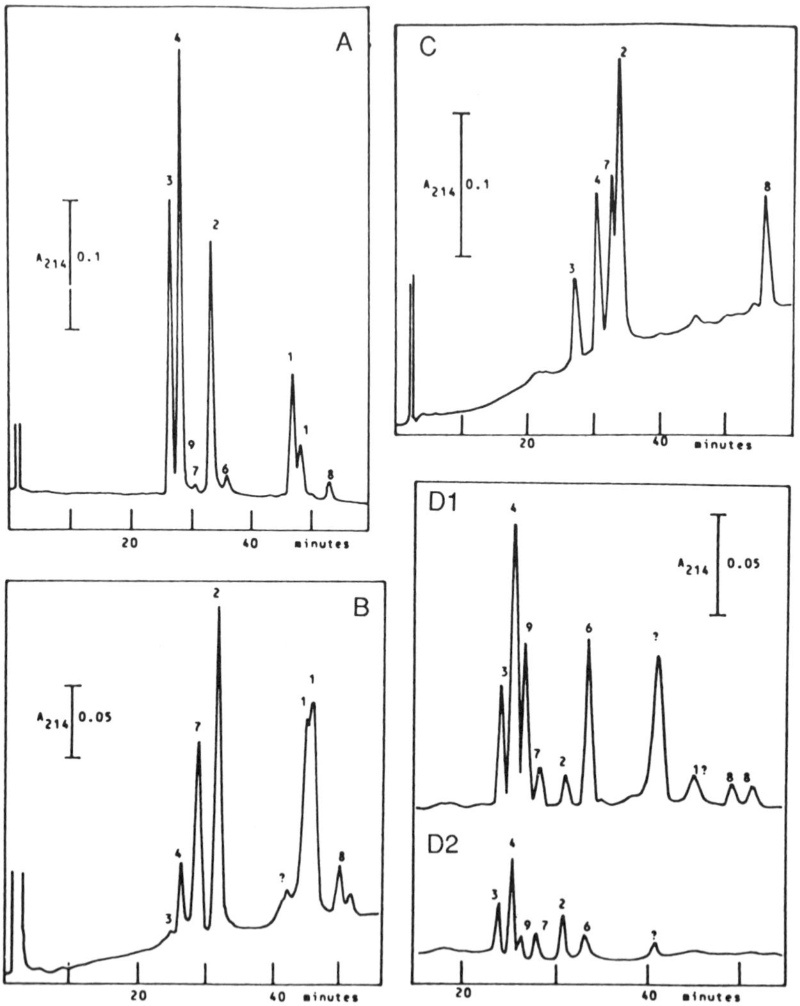

Fig. 4 HPLC analysis of GSTs. (a) liver; (b) kidney; (c) lung; (d)i spermatogenic tubules; (d)ii interstitial cells of the testis.

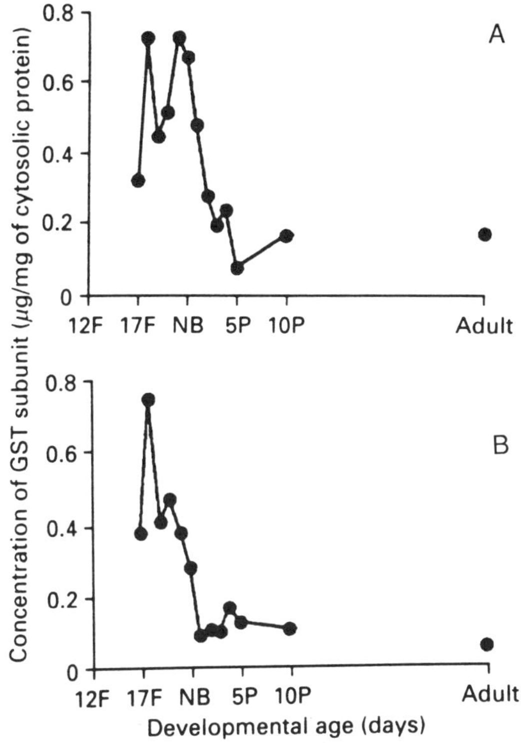

Fig. 5 Expression of GSTs 10–10 and 7–7 in the fetus and the adult (a) GST10-10; (b) GST7-7.

may be determined by sex (Meyer *et al.*, 1993b) (Table IX). (*d*) GST expression changes in tumorigenesis; for example, in rat hepatocarcinogenesis, GST subunit 7 is a marker of preneoplasia, is expressed in liver tumors and is therefore an oncofetal protein (Sato, 1989). The human homologue GSTP1 is also regarded as a marker for a number of tumors (Sato, 1989; Kantor *et al.*, 1991). (*e*) Finally, of considerable importance, a number of GSTs are induced by xenobiotics, the particular subunits affected, varying with the inducer (Meyer *et al.*, 1993b) (Fig. 6). The mechanism of this induction is in part understood. In the case of rat GST subunit 1b (what was originally defined as GST subunit 1 has been shown to comprise two forms, only one of which is inducible; Östlund-Farrants *et al.*, 1987; Hayes *et al.*, 1990), the 5′ untranslated region contains two sequences associated with induction by xenobiotics. One is the xenobiotic responsive element (XRE). This sequence was first identified in *CYP1A1*

Table IX

Differences between Male and Female Rats in the Distribution of GST Isoforms in the Liver

	μg GST/mg hepatic soluble protein	
	Male	Female
	α family	
GST1a	0.7 ± 0.3	0.4 ± 0.05
1b	0.5 ± 0.3	1.6 ± 0.01
2	2.0 ± 0.5	6.0 ± 0.5
8	0.3 ± 0.01	0.6 ± 0.08
10	0.15 ± 0.05	0.4 ± 0.15
	μ family	
3	3.0 ± 0.4	2.5 ± 0.5
4	3.2 ± 0.4	2.8 ± 0.5
6	0.18 ± 0.05	0.16 ± 0.01
11	ND	1.3 ± 0.1
	π family	
7	0.02 ± 0.05	1.0 ± 0.01

and, through the mediation of the Ah or TCDD receptor, is associated with induction by planar aromatic compounds (Paulson *et al.,* 1990), such as polycyclic aromatic hydrocarbons. However, whereas in *CYP1A1*, the XRE is multiple (Denison *et al.*, 1988), *GST1b* has only one. An additional sequence also associated with induction, namely the antioxidant responsive element (ARE), appears to play the more important role (Rushmore *et al.,* 1991).

The ARE sequence, GTGACAAAGC, contains an AP1-like element TGAC and responds to *tert*-butyl-benzoquinone (a redox compound), H_2O_2, and β-naphthoflavone. A similar structure in the *GST4* mouse alpha gene, referred to as the electrophile responsive element, contains two AP1-like sites in tandem, namely TGACATTGCTAATGGTGACAAAGC (Friling *et al.*, 1992), and is also responsive to redox compounds. Induction due to phenobarbital apparently results from the production of oxygen radicals during the course of its metabolism (Pinkus *et al.,* 1993).

The rat pi gene has an important enhancer (GPE1) that is responsive to 12-*O*-tetradecanoyl phorbol-13-acetate and the *ras* oncogene. GPEI also contains AP1-like binding motifs in a region of partial diad symmetry namely AGTCAGTCACTATGATTCAG (Sakai *et al.*, 1988; Okuda *et al.*, 1989; Muramatsu *et al.,* 1990). GST7 expression in hepatic preneoplasia

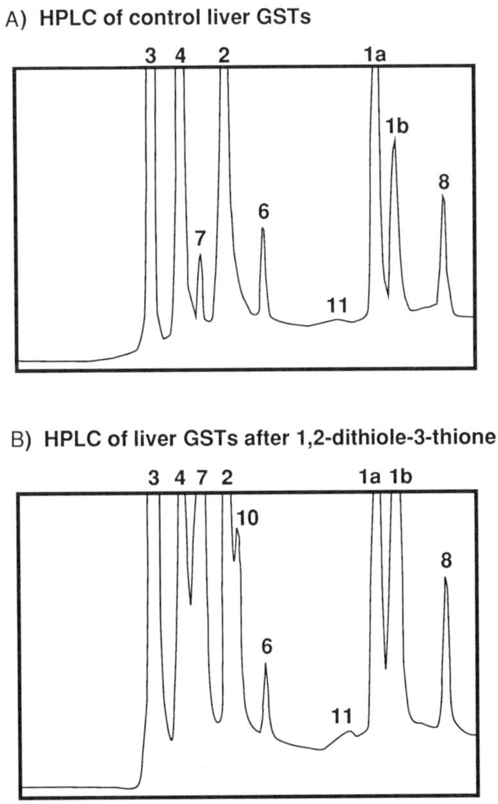

Fig. 6 Induction of rat liver GSTs by 1,2-dithiole-3-thione.

and neoplasia depends on this motif. It could be postulated that the presence of the AP1-like sequences in their associated contexts are associated with common factors responsible for the powerful induction of mouse GST subunit 4 and both rat GST subunits 1b and 7 (Daniel, 1993).

No equivalent to the GPE1 site is found in the human *GSTP1* gene; however, it does have a consensus AP1 motif that is essential for promotion. This motif is associated with retinoic acid responsiveness resulting in down-regulation of transcription (Cowell *et al.*, 1988; Dixon *et al.*, 1989; Xia *et al.*, 1991, 1993). In addition there is a sequence in the first intron that appears to be linked with the promoter and is associated with up-regulation by insulin. A lack of retinoic acid receptors (Gudas, 1992) and an increase in insulin binding sites, both of which may occur in transformed cell lines (O'Brien and Granner, 1991), might be the basis for the observed enhanced *GSTP1* expression in human malignancy. A similar AP1 site also occurs in the promoter region of rat *GST7,* but work on this gene has so far concentrated on upstream regulators, such as GPE1.

VI. Specific Examples Illustrating the Properties of GSTs

A. Aflatoxin B_1-Induced Liver Cancer in Experimental Animals and Humans

In experimental hepatocarcinogenesis, aflatoxin B_1 (AFB_1) is a poor carcinogen in the mouse, but a powerful carcinogen in the rat, provided the rat has not been treated with inducers. The carcinogenic electrophile produced by AFB_1 metabolism is now known to be AFB_1 *exo*-8,9-oxide. In humans, and to a lesser extent the rat, the isomeric *endo*-oxide is also formed but this does not appear to react significantly with DNA and may have little importance in carcinogenesis (Raney *et al.,* 1992; Iyer et al., 1994).

The AFB_1-oxides are potential substrates for GSTs. The low carcinogenicity of AFB_1 in the mouse, relative to the rat, is probably due to the 100-fold greater constitutive activity of its hepatic GSTs, which are two orders of magnitude greater than those in rat liver. The principal activity in mice is the alpha enzyme (mouse GST4-4), which has an activity of 300 $nmol \cdot min^{-1} \cdot mg^{-1}$ (Ramsdell and Eaton, 1990). Male rats exposed to appropriate inducers become much less susceptible to AFB_1-induced carcinogenesis. Kensler and colleagues (1991) have shown that when oltipraz [5-(2-(pyrazinyl)-4-methyl-1,2-dithiol-3-thione] is added to a diet containing levels of AFB_1 sufficient to produce 20% tumorigenesis DNA-bound AFB_1 and the development of preneoplastic nodules were greatly reduced and no tumors developed. Analysis of hepatic GSTs shows that administration of oltipraz and the parent compound 1,2-dithiole-3-thione induces all GSTs present in the normal male adult rat (particularly GST1b) and, in addition, brings about large inductions of GST7-7 and GST10-10, which are associated more with the fetal than the adult state (Fig. 6, Meyer *et al.,* 1993b). However, the induction that is most important in AFB_1 anticarcinogenesis is that of GST10-10, which has activity toward AFB_1-*exo*-oxide similar to that of the mouse GST4-4. GST10-10 has yet to be cloned but the sequence available shows strong structural homology to mouse GST4-4 (Hayes *et al.,* 1991; Meyer *et al.,* 1991b; Buetler and Eaton, 1992).

Hepatocarcinogenesis in man has been associated with the contamination of foodstuffs with AFB_1 and hepatitis B (HBV) infection (Palmer

Beasley *et al.*, 1981; Groopman *et al.*, 1988; Kensler *et al.*, 1991). In a prospective epidemiological study it has been shown that although AFB_1 and HBV might be independent hepatacarcinogens, they are much more powerful in combination (Ross *et al.*, 1992). In this respect it is interesting to note that inhabitants of Chongming, an island off the coast of Shanghai, who originally consumed a maize diet susceptible to *Aspergillus flavus* infestation, had the expected high incidence of hepatocellular carcinoma. When, however, wheat, which is much less readily infected, became the principle grain in the diet, the incidence of liver cancer fell rapidly, a phenomenon associated with the removal of promoting activity (Tu *et al.*, 1985). It is therefore important to know the activity of human hepatic GSTs toward AFB_1-oxides and their inducibility. Human liver contains predominantly the alpha family enzymes GSTA1 and GSTA2 and also, in about 50% of individuals the mu family enzyme, GSTM1. The activity of human liver supernatant fraction is an order of magnitude lower than that of the normal rat, and none of the enzymes present has an activity of the level associated with GST10-10 or the mouse GST4-4.

It is possible to gain some idea of the value of inducers for human anticarcinogenesis with isolated human hepatocytes obtained from biopsy material in culture. Dithiole thiones are found to increase levels of the alpha enzymes, particularly GSTA1, over those in control cells in culture and also to have some effect on GSTM1 (Morel *et al.*, 1993).

The protective effect of inducers is worth investigation in populations at risk. An epidemiological survey has been proposed to study their effect on urinary markers of AFB_1 toxication and detoxication, such as AFB_1-guanine and AFB_1-mercapturates respectively. Induction may be particularly valuable in the case of oltipraz. This compound not only induces GSTs but also cytochromes P450 catalyzing pathways of AFB_1 metabolism, alternative to epoxidation and therefore detoxifying.

B. GSTs and Human Lung Cancer—Effect of GSTM1 Polymorphism and Its Possible Mechanism

Tobacco smoke, the major cause of lung cancer, contains numerous carcinogens (the number 43 has been quoted, Hoffman and Hecht, 1990), but so far only one group has been shown to give rise to substrates for GSTs, namely the polycyclic aromatic hydrocarbons (Ketterer *et al.*, 1992). They are a diverse group, but in general have similar metabolic pathways.

Benzo[*a*]pyrene is one of the most commonly occurring and is the most studied tobacco-smoke carcinogen. The genotoxin, associated with its carcinogenesis (+)-*anti*-benzo[*a*]pyrene-7,8-diol-9,10-oxide (Harvey, 1991), which is a good substrate for GSTs M1-1 (Robertson *et al.*, 1986), M2-2

(B. Jernström and B. Mannervik, unpublished data), and M3-3 (B. Jernström, B. Ketterer, and D. J. Meyer, unpublished data) members of the human mu family and better still for GSTP1-1 of the pi family, but very poor for the alpha family (Robertson *et al.*, 1986). Unlike AFB_1-*exo*-oxide, (+)-*anti*-benzo[*a*]pyrene-7,8-diol-9,10-oxide is utilized equally well by rat and human enzymes, possibly because homology both within and across species is greater in the mu and pi families than in the alpha family.

When an epidemiological study was made comparing 125 smokers with lung cancer with 114 controls, there appeared to be a significantly greater susceptibility to lung cancer in individuals with the GSTM1 null phenotype (Seidegård *et al.*, 1986, 1990). This association supports the general belief that polycyclic aromatic hydrocarbons are important causative factors in smoking-induced lung cancer.

The GSTM1 polymorphism has been known for some years to be genetic in origin (Board, 1981), and recently Seidegård *et al.* (1986) showed that it was due to a gene deletion. At least five additional epidemiological studies have been made, involving either genotyping or phenotyping (Zhong *et al.*, 1991; Heckbert *et al.*, 1992; Brockmöller *et al.*, 1993; Nazar-Stewart *et al.*, 1993; Hirvonen *et al.*, 1993). Despite differences in experimental design the overall conclusion is (*a*) that the *GSTM1* null phenotype increases susceptibility to lung cancer, (*b*) that this is more evident when smokers with lung cancer are compared with smoking controls, and (*c*) that it is even more evident among heavy smokers with lung cancer.

When, in investigations of the origins of the effect of the GSTM1 phenotype, pulmonary GSTs were analyzed, it was shown that GST subunits A1, A2, M1, M3, and P1 were present, GSTP1 being by far the most abundant. Thus, if the detoxication of polycyclic aromatic hydrocarbon diol-epoxides derived from tobacco smoke occurred *solely in the lung*, little effect of the GSTM1 null phenotype would be expected. The liver, in terms of xenobiotic metabolism, is much more powerful than the lung (the liver is about four times larger than the lung and is much richer in drug metabolizing enzymes). Since the liver contains no GSTP1 but only GSTs A1 and A2, and M1, in GSTM1-positive individuals, it would appear to be the most likely origin of an effect of GSTM1 polymorphism on smoking-induced cancer (Ketterer *et al.*, 1992).

If the liver is to be important, a large proportion of the polycyclic aromatic hydrocarbon burden in tobacco smoke should pass unchanged through the peripheral lung to reach the liver through the systemic circulation. The transmission of any hepatic effect to the lung assumes that activated carcinogens are released into the hepatic bloodstream and transported to the lung without inactivation. There is good evidence that this occurs in the mouse, where benzo[*a*]pyrene metabolism by the liver results in the release of (+)-*anti*-benzo[*a*]pyrene-7,8-diol-9,10-oxide into the blood stream where it is transported, apparently protected, to affect other parts of the body (Ginsberg and Atherholt, 1990). That (+)-*anti*-benzo[*a*]-pyrene-7,8-diol-9,10-oxide also enters the circulation in humans is shown by its presence as a hemoglobin adduct in samples of human blood (Jankowiak *et al.*, 1990). Since GSTM1 null individuals are presumably much less able to detoxify polycyclic aromatic hydrocarbon diol epoxides by glutathione conjugation, much greater amounts of these toxins should pass into the blood and be dispatched to the lung.

Thus polycyclic aromatic hydrocarbons from tobacco smoke are dealt with by the lung in several passes. In the first pass, presumably a small amount is retained and metabolized there. Present knowledge indicates that the distribution of CYP1A1, important for their epoxidation, tends to be peripheral (Antilla *et al.*, 1992). In contrast, the GSTs tend to be bronchial in disposition and therefore encountered first (Antilla *et al.*, 1993). In the second pass, diol-epoxides originating from the liver reach the lung via the pulmonary artery and therefore enter at the proximal end of the respiratory tree and encounter the bronchial region first where GSTs P1, M1, M2, and M3 are present, but are not sufficient to eliminate the diol-epoxides completely. The load of diol-epoxide entering a lung on the second pass and therefore the potential for damage presumably depends on whether the individual has a positive or negative GSTM1 phenotype.

It is possible that the stable diol precursor of the diol-epoxide also reaches the lung from the liver. In this case its activation may be in the periphery, where CYP1A1 but not GST activity is most abundant.

The GSTM1 polymorphism is not the only one to affect smoking induced lung cancer and the coincidence of two or more should have a cumulative effect. This has already been shown in a CYP1A1 polymorphism, which, combined with the GSTM1 null phenotype, gives rise to a susceptibility 10 times that of controls (Hayashi *et al.*, 1992). A cumulative effect might also be seen with the concurrence of a GSTM1 phenotype with the debrisoquine hydroxylase "extensive metabolizer" (Caporaso *et al.*, 1990).

C. Industrial Chemicals: Toxication and Detoxication

The effects of GSTs on human disease discussed so far have been associated with diet and lifestyle. Another important area concerns disease resulting from exposure to industrial chemicals and industrial processing. Here again GSTs may be cytoprotective. Examples are GSTM1, which detoxifies styrene oxide, an intermediate in the manufacture of the polystyrene (Warholm *et al.*, 1983); GSTP1, which detoxifies acrolein, an intermediate in the manufacture of glycerol (Berhane and Mannervik, 1990); and GSTT1, which detoxifies ethylene oxide, a food sterilant, fungicide, and an intermediate in antifreeze production (Pemble *et al.*, 1994). The polycyclic aromatic hydrocarbons are also important where in large-scale combustion processes associated with traditional heavy industry.

An apparent paradox is toxication by GSTs (discussed in detail by Guengerich, this volume). This is observed with several industrial chemicals. The first example recognized was 1,2-dibromoethane (a petrol additive and soil fumigant), the glutathione conjugate of which behaves as a sulfur mustard and gives an *S*-[2(N^7-guanyl)ethyl]glutathione adduct when incubated with calf thymus DNA. This reaction is catalyzed weakly by alpha class GSTs from rat and humans (Cmarik *et al.*, 1990), but strongly by rat GST5-5. (*Salmonella typhimurium* TA1535 expressing rat GST5-5, when exposed to dibromoethane, undergoes to substantial mutagenesis compared with controls transfected with antisense constructs; Thier *et al.*, 1993.) Dichloromethane, a widely used industrial solvent, which has already been shown to be a carcinogen in the mouse, is also mutagenic in *S. typhimunium* TA1535 expressing GST5-5. Human GSTT1-1 (a homologue of rat GST5-5) has similar enzymic activity, therefore dichloromethane is a potential genotoxin in humans.

1,2-Dibromo-3-chloropropane (a nematocide and soil fumigant with many toxic properties) is another substrate for GSTs (see Table I) that gives rise to reactive episulfonium ions (Pearson *et al.*, 1990).

GSTs can also toxify by GSH depletion. This requires substrates that can be administered in large amounts, e.g., paracetamol and diethyl maleate, and that have rapid rates of reaction (for NABQI, the electrophilic metabolite of paracetamol, see Table III). These substrates cause cytosolic and presumably nuclear GSH depletion (due to nuclear pores, small molecules are in equilibrium between the free water of the cytosol and that of the nucleoplasm) (Peters, 1986), exposing both cellular compartments to oxidative damage. In the case of the cytosol and the nucleoplasm, where alpha, mu, and pi class enzymes with their low K_m values for GSH are abundant, enzymatically induced depletion can be very severe. Mitochondria, on the other hand, are protected from this by the high K_m value of the mitochondrial enzyme GST 13-13 for GSH.

D. Polymorphism in the Human Theta Class Family

Human GSTT1-1 has been isolated from liver (Meyer *et al.*, 1991a) and erythrocytes (Schröder *et al.*, 1992) and its gene, like *GSTM1*, is shown to have a null allele (Pemble *et al.*, 1994). Its substrates include dichloromethane (Meyer *et al.* 1991a; Hallier *et al.*, 1993), ethylene oxide (Hallier *et al.*, 1993), and probably (+)-benzo(*a*)pyrene-7,8-diol-9,10-oxide, which is a substrate for the rat homologue GST5-5 (B. Jernström, D. J. Meyer, and B. Ketterer, unpublished data). The GSTT1 null phenotype should reduce susceptibility to the toxicity of dichloromethane but may increase susceptibility to polycyclic aromatic hydrocarbon diol epoxides. In the etiology of tobacco-smoke-induced lung cancer, GSTT1 and GSTM1 polymophisms may be additive.

E. GSTs and the Detoxication of Xenobiotic Organic Hydroperoxides and Endogenous Lipid and DNA Hydroperoxides

Table IV shows that GSTs (particularly those of the alpha family) are relatively good enzymes for both endogenous organic hydroperoxides (polyunsaturated fatty acid hydroperoxides and thymine hydroperoxides) and exogenous organic hydroperoxides (cumene hydroperoxide; *tert*-butylhydroperoxide). Free radicals and high-energy irradiation may cause cellular lipid peroxidation. The most abundant targets are polyunsaturated fatty acyl residues in phospholipids and without detoxication, the result is a cascade of reactions leading to further radical damage, membrane decomposition, and free-radical attack on other cellular targets such as DNA. Esterified fatty acid hydroperoxides are not substrates for GSTs, but free fatty acids can be made available by the action of phospholipase A_2. It is envisaged that in peroxidized cellular membranes *in vivo* the concerted action of phospolipase A_2, GST (or Se-dependent glutathione peroxidase), and lysophosphatide acylase results in detoxication and repair of membranes (Tan *et al.*, 1984; van Kuijk *et al.*, 1987). A phospholipid hydroperoxide glutathione peroxidase, which catalyzes the reaction directly, has also been reported (Ursini *et al.*, 1985).

Free-radical attack on DNA can lead to its peroxidation and proliferating secondary DNA damage. GSTs and Se-dependent glutathione peroxidase have also been shown to reduce peroxidized DNA, an important reaction that not only leads to immediate detoxication, but also may be the first in a series of reactions eventually leading to repair. For example, 5-hydroperoxymethyluracil in DNA may be reduced to 5-hydroxymethyluracil, which can then undergo excision repair (Ketterer *et al.*, 1990; Ketterer and Coles, 1991).

Where oxidative damage to nuclear DNA is concerned, the GSTs may be particularly important since the nucleus does not contain Se-dependent glutathione peroxidase (Ketterer *et al.*, 1994), presumably because, unlike the GSTs, it is too large to diffuse through nuclear pores, which have a cut-off value for globular proteins of approximately 50,000 (Peters, 1986). Whereas GSTs have a molecular weight of approximately 50,000, Se-dependent GSH peroxidase is 84,000. The only available information concerning the distribution of Ursini's enzyme is immunohistochemical and shows that it is membrane-associated.

In addition to the above detoxication reactions, GSTs (once more notably the alpha family) catalyze a number of reactions of importance in the biosynthesis of eicosanoids involved in the inflammatory process. They can reduce the endoperoxide prostaglandin $(PG)H_2$ to its corresponding alcohol $PGF_{2\alpha}$. They can also catalyze the GSH-dependent isomerization of PGH_2 to PGE_2 and PGD_2 (Meyer and Ketterer, 1987; Ujihara *et al.*, 1988). Also mu class GSTs, particularly rat GST6-6, catalyze the conjugation of leukotriene A_4 to leukotriene C_4 (Tsuchida *et al.*, 1987). These reactions may have little physiological significance since other enzymes for which these activities are specific also catalyze these reactions (see Ketterer and Coles, 1991, for a more complete discussion). Also, the tissue distribution of the GSTs most active in these reactions (liver and kidney) does not coincide with that of cyclooxygenase, which initiates prostaglandin biosynthesis (vesicular gland, cecum, gall bladder, and lung). These reactions may, however, have a role in pathophysiology; for example, in situations of abnormal differentiation, such as malignancy, it is possible that GSTs perform some of these reactions. Such ectopic synthesis of these powerful mediators could cause considerable disturbance and be regarded as a toxic effect. For example, it has been claimed that tumors produce large quantities of PGE_2, which is a factor down-regulating the immune system and assisting carcinogenesis. The route of metabolism leading to PGE_2 in tumors has not been determined, but may involve GSTs.

VII. Conclusions

The GSTs catalyze a range of reactions of GSH with hydrophobic electrophiles. Their most established role is the glutathione conjugation of electrophiles, which would otherwise cause toxic reactions with macromolecules. These electrophiles frequently arise from the oxidation of xenobiotics by the cytochrome P450 system. This function has been illustrated by a discussion of aflatoxin B_1 carcinogenesis. It is demonstrated that the induction in the adult of a fetal GST with substantial activity toward the carcinogenic electrophile AFB_1-*exo*-oxide can prevent aflatoxin-induced hepatocarcinogenesis in the rat. The presence of a similar enzyme as a normal constituent in adult mouse liver may explain the resistance of the mouse to AFB_1-induced carcinogenesis. On the other hand the absence of such an enzyme in control or induced isolated human hepatocytes raises doubts about the importance of GSTs in anti-hepatocarcinogenesis in humans exposed to AFB_1.

The distribution of GSTs differs both qualitatively and quantitatively from tissue to tissue with the consequence that GSTs in one tissue may influence the levels of circulating substrates important for an effect in another tissue. This has been discussed in relation to tobacco smoke-induced lung cancer in humans, where greater susceptibility to this disease is associated with the null phenotype in a *GSTM1* polymorphism. This effect may originate in the inability of the livers of individuals with a GSTM1 null phenotype to detoxify polycyclic aromatic hydrocarbon diol epoxides, which then enter the circulation to damage the lungs.

A similar polymorphism has recently been discovered in the human theta enzyme *GSTT1*. It will be of interest to know whether this also has important consequences in human epidemiology. The existence of stable polymorphisms in *GSTM1* and *GSTT1* suggests that they have evolutionary advantages, but what they may be is not clear.

Occasionally glutathione conjugation results not in detoxication, but the formation of a harder electrophile. This has recently been demonstrated to occur with dichloromethane, which is mutagenic in bacteria (*S. typhimurum* TA1535) transfected with a rat *GST5* cDNA construct but not with its antisense construct. Since dichloromethane is a commonly used solvent in many countries, despite having been shown carcinogenic in the mouse, its continued use should be reexamined, now that the enzyme responsible for its mutagenicity is known.

The GST-catalyzed attack of glutathione on electrophilic oxygen in hydroperoxides, resulting in their reduction, has not been given much attention since the apparently universal Se-dependent enzyme has a much higher specific activity. However, this lack of interest may not be justified. Not only do GSTs make a major contribution to total GSH peroxidase activity in rat and human liver, but nuclei isolated by a nonaqueous method (Siebert, 1961), which ensures that all water soluble molecules remain *in situ*, shows that the rat liver nucleus does not contain Se-dependent glutathione peroxidase and therefore that GSTs offer the only enzymic protection from damage by hydroperoxide in the nucleus (S. Soboll, H. Sies, and B. Ketterer, unpublished data).

What has been described so far concerns current knowledge of function of the GSTs, which has grown steadily over many years. With the development of modern technology it has been particularly interesting to witness the recent determination of the crystal structure of proteins from three of the gene families which has enabled the mechanism of the binding and activation of glutathione to be determined. Interactions of the diverse electrophilic substrates with their binding sites are in the course of clarification, and the use of site-directed mutagenesis will enable rapid progress to be made in this area.

Knowledge of gene structure and regulation is advancing rapidly. Regulatory sequences associated with induction and malignancy have been identified and are under intensive study. The regulation of the mu family is poorly understood. Sequences downstream from the transcription start site, which are hormone responsive, have been identified, but none that have been related to induction by xenobiotics.

There is one curiosity in GST genetics that merits comment. Whereas in each of the alpha, mu, and theta families there are multiple homologous genes and evidence for gene conversion, particularly in the mu family, in the pi family there is usually only one expressed gene that is highly conserved between species.

From a practical point of view one of the most interesting areas of recent work concerns the possible use of nontoxic inducers of GSTs in preventing chemical toxicity. The use of oltipraz for the prophylaxis of AFB_1-associated liver cancer is being investigated at the moment, but such inducers may be useful for protection against any toxins which are inactivated by inducible GSTs.

Acknowledgments

This work was supported by the Cancer Research Campaign. B.K. is a Forschungspreisträger of the Alexander-Von-Humboldt Foundation, Bonn-Bad-Godesberg.

References

Aceto, A., Di Ilio, C., Lo Bello, M., Bucciarelli, T., Angelucci, S., and Federici, G. (1990). Differential activity of human, rat, mouse and bacteria glutathione transferase isoenzymes towards 4-nitroquinoline 1-oxide. *Carcinogenesis (London)* **11,** 2267–2269.

Antilla, S., Vainio, H., Hietanen, E., Camus, A.-M., Malaveille, C., Brun, G., Husgafvel-Pursiainen, K., Heikkila, L., Karjalainen, A., and Bartsch, H. (1992). Immunohistochemical detection of pulmonary cytochrome P4501A and metabolic activities associated with P4501A1 and P4501A2 isozymes in lung cancer patients. *Environ. Health Perspect.* **98,** 179–182.

Antilla, S., Hirvonen, A., Vainio, H., Husgafvel-Pursiainen, K., Hayes, J. D., and Ketterer, B. (1993). Immunohistochemical localization of glutathione S-transferases in human lung. *Cancer Res.* **53,** 5643–5648.

Armstrong, R. N. (1991). Glutathione S-transferases: Reaction mechanism, structure, and function. *Chem. Res. Toxicol.* **4,** 131–140.

Bass, N. M., Kirsch, R. E., Tuff, S. A., Marks, I., and Saunders, S. J. (1977). Ligandin heterogeneity: Evidence that the two non-identical subunits are the monomers of two distinct proteins. *Biochim. Biophys. Acta* **492,** 163–168.

Berhane, K., and Mannervik, B. (1990). Inactivation of the genotoxic aldehyde acrolein by human glutathione transferases of classes alpha, mu and pi. *Mol. Pharmacol.* **37,** 251–254.

Board, P. G. (1981). Biochemical genetics of glutathione-S-transferase in man. *Am. J. Hum. Genet.* **33,** 36–43.

Brockmöller, J., Kerb, R., Drakoulis, N., Nitz, M., and Roots, I. (1993). Genotype and phenotype of glutathione S-transferase class μ isoenzymes μ and ψ in lung cancer patients and controls. *Cancer Res.* **53,** 1004–1011.

Buetler, T. M., and Eaton, D. L. (1992). Complementary DNA cloning, messenger RNA expression and induction of α-class glutathione S-transferases in mouse tissues. *Cancer Res.* **52,** 314–318.

Caporaso, N. E., Tucker, M. A., Hoover, R. N., Hayes, R. B., Pickle, L. W., Issaq, H. J., Muschik, G. M., Green-Gallo, L., Buivys, D., Aisner, S., Resau, J. H., Trump, B. J., Tollerud, D., Weston, A., and Harris, C. C. (1990). Lung cancer and the debrisoquine metabolic phenotype. *J. Natl. Cancer Inst.* **82,** 1264–1272.

Cmarik, J. L., Inskeep, P. B., Meredith, M. J., Meyer, D. J., Ketterer, B., and Guengerich, F. P. (1990). Selectivity of rat and human glutathione S-transferases in activation of ethylene dibromide by glutathione conjugation and DNA binding and induction of unscheduled DNA synthesis in human hepatocytes. *Cancer Res.* **50,** 2747–2752.

Coles, B. (1984). Effects of modifying structure on electrophilic reactions with biological nucleophiles. *Drug Metab. Rev.* **15,** 1307–1334.

Coles, B., and Ketterer, B. (1990). The role of glutathione and glutathione transferases in chemical carcinogenesis. *CRC Crit. Rev. Biochem. Mol. Biol.* **25,** 47–70.

Coles, B., Beale, D., Miller, D., Lay, J., Kadlubar, F. F., Aitken, A., and Ketterer, B. (1987). The binding of an aminoazo dye carcinogen to a specific methionine residue in rat liver dehydrogenase *in vivo*. *Chem.-Biol. Interact.* **64,** 181–192.

Coles, B., Wilson, I., Wardman, P., Hinson, J. A., Nelson, S. D., and Ketterer, B. (1988). The spontaneous and enzymic reaction of N-acetyl-benzoquinone-imine with glutathione: A stopped-flow kinetic study. *Arch. Biochem. Biophys.* **264,** 253–260.

Cowell, I. G., Dixon, K. H., Pemble, S. E., Ketterer, B., and Taylor, J. B. (1988). The structure of the human glutathione S-transferase π gene. *Biochem. J.* **255,** 79–83.

Daniel, V. (1993). Glutathione S-transferases: Gene structure and regulation of expression. *CRC Crit. Rev. Biochem. Mol. Biol.* **28,** 173–207.

Danielson, U. H., Esterbauer, H., and Mannervik, B. (1987). Structure-activity relationships of 4-hydroxy-alkenals in the conjugation catalysed by mammalian glutathione transferases. *Biochem. J.* **247,** 707–712.

Denison, M. S., Fisher, J. M., and Whitlock, J. P., Jr. (1988). The DNA recognition site for dioxin-Ah receptor complex: Nucleotide sequence and functional analysis. *J. Biol. Chem.* **263,** 17221–17224.

Dixon, K. H., Cowell, I. G., Xia, C.-L., Pemble, S. E., Taylor, J. B., and Ketterer, B. (1989). Control of expression of the human glutathione S-transferase π gene differs from its rat orthologue. *Biochem. Biophys. Res. Commun.* **163,** 815–822.

Djuríc, Z., Coles, B., Fifer, E. K., Ketterer, B., and Beland, F. A. (1987). In vivo and in vitro formation of glutathione conjugates from the K-region epoxides of 1-nitropyrene. *Carcinogenesis (London)* **8,** 1781–1786.

Friling, R. S., Bergelson, S., and Daniel, V. (1992). Two adjacent AP1-like binding sites from the electrophile responsive element of the murine glutathione S-transferase Ya subunit gene. *Proc. Natl. Acad. Sci. U.S.A.* **89,** 668–672.

Ginsberg, G. L., and Atherholt, T. B. (1990). DNA adduct formation in mouse tissues in relation to serum levels of benzo(a)pyrene-diol-epoxide after injection of benzo(a)pyrene or the diol epoxide. *Cancer Res.* **50,** 1189–1194.

Groopman, J. D., Cain, L. G., and Kensler, T. W. (1988). Aflatoxin exposure in human populations: Measurements and relationship to cancer. *CRC Crit. Rev. Toxicol.* **19,** 113–145.

Gudas, L. J. (1992). Retinoids, retinoid-responsive genes, cell differentiation, and cancer. *Cell Growth Differ.* **3,** 655–662.

Habig, W. H., Pabst, M. J., Fleischner, G., Gatmaitan, Z., Arias, I. M., and Jakoby, W. B. (1974). The identity of glutathione S-transferase B with ligandin, a major binding protein of liver. *Proc. Natl. Acad. Sci. U.S.A.* **71,** 3879–3882.

Hallier, E., Langhof, T., Dannappel, D., Leutbecher, M., Schröder, K., Goergens, W., Müller, A., and Bolt, H. M. (1993). Polymorphism of glutathione conjugation of methyl bromide, ethylene oxide and dichloromethane in human blood: Influence on the induction of sister chromatid exchanges (SCE) in lymphocytes. *Arch. Toxicol.* **67,** 173–178.

Harris, J. M., Meyer, D. J., Coles, B., and Ketterer, B. (1991). A novel glutathione transferase (13-13) isolated from the matrix of rat liver mitochondria having structural similarity to class theta enzymes. *Biochem. J.* **278,** 137–141.

Harvey, R. G. (1991). "Polycyclic Aromatic Hydrocarbons." Cambridge Univ. Press, Cambridge, UK.

Hayashi, S., Watanabe, J., and Kawajiri, K. (1992). High susceptibility to lung cancer analysed in terms of combined genotypes of P4501A1 and Mu-class glutathione S-transferase genes. *Jpn. J. Cancer Res.* **83,** 866–870.

Hayes, J. D. (1988). Selective elution of rodent glutathione S-transferases and glyoxalase I from the S-hexylglutathione-sepharose affinity column. *Biochem. J.* **255,** 913–922.

Hayes, J. D., Judah, D. J., McLellan, L. I., Kerr, L. A., Peacock, S. D., and Neal, G. E. (1991). Ethoxyquin-induced resistance to aflatoxin B_1 in the rat is associated with the expression of a novel Alpha-class glutathione S-transferase subunit, Yc_2, which possesses high catalytic activity for aflatoxin B_1-8,9-epoxide. *Biochem. J.* **279,** 385–398.

Heckbert, S. R., Weiss, N. S., Hornung, S. K., Eaton, D. L., and Motulsky, A. G. (1992). Glutathione S-transferase and epoxide hydrolase activity in human leukocytes in relation to risk of lung cancer and other smoking-related cancers. *J. Natl. Cancer Inst.* **84,** 414–422.

Hiratsuka, A., Sebata, N., Kawashima, K., Okuda, H., Ogura, K., Watabe, T., Satoh, K., Hatayama, I., Tsuchida, S., Ishikawa, T., and Sato, K. (1990). A new class of rat glutathione S-transferase Yrs-Yrs inactivating reactive sulfate esters as metabolites of carcinogenic arylmethanols. *J. Biol. Chem.* **265,** 11973–11981.

Hirvonen, A., Husgafvel-Pursiainen, K., Antilla, S., and Vainio, H. (1993). The *GSTM1* null genotype as a potential modifier for sqamous cell carcinoma of the lung. *Carcinogenesis (London)* **14,** 1479–1481.

Hoffmann, D., and Hecht, S. S. (1990). Advances in tobacco carcinogenesis. *In* "Chemical Carcinogenesis and Mutagenesis" (C. S. Cooper and P. L. Glover, eds.), pp. 64–102. Springer-Verlag, New York.

Jakobson, I., Warholm, M., and Mannervik, B. (1979). Multiple inhibition of glutathione S-transferase A from rat liver by glutathione derivatives: kinetic analysis supporting a steady-state random sequential mechanism. *Biochem. J.* **177,** 861–868.

Jakoby, W. B., Ketterer, B., and Mannervik, B. (1984). Glutathione transferases: Nomenclature. *Biochem. Pharmacol.* **33,** 2539–2540.

Jankowiak, R., Day, B. W., Lu, P., Doxtader, M. M., Skipper, P. L., Tannenbaum, S. R., and Small, G. J. (1990). Fluorescence line-narrowing spectral analysis of in vivo human hemoglobin-benzo(a)pyrene diol epoxide adducts: Comparison to synthetic analogues. *J. Am. Chem. Soc.* **112,** 5866–5869.

Jernström, B., Martinez, M., Meyer, D. J., and Ketterer, B. (1985). Glutathione conjugation of the carcinogenic and mutagenic electrophile (±)-7β,8α-dihydroxy-9α,10α-oxy-7,8,9,10-tetrahydrobenzo(a)pyrene catalysed by purified rat liver glutathione transferases. *Carcinogenesis (London)* **6,** 85–89.

Ji, X., Zhang, P., Armstrong, R. N., and Gilliland, G. L. (1992). The three-dimensional structure of a glutathione S-transferase from the mu gene class. Structural analysis of the binary complex of isoenzyme 3-3 and glutathione at 2.2 A resolution. *Biochemistry* **31,** 10169–10184.

Kantor, R. R. S., Giardina, S. L., Bartolazzi, A., Townsend, A. J., Meyers, C. E., Cowan, K. H., Longo, D. L., and Natali, P. G. (1991). Monoclonal antibodies to glutathione S-transferase π - Immunohistochemical analysis of human tissues and cancers. *Int. J. Cancer* **47,** 193–201.

Kensler, T. W., Groopman, J. D., Curphey, T. J., and Roebuck, B. D. (1991). Protection against aflatoxin hepatocarcinogenesis by 1,2-dithiol-3-thiones mechanisms and markers. *Proc. Am. Assoc. Cancer Res.* **32,** 476–478.

Ketterer, B., and Christodoulides, L. (1969). Two specific azodye-carcinogen binding proteins of the rat liver. The identity of the amino acid residues which bind the azodye. *Chem.-Biol. Interact.* **1,** 173–183.

Ketterer, B., and Coles, B. (1991). Glutathione transferases and products of reactive oxygen. *In* "Oxidative Stress" (H. Sies, ed.), pp. 171–194. Academic Press, London.

Ketterer, B., and Meyer, D. J. (1989). Glutathione transferases: A possible role in the detoxication and repair of DNA and lipid hydroperoxides. *Mutat. Res.* **214,** 33–40.

Ketterer, B., Beale, D., and Hackney, J. F. (1971). The conjugation of aminoazo dye metabolites with a tripeptide and specific soluble proteins of the liver soluble cytoplasm of the rat. *Xenobiotica* **1,** 551–552.

Ketterer, B., Meyer, D. J., and Clark, A. G. (1988). Soluble glutathione transferase isozymes. *In* "Glutathione Conjugation: Mechanisms and Biological Significance" (H. Sies and B. Ketterer, eds.), pp. 73–135. Academic Press, London.

Ketterer, B., Meyer, D. J., Taylor, J. B., Pemble, S., Coles, B., and Fraser, G. (1990). GSTs and protection against oxidative stress. *In* "Glutathione S-Transferases and Drug Resistance" (J. D. Hayes, C. B. Pickett, and T. J. Mantle, eds.), pp. 97–109. Taylor & Frances, London.

Ketterer, B., Harris, J. M., Talaska, G., Meyer, D. J., Pemble, S. E., Taylor, J. B., Lang, N. P., and Kadlubar, F. F. (1992). The human glutathione S-transferase supergene family, its polymorphism and its effects on susceptibility to lung cancer. *Environ. Health Perspect.* **98,** 87–94.

Ketterer, B., Harris, J., Kramer, H., Kolb-Bachofen, V., Gründel, S., Schultz, B., Sies, H., and Soboll, S. (1994). Nuclear content of glutathione *S*-transferases and Se-dependent glutathione peroxidase. *Proc. Acad. Assoc. Cancer Res.* **35** (in press).

Klöne, A., Hussnätter, R., and Sies, H. (1992). Cloning, sequencing and characterization of the human alpha, glutathione S-transferase gene corresponding to the cDNA clone pGTH2. *Biochem. J.* **285,** 925–928.

Kolm, R. H., Sroga, G. E., and Mannervik, B. (1992). Participation of the phenolic hydroxyl group of Try-8 in the catalytic mechanism of human glutathione transferase P1-1. *Biochem. J.* **285,** 537–540.

Lai, H.-C. J., Qian, B., Grove, G., and Tu, C.-P. D. (1988). Gene expression of rat glutathione transferases. *J. Biol. Chem.* **263,** 11389–11395.

Liu, S., Zhang, P., Ji, X., Johnson, W. W., Gilliland, G. L., and Armstrong, R. N. (1992). Contribution of tyrosine 6 to the catalytic mechanism of isoenzyme 3-3 of glutathione transferase S-transferase. *J. Biol. Chem.* **276,** 4296–4299.

Mannervik, B., and Danielson, U. H. (1988). Glutathione transferases - structure and catalytic activity. *CRC Crit. Rev. Biochem. Mol. Biol.* **23,** 283–337.

Mannervik, B., Awasthi, Y. C., Board, P. G., Hayes, J. D., Di Ilio, C., Ketterer, B., Listowsky, I., Morgenstern, R., Muramatsu, M., Pearson, W. R., Pickett, C. B., Sato, K., Widerstern, M., and Wolf, C. R. (1992). Nomenclature for human glutathione transferases. *Biochem. J.* **282,** 305–306.

Manoharan, T. H., Gulick, A. M., Reinemer, P., Dirr, H. W., Huber, R., and Fahl, W. E. (1992). Mutational substitution of residues implicated by crystal structure in binding the substrate glutathione to human glutathione S-transferase pi. *J. Mol. Biol.* **226,** 319–322.

Meyer, D. J. (1993). The significance of an unusually low K_m for glutathione in glutathione transferase of the alpha, mu and pi classes. *Xenobiotica* **22,** 823–834.

Meyer, D. J., and Ketterer, B. (1982). 5,6-Epoxy-cholestan-3β-ol (cholesterol α-oxide): A specific substrate for rat liver glutathione transferase B. *FEBS Lett.* **150,** 499–502.

Meyer, D. J., and Ketterer, B. (1987). Prostaglandin isomerase activity of purified rat glutathione (GSH) transferases. *In* "Glutathione S-Transferases and Carcinogenesis" (T. J. Mantle, J. Hayes, and C. Pickett, eds.), pp 57–59. Taylor & Francis, London.

Meyer, D. J., Coles, B., Gilmore, K., Pemble, S., Fraser, G., and Ketterer, B. (1991a). Theta, a new class of glutathione transferases purified from rat and man. *Biochem. J.* **274,** 409–414.

Meyer, D. J., Gilmore, K. S., Coles, B., Dalton, K., Hulbert, P. B., and Ketterer, B. (1991b). Structural distinction of rat GSH transferase subunit 10. *Biochem J.* **274,** 619–623.

Meyer, D. J., Xia, C.-L., Coles, B., Chen, H., Reinemer, P., Huber, R., and Ketterer, B. (1993a). Unusual reactivity of Tyr-7 of GSH transferase P1-1. *Biochem. J.* **293,** 351–356.

Meyer, D. J., Harris, J. M., Gilmore, K. S., Coles, B., Kensler, T. W., and Ketterer, B. (1993b). Quantitation of tissue- and sex-specific induction of rat glutathione transferase subunits by dietary 1,2-dithiole-3-thiones. *Carcinogenesis* (*London*) **14,** 567–572.

Mignogna, G., Allocati, N., Aceto, A., Piccolomini, R., Di Ilio, C., Barra, D., and Martini, F. (1993). The amino acid sequence of glutathione transferase from *Proteus mirabilis,* a prototype of a new class of enzymes. *Eur. J. Biochem.* **211,** 421–425.

Morel, F., Fardel, O., Meyer, D. J., Langouet, S., Gilmore, K. S., Meunier, B., Tu, C.-P. D., Kensler, T. W., Ketterer, B., and Guillouzo, A. (1993). Preferential increase of glutathione S-transferase class α transcripts in cultured human heptocytes by phenobarbital, 3-methylcholanthrene and dithiolethiones. *Cancer Res.* **53,** 231–234.

Morrow, C. S., Cowan, K. H., and Goldsmith, M. E. (1989). Regulation of human glutathione S-transferase π gene transcription: Influence of 5′-flanking sequences and *trans*-activating factors which recognize AP-1 binding sites. *Gene* **88,** 215–225.

Morton, M. R., Bayney, R. M., and Pickett, C. B. (1990). Isolation and characterization of the rat glutathione S-transferase Yb1 subunit gene. *Arch. Biochem. Biophys.* **277,** 56–60.

Muramatsu, M., Okuda, A., Imagawa, M., and Sakai, M. (1990). Regulation of glutathione transferase P gene during hepatocarcinogenesis of the rat. *In* "Glutathione S-Transferases and Drug Resistance" (J. D. Hayes, C. B. Pickett, and T. J. Mantle, eds.), pp. 165–175. Taylor & Francis, London.

Nazar-Stewart, V., Motulsky, A. G., Eaton, D. L., White, E., Hornung, S. K., Leng, Z.-T., Stapleton, P., and Weiss, N. S. (1993). The glutathione S-transferase μ polymorphism as a marker for the susceptibility to lung carcinoma. *Cancer Res.* **53,** 2313–2318.

Nemoto, N., Gelboin, H. V., Habig, W. H., Ketley, N. J., and Jakoby, W. B. (1975). K-Region benzo(a)pyrene-4,5-oxide is conjugated by homogeneous glutathione S-transferases. *Nature (London)* **255,** 512–517.

O'Brien, R., and Granner, D. K. (1991). Regulation of gene expression by insulin. *Biochem. J.* **278,** 609–619.

Okuda, A., Sakai, M., and Muramatsu, M. (1987). The structure of the rat glutathione transferase P gene and related pseudogenes. *J. Biol. Chem.* **262,** 3858–3863.

Okuda, A., Imagawa, M., Meada, Y., Sakai, M., and Muramatsu, M. (1989). Structure and functional analysis of an enhancer GPE1 having a phorbol 12-O-tetradecanoate 13-acetate responsive element-like sequence found in the rat glutathione transferase P gene. *J. Biol. Chem.* **264,** 16919–169226.

Östlund-Farrants, A. K., Meyer, D. J., Coles, B., Southan, C., Aitken, A., Johnson, P. J., and Ketterer, B. (1987). The separation of glutathione transferase subunits by using reverse-phase high-pressure liquid chromatography. *Biochem. J.* **245,** 423–428.

Palmer Beasley, R., Hwang, L.-Y., Lin, C.-C., and Chien, C.-S. (1981). Hepatocellular carcinoma and heptatitis B virus. *Lancet* **2,** 1129–1133.

Paulson, K. E., Darnell, J. E., Rushmore, T., and Pickett, C. B. (1990). Analysis of the upstream elements of the xenobiotic compound-inducible and positionally regulated glutathione S-transferase Ya gene. *Mol. Cell. Biol.* **10,** 1841–1852.

Pearson, P. G., Soderlund, E. J., Dybing, E., and Nelson, S. D. (1990). Metabolic activation of 1,2-dibromo-3-chloropropane: Evidence for the formation of reactive episulfonium ion intermediates. *Biochemistry* **29,** 4971–4981.

Pemble, S. E., and Taylor, J. B. (1992). An evolutionary perspective on glutathione transferases inferred from class-Theta glutathione transferase cDNA sequences. *Biochem. J.* **287,** 957–963.

Pemble, S. E., Taylor, J. B. and Ketterer, B. (1993). The cloning of human glutathione transferase (GST) M5, a fifth member of the human GST mu class gene family. *Proc. Amer. Assoc. Cancer Res.* **34,** 2965.

Pemble, S., Schroeder, K. R., Spencer, S. R., Meyer, D. J., Hallier, E., Bolt, H. M., Ketterer, B., and Taylor, J. B. (1994). Human glutathione S-transferase theta (GSTT1): cDNA cloning and the characterization of a genetic polymorphism. *Biochem J.* (in press).

Peters, R. (1986). Fluorescence microphotolysis to measure nucleocytoplasmic transport and intracellular mobility. *Biochim. Biophys. Acta* **864,** 305–359.

Pinkus, R., Bergelson, S., and Daniel, V. (1993). Phenobarbital induction of AP-1 binding activity mediates activation of glutathione S-transferase and quinone reductase gene expression. *Biochem. J.* **290,** 637–640.

Ramsdell, H. S., and Eaton, D. L. (1990). Mouse liver glutathione S-transferase isoenzyme activity toward aflatoxin B_1-8,9-epoxide and benzo(a)pyrene-7,8-dihydrodiol-9,10-epoxide. *Toxicol. Appl. Pharmacol.* **105,** 216–225.

Raney, K. D., Coles, B., Guengerich, F. P., and Harris, T. M. (1992). The *endo*-8,9-epoxide of aflatoxin B_1: A new metabolite. *Chem. Res. Toxicol.* **5,** 333–335.

Reinemer, P., Dirr, H. W., Ladenstein, R., Schäffer, J., Gallay, O., and Huber, R. (1991). The three-dimensional structure of class π glutathione S-transferase in complex with glutathione sulphonate at 2.3 A resolution. *EMBO J.* **10,** 1997–2005.

Reinemer, P., Dirr, H. W., Ladenstein, R., Huber, R., Lo Bello, M., Federici, G., and Parker, M. W. (1992). Three-dimensional structure of class π glutathione S-transferase from human placenta in complex with S-hexylglutathione at 2.8 A resolution. *J. Mol. Biol.* **227,** 214–226.

Robertson, I. G. C., Guthenberg, C., Mannervik, B., and Jernström, B. (1986). Differences in stereoselectivity and catalytic efficiencies of three human glutathione transferases in the conjugation of glutathione with 7β,8α-dihydroxy-9α,10α-oxy-7,8,9,10-tetrahydrobenzo(a)pyrene. *Cancer Res.* **46,** 2220–2224.

Röhrdanz, E., Nguyen, T., and Pickett, C. B. (1992). Isolation and characterization of glutathione S-transferase A_2 subunit gene. *Arch. Biochem. Biophys.* **298,** 747–752.

Ross, R. K., Yuan, J.-M., Yu, M. C., Wogan, G. N., Qian, G.-S., Tu, J.-T., Groopman, J. D., Gao, Y.-T., and Henderson, B. E. (1992). Urinary aflatoxin biomarkers and risk of hepatocellular carcinoma. *Lancet,* **339,** 943–946.

Rozen, F., Nguyen, T., and Pickett, C. B. (1992). Isolation and characterization of a human glutathione transferase Ha_1 subunit gene. *Arch. Biochem. Biophys.* **292,** 589–593.

Rushmore, T. H., Morton, M. R., and Pickett, C. B. (1991). The antioxidant responsive element. Activation by oxidative stress and identification of the DNA consensus sequence required for functional activity. *J. Biol. Chem.* **266,** 11632–11639.

Sakai, M., Okuda, A., and Muramatsu, M. (1988). Multiple regulatory elements and phorbol 12-O-tetradecanoate 13-acetate responsiveness of the rat placental glutathione transferase gene. *Proc. Natl. Acad. Sci. U.S.A.* **85,** 9456–9460.

Sato, K. (1989). Glutathione transferases as markers of preneoplasia and neoplasia. *Adv. Cancer Res.* **52,** 205–255.

Schröeder, K. R., Hallier, E., and Bolt, H. M. (1992). Dissociation of a new glutathione S-transferase activity in human erythrocytes. *Biochem. Pharmacol.* **43,** 1671–1674.

Seidegård, J., Pero, R. W., Miller, D. G., and Beattie, E. J. (1986). A glutathione transferase in human leukocytes as a marker for the susceptibility to lung cancer. *Carcinogenesis (London)* **7,** 751–753.

Seidegård, J., Pero, R. W., Markowitz, M., Roush, G., Miller, D. G., and Beattie, E. J. (1990). Isoenzyme(s) of glutathione transferase (class Mu) as a marker for the susceptibility to lung cancer: A follow up study. *Carcinogenesis (London)* **11,** 33–36.

Siebert, G. (1961). Enzyme und substrate der glykolyse in isolierten zellkernen. *Biochem. Z.* **334,** 369–387.

Sinning, I., Kleywegt, G. J., Cowan, S. W., Reinemer, P., Dirr, H. W., Huber, R., Gilliland, G. L., Armstrong, R. N., Ji, X., Board, P. G., Olin, B., Mannervik, B., and Jones, T. A. (1993). Structure determination and refinement of human alpha class glutathione transferase A1-1 and a comparison with the mu and pi class enzymes. *J. Mol. Biol.* **232,** 192–212.

Stenberg, G., Board, P. G., and Mannervik, B. (1991). Mutation of an evolutionary conserved tyrosine residue in the active site of a human class Alpha glutathione transferase. *FEBS Lett.* **293,** 153–155.

Tan, K. H., Meyer, D. J., and Ketterer, B. (1984). Inhibition of microsomal lipid peroxidation by glutathione and glutathione transferases B and AA. Role of endogenous phospholipase A_2. *Biochem. J.* **220,** 243–252.

Taylor, J. B., Oliver, J., Sherrington, R., and Pemble, S. E. (1991). Structure of human glutathione S-transferase class Mu genes. *Biochem. J.* **274,** 587–593.

Tee, L. B. G., Gilmore, K. S., Meyer, D. J., Ketterer, B., Vandenberghe, Y., and Yeoh, G. C. T. (1992). Expression of glutathione S-transferase during rat liver development. *Biochem J.* **282,** 209–218.

te Koppele, J. M., Coles, B., Ketterer, B., and Mulder, G. J. (1988). Stereoselectivity of rat liver glutathione transferase isoenzymes for α-bromoisovaleric acid and α-bromoisovalerylurea enantiomers. *Biochem. J.* **252,** 137–142.

Telakowski-Hopkins, C. A., Rothkopt, G. S., and Pickett, C. B. (1986). Structural analysis of a rat liver glutathione transferase Ya gene. *Proc. Natl. Acad. Sci. U.S.A.* **83,** 9393–9347.

Thier, R., Pemble, S. E., Taylor, J. B., Humphreys, W. G., Persmark, M., Ketterer, B., and Guengerich, F. P. (1993). Expression of mammalian glutathione S-transferase 5-5

Salmonella typhimurium TA1535 leads to base-pair mutations upon exposure to dihalamethanes. *Proc. Natl. Acad. Sci. U.S.A.* **90,** 8576–8580.

Tsuchida, S., Izumi, T., Shimizu, T., Ishikawa, T., Hatayama, I., Satoh, K., and Sato, K. (1987). Purification of a new acidic glutathione S-transferase, GST-Yn_1Yn_1, with a high leukotriene-C_4 synthase activity from rat brain. *Eur. J. Biochem.* **170,** 159–164.

Tu, J.-T., Gao, R.-N., Zhang, D. H., and Gu, B.-C. (1985). Hepatitis B virus and primary liver cancer on Chongming island, Peoples Republic of China. *Natl. Cancer Inst. Monogr.* **69,** 213–215.

Ujihara, M., Tsuchida, S., Satoh, K., Sato, K., and Urade, Y. (1988). Biochemical and immunological demonstration of prostaglandin D_2, E_2 and $F_{2\alpha}$ formation from prostaglandin H_2 by various rat glutathione S-transferase isozymes. *Arch. Biochem. Biophys.* **264,** 428–435.

Ursini, F., Maiorino, M., and Gregolin, C. (1985). The selenoenzyme phospholipid hydroperoxide glutathione peroxidase. *Biochim. Biophys. Acta* **839,** 62–70.

van Kuijk, F. J. G. M., Sevanian, A., Handelman, G. J., and Dratz, E. A. (1987). A new role for phospolipase A_2: Protection of membranes from lipid peroxidation damage. *Trends Biochem.* **12,** 31–34.

Warholm, M., Guthenberg, C., and Mannervik, B. (1983). Molecular and catalytic properties of glutathione transferase μ from human liver: An enzyme efficiently conjugating epoxides. *Biochemistry* **22,** 3610–3617.

Xia, C. L., Cowell, I. G., Dixon, K. H., Pemble, S. E., Ketterer, B., and Taylor, J. B. (1991). Glutathione transferase π—Its minimal promoter and downstream cis-acting element. *Biochem. Biophys. Res. Commun.* **176,** 233–240.

Xia, C. L., Meyer, D. J., Chen, H., Reinemer, P., Huber, R., and Ketterer, B. (1993). Chemical modification of glutathione transferase P1-1 confirms the presence of Arg-13, Lys-44 and one carboxylate group in the GSH-binding domain of the active site. *Biochem. J.* **293,** 357–362.

Zhong, S., Howie, A. F., Ketterer, B., Taylor, J., Hayes, J. D., Beckett, G. J., Wathen, C. G., Wolf, C. R., and Spurr, N. K. (1991). Glutathione S-transferase mu locus: Use of genotyping and phenotyping assays to assess association with lung cancer susceptibility. *Carcinogenesis* (*London*) **12,** 1533–1537.

Zhong, S., Spurr, N. K., Hayes, J. D., and Wolf, R. C. (1993). Deduced amino acid sequence, gene structure and chromosomal location of a novel human class Mu glutathione S-transferase, GSTM4. *Biochem. J.* **291,** 41–50.

Enzymology of Cysteine S-Conjugate β-Lyases

Arthur J. L. Cooper

Departments of Biochemistry and Neurology
Cornell University Medical College
New York, New York 10021

I. Introduction

Certain halogenated xenobiotics have been known for over a hundred years to be converted within the body to the corresponding mercapturic acids. For example, administration of bromobenzene to experimental animals results in urinary excretion of β-bromophenylmercapturic acid [Br(Phe)$SCH_2CH(NHCOCH_3)CO_2H$] (Baumann, 1883). The mercapturate is generally more soluble than the original, unconjugated xenobiotic, and active transporters for organic acids aid in the detoxification process (Stevens and Jones, 1989). Mercapturic acid synthesis begins with the formation of a glutathione *S*-conjugate from the reaction of the xenobiotic with glutathione (GSH). This process is catalyzed by a number of glutathione *S*-transferases. With highly reactive electrophilic substances, however, substantial conjugate formation may occur nonenzymatically (Chasseaud, 1979; see also the review by Koob and Dekant, 1991), a process facilitated by the high concentration (1–10 m*M*) of glutathione in most tissues. Glutathione *S*-conjugates are converted to the corresponding cysteine *S*-conjugates by the sequential action of γ-glutamyltranspeptidase and dipeptidase; the resulting cysteine *S*-conjugates are converted to the corresponding *N*-acetylated cysteine *S*-conjugates (mercapturic acids) by *N*-acetyltransferases (Duffel and Jakoby, 1982; Jakoby, 1980; Stevens and Jones, 1989) (Fig. 1). In addition to its role in mercapturic acid formation,

Advances in Pharmacology, Volume 27

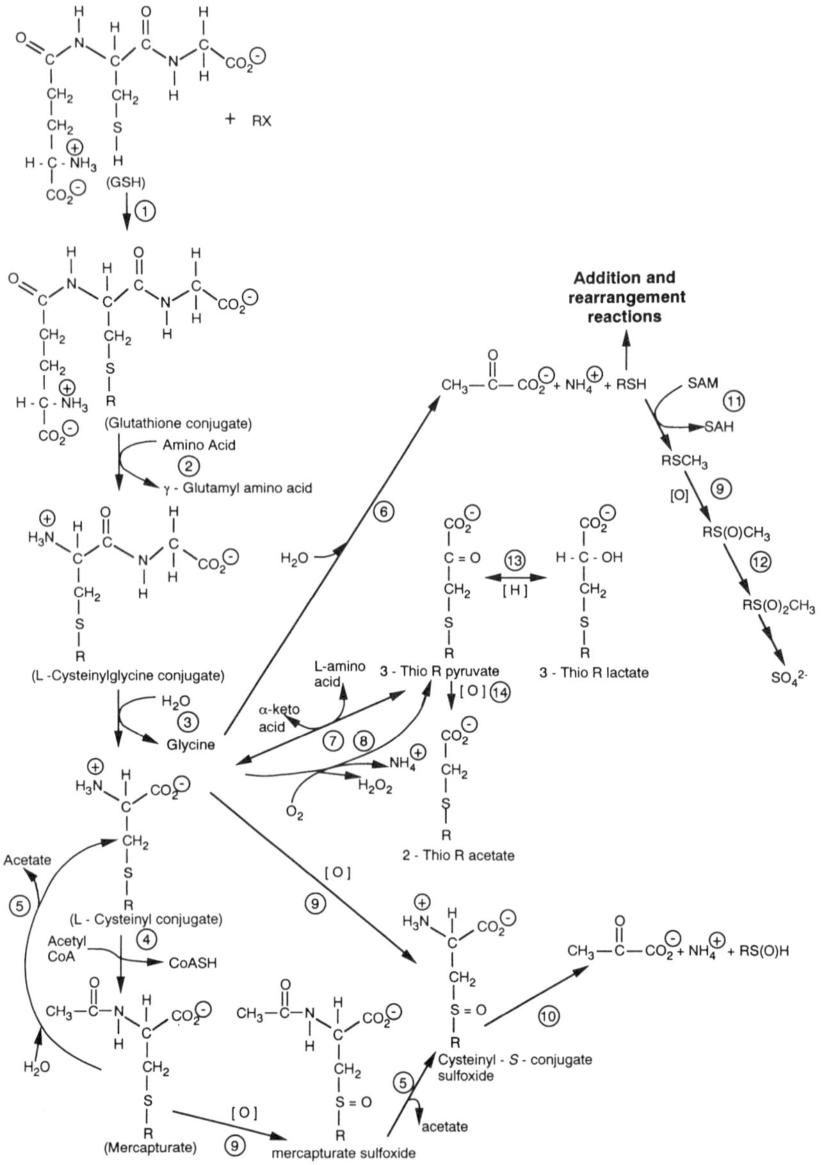

Fig. 1 Formation of mercapturates and some related reactions. (1) Glutathione *S*-transferases, (2) γ-glutamyltranspeptidase, (3) dipeptidase, (4) cysteine *S*-conjugate *N*-acetyl transferases, (5) deacylases, (6) cysteine *S*-conjugate β-lyases, (7) glutamine transaminase K, (8) L-amino acid oxidase (in the rat probably the B-(kidney) form of L-α-hydoxy acid oxidase (Hamilton, 1985)), (9) *S*-oxidase, (10) cysteine *S*-conjugate sulfoxide β-lyase (related to reaction and the alliinase reaction), (11) *S*-methyl transferase, (12) *S*-sulfoxide oxidase.

glutathione has many important biological functions including (*a*) protection against oxidative stress, (*b*) protection against free radicals, (*c*) storage of cysteine, and (*d*) enzyme regulation. Glutathione also plays a role in the maintenance of the cellular redox state, eicosanoid biology, and transport phenomena (e.g., Meister, 1989, 1992; Meister and Anderson, 1983). Thus, it is not surprising that depletion of glutathione, especially in the mitochondrial pool, leads to extensive tissue damage (e.g., Meister, 1992). Although glutathione is an important biomolecule it can, however, participate in reactions that result in bioactivation of foreign substances and in tissue damage.

The concept that the mercapturic acid pathway is an important route for the detoxification of certain drugs and other xenobiotics has been established for many years (e.g., Boyland and Chasseaud, 1969; Chasseaud, 1979; references cited in Koob and Dekant, 1991). Nevertheless, although the majority of glutathione *S*-conjugates are nontoxic and are converted to cysteine *S*-conjugates that are excreted as the mercapturic acids, there are notable exceptions, especially in the case of conjugates formed from haloalkenes. For example, the glutathione and cysteine *S*-conjugates formed from hexachloro-1,3-butadiene (Lock and Ishmael, 1979; Wolf *et al.*, 1984; Veltman *et al.*, 1988), tetrafluoroethylene (Odum and Green, 1984; Commandeur *et al.*, 1988, 1989; Boogaard *et al.*, 1989), 1,1-dichloro-2,2-difluoroethylene (Boogaard *et al.*, 1989; Commandeur *et al.*, 1988), 1,1-dibromo-2,2-difluoroethylene (Boogaard *et al.*, 1989; Commandeur *et al.*, 1988), 2-chloro-1,1,2-trifluoroethylene (Banki *et al.*, 1986; Boogaard *et al.*, 1989; Dohn *et al.*, 1985; Dekant *et al.*, 1987a), trichloroethylene (Dekant *et al.*, 1986; 1990; Elfarra *et al.*, 1986a,b), tetrachloroethylene (Green and Odum, 1985; Dekant *et al.*, 1987b; Commandeur *et al.*, 1989), hexafluoropropene (Green and Odum, 1985; Koob and Dekant, 1990), and 1,1,2-trichloro-3,3,3-trifluoro-1-propene (Vamvakas *et al.*, 1989) are all nephrotoxic in experimental animals. The haloalkane 1,2-dichloroethane and the haloalkyne dichloroacetylene are also nephrotoxic (Elfarra *et al.*, 1985). Trichloroethylene exposed to base is neurotoxic to humans (Buxton and Hayward, 1967; Schaumburg, 1992). Dichloroacetylene is directly neurotoxic to rabbits (Reichert *et al.*, 1976). Exposure to

---

GSH, glutathione; SAM, *S*-adenosylmethionine; SAH, *S*-adenosylhomocysteine. Names of compounds in the mercapturic acid pathway are shown in parentheses. The diagram is a composite of a large number of reactions occurring in many parts of the body (e.g., kidney, liver, lung, brain, gut, intestinal flora) and is meant to illustrate some of the key pathways that have been uncovered. The metabolic fate of halogenated xenobiotic (RX) depends on species, age, sex, nutritional status, and structure of R. For detailed discussions of these pathways see, for example, Stevens and Jones (1989) and Elfarra (1993).

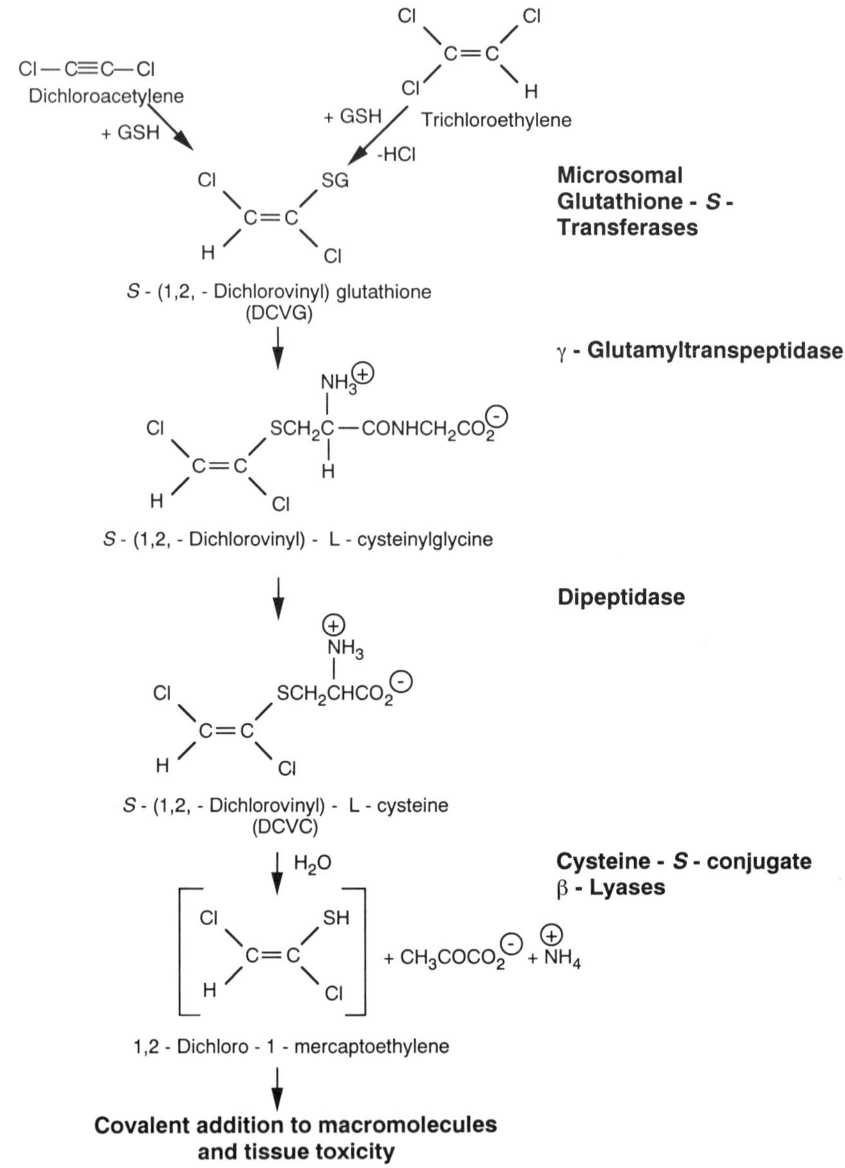

Fig. 2 Bioactivation of dichloroacetylene and trichloroethylene. In the first step, microsomal glutathione *S*-transferases catalyze the formation of *S*-(1,2-dichlorovinyl)-glutathione (DCVG) either via a direct addition reaction (with dichloroacetylene) or via an elimination reaction (-HCl; with trichloroethylene). DCVG is converted to *S*-(1,2-dichlorovinyl)-L-cysteinylglycine and thence to *S*-(1,2-dichlorovinyl)-L-cysteine (DCVC) by the sequential

base converts trichloroacetylene to dichloroacetylene; the latter is readily converted to a glutathione conjugate *in vivo* (Kanhai *et al.*, 1989). Indeed, the rate of glutathione *S*-conjugate formation with dichloroacetylene is the highest recorded for substrates of microsomal glutathione *S*-transferases thus far studied (Dekant *et al.*, 1989a).

Glutathione *S*-transferases are present in a number of tissues, such as liver, kidney, lung, and brain. In the liver, glutathione *S*-transferases account for >3% of the soluble protein (Ketterer *et al.*, 1989). The enzymes are present as cytosolic and membrane-bound forms. The cytosolic forms are dimers with subunit molecular weight of ~25,000 (Ketterer *et al.*, 1989). At least 11 distinct subunits of cytosolic glutathione *S*-transferases have been identified in the rat (Ketterer *et al.*, 1989). Interestingly, in immunohistochemical studies brain glutathione *S*-transferases have thus far only been detected in astrocytes and to a lesser extent in oligodendrocytes, but not in neurons (e.g., Cammer *et al.*, 1989; Campbell *et al.*, 1990). The enzymes may be involved in the normal functioning of the blood–brain barrier (Campbell *et al.*, 1990). Since astrocytic end feet surround the brain capillaries, glutathione *S*-transferases in the central nervous system may act as a second line of defense against potential toxins that can diffuse across the blood–brain barrier (Abramovitz *et al.*, 1988; Cammer *et al.*, 1989).

In general, glutathione *S*-transferases catalyze the attack of the nucleophilic thiol of glutathione at the electrophilic center of compounds possessing a lipophilic structure. Nucleophilic displacement (substitution), nucleophilic opening of an oxiran ring and Michael addition to polarized double bonds are all subject to subunit-specific glutathione *S*-transferase catalysis (Ketterer *et al.*, 1989). In the case of the haloalkenes, depending on the structure of the substrate, substitution or addition products, or both, may be formed (Elfarra, 1993). For example, addition occurs with chlorotrifluoroethylene and tetrafluoroethylene; substitution occurs with hexachlorobutadiene, trichloroethylene, and tetrachloroethylene; and addition and substitution occurs with hexafluoropropene (see Elfarra, 1993, and references cited therein). With the haloalkyne dichloroacetylene, an addition reaction occurs (Kanhai *et al.*, 1989). Quantitatively, the most important site of glutathione *S*-conjugate formation in the body is the liver; within the liver microsomal glutathione *S*-transferases are more effective in

---

action of γ-glutamyltranspeptidase and dipeptidase. DCVC is a substrate of cysteine *S*-conjugate β-lyases. The products of this reaction are pyruvate, ammonia, and 1,2-dichloro-1-mercaptoethylene. The toxic thiol fragment is very reactive and has not been isolated. However, much evidence suggests that its toxicity is due to covalent addition to important cellular macromolecules (see the text).

catalyzing glutathione *S*-conjugate formation with haloalkenes than are the hepatic cytosolic enzymes (Elfarra, 1993, and references quoted therein). For example, both trichloroethylene and dichloroacetylene are converted to *S*-(1,2-dichlorovinyl)glutathione selectively by hepatic microsomal glutathione *S*-transferases (Anders *et al.*, 1988; Dekant *et al.*, 1988a). The specific activity of lung microsomal glutathione *S*-transferase is about 12% that found in the liver; since the liver is four times as large as the lungs the total activity of microsomal glutathione *S*-transferases in lungs is about 3% relative to that in the liver (DePierre and Morgenstern, 1983). Nevertheless, it is conceivable that inhalation of highly reactive halogenated xenobiotics, such as dichloroacetylene, may lead to substantial glutathione *S*-conjugate formation in the lungs (Patel *et al.*, 1993). In many of the above-mentioned cases, ample evidence implicates the cysteine *S*-conjugates formed from the corresponding glutathione *S*-conjugate in the toxicity of haloalkenes and dichloroacetylene especially to kidney mitochondria. The cysteine *S*-conjugate is cleaved to pyruvate, ammonia, and a reactive thiol in reactions catalyzed by cysteine *S*-conjugate β-lyases (e.g., Elfarra and Anders, 1984; Cooper and Anders, 1990; Elfarra, 1993) (Fig. 2).

The mechanisms whereby certain glutathione and cysteine conjugates exert their toxicity have been discussed in several recent reviews (Elfarra and Anders, 1984; Anders *et al.*, 1986, 1988; Lash and Anders, 1986, 1987; Anders, 1988; Stevens and Jones, 1989; Stevens and Wallin, 1990; Koob and Dekant, 1991; Elfarra, 1993) and are further discussed elsewhere in this book. In this chapter, recent findings concerning (*a*) the identity of the cysteine *S*-conjugate β-lyases in mammals, (*b*) the reactions catalyzed by these enzymes, and (*c*) the pharmacological aspects of the cysteine *S*-conjugate-β-lyases will be reviewed.

II. Enzymes That Cleave the Thioether Bond of *S*-Substituted Cysteines

A number of enzymes are known to catalyze the cleavage of the proximal (i.e., the $S—C_3$) bond of *S*-substituted cysteines ($RSCH_2CH(NH_2)CO_2H$). In each case, the products are pyruvate (CH_3COCO_2H), ammonia (NH_3), and RSH. For example, an important enzyme in the pathway to formation of methionine in plants and microorganisms is β-cystathionase, which catalyzes the conversion of cystathionine ($R = —CH_2CH_2CH(NH_2)CO_2H$) to homocysteine, pyruvate, and ammonia (e.g., Wijesundera and Woods, 1962; Flavin and Slaughter, 1964). In another example, γ-cystathionase, which normally catalyzes a γ-elimination reaction with cystathionine to yield cysteine, ammonia, and α-ketobutyrate, can also catalyze β-elimination reactions from suitable substrates, such as cystine (R = $—SCH_2CH(NH_2)CO_2H$), djenkolic acid (R = $—CH_2SCH_2CH(NH_2)CO_2H$), and lanthionine (R = $—CH_2CH(NH_2)CO_2H$). With cystine the eliminated product is thiocysteine ($HSSCH_2CH(NH_2)CO_2H$) (see Braunstein and Goryachenkova, 1984, and references cited therein). In addition, a remarkable enzyme activity (alliinase) present in onions and garlic is able to convert simple alkyl sulfoxide analogues of cysteine ($RS(O)CH_2CH(NH_2)CO_2H$) to pyruvate, ammonia, and a sulfenic acid (RSOH); cysteine and *S*-methylcys-teine are not substrates (e.g., Schwimmer and Mazelis, 1963; Braunstein and Goryachenkova, 1984). A cysteine *S*-conjugate β-lyase isolated from an enteric bacterium and another from human liver are able to cleave cysteine conjugate sulfoxides (see below). These enzymes obviously catalyze reactions related to the alliinase reaction; however, in this chapter β-elimination reactions with cysteine *S*-conjugates will be discussed in most detail. Cysteine *S*-conjugates are defined as *S*-substituted cysteines that arise *in vivo* from reaction of xenobiotics (often, but not always, halogenated) with glutathione (Fig. 1, reactions **1, 2,** and **3**). By this definition simple *S*-alkylcysteines are *S*-cysteine conjugates because they can arise through the reaction of alkylating agents with glutathione or cysteine (e.g., Barnsley, 1964). Some cysteine *S*-conjugate β-lyases, particularly those isolated from bacteria, have broad specificity and can catalyze elimination reactions with (*a*) simple *S*-alkyl-L-cysteines (*b*) *S*-cysteine conjugates with activated R groups, such as *S*-(1,2-dichlorovinyl)-L-cysteine (DCVC), and (*c*) cystathionine and related compounds. Clearly, these cysteine *S*-conjugate β-lyases have considerable similarity and substrate overlap to the β- and γ-cystathionases. However, for the mammalian cysteine *S*-conjugate β-lyases thus far studied simple alkyl *S*-cysteine conjugates are not converted to pyruvate with elimination of HSR; for elimination to occur the —SR portion must be a good leaving group (i.e., be activated, usually by a halogen, double bond or an aromatic group).

One of the earlier experiments in which it was shown from tracer experiments that the sulfur of glutathione is incorporated into a thiol metabolite of a xenobiotic was carried out by Colucci and Buyske (1965). Benzothiazole-2-sulfonamide was shown to be converted in rats, rabbits, and dogs to *S*-(2-benzothiazolyl)glutathione and thence to the corresponding cysteine conjugate and the mercapturate; the sulfonamide portion was converted to sulfate. At about the same time, Schultze and colleagues reported that the eliminated fragment obtained from the "C—S-lyase" reaction on DCVC catalyzed by an enzyme from bovine organs (Anderson and Schultze, 1965) and an enzyme in *Escherichia coli* (Saari and Schultze,

1965) is very reactive and binds to proteins. Later, Bhattacharya and Schultze (1972) showed that ^{35}S in the active thiol released from ^{35}S-labeled DCVC by the β-lyase activity was readily bound to DNA.

Some xenobiotics have been known for some time to be readily converted in part *in vivo* to thiomethyl compounds (e.g., phenacetin, acetaminophen). At first it was assumed that a reactive ester was formed, followed by nucleophilic attack by the sulfur of methionine; the resulting sulfonium ion would then readily decompose to the methylthiol compound and homoserine (see Tateishi *et al.*, 1978b, for a discussion). However, Chatfield and Hunter (1973) provided strong evidence that the conversion of 2-acetamido-4-chloromethylthiazole to 2-acetamido-4-methylthiomethylthiazole in the rat involves the mercapturic acid pathway and a methylation reaction (Fig. 1, sequential reactions **1,2,3,6,11**). Further evidence for this pathway was obtained by Tateishi *et al.* (1978a), who investigated the methylthiolation of bromazepam in rats and identified the corresponding mercapturic acid conjugate and 6′-methylthiobromazepam in the bile. These authors also provided additional evidence for this pathway by showing that the thioether bonds of the cysteine conjugates of 2,4-dinitrobenzene and bromobenzene are readily cleaved by an enzyme present in rat liver cytosol. Incubation of *S*-(2,4-dinitrophenyl)cysteine with enzyme resulted in formation of pyruvate, ammonia, and (in this case the relatively stable thiol) 2,4-dinitrobenzenethiol. The thiol product was readily methylated with *S*-adenosylmethionine by a microsomal thiomethyl transferase. Tateishi *et al.* (1978b) were the first to use the term "cysteine conjugate β-lyase" to describe β-elimination reactions with cysteine conjugates.

III. Bacterial Cysteine *S*-Conjugate β-Lyases

Cysteine *S*-conjugate β-lyases are known to occur in enteric bacteria (Larsen and Stevens, 1986, and references cited therein). Suzuki *et al.* (1982) reported that the large intestine of rats contains a C—S lyase capable of cleaving *S*-(*p*-bromophenyl)cysteine but not the corresponding glutathione conjugate; little C—S lyase activity was found in the small intestine. Larsen and Bakke (1983) showed that the contents of the pig cecum, but not small intestine, contain a C—S lyase. These authors prepared ^{14}C-labeled glutathione, cysteine, *N*-acetylcysteine, and *N*-acetylcysteine (*S*-oxide) conjugates of propachlor (= 2-chloro-*N*-isopropylacetanilide). Each was shown to be converted to 2-mercapto-*N*-isopropylacetanilide by reactions that include scission of the C—S bond. Four bacterial species were found to catalyze the C—S lyase reaction. Later, Larsen (1985) showed that cysteine S-conjugate β-lyase activity was present in 27 of 43 intestinal bacteria studied. The β-lyase isolated from these sources exhibited broad activity toward *S*-aryl- and *S*-alkyl-linked cysteine conjugates of xenobiotics. The lyases from *Fusobacterium necrophorum* (Larsen *et al.*, 1983) and *F. varium* (Tomisawa *et al.*, 1984) have been partially purified and characterized. The enzymes have molecular weights of 228,00 and 70,000, respectively, and require pyridoxal 5′-phosphate as cofactor. The enzyme from *F. necrophorum* was purified about 50-fold and was shown to catalyze the C—S cleavage of the *S*-cysteine conjugate of propachlor (= (2-*S*-cysteinyl)-*N*-isopropylanilide), 1,2-dihydro-1-hydroxy-2-cysteinylnaphthalene, and *S*-(2-benzothiazolyl)cysteine. The propachlor cysteinyl conjugate was shown to be converted to 2-mercapto-*N*-isopropylacetanilide, pyruvate, and ammonia. The enzyme from *F. varium* was shown to be most active with aromatic cysteine *S*-conjugates (e.g., phenyl-, benzyl-, *p*-bromophenyl-) although some activity was noted with alkyl cysteine *S*-conjugates. Later, Larsen and Stevens (1986) purified a cysteine *S*-conjugate β-lyase by a factor of about 540-fold from the gastrointestinal bacterium *Eubacterium limosum*. This bacterium had the highest activity of cysteine *S*-conjugate β-lyase of the enteric bacteria studied. The enzyme has a molecular weight of about 75,000 and is composed of two subunits. The enzyme has a broad specificity and is active with the cysteine conjugate of propachlor, *trans*-9-hydroxy-10-[*S*-(L-cys-teinyl)]-9,10-dihydrophenanthrene, *S*-(benzothiazolyl)-L-cysteine, *S*-benzyl-L-cysteine, and *S*-(1,2-dichlorovinyl)-L-cysteine. The enzyme is also moderately active with the *S*-alkyl conjugate *S*-ethyl-L-cysteine. D-Conjugates are not substrates. Activity does not require added pyridoxal 5′-phosphate. The enzyme is inactivated, however, by carbonyl reagents such as hydroxylamine and cyanide, suggesting the presence of a carbonyl cofactor. Of interest is the finding that the enzyme is quite active with cystathionine, djenkolic acid, lanthionine and cystine. The *E. limosum* β-lyase is thus catalytically similar to β- and γ-cystathionases (in contrast, the *F. necrophorum* and *F. varium* β-lyases do not cleave cystathionine). More recently, Bernström *et al.* (1989) showed that rat fecal contents and purified *E. limosum* cysteine *S*-conjugate β-lyase catalyze the conversion of leukotriene E_4 to a sulfhydryl-containing fragment, 5-hydroxy-6-mercapto 7,9-*trans*-11,14-*cis*-eicosatetraenoic acid. The authors stated that the rat kidney enzyme (= glutamine transaminase K) is not active with leukotriene E_4. We agree that glutamine transaminase K is inactive with leukotriene E_4, but have found an enzyme in rat kidney that does indeed react with leukotriene E_4 (see below).

The *E. limosum* cysteine *S*-conjugate β-lyase exhibits β-lyase activity toward β-chloroalanine, but, in contrast to many pyridoxal 5′-phosphate-

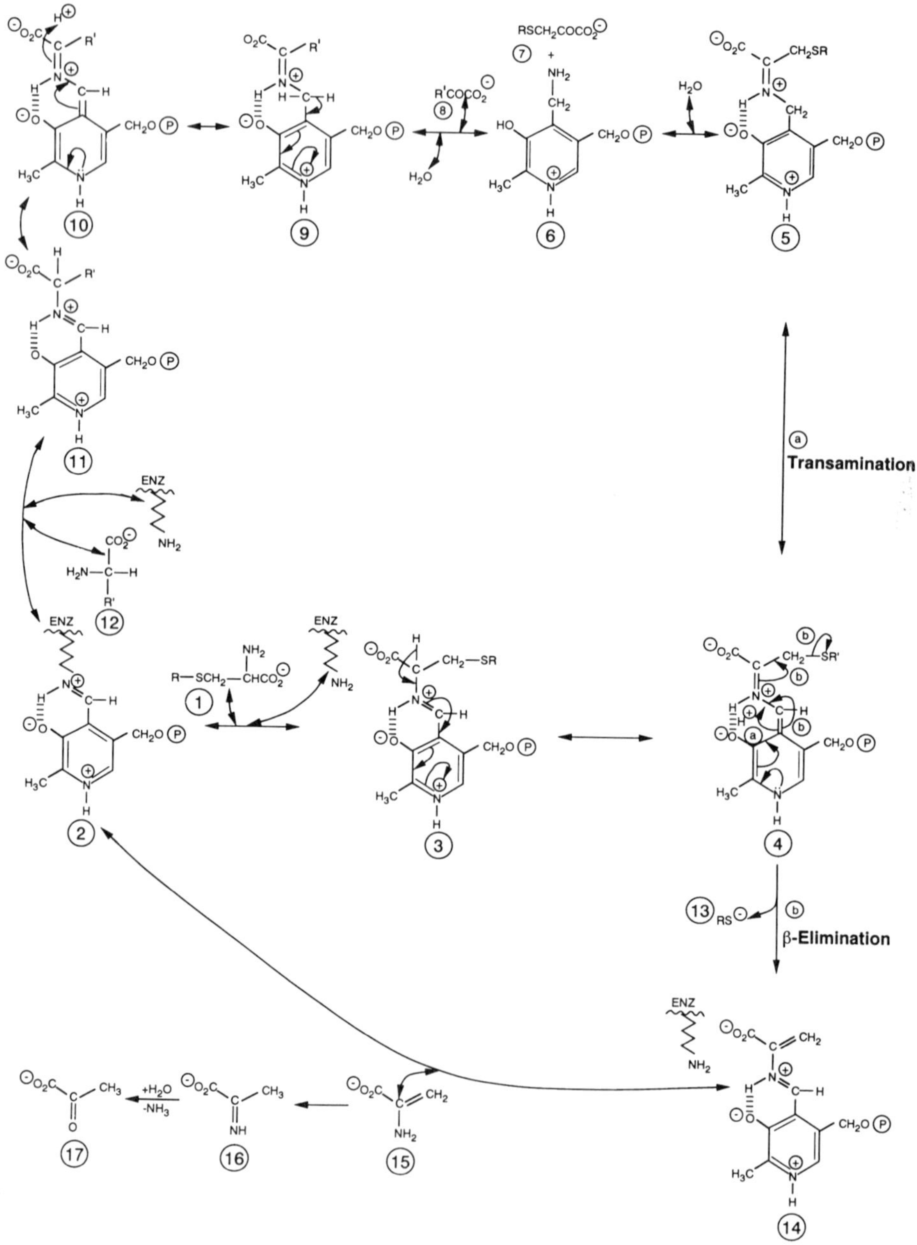
Transamination
β-Elimination
ENZ
R'COCO2
RSCH2COCO2
H2O
+H2O
-NH3
CH2O P

containing enzymes, it is not inactivated in the process. Bacterial kynureninase (Kishore, 1984), rat liver cysteine *S*-conjugate β-lyase (= kynureninase; see below) (Stevens and Jakoby, 1983), pig heart cytosolic aspartate aminotransferase (Morino and Okamoto, 1973), pig heart alanine aminotransferase (Golichowski and Jenkins, 1978), *E. coli* alanine racemase (Wang and Walsh, 1978), *Alcaligenes faecalis* L-aspartate-β-decarboxylase (Relyea *et al.*, 1974), and bacterial D-amino acid aminotransferase (Soper and Manning, 1978) can catalyze an elimination of halogen ions from β-haloalanines to yield pyruvate, but in each case the enzymes are inactivated in the process. The inactivating species is aminoacrylate bound in Schiff's-base linkage to the pyridoxal 5′-phosphate cofactor (Fig. 3, structure **14**) or free aminoacrylate (Fig. 3, structure **15**), which can (*a*) alkylate a susceptible group at the enzyme active site (Cavallini *et al.*, 1973; Soper and Manning, 1978) or (*b*) form an addition compound with the cofactor (Ueno *et al.*, 1982). In the case of L-aspartate-β-decarboxylase the inactivated enzyme was shown to contain 3-hydroxypyruvate bound in ester linkage to an active site glutamate (Relyea *et al.*, 1974). The reactions catalyzed by cysteine *S*-conjugate β-lyases probably occur via eneamine intermediates (structures **14** and **15**, Fig. 3) with both cysteine conjugate substrate and with β-chloroalanine.

Fig. 3 Proposed mechanism for the transamination and β-lyase reactions catalyzed by glutamine transaminase K. The extent to which transamination (route a) is favored over the elimination reaction (route b) is a function of the nature of the R group. When S is a simple aliphatic group, elimination is not favored. For example, *S*-methylcysteine is an excellent substrate of glutamine transaminase K (Cooper and Meister, 1981) but there is no evidence for elimination of CH_3SH. On the other hand, if R contains electron-withdrawing groups, elimination of RSH is facilitated. Thus, the ratio of α-keto-γ-methiolbutyrate-L-phenylalanine transaminase activity to lyase activity with *S*-(1,2-dichlorovinyl)-L-cysteine (R = CClH=CCl—) as substrate is about 5:1 (Stevens *et al.*, 1986b). **1,** cysteine conjugate substrate; **2,** pyridoxal 5′-phosphate bound at the active site of glutamine transaminase K through Schiff base linkage to the ε-amino group of a protein lysine; **3,** external aldimine I; **4,** quinonoid intermediate I; **5,** ketimine I; **6,** enzyme-bound pyridoxamine 5′-phosphate; **7,** α-keto acid I; **8,** α-keto acid II; **9,** ketimine II; **10,** quinonoid II; **11,** external aldimine II; **12,** amino acid substrate II; **13,** eliminated thiol-containing fragment; **14,** external aldimine III (eneamine intermediate); **15,** aminoacrylate; **16,** 2-iminobutyrate; **17,** pyruvate. The proposed mechanism is based on accepted mechanisms for pyridoxal 5′-phosphate-catalyzed transamination and β-elimination (e.g., Snell and Di Mari, 1970; Braunstein and Goryachenkova, 1984; Braunstein, 1985). The mechanism is also consistent with various nonenzymatic model systems capable of catalyzing competing transamination and β-elimination reactions (Kikuchi *et al.*, 1985; Dekant *et al.*, 1987a; 1988b,c). Note that the pathways shown depict a minimum sequence. There are likely to be many additional intermediate steps in typical enzyme-catalyzed transamination and β-lyase reactions (e.g., Braunstein, 1985). A less detailed mechanism has previously been presented by Stevens and Jones (1989).

Presumably, the *E. limosum* β-lyase is not inactivated by the product of the lyase reaction [aminoacrylate (or its Schiff's-base intermediate)] because aminoacrylate is rapidly converted to pyruvate or because no groups susceptible to alkylation are near the active center. The inactivation of kynureninases, aspartate aminotransferase, alanine aminotransferase, alanine racemase, aspartate-β-decarboxylase, and D-amino acid aminotransferase by aminoacrylate may be related to the fact that these enzymes catalyze β-elimination from 3-substituted alanines as a secondary reaction; the nonphysiological product (aminoacrylate) generated in the secondary reaction can then react with a base at the active site that is normally used as a catalyst. Also of interest is the finding that *E. limosum* β-lyase has very high activity with the *S*-oxide of the cysteine conjugate of propachlor (Larsen and Stevens, 1986). The $RS(O)^-$ group is a better leaving group than is RS^-. The product is presumably a sulfenic acid (Fig. 1, reaction **10**).

Evidently, cysteine *S*-conjugate β-lyases in the different enteric bacteria have very varied physical and catalytic properties. Earlier work showed that gut bacteria are important in catalyzing biotransformations of xenobiotics (e.g., Goldman, 1978) including the conjugate formation with glutathione (e.g., Bakke *et al.*, 1981). This earlier work and the more recent findings of the widespread presence of cysteine *S*-conjugate β-lyases in enteric bacteria and the broad specificity of these enzymes suggest that the gut flora may be important for the *in vivo* introduction of sulfur into some xenobiotics (see Larsen and Stevens, 1986). For example, propachlor does not appear to be a substrate for the mammalian cysteine *S*-conjugate β-lyases, and germ-free rats do not convert propachlor to 2-mercapto-*N*-isopropylacetanilide (Bakke *et al.*, 1981). As noted above, simple *S*-alkyl- and *S*-aryl-L-cysteine conjugates are generally substrates of the bacterial cysteine *S*-conjugate β-lyases. Of the mammalian enzymes characterized thus far (see below), none has lyase activity toward simple *S*-alkylcysteines although the human liver enzyme is very active with a number of *S*-arylcysteines. Larsen and Stevens (1986) point out that the bile is a common excretory route for intermediates of the mercapturic acid pathway and that the gut is exposed to a wide spectrum of cysteine and glutathione *S*-conjugates. Many of the thiols generated from cysteine *S*-conjugates by the action of β-lyases are relatively stable and presumably can be transported throughout the body. However, others, such as 1-mercapto-1,2-dichloroethylene generated from β-lyase activity on DCVC, are exceedingly reactive and will add to macromolecules locally (Section I). Whether the cysteine S-conjugate β-lyases in enteric bacteria can contribute to the production of toxic species from cysteine conjugates (such as DCVC) that can locally damage the gut remains to be determined.

Whether more stable thiol products generated from other cysteine *S*-conjugates in the gut can be transported to other regions to exert toxic effects also needs to be further evaluated (Larsen and Stevens, 1986).

IV. Mammalian Cysteine *S*-Conjugate β-Lyases

A. Human Liver Cysteine *S*-Conjugate β-Lyase

Tomisawa *et al.* (1986) purified a cysteine *S*-conjugate β-lyase from post-mortem human liver by about 880-fold. As noted above, the enzyme is active with *S*-arylcysteines but not with simple *S*-alkylcysteines. Of interest is the finding that the enzyme is nine times more active with *S*-phenylcysteine sulfoxide than with *S*-phenylcysteine. No activity was detected with L-cystathionine or with L-kynurenine. The enzyme was found to have a native molecular weight of 88,000, and to require pyridoxal 5′-phosphate for full activity. The enzyme exhibits a pH optimum of 8.5 and catalyzes stoichiometric formation of *p*-bromophenylmercaptan, pyruvate, and ammonia from *S*-(*p*-bromophenyl)-L-cysteine.

B. Rat Liver Cysteine *S*-Conjugate β-Lyase/Kynureninase

As noted above, Tateishi *et al.* (1978a) partially purified an enzyme from rat liver that cleaves *S*-(2,4-dinitrophenyl)-L-cysteine to pyruvate, ammonia and 2,4-dinitrophenylmercaptan. Subsequently, a β-lyase of rat liver was purified to homogeneity (Stevens and Jakoby, 1983) and was later shown to be identical to kynureninase (Stevens, 1985a). Kynureninase appears to be the major lyase of rat liver, at least when DCVC is employed as the substrate. The enzyme has a native molecular weight of about 100,000 and is composed of two identical subunits (M_r ~55,000) (Takeuchi *et al.*, 1980; Stevens, 1985a). The enzyme contains tightly bound pyridoxal 5′-phosphate. Kynureninase catalyzes the hydrolytic cleavage of kynurenine to anthranilate and alanine; the enzyme also catalyzes the hydrolytic cleavage of 3-hydroxykynurenine to 3-hydroxyanthranilate and alanine. Pyruvate is not a product. On the other hand, if a cysteine *S*-conjugate containing a good leaving group binds at the active site, electron flow from the quinonoid intermediate **4** (Fig. 3) toward the sulfur is facilitated, resulting in β-elimination of aminoacrylate; aminoacrylate undergoes spontaneous conversion to pyruvate (pathway **b,** Fig. 3) DCVC is a substrate of the rat liver enzyme but this enzyme is slowly inactivated by this amino acid and by β-chloroalanine in a pseudo-first-order fashion. In both cases, partitioning at the active site between events that lead to formation of pyruvate and events that lead to an inactivation is about 600

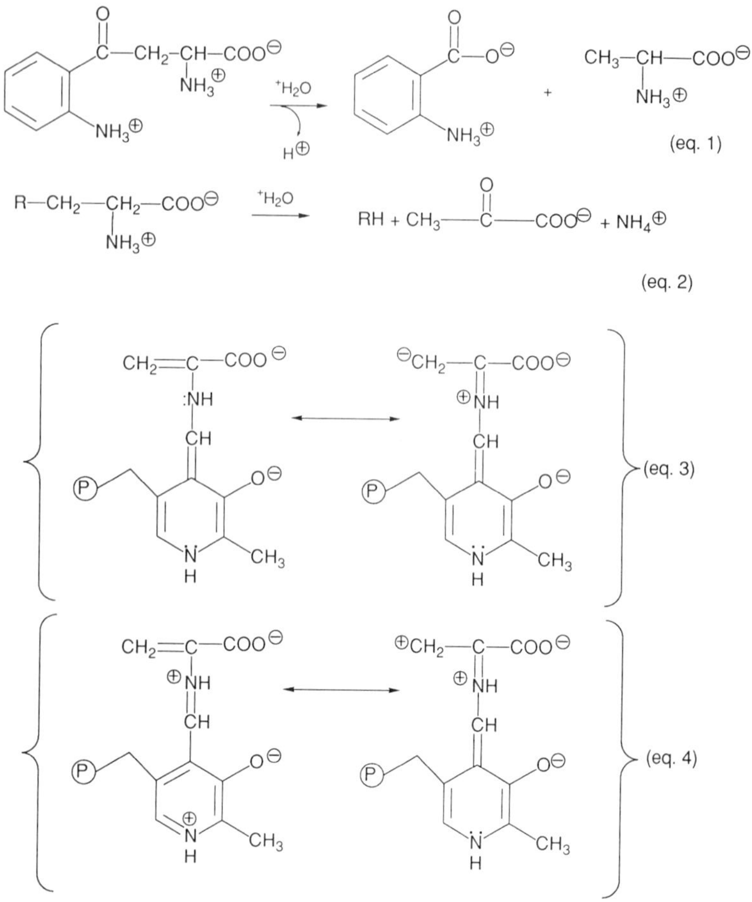

Fig. 4 Differences in the reaction mechanisms for the hydrolytic cleavage of kynurenine and cysteine *S*-conjugate β-lyase reactions catalyzed by kynureninase. Eq. (1), the "normal" biological reaction; Eq. (2), the β-lyase reaction; Eq. (3), resonance hybrid structures showing the partial carbanion nature of the eneamine intermediate resulting from kynurenine cleavage; Eq. (4), resonance hybrid structures showing the partial carbonium nature of the eneamine intermediate resulting from the β-lyase reaction. Reprinted from *J. Biol. Chem.* **260,** J. L. Stevens, Isolation and characterization of a rat liver enzyme with both cysteine conjugate β-lyase and kynureninase activity, pp. 7945–7950, © 1985, with permission from Pergamon Press, Ltd., Oxford, United Kingdom.

(Stevens and Jakoby, 1983). In this regard it is interesting to note that *Pseudomonas marginalis* kynureninase also catalyzes a β-elimination reaction with β-chloroalanine and with the cysteine conjugate *S*-(*ortho*-nitrophenol)-L-cysteine (Kishore, 1984). Similarly, for this enzyme the partition ratio between pyruvate formation and inactivation is about 500. Although the normal reaction catalyzed by kynureninase is hydrolytic cleavage at the β position of kynurenine, this reaction differs from a true β-elimination reaction. During the kynureninase reaction net loss of a proton from the α-position of the amino acid substrate does not occur and the β-carbon of the eneamine intermediate has carbanion character (Bild and Morris, 1984). After hydrolytic cleavage and protonation of the carbanionic cofactor intermediate, the resulting intermediate is alanine bound as a Schiff's base to the pyridoxal 5′-phosphate cofactor; this intermediate is subsequently hydrolyzed to free alanine (Bild and Morris, 1984; Stevens, 1985a). On the other hand, β-elimination results in net loss of the α proton and in formation of an eneamine intermediate which has carbonium character at the β-carbon (Fig. 3, structure **14**) (Snell and Di Mari, 1970; Stevens, 1985a). These differences are summarized in Fig. 4.

Kynureninase is an enzyme in the pathway by which tryptophan is converted to quinolinate. Quinolinate (via nicotinate mononucleotide) is a precursor of the important cofactor NAD^+. However, a substantial fraction of the daily requirement for this cofactor is obtained via nicotinate/nicotinamide in the diet. Clearly, exposure of the liver to cysteine *S*-conjugates will lead to diminished activity of kynureninase in that organ. Whether this loss of activity contributes to the hepatic toxicity associated with some cysteine *S*-conjugates is not clear (Stevens, 1985a).

C. Kynureninase-Type Cysteine *S*-Conjugate β-Lyase in the Brain

In the central nervous system, quinolinate is a powerful excitant and convulsant that activates the *N*-methyl-D-aspartate (NMDA) receptors (Stone and Burton, 1988). When injected intracerebroventricularly in rodents, quinolinate and other kynurenine metabolites elicit significant behavioral and convulsant effects (Lapin, 1981). Moreover, direct infusion of quinolinate into brain causes a lesion that resembles those found in Huntington's disease and temporal lobe epilepsy (Schwarcz *et al.*, 1988). Because quinolinate and other kynurenine metabolites are endogenous brain substances that are known to elicit strong excitatory responses, it has been suggested that imbalance in their production, transport, and catabolism may lead to cerebral damage and may even be involved in the pathogenesis of some neurodegenerative and psychiatric diseases (Stone and Burton, 1988; Schwarcz *et al.*, 1988; Heyes *et al.*, 1990; see also the discussion by Fukui *et al.*, 1991, and the recent commentary by Holzman, 1993). Indeed, substantial levels of quinolinate have been found in the cerebrospinal fluid and brains of patients with a broad spectrum of infections and other inflammatory neurological diseases including acquired immune deficiency syndrome (AIDS), poliovirus infection of the spinal chord, Lyme disease, and septicemia (Heyes *et al.*, 1992, and references cited therein). A strong correlation was found between the level of quinolinate in the cerebrospinal fluid and neurological impairment in human immunodeficiency virus (HIV) patients (Heyes *et al.*, 1991). Since substantial amounts of kynurenine are made in extracerebral tissues, it is possible that the brain kynurenine and its metabolites originate from elsewhere in the body. However, kynurenine formation and enzymes involved in its metabolism (including kynureninase) are now known to occur in the brain; earlier estimates suggest that perhaps 40% of brain kynurenine is synthesized endogenously (Gál and Sherman, 1980). More recent work by Fukui *et al.* (1991) suggests that kynurenine (and probably 3-hydroxykynurenine) is taken up across the blood–brain barrier at a significant rate on the large neutral amino acid (L-system) carrier; anthranilate crosses the blood–brain barrier at a significant rate by passive diffusion, whereas 3-hydroxyanthranilate, kynurenate, and quinolinate enter the brain poorly. Heyes *et al.* (1992) suggest that increased quinolinate in the pathological brains may in part be a consequence of increased indoleamine-2,3-dioxygenase activity, perhaps in the macrophages. Glutathione and cysteine *S*-conjugates are known to readily cross the blood–brain barrier (Patel *et al.*, 1993). Inactivation of kynureninase in the liver and brain by cysteine *S*-conjugates could profoundly affect the cerebral levels of kynurenine and its metabolites. In experimental animals treated with cysteine *S*-conjugates, brain kynurenine and kynurenate levels would be expected to rise, whereas the level of quinolinate would be expected to drop. The consequences of such a disturbance are unknown. However, given the interest in the quinolinate pathway, this is an area that should be worth investigating.

D. Rat Kidney Cysteine *S*-Conjugate *β*-Lyase/Glutamine Transaminase K

Stevens (1985b) showed that rat kidney possesses DCVC lyase activity in the cytosolic and mitochondrial fractions that is immunologically distinct from kynureninase of the rat liver. In LLC-PK1 kidney cells the toxicity of *S*-(1,2-dichlorovinyl)glutathione (DCVG) is dependent on its conversion to DCVC and metabolism of DCVC (Stevens *et al.*, 1986a). Moreover, the activity of cysteine *S*-conjugate *β*-lyase is stimulated in homogenates of these cells by addition of pyruvate; this stimulation is accompanied by increased covalent binding of sulfur-containing fragments from DCVC to macromolecules. Blockage of cysteine *S*-conjugate β-lyase activity by aminooxyacetate (a general inhibitor of pyridoxal 5′-phosphate-containing enzymes) prevents the formation of DCVC fragments and thereby reduces the binding of DCVC-derived metabolites to macromolecules in LLC-PK1 cells (Stevens *et al.*, 1986a). Aminooxyacetate protects *in vivo* (Elfarra *et al.*, 1986a) and *in vitro* (Lash and Anders, 1986) against the toxicity of DCVC. Aminooxyacetate protects isolated rat renal tubular cells from the toxic effects of *S*-(2-chloro,1,1,2-trifluoroethyl)glutathione and *S*-(2-chloro-1,1,2-trifluoroethyl)-L-cysteine (Dohn *et al.*, 1985). *S*-(Pentachlorobutadienyl)-L-cysteine poisons isolated rat liver mitochondria respiring on succinate; the mitochondria are protected by the addition of aminooxyacetate (Wallin *et al.*, 1987). Addition of aminooxyacetate to rat kidney homogenates completely abolished the conversion of *S*-(1,1,2,2-tetrafluoroethyl)-L-cysteine, *S*-(2,2-dichloro,1,1-difluoroethyl)-L-cysteine, and *S*-(2,2-dibromo-1,1-difluorethyl)-L-cysteine to pyruvate (Commandeur *et al.*, 1988). Pentachlorobutadienyl glutathione is toxic to isolated renal epithelial cells; the toxicity is completely prevented by addition of the pyridoxal 5′-phosphate antagonist semicarbazide (Jones *et al.*, 1986). Moreover, the α-methyl analogues of DCVC (Elfarra *et al.*, 1986b) and *S*-(1-chloro-1,2,2-trifluoroethyl)-L-cysteine (Dohn *et al.*, 1985) are not toxic. These compounds cannot form a quinonoid intermediate with the pyridoxal 5′-phosphate cofactor (Fig. 3, structure **4**) and hence cannot undergo β-elimination.

In crude rat kidney homogenates it was shown that endogenous L-amino acid oxidase (which converts cysteine *S*-conjugates to α-keto acids) is required for maximal activity; after removal of the L-amino acid oxidase activity during purification, the β-lyase activity becomes almost fully dependent on an added α-keto acid substrate (Stevens *et al.*, 1986b; see also Elfarra *et al.*, 1987). Moreover, *S*-(1,2-dichlorovinyl)-D-cysteine was shown to be almost as nephrotoxic as the L-isomer in rabbits (Wolfgang *et al.*, 1989). Since kidney contains a large amount of D-amino acid oxidase (Meister *et al.*, 1960), it is probable that the D-isomer is readily converted to the corresponding α-keto acid, which is predicted to be an excellent substrate of glutamine transaminase K. Taken together the data strongly suggested that the cytosolic kidney enzyme requires pyridoxal 5′-phosphate and is regulated by a transamination step. Subsequently, a major kidney cysteine *S*-conjugate β-lyase was highly purified from the cytosolic fraction of rat kidney homogenates and shown to be identical to cytosolic glutamine transaminase K (Stevens *et al.*, 1986b). Thus, a major cysteine *S*-conjugate β-lyase of the rat liver (= kynureninase) and another of the rat kidney (= glutamine transaminase K) are actually enzymes of known metabolic functions that are "coerced" into catalyzing a nonphysiological β-elimination reaction with cysteine *S*-conjugates containing strong electron-withdrawing groups. The α-keto acid requirement of the kidney cysteine *S*-conjugate β-lyase can be rationalized by inspection of Fig. 3. The conversion of cysteine *S*-conjugate (**1**) to pyruvate (**17**), reactive mercaptan (**13**), and ammonia via the elimination reaction (pathway **b**) results in the regeneration of pyridoxal 5′-phosphate (PLP) (actually in Schiff's-base linkage to an active site lysine residue). Thus, after one turnover through the β-elimination pathway, the enzyme is capable of productively binding another equivalent of cysteine conjugate. However, if transamination occurs (pathway **a**) the cysteine conjugate and cofactor are converted to the corresponding α-keto acid (**7**) and pyridoxamine 5′-phosphate (PMP, **6**), respectively. The pyridoxamine 5′-phosphate form of the enzyme cannot productively bind the cysteine conjugate (amino acid) substrate and the β-lyase reaction ceases. In order for the enzyme to be regenerated to the active pyridoxal 5′-phosphate form, an α-keto acid (structure **8**; e.g., phenylpyruvate, α-keto-γ-methiolbutyrate) must be supplied.

The above discussion suggests that the activities of both the liver and the kidney cysteine *S*-conjugate β-lyases may be regulated *in vivo* (see the discussion by Stevens and Jones, 1989). The liver cysteine *S*-conjugate β-lyase (kynureninase) may be regulated by inactivation with aminoacrylate, in which case restoration of enzyme activity will require new enzyme synthesis. [However, it is possible that *in vivo* the enzyme will be protected to some extent by small molecular weight thiols that can react with aminoacrylate before it has a chance to react with the enzyme (see discussion by Cavallini *et al.*, 1973).] On the other hand, the kidney enzyme (glutamine transaminase K) may be regulated by transamination and the availability of a suitable α-keto acid substrate. This process is reminiscent of the inactivation of certain amino acid decarboxylases by the occasional transamination reaction during turnover of substrate. Transamination yields the inactive pyridoxamine 5′-phosphate form of the enzyme. Activity is restored to hog brain glutamate decarboxylase by addition of pyridoxal 5′-phosphate (Spink and Martin, 1983). In the case of bacterial aspartate decarboxylase, activity is restored by addition of pyruvate, which ensures that the enzyme is converted back to the active pyridoxal 5′-phosphate form by transamination (Novogrodsky *et al.*, 1963). The extent to which the liver and kidney cysteine *S*-conjugate β-lyases are regulated *in vivo* by the above-mentioned mechanisms is not clear. Recently, it was shown, however, that activity of rat kidney cysteine *S*-conjugate β-lyase could be induced by injection of a single nonnephrotoxic dose of *N*-acetyl-*S*-(1,2,3,4,4-pentachloro-1,3-butadienyl)-L-cysteine (MacFarlane *et al.*, 1993). The enzyme and its mRNA were both elevated. On the other hand, repeated nephrotoxic doses resulted in a reduction of the lyase activity. The data suggest that at least some halogenated xenobiotics can cause induction of glutamine transaminase K/cysteine *S*-conjugate β-lyase in rat kidney, but that activity drops on destruction of kidney tissue (MacFarlane *et al.*, 1993).

E. Properties of Cytosolic Rat Glutamine Transaminase K/Cysteine *S*-Conjugate *β*-Lyase

In the late 1940s Greenstein and co-workers discovered that rat liver contains two glutaminases which they named "glutaminase I" and "glutaminase II" (e.g., Greenstein, 1949). Glutaminase I was shown to be activated by phosphate and to a lesser extent by arsenate; glutaminase II was shown to require an α-keto acid. Subsequently, Meister and colleagues (1952) showed that glutaminase II is a composite of two enzymes, namely a glutamine transaminase [Eq. (1)] and ω-amidase [Eq. (2)].

$$\text{L-Glutamine} + \alpha\text{-keto acid} \rightleftharpoons \alpha\text{-ketoglutaramate} + \text{L-amino acid} \quad (1)$$

$$\alpha\text{-Ketoglutaramate} + H_2O \rightarrow \alpha\text{-ketoglutarate} + NH_3 \quad (2)$$

The product of the transamination reaction with glutamine, i.e., α-ketoglutaramate, cyclizes to a lactam; the equilibrium favors the lactam by >99% at physiological pH (Meister, 1953; Hersh, 1971). The ω-amidase, which is of high activity in liver and kidney, however, has a strong affinity for the open-chain form of α-ketoglutaramate (Meister, 1953; Hersh, 1971). Thus, although the transaminase reaction is freely reversible, the reaction is drawn toward glutamine utilization either by cyclization of product or conversion of product to α-ketoglutarate. In a later study the rat liver glutamine transaminase was purified and characterized (Cooper and Meister, 1972). The enzyme has broad specificity toward α-keto acids but it is most active with the α-keto analogues of glutamine and methionine; the enzyme may therefore be regarded as a fully reversible glutamine–methionine aminotransferase. The enzyme, which has a native M_r of about 100,000 (two identical subunits), contains very tightly bound pyridoxal 5′-phosphate. Initial studies by Meister *et al.* (1952) suggested that highest specific activity of glutamine-pyruvate transaminase activity in the rat is in the liver with a lower specific activity in the kidney. However, by using phenylpyruvate as the α-keto acid acceptor, Kupchik and Knox (1970) concluded that the specific activity of glutamine transaminase was higher in rat kidney than in liver. Upon investigating these apparently contradictory claims, Cooper and Meister (1974) discovered that rat kidney contains a glutamine transaminase that is distinct from the predominant liver form. The two forms of glutamine transaminase were named glutamine transaminase K and L, respectively.

Glutamine transaminase K contains tightly bound pyridoxal 5′-phosphate, has an apparent M_r of ~90,000, and is composed of two identical subunits (e.g., Cooper and Meister, 1981). As with the L form, the K form has a broad substrate specificity. However, unlike the L form the K enzyme has a very high affinity for aromatic α-keto acids (K_m phenylpyruvate $<$10–20 μM). The enzyme may be regarded as fully reversible glutamine (methionine) aromatic amino acid aminotransferase. In general amino acid substrates have the following structure:

$$X(CH_2)_nCH(NH_2)CO_2H \tag{3}$$

where n = 1 or 2, and X is a large noncharged group (as in, for example, glutamine, methionine, phenylalanine, *S*-methylcysteine). Similarly, α-keto acid substrates have the following structure:

$$X(CH_2)_nCOCO_2H \tag{4}$$

(e.g., α-ketoglutaramate, phenylpyruvate, α-keto-γ-methiolbutyrate). In addition, both the cytosolic rat kidney enzyme (Cooper and Anders, 1990) and the bovine enzyme (Costa *et al.*, 1986, 1987; Ricci *et al.*, 1986) can catalyze transamination of cystine, lanthionine, homocystine, and cystathionine (see below) with a suitable α-keto acid acceptor. Thus, it is not surprising that cysteine *S*-conjugates (n = 1) and homocysteine *S*-conjugates (n = 2) (see Section V) are substrates for glutamine transaminase K. Other examples include the cysteine conjugate *S*-(1,2-dichlorovinyl)cysteine (DCVC) (n = 1 and X = —SCCl═CClH) and the homocysteine conjugate *S*-(benzothiazolyl)-L-homocysteine (n = 2, X = C_7H_4NS—) (see Section V). Unlike the rat liver cysteine *S*-conjugate β-lyase (= kynureninase), however, the rat kidney cysteine *S*-conjugate β-lyase (= glutamine transaminase K) does not appear to be inactivated by aminoacrylate (or its equivalent cofactor-bound species) following β-elimination.

A recent report by Gibson *et al.* (1993) indicates that the full-length cDNA clone for rat kidney glutamine transaminae K/cysteine *S*-conjugate β-lyase encodes for a protein subunit with a predicted M_r of 47,800. The same group has published the subunit amino acid sequence predicted from the cDNA clone (Perry *et al.*, 1993). In our laboratory we have carried out amino acid analyses on fragments of the enzyme obtained after tryptic digestion. Our (more direct) amino acid analysis is generally in agreement with the data of Perry *et al.*, with some exceptions (D. G. Abraham and A. J. L. Cooper, unpublished observations). We also note that according to Perry *et al.* (1993) β-lyase mRNA can only be detected in the rat kidney, but as noted below, glutamine transaminase K/cysteine *S*-conjugate β-lyase activity is widespread in rat tissues. The reason for the apparent discrepancy is not known but may be due to the presence of isoforms of the enzyme in rat tissues.

F. Tissue Distribution of Glutamine Transaminase K/Cysteine *S*-Conjugate β-Lyases in Mammals

In rat tissues glutamine transaminase K is most active in kidney followed by liver; activity has also been shown to be present in brain, heart, skeletal muscle, pancreas, testes, small intestine, spleen, bone marrow, and lung (Cooper, 1988; Jones *et al.*, 1988). α-Keto-γ-methiolbutyrate-stimulated cysteine *S*-conjugate β-lyase activity with DCVC as substrate in rat tissues roughly parallels the transaminase activity (Jones *et al.*, 1988).

Cavallini and colleagues have partially purified a transaminase from bovine liver (Costa *et al.*, 1986), kidney (Ricci *et al.*, 1986), and brain (Costa *et al.*, 1987) that has a very broad amino acid specificity. The enzyme is active with glutamine, methionine, and various sulfur-containing amino acids. The enzyme, as recognized by the authors, is similar to rat kidney glutamine transaminase K and has a high affinity toward phenylpyruvate. As with the rat enzyme, the bovine kidney enzyme contains pyridoxal 5′-phosphate and is a homodimer, and the M_r of the subunit is estimated to be ~47,000. The bovine glutamine transaminase purified by Costa *et al.* (1986) is the same enzyme as the cysteine *S*-conjugate β-lyase purified from bovine kidney by Lash and Anders (1988; see also Dekant *et al.*, 1987a). Lash *et al.* (1990a) have also purified and characterized cytosolic human kidney cysteine *S*-conjugate β-lyase. The human enzyme is about the same size as the rat enzyme, contains pyridoxal 5′-phosphate, and is also composed of two identical subunits. The enzyme also copurifies with glutamine transaminase K and is strongly inhibited by aminooxyacetate. Similar to that in the rat kidney (Jones *et al.*, 1988), the activity was found to be present in the cytosolic, mitochondrial, and microsomal fractions, but with the highest specific activity in the cytosol. Overall, the specific activity of the enzyme in the human kidney is about 10% that in the rat kidney; however, enough activity is present to suggest that the human kidney may be susceptible to long-term damage resulting from exposure to reactive cysteine conjugates (Lash *et al.*, 1990a). Curiously, differences in the response of the various kidney β-lyases to added α-keto acids have been reported. Thus, the β-lyase activity with DCVC is stimulated 30-fold by addition of α-keto-γ-methiolbutyrate to the purified rat kidney enzyme. Addition of this α-keto acid to purified bovine kidney enzyme (Lash and Anders, 1988) and purified human kidney enzyme (Lash *et al.*, 1990a) results in only ~2- and 1.3-fold stimulation of β-lyase activity with *S*-(2-benzothiazolyl)-L-cysteine. The lesser dependence of lyase activity on added α-keto acids noted with the human and bovine kidney enzymes is possibly due to a more favorable orientation of the quinonoid intermediate (Fig. 3, structure **4**) for β-elimination in these enzymes compared to that in the rat enzyme, or it is possible that under the conditions used 2-mercaptobenzothiazole is a better leaving group than is 1,2-dichloro-1-mercaptoethylene. *S*-(1,2-Benzothiazolyl)-L-homocysteine was also shown to be a substrate of the human and bovine enzymes (Lash and Anders, 1988; Lash *et al.*, 1990b). In this case the reaction is strongly dependent on added α-keto acid; the first step in the reaction is a transaminase reaction followed by spontaneous, nonenzymatic γ-elimination to yield 2-mercaptobenzothiazole. This reaction will be dealt with in more detail in Section V.

G. Metabolic Functions of Glutamine Transaminase K

This topic has been reviewed previously (Cooper, 1988; Cooper and Anders, 1990); therefore, only a brief outline will be given here. The major route for formation of ammonia from glutamine (amide) in the whole body is probably via the glutaminase reaction. It is possible, however, that the combined action of glutamine transaminases and ω-amidase [Eqs. (1) and (2)] may produce ammonia from glutamine in relatively large amounts in localized areas. Glutamine transaminase is particularly enriched in the S_1, S_2, and S_3 regions of the nephron of the rat kidney (Jones *et al.*, 1988; H. Endou and A. J. L. Cooper, unpublished observation) and in the choroid plexus of the brain (Cooper *et al.*, 1993) and in these regions may contribute to ammonia formation. These regions are involved in amino acid recycling, but whether glutamine transaminase K is involved in amino acid transport across cell membranes in these areas remains to be determined. Additionally, since glutamine transaminase K is present in both cytosolic and mitochondrial fractions (see below) the enzyme may be involved in amino acid transfer between the two compartments.

The α-keto analogues of a number of essential amino acids (e.g., phenylalanine, methionine) may arise *in vivo* from nonspecific transamination reactions. For example, cytosolic and mitochondrial aspartate aminotransferases (Miller and Litwack, 1971; Shrawder and Martinez-Carrion, 1972) and asparagine transaminase (Cooper, 1977) exhibit some activity with the aromatic amino acids. At least one of the branched-chain amino acid aminotransferases (Ikeda *et al.*, 1976) has significant activity with methionine. Glutamine transaminases may act normally to salvage phenylpyruvate, α-keto-γ-methiolbutyrate and other α-keto acids arising by nonspecific transamination reactions. Without such a salvage pathway a drain on essential carbon and sulfur may ensue. In addition, some α-keto acids (e.g., phenylpyruvate, α-keto-γ-methiolbutyrate) may be toxic at elevated concentrations and glutamine transaminase may play a role in maintaining these α-keto acids at a low level. Methionine may also be salvaged by an additional route involving glutamine transaminase K. Thus, 5′-methylthioadenosine (formed from *S*-adenosylhomocysteine during polyamine biosynthesis) is converted to α-keto-γ-methiolbuytrate by a complex series of reactions (Backlund *et al.*, 1982). This α-keto acid is then converted to methionine via transamination in a reaction catalyzed by glutamine transaminase K. Thus, the original sulfur atom and methyl group of methionine are retained but carbons 1 to 4 are obtained anew from four of the five ribose carbons of 5′-methylthioribose.

α-Keto-γ-mercaptobutyrate and β-mercaptopyruvate, the α-keto acid analogues of homocysteine and cysteine, respectively, are highly reactive (Cooper *et al.*, 1982; Cooper and Meister, 1985). α-Keto-γ-mercaptobutyrate may undergo spontaneous β,γ-elimination to yield potentially toxic 2-oxo-3-butenoate (vinylglyoxylate) and hydrogen sulfide. Interestingly, both α-keto acids are substrates of glutamine transaminase K (Cooper, 1988) and of kidney thiopurine *S*-methyltransferase (Donahue and Henry, 1984). The products of the thiopurine *S*-methyltransferase reaction (i.e., α-keto-γ-methiolbutyrate and β-methiolpyruvate) formed with the two sulfhydryl-containing α-keto acids are also substrates of glutamine transaminase K (Cooper, 1988) yielding L-methionine and L-*S*-methylcysteine, respectively. Thus, the kidney appears to have two ways of salvaging potentially toxic sulfur-containing α-keto acids, both of which involve glutamine transaminase K.

α-Keto-γ-methiolbutyrate is transported across the blood–brain barrier by an α-keto acid carrier (Conn and Steele, 1982). If interorgan transport of this α-keto acid occurs *in vivo* then glutamine transaminase K may play an additional role in methionine homeostasis. Finally, phenylpyruvate is a potent initiator of insulin secretion in isolated pancreatic islets; glutamine transaminase K may play a role in this process (Lenzen *et al.*, 1984).

H. Mitochondrial Glutamine Transaminase K/Cysteine *S*-Conjugate β-Lyase

Although most of the activity of glutamine transaminase K in the rat kidney is cytosolic, about 10% of the activity is present in the mitochondria (Cooper and Meister, 1974; Cooper, 1988; Stevens, 1985b; Stevens *et al.*, 1988; Jones *et al.*, 1988). In the rat liver about a third of the glutamine transaminase K activity is in the mitochondrial fraction (Cooper, 1988). Lenzen *et al.* (1984) reported that the specific activity of glutamine-phenylpyruvate transaminase is about twofold higher in mitochondria than in the cytosolic fraction of pancreatic B cells from diabetic mice. Several reports have suggested that glutamine transaminase activity is largely mitochondrial in the brains of several species (see references quoted in

Cooper *et al.*, 1993). Detailed fractionation studies showed that ~80–85% of the glutamine transaminase activity in the rat brain is in the mitochondrial fraction (Cooper, 1988). Glutamine transaminase K in rat kidney mitochondria has been reported to be present in the outer membrane (Lash *et al.*, 1986a) and within the matrix (Stevens *et al.*, 1988). The enzyme has been purified from the mitochondrial fraction of rat kidneys (Cooper, 1978; Cooper and Meister, 1981; Stevens *et al.*, 1988) and rat brain (Van Leuven, 1975, 1976; Cooper and Gross, 1977). As with the cytosolic enzyme, rat mitochondrial glutamine transaminase K contains tightly bound pyridoxal 5′-phosphate, has a M_r of ~90,000 and is composed of two identical subunits. The substrate specificity is very similar to that of the cytosolic enzyme. However, there are subtle differences in the substrate affinities between the two enzymes and in the affinity with which the two enzymes bind to hydroxylapatite and DEAE-cellulose during purification (Cooper and Gross, 1977; Stevens *et al.*, 1988). In addition the purified bovine brain (= mitochondrial?) enzyme exhibits some kinetic differences to those of the purified cytosolic bovine kidney enzyme (Costa *et al.*, 1987). Finally, we have recently shown that polyclonal antibodies raised against purified cytosolic rat kidney glutamine transaminase K in rabbits react much less readily with the rat kidney and brain mitochondrial glutamine transaminase K than with the cytosolic enzyme (Cooper *et al.*, 1993). In future work it will be interesting to determine the relationship of the two forms of the enzyme and how the mitochondrial enzyme is inserted into the mitochondria.

I. Other Cysteine *S*-Conjugate β-Lyases in the Rat Kidney

As already noted, a major target of cysteine conjugates are mitochondria (Stonard and Parker, 1971a,b; Jones *et al.*, 1986; Lash and Anders, 1986; Lash *et al.*, 1986a,b; Stevens *et al.*, 1988), and glutamine transaminase K/cysteine *S*-conjugate β-lyase is present in the rat kidney in both cytosolic and mitochondrial fractions. Stevens *et al.* (1988) noted that with DCVC as substrate the cytosolic and mitochondrial forms of glutamine transaminase K probably account for β-lyase activity. The authors also noted that *S*-(benzothiazolyl)-L-cysteine (BTC) is a relatively poor substrate of both mitochondrial and cytosolic preparations of glutamine transaminase K. However, β-lyase activity toward BTC was comparable to that with DCVC in rat kidney mitochondria, suggesting that other β-lyases beside glutamine transaminase K may be present in rat kidney mitochondria. This idea is further supported by the finding that α-keto acids that are not substrates of glutamine transaminase K stimulate the β-lyase activity of the rat kidney mitochondrial matrix (Stevens *et al.*, 1988). The characterization of this additional β-lyase in the rat kidney mitochondria must await further studies.

Abraham and Cooper (1991) recently developed activity stains for (*a*) the phenylalanine–α-keto-γ-methiolbutyrate transaminase reaction and (*b*) the cysteine *S*-conjugate β-lyase reaction with DCVC. Analysis of rat kidney homogenates on nondenaturing polyacrylamide gels revealed a major band of activity with both reactions corresponding to a M_r of 90,000. This band clearly corresponds to glutamine transaminase K and is further evidence that the two activities are catalyzed by the same enzyme. However, another band of activity was detected with an apparent M_r of ~330,000 that possesses weak transaminase activity but stronger DCVC lyase activity (Abraham and Cooper, 1991). The higher molecular weight form that is present in the cytosol of rat kidneys does not appear to have an obvious relationship to glutamine transaminase K. The higher-molecular-weight species does not react with polyclonal antibodies to rat kidney cytosolic glutamine transaminase K and is not detected in the brain (Cooper *et al.*, 1993). The relationship of this second DCVC lyase to glutamine transaminase K or to other pyridoxal 5′-phosphate enzymes remains to be established.

We have recently shown that the 330,000-Da species (but not glutamine transaminase K) can convert leukotriene E_4 (a cysteine conjugate) to pyruvate, ammonia, and a reactive sulfhydryl-containing compound; the cysteine conjugate of dopamine is also a substrate of the 330,000-Da species (D. G. Abraham and A. J. L. Cooper, unpublished observations).

J. Comments on the Distribution of Glutamine Transaminase K/Cysteine *S*-Conjugate β-Lyase in Rat Kidney and Brain

Although, as noted above, glutamine transaminase K activity is widespread in rat tissues, the regional distribution of the enzyme has thus far only been studied in two organs. Jones *et al.* (1988) prepared antibodies to purified cytosolic rat kidney glutamine transaminase K in goats. They noted immunohistochemical staining in the S_1, S_2, and S_3 regions of the proximal tubular epithelium of the rat kidney. MacFarlane *et al.* (1989) prepared antibodies to cytosolic glutamine transaminase K in sheep. They also noted immunohistochemical staining in the proximal tubules but could detect the activity only in the S_3 region. The reason for the difference between the two studies is not clear. However, we have used activity staining in microdisected regions of the kidney to show that glutamine transaminase K/cysteine *S*-conjugate β-lyase activity is indeed present in the rat kidney in the S_1, S_2, and S_3 segments of the proximal tubule (H. Endou and A. J. L. Cooper, unpublished observations).

The finding that cysteine *S*-conjugate β-lyase activity is present in the S_1, S_2, S_3, regions of the rat kidney raises interesting questions. Some damage to brain and liver may occur, but the toxicity of certain cysteine conjugates that are metabolized by the β-lyases is manifest most often in the kidney (e.g., Terracini and Parker, 1965; Potter *et al.*, 1981; Nash *et al.*, 1984; Jaffe *et al.*, 1984; Dohn *et al.*, 1985). Although some damage at higher concentrations of cysteine conjugate may occur in the S_1 and S_2 regions of the proximal tubule, the main region of damage in the kidney is in the area of the pars rectus that contains the S_3 segment. Evidently, other factors besides the localization of glutamine transaminase K/cysteine *S*-conjugate β-lyase must play a part in the selective toxicity to the S_3 segment to certain halogenated xenobiotics and their corresponding cysteine *S*-conjugates (see, for example, discussions by Stevens and Jones, 1989; Elfarra, 1993). Such factors include the distribution and specificity of (*a*) glutathione *S*-transferases (Fig. 1, reaction **1**); (*b*) γ-glutamyltranspeptidase (reaction **2**); (*c*) dipeptidases (reaction **3**); (*d*) acylases (reaction **4**); (*e*) deacylases (reaction **5**); (*f*) other cysteine *S*-conjugate β-lyases (reaction **6**); (*g*) transporters for the glutathione conjugate, cysteinylglycine conjugate, cysteine conjugate, and mercapturate; (*h*) sulfoxidation reactions; and (*i*) availability of α-keto acid substrate of glutamine transaminase K/cysteine *S*-conjugate β-lyase. A few of these points will be discussed briefly. Several studies have shown that *S*-(1,2-dichlorovinyl)glutathione (DCVG), like DCVC, is also nephrotoxic. The nephrotoxicity of DCVG may be completely blocked *in vivo* and *in vitro* by the γ-glutamyltranspeptidase inhibitor, 4,5-dihydro-5-isoxaleacetic acid (AT-125) (Dohn *et al.*, 1985; Elfarra *et al.*, 1986a; Lash and Anders, 1986). The toxicity may also be blocked by probenecid, a general inhibitor of anion transport (Dohn *et al.*, 1985; Elfarra *et al.*, 1986a,b; Lash and Anders, 1986). AT-125 blocks the nephrotoxicity of 2-bromohydroquinone glutathione conjugate but not that of *S*-(2-chloroethyl)glutathione (CEG); on the other hand, probenicid blocks the nephrotoxicity of CEG but not that of 2-bromoquinone glutathione conjugate (Kramer *et al.*, 1987). Evidently, glutathione *S*-conjugates (and *N*-acetyl cysteine *S*-conjugates) may be taken up by kidney cells by a sodium-dependent mechanism that is inhibited by probenecid but not by cysteine conjugates. Alternatively, glutathione conjugates may be metabolized by γ-glutamyl transpeptidase and aminopeptidase M, which occur at the brush borders and basolateral membranes; the cysteine conjugate is then taken up by the general sodium-dependent, organic ion transporter or by the sodium-independent L and T amino acid transporters (see Elfarra, 1993, and references quoted therein).

Sausen and Elfarra (1990a) have characterized a flavin-containing monooxygenase that converts cysteine *S*-conjugates to the corresponding sulfoxides. The sulfoxide of DCVC was found to be an even more potent nephrotoxin than DCVC itself (Sausen and Elfarra, 1990b); the sulfoxide acts as a strong Michael acceptor by reacting with glutathione (Sausen and Elfarra, 1991). Elfarra (1993) points out that this may represent a novel mechanism for activation of S-conjugates. However, as noted above, an enteric bacterial enzyme (Larsen and Stevens, 1986) and a human cysteine *S*-conjugate β-lyase (Tomisawa *et al.*, 1986) possess high activity toward cysteine *S*-conjugate sulfoxides. Thus, in addition to reaction with glutathione, the cysteine *S*-conjugate sulfoxides may provide reactive fragments via β-elimination. Since the RS— (methionine), RS(O)— (methionine sulfoxide), and RS(O)(=NH)— (methionine sulfoximine) groups are readily accommodated at the active site of glutamine transaminase K (Cooper and Meister, 1974), it seems reasonable to expect that glutamine transaminase K may catalyze β-elimination reactions with cysteine *S*-conjugate sulfoxides.

In the rat, glutamine transaminase K activity is fairly evenly distributed throughout the brain with one exception. The specific activity of the enzyme is ~6.4 times higher in the choroid plexus than in the brain as a whole (Cooper *et al.*, 1993). In addition, the specific activity of the enzyme is several times higher in cultures of mouse, rat, and chicken astrocytes than in cultures of the corresponding neurons (Makar *et al.*, 1994). As with the kidney, the localization of glutamine transaminase K/cysteine *S*-conjugate β-lyase in brain does not correlate with the regional damage induced by cysteine *S*-conjugates such as DCVC. Although the enzyme is widespread in rat (Cooper *et al.*, 1993) and human brain (Cooper and Gross, 1977), damage to the nervous system is largely confined to the sensory trigeminal nucleus and, to a lesser extent, the facial nerves, oculomotor nerves, and motor trigeminal nucleus (Buxton and Hayward, 1967; Reichert *et al.*, 1976, and references cited therein). Again, as discussed in relation to the kidney, many factors must be involved in the selective damage to nervous tissue. In this regard, it is interesting to note that Patel *et al.* (1993) recently showed that rat brain possesses carriers for DCVG and DCVC at the blood–brain barrier. The carrier for DCVC corresponds to the neutral amino acid (= L) transporter. Evidently, more work is needed to determine the selectivity of DCVC-induced damage to nervous tissue.

Other compelling reasons exist for continued study of glutamine transaminase K in nervous tissue. A recent report of the National Research

Council discusses the possibility that environmental substances may contribute to the long-term neurodegenerative diseases in humans (summarized by Stone, 1992). If so, there is a distinct possibility that glutamine transaminase K/cysteine *S*-conjugate β-lyase (and possibly kynureninase) may contribute to this process if the xenobiotics are converted *in vivo* to reactive cysteine *S*-conjugates. Halohydrocarbons (HCFCs) have been proposed as replacements for chlorofluorocarbon (CFC) propellants and refrigerants to lower the stress on the earth's ozone layer. Many of the proposed HCFCs are pentahalogenated ethanes. We have been unable to find references in the literature to suggest that these compounds are substrates of glutathione *S*-transferases. However, if the compound is metabolized *in vivo* to a halogenated ethylene then the product may be a substrate of glutathione *S*-transferases and the possibility arises that a neurotoxic cysteine *S*-conjugate will be formed. Other HCFCs suggested as replacements of CFCs include tetra- and trihalogented ethanes. Anders and associates have begun a study of the metabolism of 2,2-dichloro-1,1,1-trifluoroethane (HCFC-123) (a pentahalogenated ethane), 1,2-dichloro-1,1-difluoroethane (HCFC-132b) (a tetrahalogenated ethane), and 1,1-dichloro-1-fluoroethane (a trihalogenated ethane) (Harris and Anders, 1991a,b; Harris *et al.*, 1991). Each was shown to be metabolized in rats, but thus far, glutathione and cysteine *S*-conjugates have not been detected. Nevertheless, with undoubted increase in the production of HCFCs in the next few years and the increased human exposure to these substances it will be important to define all potential metabolic pathways of these compounds and to ensure that they are not bioactivated through pathways that include the glutamine transaminase K/cysteine *S*-conjugate β-lyase.

V. Design of Prodrugs Directed toward Glutamine Transaminase K/Cysteine *S*-Conjugate β-Lyase in the Kidney

According to a recent survey approximately 1,040,000 new cancer cases are calculated to have occurred in the United States in 1990. Of these, 24,000 are estimated to have arisen in the kidneys (Henderson *et al.*, 1991). Although in absolute numbers the incidence of kidney cancer is relatively small compared to the overall incidence of cancer in the general population, kidney cancer is nevertheless a serious problem. Metastatic renal cell carcinoma is essentially resistant to all currently used therapeutic agents (Javadpour, 1984; Fair and Cordon-Cardo, 1988). In humans, renal carcinomas appear to arise from the proximal nephron (Cordon-Cardo *et al.*, 1989, and references cited therein), and, as noted above, this region of the kidney is most susceptible to damage from cysteine *S*-conjugates in experimental animals. Because chloroalkene-derived glutathione and cysteine *S*-conjugates are mutagenic (Green and Odum, 1985; Dekant *et al.*, 1989b) and there is much evidence to link this mutagenicity to the actions of cysteine *S*-conjugate β-lyases (Vamvakas et al., 1988), it is possible that there may be a link between low-level exposure to chloroalkenes and human kidney cancer (Cooper *et al.*, 1989). In this context it is interesting to note that Henderson *et al.* 1991) list cigarette smoking as a major contributing factor to human kidney cancer; perhaps cysteine conjugates formed from xenobiotics in cigarette smoke are partially responsible. As noted in Section IV,A, a cysteine *S*-conjugate β-lyase has been purified from human kidney and may be present at high enough concentration to promote nephrotoxic damage with reactive cysteine *S*-conjugates (Lash *et al.*, 1990a). However, the detailed localization of glutamine transaminase K/cysteine *S*-conjugate β-lyase in human kidney has not yet been determined. Assuming that the enzyme activity, as in the rat, is localized to the proximal nephron, then the identity of the site of action of cysteine *S*-conjugates and site of origin of renal cancer indicate that the targeting of prodrugs of cancer therapeutic agents to the kidney may afford a useful therapeutic strategy. A promising approach to the design of prodrugs to combat kidney cancers and (perhaps) infections is that employed by Hwang and Elfarra (1989,1991), who have designed a prodrug with an anticancer agent (6-mercaptopurine) as part of a cysteine conjugate.

In contrast to nephrotoxic cysteine *S*-conjugates, which give rise to reactive fragments following cleavage with β-lyases (e.g., DCVC), other cysteine conjugates such as *S*-(benzothiazolyl)-L-cysteine and *S*-(6-purinyl)-L-cysteine give rise to relatively stable sulfhydryl-containing fragments and are not nephrotoxic (Elfarra and Hwang, 1990; Hwang and Elfarra, 1989, 1991). After treatment of rats with *S*-(benzothiazolyl)-L-cysteine kidney, liver, and plasma levels of 2-mercaptobenzothiazole and its glucuronate derivative were similar. This lack of tissue selectivity was attributed in part to the similar rates of metabolism of *S*-(benzothiazolyl)-L-cysteine by hepatic and renal cysteine-*S*-conjugate β-lyases. On the other hand, 30 min after administration of *S*-(6-purinyl)-L-cysteine the concentration of 6-mercaptopurine and its metabolites were 90- and 2.5-fold higher in kidney than in plasma and liver, respectively. The cytosolic and mitochondrial fractions of rat kidney were about equally effective in converting *S*-(6-purinyl)-L-cysteine to 6-mercaptopurine; the reaction was inhibited by aminooxyacetate (Hwang and Elfarra, 1989). The target organ selectivety of *S*-(6-purinyl)-L-cysteine appears to be due in part to the fact that the compound is a much better substrate of kidney β-lyases than of liver β-lyases (Hwang and Elfarra, 1991). As noted by Hwang and Elfarra (1989) considerable effort has gone into designing prodrugs of 6-mercaptopurine that will provide a better therapeutic index than the parent molecule drug itself. The fact that relatively large amounts of 6-mercaptopurine were found in the rat kidney after administration of *S*-(6-purinyl)-L-cysteine supports the concept that renal glutamine transaminase K/cysteine *S*-conjugate β-lyase may be exploited for the development of kidney selective prodrugs. Because of the unique substrate specificity of glutamine transaminase K, we have considered the possibility that homocysteine *S*-conjugates may be even more effective prodrugs than cysteine *S*-conjugates (Cooper and Anders, 1990). To appreciate this idea some background chemistry and additional discussion on the specificity of glutamine transaminase K will be given.

Enzymes catalyzing γ-elimination reactions from suitable amino acid substrates [Eq. (5)] are well known in biochemistry.

$$XCH_2CH_2CH(NH_2)CO_2H + H_2O \rightarrow XH + CH_3CH_2COCO_2H + NH_3 \quad (5)$$

An example, mentioned in Section II, is γ-cystathionase. In this case $X = -SCH_2CH(NH_2)CO_2H$. However, it is apparent that nonenzymatic γ-elimination reactions involving amino acids with good leaving groups in the γ-position can also occur following enzymatic oxidation at the α carbon (Cooper *et al.*, 1989) [Eq. (6)]. In this case, the resulting α-keto acid fragment is vinylglyoxylate (2-oxobutenoate).

$$\begin{aligned} XCH_2CH_2CH(NH_2)CO_2H + [O] &\rightarrow XH \\ + CH_2{=}CHCOCO_2H + NH_3 & \quad (6) \end{aligned}$$

Examples in which this type of reaction has been shown to occur include (*a*) *S*-adenosylmethionine ($X = CH_3S^+$(adenosyl)—) (Stoner and Eisenberg, 1975), L-methionine-SR-sulfoximine ($X = CH_3S({=}NH)({=}O)-$) (Cooper *et al.*, 1976), L-homocysteine (X = HS—) (Cooper and Meister, 1985), L-canavanine ($X = NH_2C({=}NH)NHO-$, or more correctly, $C(NH_2)_2{=}NO-$) (Hollander *et al.*, 1989), and homocysteine *S*-conjugates (X = RS—) (e.g., Elfarra *et al.*, 1986b). The pathways leading to γ-elimination (a retro-Michael elimination) are shown in Fig. 5. Reaction of the amino acid (**1**) with L-amino acid oxidase (AAO) leads to formation of the corresponding α-imino acid (**3**). The β-hydrogens in the imino acid are activated relative to those in the parent amino acid and γ-elimination is facilitated. γ-Elimination results in formation of 2-imino-3-butenoate (**5**) and vinylglyoxylate (**4**). Alternatively, the amino acid (**1**) is converted to the corresponding α-keto acid (**2**) by a transamination reaction that

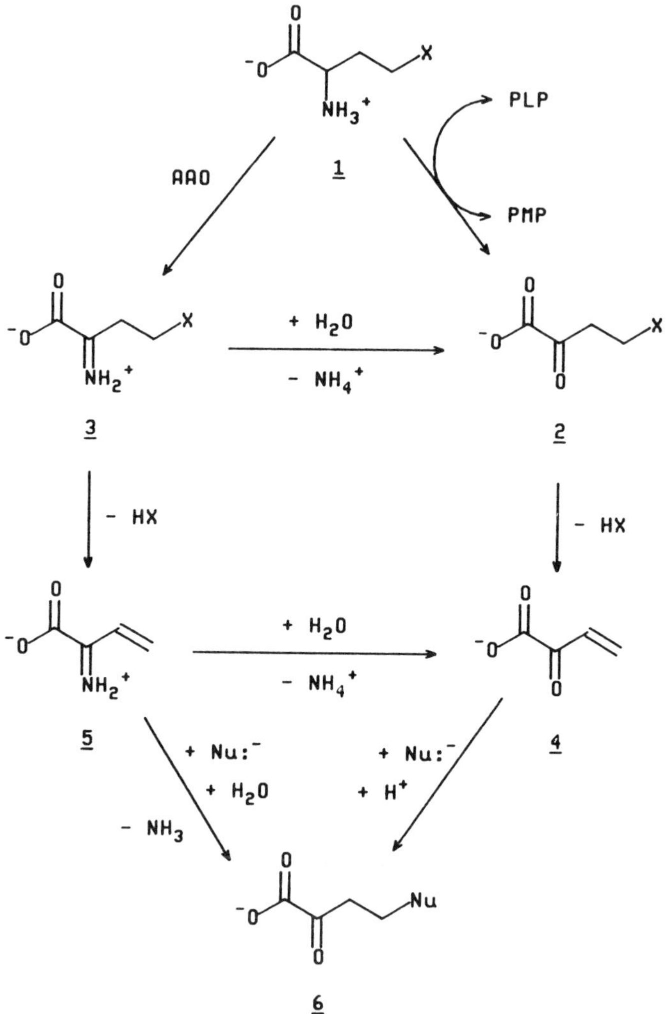

Fig. 5 Elimination and γ-addition reactions associated with α-oxidation of amino acids possessing good leaving groups (X) in the γ position. 1, amino acid with good leaving group in the γ (or 4) position; 2, 4-substituted 2-oxobutyrate; 3, 4-substituted 2-iminobutyrate; 4, vinylglyoxylate (2-oxo-3-butenoate); 5, 2-imino-3-butenoate; 6, addition product of 4 with nucleophile (Nu:$^-$). Abbreviations: AAO, L-(or D-) amino acid oxidase; PLP, pyridoxal 5′-phosphate; PMP, pyridoxamine 5′-phosphate. Pyridoxal 5′-phosphate is the cofactor of glutamine transaminase K, which participates in the first step of the conversion of some 4-substituted amino acids to vinylglyoxylate (see the text). From Cooper *et al.* (1989) with permission.

requires pyridoxal 5′-phosphate. Again, the β-hydrogens in the α-keto acid are activated relative to the amino acid and elimination to vinylglyoxylate (**4**) may occur. [However, the α-keto acid (**2**) is less reactive than the corresponding 2-imino acid (**3**) so that elimination from the α-keto acid is not as favorable as from the 2-imino acid (Hollander *et al.*, 1989).] Vinylglyoxylate has not been isolated, but it is readily trapped by Michael addition to a suitable nucleophile ($Nu:^-$), such as 2-mercaptoethanol (Cooper *et al.*, 1976) or methanethiol (Lash *et al.*, 1990b).

S-(1,2-Dichlorovinyl)-L-homocysteine (DCVHC) was shown to be even more nephrotoxic *in vivo*, in isolated renal proximal tubular cells, and in mitochondria than is DCVC (Elfarra *et al.*, 1986b; Lash *et al.*, 1986b; Lash and Anders, 1987). DCVC is strongly mutagenic in the Ames test, whereas DCVHC is less so; aminooxyacetate completely blocks the mutagenicity of both compounds (Dekant *et al.*, 1989a,b). In contrast, the α-methyl analogues of DCVHC and DCVC are not nephrotoxic or mutagenic. Aminooxyacetate and probenecid protect against the nephrotoxicity. Elimination of a reactive sulfhydryl-containing fragment was shown to occur via an enzymatic route that did not involve a γ-cystathionase-type reaction [Eq. (5)]. Rather, the elimination was stimulated by addition of α-ketobutyrate, suggesting that the product of the γ-elimination reaction is 2-oxobutenoate and not α-ketobutyrate. Taken together the data show that DCVHC is a substrate for a transaminase in the kidney. The α-keto acid product of this reaction then slowly and spontaneously undergoes nonenzymatic γ-elimination to yield vinylglyoxylate and 1-mercapto-1,2-dichloroethylene. Although it was not explicitly shown, the enzyme responsible for transamination of DCVHC in the rat kidney is almost certainly glutamine transaminase K since this amino acid fulfills the requirement for a substrate [i.e., Eq. (2), $n = 2$, X = ClHC=CCl—). In other work, Lash *et al.* (1990b) showed that *S*-(2-benzothiazolyl)-L-homocysteine is a substrate of purified bovine glutamine transaminase K/cysteine *S*-conjugate β-lyase. The authors also showed that *S*-(2-benzothiazolyl)-L-homocysteine, L-methionine sulfoximine, and L-canavanine (amino acids, which as noted above, can undergo γ-elimination upon oxidation at the α-carbon) are toxic to isolated rat kidney cells, as is the Michael acceptor methyl vinyl ketone. This finding suggests that the increased toxicity of DCVHC over DCVC is due in part to the fact that DCVC is metabolized by glutamine transaminase K/cysteine *S*-conjugate β-lyase to only one nephrotoxic species (1-mercapto-1,2-dichloroethylene), whereas DCVHC is metabolized by glutamine transaminase and/or L-amino acid oxidase to two nephrotoxic species (1-mercapto-1,2-dichloroethylene and vinylglyoxylate). Also, DCVHC is more slowly acetylated than is DCVC (Lash *et al.*, 1990b). It may be possible to exploit the fact that the homocysteine conjugates are more bioactive than the corresponding cysteine conjugates. Thus, an extension of the work of Elfarra and Hwang with *S*-(6-purinyl)-L-cysteine would be the synthesis of *S*-(6-purinyl)-L-homocysteine as a kidney-directed prodrug. Such a compound would be expected to be a substrate of human kidney glutamine transaminase K; transamination should give rise to the corresponding α-keto acid, which is predicted to spontaneously fragment to the anticancer agent 6-mercaptopurine and the strong nucleophile vinylglyoxylate (Cooper *et al.,* 1989).

VI. Conclusions

Over the last 10 to 15 years, it has become apparent that cysteine *S*-conjugate β-lyases play an important role in the bioactivation of many xenobiotics. The resulting organ damage varies with such factors as species, age, and sex, but, in general, the kidney is the most severely affected organ. In addition, the liver, gut, and nervous system may be compromised to some extent. In fact, in humans who are known to have been exposed to a reactive cysteine conjugate (i.e., DCVC, resulting from inhalation of dichloroacetylene), the major damage is to the nervous system. Even within the kidney and nervous system, damage is to relatively small defined areas. For example, damage is most pronounced within the S_3 segment of the pars rectus region of the proximal tubule of the kidney, and within the sensory trigeminal nucleus of the brain. Clearly, much work remains to be done to explain this selectivity. Evidently, the presence of cysteine *S*-conjugate β-lyases in the affected regions is important for the damage to occur, but other factors must also play a part.

Relatively little attention has been given to the role of cysteine *S*-conjugates in human health. Prolonged exposure to low levels of pollutants that can be activated through a glutathione conjugate may predispose certain individuals to risk of kidney disease and others perhaps to inexorable loss of brain function. From this epidemiological point of view the cysteine *S*-conjugate β-lyases deserve more attention. However, these enzymes are also interesting in their own right. Cysteine *S*-conjugate β-lyases thus far shown to occur in mammals belong to three classes: (*a*) those present in enteric bacteria, (*b*) a liver form (= kynureninase), and (*c*) a kidney form (= glutamine transaminase K). Of these, the last has probably received the most attention because of the nephrotoxicity associated with many cysteine *S*-conjugates. Glutamine transaminase K is indeed a fascinating enzyme with a remarkable substrate specificity. However, much remains to be determined concerning its structure, its relationship to other pyridoxal 5′-phosphate-containing enzymes, the differences between mitochondrial and cytosolic forms, and how the enzyme is targeted to cytosolic and mitochondrial regions. In addition, because of its unique substrate specificity, the enzyme may be the target of prodrugs designed to deliver anticancer drugs to the kidney. Indeed, some progress has already been made in this area.

Acknowledgments

Work carried in the author's laboratory was supported by National Institutes of Health Grant DK 16739.

NOTE ADDED IN PROOF. Elfarra and Hwang (1993) have recently shown that *S*-(6-purinyl)-L-homocysteine is indeed selectively converted to 6-mercaptopurine in rat kidney *in vivo*. The reaction is stimulated by simultaneous administration of α-keto-γ-methiolbutyrate and appears to proceed via a γ-elimination reaction as was predicted by Cooper and Anders (1990) (see Section V).

References

Abraham, D. G., and Cooper, A. J. L. (1991). Glutamine transaminase K and cysteine-*S*-conjugate β-lyase activity stains. *Anal. Biochem.* **197,** 421–427.

Abramovitz, M., Hommar, H., Ishigaki, S., Tansey, F., Cammer, W., and Listowsky, I. (1988). Characterization and localization of glutathione-*S*-transferases in rat brain and binding of hormones, neurotransmitters and drugs. *J. Neurochem.* **50,** 50–57.

Anders, M. W. (1988). Glutathione-dependent toxicity: Biosynthesis and bioactivation of cytotoxic *S*-conjugates. *ISI Atlas Sci. Pharmacol.* **2,** 99–104.

Anders, M. W., Lash, L. H., and Elfarra, A. A. (1986). Nephrotoxic amino acid and glutathione *S*-conjugates: Formation and renal activation. *Adv. Exp. Med. Biol.* **197,** 443–455.

Anders, M. W., Lash, L., Dekant, W., Elfarra, A. A., and Dohn, D. R. (1988). Biosynthesis and biotransformation of glutathione *S*-conjugates to toxic metabolites. *CRC Crit. Rev. Toxicol.* **18,** 311–341.

Anderson, P. M., and Schultze, M. O. (1965). Cleavage of *S*-(1,2-dichlorovinyl)-L-cysteine by an enzyme of bovine origin. *Arch. Biochem. Biophys.* **111,** 593–602.

Backlund, P. S., Jr., Chang, C. P., and Smith, R. A. (1982). Identification of 2-keto-4-methylthiobutyrate as an intermediate compound in methionine synthesis from 5′-methylthioadenosine. *J. Biol. Chem.* **257,** 4196–4202.

Bakke, J. E., Larsen, G. L., Aschbacher, P. W., Rafter, J. J., Gustafsson, J.-Å., and Gustafsson, B. E. (1981). Role of gut microflora in metabolism of glutathione conjugates of xenobiotics. *In* "Sulfur in Pesticide Action and Metabolism" (J. D. Rosen, P. S. Magee, and J. E. Casida, eds.), pp. 165–178. Am. Chem. Soc., Washington, DC.

Banki, K., Elfarra, A. A., Lash, L. H., and Anders, M. W. (1986). Metabolism of *S*-(2-chloro-1,1,2-trifluoroethyl)-L-cysteine to hydrogen sulfide and the role of hydrogen sulfide in *S*-(2-chloro-1,1,2-trifluoroethyl)-L-cysteine-induced mitochondrial toxicity. *Biochem. Biophys. Res. Commun.* **138,** 707–713.

Barnsley, E. A. (1964). The metabolism of *S*-methyl-L-cysteine in the rat. *Biochim. Biophys. Acta* **90,** 24–36.

Baumann, E. (1883). Über die Bildung der Mercaptursäuren in Organismus und ihre Erkennung im Harn. *Hoppe-Seyler's Physiol. Chem.* **84,** 190–197.

Bernström, K., Larsen, G. L., and Hammerström, S. (1989). Metabolism of leukotriene E_4 to 5-hydroxy-6-mercapto 7,9-*trans*-11,14-cis-eicosatetraenoic acid by microfloral cysteine conjugate β-lyase and rat cecum contents. *Arch. Biochem. Biophys.* **275,** 531–539.

Bhattacharya, R. K., and Schultze, M. O. (1972). Properties of DNA treated with *S*-(1,2-dichlorovinyl)-L-cysteine and a lyase. *Arch. Biochem. Biophys.* **153,** 105–115.

Bild, G. S., and Morris, J. C. (1984). Detection of β-carbanion formation during kynurenine hydrolysis catalyzed by *Pseudomonas marginalis* kynureninase. *Arch. Biochem. Biophys.* **235,** 41–47.

Boogaard, P. J., Commandeur, J. N. M., Mulder, G. J., Vermeulen, N. P. E., and Nagelkerke, J. F. (1989). Toxicity of the cysteine-S-conjugates and mercapturic acids of four structurally related difluoroethylenes in isolated proximal tubular cells from rat kidney. *Biochem. Pharmacol.* **38,** 3731–3741.

Boyland, E., and Chasseaud, L. F. (1969). Role of glutathione and glutathione *S*-transferases in mercapturic acid biosynthesis. *Adv. Enzymol. Relat. Areas Mol. Biol.* **32,** 163–219.

Braunstein, A. E. (1985). Introduction and historical survey. *In* "The Transaminases" (P. Christen, and D. E. Metzler, eds.), pp. 1–35. Wiley, New York.

Braunstein, A. E., and Goryachenkova, E. V. (1984). The β-replacement specific pyridoxal P-dependent lyases. *Adv. Enzymol. Relat. Areas Mol. Biol.* **56,** 1–89.

Buxton, P. H., and Hayward, M. (1967). Polyneuritis cranialis associated with industrial trichloroethylene poisoning. *J. Neurol., Neurosurg. Psychiatry* **30,** 511–518.

Cammer, W., Tansey, F., Abramovitz, M., Ishigaki, S., and Listowski, I. (1989). Differential localization of glutathione-*S*-transferase Y_p and Y_b subunits in oligodendrocytes and astrocytes of rat brain. *J. Neurochem.* **52,** 876–883.

Campbell, E., Takahashi, Y., Abramovitz, M., Peretz, M., and Listowsky, I. (1990). A distinct human testis and brain μ-class glutathione *S*-transferase. Molecular cloning and characterization of a form present even in individuals lacking hepatic type μ isoenzymes. *J. Biol. Chem.* **265,** 9188–9193.

Cavallini, D., Federici, G., Bossa, F., and Granata, F. (1973). The protective effect of thiosulfate upon the inactivation of aspartate aminotransferase by aminoacrylic-acid-producing substrates. *Eur. J. Biochem.* **39,** 301–304.

Chasseaud, L. F. (1979). The role of glutathione and glutathione *S*-transferases in the metabolism of chemical carcinogens and other electrophilic agents. *Adv. Cancer Res.* **29,** 175–274.

Chatfield, D. H., and Hunter, W. H. (1973). The metabolism of acetamidothiazoles in the rat. 2-Acetamido-4-chloromethylthiazole. *Biochem. J.* **134,** 879–884.

Colucci, D. F., and Buyske, D. A. (1965). The biotransformation of a sulfonamide to a mercaptan and to mercapturic acid and glucuronide conjugates. *Biochem. Pharmacol.* **14,** 457–466.

Commandeur, J. N. M., Brakenhoff, J. P. G., De Kanter, F. J. J., and Vermeulen, N. P. E. (1988). Nephrotoxicity of mercapturic acids of three structurally related 2,2-difluoroethylenes in the rat. Indications for different bioactivation mechanisms. *Biochem. Pharmacol.* **37,** 4495–4504.

Commandeur, J. N. M., De Kanter, F. J. J., and Vermeulen, N. P. E. (1989). Bioactivation of the cysteine-*S*-conjugate and mercapturic acid of tetrafluoroethylene to acylating reactive intermediates in the rat: Dependence of activation and deactivation on acetyl Coenzyme A availability. *Mol. Pharmacol.* **36,** 654–663.

Conn, A. R., and Steele, R. D. (1982). Transport of α-keto analogues of amino acids across the blood-brain barrier. *Am. J. Physiol.* **243,** E272–E277.

Cooper, A. J. L. (1977). Asparagine transaminase from rat liver. *J. Biol. Chem.* **252,** 2032–2038.

Cooper, A. J. L. (1978). Purification of soluble and mitochondrial glutamine transaminase K from rat kidney. Use of a sensitive assay involving transamination between L-phenylalanine and α-keto-γ-methiolbutyrate. *Anal. Biochem.* **89,** 451–460.

Cooper, A. J. L. (1988). Glutamine aminotransferases and ω-amidases. *In* "Glutamine and Glutamate in Mammals" (E. Kvamme, ed.), Vol. 1, pp. 33–52. CRC Press, Boca Raton, FL.

Cooper, A. J. L., and Anders, M. W. (1990). Glutamine transaminase K and cysteine conjugate β-lyase. *Ann. N. Y. Acad. Sci.* **585,** 118–127.

Cooper, A. J. L., and Gross, M. (1977). The glutamine transaminase–ω-amidase system in rat and human brain. *J. Neurochem.* **28,** 771–778.

Cooper, A. J. L., and Meister, A. (1972). Isolation and purification of highly purified glutamine transaminase. *Biochemistry* **11,** 661–671.

Cooper, A. J. L., and Meister, A. (1974). Isolation and purification of a new glutamine transaminase from rat kidney. *J. Biol. Chem.* **249,** 2554–2561.

Cooper, A. J. L., and Meister, A. (1981). Comparative studies of glutamine transaminases from rat tissues. *Comp. Biochem. Physiol. B* **69B,** 137–145.

Cooper, A. J. L., and Meister, A. (1985). Enzymatic oxidation of homocysteine. *Arch. Biochem. Biophys.* **239,** 556–566.

Cooper, A. J. L., Stephani, R. A., and Meister, A. (1976). Enzymatic reactions of methionine sulfoximine. Conversion to the corresponding α-imino and α-keto acids and to α-ketobutyrate and methane sulfinamide. *J. Biol. Chem.* **251,** 6674–6682.

Cooper, A. J. L., Haber, M. T., and Meister, A. (1982). On the chemistry and biochemistry of 3-mercaptopyruvic acid, the α-keto analog of cysteine. *J. Biol. Chem.* **257,** 816–826.

Cooper, A. J. L., Hollander, M. M., and Anders, M. W. (1989). Formation of highly reactive vinylglyoxylate (2-oxo-3-butenoate) from amino acids with good leaving groups in the γ-position. Toxicological implication and therapeutic potential. *Biochem. Pharmacol.* **38,** 3895–3901.

Cooper A. J. L., Abraham, D., Gelbard, A. S., Lai, J. C. K., and Petito, C. K. (1993). High activities of glutamine transaminase K (dichlorovinylcysteine β-lyase) and ω-amidase in the choroid plexus of rat brain. *J. Neurochem.* **61,** 1731–1741.

Cordon-Cardo, C., Reuter, V. E., Finstad, C. L., Sheinfeld, J., Lloyd, K. O., Fair, W. R., and Melamed, M. R. (1989). Blood-group related antigens in human kidney. Modulation of Lewis determinants in renal cell carcinoma. *Cancer Res.* **49,** 212–218.

Costa, M., Pensa, B., Fontana, M., Foppoli, C., and Cavallini, D. (1986). Transamination of L-cystathionine and related compounds by a bovine liver enzyme. Possible identification with glutamine transaminase. *Biochim. Biophys. Acta* **881,** 314–320.

Costa, M., Pensa, B., Di Costanzo, B., Coccia, R., and Cavallini, D. (1987). Transamination of L-cystathionine and related compounds by bovine brain glutamine transaminase. *Neurochem. Int.* **10,** 377–382.

Dekant, W., Metzler, M., and Henschler, D. (1986). Identification of S-1,2-dichlorovinyl-N-acetyl-cysteine as a primary metabolite of trichloroethylene: A possible explanation of its nephrocarcinogenicity. *Biochem. Pharmacol.* **35,** 2455–2458.

Dekant, W., Lash, L. H., and Anders, M. W. (1987a). Bioactivation mechanism of the cytotoxic and nephrotoxic S-conjugate S-(2-chloro-1,1,2-trifluoroethyl)-L-cysteine. *Proc. Natl. Acad. Sci. U.S.A.* **84,** 7443–7447.

Dekant, W., Martens, G., Vamvakas, S., Metzler, M., and Henschler, D. (1987b). Bioactivation of tetrachloroethylene. Role of glutathione S-transferase-catalyzed conjugation *versus* cytochrome P-450-dependent phospholipid alkylation. *Drug Metab. Dispos.* **15,** 702–709.

Dekant, W., Lash, L. H., and Anders, M. W. (1988a). Fate of glutathione conjugates and bioactivation of cysteine S-conjugates by cysteine conjugate β-lyase. *In* "Glutathione Conjugation" (H. Sies and B. Ketterer, eds.), pp. 425–447. Academic Press, London.

Dekant, W., Berthold, K., Vamvakas, S., Henschler, D., and Anders, M. W. (1988b). Thioacylating intermediates as metabolites of S-(1,2-dichlorovinyl)-L-cysteine and S-(1,2,2-trichlorovinyl)-L-cysteine formed by cysteine conjugate β-lyase. *Chem. Res. Toxicol.* **1,** 175–178.

Dekant, W., Berthold, K., Vamvakas, S., and Henschler, D. (1988c). Thioacetylating agents as ultimate intermediates in the β-lyase catalysed metabolism of S-(pentachlorobutadienyl)-L-cysteine. *Chem.-Biol. Interact.* **67,** 139–148.

Dekant, W., Vamvakas, S., and Anders, M. W. (1989a). Bioactivation of nephrotoxic haloalkanes by glutathione conjugation: Formation of toxic and mutagenic intermediates by cysteine conjugate β-lyase. *Drug Metab. Rev.* **20,** 43–83.

Dekant, W., Vamvakas, S., and Anders, M. W. (1989b). Bioactivation of nephrotoxic haloalkenes by glutathione conjugation: Formation of toxic and mutagenic intermediates by cysteine conjugate β-lyase. *Drug. Metab. Rev.* **20,** 43–83.

Dekant, W., Koob, M., and Henschler, D. (1990). Metabolism of trichloroethene—*in vivo* and *in vitro* evidence for activation by glutathione conjugation. *Chem.-Biol. Interact.* **73,** 89–101.

DePierre, J. W., and Morgenstern, R. (1983). Comparison of the distribution of microsomal and cytosolic glutathione S-transferase activities in different organs of the rat. *Biochem. Pharmacol.* **32,** 721–723.

Dohn, D. R., Leininger, J. R., Lash, L. H., Quebberman, A. J., and Anders, D. W. (1985). Nephrotoxicity of S-(2-chloro-1,1,2-trifluoroethyl)glutathione and S-(2-chloro-1,1,2,-trifluoroethyl)-L-cysteine, the glutathione and cysteine conjugates of chlorotrifluoroethene. *J. Pharmacol. Exp. Ther.* **235,** 851–857.

Donahue, J. G., and Henry, D. P. (1984). Endogenous substrates for thiopurine S-methyltransferase: Formation of S-methylcysteine and methionine *in vitro. Fed. Proc., Fed. Am. Soc. Exp. Biol.* **43,** 464 (abstr.).

Duffel, M. W., and Jakoby, W. B. (1982). Cysteine S-conjugate N-acetyltransferase from rat kidney microsomes. *Mol. Pharmacol.* **21,** 444–448.

Elfarra, A. A. (1993). Aliphatic halogenated hydrocarbons. *In* "Toxicology of the Kidney" (J. B. Hook and R. S. Goldstein, eds.), 2nd ed., pp. 387–414. Raven Press, New York.

Elfarra, A. A., and Anders, M. W. (1984). Renal processing of glutathione conjugates. Role in nephorotoxicity. *Biochem. Pharmacol.* **33,** 3729–3732.

Elfarra, A. A., and Hwang, I. Y. (1990). *In vivo* metabolites of S-(2-benzothiazolyl)-L-cysteine as markers of *in vivo* cysteine conjugate β-lyase and thiol glucuronosyl transferase activities. *Drug Metab. Dispos.* **18,** 917–922.

Elfarra, A. A., and Hwang, I. Y. (1993). Targeting of 6-mercaptopurine to the kidneys. Metabolism and kidney-selectivity of *S*-(6-purinyl)-L-cysteine analogs in rats. *Drug Metab. Dispos.* **21,** 841–845.

Elfarra, A. A., Baggs, R., and Anders, M. W. (1985). Structure-nephrotoxicity relationships of S-(2-chloroethyl)-DL-cysteine and analogs: Role for an episulfonium ion. *J. Pharmacol. Exp. Ther.* **233,** 512–516.

Elfarra, A. A., Jakobson, I., and Anders, M. W. (1986a). Mechanism of S-(1,2-dichlorovinyl)-glutathione-induced nephrotoxicity. *Biochem. Pharmacol.* **35,** 283–288.

Elfarra, A. A., and Lash, L. H., and Anders, M. W. (1986b). Metabolic activation and detoxication of nephrotoxic cysteine and homocysteine S-conjugates. *Proc. Natl. Acad. Sci. U.S.A.* **83,** 2667–2671.

Elfarra, A. A., Lash, L. H., and Anders, M. W. (1987). α-Ketoacids stimulate renal cysteine conjugate β-lyase activity and potentiate the cytotoxicity of S-(1,2-dichlorovinyl)-L-cysteine. *Mol. Pharmacol.* **31,** 208–217.

Fair, W. R., and Cordon-Cardo, C. (1988). Immunopathology—New help for the clinician. *Semin. Urol.* **6,** 31–45.

Flavin, M., and Slaughter, C. (1964). Cystathionine cleavage enzymes of *Neurospora*. *J. Biol. Chem.* **239,** 2212–2219.

Fukui, S., Schwarcz, R., Rapoport, S. I., Takada, Y., and Smith, Q. R. (1991). Blood-brain barrier transport of kynurenines: Implications for brain synthesis and metabolism. *J. Neurochem.* **56,** 2007–2017.

Gál, E. M., and Sherman, A. D. (1980). L-Kynurenine. Its synthesis and possible regulatory function in brain. *Neurochem. Res.* **5,** 223–239.

Gibson, G. G., Perry, S. J., MacFarlane, M., Schofield, M. A., King, L. J., Lock, E. A., and Goldfarb, P. S. (1993). Molecular cloning and functional expression of the cysteine conjugate β-lyase gene. *Toxicologist* **13,** 243.

Goldman, P. (1978). Biochemical pharmacology of the intestinal flora. *Annu. Rev. Pharmacol. Toxicol.* **18,** 523–539.

Golichowski, A., and Jenkins, W. T. (1978). Inactivation of pig heart alanine aminotransferase by β-chloroalanine. *Arch. Biochem. Biophys.* **189,** 109–114.

Green, T., and Odum, J. (1985). Structure/activity studies of the nephrotoxic and mutagenic action of cysteine conjugates of chloro- and fluoroalkenes. *Chem.-Biol. Interact.* **54,** 15–31.

Greenstein, J. P. (1949). Dehydropeptidases. *Adv. Enzymol. Relat. Areas Biochem.* **8,** 117–169.

Hamilton, G. A. (1985). Peroxisomal oxidases and suggestions for the mechanism of action of insulin and other hormones. *Adv. Enzymol. Relat. Areas Mol. Biol.* **57,** 85–177.

Harris, J. W., and Anders, M. W. (1991a). Metabolism of the hydrochlorofluorocarbon 1,2-dichloro-1,1-difluoroethane. *Chem. Res. Toxicol.* **4,** 180–186.

Harris, J. W., and Anders, M. W. (1991b). *In vivo* metabolism of the hydrochlorofluorocarbon 1,1-dichloro-1-fluoroethane (HCFC-141b). *Biochem. Pharmacol.* **41,** R13–R16.

Harris, J. W., Pohl, L. R., Martin, J. L., and Anders, M. W. (1991). Tissue acylation by the chlorofluorocarbon substitute 2,2-dichloro-1,1,1-trifluoroethane. *Proc. Natl. Acad. Sci. U.S.A.* **88,** 1407–1410.

Henderson, B. E., Ross, R. K., and Pike, M. C. (1991). Toward the primary prevention of cancer. *Science* **254,** 1131–1138.

Hersh, L. B. (1971). Rat liver ω-amidase: Purification and properties. *Biochemistry* **10,** 2884–2891.

Heyes, M. P., Mefford, I. N., Quearry, B. J., Dedhia, M., and Lackner, A. (1990). Increased ratio of quinolinic acid to kynurenic acid in cerebrospinal fluid of D-retrovirus-infected rhesus macaques: Relationship to clinical and viral status. *Ann. Neurol.* **27,** 666–675.

Heyes, M. P., Brew, B. J., Martin, A., Price, R. W., Salazar, A. M., Sidtis, J. J., Yergey, J. A., Mouradian, M. M., Sadler, A. E., Keilp, J., Rubinow, D., and Markey, S. P. (1991). Quinolinic acid in cerebrospinal fluid and serum in HIV-1 infection: Relationship to clinical and neurological status. *Ann. Neurol.* **29,** 202–209.

Heyes, M. P., Saito, K., and Markey, S. P. (1992). Human macrophages convert L-tryptophan into the neurotoxin quinolinic acid. *Biochem. J.* **283,** 633–635.

Hollander, M. M., Reiter, A. J., Horner, W. H., and Cooper, A. J. L. (1989). Conversion of canavanine to α-keto-γ-guanidinooxybutyrate and to vinylglyoxylate and 2-hydroxyguanidine. *Arch. Biochem. Biophys.* **270,** 698–713.

Holzman, D. (1993). Inflamed debate over neurotoxin. *Science* **259,** 25–26.

Hwang, I. Y., and Elfarra, A. A. (1989). Cysteine S-conjugates may act as kidney-selective prodrugs: Formation of 6-mercaptopurine by the renal metabolism of S-(6-purinyl)-L-cysteine and metabolism of new analogs in rats. *J. Pharmacol. Exp. Ther.* **251,** 448–454.

Hwang, I. Y., and Elfarra, A. A. (1991). Kidney selective prodrugs of 6-mercaptopurine: Biochemical basis of the kidney selectivity of S-(6-purinyl)-L-cysteine and metabolism of new analogs in rats. *J. Pharmacol. Exp. Ther.* **258,** 171–177.

Ikeda, T., Koneshi, Y., and Ichihara, A. (1976). Transaminase of branched-chain amino acids. XI. Leucine (methionine) transaminase of rat liver mitochondria. *Biochim. Biophys. Acta* **445,** 622–631.

Jaffe, D. R., Gandolfi, A. J., and Nagle, R. B. (1984). Chronic toxicity of S-(*trans*-1,2-dichlorovinyl)-L-cysteine in mice. *J. Appl. Toxicol.* **4,** 315–319.

Jakoby, W. B., ed. (1980). "Enzymatic Basis of Detoxication." Academic Press, San Diego, CA.

Javadpour, N. (1984). Overview of renal cancer. *In* "Cancer of the Kidney" (N. Javadpour, ed.), pp. 1–5. Thieme-Stratton, New York.

Jones, T. W., Wallin, A., Hjordis, T., Thor, H., Gerdes, R. G., Ormstad, K., and Orrenius, S. (1986). The mechanism of pentachlorobutadienyl-glutathione nephrotoxicity studied with isolated rat renal epithelial cells. *Arch. Biochem. Biophys.* **251,** 504–531.

Jones, T. W., Qin, C., Schaeffer, V. H., and Stevens, J. L. (1988). Immunohistochemical localization of glutamine transaminase K, a rat kidney cysteine conjugate β-lyase, and the relationship to the segment specificity of cysteine conjugate nephrotoxicity. *Mol. Pharmacol.* **34,** 621–627.

Kanhai, W., Dekant, W., and Henschler, D. (1989). Metabolism of the nephrotoxin dichloroacetylene by glutathione conjugation. *Chem. Res. Toxicol.* **2,** 51–56.

Ketterer, B., Taylor, J. B., Meyer, D. J., Coles, B., Pemble, S., Cowell, I. G., and Dixon, K. (1989). Glutathione S-transferases: Structure and function. *In* "Glutathione Centennial. Molecular Perspectives and Clinical Implications" (N. Taniguchi, T. Higashi, Y. Sakamoto, and A. Meister, eds.), pp. 241–257. Academic Press, San Diego, CA.

Kikuchi, J., Sunamoto, J., and Kondo, H. (1985). Nonenzymic transamination and β-elimination of DL-S-benzylcysteine catalyzed by a potent pyridoxal model. *J. Chem. Soc., Perkin Trans. 2,* pp. 341–345.

Kishore, G. M. (1984). Mechanism-based inactivation of bacterial kynureninase by β-substituted amino acids. *J. Biol. Chem.* **259,** 10669–10674.

Koob, M., and Dekant, W. (1990). Metabolism of hexafluoropropene. Evidence for bioactivation by glutathione conjugate formation in the kidney. *Drug. Metab. Dispos.* **18,** 911–916.

Koob, M., and Dekant, W. (1991). Bioactivation of xenobiotics by formation of toxic glutathione conjugates. *Chem.-Biol. Interact.* **77,** 107–136.

Kramer, R. A., Fouremen, G., Greene K. E., and Reed, D. J. (1987). Nephrotoxicity of S-(2-chloroethyl)glutathione in the Fischer rat: Evidence for γ-glutamyl transpeptidase-independent uptake by the kidney. *J. Pharmacol. Exp. Ther.* **242,** 741–748.

Kupchik, H. Z., and Knox, W. E. (1970). Assays of glutamine and its aminotransferase with the enol-borate of phenylpyruvate. *Arch. Biochem. Biophys.* **136,** 178–186.

Lapin, I. P. (1981). Kynurenines and seizures. *Epilepsia* **22,** 257–265.

Larsen, G. L. (1985). Distribution of cysteine conjugate β-lyase in gastrointestinal bacteria and in the environment. *Xenobiotica* **15,** 199–209.

Larsen, G. L., and Bakke, J. E. (1983). Metabolism of mercapturic acid-pathway metabolites of 2-chloro-N-isopropylacetanilide (propachlor) by gastrointestinal bacteria. *Xenobiotica* **13,** 115–126.

Larsen, G. L., and Stevens, J. L. (1986). Cysteine conjugate β-lyase in the gastrointestinal bacterium *Eubacterium limosum*. *Mol. Pharmacol.* **29,** 97–103.

Larsen, G. L., and Larson, J. D., and Gustafsson, J.-Å. (1983). Cysteine conjugate β-lyase in the gastrointestinal bacterium *Fusobacterium necrophorum*. *Xenobiotica* **13,** 689–700.

Lash, L. H., and Anders, M. W. (1986). Cytotoxicity of S-(1,2-dichlorovinyl)glutathione and S-(1,2-dichlorovinyl)-L-cysteine in isolated rat kidney cells. *J. Biol. Chem.* **261,** 13076–13081.

Lash, L. H., and Anders, M. W. (1987). Mechanism of S-(1,2-dichlorovinyl)-L-cysteine- and S-(1,2-dichlorovinyl)-L-homocysteine-induced renal mitochondrial toxicity. *Mol. Pharmacol.* **32,** 549–556.

Lash, L. H., and Anders, M. W. (1988). S-Conjugate metabolism by purified bovine kidney cytosolic cysteine conjugate β-lyase. *FASEB J.* **2,** A1059.

Lash, L. H., Elfarra, A. A., and Anders, M. W. (1986a). Renal cysteine conjugate β-lyase. Bioactivation of nephrotoxic cysteine S-conjugates in mitochondrial outer membrane. *J. Biol. Chem.* **261,** 5930–5935.

Lash, L. H., Elfarra, A. A., and Anders, M. W. (1986b). S-(1,2-Dichlorovinyl)-L-homocysteine-induced cytotoxicity in isolated rat kidney cells. *Arch. Biochem. Biophys.* **251,** 432–439.

Lash, L. H., Nelson, R. M., Van Dyke, R. A., and Anders, M. W. (1990a). Purification and characterization of human kidney cytosolic cysteine conjugate β-lyase activity. *Drug Metab. Dispos.* **18,** 50–54.

Lash, L. H., Elfarra, A. A., Rakiewicz-Nemeth, D., and Anders, M. W. (1990b). Bioactivation mechanisms of cytotoxic homocysteine S-conjugates. *Arch. Biochem. Biophys.* **276,** 322–330.

Lenzen, S., Rustenbeck, I., and Panten, U. (1984). Transamination of 3-phenylpyruvate in pancreatic B-cell mitochondria. *J. Biol. Chem.* **259,** 2043–2046.

Lock, E. A., and Ishmael, J. (1979). The acute toxic effects of hexachloro-1:3-butadiene on the rat kidney. *Arch. Toxicol.* **43,** 47–57.

MacFarlane, M., Foster, J. R., Gibson, G. G., King, L. J., and Lock, E. A. (1989). Cysteine conjugate β-lyase of rat kidney cytosol: Characterization, immunocytochemical localization, and correlation with hexachlorobutadiene nephrotoxicity. *Toxicol. Appl. Pharmacol.* **98,** 185–197.

MacFarlane, M., Schofield, M., Parker, N., Roelandt, L., David M., Lock, E. A., King L. J., Goldfarb, P. S., and Gibson, G. G. (1993). Dose-dependent induction or depression of cysteine conjugate β-lyase in rat kidney by *N*-acetyl-*S*-(1,2,3,4,4-pentachloro-1,3-butadienyl)-L-cysteine. *Toxicology* **77,** 133–144.

Makar, T. K., Nedergaard, M., Preuss, A., Hertz, L., and Cooper, A. J. L. (1994). Glutamine transaminase K and ω-amidase activities in primary cultures of astrocytes and neurons and in embryonic chick forebrain. Marked induction of brain glutamine transaminase K at time of hatching. *J. Neurochem.* (in press).

Meister, A. (1953). Preparation and enzymatic reactions of the keto analogues of asparagine and glutamine. *J. Biol. Chem.* **200,** 571–589.

Meister, A. (1989). Metabolism and function of glutathione. *In* "Glutathione: Chemical, Biochemical and Medical Aspects" (D. Dolphin, R. Poulson, and O. Avramovic, eds.), Part B, pp. 367–474. Wiley, New York.

Meister, A. (1992). On the antioxidant effects of ascorbic acid and glutathione. *Biochem. Pharmacol.* **44,** 1905–1915.

Meister, A., and Anderson, M. E. (1983). Glutathione. *Annu. Rev. Biochem.* **52,** 711–760.

Meister, A., Sober, H. A., Tice, S. V., and Fraser, P. E. (1952). Transamination and associated deamidation of asparagine and glutamine. *J. Biol. Chem.* **197,** 319–330.

Meister, A., Wellner, D., and Scott, S. J. (1960). Recent investigations of L- and D-amino acid oxidases. *J. Natl. Cancer Inst.* (*U.S.*) **24,** 31–49.

Miller, J. E., and Litwack, G. (1971). Purification, properties, and identity of liver mitochondrial tyrosine aminotransferase. *J. Biol. Chem.* **246,** 3234–3240.

Morino, Y., and Okamoto, M. (1973). Labeling of the active site of cytoplasmic aspartate aminotransferase by β-chloro-L-alanine. *Biochem. Biophys. Res. Commun.* **50,** 1061–1067.

Nash, J. A., King, L. J., Lock, E. A., and Green, T. (1984). The metabolism and disposition of hexachloro-1:3-butadiene in the rat and its relevance to nephrotoxicity. *Toxicol. Appl. Pharmacol.* **73,** 124–137.

Novogrodsky, A., Nishimura, J. S., and Meister, A. (1963). Transamination and β-decarboxylation of aspartate catalyzed by the same pyridoxal phosphate-enzyme. *J. Biol. Chem.* **238,** PC1903–PC1905.

Odum, J., and Green, T. (1984). The metabolism and nephrotoxicity of tetrafluoroethylene in the rat. *Toxicol. Appl. Pharmacol.* **76,** 306–318.

Patel, N. J., Fullone, J., and Anders, M. W. (1993). Brain uptake of S-(1,2-dichlorovinyl) glutathione and S-(1,2-dichlorovinyl)-L-cysteine, the glutathione and cysteine S-conjugates of the neurotoxin dichloroacetylene. *Mol. Brain Res.* **17,** 53–58.

Perry, S. J., Schofield, M. A., MacFarlane, M., Lock, E. A., King, L. J., Gibson, G. G., and Goldfarb, P. S. (1993). Isolation and expression of a cDNA coding for kidney cytosolic cysteine conjugate β-lyase. *Mol. Pharmacol.* **43,** 660–665.

Potter, C. L., Gandolfi, A. J., Nagle, R., and Clayton, J. W. (1981). Effects of inhaled chlorotrifluoroethylene and hexafluoropropene on the rat kidney. *Toxicol. Appl. Pharmacol.* **59,** 431–440.

Reichert, D., Liebaldt, G., and Henschler, D. (1976). Neurotoxic effects of dichloroacetylene. *Arch. Toxicol.* **37,** 23–38.

Relyea, N. M., Tate, S. S., and Meister, A. (1974). Affinity labeling of the active center of L-aspartate-β-decarboxylase with β-chloro-L-analine. *J. Biol. Chem.* **249,** 1519–1524.

Ricci, G., Nardini, M., Federici, G., and Cavallini, D. (1986). The transamination of L-cystathionine, L-cystine, and related compounds by a bovine kidney transaminase. *Eur. J. Biochem.* **157,** 57–63.

Saari, J. C., and Schultze, M. O. (1965). Cleavage of S-(1,2-dichlorovinyl)-L-cysteine by *Escherichia coli* B. *Arch. Biochem. Biophys.* **109,** 595–602.

Sausen, P. J., and Elfarra, A. A. (1990a). Cysteine conjugate S-oxidase. Characterization of a novel enzymatic activity in rat hepatic and renal microsomes. *J. Biol. Chem.* **265,** 6139–6145.

Sausen, P. J., and Elfarra, A. A. (1990b). S-(1,2-Dichlorovinyl)-L-cysteine sulfoxide, a putative metabolite of S-(1,2-dichlorovinyl)-L-cysteine, is a potent nephrotoxin. *Toxicologist* **10,** 267.

Sausen, P. J., and Elfarra, A. A. (1991). Reactivity of cysteine conjugate sulfoxides: Formation of S-[1-chloro-2-(S-glutathionyl)vinyl]-L-cysteine sulfoxide by the reaction of S-(1,2-dichlorovinyl)-L-cysteine sulfoxide with glutathione. *Chem. Res. Toxicol.* **4,** 655–660.

Schaumburg, H. H. (1992). Chemical neurotoxicity. *In* "Disorders of Central Nervous System" (A. K. Asbury, G. M. McKhann, and W. I. McDonald, eds.), 2nd ed., Vol. 2, pp. 1238–1249. Saunders, Philadelphia.

Schwarcz, R., Okuno, E., White, R. J., Bird, E. D., and Whetsell, W. O., Jr. (1988). 3-Hydroxyanthranilate oxygenase activity is increased in the brains of Huntington disease victims. *Proc. Natl. Acad. Sci. U.S.A.* **85,** 4079–4081.

Schwimmer, S., and Mazelis, M. (1963). Characterization of alliinase of *Allium cepa* (onion). *Arch. Biochem. Biophys.* **100,** 66–73.

Shrawder, E., and Martinez-Carrion, M. (1972). Evidence of phenylalanine transaminase activity in the isoenzymes of aspartate aminotransferase. *J. Biol. Chem.* **247,** 2486–2492.

Snell, E. E., and Di Mari, S. J. (1970). Schiff base intermediates in enzyme catalysis. *In* "The Enzymes." (P. D. Boyer, ed.), 3rd ed., Vol. 2, pp. 335–370. Academic Press, New York.

Soper, T. S., and Manning, J. M. (1978). β-Elimination of β-halo substrates by D-amino acid transaminase associated with inactivation of the enzyme. Trapping of a key intermediate of the reaction. *Biochemistry* **17,** 3377–3384.

Spink, D. C., and Martin, D. L. (1983). Multiple forms of glutamate decarboxylase in hog, rat, and human brain. *In* "Glutamine, Glutamate, and GABA in the Central Nervous System" (L. Hertz, E. Kvamme, E. G. McGeer, and A. Schousboe, eds.), pp. 129–143. Alan R. Liss, New York.

Stevens, J. L. (1985a). Isolation and characterization of a rat liver enzyme with both cysteine conjugate β-lyase and kynureninase activity. *J. Biol. Chem.* **260,** 7945–7950.

Stevens, J. L. (1985b). Cysteine conjugate β-lyase activities in rat kidney cortex: Subcellular localization and relationship to hepatic enzymes. *Biochem. Biophys. Res. Commun.* **129,** 499–504.

Stevens, J. L., and Jakoby, W. B. (1983). Cysteine conjugate β-lyase. *Mol. Pharmacol.* **23,** 761–765.

Stevens, J. L., and Jones, D. P. (1989). The mercapturic acid pathway: Biosynthesis, intermediary metabolism, and physiological disposition. *In* "Glutathione. Chemical, Biochemical and Medical Aspects" (D. Dolphin, R. Poulson, and O. Avramovic, eds.), Part B, pp. 45–84. Wiley, New York.

Stevens, J. L., and Wallin, A. (1990). Is the toxicity of cysteine conjugates formed during mercapturic acid biosynthesis relevant to the toxicity of covalently bound drug residues? *Drug Metab. Rev.* **22,** 617–635.

Stevens, J. L., Hayden, P., and Taylor, S. (1986a). The role of glutathione conjugate metabolism and cysteine conjugate β-lyase in the mechanism of S-cysteine toxicity in LLC-PK1 cells. *J. Biol. Chem.* **261,** 3325–3332.

Stevens, J. L., Robbins, J. D., and Byrd, R. A. (1986b). A purified cysteine conjugate β-lyase from rat kidney cytosol. Requirement for an α-keto acid or an amino acid oxidase for activity and identity with soluble glutamine transaminase K. *J. Biol. Chem.* **261,** 15529–15537.

Stevens, J. L., Ayoubi, N., and Robbins, J. D. (1988). The role of mitochondrial matrix enzymes in the metabolism and toxicity of cysteine conjugates. *J. Biol. Chem.* **263,** 3395–3401.

Stonard, M. D., and Parker, V. H. (1971a). 2-Oxoacid dehydrogenases of rat liver mitochondria as the site of action of S-(1,2-dichlorovinyl)-L-cysteine and S-(1,2-dichlorovinyl)-3-mercaptopropionic acid. *Biochem. Pharmacol.* **20,** 2417–2427.

Stonard, M. D., and Parker, V. H. (1971b). The metabolism of S-(1,2-dichlorovinyl)-L-cysteine by rat liver mitochondria. *Biochem. Pharmacol.* **20,** 2429–2437.

Stone, R. (1992). Zeroing in on brain toxins. *Science* **225,** 1063.

Stone, T. W., and Burton, N. R. (1988). NMDA receptors and ligands in the vertebrate CNS. *Prog. Neurobiol.* **30,** 333–368.

Stoner, G. L., and Eisenberg, M. A. (1975). Purification and properties of 7,8-diaminopelargonic acid aminotransferase. An enzyme in the biotin biosynthetic pathway. *J. Biol. Chem.* **250,** 4029–4036.

Suzuki, S., Tomisawa, H., Ichihara, S., Fukazawa, F., and Tateishi, M. (1982). A C-S bond cleavage enzyme of cysteine conjugates in intestinal microorganisms. *Biochem. Pharmacol.* **31,** 2137–2140.

Takeuchi, F., Otsuka, H., and Shibata, Y. (1980). Purification and properties of kynureninase from rat liver. *J. Biochem. (Tokyo)* **88,** 987–994.

Tateishi, M., Suzuki, S., and Shimizu, S. (1978a). The metabolism of bromazepam in the rat—identification of mercapturic acid and its conversion *in vitro* to methylthiobromazepam. *Biochem. Pharmacol.* **27,** 809–810.

Tateishi, M., Suzuki, S., and Shimizu, S. (1978b). Cysteine conjugate β-lyase in rat liver. A novel enzyme catalyzing formation of thiol-containing metabolites of drugs. *J. Biol. Chem.* **253,** 8854–8859.

Terracini, B., and Parker, V. H. (1965). A pathological study on the toxicity of *S*-dichlorovinyl-L-cysteine. *Food Cosmet. Toxicol.* **3,** 67–74.

Tomisawa, H., Suzuki, S., Ichihara, S., Fukazawa, H., and Tateishi, M. (1984). Purification and characterization of C—S lyase from *Fusobacterium varium*. A C—S cleavage enzyme of cysteine conjugates and some *S*-containing amino acids. *J. Biol. Chem.* **259,** 2588–2593.

Tomisawa, H., Ichihara, S., Fukazawa, H., Ichimoto, N., Tateishi, M., and Yamamoto, I. (1986). Purification and characterization of human hepatic cysteine-conjugate β-lyase. *Biochem. J.* **235,** 569–575.

Ueno, H., Likos, J. J., and Metzler, D. E. (1982). Chemistry of the inactivation of cytosolic aspartate aminotransferase by serine *O*-sulfate. *Biochemistry* **21,** 4387–4393.

Vamvakas, S., Elfarra, A. A., Dekant, W., Henschler, D., and Anders, M. W. (1988). Mutagenicity of amino acid and glutathione *S*-conjugates by the Ames test. *Mut. Res.* **206,** 83–90.

Vamvakas, S., Kremling, E., and Dekant, W. (1989). Metabolic activation of the nephrotoxic haloalkene 1,1,2-trichloro-3,3,3-trifluoro-1-propene by glutathione conjugation. *Biochem. Pharmacol.* **38,** 2297–2304.

Van Leuven, F. (1975). Highly purified glutamine transaminase from rat brain: Physical and kinetic properties. *Eur. J. Biochem.* **58,** 153–158.

Van Leuven, F. (1976). Glutamine transaminase from brain tissue: Further studies on kinetic properties and specificity of the enzyme. *Eur. J. Biochem.* **65,** 271–274.

Veltman, J. C., Dekant, W., Guengerich, F. P., and Anders, M. W. (1988). Cytotoxicity and bioactivation mechanism of benzyl 2-chloro-1,1,2-trifluoroethyl sulfide and benzyl 1,2,3,4,4-pentachlorobuta-1,3-dienyl sulfide. *Chem. Res. Toxicol.* **1,** 35–40.

Wallin, A., Jones, T. W., Vercesi, A. E., Cotgreave, I., Ormstad, K., and Orrenius, S. (1987). Toxicity of S-pentachlorobutadienyl-L-cysteine studied with isolated rat renal cortical mitochondria. *Arch. Biochem. Biophys.* **258,** 365–372.

Wang, E., and Walsh, C. (1978). Suicide substrates for the alanine racemase of *Escherichia coli* B. *Biochemistry* **17,** 1313–1321.

Wijesundera, S., and Woods, D. D. (1962). The catabolism of cystathionine by *Escherichia coli*. *J. Gen. Microbiol.* **29,** 353–366.

Wolf, C. R., Berry, P. N., Nash, J. A., Green, T., and Lock, E. A. (1984). Role of microsomal and cytosolic glutathione S-transferases in the conjugation of hexachloro-1:3-butadiene and its possible relevance to toxicity. *J. Pharmacol. Exp. Ther.* **228,** 202–208.

Wolfgang, G. H. I., Gandolfi, A. J., Stevens, J. L., and Brendel, K. (1989). *In vitro* and *in vivo* nephrotoxicity of the L and D isomers of S-(1,2-dichlorovinyl)-L-cysteine. *Toxicology* **58,** 33–42.

Formation and Fate of Nephrotoxic and Cytotoxic Glutathione S-Conjugates: Cysteine Conjugate β-Lyase Pathway

Wolfgang Dekant,* Spyridon Vamvakas,* and M. W. Anders†

**Institut für Toxikologie*
Universität Würzburg
97078 Würzburg, Germany

†Department of Pharmacology
University of Rochester
Rochester, New York 14642

I. Introduction

The glutathione *S*-transferases are multifunctional proteins that serve several cellular functions. A prominent function, which is associated with their widely recognized cytoprotective role, is the catalysis of the reaction of glutathione with electrophiles as the first enzymatic step in mercapturic acid biosynthesis. The transferases also serve as binding proteins for organic dyes in the liver. Finally, glutathione *S*-transferases may serve as targets for reactive intermediates and become covalently modified. Several volumes (Arias and Jakoby, 1976; Flohé *et al.*, 1974; Hayes *et al.*, 1990; Larsson *et al.*, 1983; Mantle *et al.*, 1987; Sakamoto *et al.*, 1983; Sies and Ketterer, 1988; Sies and Wendel, 1978; Taniguchi *et al.*, 1989) and review articles (Buetler and Eaton, 1992; Chasseaud, 1979; Coles and Ketterer, 1990; Mannervik and Danielson, 1988; Pickett and Lu, 1989; Tsuchida and Sato, 1992; Wolkoff *et al.*, 1979) that describe the chemistry, physiology, and biological functions of glutathione and the glutathione *S*-transferases have appeared.

Advances in Pharmacology, Volume 27

Although the cytoprotective function of glutathione and the glutathione *S*-transferases is well established, studies over the past decade have elaborated several mechanisms for the glutathione-dependent bioactivation and toxicity of xenobiotics (for reviews, see Anders, 1989, 1990, 1991; Anders *et al.*, 1988, 1992; Dekant *et al.*, 1992; Koob and Dekant, 1991; Monks and Lau, 1987; Vamvakas and Anders, 1990). Briefly, these mechanisms involve the glutathione- and glutathione *S*-transferase-dependent formation of direct-acting half-sulfur mustards from vicinal dihaloalkanes, the cysteine conjugate β-lyase pathway for the bioactivation of nephrotoxic haloalkenes, the glutathione-dependent bioactivation of hydroquinones and aminophenols, and the formation of glutatione *S*-conjugates that serve as transport forms; these bioactivation mechanisms are discussed elsewhere in this volume.

This chapter is concerned with the cysteine conjugate β-lyase pathway for the bioactivation of nephrotoxic and cytotoxic cysteine *S*-conjugates of haloalkenes. The β-lyase pathway includes several steps: hepatic biosynthesis of glutathione *S*-conjugates, hydrolysis of glutathione *S*-conjugates to cysteine *S*-conjugates and translocation to the kidney, and bioactivation by renal cysteine conjugate β-lyase. Several reviews of the β-lyase pathway for the bioactivation of haloalkenes have appeared (Anders, 1989, 1990, 1991; Anders *et al.*, 1986, 1988, 1990, 1992; Dekant *et al.*, 1989, 1990a,b; Koob and Dekant, 1991; Lash and Anders, 1986b; Lash *et al.*, 1988; Lock, 1987, 1988; Nagelkerke and Boogaard, 1991; Stevens and Jones, 1989; Stevens and Wallin, 1989).

II. Biosynthesis of Toxic Glutathione *S*-Conjugates

The reaction of haloalkenes with glutathione is catalyzed by the glutathione *S*-transferases (EC 2.5.1.18). Glutathione *S*-transferase activity is present in cytosolic, microsomal, and mitochondrial compartments of the liver. The cytosolic transferases are a family of homo- and heterodimeric proteins found in many organs, although the isoform distribution differs among organs (Coles and Ketterer, 1990; Ketterer and Mulder, 1990; Mannervik and Danielson, 1988). The microsomal transferase is unrelated in protein structure to the cytosolic transferases, and the functional enzyme may be a homotrimer or homotetramer (Boyer *et al.*, 1986; Lundqvist *et al.*, 1992). Microsomal glutathione *S*-transferase activity is high in the liver, although low activities are found in extrahepatic sites (DePierre and Morgenstern, 1983; Morgenstern *et al.*, 1984). (The enzymology of the cytosolic and microsomal glutathione *S*-transferases is reviewed elsewhere in this volume; see Ketterer *et al.* and Andersson *et al.*) Although glutathione *S*-transferase activity has been detected in hepatic mitochondria (Botti *et al.*, 1989; Jocelyn and Cronshaw, 1985; Kraus, 1980; Kraus and Gross, 1979; Wahlländer *et al.*, 1979), there has been little agreement about the nature of the transferases present (Ryle and Mantle, 1984). Recent studies show, however, that a glutathione *S*-transferase related to the θ-class transferases is present in rat liver mitochondria (Harris *et al.*, 1991). Microsomal glutathione *S*-transferase is present on the outer mitochondrial membrane (Morgenstern and DePierre, 1985; Morgenstern *et al.*, 1984; Nishino and Ito, 1990a) and in peroxisomal membranes (Nishino and Ito, 1990b).

The reaction of haloalkenes with glutathione is catalyzed by both the cytosolic and the microsomal glutathione *S*-transferases, but the microsomal transferase is the more efficient catalyst. This has been confirmed with chlorotrifluoroethene (Dohn and Anders, 1982a; Dohn *et al.*, 1985a; Hargus *et al.*, 1991), dichloroethyne (Kanhai *et al.*, 1989), hexafluoropropene (Koob and Dekant, 1990), and tetrafluoroethene (Odum and Green, 1984) as substrates.

The glutathione *S*-transferase-catalyzed addition of glutathione to haloalkenes may follow two reaction mechanisms: 1,1-dichloroalkenes undergo an addition–elimination reaction to give *S*-(1-chloroalkenyl)glutathione conjugates (Fig. 1), whereas 1,1-difluoroalkenes and dichloroacetylene undergo an addition reaction to afford *S*-(1,1-difluoroalkyl)glutathione conjugates (Fig. 3) or *S*-(1,2-dichlorovinyl)glutathione, respectively (Fig. 2). There are ample precedents in organic chemistry for the addition of sulfur nucleophiles to alkenes, which is termed the S_NV reaction (Bernasconi, 1989; Rappoport, 1985).

Many examples of glutathione *S*-transferase-catalyzed addition–elimination reactions of glutathione with nephrotoxic 1,1-dichloroalkenes are available: trichloroethene is biotransformed to *S*-(1,2-dichlorovinyl) glutathione (Fig. 1, X_1 = Cl, X_2 = H) (Dekant *et al.*, 1990c), hexachlorobutadiene is biotransformed to *S*-(pentachlorobutadienyl)glutathione (Fig. 1, X_1 = Cl, X_2 = ClC = CCl_2) (Dekant *et al.*, 1988a,b; Oesch and Wolf, 1989), tetrachloroethene is biotransformed to *S*-(1,2,2-trichlorovinyl) glutathione (Fig. 1, X_1 = X_2 = Cl) (Dekant *et al.*, 1987b), and 1,1,2-

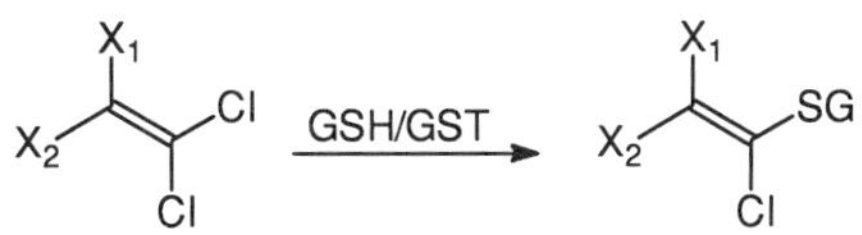

Fig. 1 Glutathione *S*-transferase-catalyzed addition–elimination reaction of glutathione with 1,1-dichloroalkenes. GSH, glutathione; GST, glutathione *S*-transferase.

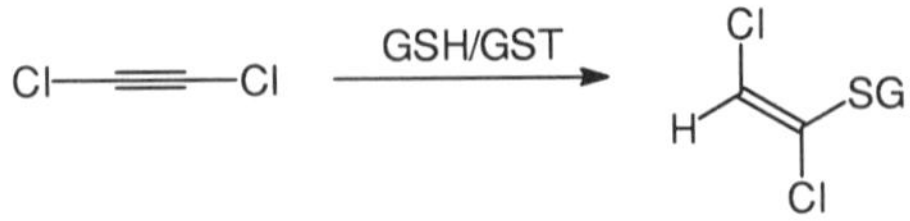

Fig. 2 Glutathione *S*-transferase-catalyzed addition reaction of glutathione with dichloroethyne. GSH, glutathione; GST, glutathione *S*-transferase.

trichloro-3,3,3-trifluoro-1-propene is biotransformed to *S*-(1,2-dichloro-3,3,3-trifluoro-1-propenyl)glutathione (Fig. 1, X_1 = Cl, X_2 = CF_3) (Vamvakas *et al.*, 1989a). The generalization that only 1,1-difluoroalkenes undergo microsomal glutathione *S*-transferase-catalyzed addition reactions does not hold for hexafluoropropene; this reactive fluoroalkene is metabolized by a cytosolic glutathione *S*-transferase-catalyzed addition–elimination reaction to give *S*-(1,2,3,3,3-pentafluoro-1-propenyl) glutathione (Fig. 1, X_1 = F, X_2 = CF_3) (Koob and Dekant, 1990). Hexachlorobutadiene is converted to a diglutathione conjugate *in vivo* and in isolated rat hepatocytes (Jones *et al.*, 1985). *S*-(Pentachlorobutadienyl)glutathione is a substrate for the cytosolic glutathione *S*-transferases, which catalyze an addition–elimination reaction to give the diglutathione conjugate 1,4-bis(glutathion-*S*-yl)-1,2,3,4-tetrachlorobuta-1,3-diene (Dekant *et al.*, 1988a). The toxicological properties of the diconjugate have not been investigated.

Dichloroethyne, which is nephrotoxic, nephrocarcinogenic, and neurotoxic (Reichert *et al.*, 1976, 1984), is the sole example of glutathione-conjugate formation with a haloalkyne. Dichloroethyne undergoes an addition reaction to give *S*-(1,2-dichlorovinyl)glutathione (Fig. 2) (Kanhai *et al.*, 1989, 1991).

Several, 1,1-difluoroalkenes are biotransformed by addition reactions to give *S*-(1,1-fluoroalkyl)glutathione conjugates. Tetrafluoroethene is biotransformed to *S*-(1,1,2,2-tetrafluoroethyl)glutathione (Fig. 3, X_1 = X_2 = F) (Odum and Green, 1984), chlorotrifluoroethene is biotransformed to *S*-(2-chloro-1,1,2-trifluoroethyl)glutathione (Fig. 3, X_1 = F, X_2 = Cl) (Dohn and Anders, 1982a; Dohn *et al.*, 1985a; Hargus *et al.*, 1991; Hassall *et al.*, 1984; Odum and Green, 1984), 2,2-dichloro-1,1-difluoroethene is

Fig. 3 Glutathione *S*-transferase-catalyzed addition reaction of glutathione with 1,1-difluoroalkenes. GSH, glutathione; GST, glutathione *S*-transferase.

biotransformed to *S*-(2,2-dichloro-1,1-difluoroethyl)glutathione (Fig. 3, $X_1 = X_2 =$ Cl) (Commandeur *et al.*, 1987), and hexafluoropropene is biotransformed to *S*-(1,1,2,3,3,3-hexafluoro-1-propyl)glutathione (Fig. 3, $X_1 =$ F, $X_2 = CF_3$) (Koob and Dekant, 1990). Indirect evidence indicates that 2-bromo-2-chloro-1,1-difluoroethene, which is a degradation product or metabolite of halothane (2-bromo-2-chloro-1,1,1-trifluoroethane), is metabolized by glutathione-conjugate formation. *N*-Acetyl-*S*-(2-bromo-2-chloro-1,1-difluoroethyl)-L-cysteine is found in the urine of children anesthetized with halothane (Wark *et al.*, 1990). This metabolite presumably arises by addition of glutathione to 2-bromo-2-chloro-1,1-difluoroethene to give *S*-(2-bromo-2-chloro-1,1-difluoroethyl)glutathione (Fig. 3, $X_1 =$ Br, $X_2 =$ Cl), which is hydrolyzed and then acetylated to afford *N*-acetyl-*S*-(2-bromo-2-chloro-1,1-difluoroethyl)-L-cysteine. This is apparently a minor pathway, because, although halothane is not nephrotoxic, both *S*-(2-bromo-2-chloro-1,1-difluoroethyl)glutathione and *S*-(2-bromo-2-chloro-1,1-difluoroethyl)-L-cysteine are nephrotoxic (Finkelstein *et al.*, 1992).

The microsomal glutathione *S*-transferase-catalyzed addition of glutathione to 1,1-difluoroethenes is regiospecific and stereoselective. With all compounds studied, addition of glutathione occurs at the electron-deficient carbon bearing two fluorines. The microsomal glutathione *S*-transferase catalyzes the stereoselective addition of glutathione to chlorotrifluoroethene to give *S*-(2-chloro-1,1,2-trifluoroethyl)glutathione, which contains a new chiral center, whereas *S*-(2-chloro-1,1,2-trifluoroethyl)glutathione formation catalyzed by the cytosolic glutathione *S*-transferases shows no stereoselectivity (Dohn *et al.*, 1985a). This observation was exploited to determine the relative contributions of the microsomal and cytosolic transferases to intracellular glutathione-conjugate formation (Hargus *et al.*, 1991). Studies with rat hepatic microsomes and purified rat hepatic microsomal glutathione *S*-transferase showed the formation of about 80% excess of the 2*S*-diastereoisomer of *S*-(2-chloro-1,1,2-trifluoroethyl) glutathione, whereas the cytosolic glutathione *S*-transferases catalyzed formation of equal amounts of the 2*S*- and 2*R*-diastereoisomers. When chlorotrifluoroethene was incubated with isolated rat hepatocytes, an 80% excess of (2*S*)-*S*-(2-chloro-1,1,2-trifluoroethyl)glutathione was formed, indicating that the microsomal transferase is largely responsible for intracellular *S*-conjugate formation.

Although extrahepatic activities of the microsomal glutathione *S*-transferase are low, extrahepatic glutathione *S*-transferase-catalyzed *S*-conjugate formation has been observed. Chlorotrifluoroethene is biotransformed to *S*-(2-chloro-1,1,2-trifluoroethyl)glutathione by rabbit renal tubules (Hassall *et al.*, 1984). Bile-duct cannulated rats exposed to hexafluoropropene excrete *N*-acetyl-*S*-(1,1,2,3,3,3-hexafluoropropyl)-L-cysteine in the urine, indicating extrahepatic *S*-conjugate formation (Koob and Dekant, 1990); this conclusion is supported by the observation that rat renal cytosol, but not microsomes, biotransform hexafluoropropene to *S*-(1,1,2,3,3,3-hexafluoropropyl)glutathione. Dichloroethyne is biotransformed to *S*-(1,2-dichlorovinyl)glutathione by lung and liver, but not by brain or kidney, microsomal fractions (N. Patel, G. Birner, W. Dekant, and M. W. Anders, unpublished observations).

In summary, the available data show that the hepatic microsomal glutathione *S*-transferase is the major catalyst for the formation glutathione conjugates of nephrotoxic and nephrocarcinogenic haloalkenes.

III. Fate of *S*-Conjugates

A. Enzymatic Processing of Toxic Glutathione *S*-Conjugates

As will be discussed below, the liver is the major site of glutathione *S*-conjugate formation with nephrotoxic haloalkenes. The glutathione *S*-conjugates are excreted in the bile and pass into the small intestine. In the bile duct and in the intestine, some of the glutathione *S*-conjugates are hydrolyzed to the corresponding L-cysteinylglycine and L-cysteine *S*-conjugates. Three enzymes catalyze the hydrolysis of glutathione *S*-conjugates: γ-glutamyltransferase, aminopeptidase M, and cysteinylglycine dipeptidase. Several reviews about the hydrolysis of glutathione and L-cysteinylglycine *S*-conjugates have appeared (Stevens and Jones, 1989; Tate, 1980, 1985, 1989).

1. γ-Glutamyltransferase (EC 2.3.2.2)

γ-Glutamyltransferase is a serine hydrolase that catalyzes the transfer of the D or L-γ-glutamyl group from γ-glutamyl di- and tripeptides, glutathione, or glutathione *S*-conjugates to acceptor amino acids or dipeptides (Fig. 4, **1** → **2** and **3**); the products of the reaction are L-cysteinylglycine or L-cysteinylglycine *S*-conjugates and L-γ-glutamyl amino acids or dipeptides (Stole *et al.*, 1990; Tate and Meister, 1974, 1985). Glutathione or glutathione *S*-conjugates may also be hydrolyzed by γ-glutamyltransferase to give L-glutamate and L-cysteinylglycine or L-cysteinylglycine *S*-conjugates, respectively, as products (Meister, 1988).

γ-Glutamyltransferase is a heterodimeric glycoprotein with subunit molecular masses of 22 kDa and 46 to 51 kDa; the heavy subunit is anchored to the lumenal surface of brush-border membranes through a hydrophobic N-terminal domain (Tate and Meister, 1985).

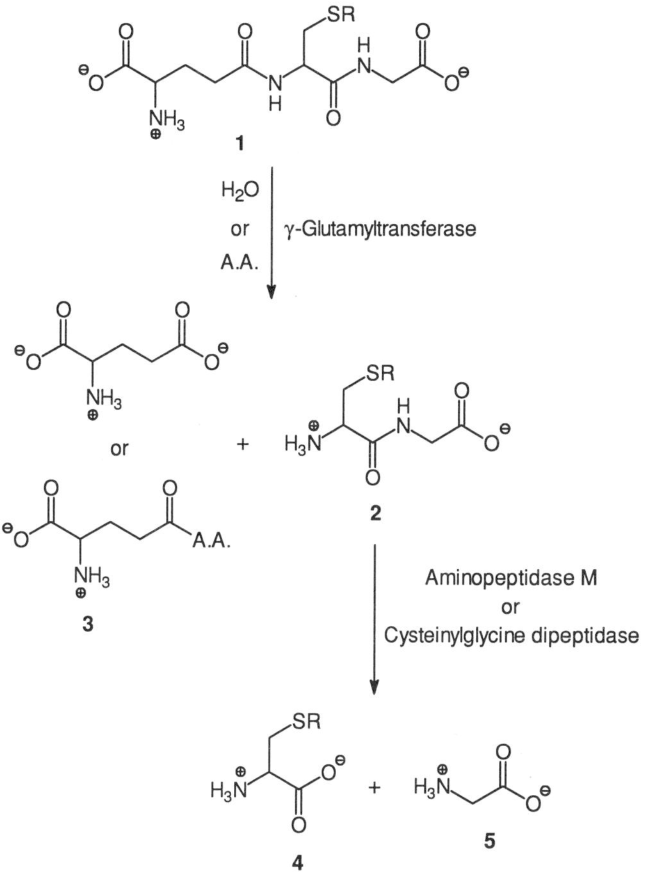

Fig. 4 Enzymatic processing of glutathione and glutathione *S*-conjugates. **1,** glutathione or glutathione *S*-conjugate; **2,** L-cysteinylglycine or L-cysteinylglycine *S*-conjugate; **3,** L-glutamic acid or L-glutamylpeptide; **4,** L-cysteine or L-cysteine *S*-conjugate; **5,** glycine; R = H, haloalkenyl, or haloalkyl groups; A.A., amino acid or dipeptide.

The enzyme is highly selective for D- or L-γ-glutamyl moieties, but tolerates a range of substituents on the α-carboxyl group (Tate and Meister, 1974). γ-Glutamyltransferase is the only known enzyme that catalyzes the hydrolysis of glutathione. Glutathione *S*-conjugates are also substrates as are a range of γ-glutamyl derivatives, including γ-glutamyl-*p*-nitroanilide, which is commonly used as a substrate to quantify γ-glutamyltransferase activity.

γ-Glutamyltransferase is an ectoenzyme present in many epithelial tissues (Kozak and Tate, 1982). Activity is present in the bile duct, in the intestinal microvilli, in the brush-border membranes of the S_2 and S_3 segments of the renal proximal tubules, and in the renal vasculature and basal–lateral membranes (Guder and Ross, 1984; Guder and Wirthensohn, 1985).

Although γ-glutamyltransferase activity is highest in renal tissue, there are significant species differences in activities among tissues (Albert *et al.*, 1961, 1964; Hinchman and Ballatori, 1990; Lau *et al.*, 1990). The whole-organ ratio of kidney/liver γ-glutamyltransferase activities amount to 142 in the rat, which is in contrast to a ratio of <5 in guinea pig, pig, macaque monkey, and humans (Hinchman and Ballatori, 1990).

Several selective inhibitors of γ-glutamyltransferase are available, and their use has helped elucidate the role of the enzyme in processing glutathione *S*-conjugates.

6-Diazo-5-oxonorleucine (DON) and *O*-diazoacetyl-L-serine (L-azaserine) are irreversible inhibitors of γ-glutamyltransferase that inactivate the enzyme by reacting covalently at the γ-glutamyl binding site (Inoue *et al.*, 1977; Tate and Meister, 1977). Treatment of mice with 6-diazo-5-oxonorleucine produced a modest glutathionuria (Griffith and Meister, 1979).

L-Serine/borate is a competitive, transition-state inhibitor of γ-glutamyltransferase (Tate and Meister, 1978). It inhibits rat kidney γ-glutamyltranferase *in vitro* (K_i = 1.45 m*M*) and the *in vivo* metabolism of D-γ-glutamyl-L-α-amino[^{14}C]butyrate to carbon dioxide, but fails to produce glutathionuria in mice (Griffith and Meister, 1979).

Both L- and D-γ-glutamyl-(*o*-carboxy)phenylhydrazide inhibit rat kidney γ-glutamyltransferase activity *in vitro* (apparent K_i = 8.2 and 22.5 *μm*, respectively), although the L-isomer is more potent than the D-isomer (Griffith and Meister, 1979). In mice, L-γ-glutamyl-(*o*-carboxy)phenylhydrazide inhibits the metabolism of D-γ-glutamyl-L-α-amino[^{14}C]butyrate to carbon dioxide and produces a profound glutathionuria.

L-(α*S*,5*S*)-α-Amino-3-chloro-4,5-dihydro-5-isoxazoleacetic acid (acivicin, AT-125) irreversibly inactivates γ-glutamyltransferase, but not cysteinylglycine dipeptidase (Allen *et al.*, 1981; Reed *et al.*, 1980). Acivicin inhibits γ-glutamyltransferase activity in a range of *in vitro* preparations, including isolated rat kidney cells, and produces glutathionuria in rats (Rankin *et al.*, 1983; Reed and Ellis, 1982).

2. Aminopeptidase M (EC 3.4.11.2, Microsomal Alanyl Aminopeptidase)

The hydrolysis of cysteinyl oligopeptides by a peptidase isolated from rat renal particulate fractions has been attributed to an aminopeptidase

(Hughey *et al.*, 1978). Aminopeptidase M catalyzes the cleavage of unblocked N-terminal amino acids from dipeptides and oligopeptides (McDonald and Barrett, 1986). Cysteinylglycine and *S*-benzyl-L-cysteine *p*-nitroanilide are hydrolyzed (Fig. 4, **2** → **4** and **5**), but cystinyl-bis-glycine is a poor substrate (McIntyre and Curthoys, 1982; Rankin *et al.*, 1980).

Aminopeptidase M activity is high in liver, kidney, and intestinal epithelium, and the enzyme is located on the lumenal surface of microvillus membranes (Kozak and Tate, 1982; Tsao and Curthoys, 1980). Activity is found in microsomal fractions because brush-border membrane fragments cosediment with microsomes (McDonald and Barrett, 1986). The enzyme is a homodimeric, zinc-containing protein with a subunit M_r of 130,000 and is inhibited by 1,10-phenanthroline and bestatin ([2*S*,3*R*]-3-amino-2-hydroxy-4-phenylbutanoyl-L-serine), but not by thiols (Kozak and Tate, 1982; Tate, 1985).

3. Cysteinylglycine Dipeptidase (EC 3.4.13.6)

The finding that epididymal membranes, which lack aminopeptidase M, hydrolyzed *S*-methylglutathione led to a search for additional glutathione-degrading enzymes and to the identification of a dipeptidase that efficiently hydrolyzes derivatives of cysteinylglycine (Kozak and Tate, 1982; McIntyre and Curthoys, 1982). A dipeptidase was purified from papain-solubilized renal brush-border membranes that hydrolyzed, for example, L-methionylglycine, L-methionyl-L-leucine, L-alanylglycine, L-leucylglycine, glycylglycine, and *S*-substituted cysteinylglycines, including *S*-methyl-L-cysteinylglycine, and cystinyl-bis-glycine (Fig. 4, **2** → **4** and **5**) (Kozak and Tate, 1982). The enzyme is a homodimeric, zinc-containing protein with a subunit M_r of 50,000. The enzyme is inhibited by 1,10-phenanthroline, thiols (L-cysteinylglycine, D- and L-penicillamine, dithiothreitol), and cilastatin, but not by bestatin.

B. Physiological Disposition of Toxic Glutathione *S*-Conjugates

The specific activity of microsomal glutathione *S*-transferases in liver is much higher than in kidney; hence, glutathione *S*-conjugate formation from nephrotoxic halogenated alkenes is considered to take place mainly in the liver (Dekant *et al.*, 1988a; DePierre and Morgenstern, 1983; Morgenstern *et al.*, 1984; Wolf *et al.*, 1984). Biotransformation of glutathione *S*-conjugates to excretable metabolites or conversion to reactive intermediates occurs elsewhere in the body. Therefore, transport systems must exist to deliver glutathione *S*-conjugates and their metabolites to these extrahepatic sites (Inoue *et al.*, 1984a; Okajima *et al.*, 1983). Hence, interorgan transport from the liver and accumulation of the *S*-conjugates in the kidney is necessary for the expression of nephrotoxicity after exposure to halogenated alkenes.

1. Hepatic Efflux of Glutathione *S*-Conjugates

Efflux of glutathione *S*-conjugates occurs mainly across the canalicular membrane into bile; sinusoidal elimination of *S*-conjugates has been observed only at high intracellular glutathione *S*-conjugate concentrations. The molecular weight of most glutathione *S*-conjugates biosynthesized in the liver is above the threshold for biliary transport in rats (Vore, 1993). In addition, the presence of polar groups in the molecule prevents passive diffusion through the membrane of the hepatocytes and requires active transport. The canalicular membrane of hepatocytes possesses an active ATP-requiring transport system that accepts many polar, high-molecular-weight compounds, including glutathione *S*-conjugates (Awasthi, 1990; Awasthi *et al.*, 1989); therefore transport into bile is the primary route of hepatic efflux of glutathione *S*-conjugates (Inoue *et al.*, 1984b,c). For example, *S*-(pentachlorobutadienyl)glutathione biosynthetized from hexachlorobutadiene in the isolated perfused liver is excreted largely into bile. Release of *S*-(pentachlorobutadienyl)glutathione into the caval perfusate occurs only after perfusion with toxic concentrations of hexachlorobutadiene (Gietl and Anders, 1991). Moreover, *S*-(2,4-dinitrophenyl)glutathione is preferentially transported across the canalicular membrane into bile; this transport is ATP-dependent and is not inhibited by glutathione concentrations up to 5 m*M* (Akerboom *et al.*, 1982, 1991; Inoue *et al.*, 1984b; Wahlländer and Sies, 1979). Sinusoidal-membrane transport of glutathione *S*-conjugates has a much lower affinity for glutathione *S*-conjugates and is inhibited by physiologic concentrations of glutathione (Inoue and Morino, 1985; Inoue *et al.*, 1984c). These observations indicate that efflux across the sinusoidal membrane occurs only at intracellular glutathione *S*-conjugate concentrations high enough to saturate biliary excretion or at hepatotoxic concentrations of the substrate for glutathione conjugation. In summary, the canalicular transport system is responsible for the *in vivo* disposition of toxic glutathione *S*-conjugates. The importance of biliary excretion is further supported by the observation that biliary cannulation protects male rats from the nephrotoxicity of hexachlorobutadiene (Nash *et al.*, 1984) and *p*-aminophenol (Gartland *et al.*, 1990), indicating that glutathione *S*-conjugate formation and biliary excretion are the first steps in the complex, multiorgan transport of nephrotoxic *S*-conjugates.

2. Intestinal Absorption

Although considerable information is available about the hepatic biosynthesis and biliary excretion of toxic glutathione *S*-conjugates, the systems involved in *S*-conjugate transport in the kidney, and the renal bioactivation of toxic glutathione and cysteine *S*-conjugates, knowledge about the sites of conversion of glutathione *S*-conjugates to cysteine *S*-conjugates and the translocation of the *S*-conjugates from the liver to the kidney is still incomplete. The role of the organs involved in the disposition of *S*-conjugates is difficult to evaluate due to the complexicity of the pathways and the lack of adequate experimental systems. Isolated cells and isolated perfused organs provide limited information, and valid *in vitro* models that simulate the complex multiorgan transport pathways and interactions are not available. The present understanding of the pathways for the processing of *S*-conjugates and their interorgan transport is presented in Fig. 5.

After secretion into bile, glutathione *S*-conjugates are either transported intact to the small intestine or degraded to the corresponding cysteine *S*-conjugates by the sequential action of γ-glutamyltransferase (EC 2.3.2.2.)

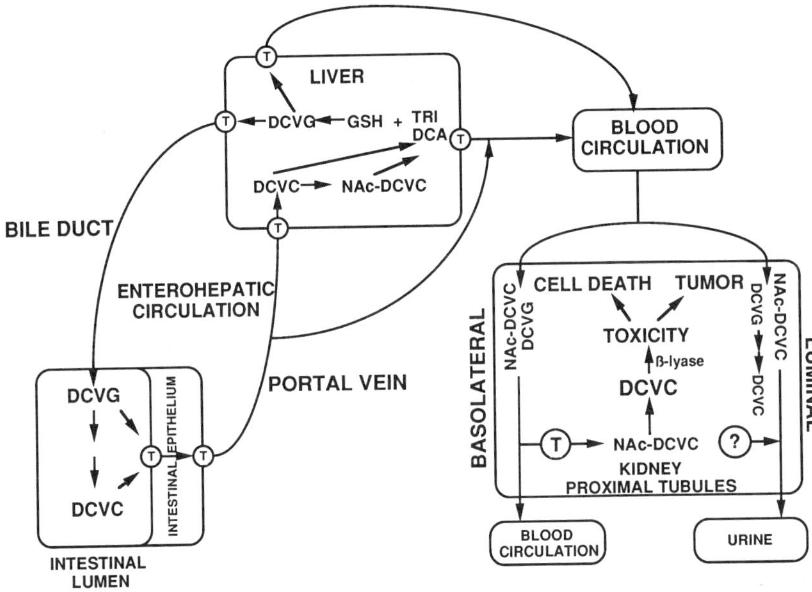

Fig. 5 The role of interorgan disposition and glutathione *S*-conjugate processing in the glutathione-dependent bioactivation of trichloroethene and dichloroethyne to *S*-(1,2-dichlorovinyl)-L-cysteine.

and cysteinylglycine dipeptidase (EC 3.4.13.6.) or aminopeptidase M (EC 3.4.11.2.), which are present in the luminal membrane of the bile duct epithelium and in the bile canalicular membrane of hepatocytes (Anderson *et al.*, 1980; Inoue *et al.*, 1983; Meister, 1988). Metabolites of glutathione *S*-conjugates biosynthesized from 1,2-dichloroethane (Rannug and Beije, 1979; Rannug *et al.*, 1978), 1,2-dibromoethane (Marchand and Reed, 1989), hexachlorobutadiene (Nash *et al.*, 1984), *p*-aminophenol (Klos *et al.*, 1992), and tetrafluoroethene (Odum and Green, 1984) have been detected in bile, indicating that metabolism of these glutathione *S*-conjugates occurs at the biliary or canalicular membrane. *S*-Conjugates translocated to the small intestine may either be reabsorbed from the gut and after passage through the liver, be translocated to the kidney, may undergo enterohepatic circulation, or may be excreted with the feces (Dekant *et al.*, 1988c). Moreover, intestinal bacteria contain cysteine conjugate β-lyase, which may cleave cysteine *S*-conjugates to produce a thiol (Bakke *et al.*, 1990; Jakoby and Stevens, 1984; Larsen and Stevens, 1985; Suzuki *et al.*, 1982; Tomisawa *et al.*, 1984). The thiol metabolite may further be converted to the methylthiol derivative by thiol *S*-methyl transferase (EC 2.1.1.9.); the methylthiol metabolite may undergo enterohepatic circulation or may be excreted with the feces. Oxidation of the thioethers thus formed to sulfoxides or sulfones has also been observed (Bakke and Gustafsson, 1984; Dulik *et al.*, 1992; Larsen and Bakke, 1983; Mulford *et al.*, 1991). Glutathione *S*-conjugates excreted in the bile may be also reabsorbed by the liver from bile after breakdown to the cysteine *S*-conjugates and further transformed to the corresponding mercapturic acid. Intrahepatic mercapturic acid formation may contribute to the disposition of *S*-conjugates in species other than the rat that have higher levels of hepatic γ-glutamyltransferase. Rats and mice have unusually low hepatic γ-glutamyltransferase activity and very high renal γ-glutamyltransferase activities when compared with guinea pigs, pigs, macaque monkeys, and humans (Hinchman and Ballatori, 1990; Hinchman *et al.*, 1991). Unfortunately, the contribution of this pathway to the disposition of nephrotoxic *S*-conjugates is not known, since only few experiments on *S*-conjugate formation and toxicity have been performed in species other than the rat.

The mechanisms of intestinal absorbtion of toxic glutathione *S*-conjugates have been studied by introducing bile containing *S*-(pentachlorobutadienyl)glutathione and *S*-(pentachlorobutadienyl)-L-cysteine directly into rat intestines via a biliary cannula and by determining the appearance of conjugates in portal blood (Gietl *et al.*, 1991). Infusion of *S*-(pentachlorobutadienyl)glutathione into the lumen of rat intestine resulted in the appearance of *S*-(pentachlorobutadienyl)glutathione and *S*-(pentachlorobutadienyl)-L-cysteine in portal blood. The average concentrations of the cysteine *S*-conjugates in blood were higher than the concentrations of the glutathione *S*-conjugates, indicating metabolism of *S*-(pentachlorobutadienyl)glutathione by γ-glutamyltransferase and dipeptidases located in the luminal surface of the enterocyte. A more efficient transport of cysteine *S*-conjugates than of glutathione *S*-conjugates through the intestinal wall is indicated by studies on the transport of *S*-(pentachlorobutadienyl)-L-cysteine. Blood concentrations of *S*-(pentachlorobutadienyl)-L-cysteine were higher after infusion of *S*-(pentachlorobutadienyl)-L-cysteine than were blood concentrations of *S*-(pentachlorobutadienyl)glutathione and *S*-(pentachlorobutadienyl)-L-cysteine after infusion of equimolar amounts of *S*-(pentachlorobutadienyl)-glutathione. γ-Glutamyltransferase-independent transport of glutathione has been observed in brush-border membrane vesicles prepared from rabbit small intestine (Vincenzini *et al.,* 1989); this mechanism may also accept glutathione *S*-conjugates. This observation indicates that different mechanisms are operative in the transport of *S*-conjugates in the enterocytes and that peptides may be transported slower than amino acids (Argiles and Lopez-Soriano, 1990).

After passage through the intestinal wall, *S*-conjugates enter the systemic circulation and thus the kidney directly or after absorption and resecretion from the hepatocytes. Glutathione *S*-conjugates absorbed by the hepatocytes at the luminal surface will be excreted with bile and thus undergo enterohepatic circulation. Due to their molecular weight, cysteine *S*-conjugates taken up by the liver may be excreted again via the sinusoidal membrane into the blood circulation. Cysteine *S*-conjugates may also be N-acetylated by hepatic *N*-acetyltransferases and may reach the systemic circulation as mercapturic acids (Duffel and Jakoby, 1982; Green and Elce, 1975).

Glutathione *S*-conjugates leaving the liver via the sinusoidal membrane and glutathione *S*-conjugates biosynthesized in organs other than the liver may be delivered intact to the kidney. Moreover, renal glutathione *S*-transferases may also synthesize toxic glutathione *S*-conjugates. For example, conjugation of 1,2-dibromoethane and 1,2-dibromo-3-chloropropane with glutathione takes place preferentially in the kidney (Kluwe *et al.,* 1981, 1982; Omichinski *et al.,* 1987, 1988; Pearson *et al.,* 1990). The formation of reactive sulfur mustards from 1,2-dibromoethane and 1,2-dibromo-3-chloropropane in the kidney is likely responsible for the nephrotoxicity of 1,2-dibromoethane and 1,2-dibromo-3-chloropropane. Intrarenal glutathione conjugation may also occur with the nephrotoxic haloalkenes perfluoropropene and chlorotrifluoroethene and with dichloroethyne. Biliary cannulation does not influence the amount of dichloroethyne (Kanhai *et al.,* 1989) and perfluoropropene (Koob and Dekant, 1990) excreted as mercapturic acids, indicating that glutathione conjugation of these reactive haloalkenes and haloalkynes may occur in the kidney. Moreover, rabbit renal tubules catalyze the formation of the toxic *S*-(1-chloro-1,2,2-trifluoroethyl)glutathione from chlorotrifluoroethene (Hassall *et al.*, 1984), and kidney microsomes from rats convert hexachlorobutadiene to *S*-(pentachlorobutadienyl)glutathione (Dekant *et al.*, 1988a; Wolf *et al.*, 1984).

Finally, *S*-conjugates present in the systemic circulation are delivered to the kidney. The high blood flow to the kidneys and their capacity to process and activate *S*-conjugates are major determinants of *S*-conjugate toxicity (Monks and Lau, 1987, 1989; Monks *et al.*, 1990; Rush *et al.*, 1984). The kidney constitutes only 0.4% of the body weight in mammals, but receives 25% of the cardiac output. The majority of the blood reaching the kidney passes through the renal cortex, which receives over 95% of the renal blood flow. Once in the kidney, *S*-conjugates may reach their target in the straight portion of the proximal tubular cells by glomerular filtration or by transport across the basolateral membrane, or both. The kidney is rich in γ-glutamyltransferase activity, dipeptidase activity, and aminoacylase activity; thus glutathione and cysteinylglycine *S*-conjugates and mercapturic acids are efficiently converted to cysteine *S*-conjugates in the proximal tubular cells (Anderson *et al.*, 1980; Guder and Ross, 1984; Guder and Wirthensohn, 1985; Hughey *et al.*, 1978). Glutathione *S*-conjugates present in blood may be filtered or may enter the peritubular circulation. The kidney removes a large portion of glutathione from the plasma; indeed, 70% of the plasma glutathione is cleared by the kidney. Glomerular filtration accounts for only 25% of the renal glutathione clearance; the balance is attributable to basolateral uptake. Studies with *S*-(pentachlorobutadienyl)glutathione in the isolated perfused rat kidney demonstrate that this glutathione *S*-conjugate is also preferentially removed by nonfiltering mechanisms (Schrenk *et al.*, 1988).

The filtered glutathione *S*-conjugates may be metabolized to the corresponding cysteine *S*-conjugates by γ-glutamyltransferase and cysteinylglycine dipeptidase or aminopeptidase M in the renal brush-border membrane (Hughey *et al.*, 1978; Jones *et al.*, 1979a,b; Moldeus *et al.*, 1978; Tsao and Curthoys, 1980). The glutamate may be transferred to an amino acid acceptor and the constituent amino acids are taken up by the cell. The glutathione *S*-conjugates in the peritubular capillaries may be metabolized to cysteine *S*-conjugates by γ-glutamyltransferase present in the renal vasculature or basolateral membrane. Transport of the intact tripeptide into renal epithelial cells by an electrogenic, sodium-coupled and probenecid-sensitive transport with a broad specificity for γ-glutamyl compounds has also been observed (Lash and Jones, 1983, 1984; Rankin and Curthoys, 1982). Glutathione *S*-conjugates transported intact (Kramer *et al.*, 1987) into the renal epithelial cell or formed there by glutathione *S*-transferase-dependent conjugation of haloalkenes may be secreted into the tubular lumen, apparently via a membrane-potential-sensitive transport system in the brush-border membrane (Schaeffer and Stevens, 1987a,b), and may then be metabolized to cysteine *S*-conjugates by γ-glutamyltransferase present on the extracellular side of the brush-border membrane. Cysteine *S*-conjugates thus produced and *S*-conjugates reaching the tubular lumen by glomerular filtration may be transported into the renal epithelial cells by sodium-dependent and sodium-independent transport systems (Lash and Anders, 1989).

The probenecid-sensitive organic-anion transporter present on the basolateral side of the proximal tubular cells seems to play the most important role in the accumulation of *S*-conjugates in proximal tubular cells and in the organ-selective toxicity. Probenecid is a selective inhibitor of the organic-anion transporter without effects on energy metabolism. Probenecid has no effect on the transport of organic cations, which are also actively accumulated by the kidney. Haloalkene-derived mercapturates have the highest affinity for the organic-anion transporter, but glutathione and cysteine *S*-conjugates with lipophilic substituents on sulfur are also substrates (Ullrich and Rumrich, 1988; Ullrich *et al.*, 1988, 1989a,b). The inhibitory effects of probenecid on the toxicity of several haloalkene cysteine *S*-conjugates and mercapturic acids support a central role for the renal organic-anion transporter in the renal accumulation of *S*-conjugates. Administration of probenecid partially blocks the *in vivo* nephrotoxicity of *S*-(1,2-dichlorovinyl)-L-cysteine and of hexachlorobutadiene, *S*-(pentachlorobutadienyl)glutathione, and *N*-acetyl-*S*-(pentachlorobutadienyl)-L-cysteine, the elimination of *S*-(pentachlorobutadienyl)-L-cysteine in the isolated perfused kidney, and the covalent binding of metabolites of [1,2-vinyl-^{14}C]*S*-(1,2-dichlorovinyl)-L-cysteine. The failure of probenecid to protect rat kidney proximal tubules cells or rabbit kidney slices from the toxicity of *S*-(1,2-dichlorovinyl)-L-cysteine may be due to the lack of N-acetylation. *In vivo,* after ip administration, *S*-(1,2-dichlorovinyl)-L-cysteine may be rapidly acetylated in the liver to *N*-acetyl-*S*-(1,2-dichlorovinyl)-L-cysteine, which is then delivered to the kidney and accumulated there. *In vitro,* acetylation before entering the kidney cell does not occur (Wolfgang *et al.*, 1989; Zhang and Stevens, 1989). Probenecid also protects isolated kidney cells from the toxicity of haloalkene- and hydroquinone-derived mercapturic acids (Koob *et al.*, 1990).

Inside the renal cell, cysteine *S*-conjugates may have three fates: they may be secreted unchanged into the plasma and then returned to the liver for further metabolism (renal–hepatic circulation); they may be N-acetylated to form the corresponding mercapturic acids and excreted finally with urine (Heuner *et al.*, 1991); or they may undergo β-lyase-catalyzed metabolism to yield reactive intermediates. Cysteine *S*-conjugates taken up by the liver are generally thought to be metabolized to mercapturic acids, which are secreted into plasma and returned to the kidneys to be excreted in urine. For example, iv administration of *S*-carbamido-[^{14}C]methylcysteine to mice leads to initial accumulation of ^{14}C in the liver, followed by excretion of ^{14}C in the urine as the mercapturic acid (Inoue *et al.*, 1981, 1984a, 1987). Similarly, Inoue *et al.* (1984a) showed conversion of *S*-benzyl-L-[U-^{14}C]cysteine to the corresponding mercapturic acid in isolated hepatocytes. Probenecid-sensitive renal transport systems have been identified for several mercapturic acids and are thought to function in urinary excretion of these compounds.

Mercapturic acids in the kidney may be cleaved to the corresponding cysteine *S*-conjugates by aminoacylases present in renal cells (see Anders and Dekant, this volume).

3. Determinants of Organ- and Cell-Selective Toxicity of *S*-Conjugates

The high renal activities of γ-glutamyltransferase, dipeptidase, β-lyase, and acylase, the ability of the kidney to accumulate amino-acid derivatives, and the high renal blood flow may explain the selective toxicity of *S*-conjugates to the proximal tubules. The mechanisms for site-selective haloalkene toxicity in the kidney are still only partially understood. Hexachlorobutadiene and derived *S*-conjugates produce cell damage predominantly in the S_3 segment (pars recta) of the renal proximal tubule of the rat (Hook *et al.*, 1982; Ishmael *et al.*, 1982; Lock and Ishmael, 1979). Similarly, *S*-(2-chloro-1,1,2-trifluoroethyl)glutathione and *S*-2(-chloro-1,1,2-trifluoroethyl)-L-cysteine (Dohn *et al.*, 1985b), as well as the precursor haloalkene chlorotrifluoroethene (Potter *et al.*, 1981), produce damage to the pars recta of the rat renal proximal tubule. The selectivity of *S*-conjugates for the pars recta of the kidney may be associated with the location of the β-lyase in the renal proximal tubular epithelium (Jones *et al.*, 1988; MacFarlane *et al.*, 1989); immunohistochemical studies indicate, however, that the β-lyase is distributed along the entire proximal tubule and is not concentrated in the S_3 segment where damage is seen. The activity of the organic-anion transport system is also higher in the S_2 section of the tubule. However, γ-glutamyltransferase is concentrated in the S_3 segment of the proximal tubule.

The selectivity of *S*-conjugates for the S_3 segment may reflect differential sensitivities of the different segments of the proximal tubule to S-conjugate-induced damage or may result from the distribution of aminoacylase activity along the proximal tubule.

Differences in the sensitivity of the cells along the nephron may also contribute to the observed site-selective toxicity. Mitochondrial toxicity is an important event in the toxic effects of *S*-conjugates to the proximal tubules cells. It is well known that these cells are highly dependent on mitochondria to maintain their energy homeostasis, since the cytosolic glycolytic enzymes have a very low activity in this part of the nephron. Because the number of mitochondria is lowest in the S_3 segment, this region will generally be most susceptible to mitochondrial toxicity, thus offering a possible reason for the strict site-selectivity of the *S*-conjugate toxicity.

IV. Bioactivation of Cytotoxic and Nephrotoxic Cysteine *S*-Conjugates

Several pyridoxal phosphate-dependent enzymes catalyze the elimination of leaving groups from the β-carbon of amino acids. Enzymatic metabolism of cysteine S-conjugates to pyruvate, ammonia, and thiols has been observed in several strains of bacteria, in rat, rabbit, and human liver and kidney. Enzymatic cleavage of cysteine *S*-conjugates was first reported by Anderson and Schultze in their investigations on the toxic factor of trichloroethene-extracted soybean meal (Anderson and Schultze, 1965). *S*-(1,2-Dichlorovinyl)-L-cysteine was identified as the toxic factor (McKinney *et al.*, 1959). *S*-(1,2-Dichlorovinyl)-L-cysteine causes aplastic anemia in calves (McKinney *et al.*, 1957), but is nephrotoxic in all other species studied (Terracini and Parker, 1965). Work in the 1960s demonstrated that *S*-(1,2-dichlorovinyl)-L-cysteine is cleaved by a C—S lyase present in bovine liver and kidney (Bhattacharya and Schultze, 1967). A sulfur-containing metabolite of *S*-(1,2-dichlorovinyl)-L-cysteine was covalently bound to proteins and DNA (Bhattacharya and Schultze, 1971a,b, 1972, 1973a,b). The C—S lyase activity was purified from rat liver and kidney and termed cysteine conjugate β-lyase (Nelson *et al.*, 1988; Stevens, 1985; Stevens and Jakoby, 1983; Stevens *et al.*, 1986a; Tateishi *et al.*, 1978; Tomisawa *et al.*, 1986). Subsequent studies demonstrated that hepatic β-lyase is attributable to kynureine aminotransferase and that renal cytosolic β-lyase activity is identical with glutamine transaminase K. Hepatic β-lyase does not cross-react with an antibody to the renal enzyme, and only a small fraction of renal β-lyase activity can be accounted for as kynureinase. β-Lyase activity is also present in the mitochondrial outer membrane and matrix fractions and this enzyme is similar to the renal cytosolic glutamine transaminase K (Lash *et al.*, 1986; Stevens *et al.*, 1988).

β-Lyase is a pyridoxal phosphate-dependent enzyme and catalyzes both β-elimination and transamination reactions with cysteine *S*-conjugates (Cooper and Anders, 1990; Elfarra *et al.*, 1987; Stevens *et al.*, 1989) (Fig. 6). *S*-(1,2-Dichlorovinyl)-L-cysteine is transformed by transamination to *S*-(1,2-dichlorovinyl)-2-oxo-3-mercaptopropionic acid. Cysteine conjugate β-lyase activity is dependent on amino acid oxidase, which converts amino acids to α-keto acids (Stevens *et al.*, 1986a). During purification of cysteine conjugate β-lyase, copurification of an α-hydroxy acid oxidase is necessary for detection of β-elimination from cysteine *S*-conjugates. During transamination catalyzed by cysteine conjugate β-lyase, the pyridoxal-phosphate cofactor is transformed to pyridoxamine phosphate, which is not competent to catalyze elimination reactions. The pyridoxal-phosphate form of the enzyme is regenerated by transfer of the amino group of pyridoxamine phosphate to an acceptor α-keto acid. Amino acid oxidase biotransforms *S*-(1,2-dichlorovinyl)-L-cysteine to its corresponding α-keto acid, *S*-(1,2-dichlorovinyl)-L-2-oxo-3-merceptopropionate, which can serve as an amino-group acceptor. This explains the dependence of β-lyase activity on amino acid oxidase and the potentiation of cysteine *S*-conjugate toxicity by α-keto acids (Elfarra *et al.*, 1987; Lash *et al.*, 1986).

The role of metabolism of cysteine *S*-conjugates in *S*-conjugate-induced toxicity has been established by the use of aminooxyacetic acid, an inhibitor of pyridoxal phosphate-dependent enzymes, and by α-keto acids such

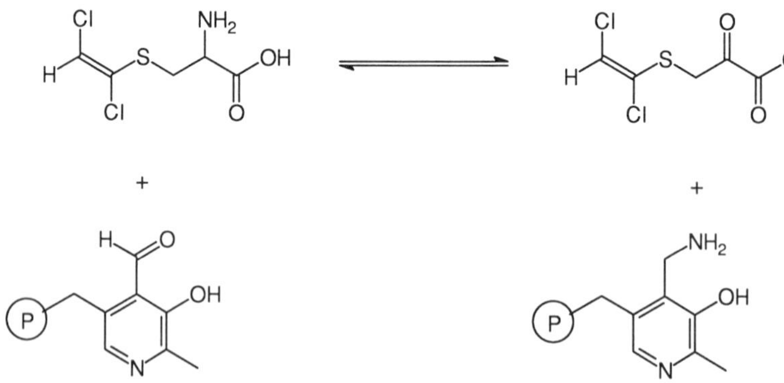

Fig. 6 Cysteine conjugate β-lyase-catalyzed transamination of cysteine *S*-conjugates.

as α-keto-γ-methiolbutyrate, which enhance both renal and cytosolic β-lyase activities (Elfarra *et al.*, 1987; Lash *et al.*, 1986). Moreover, α-methyl analogs of nephrotoxic cysteine *S*-conjugates, which cannot be cleaved by β-lyase, are not toxic (Elfarra and Anders, 1985; Elfarra *et al.*, 1986). Abstraction of the α-proton in cysteine *S*-conjugates occurs readily after formation of the Schiff base with pyridoxal phosphate and results in β-elimination (Anders *et al.*, 1988); therefore, substitution of the α-proton by a methyl group makes the molecule nonmetabolizable by cysteine conjugate β-lyase.

These results suggest that the thiol metabolites formed by enzymatic cleavage of cysteine *S*-conjugates and their interaction with cellular macromolecules are responsible for *S*-conjugate toxicity. Thus, the chemical structure and, hence, the chemical reactivity of the thiols formed may control the expression of cysteine *S*-conjugate toxicity. Three different types of thiols are formed by cysteine conjugate β-lyase-dependent metabolism, and the thiols formed differ markedly in chemical reactivity and biological effects.

A. Unstable Thiols as Products of β-Lyase-Catalyzed Metabolism

Cysteine *S*-conjugates biosynthesized from haloalkenes are cleaved to unstable thiols by cysteine conjugate β-lyase. Different types of reactive intermediates are formed from *S*-(1,1-difluoroalkyl)-L-cysteine derivatives and from *S*-(1-chloroalkenyl)-L-cytsteine derivatives.

α-Fluoroethanethiols are products of the cysteine conjugate β-lyase metabolism of *S*-(2-chloro-1,1,2-trifluoroethyl)-L-cysteine and of *S*-(1,1,2,2-tetrafluoroethyl)-L-cysteine (Dekant *et al.*, 1987a; Green and Odum, 1985) (Fig. 7). These fluoroethanethiols can be trapped in model systems mimicking β-lyase activity, but rapidly eliminate hydrogen fluoride to yield thionoacyl fluorides. The concept of intermediate thionoacyl fluoride formation is further supported by the following observations: (a) Incubation of *S*-(2-chloro-1,1,2-trifluoroethyl)-L-cysteine with a pyridoxal model system or with purified bovine kidney β-lyase preparations yields chlorofluoroacetic acid and inorganic fluoride as terminal products. Incubation with the pyridoxal model system in the presence of the model nucleophile diethylamine results in the formation of *N*,*N*-diethylchlorofluorothioacetamide (Dekant *et al.*, 1987a). (b) Studies on the chemical reactivity of synthetic 2-chloro-1,1,2-trifluoroethanethiol and 1,1,2,2-tetrafluoroethanethiol indicated the formation of thioamides and thiono acids in the presence of amines and water, respectively (Fokin *et al.*, 1961). (c) α-Fluoroalkyl disulfides have been developed as proreactive intermedi-

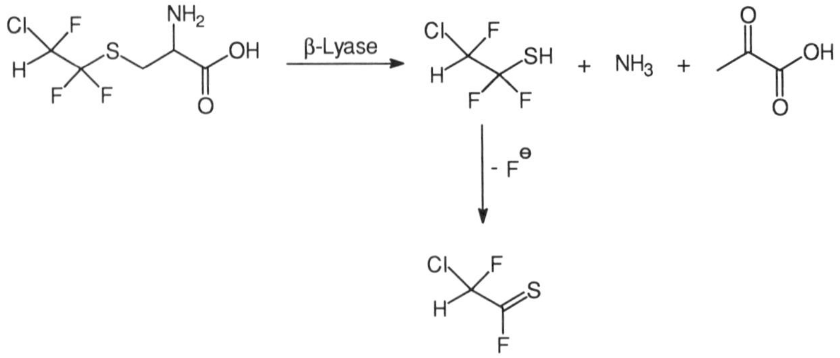

Fig. 7 Cysteine conjugate β-lyase-catalyzed biotransformation of *S*-(1-chloro-1,2,3-trifluoroalkyl)-L-cysteine to chlorofluorothionoacetyl fluoride.

ates to characterize the reactivity of α-haloalkyl- and α-haloalkenylthiols. They can be cleaved in inert solvents to yield the corresponding thiols and their transformation products, which can then be trapped and characterized. When 2-chloro-1,1,2-trifluoroethyl 2-nitrophenyl disulfide was cleaved in the presence of cyclopentadiene, which undergoes a cycloaddition reaction with carbon—sulfur double bonds, four isomers of thianorbornene were identified as stable products formed from the thiolate. These observations strongly indicate that thioacyl fluorides are formed as reactive intermediates from fluoroalkyl-L-cysteine *S*-conjugates (Dekant *et al.*, 1991; Müller *et al.*, 1991).

The *S*-conjugates *S*-(1,2-dichlorovinyl)-L-cysteine, *S*-(1,2,2-trichlorovinyl)-L-cysteine, and *S*-(pentachlorobutadienyl)-L-cysteine are transformed by cysteine conjugate β-lyase to yield halovinylthiols, and these vinylthiols give rise to electrophilic species whose interaction with cellular marcomolecules is responsible for the cytotoxicity and mutagenicity of *S*-(1,2-dichlorovinyl)-L-cysteine, *S*-(1,1,2-trichlorovinyl)-L-cysteine, and *S*-(pentachlorobutadienyl)-L-cysteine. Incubation of these *S*-conjugates with a β-lyase model system or with partially purified bacterial β-lyase resulted in the formation of chloroacetic acid and chlorothionoacetic acid from *S*-(1,2-dichlorovinyl)-L-cysteine, dichloroacetic acid from *S*-(1,2,2-trichlorovinyl)-L-cysteine, and 2,3,4,4,-tetrachlorobutenoic acid and 2,3,4,4-tetrachlorothionobutenoic acid from *S*-(pentachlorobutadienyl)-L-cysteine (Dekant *et al.*, 1988d,e). Incubation with the β-lyase model system in the presence of diethylamine gave the corresponding thioamides, indicating that thionoacylating agents are formed as reactive intermediates. Two different types of reactive intermediates may be formed: (a) Thionoacyl chlorides are one type; they are tautomers of α-chloro-substituted thiols that are not in equilibrium with the thiols (Mayer and Scheithauer, 1976). The thionoacyl chloride is the more stable configuration, indicating that the vinylic thiols may rapidly tautomerize to the more stable thionoacyl chlorides. (b) As an alternative to tautomerization, α-chlorovinyl thiols may eliminate hydrochloric acid to yield thioketenes. Thioketenes are known thionoacylating agents and are highly reactive with nitrogen nucleophiles (Adiwidjaja *et al.*, 1991; Raasch, 1970, 1972). To distinguish between thionoacyl chlorides and thioketenes, α-chlorovinyl thiols were generated from cysteine *S*-conjugates and from the corresponding α-chlorovinyl 2-nitrophenyl disulfides in the presence of cyclopentadiene. Cyclopentadiene undergoes a [4 + 2]cycloaddition reaction with the carbon—sulfur double bond of thioketenes and gives products different in their mass spectra and chromatographic properties from thionoacyl chlorides and thioketenes (Dekant *et al.*, 1991). In these experiments, only thianobornenes indicative of the formation of thioketenes as reactive intermediates were observed, indicating that thioketenes are the exclusive reactive intermediates formed from α-halovinyl cysteine *S*-conjugates (Fig. 8).

Both thioketenes and thioacyl fluorides are reactive, acylating agents and may react with amino and hydroxyl groups in proteins and lipids. N^{ε}-(Difluorothioacetyl)-L-lysine and N^{ε}-(chlorofluorothioacetyl)-L-lysine have been identified as adducts in proteins of subcellular fractions incubated with *S*-(2-chloro-1,1,2-trifluoroethyl)-L-cysteine and of *S*-(1,1,2,2-tetrafluoroethyl)-L-cysteine (Hargus and Anders, 1991; Hayden and Stevens, 1990; Hayden *et al.*, 1991a,b, 1992). Difluorothionoacetamido–lipid adducts have also been identified in mitochondria. Moreover, N^{ε}-(chlorofluoroacetyl)-L-lysine is formed in renal proteins of rats treated with *S*-(2-chloro-1,1,2-trifluoroethyl)-L-cysteine (Harris *et al.*, 1992). Covalent

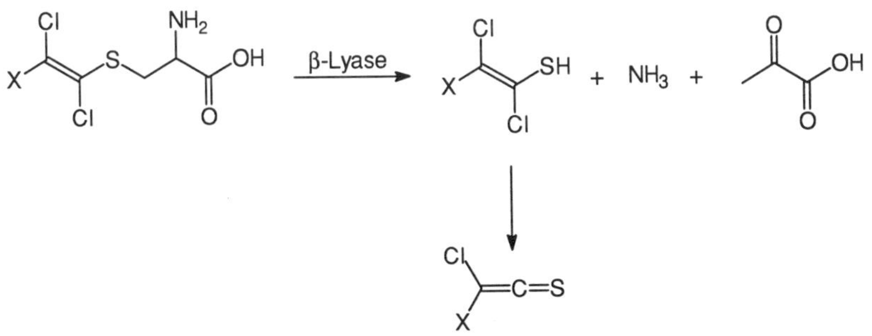

Fig. 8 Cysteine conjugate β-lyase-catalyzed biotransformation of *S*-(1-chloroalkenyl)-L-cysteine *S*-conjugates to thioketenes. X = Cl, *S*-(trichlorovinyl)-L-cysteine; X = C_2Cl_3, *S*-(pentachlorobutadienyl)-L-cysteine.

binding of the thioketenes or thioacyl fluorides to cellular macromolecules is associated with *S*-conjugate-induced toxicity and mutagenicity (Darnerud *et al.*, 1988; Hayden and Stevens, 1990). In addition to protein binding, several studies have shown that cysteine conjugate β-lyase-derived metabolites may bind to DNA.

The pioneering work of Schultze and co-workers has demonstrated that a reactive metabolite formed by the β-lyase-catalyzed cleavage of *S*-(1,2-dichlorovinyl)-L-cysteine binds to DNA and results in structural modifications that might influence gene expression (Bhattacharya and Schultze, 1971a,b, 1972, 1973a,b).

Hexachlorobutadiene is also metabolized by *S*-conjugate formation in mice to metabolites that bind to renal DNA. DNA binding in kidneys is much higher than in liver, and adducts are almost exclusively found in mitochondrial DNA (Schrenk and Dekant, 1989). Although the structure of the hexachlorobutadiene-DNA adducts has not been elucidated, indirect evidence indicates that they are formed by the β-lyase-catalyzed metabolism of *S*-(1,2,3,4,4-pentachlorobutadienyl)-L-cysteine. Adducts formed from this *S*-conjugate in bacteria and in renal proximal tubular cells were identical with the hexachlorobutadiene-DNA adducts in their retention time in HPLC under different separation conditons (Müller *et al.*, 1991; Vamvakas *et al.*, 1988a).

Phosphorus-32 postlabeling techniques demonstrated that only mutagenic *S*-conjugates, which form thioketenes, modify DNA; metabolically generated thionoacylfluorides did not result in detectable DNA modification (Müller *et al.*, 1991).

B. Chemically Stable Thiols as Products of β-Lyase-Catalyzed Metabolism

Cysteine *S*-conjugates of aromatic, heteroaromatic, and alkylaromatic compounds are substrates for β-lyase from a variety of sources, and chemically stable thiols are formed as products (Bakke and Gustafsson, 1984; Bakke *et al.*, 1980, 1981, 1982). The possible toxic effects of these cysteine conjugates and of the thiols formed as metabolites have not been investigated in detail. These thiols may be oxidized by flavin-dependent monooxygenases to sulfenic acids, which are highly reactive with glutathione and other thiols, and form disulfides (Bakke and Gustafsson, 1984; Bakke *et al.*, 1980, 1981, 1982). Protein–glutathione mixed disulfide and glutathione disulfide formation after oxidation of thiols to sulfenic acids has been implicated in the mechanisms of toxicity of several thiols. Renal damage, the target organ for *S*-conjugate-induced toxicity, has not been observed when *N*-acetyl-*S*-(pentachlorophenyl)-L-cysteine was given to rats; this compound is metabolized by aminoacylase and β-lyase to pentachlorothiophenol, which may be excreted in the urine or may undergo *S*-methylation to yield pentachlorothioanisole (Renner, 1983; Smith and Francis, 1983). Moreover, *S*-(2-benzothiazolyl)-L-cysteine, which is not toxic to isolated rat renal proximal tubular cells, is also metabolized by β-lyase to 2-mercaptobenzothiazole; determination of the stable thiol by UV spectroscopy is used for the determination of β-lyase activity (Dohn and Anders, 1982b). These limited results indicate that chemically stable thiols, which are formed as metabolites of certain cysteine *S*-conjugates, exert little toxicity. Stable thiols formed by β-lyase *in vivo* may be excreted unchanged or may serve as precursors of a variety of sulfur-containing metabolites, such as thiomethyl derivatives, sulfinates, sulfones, and *S*-glucuronides. Structure/activity studies on the mutagenicity and cytotoxicity of cysteine *S*-conjugates indicate that their mutagenicity is dependent on the formation of unstable α-chlorinated thiols. *S*-(Pentachlorophenyl)-L-cysteine and *S*-(benzyl)-L-cysteine, which yield the stable thiols pentachlorothiophenol and benzyl thiol, are not mutagenic and cytotoxic, whereas *S*-conjugates yielding unstable α-chloroenethiols are potent mutagens in the Ames test and are highly cytotoxic in renal epithelial cells (Vamvakas *et al.*, 1988b,c, 1989b).

V. Cellular Effects of Cytotoxic and Nephrotoxic Cysteine *S*-Conjugates

The first report on toxic cysteine *S*-conjugates derived from haloalkenes appeared in 1959, when *S*-(1,2-dichlorovinyl)-L-cysteine was identified as the toxic factor in soybean meal extracted with trichloroethene (McKinney *et al.*, 1959). *S*-(1,2,-Dichlorovinyl)-L-cysteine produced aplastic anemia and biochemical alterations in the DNA structure in bone marrow, lymph nodes, and thymus in cattle (Bhattacharya and Schultze, 1971a,b, 1972; Schultze *et al.*, 1959). The hematopoietic toxicity of *S*-(1,2-dichlorovinyl)-L-cysteine in cattle is a unique observation. In all other species studied, *S*-(1,2-dichlorovinyl)-L-cysteine was nephrotoxic, not hematotoxic.

A. Nephrotoxicity *in Vivo*

Mice, rats, guinea pigs, and dogs displayed toxic alterations and necrosis of the renal tubular epithelium 24 h after administration of *S*-(1,2-dichlorovinyl)-L-cysteine (Jaffe *et al.*, 1984; Koechel *et al.*, 1991; Terracini and Parker, 1965). The histopathological changes in the proximal tubules were confirmed by an increase in blood urea nitrogen concentrations and urinary glucose excretion. Several other haloalkenyl cysteine *S*-conjugates, i.e., *S*-(pentachlorobutadienyl)-L-cysteine, and *S*-(2-chloro-

1,1,2-trifluoroethyl)-L--cysteine, *S*-(2-bromo-2-chloro-1,1-difluoroethyl)-L-cysteine, and *S*-(1,1,2,2-tetrafluoroethyl)-L-cysteine, caused identical toxicity to the proximal tubules after administration to rats and mice (Dohn *et al.*, 1985b; Green and Odum, 1985; Odum and Green, 1984). Several strategies, based on the knowledge that β-lyases are pyridoxal phosphate-dependent enzymes, have been exploited to test the hypothesis that β-lyase-catalyzed cleavage is required for the expression of *S*-conjugate-induced nephrotoxicity. The β-lyase inhibitor aminooxyacetic acid blocks *S*-conjugate-induced proximal tubule damage *in vivo*. Also, the α-methyl derivatives *S*-(1,2-dichlorovinyl)-α-methyl-L-cysteine and *S*-(1-chloro-1,2,2-trifluoro)-α-methyl-L-cysteine are not nephrotoxic. These *S*-conjugates cannot be cleaved by β-lyase; cleavage of the carbon—sulfur bond does not occur and reactive intermediates are not formed (Dohn *et al.*, 1985b; Elfarra *et al.*, 1986; Finkelstein *et al.*, 1992; Jaffe *et al.*, 1983; Nash *et al.*, 1984; Odum and Green, 1984).

In initiation–promotion experiments, *S*-(1,2-dichlorovinyl)-L-cysteine showed weak promoting activity on *N*-nitrosodimethyl amine-initiated renal tumors in mice (Meadows *et al.*, 1988). *S*-(1,2,-Dichlorovinyl)-L-cysteine, *S*-(1,1,2-trichlorovinyl)-L-cysteine, and *S*-(pentachlorobutadienyl)-L-cysteine are the key metabolites responsible not only for the nephrotoxicity but, likely, also for the nephrocarcinogenicity of the parent haloalkenes trichloroethene, tetrachloroethene, and hexachlorobutadiene, and of dichloroethyne. This indicates that the *S*-conjugates may be nephrocarcinogenic. However, long-term carcinogenicity experiments have not been carried out with the cysteine *S*-conjugates.

The glutathione *S*-conjugates *S*-(1,2,-dichlorovinyl)glutathione, *S*-(pentachlorobutadienyl)glutathione, and *S*-(2,-chloro-1,1,2-trifluoroethyl) glutathione are also nephrotoxic in rats. Metabolism to the corresponding cysteine *S*-conjugates is essential for the nephrotoxic effects: inhibition of γ-glutamyltransferase by the irreversible inhibitor acivicin blocked *S*-(1,2-dichlorovinyl)glutathione- and *S*-(2-chloro-1,1,2-trifluoroethyl)glutathione-induced nephrotoxicity *in vivo*. Aminooxyacetic acid also blocked the toxicity of the glutathione *S*-conjugates. Hence, available data from *in vivo* experiments are in agreement with the proposed mechanism (Dohn *et al.*, 1985b; Elfarra *et al.*, 1986; Nash *et al.*, 1984).

B. Cytotoxicity *in Vitro*

The toxicity of a number of haloalkenyl *S*-conjugates has been studied *in vitro* in renal slices, in isolated proximal tubules, in freshly isolated rat proximal tubule cells, and in LLC-PK_1 cells, a cultured line of porcine kidney cells. In isolated rat renal cortical slices, *S*-conjugates increased the production of the β-lyase cleavage products pyruvate and ammonia and generated a reactive moiety that inhibited the activity of the renal *p*-aminohippurate and tetraethylammonium transport systems (Green and Odum, 1985). Exposure of rabbit renal proximal tubules to *S*-(1,2-dichlorovinyl)-L-cysteine, *S*-(pentachlorobutadienyl)-L-cysteine, *S*-(1,1,2,2-tetrafluoroethyl)-L-cysteine, and *S*-(2-chloro-1,1,2-trifluoroethyl)-L-cysteine at a concentration of 25 μM resulted in a time-dependent loss of tubule cell viability (Groves *et al.*, 1991). Freshly isolated rat kidney proximal tubules cells have been widely used as a model to study *S*-conjugate cytotoxicity. Cell viability, assessed by trypan blue exclusion and lactate dehydrogenase leakage, decreased in a time- and dose-dependent manner when the cells were incubated with *S*-(1,2-dichlorovinyl)-L-cysteine, *S*-(1,1,2-trichlorovinyl)-L-cysteine, *S*-(pentachlorobutadienyl)-L-cysteine, and *S*-(2-chloro-1,1,2-trifluoroethyl)-L-cysteine. Cytotoxicity was reduced in presence of the β-lyase inhibitor aminooxyacetic acid. In addition, the α-methyl analog of *S*-(1,2-dichlorovinyl)-L-cysteine was not cytotoxic to isolated renal cells, indicating that cleavage by the pyridoxal-phosphate-dependent β-lyase is required for the expression of toxicity. The glutathione *S*-conjugates of tri- and tetrachloroethene, hexachlorobutadiene, and chlorotrifluoroethene also reduced renal tubule cell viability. Acivicin, 1, 10-phenanthroline, and phenylalanylglycine, inhibitors of γ-glutamyltransferase, aminopeptidase M, and cysteinylglycine dipeptidase, respectively, blocked the toxicity of the glutathione conjugates, indicating that metabolism to the corresponding cysteine conjugates is required for the glutathione conjugates to cause toxicity (Dohn *et al.*, 1985b; Jaffe *et al.*, 1983; Jones *et al.*, 1986; Lash and Anders, 1986a; Schnellmann *et al.*, 1987; Vamvakas *et al.*, 1989c).

The renal eptihelial cell line LLC-PK_1, grown on a solid substrate, possesses many morphological and biochemical characteristics of the proximal tubular epithelium, the target site of *S*-conjugate toxicity (Gstraunthaler *et al.*, 1985; Hull *et al.*, 1976). LLC-PK_1 cells retain the degradative enzymes of the mercapturate pathway, such as γ-glutamyltransferase, and both *S*-(1,2-dichlorovinyl)glutathione and *S*-(1,2-dichlorovinyl)-L-cysteine are cytotoxic in LLC-PK_1 monolayers. Acivivin, an inhibitor of γ-glutamyltransferase, blocked the toxicity of the glutathione *S*-conjugates, and aminooxyacetic acid blocked the toxicity of both *S*-(1,2-dichlorovinyl)glutathione and *S*-(1,2-dichlorovinyl)-L-cysteine and the metabolism of *S*-(1,2-dichlorovinyl)-L-cysteine to pyruvate in these cells. The glutathione and cysteine *S*-conjugates of tetrachloroethene, hexachlorobutadiene, and chlorotrifluoethene were also toxic to LLC-PK_1 monolayers (Mertens *et al.*, 1988, 1990). Similar results were obtained when *S*-conjugates were incubated with proximal tubule cells of human origin (J. C. Chen *et al.*, 1990; Zhang and Stevens, 1989). Furthermore, incubation of the cells with [^{35}S]*S*-(1,2-dichlorovinyl)-L-cysteine resulted in β-lyase-dependent binding of ^{35}S-containing metabolites to cellular marcomolecules, and the binding was proportional to toxicity. The data provide direct evidence that in LLC-PK_1 *S*-conjugates are metabolized to reactive species that covalently bind to cellular marcomolecules causing toxicity and cell death (Stevens *et al.*, 1986b).

C. Biochemical Mechanisms of *S*-Conjugate Cytotoxicity

The mitochondria are the primary subcellular targets for *S*-conjugate-induced cytotoxicity in freshly isolated renal proximal tubular cells. For example, cellular respiration and ATP content were markedly reduced by *S*-(1,2,3,4,4-pentachlorobutadienyl)-L-cysteine and *S*-(1,2-dichlorovinyl)-L-cysteine (Lash and Anders, 1987). The disruption of mitochondrial energy metabolism may contribute to glutathione depletion, reduced organic anion and cation transport, and the elevation in cytosolic Ca^{2+} concentrations due to impaired function of the Ca^{2+}-ATPases of the plasma membrane and the endoplasmatic reticulum (Jones *et al.*, 1986; Wallin *et al.*, 1987). All of these processes require energy in the form of ATP.

The correlation between covalent binding to mitochondria, mitochondrial toxicity, and *S*-conjugate-induced cell death is only partly understood. *S*-(1,2-Dichlorovinyl)-L-cysteine inhibited succinate-linked state 3 respiration, altered the concentration of several citric acid cycle intermediates, and impaired the ability of mitochondria to generate a membrane potential; *S*-(1,2,3,4,4-pentachlorobutadienyl)-L-cysteine was also found to uncouple oxidative phosphorylation (Schnellmann *et al.*, 1989). Binding of radioactivity from several ^{35}S-labeled *S*-conjugates to macromolecules in mitochondria was observed. However, the relative extent of metabolism and binding of ^{35}S-containing metabolites did not correlate with their mitochondrial toxicity (Hayden and Stevens, 1990).

Covalent binding of *S*-conjugates metabolites is associated with mitochondrial dysfunction and ATP depletion, but the contribution of mitochondrial toxicity to cell death is unclear. Antioxidants protect isolated rat kidney cells from *S*-conjugate-induced cell death without inhibiting covalent binding of radioactivity to mitochondrial macromolecules (Q. Chen *et al.*, 1990). Interestingly, antioxidant protection effectively blocked cell death even when the antioxidant was given several hours after the toxic *S*-conjugates. Moreover, mitochondrial glutathione concentrations decline rapidly in kidney cells incubated with *S*-conjugates; the glutathione loss is nearly quantitatively recovered as glutathione disulfide. These observations indicate that mitochondrial dysfunction and cell death can be dissociated and that a peroxidative mechanism may be important in the events resulting in cell death (Lash and Anders, 1987). However, mechanisms other than peroxidative damage have also been proposed to play a major role in *S*-conjugate-induced toxicity.

After exposure to nephrotoxic *S*-conjugates, a number of cellular functions are disturbed prior to cell death. Numerous studies on the mechanisms of *S*-conjugate-induced cytotoxicity have been published in the last 10 years. However, most of these reports monitored changes of single biochemical parameters indicative of toxicity but did not establish links and cause–effect relationships between these events. Hence, although abundant data are available on toxic changes induced by *S*-conjugates, the molecular mechanisms or the toxic cascades responsible for cell death cannot be precisely delineated.

Perturbations in intracellular Ca^{2+} homeostasis may be a central event in *S*-conjugate-induced toxicity (Jones *et al.*, 1986; Lash and Anders, 1986b). Studies with fluorescence digital imaging microscopy showed that *S*-(1,2-dichlorovinyl)-L-cysteine increased cytosolic Ca^{2+} concentrations prior to the onset of cell death. The increased cytosolic Ca^{2+} concentrations were associated with impaired ability of the mitochondria to sequester cytosolic Ca^{2+} and preceded severe perturbations of the mitochondrial membrane potential (Vamvakas *et al.*, 1990).

A similar selective impairment of mitochondrial Ca^{2+} sequestration was observed with several prooxidants that induce increased oxidation and hydrolysis of mitochondrial pyridine nucleotides (Richter and Kass, 1991). The products of this reaction are nicotinamide and ADP-ribose; hence the key step in the induction of mitochondrial Ca^{2+} efflux is probably modification of mitochondrial membrane proteins by ADP-ribose moieties. In freshly isolated mitochondria from the pig cortex, *S*-(1,2-dichlorovinyl)-L-cysteine induces Ca^{2+} release. The release was preceded by increased oxidation and hydrolysis of mitochondrial pyridine nucleotides, whereas the mitochondrial membrane potential remianed unchanged over the entire incubation time (Vamvakas *et al.*, 1992). Moreover, inhibition of the hydrolysis of oxidized pyridine nucleotides and of the mono (ADP-ribosyl)ation of mitochondrial membrane proteins blocked *S*-(1,2-dichlorovinyl)-L-cysteine-induced Ca^{2+}release, indicating that *S*-conjugates may cause selective depletion of mitochondrial Ca^{2+} by modification of mitochondrial membrane proteins by the ADP-ribose moieties formed from the hydrolysis of the oxidized pyridine nucleotides. Since this cascade is induced by several prooxidants, the results are in agreement with the perturbations in the intracellular redox potential observed in LLC-PK_1 cells treated with *S*-(1,2-dichlorovinyl)-L-cysteine (Q. Chen *et al.*, 1990) and also with the finding that *S*-(1,2-dichlorovinyl)-L-cysteine inhibits sulfhydryl-sensitive dehydrogenases in isolated rat kidney mitochondira (Lash and Anders, 1987).

The specific interference with the Ca^{2+}-buffering capacity of the mitochondria and the increase in the Ca^{2+} concentrations may induce cell death by activation of Ca^{2+}-dependent degradative enzymes, such as proteases and phospholipases that induce damage of the plasma membrane and the cytoskeleton (Boobis *et al.*, 1989; Orrenius *et al.*, 1989). In addition, activation of Ca^{2+}-dependent endonucleases may result in increased formation of DNA double-strand breaks followed by the induction of poly(ADP-ribosyl)ation of nuclear proteins (Cantoni *et al.*, 1989; McConkey *et al.*, 1988). High levels of poly(ADP-ribosyl)ation may cause cell death by depletion of NAD^+ and ATP, whereas the cell may survive moderate increases, that result in altered chromatin structure and function, and thus altered gene expression (Boulikas, 1993; Cerutti, 1989). These effects may provide a link between acute and toxicity and a possible mechanism of carcinogenicity for haloalkenes.

D. Mutagenicity of Cysteine and Glutathione *S*-Conjugates in Bacteria

The demonstration of high β-lyase activities in *S. typhimurium* strains used in the mutagenicity tests orginally described by Ames was important in the elucidation of the bacterial mutagenicity and metabolism of haloalkenyl and haloalkyl cysteine and glutathione *S*-conjugates. *S*-(1,2-Dichlorovinyl)-L-cysteine, *S*-(1,1,2-trichlorovinyl)-L-cysteine, and *S*-(pentachlorobutadienyl)-L-cysteine were mutagenic in the Ames preincubation assay without the addition of an exogenous activating system (Dekant *et al.*, 1986). Mutagenicity of these *S*-conjugates was also observed in the presence of rat renal cytosol (Green and Odum, 1985). Incubation of halovinyl *S*-conjugates with *S. typhimurium* homogenates resulted in time- and concentration-dependent production of pyruvate, which is formed by β-lyase-catalyzed cleavage of the *S*-conjugates in equimolar amounts with the presumed mutagenic intermediates. Both mutagenicity and pyruvate production were decreased in the presence of the β-lyase inhibitor aminooxyacetic acid. Moreover, the α-methyl derivative of *S*-(1,2-dichlorovinyl)-L-cysteine *S*-(1,2-dichlorovinyl)-DL-α-methylcysteine, which cannot be cleaved by β-lyase, was not mutagenic and did not produce pyruvate (Vamvakas *et al.*, 1988c). The corresponding mercapturic acids were also potent bacterial mutagens in the presence of soluble aminoacylases from rat kidney cytosol, and ^{14}C-labeled *N*-acetyl-*S*-(1,2,3,4,4-pentachlorobutadienyl)-L-cysteine was metabolized to covalently bound metabolites in the presence of rat aminoacylases (Vamvakas *et al.*, 1987).

Rat kidney microsomes contain high activities of γ-glutamyltransferase and dipeptidases. Accordingly, *S*-(1,2-dichlorovinyl)glutathione, *S*-(1,1,2-trichlorovinyl)glutathione, and *S*-(pentachlorobutadienyl)glutathione were potent mutagens in the presence of rat kidney particulate fractions, probably because the corresponding cysteine *S*-conjugates formed can be cleaved by bacterial β-lyase. This was confirmed with the γ-glutamyltransferase inhibitor serine borate, which diminished the mutagenicity of the glutathione *S*-conjugates. In absence of subcellular fractions catalyzing glutathione conjugate formation, hexachlorobutadiene and tetrachloroethene failed to increase the number of revertants above control values (Vamvakas *et al.*, 1988d, 1989d).

A mutagenic effect was, however, obtained with the parent haloalkenes hexachlorobutadiene and tetrachloroethene when the mercapturic acid pathway was simulated in the Ames preincubation assay by sequential addition of rat liver microsomes and glutathione, followed by rat kidney microsomes. Inhibition of γ-glutamyltransferase and of β-lyase present in these organ fractions diminished the mutagenicity of hexachlorobutadiene and tetrachloroethene (Vamvakas *et al.*, 1988d, 1989d). In summary, glutathione conjugation followed by β-lyase-catalyzed cleavage of the corresponding cysteine *S*-conjugates is required for the mutagenicity of the nephrocarcinogenic chloroalkenes.

E. Structure–Mutagenicity and Structure–Cytotoxicity

All α-chlorovinyl cysteine *S*-conjugates studied are mutagenic in *S. typhimurium*. Haloethyl cysteine *S*-conjugates containing at least one bromine atom [*S*-(2-bromo-2-chloro-1,1-difluoroethy)-L-cysteine, *S*-(2-bromo-1,1,2-trifluoroethyl)-L-cysteine, and *S*-(2,2-dibromo-1,1-difluoroethyl)-L-cysteine] are also mutagenic in *S. typhimurium* (Finkelstein *et al.*, 1993). However, both the absolute numbers of induced revertants and the relative mutgenic potency (number of revertants/nmol*S*-conjugate) of these haloethyl cysteine *S*-conjugates were approximately 10-fold lower compared to the values obtained with the halovinyl cysteine *S*-conjugates. The β-lyase inhibitor aminooxyacetic acid blocked the mutagenicity of the brominated haloethyl cysteine *S*-conjugates. Haloethylcysteine *S*-conjugates lacking a bromine substitutent, i.e., *S*-(1,2-dichloro-1,1-difluoroethyl)-L-cysteine and *S*-(2-chloro-1,1,2-trifluoroethyl)-L-cysteine failed to increase the number of revertants per plate above the control values in all *S. typhimurium* strains tested (TA 2638, TA 100, TA 102, and TA98).

In contrast to the differences in mutagenicity, all haloethyl cysteine *S*-conjugates induced Ca^{2+} release in incubations with freshly isolated pig kidney mitochondria in a time-dependent manner.

F. Genotoxicity in Cultured Kidney Cells

Cysteine and glutathione *S*-conjugates derived from nephrocarcinogenic chloroalkenes were genotoxic in mammalian cells and induced γ-glutamyltransferase- and β-lyase-dependent DNA repair in the porcine kidney cell line LLC-PK_1 (Vamvakas *et al.*, 1989c,e). Although DNA repair was observed with concentrations of *S*-conjugates that did not cause severe toxicity, the concentration range over which DNA repair occurred in the absence of cell death was very small. Moreover, the extent of DNA repair induced by the *S*-conjugates was very low compared with the effects of known DNA-alkylating agents, such as streptozotocin or 4-nitroquinoline-1-oxide.

The weak genotoxic potency of the *S*-conjugates in renal cells in culture and the predominance of cytotoxicity and cell death are in agreement with the carcinogenicity studies in rats and mice. In these *in vivo* long-term studies, chloroalkenes induced renal cell tumors only with dose regimens that cause marked nephrotoxicity. All carcinogenicity studies resulting in renal tubular carcinomas also revealed concomitant severe inflammatory changes in the proximal tubules, the site of origin of the tumors. Nephrotoxicity was dose-dependent and quantitatively correlated with tumor incidences (Kociba *et al.*, 1977; National Cancer Institute, 1986; Reichert *et al.*, 1984).

In an attempt to establish a link between the extranuclear cytotoxic effects of the chlorovinyl *S*-conjugates and the formation of renal tumors observed with the parent chloroalkenes, the effects of mitochondrial Ca^{2+} release and increases in cytosolic Ca^{2+} concentrations were investigated in cultured renal cells. Among the Ca^{2+}-dependent degradative pathways, induction of DNA double-strand breaks by activation of Ca^{2+}- and Mg^{2+}-dependent endonucleases is of particular interest in terms of tumor formation. In LLC-PK_1 cells, *S*-(1,2-dichlorovinyl)-L-cysteine induced DNA fragmentation at concentrations that produced little or no impairment of cell growth. The Ca^{2+}-dependent DNA damage could be inhibited by blocking the mitochondrial Ca^{2+} release and was followed by increased poly(ADP-ribosyl)ation of nuclear proteins (Vamvakas *et al.*, 1992). A nonlethal increase in the amount of nuclear poly(ADP-ribosyl)conjugates may change structure and function of nuclear proteins and thus alter the expression of genes involved in cell proliferation and differentiation and possibly in tumor formation (Boulikas, 1993).

G. Genotoxicity *In Vivo*

Administration of *S*-(1,2-dichlorovinyl)-L-cysteine to rats resulted in increased formation of DNA single- and double-strand breaks (Jaffe *et al.*, 1985; McLaren *et al.*, 1993). Gel electrophoretic analysis of the DNA isolated from the renal cortex of rats treated with *S*-(1,2-dichlorovinyl)-L-cysteine revealed the oligonucleosomal ladder pattern typical of DNA fragmentation induced by Ca^{2+}- and Mg^{2+}-dependent endonucleases (D. Bittner and S. Vamvakas, unpublished data). Moreover, the induction of DNA double-strand breaks was followed by increased poly(ADP-ribosyl)ation of nuclear proteins, estimated by the incorporation of $[^{32}P]NAD^+$ into isolated renal cortex nuclei from treated rats.

H. Induction of Protooncogene Expression

The protooncogenes c-*fos* and c-*myc* play an important role in growth and differentiation of renal tissue. They are highly expressed during embryogenesis in the mitotically active tubular epithelium, whereas in terminally differentiated tubule cells of the kidney no expression is seen (Mugrauer and Ekblom, 1991; Mugrauer *et al.*, 1988). Furthermore, enhanced expression of c-*fos* and c-*myc* has been demonstrated in renal tumors in both humans and experimental animals (Shore *et al.*, 1988; Yao *et al.*, 1988).

In a recent study, the effects of *S*-(1,2-dichlorovinyl)-L-cysteine on the expression of these protooncogenes were investigated in LLC-PK_1 cells. Exponentially growing LLC-PK_1 monolayers were synchronized at the G_0 phase of the cell cycle by incubation for 24 h in medium containing only 0.2% fetal calf serum (instead of the 10% fetal calf serum normally used). Neither c-*fos* nor c-*myc* mRNA was detectable by Northern-blot analysis in these serum-deprived cells (Vamvakas and Köster, 1993; Vamvakas *et al.*, 1993). Monolayers exposed to 10% fetal calf serum after the 24-h starvation period revealed weak, but clearly detectable, c-*fos* and c-*myc* transcripts. Addition of 500 μM *S*-(1,2-dichlorovinyl)-L-cysteine to the serum-supplemented medium induced an additional twofold increase in the levels of c-*myc* and c-*fos* mRNA. The increase was sustained over the entire 5-h incubation.

As described above, *S*-(1,2-dichlorovinyl)-L-cysteine may induce both point mutations and also Ca^{2+}-mediated DNA double-strand breaks and increased poly(ADP-ribosyl)ation of nuclear proteins. If the induction of gene expression results from the formation of DNA adducts, which are converted into heritable mutations, the effects should persist after removal of *S*-(1,2-dichlorovinyl)-L-cysteine. To investigate the persistency of the induction of gene expression, LLC-PK_1 cells were pulse-treated with *S*-(1,2-dichlorovinyl)-L-cysteine for 5 h, and then incubated in *S*-(1,2-dichlorovinyl)-L-cysteine-free medium for 24, 48, and 72 h. Northern-blot analysis of the mRNA isolated at these time points did not reveal detectable c-*fos* and c-*myc* transcripts. Hence, *S*-(1,2-dichlorovinyl)-L-cysteine-induced DNA double-strand breaks, and poly(ADP-ribosyl)ation of nuclear proteins as a consequence of disturbed Ca^{2+} homeostasis may be more important for the observed changes in oncogene expression than mutational events. This is supported by the induced c-*fos* mRNA accumulation of LLC-PK_1 cells treated with the Ca^{2+} ionophore A23187 or with the tumor promoter thapsigargin, which increases cytosolic Ca^{2+} concentrations by inhibiting the Ca^{2+} ATPase of the endoplasmic reticulum (Vamvakas and Köster, 1993). Ca^{2+} is also involved in the induction of c-*fos* expression in renal cells exposed to oxidative stress generated by xanthine/xanthine oxidase, as demonstrated by reduction of the c-*fos* mRNA transcript levels in presence of the Ca^{2+} chelator EGTA (Maki *et al.*, 1992).

I. Possible Mechanisms in the Nephrocarcinogenicity of Haloalkenes

S-Conjugates from dichloroethyne, trichloroethene, perchloroethene, and hexachlorobutadiene may induce renal cell tumors by three interacting pathways. One, the thioketenes produced upon β-lyase cleavage may form DNA-adducts and induce point mutations. Two, *S*-conjugates may indirectly alter gene expression by induction of Ca^{2+}-dependent DNA double-strand breaks and increased poly(ADP-ribosyl)ation of nuclear proteins. This post-translational modification of structure and function of nuclear proteins may be an important epigenetic mechanism for modulation of gene expression (Boulikas, 1993; Cerutti, 1985). Several tumor promotors enhance the poly(ADP-ribose)polymerase activity and inhibition of this enzyme blocks the promoting effects of 12-*O*-tetradecanoylphorbol-13-acetate *in vitro* and of phenobarbital *in vivo* (Borek and Cleaver, 1986; Singh, 1990a,b; Tsujiuchi *et al.*, 1990). Furthermore, increased levels of poly(ADP-ribosyl)ation were found in rapidly proliferating tissues as compared with terminally differentiated, quiescent tissues. In particular, this last finding provides a link to the third aspect of the nephrocarcinogenicity of dichloroethyne, trichloroethene, tetrachloroethene, and hexachlorobutadiene: these compounds increase the incidence of renal tumors in rats only at doses that also induce severe nephrotoxicity and forced cell proliferation in the proximal tubules. The nephrotoxicity is quantitatively correlated with the formation of renal cell tumors. The increased cell turnover enhances the frequency of spontaneous mutations, and the probability of DNA damage to be converted into heritable mutations before DNA repair can take place (Goldworthy *et al.*, 1990). The concept of the interplay between immediate DNA damage and extranuclear effects in renal tumorigenicity is supported by a recent study on the morphological and biochemical features of preneoplastic lesions in rodent kidney (Dietrich and Swenberg, 1991). This study demonstrated that the preneoplastic lesions observed in long-term studies with genotoxic nephrocarcinogens are identical with the preneoplastic lesions induced in the kidneys of animals treated with nongenotoxic carcinogens.

With the currently available data, it is not possible to estimate the quantitative contributions of direct DNA damage, epigenetic alterations of gene expression, and enhanced cell proliferation to the nephrocarcinogenic effects of dichloroethyne, trichloroethene, tetrachloroethene, and hexachlorobutadiene. To assess the importance of these three pathways, *in vivo* investigations on the induction of cell proliferation, alteration of gene expression by poly(ADP-ribosyl)ation, and formation of DNA adducts should be carried out concomitantly, i.e., in preneoplastic lesions and renal cell tumors formed by chloroalkenes or their *S*-conjugates.

Acknowledgments

Work in the authors laboratories was supported by the Deutsche Forschungsgemeinschaft (W.D., S.V.), the Doktor-Robert-Pfleger-Stiftung (W.D.), the Nato Collaborative Research Grant (W.D., M.W.A.), and National Institute of Environmental Health Sciences Grant ES03127 (M.W.A.).

References

Adiwidjaja, G., Kirsch, C., Pedersen, F., Schaumann, E., Schmerse, G. C., and Senning, A. (1991). Generation and trapping of dichlorothioketene. *Chem. Ber.* **124,** 1485–1487.

Akerboom, T. P. M., Bilzer, M., and Sies, H. (1982). Competition between transport of glutathione disulfide (GSSG) and glutathione S-conjugates from perfused rat liver into bile. *FEBS Lett.* **140,** 73–76.

Akerboom, T. P. M., Narayanaswami, V., Kunst, M., and Sies, H. (1991). ATP-dependent S-(2,4-dinitrophenyl)glutathione transport in canalicular plasma membrane vesicles from rat liver. *J. Biol. Chem.* **266,** 13147–13152.

Albert, Z., Orlowski, M., and Szewczuk, A. (1961). Histochemical demonstration of gamma-glutamyl transpeptidase. *Nature (London)* **181,** 767–768.

Albert, Z., Orlowska, J., Orlowski, M., and Szewczuk, A. (1964). Histochemical and biochemical investigations of gamma-glutamyltranspeptidase in the tissues of man and laboratory rodents. *Acta Histochem.* **18,** 78–89.

Allen, L. M., Corrigan, M. V., and Meinking, T. (1981). Interaction of AT-125, (αS,5S)-amino-3-chloro-4,5-dihydro-isoxazoleacetic acid, with bovine kidney γ-glutamyl transpeptidase. *Chem. Biol. Interact.* **33,** 361–365.

Anders, M. W. (1989). Biotransformation and bioactivation of xenobiotics by the kidney. *In* "Intermediary Xenobiotic Metabolism in Animals: Methodology, Mechanisms and Significance" (D. H. Hutson, J. Caldwell, and G. D. Paulson, eds.), pp. 81–97. Taylor & Francis, London.

Anders, M. W. (1990). Glutathione-dependent bioactivation of xenobiotics. "New Horizons in Molecular Toxicology." Federation of American Societies for Experimental Biology, Indianapolis, IN.

Anders, M. W. (1991). Glutathione-dependent bioactivation of xenobiotics: Implications for mutagenicity and carcinogenicity. *In* "Xenobiotics and Cancer" (L. Ernster, ed.), pp. 89–99. Jpn. Sci. Soc. Press, Tokyo.

Anders, M. W., Lash, L. H., and Elfarra, A. A. (1986). Nephrotoxic amino acid and glutathione S-conjugates: Formation and renal activation. *Adv. Exp. Med. Biol.* **197,** 443–455.

Anders, M. W., Lash, L., Dekant, W., Elfarra, A. A., and Dohn, D. R. (1988). Biosynthesis and biotransformation of glutathione S-conjugates to toxic metabolites. *CRC Crit. Rev. Toxicol.* **18,** 311–342.

Anders, M. W., Vamakas, S., and Dekant, W. (1990). Bioactivation of haloalkenes through glutathione conjugation. *In* "Glutathione S-Transferases and Drug Resistance" (J. D. Hayes, C. B. Pickett, and T. J. Mantle, eds.), pp. 121–130. Taylor & Francis, London.

Anders, M. W., Dekant, W., and Vamvakas, S. (1992). Glutathione-dependent toxicity. *Xenobiotica* **22,** 1135–1145.

Anderson, M. E., Bridges, R. J., and Meister, A. (1980). Direct evidence for inter-organ transport of glutathione and that the non-filtration renal mechanism for glutathione utilization involves γ-glutamyl transpeptidase. *Biochem. Biophys. Res. Commun.* **96,** 848–853.

Anderson, P. M., and Schultze, M. O. (1965). Cleavage of S-(1,2-dichlorovinyl)-L-cysteine by an enzyme of bovine origin. *Arch. Biochem. Biophys.* **111,** 593–602.

Argiles, J. M., and Lopez-Soriano, F. J. (1990). Intestinal amino acid transport: An overview. *Int. J. Biochem.* **22,** 931–937.

Arias, I. M., and Jakoby, W. B., eds. (1976). "Glutathione: Metabolism and Function." Raven Press, New York.

Awasthi, Y. C. (1990). The interrelationship between p-glycoprotein and glutathione S-conjugate transporter(s). *Trends Biochem. Sci.* **15,** 376–377.

Awasthi, Y. C., Singh, S. V., Ahmad, H., Wronski, L. W., Srivastava, S. K., and Labelle, E. F. (1989). ATP dependent primary active transport of xenobiotic-glutathione conjugates by human erythrocyte membrane. *Cell Biochem.* **91,** 131–136.

Bakke, J., and Gustafsson, J.-A. (1984). Mercapturic acid pathway metabolites of xenobiotics: Generation of potentially toxic metabolites during enterohepatic circulation. *Trends Pharmacol. Sci., Dec.,* 517–521.

Bakke, J. E., Gustafsson, J. A., and Gustafsson, B. E. (1980). Metabolism of propachlor by the germfree rat. *Science* **210,** 433–435.

Bakke, J. E., Rafter, J. J., Lindeskog, P., Feil, V. J., Gustafsson, J. A., and Gustafsson, B. E. (1981). Metabolism of 2-acetamido-4-(chloromethyl)thiazole in germfree and conventional rats. *Biochem. Pharmacol.* **30,** 1839–1844.

Bakke, J. E., Bergman, A. L., and Larsen, G. L. (1982). Metabolism of 2,4′,5-trichlorobiphenyl by the mercapturic acid pathway. *Science* **217,** 645–647.

Bakke, J. E., Feil, V. J., and Mulford, D. J. (1990). Biliary excretion and intestinal metabolism in the intermediary metabolism of pentachlorothioanisole. *Xenobiotica* **20,** 601–605.

Bernasconi, C. F. (1989). Nucleophilic addition to olefins. Kinetics and mechanism. *Tetrahedron* **45,** 4017–4090.

Bhattacharya, R. K., and Schultze, M. O. (1967). Enzymes from bovine and turkey kidneys which cleaved S-(1,2-dichlorovinyl)-L-cysteine. *Comp. Biochem. Physiol.* **22,** 723–735.

Bhattacharya, R. K., and Schultze, M. O. (1971a). Properties of DNA isolated from organs of calves treated with S-(1,2-dichlorovinyl)-L-cysteine. III. Template activity for bacterial RNA polymerase. *Biochem. Biophys. Res. Commun.* **45,** 1526–1532.

Bhattacharya, R. K., and Schultze, M. O. (1971b). Properties of DNA isolated from tissues of calves treated with S-(1,2-dichlorovinyl) L cysteine. II. Primer-template activity for bacterial DNA polymerases. *Arch. Biochem. Biophys.* **145,** 575–582.

Bhattacharya, R. K., and Schultze, M. O. (1972). Properties of DNA treated with S-(1,2-dichlorovinyl)-L-cysteine and a lyase. *Arch. Biochem. Biophys.* **153,** 105–115.

Bhattacharya, R. K., and Schultze, M. O. (1973a). Modification of polynucleotides by a fragment produced by enzymatic cleavage of S-(1,2-dichlorovinyl)-L-cysteine. *Biochem. Biophys. Res. Commun.* **53,** 172–181.

Bhattacharya, R. K., and Schultze, M. O. (1973b). Hybridization of DNA modified by interaction with a metabolic fragment from S-(1,2-dichlorovinyl)-L-cysteine. *Biochem. Biophys. Res. Commun.* **54,** 538–543.

Boobis, A. R., Fawthrop, D. J., and Davies, D. S. (1989). Mechanisms of cell death. *Trends Pharmacol. Sci.* **10,** 275–280.

Borek, C., and Cleaver, J. E. (1986). Antagonistic action of a tumor promotor and a poly(adenosine diphosphoribose) synthesis inhibitor in radiation induced transformation *in vitro*. *Biochem. Biophys. Res. Commun.* **134,** 1334–1341.

Botti, B., Ceccarelli, D., Tomasi, A., Muscatello, U., and Masini, A. (1989). Biochemical mechanism of GSH depletion induced by 1,2-dibromoethane in isolated rat liver mitochondria. Evidence of a GSH conjugation process. *Biochim. Biophys. Acta* **992,** 327–332.

Boulikas, T. (1993). Poly(ADP-ribosyl)ation, DNA strand breaks, chromatin and cancer. *Curr. Persp. Molec. Cell Oncol.* **1,** 1–18.

Boyer, T. D., Vessey, D. A., and Kempner, E. (1986). Radiation inactivation of microsomal glutathione transferase. *J. Biol. Chem.* **261,** 16963–16968.

Buetler, T. M., and Eaton, D. L. (1992). Glutathione S-transferases: Amino acid sequence comparison, classification and phylogenetic relationship. *Environ. Carcinog. Ecotoxicol. Rev.* **C10,** 181–203.

Cantoni, O., Sestili, P., Cattabeni, F., Bellomo, G., Pou, S., Cohen, M., and Cerutti, P. (1989). Calcium chelator Quin 2 prevents hydrogen-peroxide-induced DNA breakage and cytotoxicity. *Eur. J. Biochem.* **182,** 209–212.

Cerutti, P. A. (1985). Prooxidant states and tumor promotion. *Science* **227,** 375–381.

Cerutti, P A. (1989). Response modification in carcinogenesis. *Environ. Health Perspect.* **81,** 39–43.

Chasseaud, L. F. (1979). The role of glutathione and glutathione S-transferases in the metabolism of chemical carcinogens and other electrophilic agents. *Adv. Cancer Res.* **29,** 175–274.

Chen, J. C., Stevens, J. L., Trifillis, A. L., and Jones, T. W. (1990). Renal cysteine conjugate β-lyase-mediated toxicity studied with primary cultures of human proximal tubular cells. *Toxicol. Appl. Pharmacol.* **103,** 463–473.

Chen, Q., Jones, T. W., Brown, P. C., and Stevens, J. L. (1990). The mechanism of cysteine conjugate cytotoxicity in renal epithelial cells. *J. Biol. Chem.* **265,** 21603–21611.

Coles, B., and Ketterer, B. (1990). Role of glutathione and glutathione transferases in chemical carcinogenesis. *CRC Crit. Rev. Biochem. Mol. Biol.* **25,** 47–70.

Commanduer, J. N. M., Oostendorp, R. A. J., Schoofs, P. R., Xu, B., and Vermeulen, N. P. E. (1987). Nephrotoxicity and hepatotoxicity of 1,1-dichloro-2,2-difluoroethylene in the rat. *Biochem. Pharmacol.* **36,** 4229–4237.

Cooper, A. J. L., and Anders, M. W. (1990). Glutamine transaminase K/cysteine conjugate β-lyase. *Ann. N. Y. Acad. Sci.* **585,** 118–127.

Darnerud, P. O., Brandt, I., Feil, V. J., and Bakke, J. E. (1988). S-(1,2-Dichloro-(^{14}C)vinyl)-L-cysteine (DCVC) in the mouse kidney: Correlation between tissue-binding and toxicity. *Toxicol. Appl. Pharmacol.* **95,** 423–434.

Dekant, W., Vamvakas, S., Berthold, K., Schmidt, S., Wild, D., and Henschler, D. (1986). Bacterial β-lyase mediated cleavage and mutagenicity of cysteine conjugates derived from the nephrocarcinogenic alkenes trichloroethylene, tetrachloroethylene and hexachlorobutadiene. *Chem.-Biol. Interact.* **60,** 31–45.

Dekant, W., Lash, L. H., and Anders, M. W. (1987a). Bioactivation mechanism of the cytotoxic and nephrotoxic S-conjugate S-(2-chloro-1,2-trifluoroethyl)-L-cysteine. *Proc. Natl. Acad. Sci. U.S.A.* **84,** 7443–7447.

Dekant, W., Martens, G., Vamvakas, S., Metzler, M., and Henschler, D. (1987b). Bioactivation of tetrachloroethylene: Role of glutathione S-transferase-catalyzed conjugation versus cytochrome P-450-dependent phospholipid alkylation. *Drug Metab. Dispos.* **15,** 702–709.

Dekant, W., Vamvakas, S., Henschler, D., and Anders, M. W. (1988a). Enzymatic conjugation of hexachloro-1,3-butadiene with glutathione: Formation of 1-(glutathion-S-yl)-1,2,3,4,4-pentachlorobuta-1,3-diene and 1,4-bis(glutathion-S-yl)-1,2,3,4-tetrachlorobuta-1,3-diene. *Drug Metab. Dispos.* **16,** 701–706.

Dekant, W., Schrenk, D., Vamvakas, S., and Henschler, D. (1988b). Metabolism of hexachloro-1,3-butadiene in mice: *In vivo* and *in vitro* evidence for activation by glutathione conjugation. *Xenobiotica* **18,** 803–816.

Dekant, W., Lash, L. H., and Anders, M. W. (1988c). Fate of glutathione conjugates and bioactivation of cysteine S-conjugates by cysteine conjugate β-lyase. *In* "Glutathione Conjugation: Its Mechanism and Biological Significance" (H. Sies and B. Ketterer, eds.), pp. 415–447. Academic Press, Orlando, FL.

Dekant, W., Berthold, K., Vamvakas, S., and Henschler, D. (1988d). Thioacylating agents as ultimate intermediates in the β-lyase catalyzed metabolism of S-(pentachlorobutadienyl)-L-cysteine. *Chem.-Biol. Interact.* **67,** 139–148.

Dekant, W., Berthold, K., Vamvakas, S., Henschler, D., and Anders, M. W. (1988e). Thioacylating intermediates as metabolites of S-(1,2-dichlorovinyl)-L-cysteine and S-(1,2,2-trichlorovinyl)-L-cysteine formed by cysteine conjugate β-lyase. *Chem. Res. Toxicol.* **1,** 175–178.

Dekant, W., Vamvakas, S., and Anders, M. W. (1989). Bioactivation of nephrotoxic haloalkenes by glutathione conjugation: Formation of toxic and mutagenic intermediates by cysteine conjugate β-lyase. *Drug Metab. Rev.* **20,** 43–83.

Dekant, W., Vamvakas, S., and Anders, M. W. (1990a). Bioactivation of hexachlorobutadiene by glutathione conjugation. *Food Chem. Toxicol.* **28,** 285–293.

Dekant, W., Vamvakas, S., Koob, M., Köchling, A., Kanhai, W., Müller, D., and Henschler, D. (1990b). A mechanism of haloalkene-induced renal carcinogenesis. *Environ. Health Perspect.* **88,** 107–110.

Dekant, W., Koob, M., and Henschler, D. (1990c). Metabolism of trichloroethene—*In vivo* and *in vitro* evidence for activation by glutathione conjugation. *Chem.-Biol. Interact.* **73,** 89–101.

Dekant, W., Urban, G., Görsman, C., and Anders, M. W. (1991). Thioketene formation from α-haloalkenyl 2-nitrophenyl disulfides: Models for biological reactive intermediates of cytotoxic S-conjugates. *J. Am. Chem. Soc.* **113,** 5120–5122.

Dekant, W., Vamvakas, S., and Anders, M. W. (1992). The kidney as a target organ for xenobiotics bioactivated by glutathione conjugation. *In* "Tissue-Specific Toxicity. Biochemical Mechanisms" (W. Dekant and H. G. Neumann, eds.), pp. 163–194. Academic Press, London.

DePierre, J. W., and Morgenstern, R. (1983). Comparison of the distribution of microsomal and cytosolic glutathione S-transferase activities in different organs of the rat. *Biochem. Pharmacol.* **32,** 721–723.

Dietrich, D. R., and Swenberg, J. A. (1991). Preneoplastic lesions in rodent kidney induced spontaneously or by non-genotoxic agents: Predictive nature and comparison to lesions induced by genotoxic carcinogens. *Mutat. Res.* **248,** 239–260.

Dohn, D. R., and Anders, M. W. (1982a). The enzymatic reaction of chlorotrifluoroethylene with glutathione. *Biochem. Biophys. Res. Commun.* **109,** 1339–1345.

Dohn, D. R., and Anders, M. W. (1982b). A simple assay for cysteine conjugate β-lyase activity with S-(2-benzothiazolyl)cysteine as the substrate. *Anal. Biochem.* **120,** 379–386.

Dohn, D. R., Quebbemann, A. J., Borch, R. F., and Anders, M. W. (1985a). Enzymatic reaction of chlorotrifluoroethene with glutathione: ^{19}F NMR evidence for stereochemical control of the reaction. *Biochemistry* **24,** 5137–5143.

Dohn, D. R., Leininger, J. R., Lash, L. H., Quebbemann, A. J., and Anders, M. W. (1985b). Nephrotoxicity of S-(2-chloro-1,1,2-trifluoroethyl)glutathione and S-(2-chloro-1,1,2-trifluoroethyl)-L-cysteine, the glutathione and cysteine conjugates of chlorotrifluoroethene. *J. Pharmacol. Exp. Ther.* **235,** 851–857.

Duffel, M. W., and Jakoby, W. B. (1982). Cysteine S-conjugate N-acetyltransferase from rat kidney microsomes. *Mol. Pharmacol.* **21,** 444–448.

Dulik, D. M., Huwe, J. K., Bakke, J. E., Connors, M. S., and Fenselau, C. (1992). Conjugation of polychlorinated agrochemical sulphoxides and sulphones by glutathione. *Xenobiotica* **22,** 325–334.

Elfarra, A. A., and Anders, M. W. (1985). S-(1,2-Dichlorovinyl)-L-homocysteine (DCVHC), and analogue of the renal toxin S-(1,2-dichlorovinyl)-L-cysteine, is a potent nephrotoxin. *Fed. Proc. Fed. Am. Soc. Exp. Biol.* **44,** 1624. Abstract.

Elfarra, A. A., Jakobson, I., and Anders, M. W. (1986). Mechanism of S-(1,2-dichlorovinyl)-glutathione-induced nephrotoxicity. *Biochem. Pharmacol.* **35,** 283–288.

Elfarra, A. A., Lash, L. H., and Anders, M. W. (1987). α-Ketoacids stimulate renal cysteine conjugate β-lyase activity and potentiate the cytotoxicity of S-(1,2-dichlorovinyl)-L-cysteine. *Mol. Pharmacol.* **31,** 208–212.

Finkelstein, M. B., Baggs, R. B., and Anders, M. W. (1992). Nephrotoxicity of the glutathione and cysteine conjugates of 2-bromo-2-chloro-1,1-difluoroethene. *J. Pharmacol. Exp. Ther.* **261,** 1248–1252.

Finkelstein, M. B., Vamvakas, S., Bittner, D., and Anders, M. W. (1993). Structure-mutagenicity and structure-cytotoxicity studies on bromine-containing cysteine *S*-conjugates and related compounds. *Chem. Res. Toxicol.* In press.

Flohé, L., Benöhr, H. C., Sies, H., Waller, H. D., and Wendel, A., eds. (1974). "Glutathione." Academic Press, New York.

Fokin, A. V., Skladnev, A. A., and Knunyants, I. L. (1961). Reaction of fluoroolefins with hydrogen sulfide. *Dokl. Akad. Nauk SSSR* **138,** 1132–1135.

Gartland, K. P. R., Eason, C. T., Bonner, F. W., and Nicholson, J. K. (1990). Effects of biliary cannulation and buthionine sulphoximine pretreatment on the nephrotoxicity of para-aminophenol in the Fisher 334 rat. *Arch. Toxicol.* **64,** 14–25.

Gietl, Y. S., and Anders, M. W. (1991). Biosynthesis and biliary excretion of S-conjugates of hexachlorobuta-1,3-diene in the perfused rat liver. *Drug. Metab. Dispos.* **19,** 274–277.

Gietl, Y. S., Vamvakas, S., and Anders, M. W. (1991). Intestinal absorbtion of S-(pentachlorbutandienyl)glutathione and S-(pentachlorbutandienyl)-L-cysteine S-conjugates of hexachlorobuta-1,3-diene. *Drug. Metab. Dispos.* **19,** 703–707.

Goldworthy, T. L., Morgan, K. P., Popp, J. A., and Butterworth, B. E. (1990). Chemically-induced cell proliferation. *Chem. Ind. Inst. Toxicol. Act.* **10,** 1–12.

Green, R. M., and Elce, J. S. (1975). Acetylation of S-substituted cysteines by rat liver and kidney microsomal N-acetyltransferase. *Biochem. J.* **147,** 283.

Green, T., and Odum, J. (1985). Structure/activity studies of the nephrotoxic and mutagenic action of cysteine conjugates of chloro- and fluoroalkenes. *Chem.-Biol. Interact.* **54,** 15–31.

Griffith, O. W., and Meister, A. (1979). Translocation of intracellular glutathione to membrane-bound γ-glutamyl transpeptidase as a discrete step in the γ-glutamyl cycle: Glutathionuria after inhibition of transpeptidase. *Proc. Natl. Acad. Sci. U.S.A.* **76,** 268–272.

Groves, C. E., Lock, E. A., and Schnellmann, R. G. (1991). Role of lipid peroxidation in renal proximal tubule cell death induced by haloalkene cysteine conjugates. *Toxicol. Appl. Pharmacol.* **107,** 54–62.

Gstraunthaler, G., Pfaller, W., and Kotanko, P. (1985). Biochemical characterization of renal epithelial cell cultures (LLC-PK_1 and MDCK). *Am. J. Physiol.* **248,** F536–F544.

Guder, W. G., and Ross, B. D. (1984). Enzyme distribution along the nephron. *Kidney Int.* **26,** 101–111.

Guder, W. G., and Wirthensohn, G. (1985). Enzyme distribution and unique biochemical pathways in specific cells along the nephron. *In* "Renal Heterogenicity and Target Cell Toxicity" (P. H. Bach and E. A. Lock, eds.), pp. 195–198. Wiley, Chichester.

Hargus, S. J., and Anders, M. W. (1991). Immunochemical detection of covalently modified kidney proteins in S-(1,1,2,2-tetrafluorethyl)-L-cystein-treated rats. *Biochem. Pharmacol.* **42,** R17–R20.

Hargus, S. J., Fitzsimmons, M. E., Aniya, Y., and Anders, M. W. (1991). Stereochemistry of the microsomal glutathione S-transferase catalyzed addition of glutathione to chlorotrifluoroethene. *Biochemistry* **30,** 717–721.

Harris, J. M., Meyer, D. J., Coles, B., and Ketterer, B. (1991). A novel glutathione transferase (13-13) isolated from the matrix of rat liver mitochondria having structural similarity to class Theta enzymes. *Biochem. J.* **278,** 137–141.

Harris, J. W., Dekant, W., and Anders, M. W. (1992). *In vivo* detection and characterization of protein adducts resulting from bioactivation of haloethene cysteine S-conjugates by ^{19}F NMR: chlorotrifluoroethene and tetrafluoroethene. *Chem. Res. Toxicol.* **5,** 34–41.

Hassall, C. D., Gandolfi, A. J., Duhamel, R. C., and Brendel, K. (1984). The formation and biotransformation of cysteine conjugates of halogenated ethylenes by rabbit renal tubules. *Chem.-Biol. Interact.* **49,** 283–297.

Hayden, P. J., and Stevens, J. L. (1990). Cysteine conjugate toxicity, metabolism, and binding to macromolecules in isolated rat kidney mitochondria. *Mol. Pharmacol.* **37,** 468–476.

Hayden, P. J., Ichimura, T., McCann, D. J., Pohl, L. R., and Stevens, J. L. (1991a). Detection of cysteine conjugate metabolite adduct formation with specific mitochondrial proteins using antibodies raised against halothane metabolite adducts. *J. Biol. Chem.* **266,** 18415–18418.

Hayden, P. J., Yang, Y., Ward, A. J. I., Dulik, D. M., McCann, D. J., and Stevens, J. L. (1991b). Formation of difluorothionoacetyl-protein adducts by S-(1,1,2,2-tetrafluoroethyl)-L-cysteine metabolites: Nucleophilic catalysis of stable adduct formation by histidine and tyrosine. *Biochemistry* **30,** 5935–5943.

Hayden, P. J., Welsh, C. J., Yang, Y., Schaefer, W. H., Ward, A. J. I., and Stevens, J. L. (1992). Formation of mitochondrial phospholipid adducts by nephrotoxic cysteine conjugate metabolites. *Chem. Res. Toxicol.* **5,** 231–237.

Hayes, J. D., Pickett, C. B., and Mantle, T. J., eds. (1990). "Glutathione S-Transferase and Drug Resistance." Taylor & Francis, London.

Heuner, A., Dekant, W., Schwegler, J. S., and Silbernagl, S. (1991). Localization and capacity of the last step of mercapturic acid biosynthesis and the reabsorption and acetylation of cysteine S-conjugates in the rat kidney. *Eur. J. Physiol.* **417,** 523–527.

Hinchman, C. A., and Ballatori, N. (1990). Glutathione-degrading capacities of liver and kidney in different species. *Biochem. Pharmacol.* **40,** 1131–1135.

Hinchman, C. A., Matsumoto, H., Simmons, T. W., and Ballatori, N. (1991). Intrahepatic conversion of a glutathione conjugate to its mercapturiv acid. *J. Biol. Chem.* **266,** 22179–22185.

Hook, J. B., Rose, M. S., and Lock, E. A. (1982). The nephrotoxicity of hexacloro-1 : 3-butadiene in the rat: Studies of organic anion and cation transport in renal slices and the effects of monooxygenase inducers. *Toxicol. Appl. Pharmacol.* **65,** 373–382.

Hughey, R. P., Rankin, B. B., Elce, J. S., and Curthoys, N. P. (1978). Specificity of a particulate rat renal peptidase and its localization along with other enzymes of mercapturic acid synthesis. *Arch. Biochem. Biophys.* **186,** 211–217.

Hull, R. N., Cherry, W. R., and Weaver, G. W. (1976). The origin and characteristics of a pig kidney cell strain LLC-PK_1. *In Vitro Cell. Dev. Biol.* **12,** 670–677.

Inoue, M., and Morino, Y. (1985). Direct evidence for the role of the membrane potential in glutathione transport by renal brush-border membranes. *J. Biol. Chem.* **260,** 326–331.

Inoue, M., Horiuchi, S., and Morino, Y. (1977). Affinity labeliing of rat kidney γ-glutamyl transpeptidase. *Eur. J. Biochem.* **73,** 335–342.

Inoue, M., Okajima, K., and Morino, Y. (1981). Renal transtubular transport of mercapturic acid *in vivo*. *Biochim. Biophys. Acta* **641,** 122–128.

Inoue, M., Kinne, R., Tran, T., Biempica, L., and Arias, I. M. (1983). Rat liver canalicular membrane vesicles - Isolation and topological characterization. *J. Biol. Chem.* **258,** 5183–5188.

Inoue, M., Okajima, K., and Morino, Y. (1984a). Hepato-renal cooperation in biotransformation, membrane transport, and elimination of cysteine S-conjugates of xenobiotics. *J. Biochem.* (*Tokyo*) **95,** 247–254.

Inoue, M., Akerboom, T. P., Sies, H., Kinne, R., Thoa, T., and Arias, I. M. (1984b). Biliary transport of glutathione S-conjugate by rat liver canalicular membrane vesicles. *J. Biol. Chem.* **259,** 4998–5002.

Inoue, M., Kinne, R., Tran, T., and Arias, I. M. (1984c). Glutathione transport across hepatocyte plasma membranes. Analysis using isolated rat-liver sinusoidal-membrane vesicles. *Eur. J. Biochem.* **138,** 491–495.

Inoue, M., Okajima, K., Nagase, S., and Morino, Y. (1987). Inter-organ metabolism and transport of a cysteine-S-conjugate of xenobiotics in normal and mutant analbuminemic rats. *Biochem. Pharmacol.* **36,** 2145–2150.

Ishmael, J., Pratt, I., and Lock, E. A. (1982). Necrosis of the pars recta (S3 segment) of the rat kidney produced by hexachloro-1 : 3-butadiene. *J. Pathol.* **138,** 99–113.

Jaffe, D. R., Hassall, C. D., Brendel, K., and Gandolfi, A. J. (1983). *In vivo* and *in vitro* nephrotoxicity of the cysteine conjugate of hexachlorobutadiene. *J. Toxicol. Environ. Health* **11,** 857–867.

Jaffe, D. R., Gandolfi, A. J., and Nagle, R. B. (1984). Chronic toxicity of S-(trans-1,2-dichlorovinyl)-L-cysteine in mice. *J. Appl. Toxicol.* **4,** 315–319.

Jaffe, D. R., Hassall, C. D., Gandolfi, A. J., and Brendel, K. (1985). Production of DNA single strand breaks in renal tissue after exposure to 1,2-dichlorovinylcysteine. *Toxicology* **35,** 25–33.

Jakoby, W. B., and Stevens, J. (1984). Cysteine conjugate β-lyase and the thiomethyl shunt. *Biochem. Soc. Trans.* **12,** 33–35.

Jocelyn, P. C., and Cronshaw, A. (1985). Properties of mitochondria treated with 1-chloro-2,4-dinitrobenzene. *Biochem. Pharmacol.* **34,** 1588–1590.

Jones, D. P., Moldeus, P., Stead, A. H., Ormstad, K., Jörnvall, H., and Orrenius, S. (1979a). Metabolism of glutathione and a glutathione conjugate by isolated kidney cells. *J. Biol. Chem.* **254,** 2787–2792.

Jones, D. P., Sundby, G.-B., Ormstad, K., and Orrenius, S. (1979b). Use of isolated kidney cells for study of drug metabolism. *Biochem. Pharmacol.* **28,** 929–935.

Jones, T. W., Gerdes, R. G., Ormstad, K., and Orrenius, S. (1985). The formation of both a mono- and a bis-substituted glutathione conjugate of hexachlorobutadiene by isolated hepatocytes and following *in vivo* administration to the rat. *Chem.-Biol. Interact.* **56,** 251–267.

Jones, T. W., Wallin, A., Thor, H., Gerdes, R. G., Ormstad, K., and Orrenius, S. (1986). The mechanism of pentachlorobutadienyl-glutathione nephrotoxicity studied with isolated rat renal epithelial cells. Arch. Biochem. Biophys. **251,** 504–513.

Jones, T. W., Qin, C., Schaeffer, V. H., and Stevens, J. L. (1988). Immunohistochemical localization of glutamine transaminase K, a rat kidney cysteine conjugate β-lyase, and the relationship to the segment specificity of cysteine conjugate nephrotoxicity. *Mol. Pharmacol.* **34,** 621–627.

Kanhai, W., Dekant, W., and Henschler, D. (1989). Metabolism of the nephrotoxin dichloroacetylene by glutathione conjugation. *Chem. Res. Toxicol.* **2,** 51–56.

Kanhai, W., Koob, M., Dekant, W., and Henschler, D. (1991). Metabolism of ^{14}C-dichloroetyne in rats. *Xenobiotica* **21,** 905–916.

Ketterer, B., and Mulder, G. J. (1990). Glutathione conjugation. *In* "Conjugation Reactions in Drug Metabolism: An Integrated Approach" (G. J. Mulder, ed.), pp. 307–364. Taylor & Francis, London.

Klos, C., Koob, M., Kramer, C., and Dekant, W. (1992). p-Aminophenol nephrotoxicity: Biosynthesis of toxic glutathione conjugates. *Toxicol. Appl. Pharmacol.* **115,** 98–106.

Kluwe, W. M., McNish, R., and Hook, J. B. (1981). Depletion by 1,2-dibromoethane, 1,2-dibromo-3-chloropropane, tris(2,3-dibromopropyl)phosphate, and hexachloro-1,3-butadiene of reduced non-protein sulfhydryl groups in target and non-target organs. *Biochem. Pharmacol.* **30,** 2265–2271.

Kluwe, W. M., Harrington, F. W., and Cooper, S. E. (1982). Toxic effects of organohalide compounds on renal tubular cells *in vivo* and *in vitro*. *J. Pharmacol. Exp. Ther.* **220,** 597–603.

Kociba, R. J., Keyes, D. G., Jersey, G. C., Ballard, J. J., Dittenber, D. A., Quast, J. F., Wade, L. E., Humiston, C. G., and Schwetz, B. A. (1977). Results of a two year chronic toxicity study with hexachlorobutadiene in rats. *Am. Ind. Hyg. Assoc. J.* **38,** 589–602.

Koechel, D. A., Krejci, M. E., and Ridgewell, R. E. (1991). The accute effects of S-(1,2-dichlorovinyl)-L-cysteine and related chemicals on renal function and ultrastructure in the pentobarbital-anesthetized dog: Structure-activity relationships, biotransformation, and unique site-specific nephrotoxicity. *Fundam. Appl. Toxicol.* **17,** 17–33.

Koob, M., and Dekant, W. (1990). Metabolism of hexafluoroproene—Evidence for bioactivation by glutathione conjugate formation in the kidney. *Drug Metab. Dispos.* **18,** 911–916.

Koob, M., and Dekant, W. (1991). Bioactivation of xenobiotics by formation of toxic glutathione conjugates. *Chem.-Biol. Interact.* **77,** 107–136.

Koob, M., Köchling, A., Weber, S., Lau, S. S., Monks, T. J., and Dekant, W. (1990). Cytotoxicity of 2-bromo-(N-acetylcystein-S-yl)hydroquinone isomers in rat renal tubular epithelial cells. *Toxicol. Lett.* **53,** 143–145.

Kozak, E. M., and Tate, S. S. (1982). Glutathione-degrading enzymes of microvillus membranes. *J. Biol. Chem.* **257,** 6322–6327.

Kramer, R. A., Foureman, G., Greene, K. E., and Reed, D. J. (1987). Nephrotoxicity of S-(2-chloroethyl)glutathione in the Fischer rat: Evidence for γ-glutamyltranspeptidase-independent uptake by the kidney. *J. Pharmacol. Exp. Ther.* **242,** 741–748.

Kraus, P. (1980). Resolution, purification and some properties of three glutathione transferases from rat liver mitochondria. *Hoppe-Seyler's Z. Physiol. Chem.* **361,** 9–15.

Kraus, P., and Gross, B. (1979). Particle-bound glutathione-S-transferases. *Enzyme* **24,** 205–208.

Larsen, G. L., and Bakke, J. E. (1983). Metabolism of mercapturic acid-pathway metabolites of 2-chloro-N-isopropylacetanilide (propachlor) by gastrointestinal bacteria. *Xenobiotica* **13,** 115–126.

Larsen, G. L., and Stevens, J. L. (1985). Cysteine conjugate β-lyase in the gastrointestinal bacterium *Eubacterium limosum*. *Mol. Pharmacol.* **29,** 97–103.

Larsson, A., Orrenius, S., Holmgren, A., and Mannervik, B., eds. (1983). "Functions of Glutathione: Biochemical, Physiological, Toxicological, and Clinical Aspects." Raven Press, New York.

Lash, L. H., and Anders, M. W. (1986a). Cytotoxicity of S-(1,2-dichlorovinyl)glutathione and S-(1,2-dichlorovinyl)-L-cysteine in isolated rat kidney cells. *J. Biol. Chem.* **261,** 13076–13081.

Lash, L. H., and Anders, M. W. (1986b). Bioactivation and cytotoxicity of nephrotoxic amino acid and glutathione S-conjugates. *Comments Toxicol.* **1,** 87–107.

Lash, L. H., and Anders, M. W. (1987). Mechanism of S-(1,2-dichlorovinyl)-L-cysteine- and S-(1,2-dichlorovinyl)-L-homocysteine-induced renal mitochondrial toxicity. *Mol. Pharmacol.* **32,** 549–556.

Lash, L. H., and Anders, M. W. (1989). Uptake of nephrotoxic S-conjugates by isolated rat renal proximal tubular cells. *J. Pharmacol. Exp. Ther.* **248,** 531–537.

Lash, L. H., and Jones, D. P. (1983). Transport of glutathione by renal basal-lateral membrane vesicles. *Biochem. Biophys. Res. Commun.* **112,** 55–60.

Lash, L. H., and Jones, D. P. (1984). Renal glutathione transport. Characteristics of the sodium-dependent system in the basal-lateral membrane. *J. Biol. Chem.* **259,** 14508–14514.

Lash, L. H., Elfarra, A. A., and Anders, M. W. (1986). Renal cysteine conjugate β-lyase: Bioactivation of nephrotoxic cysteine S-conjugates in mitochondrial outer membrane. *J. Biol. Chem.* **261,** 5930–5935.

Lash, L. H., Jones, D. P., and Anders, M. W. (1988). Glutathione homeostasis and glutathione S-conjugate toxicity in the kidney. *Rev. Biochem. Toxicol.* **9,** 29–67.

Lau, S. S., Jones, T. W., Sioco, R., Hill, B. A., Pinon, R. K., and Monks, T. J. (1990). Species differences in renal γ-glutamyl transpeptidase activity do not correlate with susceptibility to 2-bromo-(diglutathion-S-yl)-hydroquinone nephrotoxicity. *Toxicology* **64,** 291–311.

Lock, E. A. (1987). The nephrotoxicity of haloalkane and haloalkene glutathione conjugates. *In* "Selectivity and Molecular Mechanisms of Toxicity" (F. DeMatteis and E. A. Lock, eds.), pp. 59–83. Macmillan, New York.

Lock, E. A. (1988). Studies on the mechanism of nephrotoxicity and nephrocarcinogenicity of halogenated alkenes. *CRC Crit. Rev. Toxicol.* **19,** 23–42.

Lock, E. A., and Ishmael, J. (1979). The acute toxic effects of hexachloro-1:3-butadiene on the rat kidney. *Arch. Toxicol.* **43,** 47–57.

Lundqvist, G., Yucel-Lindberg, T., and Morgenstern, R. (1992). The oligomeric structure of rat liver microsomal glutathione transferase studied by chemical cross-linking. *Biochim. Biophys. Acta* **1159,** 103–108.

MacFarlane, M., Foster, J. R., Gibson, G. G., King, L. J., and Lock, E. A. (1989). Cysteine conjugate β-lyase of rat kidney cytosol: Characterization, immunocytochemical localization, and correlation with hexachlorobutadiene nephrotoxicity. *Toxicol. Appl. Pharmacol.* **98,** 185–197.

Maki, A., Berezesky, I. K., Fargnoli, J., Holbrook, N. J., and Trump, B. F. (1992). Role of $[CA^{2+}]$i in induction of c-fos, c-jun, and c-myc mRNA in rat PTE after oxidative stress. *FASEB J.* **6,** 919–924.

Mannervik, B., and Danielson, U. H. (1988). Glutathione transferases—Structure and catalytic activity. *CRC Crit. Rev. Biochem.* **23,** 281–334.

Mantle, J. C., Pickett, C. B., and Hayes, J. D., eds. (1990). "Glutathione S-Transferases and Carcinogenesis." Taylor & Francis, London.

Marchand, D. H., and Reed, D. J. (1989). Identification of the reactive glutathione conjugate S-(2-chloroethyl)glutathione in the bile of 1-bromo-2-chlorethane-treated rats by high-pressure liquid chromatography and precolumn derivatization with o-phthaladehyde. *Chem. Res. Toxicol.* **2,** 449–454.

Mayer, R., and Scheithauer, S. (1976). Thiocarbonsäure-halogenide. *In* "Methoden der Chemie (Houben-Wey) (E. Mueller, ed.) pp. 615–627.

McConkey, S. J., Hartzell, P., Nicotera, P., Wyllic, A. H., and Orrenius, S. (1988). Stimulation of endogenous endonuclease activity in hepatocytes exposed to oxidative stress. *Toxicol. Lett.* **42,** 123–130.

McDonald, J. K., and Barrett, A. J. (1986). "Mammalian Proteases: A Glossary and Bibliography," Vol. 2, pp. 59–71. Academic Press, London.

McIntyre, T., and Curthoys, N. P. (1982). Renal catabolism of glutathione. Characterization of a particulate rat renal dipeptidase that catalyzes the hydrolysis of cysteinylglycine. *J. Biol. Chem.* **257,** 11915–11921.

McKinney, L. L., Weakley, F. B., Eldridge, A. C., Campbell, R. E., Cowan, J. C., Picken, J. C., Jr., and Biester, H. E. (1957). S-(Dichlorovinyl)-L-cysteine: An agent causing fatal aplastic anemia in calves. *J. Am. Chem. Soc.* **79,** 3932–3933.

McKinney, L. L., Picken, J. C., Jr., Weakley, F. B., Eldridge, A. C., Campbell, R. E., Cowan, J. C., and Biester, H. E. (1959). Possible toxic factor of trichloroethylene-extracted soybean oil meal. *J. Am. Chem. Soc.* **81,** 909–915.

McLaren, J., Boulikas, T., and Vamvakas, S. (1993). Modification of nuclear proteins by poly(ADP-ribosyl)ation: Possible involvement in renal carcinogenesis. *Toxicology* In press.

Meadows, S. D., Gandolfi, A. J., Nagle, R. B., and Shively, J. W. (1988). Enhancement of DMN-induced kidney tumors by 1,2-dichlorovinylcysteine in Swiss-Webster mice. *Drug Chem. Toxicol.* **11,** 307–318.

Meister, A. (1988). Glutathione metabolism and its selective modification. *J. Biol. Chem.* **263,** 17205–17208.

Mertens, J. J. W. M., Weijnen, J. G. J., Doorn, W. J., Spenkelink, B., Temmink, J. H. M., and van Bladeren, P. J. (1988). Differential toxicity as a result of apical and basolateral treatment of LLC-PK_1 monolayers with S-(1,2,3,4,4-pentachlorobutadienyl)-glutathione and N-acetyl-S-(1,2,3,4,4-pentachlorobutadienyl)-L-cysteine. *Chem.-Biol. Interact.* **65,** 283–293.

Mertens, J. J. W. M., Keukens, E. A. J., Appel, M., Spenkelink, B., Temmink, J. H. M., and van Bladeren, P. J. (1990). Cytotoxicity of S-(1,2,3,4,4-pentachlorobutadienyl)-L-cysteine after apical and basolateral exposure of LLC-PK_1 monolayers, involvement of an amino acid transport system. *Chem.-Biol. Interact.* **75,** 119–130.

Moldéus, P., Jones, D. P., Ormstad, K., and Orrenius, S. (1978). Formation and metabolism of a glutathione-S-conjugate in isolated rat liver and kidney cells. *Biochem. Biophys. Res. Commun.* **83,** 195–200.

Monks, T. J., and Lau, S. S. (1987). Renal transport processes and glutathione conjugate-mediated nephrotoxicity. *Drug Metab. Dispos.* **15,** 437–441.

Monks, T. J., and Lau, S. S. (1989). Sulphur conjugate-mediated toxicity. *Rev. Biochem. Toxicol.* **10,** 41–90.

Monks, T. J., Anders, M. W., Dekant, W., Stevens, J. L., Lau, S. S., and van Bladeren, P. J. (1990). Glutathione conjugate mediated toxicities. *Toxicol. Appl. Pharmacol.* **106, 1–19.**

Morgenstern, R., and DePierre, J. W. (1985). Microsomal glutathione transferase. *Rev. Biochem. Toxicol.* **7,** 67–104.

Morgenstern, R., Lundqvist, G., Andersson, G., Balk, L., and DePierre, J. W. (1984). The distribution of microsomal glutathione transferase among different organelles, different organs, and different organisms. *Biochem. Pharmacol.* **33,** 3609–3614.

Mugrauer, G., and Ekblom, P. (1991). Contrasting expression patterns of three members of the myc family of protooncogenes in the developing and adult mouse kidney. *J. Cell Biol.* **112,** 13–25.

Mugrauer, G., Alt, R. W., and Ekblom, P. (1988). N-myc proto-oncogene expression during organogenesis in the developing mouse as revealed by in situ hybridization. *J. Cell Biol.* **107,** 1325–1335.

Mulford, D. J., Bakke, J. E., and Feil, V. J. (1991). Glutathione-S-transferase-mediated methylthio displacement and turnover in the metabolism of pentachlorothioanisole by rats. *Xenobiotica* **21,** 597–603.

Müller, D. A., Urban, G., and Dekant, W. (1991). Sulfur containing proreactive intermediates: Hydrolysis and mutagenicity of halovinyl 2-nitrophenyl disulfides. *Chem.-Biol. Interact.* **77,** 159–172.

Nagelkerke, J. F., and Boogaard, P. J. (1991). Nephrotoxicity of halogenated alkenyl cysteine-S-conjugates. *Life Sci.* **49,** 1769–1776.

Nash, J. A., King, L. J., Lock, E. A., and Green, T. (1984). The metabolism and disposition of hexachloro-1 : 3-butadiene in the rat and its relevance to nephrotoxicity. *Toxicol. Appl. Pharmacol.* **73,** 124–137.

National Cancer Institute (1986). Carcinogenesis bioassay of tetrachloroethylene. *Natl. Toxicol. Program Techl. Rep. Ser.* **232.**

Nelson, R., Van Dyke, R., Powis, G., Lash, L. H., and Anders, M. W. (1988). Characterization of human liver and kidney cysteine conjugate β-lyase activity. *FASEB J.* **2,** A1355.

Nishino, H., and Ito, A. (1990a). Purification and properties of glutathione S-transferase from outer mitochondrial membrane of rat liver. *Biochem. Int.* **20,** 1059–1066.

Nishino, H., and Ito, A. (1990b). Subcellular distribution of N-ethylmaleimide-stimulatable glutathione S-transferase activity in rat liver. Evidence of localization of glutathione S-transferase in peroxisomal membrane. *Biochem. Int.* **21,** 331–340.

Odum, J., and Green, T. (1984). The metabolism and nephrotoxicity of tetrafluoroethylene in the rat. *Toxicol. Appl. Pharmacol.* **76,** 306–318.

Oesch, F., and Wolf, C. R. (1989). Properties of the microsomal and cytosolic glutathione transferases involved in hexachloro-1 : 3-butadiene conjugation. *Biochem. Pharmacol.* **38,** 353–359.

Okajima, K., Inoue, M., Itoh, K., Horiuchi, S., and Morino, Y. (1983). Interorgan cooperation in enzymic processing. *In* "Glutathione: Storage, Transport and Turnover in Mammals" (Y. Sakamoto, T. Higashi, and N. Tateishi, eds.), pp. 129–144. *J. Sci. Soc. Press, Tokyo.*

Omichinski, J. G., Brunborg, G., Soderlund, E. J., Dahl, J. E., Bausano, J. A., Holme, J. A., Nelson, S. D., and Dybing, E. (1987). Renal necrosis and DNA damage caused by selectively deuterated and methylated analogs of 1,2-dibromo-3-chloropropane in the rat. Toxicol. Appl. Pharmacol. **91,** 358–370.

Omichinski, J. G., Brunborg, G., Holme, J. A., Soderlund, E. J., Nelson, S. D., and Dybing, E. (1988). The role of oxidative and conjugative pathways in the activation of 1,2-dibromo-3-chloropropane to DNA-damaging products in rat testicular cells. *Mol. Pharmacol.* **34,** 74–79.

Orrenius, S., McConkey, D. J., Bellomo, G., and Nicotera, P. (1989). Role of Ca^{2+} in toxic cell killing. *Trends Pharmacol. Sci.* **10,** 281–285.

Pearson, P. G., Soderlund, E. J., Dybing, E., and Nelson, S. D. (1990). Metabolic activation of 1,2-dibromo-3-chloropropane: Evidence for the formation of reactive episulfonium ion intermediates. *Biochemistry* **29,** 4971–4977.

Pickett, C. B., and Lu, A. Y. H. (1989). Glutathione S-transferases: Gene structure, regulation, and biological function. *Annu. Rev. Biochem.* **58,** 743–764.

Potter, C. L., Gandolfi, A. J., Nagle, R., and Clayton, J. W. (1981). Effects of inhaled chlorotrifluoroethylene and hexafluoropropene on the rat kidney. *Toxicol. Appl. Pharmacol.* **59,** 431–440.

Raasch, M. S. (1970). Bis(trifluoromethyl)thioketene. I. Synthesis and cycloaddition reactions. *J. Org. Chem.* **35,** 3470–3483.

Raasch, M. S. (1972). Bis(trifluoromethyl)thioketene. II. Acyclic derivatives. *J. Org. Chem.* **37,** 1347–1355.

Rankin, B. B., and Curthoys, N. P. (1982). Evidence for the renal paratubular transport of glutathione. *FEBS Lett.* **147,** 193–196.

Rankin, B. B., McIntyre, T. M., and Curthoys, N. P. (1980). Brush border membrane hydrolysis of S-benzyl-L-cysteine-p-nitroanilide, an activity of aminopeptidase M. *Biochem. Biophys. Res. Commun.* **96,** 991–996.

Rankin, B. B., McIntyre, T. M., and Curthoys, N. P. (1983). Role of the kidney in the interorgan metabolism of glutathione. *In* "Functions of Glutathione: Biochemical, Physiological, Toxicological, and Clinical Aspects" (A. Larsson, S. Orrenius, A. Holmgren, and B. Mannervik, eds.), pp. 31–38. Raven Press, New York.

Rannug, U., and Beije, B. (1979). The mutagenic effect of 1,2-dichloroethane on *Salmonella typhimurium*. II. Activation by the isolated perfused rat liver. *Chem.-Biol. Interact.* **24,** 265–285.

Rannug, U., Sundvall, A., and Ramel, C. (1978). The mutagenic effect of 1,2-dichloroethane on *Salmonella typhimurium*. I. Activation through conjugation with glutathione *in vitro*. *Chem.-Biol. Interact.* **20,** 1–16.

Rappoport, Z. (1985). The rich mechanistic world of nucleophilic vinylic (S_NV) substitution. *Recl. Trav. Chim. Pays-Bas* **104,** 309–349.

Reed, D. J., and Ellis, W. W. (1982). Influence of γ-glutamyl transpeptidase inactivation on the status of extracellular glutathione and glutathione conjugates. *Adv. Exp. Med. Biol.* **136A,** 75–86.

Reed, D. J., Ellis, W. W., and Meck, R. A. (1980). The inhibition of γ-glutamyl transpeptidase and glutathione metabolism of isolated rat kidney cells by L-(αS,5S)-α-amino-3-chloro-4,5-dihydro-5-isoxazoleacetic acid (AT-125; NSC-163501). *Biochem. Biophys. Res. Commun.* **94,** 1273–1277.

Reichert, D., Liebaldt, G., and Henschler, D. (1976). Neurotoxic effects of dichloroacetylene. *Arch. Toxicol.* **37,** 23–38.

Reichert, D., Spengler, U., Romen, W., and Henschler, D. (1984). Carcinogenicity of dichloroacetylene: An inhalation study. *Carcinogenesis* (*London*) **5,** 1411–1420.

Renner, G. (1983). Formation of pentachlorothioanisole from N-acetyl-S-(pentachlorophenyl) cysteine in blood and liver of rats *in vivo*. *Life Sci.* **33,** 1427–1431.

Richter, C., and Kass, G. E. N. (1991). Oxidative stress in mitochondria: Its relationship to cellular Ca^{2+} homeostasis, cell death, proliferation, and differentiation. *Chem.-Biol. Interact.* **777,** 1–23.

Rush, G. F., Smith, J. H., Newton, J. F., and Hook, J. B. (1984). Chemically induced nephrotoxicity: Role of metabolic activation. *CRC Crit. Rev. Toxicol.* **13,** 99–160.

Ryle, C. M., and Mantle, T. J. (1984). Studies on the glutathione S-transferase activity associated with rat liver mitochondria. *Biochem. J.* **222,** 553–556.

Sakamoto, Y., Higashi, T., and Tateishi, N., eds. (1983). "Glutathione: Storage, Transport and Turnover in Mammals." J. Sci. Soc. Press, Tokyo.

Schaeffer, V. H., and Stevens, J. L. (1987a). The transport of S-cysteine conjugates in LLC-PK_1 cells and its role in toxicity. *Mol. Pharmacol.* **31,** 506–512.

Schaeffer, V. H., and Stevens, J. L. (1987b). Mechanism of transport for toxic cysteine conjugates in rat kidney cortex membrane vesicles. *Mol. Pharmacol.* **32,** 293–298.

Schnellmann, R. G., Lock, E. A., and Mandel, L. J. (1987). A mechanism of S-(1,2,3,4,4-pentachloro-1,3-butadienyl)-L-cysteine toxicity to rabbit renal proximal tubules. *Toxicol. Appl. Pharmacol.* **90,** 513–521.

Schnellmann, R. G., Cross, T. J., and Lock, E. A. (1989). Pentachlorobutadienyl-L-cysteine uncouples oxidative phosphorylation by dissipating the proton gradient. *Toxicol. Appl. Pharmacol.* **100,** 498–505.

Schrenk, D., and Dekant, W. (1989). Covalent binding of hexachlorobutadiene metabolites in mice: Binding of reactive intermediates to renal mitochondrial DNA. *Carcinogenesis (London)* **10,** 1139–1141.

Schrenk, D., Dekant, W., and Henschler, D. (1988). Metabolism and excretion of S-conjugates derived from hexachlorobutadiene in the isolated perfused rat kidney. *Mol. Pharmacol.* **34,** 407–412.

Schultze, M. O., Klubes, P., Perman, V., Mizuno, N. S., Bates, F. W., and Sautter, J. H. (1959). Blood dyscrasia in calves induced by S-(dichlorovinyl)-L-cysteine. *Blood* **14,** 1015–1025.

Shore, S. K., Hard, G. C., and Magee, P. N. (1988). Deregulation and overexpression of c-fos proto-oncogene in rat renal cell-lines and primary tumors induced by dimethylnitrosamine. *Oncogene Res.* **3,** 567–570.

Sies, H., and Ketterer, B., eds. (1988). "Glutathione Conjugation: Mechanisms and Biological Significance." Academic Press, London.

Sies, H., and Wendel, A., eds. (1978). "Functions of Glutathione in Liver and Kidney." Springer-Verlag, Berlin.

Singh, N. (1990a). A comparative study on the effect of tumor promoters on poly-ADP-ribosylation in A431 cells. *Int. J. Cancer* **46,** 648–651.

Singh, N. (1990b). Effect of tumour promoters on poly-ADP-ribosylation in human epidermoid carcinoma HeP2 cells. *Int. J. Exp. Pathol.* **71,** 809–902.

Smith, A. G., and Francis, J. E. (1983). Evidence for the active renal secretion of S-pentachlorophenyl-N-acetyl-L-cysteine by female rats. *Biochem. Pharmacol.* **32,** 3797–3801.

Stevens, J. L. (1985). Isolation and characterization of a rat liver enzyme with both cysteine conjugate β-lyase and kynureninase activity. *J. Biol. Chem.* **260,** 7945–7950.

Stevens, J. L., and Jakoby, W. B. (1983). Cysteine conjugate β-lyase. *Mol. Pharmacol.* **23,** 761–765.

Stevens, J. L., and Jones, D. P. (1989). The mercapturic acid pathway: Biosynthesis, intermediary metabolism, and physiological disposition. *In* "Glutathione: Chemical, Biochemical, and Medical Aspects" (D. Dolphin, R. Poulson, and O. Avramovic, eds.), Part B, pp. 46–84. Wiley, New York.

Stevens, J. L., and Wallin, A. (1989). Is the toxicity of cysteine conjugates formed during mercapturic acid biosynthesis relevant to the toxicity of covalently bound drug residues? *Drug Metab. Rev.* **22,** 617–635.

Stevens, J. L., Robbins, J. D., and Byrd, R. A. (1986a). A purified cysteine conjugate β-lyase from rat kidney cytosol. *J. Biol. Chem.* **261,** 15529–15537.

Stevens, J. L., Hayden, P., and Taylor, G. (1986b). The role of glutathione conjugate metabolism and cysteine conjugate β-lyase in the mechanism of S-cysteine conjugate toxicity in LLC-PK1 cells. *J. Biol. Chem.* **261,** 3325–3332.

Stevens, J. L., Ayoubi, N., and Robbins, J. D. (1988). The role of mitochondrial matrix enzymes in the metabolism and toxicity of cysteine conjugates. *J. Biol. Chem.* **263,** 3395–3401.

Stevens, J. L., Hatzinger, P. B., and Hayden, P. J. (1989). Quantitation of multiple pathways for the metabolism of nephrotoxic cysteine conjugates using selective inhibitors of L-α-hydroxy acid oxidase (L-amino acid oxidase) and cysteine conjugate β-lyase. *Drug Metab. Dispos.* **17,** 297–303.

Stole, E., Seddon, A. P., Wellner, D., and Meister, A. (1990). Identification of a highly reactive threonine residue at the active site of γ-glutamyltranspeptidase. *Proc. Natl. Acad. Sci. U.S.A.* **87,** 1706–1709.

Suzuki, S., Tomisawa, H., Ichihara, S., Fukazawa, H., and Tateishi, M. (1982). A C-S bond cleavage enzyme of cysteine conjugates in intestinal microorganisms. *Biochem. Pharmacol.* **31,** 2137–2140.

Taniguchi, N., Higashi, T., Sakamoto, Y., and Meister, A., eds. (1989). "Glutathione Centennial: Molecular Perspectives and Clinical Implications." Academic Press, San Diego, CA.

Tate, S. S. (1980). Enzymes of mercapturic acid formation. *In* "Enzymatic Basis of Detoxication" (W. B. Jakoby, ed.), Vol. 2, pp. 95–120. Academic Press, New York.

Tate, S. S. (1985). Microvillus membrane peptidases that catalyze hydrolysis of cysteinylglycine and its derivatives. *In* "Methods in Enzymology" (A. Meister, ed.), Vol. 113, pp. 471–484. Academic Press, Orlando, FL.

Tate, S. S. (1989). Glutathione-degrading enzymes of brush border: Molecular and functional aspects. *In* "Glutathione Centenial: Molecular Perspectives and Clinical Implications" (N. Taniguchi, T. Highashi, Y. Sakamoto, and A. Meister, eds.), pp. 189–208. Academic Press, San Diego, CA.

Tate, S. S., and Meister, A. (1974). Interaction of γ-glutamyl transpeptidase with amino acids, dipeptides, and derivatives and analogs of glutathione. *J. Biol. Chem.* **249,** 7593–7602.

Tate, S. S., and Meister, A. (1977). Affinity labeling of γ-glutamyl transpeptidase and location of the γ-glutamyl binding site on the light subunit. *Proc. Natl. Acad. Sci. U.S.A.* **74,** 931–935.

Tate, S. S., and Meister, A. (1978). Serine-borate complex as a transition-state inhibitor of γ-glutamyl transpeptidase. *Proc. Natl. Acad. Sci. U.S.A.* **75,** 4806–4809.

Tate, S. S., and Meister, A. (1985). γ-Glutamyl transpeptidase from kidney. *In* "Methods in Enzymology" (A. Meister, ed.), Vol. 113, pp. 400–419. Academic Press, Orlando, FL.

Tateishi, M., Suzuki, S., and Shimizu, H. (1978). Cysteine conjugate β-lyase in rat liver. *J. Biol. Chem.* **253,** 8854–8859.

Terracini, B., and Parker, V. H. (1965). A pathological study on the toxicity of S-dichlorovinyl-L-cysteine. *Food Cosmet. Toxicol.* **3,** 67–74.

Tomisawa, H., Suzuki, S., Ichihara, S., Fukazawa, H., and Tateishi, M. (1984). Purification and characterization of C-S lyase from Fusobacterium varium. A C-S cleavage enzyme of cysteine conjugates and some S-containing amino acids. *J. Biol. Chem.* **259,** 2588–2593.

Tomisawa, H., Ichihara, S., Fukazawa, H., Ichimoto, N., Tateishi, M., and Yamamoto, I. (1986). Purification and characterization of human hepatic cysteine-conjugate β-lyase. *Biochem. J.* **235,** 569–575.

Tsao, B., and Curthoys, N. P. (1980). The absolute asymetry of orientation of γ-glutamyltranspeptidase and aminopeptidase on the external surface of the rat renal brush border membrane. *J. Biol. Chem.* **255,** 7708–7711.

Tsuchida, S., and Sato, K. (1992). Glutathione transferases and cancer. *CRC Crit. Rev. Biochem. Mol. Biol.* **27,** 337–384.

Tsujiuchi, T., Tsutsumi, M., Denda, A., Kondoh, S., Nakae, D., Maruyama, H., and Konishi, Y. (1990). Possible involvement of poly ADP-ribosylation in phenobarbital promotion of rat hepatocarcinogenesis. *Carcinogenesis (London)* **11,** 1783–1787.

Ullrich, K. J., and Rumrich, G. (1988). Contraluminal transport systems in the proximal renal tubule involved in secretion of organic anions. *Am. J. Physiol.* **254,** 453–462.

Ullrich, K. J., Rumrich, G., and Kloess, S. (1988). Contraluminal para-aminohippurate (PAH) transport in the proximal tubule of the rat kidney. *Pfluegers Arch.* **413,** 134–146.

Ullrich, K. J., Rumrich, G., and Klöss, S. (1989a). Contraluminal organic anion and cation transport in the proximal renal tubule. V. Interaction with sulfamoyl- and phenoxy diuretics, and with β-lactam antibiotics. *Kidney Int.* **36,** 78–88.

Ullrich, K. J., Rumrich, G., Wieland, T., and Dekant, W. (1989b). Contraluminal para-aminohippurate (PAH) transport in the proximal tubule of the rat kidney. *Eur. J. Physiol.* **415,** 342–350.

Vamvakas, S., and Anders, M. W. (1990). Formation of reactive intermediates by Phase II enzymes: Glutathione-dependent bioactivation reactions. *Adv. Exp. Med. Biol.* **283,** 13–24.

Vamvakas, S., and Köster, U. (1993). The nephrotoxin dichlorovinylcysteine induces expression of the protooncogenes c-fos and c-myc in LLC-PK_1 cells—a comparative investigation with growth factors and 12-O-tetradecanoylphorbolacetate. *Cell Biol. Toxicol.* **9,** 1–13.

Vamvakas, S., Dekant, W., Berthold, K., Schmidt, S., Wild, D., and Henschler, D. (1987). Enzymatic transformation of mercapturic acids derived from halogenated alkenes to reactive and mutagenic intermediates. *Biochem. Pharmacol.* **36,** 2741–2748.

Vamvakas, S. Müller, D. A., Dekant, W., and Henschler, D. (1988a). DNA-Binding of sulfur-containing metabolites from ^{35}S-(pentachlorobutadienyl)-L-cysteine in bacteria and isolated renal tubular cells. *Drug Metab. Drug Interact.* **6,** 349–358.

Vamvakas, S., Berthold, K., Dekant, W., and Henschler, D. (1988b). Bacterial cysteine conjugate β-lyase and the metabolism of cysteine S-conjugates: Structural requirements for the cleavage of S-conjugates and the formation of reactive intermediates. *Chem.-Biol. Interact.* **65,** 59–71.

Vamvakas, S., Elfarra, A. A., Dekant, W., Henschler, D., and Anders, M. W. (1988c). Mutagenicity of amino acid and glutathione S-conjugates in the Ames test. *Mutat. Res.* **206,** 83–90.

Vamvakas, S., Kordowich, F. J., Dekant, W., Neudecker, T., and Henschler, D. (1988d). Mutagenicity of hexachloro-1,3-butadiene and its S-conjugates in the Ames test - Role of activation by the mercapturic acid pathway in its nephrocarcinogenicity. *Carcinogenesis (London)* **9,** 907–910.

Vamvakas, S., Kremling, E., and Dekant, W. (1989a). Metabolic activation of the nephrotoxic haloalkene 1,1,2-trichloro-3,3,3-trifluoro-1-propene by glutathione conjugation. *Biochem. Pharmacol.* **38,** 2297–2304.

Vamvakas, S., Köchling, A., Berthold, K., and Dekant, W. (1989b). Cytotoxicity of cysteine S-conjugates: structure-activity relationships. *Chem.-Biol. Interact.* **71,** 79–90.

Vamvakas, S., Dekant, W., and Henschler, D. (1989c). Genotoxicity of haloalkene and haloalkane glutathione S-conjugates in porcine kidney cells. *Toxicol. In Vitro* **3,** 151–156.

Vamvakas, S., Herkenhoff, M., Dekant, W., and Henschler, D. (1989d). Mutagenicity of tetrachloroethylene in the Ames-test—Metabolic activation by conjugation with glutathione. *J. Biochem. Toxicol.* **4,** 21–27.

Vamvakas, S., Dekant, W., and Henschler, D. (1989e). Assessment of unscheduled DNA synthesis in a cultured line of renal epithelial cells exposed to cysteine S-conjugates of haloalkenes and haloalkanes. *Mutat. Res.* **222,** 329–335.

Vamvakas, S., Sharma, V. K., Shen, S.-S., and Anders, M. W. (1990). Perturbations of intracellular calcium distribution in kidney cells by nephrotoxic haloalkenyl cysteine S-conjugates. *Mol. Pharmacol.* **38,** 455–461.

Vamvakas, S., Bittner, D., Dekant, W., and Anders, M. W. (1992). Events that precede and that follow S-(1,2-dichlorovinyl)-L-cysteine-induced release of mitochondrial Ca^{2+} and their association with cytotoxicity to renal cells. *Biochem. Pharmacol.* **44,** 1131–1138.

Vamvakas, S., Bittner, D., and Köster, U. (1993). Enhanced expression of the protooncogenes c-myc and c-fos in normal and malignant growth. *Toxicol. Lett.* **67,** 161–172.

Vincenzini, M. T., Iantomasi, T., and Favilli, F. (1989). Glutathione transport across intestinal brushborder membranes: Effects of ions, pH and inhibitors. *Biochim. Biophys. Acta* **987,** 29–37.

Vore, M. (1993). Canalicular transport: Discovery of ATP-dependent mechanisms. *Toxicol. Appl. Pharmacol.* **118,** 2–7.

Wahlländer, A., and Sies, H. (1979). Glutathione S-conjugate formation from 1-chloro-2,4-dinitrobenzene and biliary S-conjugate excretion in the perfused rat liver. *Eur. J. Biochem.* **96,** 441–446.

Wahlländer, A., Soboll, S., and Sies, H. (1979). Hepatic mitochondrial and cytosolic glutathione content and the subcellular distribution of GSH-S-transferases. *FEBS Lett.* **97,** 138–140.

Wallin, A., Jones, T. W., Vercesi, A. E., Cotgreave, I., Ormstad, K., and Orrenius, S. (1987). Toxicity of S-pentachlorobutadienyl-L-cysteine studies with isolated rat renal cortical mitochondria. *Arch. Biochem. Biophys.* **258,** 365–372.

Wark, H., Earl, J., Chau, D. D., Overton, J., and Cheung, H. T. A. (1990). A urinary cysteine-halothane metabolite: Validation and measurement in children. *Br. J. Anaesth.* **64,** 469–473.

Wolf, C. R., Berry, P. N., Nash, J. A., Green, T., and Lock, E. A. (1984). Role of microsomal and cytosolic glutathione S-transferases in the conjugation of hexachloro-1 : 3-butadiene and its possible relevance to toxicity. *J. Pharmacol. Exp. Ther.* **228,** 202–208.

Wolfgang, G. H. I., Gandolfi, A. J., Stevens, J. L., and Brendel, K. (1989). N-Acetyl-S-(1,2-dichlorovinyl)-L-cysteine produces a similar toxicity to S-(1,2-dichlorovinyl)-L-cysteine in rabbit renal slices: Differential transport and metabolism. *Toxicol. Appl. Pharmacol.* **101,** 205–219.

Wolkoff, A. W., Weisiger, R. A., and Jakoby, W. B. (1979). The multiple roles of the glutathione transferases (ligandins). *Prog. Liver Dis.* **6,** 213–224.

Yao, M., Shuin, T., Misaki, H., and Kubota, Y. (1988). Enhanced expression of c-myc and epidermal growth factor (C-erb-1) genes in primary human renal cancer. *Cancer Res.* **48,** 6753–6757.

Zhang, G., and Stevens, J. L. (1989). Transport and activation of S-(1,2-dichlorovinyl)-L-cysteine and N-acetyl-S-(1,2-dichlorovinyl)-L-cysteine in rat kidney proximal tubules. *Toxicol. Appl. Pharmacol.* **100,** 51–61.

Reversibility in Glutathione-Conjugate Formation

Thomas A. Baillie and Kelem Kassahun

Department of Medicinal Chemistry
School of Pharmacy
University of Washington
Seattle, Washington 98195

I. Introduction

Traditionally, the conjugation of electrophilic xenobiotic metabolites with glutathione (γ-L-glutamyl-L-cysteinylglycine; GSH; Fig. 1) has been viewed as a detoxification pathway leading to the formation of polar, water-soluble products, which are excreted mainly into bile (Chasseuad, 1979). Although this appears to hold true in the majority of cases, an increasing number of examples have come to light in recent years where GSH conjugates (or the cysteine adducts to which they are degraded) exhibit comparable or even greater toxic potential than that of the original xenobiotic (Anders *et al.*, 1988, 1992; van Bladeren, 1988; Monks and Lau, 1989; Koob and Dekant, 1991). These cases have been the focus of considerable research interest, as a consequence of which much has been learned of the underlying biochemical mechanisms (Monks *et al.*, 1990). This chapter will deal with one mechanism of GSH conjugate-mediated toxicity, namely reversibility in adduct formation, whereby the xenobiotic (or xenobiotic metabolite) is regenerated upon spontaneous reversal of the initial conjugation reaction. To the extent that S-linked conjugates of GSH can be formed in one organ and transported to some distant site before collapsing to release the original electrophilic species, GSH may be considered to be acting in this situation as a moelcular vehicle for the

Advances in Pharmacology, Volume 27

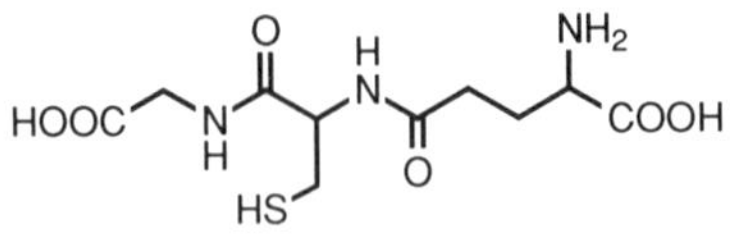

Glutathione (GSH)

Fig. 1 Structure of glutathione (GSH).

transport of chemically reactive metabolites *in vivo* (Baillie and Slatter, 1991). From the limited data available to date, it appears that reversibility in GSH conjugation is important only for certain types of functional groups, as outlined below. Chemical aspects of these reversible conjugation reactions have been discussed elsewhere (Baillie and Slatter, 1991) and therefore will not be presented in any depth in this chapter.

II. Classes of Compounds That Undergo Reversible Conjugation with Glutathione

A. Isothiocyanates

Compounds bearing the reactive, electrophilic isothiocyanate moiety (R—N=C=S) are of biological interest in that several members of this family occur as natural products (notably in cruciferous vegetables) where they are present as heteroglycoside conjugates known as glucosinolates. When such plants are degraded, e.g., during the preparation of salads for human consumption, the free isothiocyanates are released by the action of the enzyme myrosinase (thioglucoside glucohydrolase, EC 3.2.3.1), which is present in the host plant. As a consequence, humans can be exposed to isothiocyanates as dietary constituents, and over 50 glucosinolate precursors have been identified from a variety of plant species (Drobnica *et al.*, 1977). The most extensively studied isothiocyanates to date have been benzylisothiocyanate (derived from phenylalanine), allylisothiocyanate (derived from methionine), and phenethylisothiocyanate (derived from 2-amino-4-phenylbutyric acid), each of which exhibits interesting biological activities (Fig. 2). Thus, benzylisothiocyanate is an effective antibiotic and is marketed in Europe under the trade name Tromacaps for the treatment of respiratory and urinary tract infections (Mennicke *et al.*, 1988). Allylisothiocyanate has been found to be a bladder carcinogen in male rats (Dunnick *et al.*, 1982), whereas both phenethyl- and benzylisothiocyanate appear to have cancer chemoprotective properties (Zhang

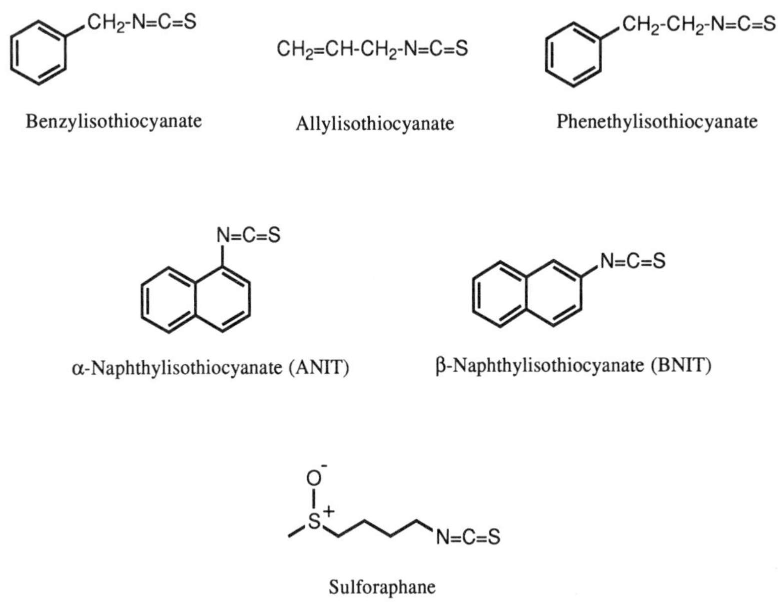

Fig. 2 Structures of isothiocyanates referred to in the text.

et al., 1992, and references therein), the mechanism of which currently is the subject of considerable research interest.

In view of the diverse biological properties of isothiocyanates, a number of studies have been carried out in both animals and human volunteers to determine the biological fate of this class of natural products. These investigations revealed that the isothiocyanate moiety of the parent compound reacts readily with GSH to afford the thiocarbamate thioester conjugate (a dithiocarbamidic ester) and that the reaction takes place both spontaneously and (at an accelerated rate) in the presence of glutathione *S*-transferase enzymes (Brüsewitz *et al.*, 1977; Mennicke *et al.*, 1983, 1988). The resulting isothiocyanate–GSH conjugates are excreted into bile, and the corresponding *N*-acetylcysteine adducts normally are found as prominent metabolites in urine where they can account for 50% or more of the administered dose (Brüsewitz *et al.*, 1977; Mennicke *et al.*, 1983, 1987, 1988; Eklind *et al.*, 1990). Interestingly, the cysteine conjugate formed as an intermediate in the degradation of the phenethylisothiocyanate–GSH adduct was shown recently to undergo oxidative deamination in the mouse to afford the corresponding mercaptopyruvic acid; the latter species proved to be the major metabolite in urine, where it was present largely in a cyclic tautomeric form (Eklind *et al.*, 1990). From these studies, therefore, it may be concluded that conjugation with GSH represents a quantitatively important pathway for the metabolism of isothiocyanates in mammalian species.

In an early report on the metabolism of benzylisothiocyanate, Brüsewitz *et al.* (1977) noted that S-linked conjugates of this compound underwent pH-dependent elimination to regenerate the free isothiocyanate, the rate of the reaction being accelerated markedly at pH values above 6. Indeed, this base-lability of thiocarbamate thioesters has led to the design of specific analytical protocols for the assay of isothiocyanate-*N*-acetylcysteine conjugates in urine based on their conversion to stable thiourea derivatives (Mennicke *et al.*, 1987). However, the potential biological significance of this phenomenon, i.e., reversibility in the formation of S-linked conjugates from isothiocyanates, was first recognized by van Bladeren and coworkers (Bruggemen *et al.*, 1986), who demonstrated that free benzyl- and allylisothiocyanate were released from their respective cysteine and GSH conjugates in aqueous phosphate buffer at pH 7.4 and 37°C. These workers also showed that when a solution of the cysteine conjugate of benzylisothiocyanate was treated with GSH, the corresponding GSH adduct was formed in a time-dependent fashion such that the two conjugates reached equilibrium after 90 min. A similar exchange phenomenon was observed when the GSH conjugate of benzylisothiocyanate was allowed to react with cysteine under simulated physiological conditions. These findings indicated that S-linked conjugates of isothiocyanates may serve as latent forms of the parent isothiocyanate *in vivo* and that conjugation with GSH might not fulfill a conventional "detoxification" role for this class of compounds. In a series of seminal experiments designed to test this hypothesis, Bruggeman *et al.* (1986) compared the cytotoxic properties of benzyl- and allylisothiocyanate with those of the corresponding GSH and L-cysteine conjugates toward isolated rat hepatocytes *in vitro*. In all cases, the S-linked conjugates proved to be cytotoxic, although addition to the media of excess GSH or L-cysteine protected the hepatocytes against injury. The latter observation is important from a mechanistic standpoint since it is consistent with the free isothiocyanate (R—N=C=S) acting as the toxic species, the release of which is suppressed by added thiols (R′SH) according to the equilibrium:

$$\text{R—NH—CS—SR}' \rightleftharpoons \text{R—N=C=S} + \text{R}'\text{SH}$$

Based on the results of these studies, the authors concluded that "glutathione and cysteine can be regarded as transporting agents for the isothiocyanates through the body. Initial detoxification can be followed by release of the reactive compound at some other site" (Bruggeman *et al.*, 1986).

The concept of reversible GSH conjugation as a key element in the adverse effects of toxic isothiocyanates was developed further by Roth, Reed, and co-workers, who examined α-napthylisothiocyanate (ANIT), a model compound that causes intrahepatic cholestasis as a result of injury to bile duct epithelium and parenchymal cells. When freshly isolated rat hepatocytes were incubated with ANIT, a GSH conjugate was formed and was identified as the expected thiocarbamate thioester derivative (Carpenter-Deyo *et al.*, 1991). This conjugate was exported from hepatocytes into the medium where it underwent facile degradation to regenerate the parent isothiocyanate. As a result, ANIT depleted GSH from hepatocytes and caused cytotoxicity, as reflected by leakage from cells of lactate dehydrogenase. The GSH conjugate also was cytotoxic under these conditions, apparently by acting as a latent form of ANIT, which, in turn, facilitated the efflux of GSH from cells. Stability studies with ANIT–GSH revealed that the position of the above equilibrium between the conjugate and the free isothiocyanate lay much further to the right than with the GSH conjugates of benzyl- or allylisothiocyanate, such that when dissolved in aqueous solutions at neutral pH in the absence of free GSH, 95% of the ANIT–GSH adduct dissociated to free ANIT within 5 min. This high degree of instability of S-linked conjugates of aromatic isothiocyanates probably accounts for the failure of previous *in vivo* studies to detect the *N*-acetylcysteine derivatives of ANIT, β-naphthylisothiocyanate (BNIT), and phenylisothiocyanate in the urine of rats dosed with the parent isothiocyanates (Mennicke *et al.*, 1983). Interestingly, recent *in vivo* experiments with ANIT have shown that pretreatment of rats with agents that deplete hepatic GSH *protects* animals against the hepatotoxic effects of ANIT, suggesting that "GSH plays a causal or permissive role in the liver injury caused by ANIT" (Dahm and Roth, 1991; Dahm *et al.*, 1991). In light of these findings, it will be interesting to determine whether the metabolism of ANIT differs in the GSH-depleted animal, such that pathways other than GSH conjugation now predominate and the bile duct epithelium is no longer exposed to high levels of the toxic ANIT–GSH conjugate. In contrast to these results, however, modulation of intracellular GSH levels had no effect on the cytotoxicity of allylisothiocyanate toward rat hepatocytes *in vitro* (Bruggeman *et al.*, 1988). Clearly, further studies will be needed to define the mechanisms of isothiocyanate-dependent toxicities, although it seems likely that reversible GSH conjugation is playing an important role in these processes.

As noted earlier, certain isothiocyanates appear to have anticarcinogenic properties, which may stem from their effects on enzyme systems involved in the metabolism of carcinogens. For example, phenethylisothiocyanate has been shown to be relatively specific inhibitor of CYP2E1, the principal isozyme of cytochrome P450 that catalyzes the conversion of *N*-nitrosodimethylamine and a number of other low-molecular-weight carcinogens to highly reactive electrophilic intermediates that alkylate DNA (Ishizaki *et al.*, 1990; Guengerich *et al.*, 1991). It is possible that labile S-linked conjugates of phenethylisothiocyanate play a role as transport forms of the inhibitory isothiocyanate or that they might even inhibit the enzyme per se, although no information is yet available on this topic. An alternative mechanism that has been proposed to account for the anticarcinogenic effects of isothiocyanates is that they act as inducers of glutathione *S*-transferases and quinone reductase enzymes, which catalyze the detoxification of electrophilic alkylating agents. Sulforaphane, a naturally occurring isothiocyanate found in broccoli, may act by this mechanism (Zhang *et al.*, 1992), and it will be interesting to determine whether a labile GSH conjugate of this compound is formed and contributes to its enzyme-inducing properties. Recently, the cysteine conjugates of benzyl- and 3-phenylpropyl-isothiocyanate were found to induce glutathione *S*-transferases in several tissues of A/J mice (Zheng *et al.*, 1992). In the bladder, the conjugates proved to be more potent inducing agents than their respective free isothiocyanates and also to be less toxic. It was proposed, therefore, that these cysteine conjugates may serve as bladder-directed chemical delivery systems for chemopreventative isothiocyanates, and thus may have utility in protecting against nitrosamine-associated bladder cancers.

Although isothiocyanates may be expected to form adducts with a variety of nucleophilic residues on proteins, reaction with cysteinyl —SH groups represents the kinetically favored process (Drobnica *et al.*, 1977). It is not surprising, therefore, that isothiocyanates are known to bind to, and thereby inhibit, sulfhydryl-dependent enzymes, e.g., glyceraldehyde-3-phosphate dehydrogenase, hexokinase, glutamate dehydrogenase, alcohol dehydrogenase, glutamate-oxaloacetate transaminase, glutathione reductase, succinate dehydrogenase, and papain (reviewed by Drobnica *et al.*, 1977). In recent studies performed in our laboratory, we found that both benzylisothiocyanate and the cysteine conjugate of benzylisothiocyanate inhibited rat liver glutathione reductase *in vitro*. Thus, following incubation with the conjugate (20 μM) with freshly isolated liver cells at pH 7.4 and 37°C, 20% of the enzyme activity was lost over a period of 5 h, compared with 25% inhibition when the free isothiocyanate was incubated under similar conditions (K. Kassahun and T. A. Baillie, unpublished observations). These results indicated that the cysteine conjugate of benzylisothiocyanate served as a depot form of the free isothiocyanate, which, in turn, carbamoylated and thereby inhibited the enzyme.

Based on the reversibility of the reaction between isothiocyanates and GSH, it may be expected that thiocarbamate thioesters formed by attack of isothiocyanates at protein-SH groups may be labile and that the parent isothiocyanate moiety may either dissociate from the protein or undergo rearrangement to form a more stable adduct with proximal —OH or —NH_2 groups. Although little attention appears to have been focused on the nature and properties of isothiocyanate–protein adducts, a recent study on protein and peptide conjugates of CS_2 provided evidence for the formation of protein–N═C═S species as products of the reaction of this compound with the ε-NH_2 groups of lysine residues (DeCaprio *et al.*, 1992). The authors of this study proposed that subsequent nucleophilic addition to the protein-bound isothiocyanate by a neighboring —SH or —NH_2 functionality would be responsible for the formation of intramolecular protein cross-links. Thus, an understanding of the chemistry of the reversible reaction between isothiocyanates and small peptides, e.g., GSH, should aid in rationalizing the nature of protein-bound adducts of isothiocyanates and related structures.

B. Isocyanates

Organic isocyanates (R—N═C═O) are considerably more reactive toward nucleophiles than their corresponding isothiocyanates (Drobnica *et al.*, 1977) and readily form adducts with peptides, proteins, and nucleic acids (Lee, 1992, and references therein). As a result, isocyanates often are highly toxic, as highlighted by the catastrophe in Bhopal, India, in 1984 when 30–40 tons of methylisocyanate (MIC; Fig. 3) were released into the atmosphere, leading to the death of some 3500 inhabitants (Heylin, 1985). Aside from this industrial disaster, humans may be exposed routinely to isocyanates from a number of sources. Isocyanates are used widely in the manufacture of pesticides, polyurethanes, and paints and, therefore, pose a hazard from an occupational health standpoint. Simple alkylformamides, such as the commonly used solvent dimethylformamide (DMF) and its monomethyl congener NMF (an experimental antitumor agent), undergo metabolic oxidation to yield isocyanates, which cause liver injury in animals and humans (Cross *et al.*, 1990; Hyland *et al.*, 1992). Also, several 2-chloroethylnitrosoureas employed clinically as chemotherapeutic alkylating agents decompose *in vivo* to yield isocyanates, the formation of which is believed to be associated with certain serious adverse reactions to this class of drugs (Smith, 1989). Recently, it was reported that 2-naphthylisothiocyanate (BNIT) undergoes cytochrome P450-dependent metabolism *in vitro* to afford the corresponding isocyanate

CH_3-N=C=O

Methylisocyanate (MIC)

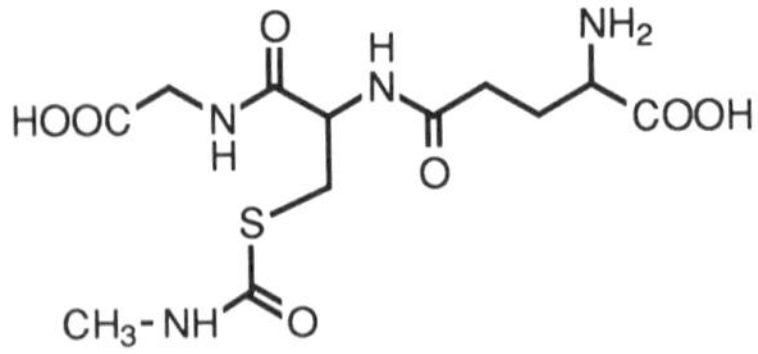

S-(*N*-Methylcarbamoyl)glutathione (SMG)

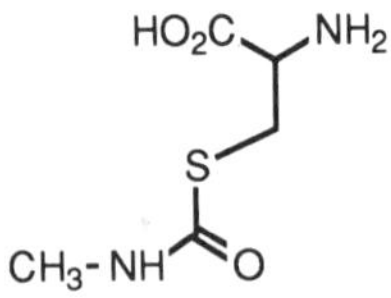

S-(*N*-Methylcarbamoyl)cysteine (SMC)

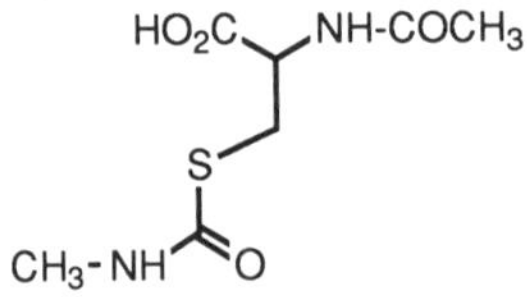

N-Acetyl-*S*-(*N*-methylcarbamoyl)cysteine (AMCC)

n-Propylisocyanate

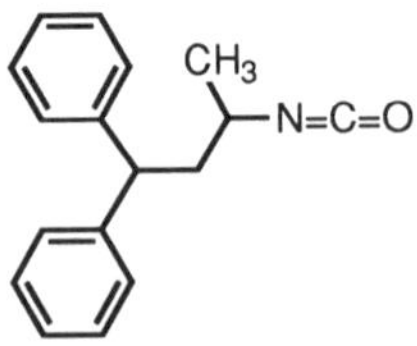

N-(1-Methyl-3,3-diphenylpropyl)isocyanate

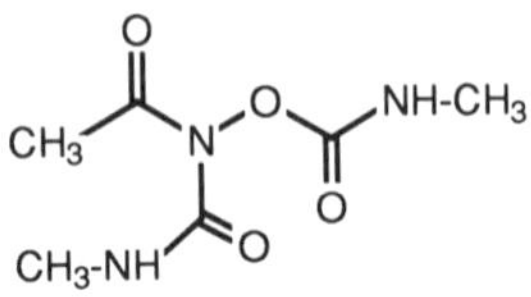

Caracemide

Fig. 3 Structures of isocyanates and S-linked isocyanate conjugates referred to in the text. The structure of caracemide, a metabolic precursor of MIC, is also shown.

(Lee, 1992), and it may be speculated, therefore, that many of the dietary isothiocyanates discussed in the preceding section may be transformed in part to more toxic isocyanates *in vivo*.

On the basis of the finding that isothiocyanates undergo reversible conjugation with GSH, we explored the possibility that isocyanates may behave similarly and form carbamate thioester conjugates with GSH that could serve as "carriers" of the reactive isocyanate moiety, according to the following process:

$$R\text{—}NH\text{—}CO\text{—}SR' \rightleftharpoons R\text{—}N\text{=}C\text{=}O + R'SH$$

Studies with MIC demonstrated that this, indeed, was the case, since administration of the isocyanate to rats by ip injection led to the excretion in bile of *S*-(*N*-methylcarbamoyl)glutathione (SMG) and to the elimination in urine of the corresponding *N*-acetylcysteine adduct (AMCC), which accounted for some 25% of the dose (Pearson *et al.*, 1990; Slatter *et al.*, 1991) (Fig. 3). These S-linked conjugates, together with the corresponding cysteine derivative *S*-(*N*-methylcarbamoyl)cysteine (SMC), proved to be unstable *in vitro* under simulated physiological conditions, where they reverted to their respective thiols and, by inference, to MIC. Incubation of SMG with free cysteine led to the formation of SMC (consistent with the liberation of MIC as an intermediate), whereas the complementary reaction of SMC with free GSH afforded SMG (Pearson *et al.*, 1990). Further *in vitro* studies with SMG and SMC showed that both conjugates were effective carbamoylating agents toward model peptides and proteins, in which cysteinyl-SH groups were the primary sites of modification (Pearson *et al.*, 1991). These results have important implications with respect to the toxicological properties of isocyanates, since reversible conjugation with GSH will serve to extend the effective biological life-time of reactive isocyanates *in vivo* and, thereby, influence their tissue distribution. For example, it has been estimated that free MIC has a half-life of approximately 2–5 min in neutral aqueous solution (Brown *et al.*, 1987), whereas the half-life of SMG is closer to 1 h at pH 7.4 and 37°C (Pearson *et al.*, 1990). (The latter value will depend on the concentration of free thiols in solution, which influence the position of the isocyanate/carbamate thioester equilibrium.) SMG has been shown to be cytotoxic toward isolated mouse hepatocytes (Han *et al.*, 1990), to inhibit the growth of murine TLX5 lymphoma cells in culture (Han *et al.*, 1990), and to be toxic to mouse embryos in culture (Guest *et al.*, 1992), probably as a result of its ability to release MIC at cell surfaces. It has been proposed, therefore, that SMG may be viewed as a "transport" form of MIC, which serves to mediate some of the toxic effects of MIC *in vivo* (Baillie and Slatter, 1991). Interestingly, SMG was first identified as a biliary metabolite of the experimental antitumor agent NMF (Threadgill *et al.*, 1987), which undergoes oxidation by CYP2E1 in liver tissue to MIC (Hyland *et al.*, 1992). This finding led to the hypothesis that both the hepatotoxic properties and the antitumor effects of NMF may be mediated by MIC, the isocyanate being delivered to extrahepatic tumor cells in the form of the labile carbamate thioester conjugate SMG (Han *et al.*, 1990). Moreover, it seems likely that a group of experimental antineoplastic drugs, which are carbamate thioester derivatives of cysteine, may be acting similarly to kill tumor cells through their ability to release toxic isocyanates *in vivo* (Németh *et al.*, 1978; Jayaram *et al.*, 1990) and that the aldehyde dehydrogenase-inhibiting effects of chlorpropamide, a sulfonylurea derivative, may be due to *n*-propylisocyanate, whose GSH conjugate has been found to be an even more potent inhibitor of the enzyme than chlorpropamide itself (Shirota *et al.*, 1990).

With the awareness that reversible conjugation with GSH represents a quantitatively important route of metabolism of isocyanates *in vivo* and that the corresponding *N*-acetylcysteine derivatives (which can be analyzed effectively by LC-MS techniques) are excreted in urine as end-products of this pathway (Slatter *et al.*, 1991), several other compounds have now been shown to serve as either chemical or metabolic precursors of isocyanates *in vivo*. Thus, the monomethyl metabolite of the bronchodilating drug bambuterol decomposes at neutral pH to yield MIC (Rashed *et al.*, 1989), and the novel anticancer agent caracemide appears to act as an efficient source of MIC when administered to rats since both the *O*-methyl- and the *N*-methylcarbamoyl side chains of the drug were found to release MIC *in vivo* (Slatter *et al.*, 1993). The secondary formamide *N*-(1-methyl-3,3-diphenylpropyl)formamide undergoes metabolic oxidation in rats to yield the corresponding isocyanate, which, in turn, is metabolized via the GSH pathway (Mutlib *et al.*, 1990). Interestingly, it was shown in this case that treatment of both the GSH and the *N*-acetylcysteine adducts with mild alkali led to release of the intermediate isocyanate, which was sufficiently stable to be characterized. It has been known for many years that carmustine (BCNU), one of the most frequently used chemotherapeutic agents in the treatment of malignant brain tumors, decomposes in aqueous media to form an alkylating intermediate (probably the chloroethyl carbonium ion) and the carbamoylating species 2-chloroethylisocyanate (Kann, 1981). It was predicted, therefore, that BCNU would give rise to the GSH conjugate of 2-chloroethylisocyanate *in vivo*, and a targeted search for this metabolite (SCG, Fig 4) in the bile of BCNU-dosed rats led to its identification by LC-MS/MS techniques (Davis *et al.*, 1993). Once again, SCG proved to be unstable in aqueous media at pH 7.4, almost certainly as a result of its decomposition to regenerate 2-chloroethylisocyanate since free GSH was identified as a by-product of the reaction. Support for this view was obtained from *in vitro* experiments in which synthetic SCG was shown to inhibit rat liver glutathione reductase, a sulfhydryl-dependent enzyme that is susceptible to inhibition by isocyanates. The corresponding mercapturic acid derivative

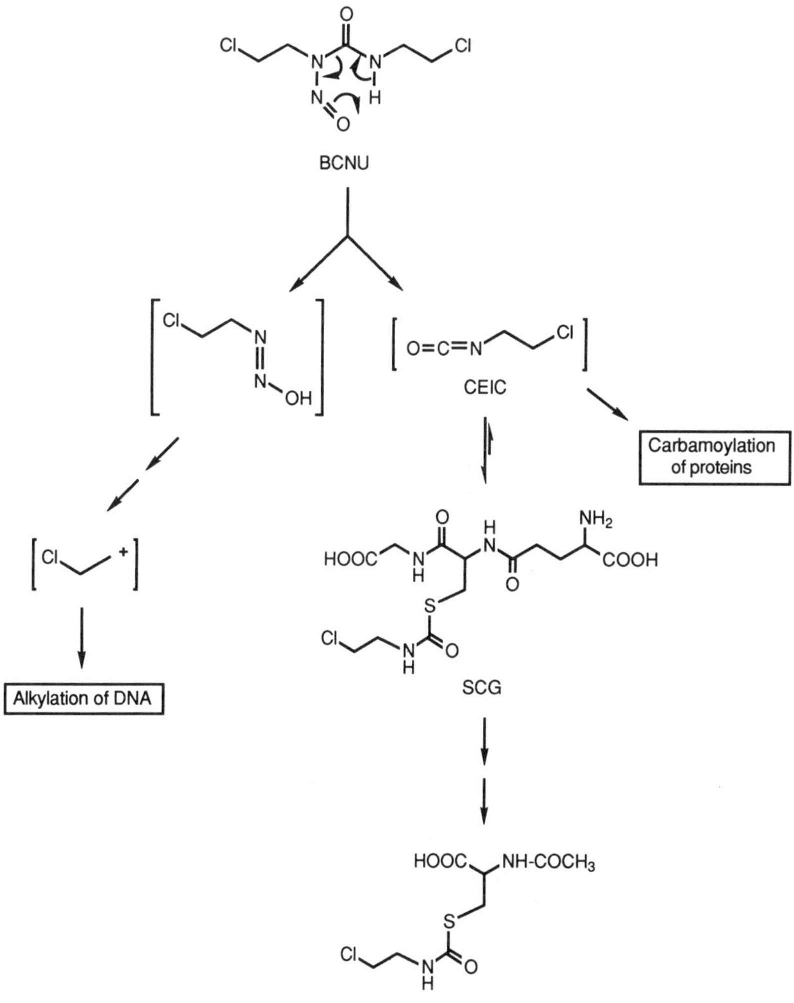

Fig. 4 Decomposition of BCNU in aqueous media to the chloroethyl carbonium ion and chloroethylisocyanate (CEIC) and proposed GSH-dependent metabolism of CEIC to the S-linked conjugates SCG and ACCC. Intermediates shown in brackets have not been isolated. (Reproduced from Davis *et al.*, 1993.)

N-acetyl-*S*-(*N*-[2-chloroethyl]carbamyol)cysteine (ACCC; Fig. 4) had a similar inhibitory effect on the enzyme (Fig. 5). Interestingly, SCG also serves as an alkylating agent toward guanosine, where an aminoethyl group is transferred from the conjugate to the N-7 position of the base

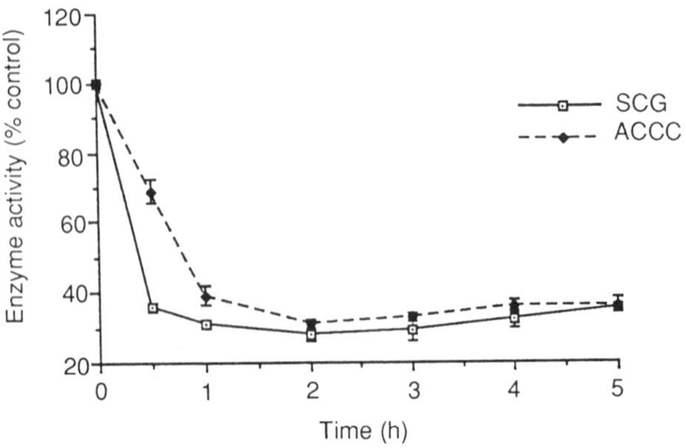

Fig. 5 Inhibition of rat liver glutathione reductase *in vitro* by S-linked conjugates of CEIC. Freshly isolated rat hepatocytes were incubated with 0.1 m*M* SCG or ACCC, and enzyme activity was determined at the intervals shown. Results are expressed as percentage activity in matched control samples incubated in the absence of inhibitor and represent means ± SE ($n = 3$). All values beyond time zero were significantly different from control samples ($P < 0.05$, Student's t test). (Reproduced from Davis *et al.*, 1993.)

(Stahl *et al.*, 1992). Therefore, SCG possesses both alkylating and carbamoylating activities and, as such, may contribute to the antitumor effects of BCNU.

Based on the above discussion, it seems likely that many xenobiotics with carbamate, formamide, urea, or sulfonylurea functionalities will undergo metabolic transformation to reactive isocyanates and that reversible conjugation of the latter short-lived, potentially toxic intermediates with GSH may play an important role in mediating the disposition and adverse effects of the parent compounds *in vivo*.

C. α,β-Unsaturated Carbonyl Compounds

A number of α,β-unsaturated aldehydes have been identified as natural products, air pollutants, xenobiotic metabolites, and products of lipid peroxidation (Witz, 1989). Since compounds of this type are electrophilic alkylating agents, they are potentially toxic to mammalian systems as a result of their propensity to generate covalent adducts with biological macromolecules. GSH is known to form S-linked conjugates with α,β-unsaturated aldehydes via 1,4-Michael addition reactions, as exemplified by the simplest member of the series, acrolein (Fig. 6). Although such thiol additions are, in principle, chemically reversible (Esterbauer *et al.*, 1975), GSH conjugates of α,β-unsaturated carbonyl compounds appear, in general, to be relatively stable compounds, and the rates of the *retro-*

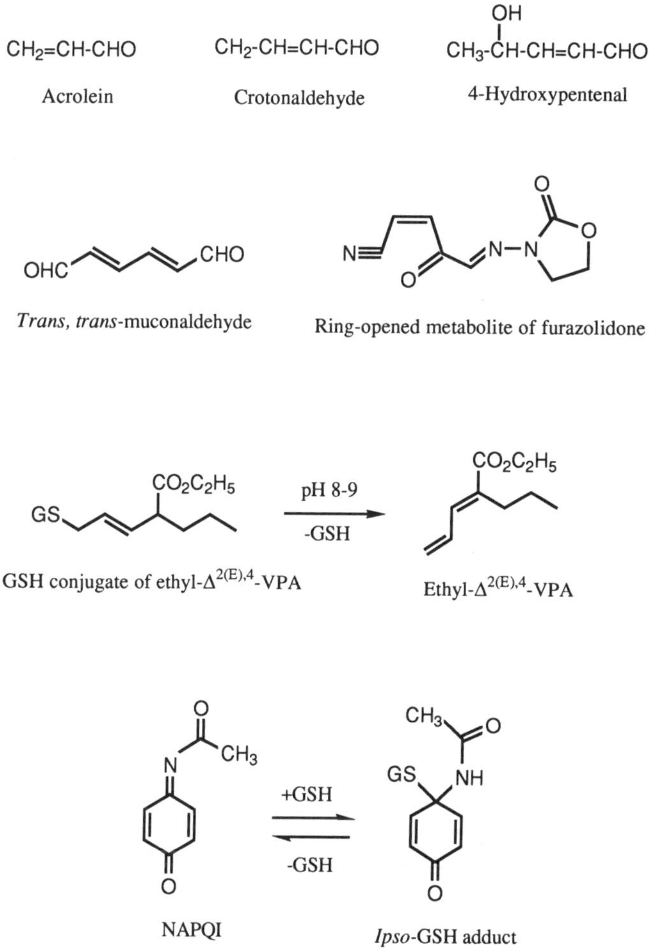

Fig. 6 Structures of α,β-unsaturated carbonyl compounds referred to in the text.

Michael reactions typically are very slow. Recently, however, evidence has been obtained which suggests that metabolic processing of such GSH conjugates along the mercapturic acid pathway may yield products that do revert more readily to their respective parent α,β-unsaturated aldehydes. Thus, the GSH conjugate of acrolein, which is known to be nephrotoxic in rats, failed to elicit kidney damage when animals were pretreated with acivicin, a specific inhibitor of γ-glutamyltranspeptidase (Horvath *et al.*, 1992). Furthermore, *in vitro* experiments with cultured LLC-PK1 cells and isolated rat renal proximal tubular cells showed that both the *N*-

acetylcysteine conjugate of acrolein and its *S*-oxide derivative were cytotoxic, although the cytotoxicity of the *N*-acetylcysteine adduct, but not the *S*-oxide, was reduced when methimazole (an inhibitor of the flavin-containing monooxygenase) was added to incubation media (Hashmi *et al.*, 1992). Collectively, these findings support the view that the nephrotoxic effects of the acrolein–GSH conjugate may be mediated by the release of free acrolein, not from the GSH adduct, however, but from the corresponding *N*-acetylcysteine conjugate *S*-oxide formed as a product of biotransformation in kidney cells (Hashmi *et al.*, 1992). Theoretically, the *S*-oxide would be expected to undergo β-elimination to yield acrolein more readily than the corresponding *N*-acetylcysteine derivative. Whether similar metabolic activation is responsible for the weak antitumor properties of the cysteine conjugates of acrolein and the related α,β-unsaturated aldehydes crotonaldehyde and 4-hydroxypentenal remains to be determined (Tillian *et al.*, 1976, 1978). Also, it will be important to establish whether reversible GSH conjugation plays a role in the "transport" of *trans, trans*-muconaldehyde, a reactive ring-opened α,β-unsaturated dialdehyde metabolite of benzene, from liver to bone marrow, which is a target tissue for benzene toxicity (Latriano *et al.*, 1986; Witz, 1989).

Reversible conjugation of an α,β-unsaturated carbonyl system with GSH has been shown to occur with a substituted acrylonitrile metabolite of furazolidone, a drug used in veterinary medicine for the treatment of gastrointestinal infections (Vroomen *et al.*, 1988, 1990). *In vitro*, the GSH adduct underwent exchange with mercaptoethanol and with microsomal protein, suggesting that the *retro*-Michael reaction occurred readily with this metabolite. In recent work in our own laboratory, we have studied the GSH conjugate of 2-*n*-propyl-2(*E*),4-pentadienoic acid (Kassahun *et al.*, 1991), a hepatotoxic metabolite of the antiepileptic drug valproic acid (Kesterson *et al.*, 1984). In the course of preparing an authentic specimen of this GSH adduct by synthesis, we found that attempts to hydrolyze the corresponding ethyl ester under very mild alkaline conditions (pH 8-9) led to facile *retro*-Michael cleavage with formation of ethyl 2-*n*-propyl-2(*E*),4-pentadienoate (K. Kassahun, L. Jin, and T. A. Baillie, unpublished observations). In light of the hypothesis that the biologically derived conjugate is formed by a 1,6-addition of GSH to the coenzyme A thioester of 2-*n*-propyl-2(*E*),4-pentadienoic acid (Kassahun *et al.*, 1991), these chemical studies with the ethyl ester derivative may have implications for the behavior of the putative GSH–coenzyme A double conjugate of this diene.

Benzoquinone and benzoquinone imine derivatives form stable thioether adducts with GSH when conjugation occurs via a Michael-type addition reaction to the β-carbon. However, it is known from chemical model systems that attack by GSH also may take place at the carbonyl (or imine) carbon atom to yield an unstable *ipso* adduct (Meisenheimer complex). Although speculative at present, such *ipso* adducts with GSH may serve as latent forms of their respective quinones, many of which (e.g., *N*-acetyl-*p*-benzoquinone imine; NAPQI; Fig. 6) are toxic in biological systems (reviewed in Baillie and Slatter, 1991).

D. Miscellaneous

In addition to the classes of compound discussed above, there is evidence that GSH conjugates involving other functional groups may exhibit reversibility under physiological conditions. For example, simple aldehydes form labile hemithioacetals (R—CH[OH]—SR′) when allowed to react with sulfhydryl-containing compounds in aqueous media, and the hemithioacetal generated from the spontaneous reaction between formaldehyde and GSH (*S*-hydroxymethylglutathione) is believed to serve as the physiological substrate for formaldehyde dehydrogenase enzymes (Mason *et al.*, 1986). Interestingly, this conjugate has been shown to exist in equilibrium not only with free formaldehyde, but with several cyclized forms of the adduct (Naylor *et al.*, 1988). Therefore, since formaldehyde can react with nucleic acids to yield cross-linked products, reversibility in GSH conjugation may be important in mediating the genotoxic effects of formaldehyde and related simple aldehydes. Similarly, nitrosoarenes (AR—N=O), formed as intermediates in the metabolic reduction of nitroaromatics, afford unstable conjugates with GSH that can either revert to the parent nitrosoarene or rearrange to the more stable sulfinamide derivative (Saito and Kato, 1984). However, the potential role of GSH as a vehicle for the transport of these toxic nitroso intermediates *in vivo* remains to be explored.

III. Conclusions

From the foregoing discussion, it is evident that reversibility in GSH conjugate formation may represent a toxicologically important phenomenon largely for compounds bearing preformed isothiocyanate, isocyanate, or α,β-unsaturated carbonyl functionalities and for foreign compounds that can be metabolized to such electrophilic species. However, it is to be expected that new examples of reversible GSH conjugation will emerge from metabolic studies on the current generation of drug candidates, which are characterized by increasingly complex structures and nontraditional functional groups. Fortunately, analytical methods for the detection and identification of labile GSH conjugates in complex biological matrices (e.g., bile) have become significantly more sensitive, selective, and versatile in recent years, notably with the advent of commercial instrumentation for on-line liquid chromatography/electrospray ionization mass spectrometry and tandem mass spectrometry (Baillie and Davis, 1993). In addition, developments in high-field NMR techniques for the direct analysis of crude biological specimens will facilitate the study of GSH adducts that can revert spontaneously to their respective precursors in aqueous solution. With access to these powerful new analytical tools, we should be in a strong position to investigate the contribution of reversible GSH conjugation to the toxicology of foreign compounds.

Acknowledgments

We thank Ms. Cindy Chernoff for assistance in manuscript preparation. The authors' research in the area of reversible GSH conjugation has been supported by a research grant from the National Institutes of Health (ES05500), which is gratefully acknowledged.

References

Anders, M. W., Lash, L., Dekant, W., Elfarra, A. A., and Dohn, D. R. (1988). Biosynthesis and biotransformation of glutathione S-conjugates to toxic metabolites. *CRC Crit. Rev. Toxicol.* **18,** 311–341.

Anders, M. W., Dekant, W., and Vamvakas, S. (1992). Glutathione-dependent toxicity. *Xenobiotica* **22,** 1135–1145.

Baillie, T. A., and Davis, M. R. (1993). Mass spectrometry in the analysis of glutathione conjugates. *Biol. Mass Spectrom* **22,** 319–325.

Baillie, T. A., and Slatter, J. G. (1991). Glutathione: A vehicle for the transport of chemically reactive metabolites in vivo. *Acc. Chem. Res.* **24,** 264–270.

Brown, W. E., Green, A. H., Cedel, T. E., and Cairns, J. (1987). Biochemistry of protein-isocyanate interactions: A comparison of the effects of aryl *versus* alkyl isocyanates. *Environ. Health Perspect.* **72,** 5–11.

Bruggeman, I. M., Temmink, J. H. M., and van Bladeren, P. J. (1986). Glutathione- and cysteine-mediated cytotoxicity of allyl and benzyl isothiocyanate. *Toxicol. Appl. Pharmacol.* **83,** 349–359.

Bruggeman, I. M., Spenkelink, A., Temmink, J. H. M., and van Bladeren, P. J. (1988). Differential effects of raising and lowering intracellular glutathione levels on the cytotoxicity of allyl isothiocyanate, *tert*-butylhydroperoxide and chlorodinitrobenzene. *Toxicol. In Vitro* **2,** 31–35.

Brüsewitz, G., Cameron, B. D., Chasseaud, L. F., Görler, K., Hawkins, D. R., Koch, H., and Mennicke, W. H. (1977). The metabolism of benzyl-isothiocyanate and its cysteine conjugate. *Biochem. J.* **162,** 99–107.

Carpenter-Deyo, L., Marchand, D. H., Jean, P. A., Roth, R. A., and Reed, D. J. (1991). Involvement of glutathione in 1-naphthylisothiocyanate (ANIT) metabolism and toxicity to isolated hepatocytes. *Biochem. Pharmacol.* **42,** 2171–2180.

Chasseaud, L. F. (1979). The role of glutathione and glutathione S-transferases in the metabolism of chemical carcinogens and other electrophilic agents. *Adv. Cancer Res.* **29,** 175–274.

Cross, H., Dayal, R., Hyland, R., and Gescher, A. (1990). N-Alkylformamides are metabolized to N-alkylcarbamoylating species by hepatic microsomes from rodents and humans. *Chem. Res. Toxicol.* **3,** 357–362.

Dahm, L. J., and Roth, R. A. (1991). Protection against α-naphthylisothiocyanate-induced liver injury by decreased hepatic non-protein sulfhydryl content. *Biochem. Pharmacol.* **42,** 1181–1188.

Dahm, L. J., Bailie, M. B., and Roth, R. A. (1991). Relationship between α-naphthylisothiocyanate-induced liver injury and elevations in hepatic non-protein sulfhydryl content. *Biochem. Pharmacol.* **42,** 1189–1194.

Davis, M. R., Kassahun, K., Jochheim, C. M., Brandt, K. M., and Baillie, T. A. (1993). Glutathione and N-acetylceysteine conjugates of 2-chloroethyl ioscyanate. Identification as metabolites of N,N′-*bis*(2-chloroethyl)-N-nitrosourea in the rat and inhibitory properties towards glutathione reductase in vitro. *Chem. Res. Toxicol.* **6,** 376–383.

DeCaprio, A. P., Spink, D. C., Chen, X., Fowke, J. H., Zhu, M., and Bank, S. (1992). Characterization of isothiocyanates, thioureas, and other lysine adduction produts in carbon disulfide-treated peptides and protein. *Chem. Res. Toxicol.* **5,** 496–504.

Drobnica, L., Kristian, P., and Augustin, J. (1977). The chemistry of the -NCS group. *In* "The chemistry of Cyanates and their Thio Derivatives" (S. Patai, ed.), Part 2, pp. 1003–1221. Wiley, New York.

Dunnick, J. K., Prejean, J. D., Haseman, J., Thompson, R. B., Giles, H. D., and McConnell, E. E. (1982). Carcinogenesis bioassay of allyl isothiocyanate. *Fundam. Appl. Toxicol.* **2,** 114–120.

Eklind, K. I., Morse, M. A., and Chung, F.-L. (1990). Distribution and metabolism of the natural anticarcinogen phenethylisothiocyanate in A/J mice. *Carcinogenesis* (*London*) **11,** 2033–2036.

Esterbauer, H., Zollner, H., and Scholz, N. Z. (1975). Reaction of glutathione with conjugated carbonyls. *Z. Naturforsch.* **30,** 466–473.

Guengerich, F. P., Kim, D.-H., and Iwasaki, M. (1991). Role of human cytochrome P-450 IIE1 in the oxidation of many low molecular weight cancer suspects. *Chem. Res. Toxicol.* **4,** 168–179.

Guest, I., Baillie, T. A., and Varma, D. R. (1992). Toxicity of the methyl isocyanate metabolite S-(N-methylcarbamoyl)GSH on mouse embryos in culture. *Teratology* **46,** 61–67.

Han, D.-H., Pearson, P. G., Baillie, T. A., Daya, R., Tsang, L. H., and Gescher, A. (1990). Chemical synthesis and cytotoxic properties of N-alkylcarbamic acid thioesters, metabolites of hepatotoxic formamides. *Chem. Res. Toxicol.* **3,** 118–124.

Hashmi, M., Vamvakas, S., and Anders, M. W. (1992). Bioactivation mechanism of S-(3-oxopropyl)-N-acetyl-L-cysteine, the mercapturic acid of acrolein. *Chem. Res. Toxicol.* **5,** 360–365.

Heylin, M. (1985). Bhopal. *Chem. Eng. News* **63,** 14–15.

Horvath, J. H., Witmer, C. M., and Witz, G. (1992). Nephrotoxicity of the 1 : 1 acrolein-glutathione adduct in the rat. *Toxicol. Appl. Pharmacol.* **117,** 200–207.

Hyland, R., Gescher, A., Thummel, K., Schiller, C., Jheeta, P., Mynett, K., Smith, A. W., and Mráz, J. (1992). Metabolic oxidation and toxification of N-methylformamide catalyzed by the cytochrome P450 isoenzyme CYP2E1. *Mol. Pharmacol.* **41,** 259–266.

Ishizaki, H., Brady, J. F., Ning, S. M., and Yang, C. S. (1990). Effect of phenethyl isothiocyanate on microsomal N-nitrosodimethylamine metabolism and other monooxygenase activities. *Xenobiotica* **20,** 255–264.

Jayaram, H. N., Lui, M. S., Plowman, J., Pillwein, K., Reardon, M. A., Elliott, W. L., and Weber, G. (1990). Oncolytic activity and mechanism of action of a novel L-cysteine derivative, L-cysteine, ethyl ester, S-(N-methylcarbamate)monohydrochloride. *Cancer Chemother. Pharmacol.* **26,** 88–92.

Kann, H. E. (1981). Carbamoylating activity of nitrosoureas. *In* "Nitrosoureas" (A. W. Prestakyo, S. T. Crooke, S. K. Baker, and P. S. Schein, eds.), pp. 95–105. Academic Press, New York.

Kassahun, K., Farrell, K., and Abbott, F. (1991). Identification and characterization of the glutathione and N-acetylcysteine conjugates of 2,4-diene-VPA, a toxic metabolite of valproic acid, in rats and humans. *Drug Metab. Dispos.* **19,** 525–535.

Kesterson, J. W., Granneman, G. R., and Machinist, J. M. (1984). The hepatotoxicity of valproic acid and its metabolites in rats. I. Toxicologic, biochemical and histopathologic studies. *Hepatology* **4,** 1143–1152.

Koob, M., and Dekant, W. (1991). Bioactivation of xenobiotics by formation of toxic glutathione conjugates. *Chem.-Biol. Interact.* **77,** 107–136.

Latriano, L., Goldstein, B. D., and Witz, G. (1986). Formation of muconaldehyde, an open-ring metabolite of benzene in mouse liver microsomes: A novel pathway for toxic metabolites. *Proc. Natl. Acad. Sci. U.S.A.* **83,** 8356–8360.

Lee, M.-S. (1992). Oxidative conversion by rat liver microsomes of 2-naphthyl isothiocyanate to 2-naphthyl isocyanate, a genotoxicant. *Chem. Res. Toxicol.* **5,** 791–796.

Mason, R. P., Sanders, J. K. M., Crawford, A., and Hunter, B. K. (1986). Formaldehyde metabolism by *Escherichia coli*. Detection by *in vivo* ^{13}C NMR spectroscopy of S-(hydroxymethyl)glutathione as a transient intracellular intermediate. *Biochemistry* **25,** 4504–4507.

Mennicke, W. H, Görler, K., and Krumbiegel, G. (1983). Metabolism of some naturally occurring isothiocyanates in the rat. *Xenobiotica* **13,** 203–207.

Mennicke, W. H., Kral, T., Krumbiegel, G., and Rittmann, N. (1987). Determination of N-acetyl-S-(N-alkylthiocarbamoyl)-L-cysteine, a principal metabolite of alkylisothiocyanates, in rat urine. *J. Chromatogr.* **414,** 19–24.

Mennicke, W. H., Görler, K., Krumbiegel, G., Lorenz, D., and Rittmann, N. (1988). Studies on the metabolism and excretion of benzyl isothiocyanate in man. *Xenobiotica* **18,** 441–447.

Monks, T. J., and Lau, S. S. (1989). Sulfur conjugate-mediated toxicity. *Rev. Biochem. Toxicol.* **10,** 41–90.

Monks, T. J., Anders, M. W., Dekant, W., Stevens, J. L., Lau, S. S., and van Bladeren, P. J. (1990). Glutathione conjugate mediated toxicities. *Toxicol. Appl. Pharmacol.* **106,** 1–19.

Mutlib, A. E., Talaat, R. E., Slatter, J. G., and Abbott, F. S. (1990). Formation and reversibility of S-linked conjugates of N-(1-methyl-3,3-diphenylpropyl)-isocyanate, an *in vivo* metabolite of N-(1-methyl-3,3-diphenylpropyl)-formamide, in rats. *Drug Metab. Dispos.* **18,** 1038–1045.

Naylor, S., Mason, R. P., Sanders, J. K. M., Williams, D. H., and Moneti, G. (1988). Formaldehyde adducts of glutathione. Structure elucidation by two-dimensional NMR spectroscopy and fast atom bombardment mass spectrometry. *Biochem. J.* **249,** 573–579.

Németh, L., Somfai-Relle, S., Kellner, B., Sugár, J., Bognár, Fárkas, J., Bálint, J., Pályi, I., Tóth, K., Szentirmay, Z., Somosy, Z., and Pokorny, E. (1978). Study of the antitumoral activity of S-carbamoyl-L-cysteine derivatives in animal experiments. *Arzneim.-Forsch.* **28,** 1119–1123.

Pearson, P. G., Slatter, J. G., Rashed, M. S., Han, D.-H., Grillo, M. P., and Baillie, T. A. (1990). S-(N-Methylcarbamoy)glutathione: A reactive S-linked metabolite of methyl isocyanate. *Biochem. Biophys. Res. Commun.* **166,** 245–250.

Pearson, P. G., Slatter, J. G., Rashed, M. S., Han, D.-H., and Baillie, T. A. (1991). Carbamoylation of peptides and proteins in vitro by S-(N-methylcarbamoyl)glutathione and S-(N-methylcarbamoyl)cysteine, two electrophilic S-linked conjugates of methyl isocyanate. *Chem. Res. Toxicol.* **4,** 436–444.

Rashed, M. S., Pearson, P. G., Han, D.-H., and Baillie, T. A. (1989). Application of liquid chromatography/thermospray mass spectrometry to studies on the formation of glutathione and cysteine conjugates from monomethylcarbamate metabolites of bambuterol. *Rapid Commun. Mass Spectrom.* **3,** 360–363.

Saito, K., and Kato, R. (1984). Glutathione conjugation of an arylnitroso compound: Detection and monitoring labile intermediates *in situ* inside a fast atom bombardment mass spectrometer. *Biochem. Biophys. Res. Commun.* **124,** 1–5.

Shirota, F. N., Elberling, J. A., Nagasawa, H. T., and DeMaster, E. G. (1990). Prodrugs of cysteine and glutathione do not reverse the action of the reactive intermediate from chlorpropamide. *Pharmacologist* **32,** 149.

Slatter, J. G., Rashed, M. S., Pearson, P. G., Han, D.-H., and Baillie, T. A. (1991). Biotransformation of methyl isocyanate in the rat. Evidence for glutathione conjugation as a major pathway of metabolism and implications for isocyanate-mediated toxicities. *Chem. Res. Toxicol.* **4,** 157–161.

Slatter, J. G., Davis, M. R., Han, D.-H., Pearson, P. G., and Baillie, T. A. (1993). Studies on the metabolic fate of caracemide, an experimental antitumor agent, in the rat. Evidence for the release of methyl isocyanate in vivo. *Chem. Res. Toxicol.* **6,** 335–340.

Smith, A. C. (1989). The pulmonary toxicity of the nitrosoureas. *Pharmacol. Ther.* **41,** 443–460.

Stahl, W., Lenhardt, S., Przybylski, M., and Eisenbrand, G. (1992). Mechanism of glutathione-mediated DNA damage by the antineoplastic agent 1,3-*bis*(2-chloroethyl)-N-nitrosourea. *Chem. Res. Toxicol.* **5,** 106–109.

Threadgill, M. D., Axworthy, D. B., Baillie, T. A., Farmer, P. B., Farrow, K. C., Gescher, A., Kestell, P., Pearson, P. G., and Shaw, A. J. (1987). Metabolism of N-methylformamide in mice: primary kinetic deuterium isotope effect and identification of S-(N-methylcarbamoyl)glutathione as a metabolite. *J. Pharmacol. Exp. Ther.* **242,** 312–319.

Tillian, H. M., Schauenstein, E., Ertl, A., and Esterbauer, H. (1976). Therapeutic effects of cysteine adducts of α,β-unsaturated aldehydes on Ehrlich ascites tumor of mice. *Eur. J. Cancer* **12,** 989–993.

Tillian, H. M., Schauenstein, E., and Esterbauer, H. (1978). Therapeutic effects of cysteine adducts of α,β-unsaturated aldehydes on Ehrlich acites tumor of mice. II. *Eur. J. Cancer* **14,** 533–536.

van Bladeren, P. J. (1988). Formation of toxic metabolites from drugs and other xenobiotics by glutathione conjugation. *Trends Pharmacol. Sci.* **9,** 295–299.

Vroomen, L. H. M., Berghmans, M. C. J., Groten, J. P., Koeman, J. H., and van Bladeren, P. J. (1988). Reversible interaction of a reactive intermediate derived from furazolidone with glutathione and protein. *Toxicol. Appl. Pharmacol.* **95,** 53–60.

Vroomen, L. H. M., Berghams, M. C. J., van Bladeren, P. J., Groten, J. P., Wissink, C. J., and Kuiper, H. A. (1990). *In vivo* and *in vitro* metabolic studies of furazolidone: A risk evaluation. *Drug Metab. Rev.* **22,** 663–676.

Witz, G. (1989). Biological interactions of α,β-unsaturated aldehydes. *Free Radical Biol. Med.* **7,** 333–349.

Zhang, Y., Talalay, P., Cho, C.-G., and Posner, G. H. (1992). A major inducer of anticarcinogenic protective enzymes from broccoli: Isolation and elucidation of structure. *Proc. Natl. Acad. Sci. U.S.A.* **89,** 2399–2403.

Zheng, G., Kenney, P. M., and Lam. L. (1992). Phenylalkyl isothiocyanate-cysteine conjugates as glutathione S-transferase stimulating agents. *J. Med. Chem.* **35,** 185–188.

Glutathione Conjugation as a Mechanism for the Transport of Reactive Metabolites

Terrence J. Monks and Serrine S. Lau
Division of Pharmacology and Toxicology
College of Pharmacy
University of Texas at Austin
Austin, Texas 78712

1. Introduction

Glutathione (γ-glutamylcysteinylglycine; GSH) is the predominant intracellular nonprotein sulfhydryl present in the cytosol of animal and plant cells and participates in a variety of biologically important reactions (Sies and Ketterer, 1988; Taniguchi *et al.*, 1989; Vinna, 1990). In its reduced form GSH is a strong nucleophile and can react with electrophiles by a direct S_N2 mechanism. These reactions may also be catalyzed by one or more of the microsomal (see Andersson *et al.*, this volume) or cytosolic (see Ketterer and Christodoulides, this volume) GSH *S*-transferase isoenzymes. Most compounds that are conjugated with GSH are ultimately excreted in urine as the corresponding mercapturic acids, which are *S*-conjugates of *N*-acetylcysteine. Conjugation with GSH has usually been considered a detoxication reaction. The increased water solubility and the active secretion of organic acids by renal tubules greatly facilitate mercapturic acid excretion. However, conjugation with GSH has now been implicated in the bioactivation of a variety of chemicals to mutagenic, carcinogenic (see Guengerich, this volume), and cytotoxic metabolites. In particular, the bioactivation of a variety of halogenated alkanes and alkenes by cysteine conjugate β-lyase is now well documented (see Dekant *et al.*, this volume). The focus of this chapter however will be on those

Advances in Pharmacology, Volume 27

conjugations of GSH that result in only a masking of the biological reactivity of the conjugated electrophile rather than in complete detoxication. The concealed reactivity of these conjugates can subsequently be exposed after further metabolism, usually at sites distal to their site of formation. These GSH conjugates can therefore be regarded as transport forms of the reactive metabolite (Monks and Lau, 1989; Monks *et al.*, 1990a; Lau and Monks, 1991). In this way GSH conjugation might therefore be regarded as the Trojan Horse of Toxicology.

II. Glutathione Conjugation as a Mechanism for the Transport of Redox Active Compounds

The reactivity of quinones resides in their ability to undergo "redox-cycling" and to create an oxidative stress (Smith *et al.*, 1985) or to react directly with cellular nucleophiles, such as protein and nonprotein sulfhydryls (Jocelyn, 1972; Finley, 1974), or both. GSH is the major nonprotein sulfhydryl present in cells (Reed and Meredith, 1985). The addition of a thiol to the double bond of a quinone represents nucleophilic addition to an α,β-unsaturated carbonyl. Although there are several studies on the addition of sulfur nucleophiles to quinones, there is little information available on the biological consequences of these reactions. Recent evidence indicates that the conjugation of quinones with GSH does not eliminate their redox properties (Wefers and Sies, 1983; Monks and Lau, 1990) and the conjugates exhibit diverse biological activities (Monks and Lau, 1992). Since the conjugates retain their redox properties, they retain the potential to generate reactive oxygen species and arylate tissue macromolecules. A combination of physiological, biochemical, and electrochemical factors appear to determine the reactivity of these conjugates (Monks *et al.*, 1988b) and the sites at which they express their reactivity. In particular, the nephrotoxicity of redox-active GSH conjugates has been investigated in some detail and is discussed below.

A. Targeting of Quinones to the Kidney by Conjugation with Glutathione

The physiological function of the kidney and its biochemical profile appear to predispose this organ to the toxicity of metabolites derived from the mercapturic acid biosynthetic pathway. Thus, kidneys possess relatively high activities of the enzymes involved in mercapturic acid biosynthesis; γ-glutamyl transpeptidase (γ-GT), cysteinylglycine dipeptidase(s), and *N*-acetyl transferases (Goldstein, 1993). The first step in the metabolism of

GSH conjugates involves either hydrolysis or transamination by γ-GT and transfer of the γ-glutamyl group to an appropriate acceptor (Meister and Tate, 1980). γ-GT is an ubiquitous membrane-bound enzyme (Tate, 1980), the active site of which is oriented on the outer surface of the cell (Horiuchi *et al.*, 1978). Kidneys possess the highest level of γ-GT activity followed by the pancreas, which in the rat has 20% of the level of the kidney (Goldbag *et al.*, 1960). Although most other tissues have less than 1% of the activity found in the kidney, it is important to note that γ-GT activity is usually localized to one cell type within an organ and to one area of the membrane, where it may be highly concentrated. The product of the γ-GT-catalyzed reaction is the *S*-substituted cysteinyl-glycine dipeptide, which, in turn, is subject to hydrolytic cleavage of the glycine residue. Several membrane-bound peptidases may catalyze this reaction (Tate, 1980). As discussed below, the high renal activity of γ-GT plays a major role in the target-organ toxicity of a variety of GSH conjugates.

1. 2-Bromo-(diglutathion-*S*-yl)hydroquinone

As little as 10 μmol/kg of 2-bromo-(diglutathion-*S*-yl)hydroquinone, administered by tail-vein injection, was sufficient to cause glucosuria, enzymuria, and renal proximal tubular cell necrosis in male Sprague–Dawley rats (Monks *et al.*, 1988b) with significant elevations in blood urea nitrogen at a dose of 15 μmol/kg (Monks *et al.*, 1985). GSH conjugates of 1,4-benzoquinone are also nephrotoxic. For example, 2,3,5-(triglutathion-*S*-yl)hydroquinone (10–20 μmol/kg) caused severe renal proximal tubular cell necrosis in rats (Lau *et al.*, 1988). An important question arises of how and in what form are the conjugates delivered to the kidney after their *in situ* formation from the respective hydroquinones. After formation in the liver, the consensus is that GSH conjugates are excreted into bile, followed by enterohepatic recirculation, and formation in the liver of the corresponding mercapturate, which is then delivered via the circulation to the kidney. In support of this scenario, the renal toxicity of hexachloro-1,3-butadiene was inhibited by pretreatment of animals with probenecid (Lock and Ishmael, 1985), an organic anion transport inhibitor. In contrast, pretreatment of animals with acivicin (AT-125) to inhibit γ-GT did not prevent the nephrotoxicity of hexachloro-1,3-butadiene (Davis, 1988). These data support the view that hexachloro-1,3-butadiene is delivered to the kidney as the mercapturate, rather than as the GSH conjugate. The tissue and cell selectivity of the damage is therefore probably determined by a combination of the relative distribution of the organic anion transporter and of the acylases, which are required to produce the cysteine conjugate for final bioactivation by β-lyase. In contrast, the tissue selectivity of 2-bromo-(diglutathion-*S*-yl)hydroquinone and 2,3,5-(triglutathion-*S*-yl)hydroquinone appears to be a consequence of their targeting to renal proximal tubule cells by brush-border γ-GT. Thus, inhibition of γ-GT by pretreatment of animals with AT-125 protected them against both 2-bromo-(diglutathion-*S*-yl)hydroquinone- and 2,3,5-(triglutathion-*S*-yl) hydroquinone-mediated nephrotoxicity, whereas probenecid did not protect animals against either conjugate (Monks *et al.*, 1988b; Lau *et al.*, 1988). Taken together, the data suggested that hydroquinone and 2-bromo-hydroquinone are delivered to kidneys in the form of the intact GSH conjugate rather than as the corresponding mercapturic acids.

Studies on the relative toxicity of 2-bromo-(dicystein-*S*-yl)hydroquinone and 2-bromo-(di-*N*-acetylcystein-*S*-yl)hydroquinone provided further evidence against a role for mercapturic acid formation in 2-bromo-(diglutathion-*S*-yl)hydroquinone-mediated nephrotoxicity (Monks *et al.*, 1991). Thus, although both 2-bromo-(dicystein-*S*-yl)hydroquinone and 2-bromo-(di-*N*-acetylcystein-*S*-yl)hydroquinone caused renal proximal tubular necrosis in male rats, only the toxicity of the mercapturic acid was inhibited by pretreatment of animals with either probenecid or aminooxyacetic acid, the latter used to inhibit the activity of cysteine conjugate β-lyase. Since neither of these protocols prevented the toxicity of 2-bromo-(diglutathion-*S*-yl)hydroquinone, the data were consistent with the view that metabolism to the mercapturate is a minor metabolic pathway of 2-bromo-(diglutathion-*S*-yl)hydroquinone. It is important to reiterate that, in contrast to the toxicity of halogenated alkanes and alkenes (see Dekant *et al.*, this volume), cysteine conjugate β-lyase does not play a major role in either 2-bromo-(diglutathion-*S*-yl)hydroquinone- or 2,3,5-(triglutathion-*S*-yl)hydroquinone-mediated nephrotoxicity. Thus, the nephrotoxicity of 6-bromo-2,5-dihydroxythiophenol, a putative β-lyase-catalyzed metabolite of 2-bromo-3-(glutathion-*S*-yl)hydroquinone, was dependent upon the presence of the quinone function, rather than the thiol function (Monks *et al.*, 1988a). In addition, although 2-bromo-(diglutathion-*S*-yl)-hydroquinone has been identified as an *in vivo* metabolite of 2-bromohydroquinone, no evidence for mercapturic acid formation has been obtained (Lau and Monks, 1990). Mercapturic acid formation from quinone–GSH conjugates may be limited by the ability of either the cysteinyl-glycine or the cysteine conjugate, or both, to undergo an oxidative cyclization reaction that results in the formation of a 1,4-benzothiazine (Fig. 1). This reaction can therefore channel the products of the γ-GT-catalyzed metabolism of quinone–GSH conjugates away from the classic mercapturic acid pathway (Monks *et al.*, 1990b).

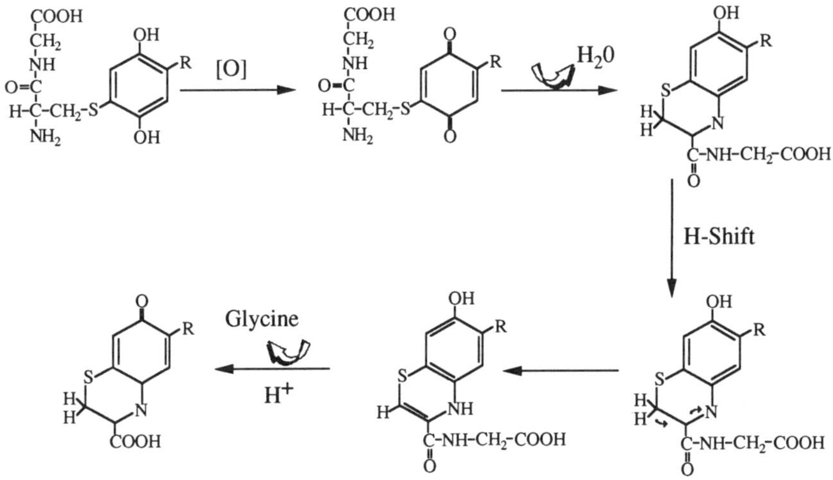

Fig. 1 Oxidative cyclization and 1,4-benzothiazine formation from cysteinyl-glycine-conjugated quinones. An identical cyclization reaction can take place with the cysteine conjugate.

2. Modulation of Glutathione-Conjugated Quinone Toxicity by γ-Glutamyl Transpeptidase and 1,4-Benzothiazine Formation

Although experiments on the inhibition of γ-GT provided evidence that metabolism of the GSH-conjugated hydroquinones was necessary for the expression of toxicity, they did not discern whether uptake of the resulting cysteinylglycine or cysteine conjugates was a prerequisite for toxicity. That is, was the toxicity mediated extracellularly or intracellularly? The possibility that the products of the γ-GT-catalyzed metabolism of GSH-conjugated hydroquinones exert their action at the surface of the brush-border membrane was suggested by the finding that such metabolites are inherently easier to oxidize to the reactive quinone than the GSH conjugate (Monks and Lau, 1990). Thus, 2-bromo-(dicystein-*S*-yl)hydroquinone exhibits a lower oxidation potential than 2-bromo-(diglutathion-*S*-yl)hydroquinone. Moreover, the covalent binding of 2-bromo-(diglutathion-*S*-yl)-[^{14}C]-hydroquinone was significantly inhibited by both ascorbic acid and by inhibition of γ-GT (Monks *et al.*, 1988b). The corresponding quinones might therefore interact with membrane nucleophiles during passage through the brush-border membrane. Quinones are also more hydrophobic than the corresponding hydroquinones, which may further contribute to the partitioning of the cysteinyl-quinone to the lipid membrane.

To examine this question in more detail, the GSH conjugates of 2,5-dichloro-1,4-benzoquinone and tetrachloro-1,4-benzoquinone were synthesized (Mertens *et al.*, 1991). These conjugates exhibit certain similarities to the hydroquinone conjugates, the major difference being that they reside in the quinone form. However, when administered by intravenous injection to rats, 2,5-dichloro-3-(glutathion-*S*-yl)-1,4-benzoquinone and 2,5,6-trichloro-3-(glutathion-*S*-yl)-1,4-benzoquinone were less potent nephrotoxicants than the corresponding reduced conjugates (Mertens *et al.*, 1991). To explain this result, it was proposed that the quinone conjugates probably reacted extensively with extrarenal nucleophiles, including plasma proteins, thereby decreasing the effective dose delivered to the kidney. Surprisingly, however, the nephrotoxicity of both 2,5-dichloro-3-(glutathion-*S*-yl)hydroquinone and 2,5,6-trichloro-3-(glutathion-*S*-yl) hydroquinone was potentiated by inhibiton of γ-GT with AT-125 (Mertens *et al.*, 1991). This result suggested that for these conjugates, metabolism by γ-GT may actually be a detoxication reaction, probably by enhancing 1,4-benzothiazine formation, a reaction that also eliminates the reactive quinone function from the molecule (Fig. 1). Whether the toxicity of 2,5-dichloro-3-(glutathion-*S*-yl)hydroquinone and 2,5,6-trichloro-3-(glutathion-*S*-yl)hydroquinone, which occurs in the presence of decreased γ-GT activity, is a consequence of their increased extracellular (intraluminal) oxidation cannot yet be answered. However, if this is indeed the case, then this raises the question, which factors determine the site selectivity of these conjugates?

3. Intracellularly or Extracellularly Mediated Toxicity?

2-Methyl-3-(*N*-acetylcystein-*S*-yl)-1,4-naphthoquinone, but not 2-methyl-3-(glutathion-*S*-yl)-1,4-naphthoquinone, produced renal proximal tubular necrosis when administered to male Sprague–Dawley rats (Lau *et al.*, 1990). Consistent with these observations, only the mercapturate and not the GSH conjugate of menadione was cytotoxic when incubated with isolated rat kidney cortical epithelial cells (Brown *et al.*, 1991). The differences between the mercapturate and the GSH conjugate of menadione are most likely related to the ability of the latter to undergo γ-GT-catalyzed oxidative cyclization and 1,4-benzothiazine formation (Brown *et al.*, 1991). Since this reaction eliminates the reactive quinone function from the molecule, it effectively prevents redox cycling of the thioether. In contrast, the presence of the *N*-acetyl group in the mercapturate hinders condensation of the cysteinyl amino group with the quinone carbonyl group. Consequently, 2-methyl-3-(*N*-acetylcystein-*S*-yl)1,4-naphthoquinone retains the ability to redox cycle with the concomitant formation of reactive oxygen species (Brown *et al.*, 1991). Consistent with an extracellular mode of activity, basolateral exposure of rat renal proximal tubular cells, grown on porous supports, to 2-methyl-3-(*N*-acetylcystein-*S*-yl)-1,4-naphthoquinone in the presence of probenecid potentiated toxicity (Haenen *et al.*, 1993). However, inhibition of the intracellular *N*-deacetylation of the mercapturate with paraoxon also potentiated the toxicity, indicating that perhaps both extra- and intracellular events play a role in the toxicity. In addiiton, the GSH conjugate was only toxic to cells after basolateral exposure, indicating apical detoxication by brush-border γ-GT. Consistent with this view, 2-methyl-3-(glutathion-*S*-yl)-1,4-naphthoquinone was only toxic to proximal tubular cells after apical exposure, when γ-GT was inhibited. The GSH conjugate of menadione (600 μM) has been shown to cause toxicity in the isolated perfused rat kidney (Redegeld *et al.*, 1989). The reasons for the model-dependent differences in the toxicity of 2-methyl-3-(glutathion-*S*-yl)-1,4-naphthoquinone are not known, although the major variable in the model that reports toxicity (Redegeld *et al.*, 1989) is the extremely high concentration of the conjugate used to perfuse the kidney.

4. 4-Aminophenol

4-Aminophenol causes acute renal proximal tubular necrosis after administration to rats (Green *et al.*, 1969; Newton *et al.*, 1982). Gartland *et al.* (1989) demonstrated that either depletion of hepatic GSH by pretreatment of animals with buthionine sulfoximine or cannulation of the bile duct to decrease the delivery of hepatic metabolites to the kidney afforded partial protection against 4-aminophenol nephrotoxicity, suggesting a role for GSH conjugation in 4-aminophenol nephrotoxicity. Indeed, oxidation of 4-aminophenol to the quinoneimine and reaction with GSH gave rise to several isomeric multisubstituted conjugates (Eckert *et al.*, 1990). Subsequently, Fowler *et al.* (1991) demonstrated that 4-amino-3-(glutathion-*S*-yl)-phenol reproduced 4-aminophenol nephrotoxicity in male Fischer-344 rats at doses three- to four-fold lower than that of 4-aminophenol. Klos *et al.* (1992) subsequently identified 4-amino-2-(glutathion-*S*-yl)-phenol, 4-amino-3-(glutathion-*S*-yl)phenol, 4-amino-2,5-(di-glutathion-*S*-yl)phenol and 4-amino-2,3,5 (or 6)-(tri-glutathion-*S*-yl)phenol in the bile of Wistar rats after administration of 4-aminophenol (100 mg/kg; ip). The latter three conjugates were all capable of causing cytotoxicity when incubated with rat kidney cortical cells, and the toxicity was prevented by inhibition of γ-GT. Interestingly, pretreatment of rats with AT-125 slightly potentiated the nephrotoxicity of 4-aminophenol (458 μmol/kg) (Fowler *et al.*, 1993), whereas probenecid had little or no effect. Coadministration of ascorbic acid with 4-aminophenol significantly protected against the toxicity with a concomitant reduction in the concentration of both total and covalently bound radiolabel in the kidney (Fowler *et al.*, 1993), indicating that intrarenal oxidation of 4-aminophenol or its metabolites plays an important role in the mechanism of toxicity.

III. Transport of Glutathione-Conjugated Redox Compounds to the Kidney as a Common Factor in Chemical-Induced Nephrocarcinogenicity

A. Hydroquinone

The National Toxicology Program recently described renal tubular cell degeneration in the renal cortex of male and female rats (F344/N) receiving 1.82 mmol/kg hydroquinone (HQ) (13-week gavage studies) (Kari, 1989). In addition, the results from long-term studies demonstrated that HQ exhibited "evidence of carcinogenic activity" in male, but not female, Fischer 344/N rats (nor mice) as shown by marked increases in tubular cell adenomas of the kidneys (Kari, 1989). These findings were confirmed by Shibata *et al.* (1991), who reported the induction of renal cell tumors in rats and mice after exposure to hydroquinone in the diet (0.8%) for 2 years. Although the mechanism of hydroquinone-mediated nephrotoxicity and nephrocarcinogenicity in male Fischer 344/N rats is not known, GSH conjugates of hydroquinone may play an important role. For example, 2-(*N*-acetyl-cystein-*S*-yl)hydroquinone and 2-(glutathion-*S*-yl)hydroquinone have been identified as *in vivo* and *in vitro* metabolites of hydroquinone (Tunek *et al.*, 1980; Lunte and Kissinger, 1983; Sawahata and Neal, 1983; Nerland and Pierce, 1990), and chemical reaction of GSH with 1,4-benzoquinone gives rise to 2-(glutathion-*S*-yl)hydroquinone and all of the possible multisubstituted GSH adducts, including 2,3-(di-glutathion-*S*-yl)hydroquinone, 2,5-(diglutathion-*S*-yl)hydroquinone, 2,6-(diglutathion-*S*-yl)hydroquinone, 2,3,5-(triglutathion-*S*-yl)-hydroquinone, and 2,3,5,6-(tetraglutathion-*S*-yl)hydroquinone (Lau *et al.*, 1988; Eckert *et al.*, 1990). Administration of each of these conjugates to rats demonstrated a direct correlation between the increasing degree of GSH substitution and the extent of proximal tubular necrosis, as determined by elevations in blood urea nitrogen. In particular, 2,3,5-(triglutathion-*S*-yl)-hydroquinone (10-20 μmol/kg, iv) caused severe renal proximal tubular necrosis in male Sprague–Dawley rats (Lau *et al.*, 1988). Histological examination of the kidneys from 2,3,5-(triglutathion-*S*-yl)hydroquinone-treated rats (20 μmol/kg, iv) revealed that severe necrosis was localized to the S_3 segment of the proximal tubules within the cortico-medullary junction. Thus, GSH conjugates of hydroquinone, in particular, 2,3,5-(triglutathion-*S*-yl)hydroquinone, may contribute to hydroquinone-mediated nephrocarcinogenicity. In support of this view, 2-(*N*-acetylcystein-*S*-yl)hydroquinone has been reported to be an *in vivo* metabolite of benzene, phenol, and hydroquinone (Nerland and Pierce, 1990). Consistent with the species-dependent nephrocarcinogenicity of hydroquinone, rats, but not mice, were susceptible to 2,3,5-(triglutathion-*S*-yl)hydroquinone-mediated nephrotoxicity (Kleiner *et al.*, 1993). Such conjugates may therefore play an important role in hydroquinone-mediated nephrocarcinogenicity. The recent identification of 2,3,5-(triglutathion-*S*-yl)hydroquinone and other multi-GSH conjugates as *in vivo* metabolites of hydroquinone in the rat (Hill *et al.*, 1993) lends further support to this hypothesis. In addition, GSH conjugates of hydroquinone have been shown to catalyze 8-hydroxy-deoxyguanosine formation in calf thymus DNA (Canales *et al.*, 1993). Such oxidative DNA damage may be important in addressing the question whether hydroquinone is a genotoxic or nongenotoxic carcinogen.

B. 3-*tert*-Butyl-4-hydroxyanisole

Long-term administration of 3-*tert*-butyl-4-hydroxyanisole increased the formation of preneoplastic and neoplastic foci in rat kideny (Tsuda *et al.*, 1984), and thioether metabolites have been reported in rat urine (Tajima *et al.*, 1991). 2-*tert*-Butylhydroquinone, a metabolite of 3-*tert*-butyl-4-hydroxyanisole, is oxidized in rat liver microsomes and in the presence of GSH forms 2-*tert*-butyl-5-(glutathion-*S*-yl)hydroquinone and 2-*tert*-butyl-6-(glutathion-*S*-yl)hydroquinone (Tajima *et al.*, 1991). Whether the GSH conjugates are involved in the carcinogenic effects of 3-*tert*-butyl-4-hydroxyanisole remains to be determined. However, exposure of male, Fischer-344 rats to 2-*tert*-butyl-3,6-(di-glutathion-*S*-yl)hydroquinone (200 μmol/kg) caused mild damage to renal proximal tubules, as observed histologically and by the significant increases in the urinary excretion of γ-GT, alkaline phosphatase, LDH, and glucose (T. J. Monks *et al.*, unpublished data). Long-term exposure to 3-*tert*-butyl-4-hydroxyanisole may therefore result in the continuous delivery of redox-active GSH conjugates to renal proximal tubules, causing cell damage and a sustained regenerative response that may be sufficient to promote the formation of neoplasia. The role of GSH in 3-*tert*-butyl-4-hydroxyanisole-induced cell proliferation is further supported by the report that depletion of GSH with diethyl maleate prevented the forestomach hyperplasia associated with the injestion of this antioxidant (Hirose *et al.*, 1987). Recent studies have shown that 2-*tert*-butylhydroquinone is also capable of inducing oxidative DNA damage in the presence of prostaglandin H synthase (Schilderman *et al.*, 1993). It is not yet known whether the GSH conjugates can also catalyze such DNA damage, but would appear likely in view of the effects of GSH conjugation on the redox properties of hydroquinones and on the ability of the GSH conjugates of hydroquinone to catalyze 8-hydroxydeoxyguanosine formation as described above.

C. Catechol Estrogens

Despite intensive research, the mechanism of both natural and synthetic estrogen-mediated carcinogenesis remains unclear. Estrogens, which are known to promote tumors, have been regarded as acting by "epigenetic" pathways, that is, by inducing cancer without direct interaction with DNA (Wanless and Medline, 1982; Siegfried *et al.*, 1984). Conversely, a mechanism of tumorigenesis by metabolic activation and subsequent damage to cellular macromolecules (including DNA) was suggested by studies that demonstrated diethylstilbestrol-induced sister chromatid exchange (Rudiger *et al.*, 1979; Hill and Wolff, 1983), unscheduled DNA synthesis (Tsutsui *et al.*, 1984), and aneupoidy in Syrian hamster embryo cells (Tsutsui *et al.*, 1983). In addition, the induction of renal carcinomas in hamsters by a series of structural analogs did not directly correlate with estrogenic potency (Liehr, 1983; Li *et al.*, 1983). Indeed, a key issue is the potential role that metabolism plays in hormonal carcinogenesis. The nephrocarcinogenicity of natural and synthetic estrogens in the Golden Syrian hamster has been suggested to be a consequence of catechol–estrogen formation (Li *et al.*, 1983; Li *et al.*, 1985; Liehr, 1983) and GSH conjugates of the catechol estrogens have been identified as major *in vivo* and *in vitro* metabolites of estradiol (Marks and Hecker, 1969; Kuss, 1969, 1971; Hoppen *et al.*, 1974; Elce, 1970, 1972; Elce and Harris, 1971; Elce and Chandra, 1973; Ball *et al.*, 1983).

The catechol estrogens are a major group of reactive estrogen metabolites formed by aromatic hydroxylation of the primary estrogens at either the C-2 or the C-4 positions, and there is general consensus that this is a major route of estrogen metabolism in humans and other mammals (Knuppen *et al.*, 1983). In terms of urinary excretion, the 2-hydroxy estrogens are the dominant metabolites of estradiol in humans (Ball and Knuppen, 1980). Aromatic hydroxylation of estradiol also occurs at the isomeric C-4 *ortho*-position (Williams *et al.*, 1974), although the activity of this pathway is smaller by almost an order of magnitude relative to that at C-2. The catechol estrogens are relatively ineffective estrogen agonists, although the catechol estrogens can interact with enzyme systems known to regulate key endocrine functions (Ball and Knuppen, 1980). In contrast to the wide agreement on these findings, reports dealing with the abundance and the physiological significance of catechol estrogens are contradictory. However, it is the nonestrogenic activities of the catechol estrogens that are of the greatest interest, although little is known of the biological significance of these compounds and of their role in estrogen physiology. Moreover, an important evolving concept in connection with the mechanism of action of steroids is the extent to which steroid hormones may modulate cellular functions through mechanisms other than the classical steroid receptors (Weisz, 1983).

Estrogen 2/4-hydroxylase activity exists in nearly all rat tissues, but by far the most important tissue, on a quantitative basis, is the liver (Ball *et al.*, 1978; Ball and Knuppen, 1978; Paul *et al.*, 1977). In contrast, estrogen 2- and 4-hydroxylase activity has been shown to be significantly greater in hamster kidney than in the corresponding rat tissue (Li *et al.*, 1985). However, catechol-estrogen formation from estrone was essentially the same in hamster kidney and liver microsomes, and catechol–estrogen formation with 17-β-estradiol as substrate was substantially higher in hamster liver microsomes than in the kidney. Moreover, estrone was a better substrate for hamster kidney estrogen hydroxylase, yet was not as potent as 17-β-estradiol in inducing renal tumors. In addition, substantial covalent binding to hamster liver microsomes with [4-^{14}C]-17-β-estradiol could be demonstrated, whereas covalent binding to kidney microsomes was negligible (Haaf *et al.*, 1987). Thus, are levels of catechol estrogens formed within the kidney sufficient to play a role in estrogen-mediated nephrocarcinogenicity? It is possible that extrarenal sites, in particular the liver, may play a substantial role in the generation of catechol estrogens for transport to the kidney. However, plasma levels of both estrogens and of "free" catechol estrogens are extremely low, and these two factors raise the questions how and in what form would the catechol estrogens be transported to the kidney?

Catechols are readily oxidized to the corresponding *ortho*-quinones, and thioether conjugates of the catechol estrogens have been identified as major *in vivo* and *in vitro* metabolites. The relative importance of catechol–estrogen thioether formation to the overall metabolism of estrogens, however, is not known. Ball *et al.* (1983) noted that in the liver estradiol was metabolized to lipophilic, water-soluble, or protein-bound products (30 : 65 : 5). The major metabolites were catechol–estrogens, which were present mainly in the water-soluble fraction. The authors stated, "As the conjugate fraction is the dominant one, the overall importance of catechol estrogen monomethyl ethers should not be overestimated . . . the major importance of conjugates and minor importance of methyl ethers in vivo should be stressed.'' The significance of GSH conjugation may have been underestimated as a consequence of several factors. In particular, Li *et al.* (1989) showed that catechol-*O*-methyl transferase (COMT) activity was low in hamster tissues relative to other rodent species; activity in hamster liver, for example, was 10-fold less than in mouse liver, and 100-fold less than in rat liver. The fraction of catechol estrogen available for oxidation and thioether formation is thus greater in the hamster than in other rodent species. Both GSH and cysteine conjugates are also obtained from *in vitro* incubations of 17β-estradiol with human liver homogenates in the presence of these thiols, but no estrogen mercapturic acid has yet been demonstrated in human urine (Ball *et al.*, 1976). It was suggested that in contrast to the rat, humans do not synthesize estrogen–GSH conjugates (Elce *et al.*, 1973). Alternatively, catechol–estrogen thioether conjugates may be formed in the human (as evidenced by the *in vitro* data), but are metabolized to products that are either not readily excreted or become bound to tissue macromolecules, or both. After the administration of radiolabeled 2-hydroxy-1-(glutathion-*S*-yl)-estradiol to rats, only 15% of the dose was recovered in the urine and 5% in the feces (Elce, 1972). Only the *N*-acetylcysteine derivative was identified in the urine. The authors noted that ''Oestrogen-glutathione conjugates formed in the intact rat may be excreted in an apparently non-steroidal possibly protein-bound form.'' Cysteinylglycine- and cysteine-conjugated *para*-quinones can undergo oxidative cyclization and 1,4-benzothiazine formation, as described above. Subsequent poly(di)merization yields insoluble pigments, which exhibit pH indicator properties (Monks *et al.*, 1990b) similar to those of the trichochromes formed during melanin synthesis from cysteinly-DOPA (Prota, 1989). Thus, both *ortho*- and *para*-quinones undergo the cyclization reaction. Although the oxidative cyclization reaction removes the reactive quinone moiety from the molecule and might be considered an intramolecular detoxication reaction, the intracellular or intralumenal formation of insoluble polymers might have important biological consequences. First, the reaction offers an explanation for the inability to recover 2-hydroxy-1-(glutathion-*S*-yl)-estradiol after *in vivo* administration (see above). Second, and perhaps more important the presence of insoluble polymers within renal tubules (the putative origin of estrogen induced tumors) will provide a constant irritant that might well be the proliferative stimulus necessary for estradiol-mediated nephrocarcinogenicity. Indeed, it has been noted that estrogen-induced renal tumors are sometimes initiated at sites of deliberate physical injury to the tissue (Kirkman, 1959). The deposition of a pigment in the cells of proximal convoluted tubules of rat kidney after long-term treatment with estrogen has been reported (Harris, 1966), and the author speculated that ''the overload of metabolite overwhelms the clearing mechanism, resulting in the deposition of pigment'' and it was suggested that the pigment "be classified in the category of pseudomelanin's.'' The ability of catechol–estrogen thioethers to undergo redox cycling, with the concomitant generation of reactive oxygen species, in combination with the possible formation of insoluble polymer, may play an important role in estrogen mediated carcinogenicity (Fig. 2). GSH conjugation of catechol estrogens therefore provides an effective mechanism of transporting them to the kidney, where they can be selectively accumulated by renal proximal tubular cells by the activity of γ-GT. In humans, renal cell carcinomas are considered to

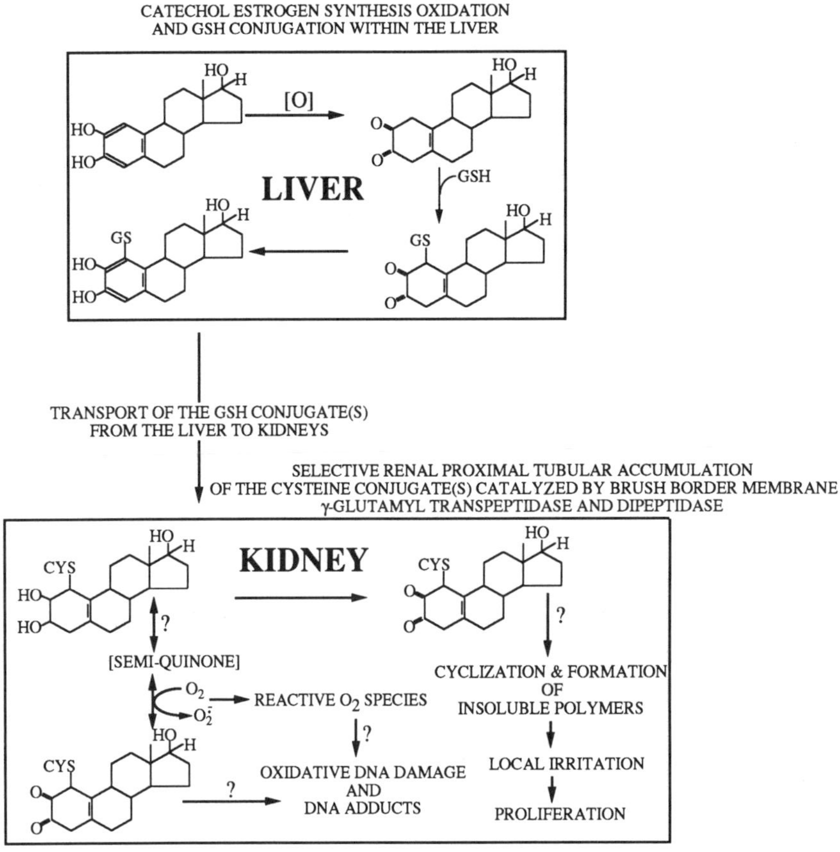

Fig. 2 Hypothetical involvement of catechol–estrogen glutathione conjugates in estradiol-mediated nephrocarcinogenicity (see text for description). Although the figure illustrates the kidney as the target for the conjugates, other tissues expressing γ-glutamyltranspeptidase and the appropriate amino acid transporters could also be potential targets for *ortho*- or *para*-quinone–glutathione conjugates.

originate from the epithelium of the proximal tubule (deKernion, 1986; Tannenbaum, 1971). It has recently been reported that estrogen-induced hamster renal cancers arise from microcarcinomas within the proximal tubules, and it is postulated that these tumors are mesenchymal–epithelial in nature (Gonzalez *et al.*, 1989).

IV. γ-Glutamyltranspeptidase and the Toxicity of Glutathione-Conjugated Redox Compounds

GSH conjugation to redox cycling compounds followed by selective (γ-GT-mediated?) uptake may be a common mechanism of toxicity of these compounds. The potential for quinone-thioethers to cause toxicity to cells other than those of the renal proximal tubules and to cells that express high γ-GT activity has been discussed (Monks and Lau, 1989, 1992). Thus, in addition to the kidney, other tissues that express high γ-GT activity, such as the pancreas, spleen, and seminal vesicles, might also be susceptible to the toxic effects of quinone-linked GSH conjugates. In this respect GSH might be envisioned as a carrier of redox cycling compounds through the body in which the initial conjugation can be followed by either tissue- or cell-selective uptake and release of the reactive compound at the target site (Fig. 2). However, it is also important to note that the toxicity of redox-active GSH conjugates will be modulated by a variety of factors, including the relative activities of cysteine conjugate *N*-acetyl transferase and the corresponding *N*-deacetylase, the relative activities of one- and two electron reductases and the availability of various antioxidants, such as ascorbic acid, GSH, and NAD(P)H. The combination of all these factors will determine both species and tissue susceptibility of γ-GT containing cells to the toxicity of quinone-thioethers. However, it is clear that quinone thioethers possess a variety of biological and toxicological activity (Monks and Lau, 1992). The ubiquitous nature of quinones and the high concentrations of GSH within cells virtually guarantees that humans will be exposed to the potential adverse effects of the resulting quinone thioethers. Knowledge of the disposition of quinone thioethers will therefore be an important prerequiste to understanding their mechanism of action.

V. Targeting of Glutathione Conjugates to Neoplastic Tissues Expressing γ-Glutamyltranspeptidase

The ability of γ-GT to predispose certain cells to the adverse effects of quinone thioethers might have some potential chemotherapeutic applications. For example, γ-GT is widely used as a marker of preneoplastic lesions in the liver during chemical carcinogenesis (Peraino *et al.,* 1983), and abnormally high levels of γ-GT were also observed in tumors of a variety of tissues, including hepatocellular carcinomas (Fiala *et al.,* 1976; Williams *et al.,* 1980), malignant squamous carcinomas of the skin (DeYoung *et al.,* 1978), squamous cell carcinomas of the buccal pouch epithelium (Solt, 1981), adenocarcinomas of the lungs (Dempo *et al.,* 1981), and some mammary tumors (Jaken and Mason, 1978). The high concentrations of γ-GT in tumors, in addition to its specific localization within the brush-border membrane of the renal proximal tubules, stimulated interest in the possibility that this enzyme might be exploited for the site-selective delivery of prodrugs (Magnan *et al.,*1982; Minard *et al.,* 1980; see Vamvakas and Anders, this volume). This concept was applied to dopamine by converting it to its γ-glutamyl derivatives thereby producing kidney-specific dopamine precursors with selective renal vasodilator activities (Wilk *et al.,* 1978). In addition, a series of *N*-acetylated-γ-glutamyl derivatives of sulfamethoxazole were designed as kidney specific prodrugs (Orlowski *et al.,* 1980). The γ-glutamyl adduct of phenylenediamine mustard was investigated for its potential use as an anti-hepatoma drug (Manson *et al.,* 1981; Smith *et al.,* 1984), and the *N*-γ-glutamyl derivative of 4-aminophenol, a naturally occurring substance isolated from mushrooms (Weaver *et al.,* 1971) and its synthetic isomer *N*-γ-glutamyl-3-aminophenol showed antitumor activity against B-16 melanoma (Rosowsky *et al.,* 1979). If the renal toxicity of redox-active GSH conjugates could be circumvented, for example, either by the selective inhibition of renal γ-GT or by localized delivery, then such compounds might prove useful directed against γ-GT-containing neoplastic cells, particulary in the liver, where nontransformed hepatocytes express little γ-GT.

VI. Developmental Toxicity of Glutathione-Conjugated Redox Compounds

γ-GT is present in the embryonic yolk sac endoderm (Miki and Kugler, 1984) and enzyme activity increases with gestational age (Andrews *et al.,* 1992), suggesting that the developing yolk sac might be at risk from maternal exposure to redox active GSH conjugates (and perhaps, other GSH conjugates that are processed to reactive metabolites, as discussed elsewhere in this volume). During organogenesis, the postimplantation rat embryo relies exclusively on the visceral yolk sac for the uptake and transport of macromolecules (Lloyd, 1990), and the activity of γ-GT appears to be essential for normal development since inhibition of this enzyme causes severe developmental abnormalities in both the embryo and the yolk sac (Stark *et al.*, 1987). Andrews *et al.* (1993) reported that *in vitro* exposure of embryos with intact visceral yolk sacs to 2-bromo-6-(glutathion-*S*-yl)hydroquinone (120 μM) caused significant decreases in yolk sac diameter, crown-rump length, head length, total protein, developmental score, and somite number, and a significant increase in the percentage of abnormal embryos. Embryos exposed to 2-bromo-(diglutathion-*S*-yl)hydroquinone (80 μM) had significantly lower developmental scores, and after exposure to 10, 25, and 120 μM, embryos also exhibited significantly lower yolk sac diameters with a higher percentage of abnormalities. Studies on the potential *in vivo* developmental toxicity of 2-bromo-(diglutathion-*S*-yl)hydroquinone were complicated by maternal toxicity (Andrews *et al.*, 1993). However, the potential developmental toxicity of redox active GSH conjugates should not be underestimated since both 2-bromo-6-(glutathion-*S*-yl)hydroquinone and 2-bromo-(diglutathion-*S*-yl)hydroquinone were more potent *in vitro* embryotoxicants than 9 of 13 recently evaluated substituted phenols when compared on a molar basis (Oglesby *et al.*, 1992).

VII. Transport of Reversible Glutathione Conjugates

Recently, a novel class of toxic GSH conjugates has been described in which the parent compound is in equilibrium with its GSH conjugate. These reactions are of particular toxicological significance because they provide a means by which a reactive molecule may initially be detoxified by conjugation with GSH, only for release of the reactive moiety to occur distal to its site of formation. In contrast to the conjugation of quinones with GSH, as described above, metabolism of GSH conjugates that are chemically reversible is not required; their activation, as defined by release of the participating reactant in its unchanged form, is largely controlled by physicochemical conditions, such as pH and the relative concentration of the reactants. A succint and insightful review of these reactions has been published (Baillie and Slatter, 1991; see also Baillie and Kassahun, this volume).

A. Glutathione and the Transport of Heavy Metals

GSH may play a role in the delivery of some heavy metals to the kidney (see Ballatori, this volume). Both methyl and inorganic mercury are carried in blood to the kidney as complexes with GSH (Naganuma *et al.*, 1988; Tanaka *et al.*, 1990). Renal methyl mercury concentrations were reduced and urinary methyl mercury excretion increased in animals treated with AT-125, implicating a role for γ-GT in renal methyl mercury accumulation (Naganuma *et al.*, 1988). In addition, renal methyl mercury uptake can occur through the basolateral membrane, by the inorganic anion transport system (Tanaka *et al.*, 1992). In contrast, cadmium is believed to be delivered to the kidney as a complex with metallothionein (Foulkes, 1978), although renal uptake of cadmium can be demonstrated in the presence of cysteine and GSH (Felley-Bosco and Diezi, 1987; Foulkes and Blanck, 1990). The relative importance of the metallothionein and GSH delivery systems in heavy metal-mediated nephrotoxicity requires clarification.

B. Isothiocyanates

Isothiocyanates are naturally occurring compounds (Fenwick *et al.*, 1982) that are of toxicological interest; amyl isothiocyanate has been shown to be a bladder carcinogen in rats (Dunnick *et al.*, 1982), whereas benzyl isothiocyanate exhibits anticarcinogenic activity (Wattenberg, 1981). These compounds are extensively metabolized by conjugation with GSH since most of the urinary metabolites are derived from the mercapturate pathway (Brusewitz *et al.*, 1977; Mennicke *et al.*, 1983). Isothiocyanates are relatively unstable and react readily with a variety of nucleophiles. The reaction with thiols is reversible (Drobnica *et al.*, 1977), the dithiocarbamate being in equilibrium with the parent compound. For example, the mercapturate derived from benzyl isothiocyanate is unstable, and free benzyl isothiocyanate is released in urine containing this metabolite (Brusewitz *et al.*, 1977). The thiol conjugates of amyl and benzyl isothiocyanate were subsequently demonstrated to exhibit toxicity almost identical to that of the parent compounds (Bruggeman *et al.*, 1986). Since the position of the dithiocarbamate : isothiocyanate equilibrium is influenced by pH and the concentration of the reactants, it is possible that the isothiocyanates become conjugated with GSH at one site, and where conditions of pH and GSH concentrations are favorable, are released at a second site. Thus, amyl isothiocyanate exerts its carcinogenic activity in the bladder of the rat (Dunnick *et al.*, 1982) where the pH of urine is usually basic, favoring release of free isothiocyanate from its mercapturic acid pathway metabolites, which are readily excreted in urine.

C. Acrolein and Other α,β-Unsaturated Aldehydes

Temmink *et al.* (1986) suggested that a similar role for GSH could occur with other chemicals, such as α,β-unsaturated carbonyls, where a reversible Michael addition can be proposed. Although such reactions are theoretically reversible, most such adducts are quite stable. However, thiol adducts of several α,β-unsaturated aldehydes are cytotoxic. For example, the crotonaldehyde–cysteine and *trans*-4-hydroxypentenal–cysteine adducts are carcinostatic, probably because of release of the active aldehyde (Tillian *et al.*, 1976, 1978, Conroy *et al.*, 1977). *S*-(3-Oxopropyl)GSH, the GSH conjugate of acrolein, is nephrotoxic when administered to male Sprague–Dawley rats (Horvath *et al.*, 1992). Whether the nephrotoxicity is due to release of acrolein is not known, but appears likely in view of the recently proposed bioactivation mechanism for *S*-(3-oxopropyl)-*N*-acetyl-L-cysteine (Hashmi *et al.*, 1992). Thus, although base-catalyzed release of acrolein could not be demonstrated from *S*-(3-oxopropyl)-*N*-acetyl-L-cysteine, the corresponding *S*-oxide readily eliminated acrolein. Both the mercapturate and the *S*-oxide were cytotoxic to LLC-PK_1 cells and in isolated renal proximal tubular cells. However, only the toxicity of the mercapturate was inhibited by the inclusion of methimazole, an

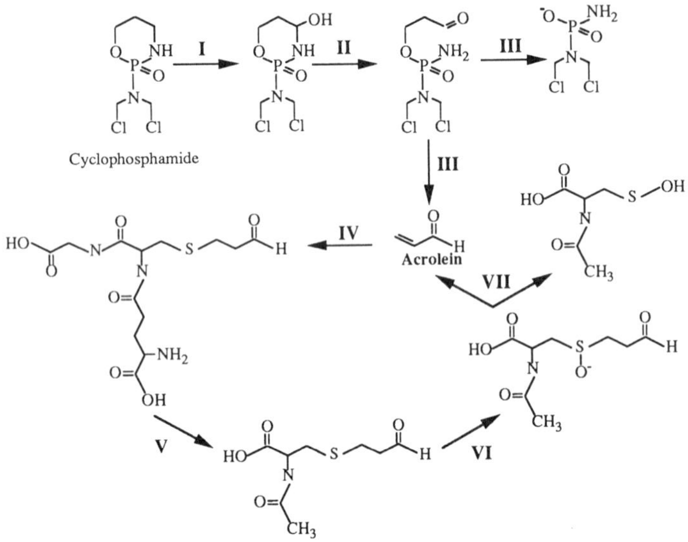

Fig. 3 Metabolism of cyclophosphamide to *S*-(3-oxopropyl)-*N*-acetyl-L-cystein-*S*-oxide, a putative bladder toxic and nephrotoxic metabolite. **(I)** Cyclophosphamide is metabolized **(I)** by either cytochromes P450 or prostaglandin H synthase to 4-hydroxycyclophosphamide, which may **(II)** tautomerize to aldophosphamide, an unstable product that rearranges **(III)** to phosphoramide mustard and acrolein. Acrolein, an α,β-unsaturated aldehyde, subsequently reacts **(IV)** with GSH to form *S*-(3-oxopropyl)GSH, which is then metabolized **(V)** to the corresponding mercapturic acid, *S*-(3-oxopropyl)-*N*-acetyl-L-cysteine. Sulfoxidation **(VI)** of the mercapturate by renal flavin-containing monooxygenases gives the *S*-oxide, which can then undergo a base-catalyzed elimination that releases acrolein and *N*-acetyl-L-cysteine sulfinic acid.

inhibitor of flavin-containing monooxygenases. Thus, the toxicity of *S*-(3-oxopropyl)-*N*-acetyl-L-cysteine appears to involve flavin-containing monooxygenase mediated sulfoxidation followed by β-elimination of the cytotoxic acrolein (Hashmi *et al.*, 1992). Acrolein has also been implicated in the bladder toxicity associated with cyclophosphamide chemotherapy (Stillwell and Benson, 1988). Subsequently, acrolein found in urine after cyclophosphamide administration appeared to be present in a "bound" form (Fraiser and Kehrer, 1992). Fraiser *et al.* (1993) then provided evidence that the bladder toxicity may involve *S*-oxidation of *S*-(3-oxopropyl)-*N*-acetyl-L-cysteine (Fig. 3). Thus, the *S*-oxide released significant amounts of acrolein *in vitro,* and instillation of the *S*-oxide into the bladder resulted in a hemorrhagic cystisis similar to that seen after either cyclophosphamide or acrolein administration.

VIII. Glutathione, Nitric Oxide, and Endothelial-Derived Relaxing Factor

The roles of nitric oxide as a mediator of vascular tone, as a retrograde messenger in the central nervous system, and in macrophage cytotoxicity are well documented, and there is widespread acceptance of the assertion that endothelial-derived releasing factor (EDRF) is nitric oxide. However, nitric oxide is a free radical, reactive toward molecular oxygen, superoxide, and a variety of metalloproteins and exhibits limited solubility in water. Consequently, there are several reports that challenge the role of nitric oxide as EDRF. As early as 1987, Long *et al.* discriminated, on a chromatographic basis, between EDRF and nitric oxide in smooth muscle. Rosenblum (1992) has summarized the available evidence indicating that at least in the cerebellum EDRF is not nitric oxide. If EDRF is not nitric oxide, then what is? The most popular alternative is that either nitric oxide is produced from a precursor, most likely a *S*-nitroso derivative, or that a nitrosothiol itself is EDRF, a view supported by the observation that nitrosothiols are potent vasodilators (Meyers *et al.*, 1990; Rubanyi *et al.*, 1991). Organic nitrates require metabolism in order to express their vasodilatory activity, and *S*-nitroso-GSH and the cysteine analog have been suggested to be the proximal activators of guanyl cyclase, resulting in the relaxation of vascular smooth muscle (Ignarro *et al.*, 1981; Yeates *et al.*, 1985) and inhibition of platelet aggregation (Mellion *et al.*, 1983). Clancy and Abramson (1992) discussed some of the functional properties of *S*-nitroso-GSH, which is capable of (*a*) inhibiting platelet aggregation, (*b*) promoting ADP ribosylation of a cytosolic 32-kDa protein in neutrophils, and (*c*) increasing cGMP synthesis in cultured lymphocytes. Nitric oxide and *S*-nitrosothiol derivatives also inhibited the formation of superoxide anion, adhesion of neutrophils to vascular endothelium and leukotriene B_4 synthesis in activated neutrophils (Ney *et al.*, 1990; Kubes *et al.*, 1991; Clancy and Abramson, 1992). *S*-nitrosation and the subsequent release of nitric oxide from the *S*-nitrosothiol thereby provide a suitable mechanism for the storage, transport, and release of nitric oxide.

IX. Summary

From this and other chapters in this volume, it should be clear that GSH conjugation no longer represents a mechanism for the detoxication of xenobiotics or their metabolites. Although the majority of conjugations with GSH do facilitate the efficient excretion of xenobiotics from the body, many examples now exist where this process results in enhanced biological reactivity (Monks *et al.*, 1990a; Monks and Lau, 1992, 1994). The number of examples in which GSH conjugation plays an important role in the generation of biologically reactive intermediates is expanding rapidly and GSH-dependent toxicity is manifested in many diverse ways. As emphasized in this chapter, GSH can act as a transport form for reactive metabolites, permitting the delivery of such metabolites to target tissues distal to the site of the initial conjugation. This type of GSH conjugate may be important in the mutagenic, carcinogenic, nephrotoxic, embryotoxic, cataractogenic, methemoglobinemic, and neurotoxic properties of a variety of redox active compounds (Monks and Lau, 1992).

References

Andrews, J. E., Ebron-McCoy, M., Zucker, R. M., Elstein, K. H., and Rogers, J. M. (1992). *In vitro/in vivo* comparison of yolk sac function and embryo development. *Toxicol. In Vitro* **6,** 1–6.

Andrews, J. E., Rogers, J. M., Ebron-McCoy, M., Logsdon, T. R., Monks, T. J., and Lau, S. S. (1993). Developmental toxicity of bromohydroquinone and bromohydroquinone-glutathione conjugates *in vivo* and in whole embryo culture. *Toxicol. Appl. Pharmacol.* **120,** 1–7.

Baille, T. A., and Slatter, J. G. (1991). Glutathione: A vehicle for the transport of chemically reactive metabolites in vivo. *Acc. Chem. Res.* **24,** 264–270.

Ball, P., and Knuppen, R. (1978). Formation of 2- and 4-hydroxyestrogens by brain, pituitary and the liver of the human fetus. *J. Clin. Endocrinol. Metab.* **47,** 732–737.

Ball, P., and Knuppen, R. (1980). Catechol estrogens: Chemistry; biogenesis, metabolism occurrence and physiological significance. *Acta Endocrinol.* (*Copenhagen*) **232,** 1–127.

Ball, P., Farthmann, E., and Knuppen, R. (1976). Comparative studies on the metabolism of oestradiol-17β and 2-hydroxyoestradiol-17β in man *in vitro* and *in vivo*. *J. Steroid Biochem.* **7,** 139–143.

Ball, P., Haupt, M., and Knuppen, R. (1978). Comparative studies on the metabolism of estradiol in the brain, the pituitary and the liver of the rat. *Acta Endocrinol.* (*Copenhagen*) **87,** 1–11.

Ball, P., Haupt, M., and Knuppen, R. (1983). Biogenesis and metabolism of catechol estrogens *in vitro*. *In* "Catechol Estrogens" (G. R. Merriam and M. B. Lipsett, eds.), pp. 91–103. Raven Press, New York.

Brown, P. C., Dulik, D. M., and Jones, T. W. (1991). The toxicity of menadione (2-methyl-1,4-naphthoquinone) and two thioethers conjugates studied with isolated renal epithelial cells. *Arch. Biochem. Biophys.* **285,** 187–196.

Bruggeman, J. M., Temmink, J. M. H., and van Bladeren, P. J. (1986). Glutathione- and cysteine-mediated cytotoxicity of allyl and benzyl isothiocyanate. *Toxicol. Appl. Pharmacol.* **83,** 349–359.

Brusewitz, G., Cameron, B. D., Chasseaud, L. F., Görler, K., Hawkins, D. R., Kock, H., and Mennicke, W. H. (1977). The metabolism of benzyl isothiocyanate and its cysteine conjugate. *Biochem. J.* **162,** 99–107.

Canales, P. L., Kleiner, H. E., Monks, T. J., and Lau, S. S. (1993). Formation of 8-hydroxydeoxyguanosine by quinol-thioethers. *Toxicologist* **13,** 202.

Clancy, R. M., and Abramson, S. B. (1992). Novel synthesis of *S*-nitrosoglutathione and degradation by human neutrophils. *Anal. Biochem.* **204,** 365–371.

Conroy, P. J., Nodes, J. T., Slater, T. F., and White, G. W. (1977). The inhibitor effects of a 4-hydroxypentenal: cysteine adduct against sarcoma 180 cells in mice. *Eur. J. Cancer* **13,** 55–63.

Davis, M. E. (1988). Effects of AT-125 on the nephrotoxicity of hexachloro-1,3-butadiene in rats. *Toxicol. Appl. Pharmacol.* **95,** 44–52.

deKernion, J. B. (1986). Renal tumors. *In* "Campbell's Urology" (P. C. Walsh, R. F. Gittes, A. D. Perlmutter, and T. A. Stamey, eds.), 5th ed., pp. 1294–1342. Saunders, Philadelphia.

Dempo, K., Elliot, K. A., Demond, W., and Fishman, W. H. (1981). Demonstration of gamma glutamyl transpeptidase, alkaline phosphatase, CEA and HGG in human lung cancer. *Oncodev. Biol. Med.* **2,** 21–37.

DeYoung, L. M., Richards, W. L., Bonzelet, W., Tsai, L. L., and Boutwell, R. K. (1978). Localization and significance of γ-glutamyl transpeptidase in normal and neoplastic mouse skin. *Cancer Res.* **38,** 3697–3701.

Drobnica, L., Kristian, P., and Augustin, J. (1977). The chemistry of the isothiocyanate group. *In* "The Chemistry of Cyanates and Their Thio-Derivatives" (S. Patai, ed.), pp. 1003–1221, Wiley, Chichester.

Dunnick, J. K., Prejean, J. D., Haseman, J., Thompson, R. B., Giles, H. D., and McConnell, E. E. (1982). Carcinogenesis bioassay of allyl isothiocyanate. *Fundam. Appl. Toxicol.* **2,** 114–120.

Eckert, K.-G., Eyer, P., Sonnenbichler, J., and Zetl, I. (1990). Activation and detoxication of aminophenols. II. Synthesis and structural elucidation of various thiol addition products of 1,4-benzoquinoneimine and *N*-acetyl-1,4-benzoquinoneimine. *Xenobiotica* **20,** 333–350.

Elce, J. S. (1970). Metabolism of a glutathione conjugate of 2-hydroxyoestradiol by rat liver and kidney preparations *in vitro*. *Biochem. J.* **116,** 913–917.

Elce, J. S. (1972). Metabolism of a glutathione conjugate of 2-hydroxyoestradiol-17β in the adult male rat. *Biochem. J.* **126,** 1067–1071.

Elce, J. S., and Chandra, J. (1973). Estrogen mercapturic acids in the adult male rat. *Steroids* **22,** 699–705.

Elce, J. S., and Harris, J. (1971). Conjugation of 2-hydroxy-estradiol-17β [1,3,5,(10)-estratriene-2,3,17β-triol] with glutathione in the rat. *Steroids* **18,** 583–591.

Elce, J. S., Bird, C. E., and Chandra, J. (1973). Water soluble products of the estrogens. Bile metabolites of 2-hydroxyestradiol-17β in man. *J. Clin. Endocrinol. Metab.* **36,** 1027–1030.

Felley-Bosco, E., and Diezi, J. (1987). Fate of cadmium in rat renal tubules: A microinjection study. *Toxicol. Appl. Pharmacol.* **91,** 204–211.

Fenwick,, G. R., Heany, R. K., and Mullin, W. J. (1982). Glucosinulates and their breakdown products in food and food plants. *CRC Crit. Rev. Food. Sci. Nutr.* **181,** 123–201.

Fiala, S., Mohindru, A., Kettering, W. G., Fiala, A. E., and Morris, H. P. (1976). Glutathione and gamma glutamyl transpeptidase in rat liver during chemical carcinogenesis. *J. Natl. Cancer Inst.* (*U.S.*) **57,** 591–598.

Finley, K. T. (1974). The addition and substitution chemistry of quinones. *In* "The Chemistry of the Quinonoid Compounds" (S. Patai, ed.), Part 2, pp. 877–1144. Wiley, New York.

Foulkes, E. C. (1978). Renal tubular transport of cadmium metallothionein. *Toxicol. Appl. Pharmacol.* **45,** 505–512.

Foulkes, E. C., and Blanck, S. (1990). Acute cadmium uptake by rabbit kidneys: Mechanism and effect. *Toxicol. Appl. Pharmacol.* **102,** 464–473.

Fowler, L. M., Moore, R. D., Foster, J. R., and Lock, E. A. (1991). Nephrotoxicity of 4-aminophenol glutathione conjugate. *Hum. Exp. Toxicol.* **10,** 451–459.

Fowler, L. M., Foster, J. R., and Lock, E. A. (1993). Effect of ascorbic acid, acivicin and probenecid on the nephrotoxicity of 4-aminophenol in the Fischer 344 rat. *Arch. Toxicol.* **67,** 613–621.

Fraiser, L., and Kehrer, J. P. (1992). Murine strain differences in metabolism and bladder toxicity of cyclophosphamide. *Toxicology* **75,** 257–273.

Fraiser, L., Skarovsky, C., and Kehrer, J. P. (1993). Bioactivation of a glutathione-acrolein conjugate to bladder toxic species. *Toxicologist* **13,** 226.

Gartland, K. P. R., Bonner, F. W., Timbrell, J. A., and Nicholson, J. K. (1989). Biochemical characterization of 4-aminophenol-induced nephrotoxic lesions in the F344 rat. *Arch. Toxicol.* **63,** 97–106.

Goldbag, J. A., Friedman, O. M., Pineda, E. P., Smith, E. E., Chatterji, R. R., Stein, E. H., and Rutenberg, A. M. (1960). The colorimetric determination of γ-glutamyl transpeptidase with a synthetic substrate. *Arch. Biochem. Biophys.* **91,** 61–70.

Goldstein, R. S. (1993). Biochemical heterogeneity and site-specific tubular injury. *In* "Toxicology of the Kidney" (J. R. Hook and R. S. Goldstein, eds.), 2nd ed., pp. 201–247. Raven Press, New York.

Gonzalez, A., Oberley, T. O., and Li, J. J. (1989). Morphological and immunohistochemical studies of the estrogen-induced Syrian hamster renal tumor: Probable cell of origin. *Cancer Res.* **49,** 1020–1028.

Green, C. R., Ham, K. N., and Tange, J. D. (1969). Kidney lesions induced in rats by 4-aminophenol. *Br. Med. J.* **1,** 162–164.

Haaf, H., Li, S. A., and Li, J. L. (1987). Covalent binding of estrogen metabolites to hamster liver microsomal proteins: Inhibition by ascorbic acid and catechol-O-methyl transferase. *Carcinogenesis* (*London*) **8,** 209–215.

Haenen, H. E. M. G., Temmink, J. H. M., and van Bladeren, P. J. (1993). Differential toxicity of 2-methyl-3-(glutathion-*S*-yl)-1,4-naphthoquinone and 2-methyl-3-(*N*-acetyl-cystein-*S*-yl)-1,4-naphthoquinone after apical or basolateral exposure to rat renal proximal tubular cells. *Toxicologist* **13,** 242.

Harris, C. (1966). A lipofuscin-like pigment in the kidneys of estrogen-treated rats. *Arch. Pathol.* **82,** 353–355.

Hashmi, M., Vamvakas, S., and Anders, M. W. (1992). Bioactivation mechanisms of *S*-(3-oxopropyl)-*N*-acetyl-L-cysteine, the mercapturic acid of acrolein. *Chem. Res. Toxicol.* **5,** 360–365.

Hill, A., and Wolff, S. (1983). Increased induction of sister chromatid exchange by diethylstilbestrol in lymphocytes from pregnant and premenopausal women. *Cancer Res.* **43,** 4114–4118.

Hill, B. A., Kleiner, H. E., Ryan, E. A., Dulik, D. M., Monks, T. J., and Lau, S. S. (1993). Identification of multi-S-substituted conjugates of hydroquinone by HPLC-coulometric electrode array analysis and mass spectroscopy. *Chem. Res. Toxicol.* **6,** 459–469.

Hirose, M., Inoue, T., Masuda, A., Tsuda, H., and Ito, N. (1987). Effects of simultaneous treatment with various chemicals on BHA-induced development of rat forestomatch hyperplasia-complete inhibition by diethylmaleate in a 5-week feeding study. *Carcinogenesis (London)* **8,** 1555–1558.

Hoppen, H.-O., Siekmann, L., and Breuer, H. (1974). Metabolism of oestrone in the microsomal fraction of rat liver. *Hoppe-Seyler's Z. Physiol. Chem.* **355,** 1305–1315.

Horiuchi, S., Inoue, M., and Morino, Y. (1978). γ-Glutamyl transpeptidase. Sidedness of its active site on renal brush-border membrane. *Eur. J. Biochem.* **87,** 429–437.

Horvath, J. J., Witmer, C. M., and Witz, G. (1992). Nephrotoxicity of the 1 : 1 acrolein-glutathione adduct in the rat. *Toxicol. Appl. Pharmacol.* **117,** 200–207.

Ignarro, L. J., Lippton, H., Edwards, J. C., Baricos, W. H., Hyman, A. L., Kadowitz, P. J., and Gruetter, C. A. (1981). Mechanism of vascular smooth muscle relaxation by organic nitrates, nitroprusside and nitric oxide: Evidence for the involvement of *S*-nitrosothiols as active intermediates. *J. Pharmacol. Exp. Ther.* **218,** 739–749.

Jaken, S., and Mason, M. (1978). Differences in the isoelectric focussing patterns of γ-glutamyl transpeptidase from normal and cancerous rat mammary tissue. *Proc. Natl. Acad. Sci. U.S.A.* **75,** 1750–1753.

Jocelyn, P. C. (1972). "Biochemistry of the SH Group," Academic Press, London.

Kari, F. W. (1989). Toxicology and carcinogenesis studies of hydroquinoine in F344/N rats and B6C3F1 mice (Gavage studies). *Nat. Toxicol. Program Tech. Rep. Ser.* **366.**

Kirkman, H. (1959). Estrogen-induced tumors of the kidney. III. Growth characteristics in the Syrian hamster. *Natl. Cancer Inst. Monogr.* **1,** 1–57.

Kleiner, H. E., Monks, T. J., and Lau, S. S. (1993). N-Acetylation/N-deacetylation as a determinant of 2,3,5-triglutathion-*S*-yl)-hydroquinone mediated nephrotoxicity. *Toxicologist* **13,** 241.

Klos, C., Koob, M., Kramer, C., and Dekant, W. (1992). p-Aminophenol nephrotoxicity: Biosynthesis of toxic glutathione conjugates. *Toxicol. Appl. Pharmacol.* **115,** 98–106.

Knuppen, R., Ball, P., Haupt, O., Hollandt, H., and Krass, F. (1983). Synthesis, properties and isolation of catechol estrogens. *In* "Catechol Estrogens" (G. R. Merrian and M. B. Lipsett, eds.), pp. 5–17. Raven Press, New York.

Kubes, P., Suzuke, M., and Granger, D. N. (1991). Nitric oxide: An endogenous modulator of leukocyte adhesion. *Proc. Natl Acad. Sci. U.S.A.* **88,** 4651–4655.

Kuss, E. (1969). Water-soluble metabolites of oestradiol-17β. III. Separation and identification of 1- and 4-glutathione thioethers of 2,3-dihydroxy-oestratrienes. *Hoppe-Seyler's Z. Physiol. Chem.* **350,** 95–97.

Kuss, E. (1971). Microsomal oxidation of estradiol-17β. 2-Hydroxylation and 1- or 4-thioether formation with and without 17-hydroxyl dehydrogenation. *Hoppe-Seyler's Z. Physiol. Chem.* **352,** 817–836.

Lau, S. S., and Monks, T. J. (1990). The *in vivo* disposition of 2-bromo-[^{14}C]hydroquinone and the effect of γ-glutamyl transpeptidase inhibition. *Toxicol. Appl. Pharmacol.* **103,** 121–132.

Lau, S. S., and Monks, T. J. (1991). Glutathione conjugation as a mechanism of targeting latent quinones to the kidney. *In* "Biological Reactive Intermediates. IV. Molecular and

Cellular Effects and Human Impact'' (C. M. Witmer, R. R. Snyder, D. J. Jollow, G. F. Kalf, J. J. Kocsis, and I. G. Sipes, eds.), pp. 457–464, Plenum, New York.

Lau, S. S. Hill, B. A., Highet, R. J., and Monks, T. J. (1988). Sequential oxidation and glutathione addition to 1,4-benzoquinone: Correlation of toxicity with increased glutathione substitution. *Mol. Pharmacol.* **34,** 829–836.

Lau, S. S., Jones, T. W., Highet, R. J., Hill, B. A., and Monks, T. J. (1990). Differences in the localization and extent of the renal proximal tubular necrosis caused by mercapturic acid and glutathione conjugates of menadione and 1,4-naphthoquinone. *Toxicol. Appl. Pharmacol.* **104,** 334–350.

Li, J. J., Li, S. A., Klicka, J. K., Parson, J. A., and Lam, L. K. T. (1983). Relative carcinogenic activity of various synthetic and natural estrogens in the Syrian hamster kidney. *Cancer Res.* **43,** 5200–5204.

Li, S. A., Klicka, J. K., and Li, J. J. (1985). Estrogen 2-4-hydroxylase activity and catechol estrogen formation in the hamster kidney: Implications for estrogen carcinogenesis. *Cancer Res.* **45,** 181–185.

Li, S. I., Purdy, R. H., and Li, S. L. (1989). Variations in catechol O-methyl transferase activity in rodent tissues: Possible role in estrogen carcinogenicity. *Carcinogenesis* (*London*) **10,** 63–67.

Liehr, J. G. (1983). 2-Fluoroestradiol: Separation of estrogenicity from carcinogenicity. *Mol. Pharmacol.* **23,** 278–281.

Lloyd, J. B. (1990). Cell physiology of the rat visceral yolk sac: A study of pinocytosis and lysosome function. *Teratology* **41,** 383–392.

Lock, E. A., and Ishmael, J. (1985). Effect of the organic acid transport inhibitor probencid on renal cortical uptake and proximal tubular toxicity of hexachloro-1,3-butadiene and its conjugates. *Toxicol. Appl. Pharmacol.* **81,** 32–42.

Long, C. J., Shikano, K., and Berkowitz, B. A. (1987). Anion exchange resins discriminate between nitric oxide and EDRF. *Eur. J. Pharmacol.* **142,** 317–318.

Lunte, S. M., and Kissinger, P. T. (1983). Detection and identification of sulfhydryl conjugates of *p*-benzoquinone in microsomal incubations of benzene and phenol. *Chem. Biol. Interact.* **47,** 195–212.

Magnan, S. D. J., Shirota, F. N., and Nahasawa, H. T. (1982). Drug latentiation by γ-glutamyl transpeptidase. *J. Med. Chem.* **20,** 1018–1021.

Manson, M. M., Legg, R. F., Watson, J. V., Green, J. A., and Neal, G. E. (1981). An examination of the relative resistances to aflatoxin B_1 and susceptibilities to γ-glutamyl *p*-phenylene diamine mustard of γ-glutamyl transferase negative and positive cell lines. *Carcinogenesis* (*London*) **2,** 661–670.

Marks, F., and Hecker, E. (1969). Metabolism and mechanism of action of estrogens. XII. Structure and mechanism of formation of water-soluble and protein-bound metabolites of estrone in rat liver microsomes *in vitro* and *in vivo*. *Biochim. Biophys. Acta* **187,** 250–265.

Meister, A., and Tate, S. S. (1980). Glutathione and related γ-glutamyl compounds. Biosynthesis and utilization. *Annu. Rev. Biochem.* **45,** 559–604.

Mellion, B. T., Ignarro, L. J., Myers, C. B., Ohlstein, E. H., Ballot, B. A., Hyman, A. L., and Kadowitz, P. J. (1983). Inhibition of human platelet aggregation by *S*-nitrothiols. Heme-dependent activation of soluble guanylate cyclase and stimulation of cyclic GMP accumulation. *Mol. Pharmacol.* **23,** 653–664.

Mennicke, W. H., Görler, K., and Krumbiegel, G. (1983). Metabolism of some naturally occurring isothiocyanates in the rat. *Xenobiotica* **13,** 203–207.

Mertens, J. J. W. M., Temmink, J. H. M., van Bladeren, P. J., Jones, T. W., Lo, H.-H., Lau, S. S., and Monks, T. J. (1991). Inhibition of γ-glutamyl transferase potentiates the nephrotoxicity of glutathione conjugated chlorohydroquinones. *Toxicol. Appl. Pharmacol.* **110,** 45–60.

Meyers, P. R., Minor, R. L. J., Guerra, R., Jr., Bates, J. N., and Harrison, D. G. (1990). Vasorelaxant properties of the endothelium-derived factors more closely resemble *S*-nitroso-cysteine than nitric oxide. *Nature* (*London*) **345,** 161–163.

Miki, A., and Kugler, P. (1984). Comparative enzyme histochemical study on the visceral yolf sac endoderm in the rat in vivo and in vitro. *Histochemistry* **81,** 409–415.

Minard, F. N., Grant, D. S., Cain, J. C., Jones, P. H., and Knycl, J. J. (1980). Metabolism of γ-glutamyl dopamide and its carboxylic acid esters. *Biochem. Pharmacol.* **29,** 69–75.

Monks, T. J., and Lau, S. S. (1989). Sulfur conjugate-mediated toxicities. *Rev. Biochem. Toxicol.* **10,** 41–90.

Monks, T. J., and Lau, S. S. (1990). Glutathione, γ-glutamyl transpeptidase and the mercapturic acid pathway as modulators of 2-bromohydroquinone oxidation. *Toxicol. Appl. Pharmacol.* **103,** 557–563.

Monks, T. J., and Lau, S. S. (1992). Toxicology of quinone-thioethers. *CRC Crit. Rev. Toxicol.* **22,** 243–270.

Monks, T. J., and Lau, S. S. (1994). Glutathione conjugate mediated toxicities. *In* "Handbook of Experimental Pharmacology", Vol. 112 (F. C. Kauffman, ed.), Springer Verlag Press, Berlin (in press).

Monks, T. J., Lau, S. S., Highet, R. J., and Gillette, J. R. (1985). Glutathione conjugates of 2-bromohydroquinone are nephrotoxic. *Drug Metab. Dispos.* **13,** 553–559.

Monks, T. J., Highet, R. J., Chu, P. S., and Lau, S. S. (1988a). Synthesis and nephrotoxicity of 6-bromo-2,5-dihydroxythiophenol. *Mol. Pharmacol.* **34,** 15–22.

Monks, T. J., Highet, R. J., and Lau, S. S. (1988b). 2-Bromo-(diglutathion-*S*-yl)hydroquinone nephrotoxicity: Physiological, biochemical and electrochemical determinants. *Mol. Pharmacol.* **34,** 492–500.

Monks, T. J., Anders, M. W., Dekant, W., Stevens, J. L., Lau, S. S., and van Bladeren, P. J. (1990a). Glutathione conjugate mediated toxicities. *Toxicol. Appl. Pharmacol.* **106,** 1–19.

Monks, T. J., Highet, R. J., and Lau, S. S. (1990b). Oxidative cyclization, 1, 4-benzothiazine formation and dimerization of 2-bromo-3-(glutathion-*S*-yl)hydroquinone. *Mol. Pharmacol.* **38,** 121–127.

Monks, T. J., Jones, T. W., Hill, B. A., and Lau, S. S. (1991). Nephrotoxicity of 2-bromo-(cystein-*S*-yl)hydroquinone and 2-bromo-(*N*-acetylcystein-*S*-yl)hydroquinone thioethers. *Toxicol. Appl. Pharmacol.* **111,** 279–298.

Naganuma, A., Oda-Urano, N., Tanaka, T., and Imura, N. (1988). Possible role of hepatic glutathione in transport of methylmercury into mouse kidney. *Biochem. Pharmacol.* **37,** 291–296.

Nerland, D. E., and Pierce, W. M. (1990). Identification of *N*-acetyl-*S*-(2,5-dihydroxyphenyl)-L-cysteine as a urinary metabolite of benzene, phenol, and hydroquinone. *Drug. Metab. Dispos.* **18,** 958–961.

Newton, J. F., Kuo, C.-H., Gemborys, M. W., Mudge, G. H., and Hook, J. B. (1982). Nephrotoxicity of 4-aminophenol, a metabolite of acetaminophen in the F344 rat. *Toxicol. Appl. Pharmacol.* **65,** 336–344.

Ney, P., Schroder, H., and Schror, K. (1990). Nitrovasodilator-induced inhibition of LTB4 release from human PMN may be mediated by cyclic GMP. *Eicosanoids* **3,** 243–245.

Oglesby, L. A., Ebron-McCoy, M. T., Logsdon, T. R., Copeland, F., Beyer, P. E., and Kavlock, R. J. (1992). *In vitro* embryotoxicity of a series of *para*-substituted phenols: Structure, activity and correlation with *in vivo* data. *Teratology* **45,** 11–33.

Orlowski, M., Mizoguchi, H., and Wilk, S. (1980). N-Acyl-γ-glutamyl derivatives of sulfamethoxazole as models of kidney-selective prodrugs. *J. Pharmacol. Ther.* **212,** 167–172.

Paul, S. M., Axelrod, J., and Diliberb, E. J. (1977). Catechol-estrogen forming enzyme of brain: Demonstration of a cytochrome P-450 monooxygenase. *Endocrinology* (*Baltimore*) **101,** 1604–1610.

Peraino, C., Richards, W. L., and Stevens, F. J. (1983). Multi-stage-hepatocarcinogenesis. *In* "Mechanisms of Tumor Promotion" (T. J. Slaga, ed.), Vol. 1, pp. 1–53. CRC Press, Boca Raton, FL.

Prota, G. (1988). Progress in the chemistry of melanins and related metabolites. *Med. Res. Rev.* **8,** 525–556.

Redegeld, F. A. M., Hofman, G. A., van der Loo, P. G. F., Koster, A. S., and Noordhoek, J. (1989). Nephrotoxicity of the GSH conjugate of menadione (2-methyl-1,4-naphthoquinone) in the isolated perfused rat kidney. Role of metabolism by γ-glutamyl transpeptidase and probenecid-sensitive transport. *J. Pharmacol. Exp. Ther.* **256,** 665–669.

Reed, D. J., and Meredith, M. J. (1985). Cellular defense mechanisms against reactive metabolites. *In* "Bioactivation of Foreign Compounds" (M. W. Anders, eds.), pp. 71–108. Academic Press, Orlando, FL.

Rosenblum, W. I. (1992). Endothelium-derived relaxing factor in brain blood vessels is not nitric oxide. *Stroke* **23,** 1527–1532.

Rosowsky, A., Wick, M. M., and Kim, S. H. (1979). Structural analogues of L-glutamic acid, γ-(4-hydroxyanilide) and γ-(3,4-dihydroxyanilide) as potential agents against melanoma. *J. Med. Chem.* **22,** 1034–1037.

Rubanyi, G. M., John, W. D., Bates, F. N., and Harrison, D. (1991). Evidence that a *S*-nitrosothiol but not nitric oxide, may be identical with endothelium derived relaxing factor. *J. Cardiovasc. Pharmacol.* **17,** S41–S45.

Rudiger, H. W., Haenisch, F., Metzler, M., Oesch, F., and Glatt, H. R. (1979). Metabolites of diethylstilbestrol induce sister chromatid exchange in human cultured fibroblasts. *Nature* (*London*) **281,** 392–394.

Sawahata, T., and Neal, R. A. (1983). Biotransformation of phenol to hydroquinone and catechol by rat liver microsomes. *Mol. Pharmacol.* **23,** 453–460.

Schilderman, P., ten Vaarwerk, F. J., van Maanen, J. M. S., Lafleur, M. V. M., Westmijze, E. J., ten Hoor, F., and Kleinjans, J. C. S. (1993). Induction of oxidative DNA-damage during prostaglandin H synthase-mediated biotransformation of butylated hydroxyanisole. *Hum. Exp. Toxicol.* **12,** 350.

Shibata, M.-A., Hirose, M., Tanaka, H., Asakawa, E., Shirai, T., and Ito, N. (1991). Induction of renal cell tumors in rats and mice, and enhancement of hepatocellular tumor development in mice after long-term hydroquinone treatment. *Jpn. J. Cancer Res.* **82,** 1211–1219.

Siegfried, J. M., Nelson, K. G., Martin, J. L., and Kaufman, D. G. (1984). Promotional effect of diethylstilbestrol on human endometrial stromal cells pretreated with a direct-acting carcinogen. *Carcinogenesis* (*London*) **5,** 641–646.

Sies, H., and Ketterer, B., eds. (1988). "Glutathione Conjugation: Mechanisms and Biological Significance." Academic Press, San Diego, CA.

Smith, G. D., Charkravarty, P. K., Connors, T. A., and Peters, T. J. (1984). Synthesis and preliminary characterization of a novel substrate for γ-glutamyl transferase. A potential anti-hepatoma drug. *Biochem. Pharmacol.* **33,** 527–529.

Smith, M. T., Evans, C. G., Thor, H., and Orrenius, S. (1985). "Quinone-induced oxidative injury to cells and tissues. *In* "Oxidative Stress (H. Sies, ed.), pp. 91–113. Academic Press, London.

Solt, D. B. (1981). Localization of gamma glutamyl transpeptidase in hamster buccal pouch epithelium treated with 7,12-dimethylbenz[*a*]anthracene. *J. Natl. Cancer. Inst.* **67,** 193–197.

Stark, K. L., Harris, C., and Juchau, M. R. (1987). Embryotoxicity elicited by inhibition of γ-glutamyltransferase by acivicin and transferase antibodies in cultured rat embryos. *Toxicol. Appl. Pharamacol.* **89,** 88–96.

Stillwell, T. J., and Benson, R. C. (1988). Cyclophosphamide-induced hemorrhagic cystisis. *Cancer (Philadelphia)* **61,** 451–457.

Tajima, K., Hashizaki M., Yamamoto K., and Mizutani T. (1991). Identification and structure characterization of *S*-containing metabolites of 3-*tert*-butyl-4-hydroxyanisole in rat urine and liver microsomes. *Drug Metab. Dispos.* **19,** 1028–1033.

Tanaka, T. A., Naganuma, A., and Imura, N. (1990). Role of γ-glutamyltranspeptidase in renal uptake and toxicity of inorganic mercury in mice. *Toxicology* **60,** 187–198.

Tanaka, T. A., Naganuma, A., and Imura, N. (1992). Routes for renal transport of methylmercury in mice. *Eur. J. Pharmacol.* **228,** 9–14.

Taniguchi, N., Higashi, T., Sakamoto, Y., and Meister, A., eds. (1989). "Glutathione Centennial: Molecular Perspectives and Clinical Implications." Academic Press, San Diego, CA.

Tannenbaum, M. (1971). Ultrastructural pathology of human renal cell tumors. *Pathol. Ann.* **6,** 249–277.

Tate, S. S. (1980). Enzymes of mercapturic acid formation. *In* "Enzymatic Basis of Detoxification" (W. B. Jakoby, ed.), Vol. 2, pp. 95–120. Academic Press, Orlando, FL.

Temmink, J. H. M., Bruggeman, I. M., and van Bladeren, P. J. (1986). Cytomorphological changes in liver cells exposed to allyl and benzyl isothocyanate and their cysteine and glutathione conjugates. *Arch. Toxicol.* **59,** 103–110.

Tillian, H., Schauenstein, E., Ertyl, A., and Esterbauer, H. (1976). Therapeutic effects of cysteine adducts of α,β-unsaturated aldehydes on Ehrlich ascites tumor of mice. I. *Eur. J. Cancer* **12,** 989–993.

Tillian, H., Schauenstein, E., Ertyl, A., and Esterbauer, H. (1978). Therapeutic effects of cysteine adducts of α,β-unsaturated aldehydes on Ehrlich ascites tumor of mice. II. *Eur. J. Cancer* **14,** 533–536.

Tsuda, H., Fukushima S., Imaida K., Sakata T., and Ito, N. (1984). Modification of carcinogenesis by antioxidants and other compounds. *Acta Pharmacol. Toxicol.* **55,** 125–143.

Tsutsui, T., Maizumi, H., McLachlan, J. A., and Barrett, J. C. (1983). Aneuploidy induction and cell transformation by diethylstilbestrol: A possible chromosomal mechanism in carcinogenesis. *Cancer Res.,* **43,** 3814–3821.

Tsutsui, T., Degen, G. H., Schiffmann, D., Wong, A., Maizumi, H., McLachlan, J. A., and Barrett, J. C. (1984). Dependence on exogenous metabolic activation for induction of unscheduled DNA synthesis in Syrian hamster embryo cells by diethylstilbestrol and related compounds. *Cancer Res.* **44,** 184–189.

Tunek, A., Platt, K. L., Pryzbylski, M., and Oesch, F. (1980). Multi-step metabolic activation of benzene. Effect of superoxide dismutase on covalent binding to microsomal macromolecules, and identification of glutathione conjugates using high pressure liquid chromatography and field desorption mass spectrometry. *Chem. Biol. Interact.* **33,** 1–17.

Viña, J., ed. (1990). "Glutathione: Metabolism and Physiological Functions." CRC Press, Boca Raton, FL.

Wanless, I. R., and Medline, A. (1982). Role of estrogens as promoters of hepatic neoplasia. *Lab. Invest.* **46,** 313–320.

Wattenberg, L. W. (1981). Inhibition of carcinogen-induced neoplasia by sodium cyanate, tertbutyl isocyanate and benzyl isothiacyanate administered subsequent to carcinogen exposure. *Cancer Res.* **41,** 2991–2994.

Weaver, R. F., Rajagopalan, K. V., Handler P., Rosenthal D., and Jeffs, P. W. (1971). Isolation from the mushroom *Agaricus bisporus* and chemical synthesis of γ-L-glutamyl-4-hydroxybenzene. *J. Biol. Chem.* **246,** 2010–2014.

Wefers, H., and Sies, H. (1983). Hepatic low-level chemiluminesence during redox cycling of menadione and the menadione-glutathione conjugate: Relation to glutathione and NAD(P)H: quinone reductase (DT-diaphorase) activity. *Arch. Biochem. Biophys.* **224,** 568–578.

Weisz, J. (1983). Summary of discussion: Biologic activities of catechol estrogens. *In* "Catechol Estrogens" (G. R. Merrian and M. B. Lipsett, eds.), pp. 215–223. Raven Press, New York.

Wilk, S., Mizoguchi, H., and Orlowski, M. (1978). γ-glutamyl dopa. A kidney-specific dopamine precursor. *J. Pharmacol. Ther.* **206,** 227–232.

Williams, G., Ohmori, T., Katayama, S., and Rice, J. (1980). Alteration by phenobarbital of membrane-associated enzymes including gamma glutamyl transpeptidase. *Carcinogenesis (N.Y.)* **1,** 813–818.

Williams, J. G., Longcope, C., and Williams, K. I. H. (1974). 4-Hydroxyestrone: A new metabolite of estradiol-17β from humans. *Steroids* **24,** 687–701.

Yeates, R. A., Laufen, H., and Leitold, M. (1985). The reaction between organic nitrates and sulfhydryl compounds. A possible model system for the activation of organic nitrates. *Mol. Pharmacol.* **28,** 555–559.

Metabolism and Genotoxicity of Dihaloalkanes

F. Peter Guengerich
Department of Biochemistry and Center in Molecular Toxicology
Vanderbilt University School of Medicine
Nashville, Tennessee 37232

I. Introduction

Dihaloalkanes are of interest because of their demonstrated toxicity and carcinogenicity and their large-scale use in the chemical industry. Most of this discussion will deal with the ethylene dihalides (1,2-dihaloalkanes), particularly ethylene dibromide (1,2-dibromoethane) and ethylene dichloride (1,2-dichloroethane). Ethylene dibromide has been used extensively in the past as a pesticide and an anti-knock agent in leaded gasoline. It produces several kinds of tumors in rodents (National Institute of Occupational Safety and Health, 1981; National Toxicology Program, 1982; Huff, 1983b; Olson *et al.*, 1973; Moslen *et al.*, 1985), although there is no clear epidemiological evidence that it is a human carcinogen. Deaths of humans have been reported in some cases of high exposure (Letz *et al.*, 1984). The use of ethylene dibromide was considerably curtailed in 1983 (Sun, 1984). Ethylene dichloride is still widely used, to the extent of $\sim 5 \times 10^9$ kg per year in the United States, particularly as a precursor of the monomer vinyl chloride in the plastics industry. It produced mouse liver tumors when administered by gavage but not by inhalation (Agency for Toxic Substances and Disease Registry, U. S. Public Health Service, and U. S. Environmental Protection Agency, 1989). Reitz *et al.* (1982) showed that the difference between the two routes might be accounted for by the fivefold higher peak blood levels attained by the gavage route. Thus, the

Advances in Pharmacology, Volume 27

metabolism of the chemicals appears to be critical to their biological action, and most of this chapter will be devoted to mechanisms of bioactivation and detoxication and the characterization of DNA and protein adducts.

The role of glutathione (GSH) conjugation in pathways of metabolism may be traced back to the report of excretion of thioglycolic acid and other related sulfur compounds following ethylene dibromide administration to rats by Nachtomi (1970). However, the general feeling was that this was a pathway of detoxication. Interest in ethylene dihalides increased in the late 1970s with reports of the carcinogenic activity of these compounds (Ames *et al.*, 1980; see this monograph for an excellent review of the earlier literature and controversies in the field.)

A seminal contribution was the report by Rannug *et al.* (1978) that rat liver cytosol was capable of converting ethylene dichloride to a product mutagenic to *Salmonella typhimurium* TA 1535, a base-pair sensitive tester strain, in the presence of GSH. NADPH-fortified liver microsomes were inactive in this regard. This group also demonstrated that mutagenic compounds could be detected in the perfusate when ethylene dichloride was introduced into a liver perfusion system (Rannug and Beije, 1979). This work led to postulation that GSH reacted enzymatically with ethylene dichloride to form a half-mustard—*S*-(2-chloroethyl)glutathione—capable of reacting with DNA; some glutathione *S*-transferase preparations were shown to be active in this regard (Rannug *et al.*, 1978; Rannug, 1980). An episulfonium ion was postulated to be an intermediate in the process of reaction of the half-mustard.

Similar studies were also reported by van Bladeren *et al.* (1980a, 1981a,b), who also showed that *cis*-1,2-dichlorocychohexane was converted to a mutagen (van Bladeren *et al.*, 1979; Buijs *et al.*, 1984). The activity of the *cis* isomer is consistent with an initial S_N2 conjugation reaction and then elimination of chloride to form an episulfonium ion. van Bladeren *et al.* (1980a) also demonstrated the mutagenicity of the methyl ester of *N*-acetyl-*S*-(2-bromoethyl)cysteine and showed that the number of *S. typhimurium* revertants could be attenuated by the addition of GSH. The expected halide order Br > Cl was seen, and addition of substituents to the basic 1,2-dihaloethane moiety decreased the reactivity (van Bladeren *et al.*, 1981a).

Several other lines of evidence argue strongly for the role of GSH conjugation in mutagenicity. DNA strand breaks are attenuated by compounds that deplete GSH (White *et al.*, 1981) and deuterium substitution leads to enhanced toxicity, arguing against a role for oxidation (White *et al.*, 1983, 1984). *S. typhimurium* TA100 mutants having attenuated levels of GSH show low levels of mutation with ethylene dibromide (Kerklaan *et al.,* 1985). The greater mutagenicity of ethylene dihalides in AHH-1 compared to TK6 human lymphoblastoid cells can be attributed to a fivefold difference in levels of glutathione *S*-transferase (Crespi *et al.,* 1985). The formation of DNA adducts from radioactive ethylene dichloride was greater with GSH-fortified rat liver cytosol than with NADPH-fortified microsomes (Guengerich *et al.,* 1980). Ethylene dibromide-dependent unscheduled DNA synthesis can be demonstrated *in vitro* with rat hepatocytes and spermocytes (although only in the hepatocytes was unscheduled DNA synthesis demonstrated after administration of ethylene dibromide *in vivo*) (Working *et al.,* 1986). Modulation of cytochrome P450 and glutathione *S*-transferase activities in the hepatocytes yielded patterns consistent with the view that the GSH conjugation pathway was more important. Unscheduled DNA synthesis has also been demonstrated in cultured human hepatocytes and can be inhibited by depletion of GSH levels (Cmarik *et al.,* 1990). Finally, the increased hepatocarcinogenicity of ethylene dibromide following administration of disulfiram (Wong *et al.,* 1982) can be attributed to the inhibition of oxidation of ethylene dihalides by cytochrome P450 2E1 (Igwe *et al.,* 1986; Kim and Guengerich, 1990). Thus, there is considerable evidence to support the view that GSH conjugation of ethylene dihalides may be critical to their mutagenic and carcinogenic effects.

II. Adducts Derived from 1,2-Dihaloethanes

A. Characterization of DNA Adducts Derived from *vic*-Dihaloethanes

After the results of Rannug's studies became available (Rannug *et al.,* 1978; Rannug and Beije, 1979; Rannug, 1980), this laboratory was able to demonstrate that the covalent attachment of label from ethylene dichloride to protein was catalyzed by rat liver microsomes in the presence of NADPH and that the covalent attachment of label to DNA was catalyzed by rat liver cytosol in the presence of GSH (Guengerich *et al.,* 1980) (Fig. 3). The binding of ethylene dihalides to proteins via oxidation has also been noted by others (Hill *et al.,* 1978; Banerjee and Van Duuren, 1979). The conclusions of Banerjee *et al.* (1980) on the roles of microsomal and cytosolic enzymes in the formation of DNA adducts from ethylene dichloride appear to be in error. Oxidation involves the formation of a *gem*-halohydrin and spontaneous dehydrohalogenation to form 2-chloroacetaldehyde, which is a reactive soft electrophile capable of reacting rapidly with soft nucleophiles such as GSH and protein thiols. 2-

Haloacetaldehydes are also capable of reacting with DNA to form etheno adducts (with adenyl, guanyl, and cytosinyl moieties) (Leonard, 1984) but the kinetics are rather slow (Guengerich *et al.*, 1981). The major enzyme involved in the oxidation of ethylene dichloride and ethylene dibromide is P450 2E1 (Guengerich *et al.*, 1991). Suggestions of the interaction of oxidative and conjugative reactions in the formation of DNA adducts from ethylene dichloride (Guengerich *et al.*, 1980) were the results of artifacts associated with the reversible binding of GSH conjugates (presumably $GSCH_2CHO$) to DNA (Koga *et al.*, 1986). Also, a reaction involving a haloso intermediate was suggested by some of the early results (Guengerich *et al.*, 1980) but later work with ^2H and ^{18}O isotopes is only consistent with hydroxylation at the methylene to form the *gem*-halohydrin (Guengerich *et al.*, 1986).

The GSH conjugation pathway responsible for the formation of DNA adducts and bacterial mutagenicity has been studied in considerable detail. The pathway involving a GSH half-mustard suggested by Rannug (Rannug and Beije, 1979; Rannug, 1980) predicts that radiolabels from both the ethylene dihalide and GSH should become attached to DNA, and this equimolar binding was demonstrated in this laboratory with a partially purified glutathione *S*-transferase preparation and with rat hepatocytes (Ozawa and Guengerich, 1983). The structure *S*-[2-(N^7-guanyl)ethyl]glutathione (Figs. 1 and 2) was proposed for the major DNA adduct after N^7-ethylguanine was identified following modified Raney nickel reduction (Ozawa and Guengerich, 1983). Later the intact DNA adduct was isolated from DNA by neutral thermal hydrolysis and reversed-phase and ion-exchange HPLC and definitively characterized as *S*-[2-(N^7-guanyl)ethyl]-

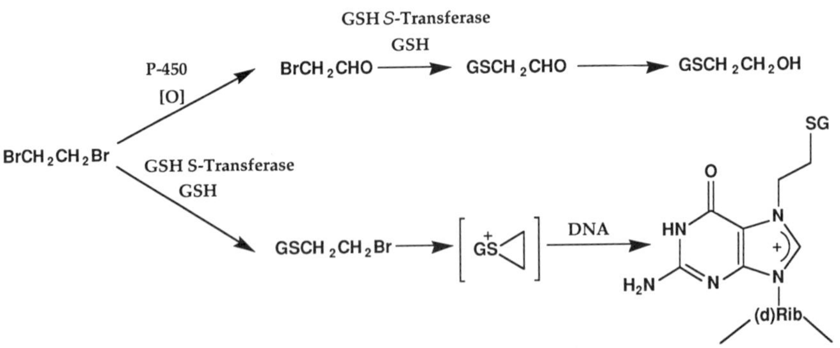

Fig. 1 Possible mechanisms for activation of ethylene dibromide and other 1,2-dihaloalkanes through oxidation (upper pathway) and GSH conjugation (lower pathway).

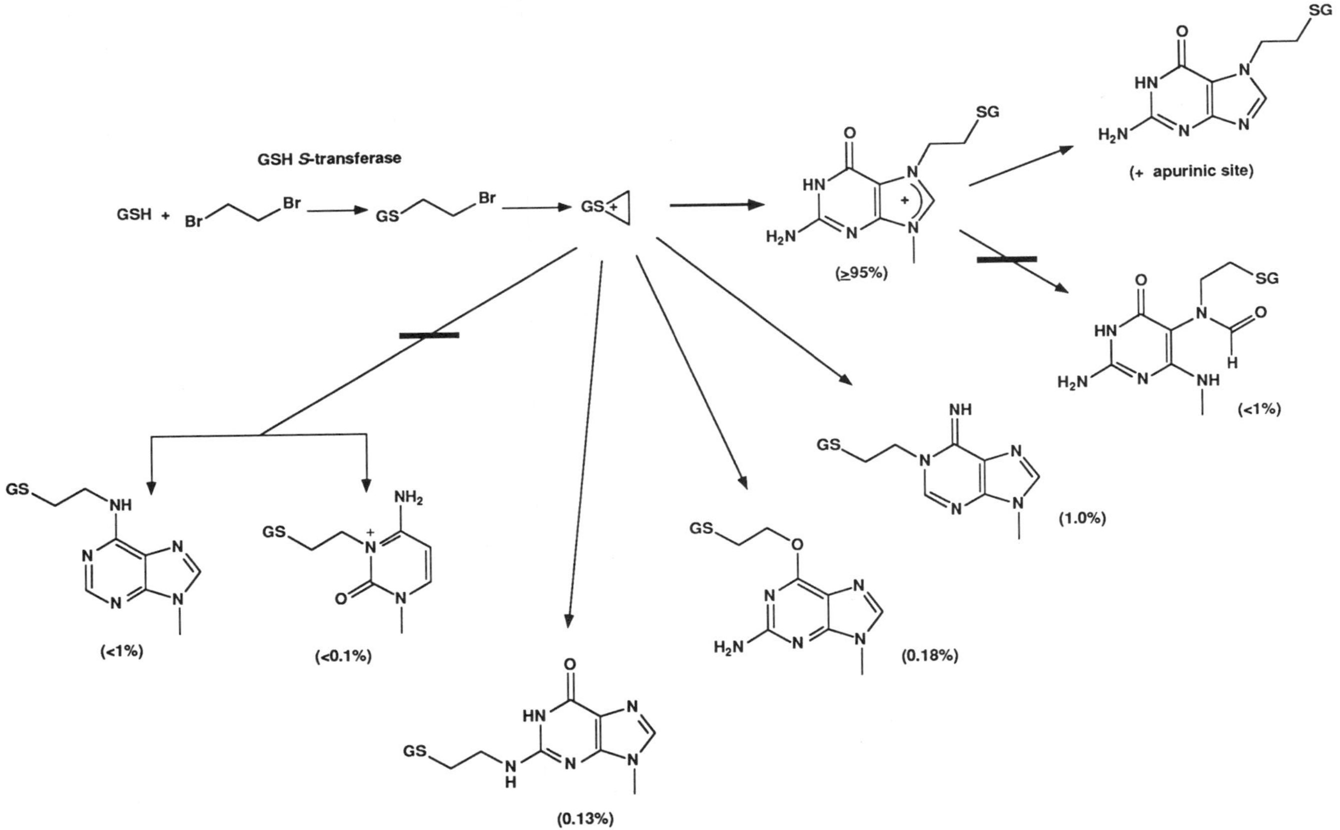

Fig. 2 DNA adducts shown to be derived from ethylene dibromide. The percentages in parentheses show the approximate yields in typical experiments—see Cmarik *et al.* (1992). Three of the postulated adducts could not be detected at the indicated levels when the authentic compounds were used as standards.

glutathione by NMR and mass spectroscopy (Koga *et al.*, 1986). This adduct has consistently accounted for >95% of the DNA or RNA adducts formed *in vitro* or *in vivo* in this laboratory over a number of years (Ozawa and Guengerich, 1983; Inskeep *et al.*, 1986; Koga *et al.*, 1986; Peterson *et al.*, 1988; and Kim and Guengerich, 1989, 1990; Humphreys *et al.*, 1990; Oida *et al.*, 1991; Cmarik *et al.*, 1992). The level of this adduct is relatively high with regard to other DNA adducts; levels of 200–500 pmol adduct (mg DNA)$^{-1}$ can be routinely obtained *in vitro* or *in vivo*. The level of this adduct formed from ethylene dichloride is at least two orders of magnitude lower and has not been rigorously quantified (Ozawa and Guengerich, 1983; Inskeep *et al.*, 1986).

Minor adducts have been detected following mild acid hydrolysis of DNA (Fig. 2). *S*-[2(N^1-Adenyl)ethyl]glutathione was isolated from RNA and identified by its NMR, mass, and UV spectra and by the formation of N^1-ethyladenine following modified Raney nickel reduction (Kim *et al.*, 1990). The extent of formation is ~1–2%, with the level being somewhat higher in RNA rather than DNA because of the exposure of the base pairing region in single-stranded material (Kim *et al.*, 1990). In the same report, *S*-[2-(N^7-guanyl)ethyl]cysteinylglycine was isolated but shown to be an artifact of the acid treatment. These and previous studies (Inskeep *et al.*, 1986) indicate that cleavage of the γ-glutamyl amide bond is not required for DNA binding nor does it occur readily in the case of ethylene dibromide (even in kidney, where γ-glutamyltranspeptidase is present). However, *S*-[2-(N^7-guanyl)ethyl]cysteinylglycine may be an adduct formed from ethylene dichloride (Inskeep *et al.*, 1986) because of the greater stability of *S*-(2-chloroethyl)glutathione relative to *S*-(2-bromoethyl)glutathione (Marchand and Reed, 1989). Two other DNA adducts have been identified (by cochromotography with synthetic materials) in DNA treated with (2-chloroethyl)glutathione- *S*-[2-(N^2-guanyl)ethyl]glutathione and *S*-[2-(O^6-guanyl)ethyl]glutathione were present at levels of 0.1–0.2% of the total adducts (Cmarik *et al.*, 1992). The finding of traces of the O^6-guanyl derivative is consistent with the reported formation of some O^6-guanyl adduct from chloroethyl sulfide (Ludlum *et al.*, 1986).

Several characterized adducts have been shown not to be present at detectable levels in DNA modified with ethylene dibromide or *S*-(2-chloroethyl)glutathione. The list includes *S*-[2-(N^4-cytosinylethyl]glutathione, *S*-[2-(N^6-adenyl)ethyl]glutathione, and the guanyl imidazole ring-opened formamidopyrimidine product, *S*-[2-[*N*-formyl-*N*-(2,6-diamino-4-oxo-3,4-dihydropyrimidin-5-yl)amino]ethyl]glutathione (Kim *et al.*, 1990; Cmarik *et al.*, 1992) (Fig. 2).

Shih and Hill (1981) reported the binding of radiolabel from ethylene dibromide to polycytidylic acid in the presence of rat liver cytosol and

GSH, although no adducts were characterized and a later search for *S*-[2-(N^3-cytosinyl)ethyl]glutathione was unsuccessful (Cmarik *et al.*, 1992).

B. Mechanism of Enzymatic Formation of DNA Adducts

The formation of DNA adducts bearing ethyl-glutathione moieties involves enzymatic formation of the half-mustard *S*-(2-haloethyl)glutathione as the initial step and this is the only enzyme-catalyzed reaction. In our studies on the catalytic specificity of purified cytosolic glutathione *S*-transferases, rat enzymes 2-2 and 3-3 and human enzymes 1A1-1A1 and 1A2-1A2 were most active of those tested (Cmarik *et al.*, 1990). Recently a rat mitochondrial θ family glutathione S-transferase has also been shown to have high activity (B. Ketterer, personal communication).

An episulfonium, or thiiranium, ion is implicated in the reaction of the half-mustard with DNA as well as other nucleophiles, as clearly demonstrated in studies with stereospecifically labeled ethylene dibromide (Peterson *et al.*, 1988) and other studies. Incubation of [*erythro*-$^2H_2$]- or [*threo*-$^2H_2$]ethylene dibromide with rat liver cytosol, GSH, and DNA led to essentially complete inversion of stereochemistry (Fig. 3).

Anders and Livesey (Livesey and Anders, 1979; Anders and Livesey, 1980; Livesey *et al.*, 1982) found that model half-mustards generate ethylene through an elimination reaction; presumably this occurs with the GSH half-mustards as well but has not been examined. The generation of *trans*-2-butene from *meso*-2,3-dibromobutane and that of *cis*-2-butene from *racemic*-2,3-dibromobutene are indicative of an E_2 mechanism involving the episulfonium ion and consistent with the results of the adduct studies (Peterson *et al.*, 1988; Livesey and Anders, 1979; Livesey *et al.*, 1982; Anders and Livesey, 1980).

In addition, other model studies support the role of episulfonium ions in reactions of these half-mustards. Dohn and Casida (1987) demonstrated the scrambling of methylene deuterium label between the two methylenes in the halogenation of *S*-(2-hydroxyethyl)cysteine, and we found similar scrambling in all of our own efforts to prepare half-mustards or *S*-(2-ethyl)cysteine derivatives bearing good cleaving groups (Peterson *et al.*, 1988). Dohn and Casida (1987) also observed an NMR spectrum consistent with that of an episulfonium ion when they treated *S*-(2-hydroxyethyl)cysteine with trifluoromethylsulfonic acid. On the basis of kinetic studies with *S*-(2-chloroethyl)cysteine, Schasteen and Reed (1983) suggested that a thiomorpholine derivative might be a reactive intermediate instead of an episulfonium ion. However, the pH dependence of the *S*-(2-chloroethyl)glutathione reaction can be interpreted in light of the removal of the inhibitory effect of the protonated amino group of *S*-

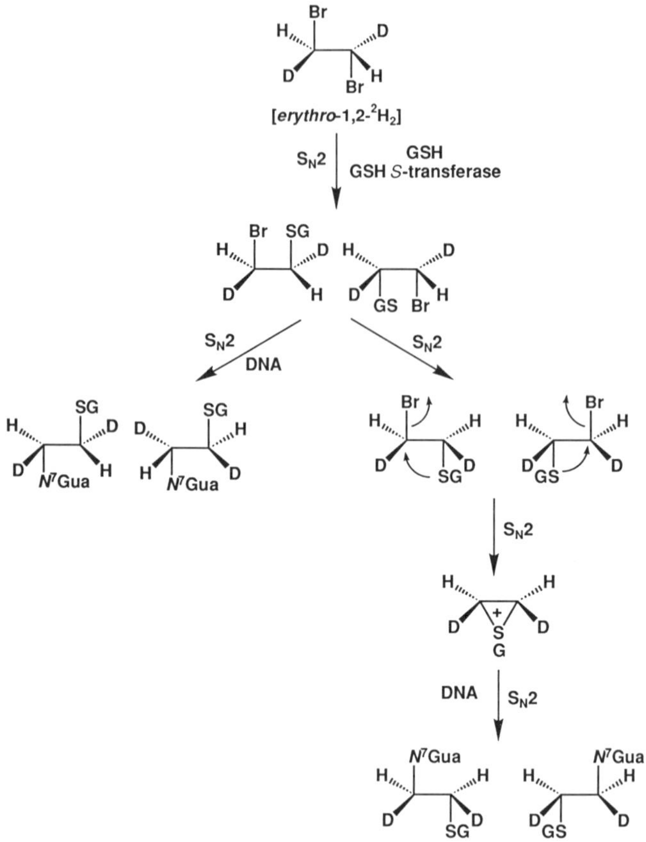

Fig. 3 Possible mechanisms of reaction of *S*-(2-bromoethyl)glutathione generated from [*erythro*-1,2-$^2H_2$]ethylene dibromide with nucleophiles and stereochemical consequences (Peterson *et al.*, 1988). The difference between an even and odd number of inversion (S_N2) steps gives rise to different stereochemistry.

(2-chloroethyl)glutathione at high pH (Peterson *et al.*, 1988). Vadi *et al.* (1985) demonstrated the nicking of pBR322 plasmid DNA by *S*-(2-chloroethyl)glutathione but not *S*-(2-chloroethyl)glutathione and also suggested the lack of involvement of an episulfonium ion on this basis. The differential cleavage of plasmid DNA by the two half-mustards has been repeated (Humphreys *et al.*, 1991) but the differences clearly cannot be accounted for in terms of the extent of formation of DNA adducts (Humphreys *et al.*, 1990).

Other work with model half-mustards indicates that reaction with the model nucleophile 4-(4′-nitrobenzyl)pyridine is essentially pH independent except when the cysteinyl α-amino group is free and protonated (Peterson *et al.*, 1988). This result is interpreted to mean that the protonated amino group is in a position to inhibit development of a transition state leading to the positively charged episulfonium ion; extension of the carbon chain by one methylene group [*S*-(2-chloro)homocysteine] removed the inhibition (Peterson *et al.*, 1988). Nonvicinal, α,ω-dihaloalkanes are good substrates for glutathione *S*-transferase but do not yield DNA adducts because of the lack of anchimeric assistance (i.e., inability to generate episulfonium ions) (Buijs *et al.*, 1984; Inskeep and Guengerich, 1984).

Half-mustards and episulfonium ions react not only with nucleic acids but also with GSH (Ozawa and Guengerich, 1983; Cmarik *et al.*, 1990), protein sulfhydryls, and H_2O (Ozawa and Guengerich, 1983); *S*-(2-hydroxyethyl)glutathione is a major product (Ozawa and Guengerich, 1983). Conceivably sulfation of the hydroxyl moiety would form a good leaving group and regenerate the episulfonium ion. However, evidence against such a reaction has been presented (Kim and Guengerich, 1990).

Lin *et al.* (1985) reported the formation of etheno adducts in polyadenylic acid and polycytidylic acid incubated in the presence of ethylene dichloride and NADPH-fortified microsomes; indeed the formation of 1,N^6-ethenoadenosine from adenosine can be demonstrated in such reactions because of the slow reaction of 2-chloroacetaldehyde with DNA (Guengerich *et al.*, 1981). However, the level of mutations is very low (Rannug *et al.*, 1976; Rannug, 1980; van Bladeren *et al.*, 1981a; Guengerich *et al.*, 1980). Lin *et al.* (1985) were unable to detect DNA adducts in GSH-supported reactions, but it is not clear that the separation methods were adequate for dealing with such a hydrophilic residue (Ozawa and Guengerich, 1983; Koga *et al.*, 1986). Only low levels of hepatic DNA adducts are formed from ethylene dichloride *in vivo* (Inskeep *et al.*, 1986) and one of these migrated with authentic *S*-[2-(N^7-guanyl)ethyl]glutathione on HPLC. Another adduct may be the derivative in which the γ-glutamyl residue has been removed but further evidence is lacking. An important consideration with ethylene dichloride is the greater stability of *S*-(2-chloroethyl)glutathione than *S*-(2-bromoethyl)glutathione and the finding of a finite level of the latter half-mustard in bile (Marchand and Reed, 1989). Thus *S*-(2-chloroethyl)glutathione may also be able to migrate to the kidney, although the DNA adducts formed there (Kim and Guengerich, 1989; Inskeep *et al.*, 1986) could also have been formed within that tissue.

C. Modulation of DNA Adduct Levels

The only studies on adduct levels have involved the major adduct, *S*-[2-(N^7-guanyl)ethyl]glutathione. The level of this adduct found in both liver and kidney DNA is directly proportional to the dose, at least in the dose range 2–37 mg kg^{-1} (Kim and Guengerich, 1989). GSH is required for generation of the adduct but a second GSH can react with the half-mustard to form the bis conjugate (Ozawa and Guengerich, 1983; Cmarik *et al.*, 1990):

$$GSH + BrCH_2CH_2Br \rightarrow GSCH_2CH_2Br \xrightarrow{GSH} GSCH_2CH_2SG$$

In *in vitro* experiments the optimal GSH concentration for DNA adduct formation of DNA adducts is ~10 m*M*, which is near the normal physiological level. The effect of *in vivo* modulation of GSH and enzyme levels was considered (Kim and Guengerich, 1990). Depletion of GSH levels by treatment of rats with either diethylmaleate or buthionine sulfoximine lowered the formation of hepatic *S*-[2-(N^7-guanyl)ethyl]glutathione adducts. The levels of adducts could be increased by treatment of the animals with butylated hydroxytoluene, a known inducer of glutathione *S*-transferases. The level of adduct formation could also be induced by treatment of the animals with disulfiram. Such treatment did not affect the level of *in vitro* glutathione *S*-transferase activity seen with ethylene dibromide (Kim and Guengerich, 1990) but is known to inhibit cytochrome P450 2E1 (Guengerich *et al.*, 1991; Brady *et al.*, 1991). Such inhibition appears to block the oxidation of ethylene dibromide, effectively a detoxication reaction, and make more substrate available for activation via the GSH conjugation pathway (Kim and Guengerich, 1990). Indeed, the effect of disulfiram on the pharmacokinetics of ethylene dichloride has been demonstrated (Nachtomi, 1981; van Bladeren *et al.*, 1981c; Milks *et al.*, 1982; Igwe *et al.*, 1986; Cheever *et al.*, 1990). This finding is significant in light of the demonstrated strong cocarcinogenic effect of disulfiram in induction of liver tumors by ethylene dibromide (Wong *et al.*, 1982) and ethylene dichloride (Cheever *et al.*, 1990), and the increase in both the level of GSH-derived DNA adduct and tumorigenicity is consistent with the view that this pathway is involved in tumorigenesis.

D. Fates of DNA Adducts

The major fate of *S*-[2-(N^7-guanyl)ethyl]glutathione adducts in DNA is depurination (Inskeep and Guengerich, 1984) (Fig. 2). Apparently no opening of the imidazole ring occurs (Kim *et al.*, 1990). *S*-[2-(N^7-Guanyl)ethyl]-glutathione is released and degraded to the mercapturic acid *N*-acetyl-*S*-[2-(N^7-guanyl)ethyl]cysteine, which is excreted in urine (Kim and Guengerich, 1989). In principle this product might be a useful marker for biologically effective doses of ethylene dihalides, since the amount excreted appears to parallel DNA adduct levels (Kim *et al.*, 1990). However, considerable increases in sensitivity of the methodology would be required before application.

The $t_{1/2}$ for release of *S*-[2-(N^7-guanyl)ethyl]glutathione from calf thymus DNA at 37°C and pH 7.0 is ~150 h (Inskeep *et al.*, 1986). However, *in vivo* the $t_{1/2}$ is ~70–100 h (Inskeep *et al.*, 1986; Kim *et al.*, 1990). A similar $t_{1/2}$ was found for *S*-[2-(N^1-adenyl)ethyl]glutathione and a minor, unidentified DNA adduct (Kim *et al.*, 1990). These results suggest that there might be some enzymatic repair of the *S*-[2-(N^7-guanyl)ethyl]glutathione lesion in DNA. In *Escherichia coli,* the mutation frequency of *S*-(2-chloroethyl)glutathione was enhanced in strains devoid of the uvr_{ABC} repair system (Cmarik *et al.*, 1992), suggesting a role for glycosylases. Further, ethylene dibromide has been shown to induce GSH-dependent unscheduled DNA synthesis in cultured hepatocytes and spermocytes (Cmarik *et al.*, 1990; Working *et al.*, 1986). These findings may be interpreted to mean that repair systems are operative.

E. Protein and Glutathione Adducts and Toxicity

Half-mustards are toxic in isolated hepatocytes and this toxicity is probably related to alkylation of proteins, not DNA. The reaction of ethylene dihalides to form bis(ethylene)glutathione adducts is well known (Ozawa and Guengerich, 1983; Nachtomi, 1970; Jean and Reed, 1989, 1992; Webb *et al.*, 1987). Treatment of isolated hepatocytes with *S*-(2-chloroethyl)cysteine decreased GSH levels, Ca^{2+} transport, Ca^{2+}-dependent ATPase activity, and cell viability (Webb *et al.*, 1987). Lipid peroxidation has been postulated (Tomasi *et al.*, 1983) but did not appear to be involved in the studies of Webb *et al.* (1987); indeed there is little solid evidence that reactions involving free radicals are of significance in the metabolism of any dihaloalkanes. Nevertheless, there does appear to be a considerable protective effect of vitamin E against cell toxicity (Webb *et al.*, 1987; Warren *et al.*, 1991; Warren and Reed, 1991).

Studies with liver microsomes and cytosolic fractions suggest that protein adducts are more readily formed by microsomal oxidation of ethylene dihalides to 2-haloacetaldehydes and reaction with thiols, as opposed to the cytosolic GSH conjugation pathway (Guengerich *et al.*, 1980, 1981). Jean and Reed (1989) have reported that *S*-(2-chloroethyl)glutathione reacts 50 times more rapidly with thiols than with DNA guanyl residues (or His). They later published evidence that, in rat hepatocytes, most of the formation of *S*-(2-hydroxyethyl)glutathione was due to the hydrolysis of *S*-(2-haloethyl)glutathione rather than conjugation of GSH with 2-haloacetaldehydes formed in oxidation (Jean and Reed, 1992). However, this conclusion appears to be inconsistent with the *in vivo* work of van Bladeren *et al.* (1981b). In that study more than 50% of the dose of ethylene dibromide given to rats could be accounted for as *N*-acetyl-*S*-(2-hydroxyethyl)cysteine recovered in urine. In further studies with deuterium labeling, the found that >80% of the mercapturic acid was derived from the oxidative pathway, since our deuterium atom had been lost from the perdeuterated ethylene dibromide. Indeed, in light of the kinetic isotope effect (Hales *et al.*, 1987; Guengerich *et al.*, 1986), this value is probably an underestimate. Jean and Reed (1992) also suggested that most of the protein adducts formed in hepatocytes are the result of the GSH half-mustard pathway, although this conclusion is only based upon the kinetics of binding relative to GSH depletion. No studies have yet been reported that distinguish the protein adducts derived from the two pathways (presumably $CySCH_2CH_2OH$ or $CySCH_2CO_2H$ vs $CySCH_2CH_2GSH$).

III. Other Dihaloalkanes

Most of this discussion has dealt with ethylene dibromide and, to a lesser extent, ethylene dichloride and 1-bromo-2-chloroethane. The addition of functional moieties to the basic 1,2-dihaloalkane structure tends to lower the extent of DNA adduct formation (Inskeep and Guengerich, 1984; van Bladeren *et al.*, 1981a), either by retarding GSH conjugation, episulfonium ion formation, or subsequent reactions. However, several other compounds in this category are of particular interest in light of their use as pesticides and solvents.

A. 1,2-Dibromo-3-chloropropane

1,2-Dibromo-3-chloropropane (DBCP) is of interest because of its past use as a pesticide and its toxicity in kidney and testis (Huff, 1983a; Rao *et al.*, 1983; Omichinski *et al.*, 1988). It can also produce rodent tumors at several sites (Huff, 1983a) and its use has been considerably restricted because of these concerns (Anonymous, 1983). There is considerable evidence that GSH conjugation is involved in the activation of DBCP. The characterized GSH conjugates support such a view (Pearson *et al.*, 1990; Humphreys *et al.*, 1991). In addition, N^7-guanyl DNA adducts have also been formed *in vitro* and characterized (Fig. 4) (Humphreys *et al.*, 1991). These conjugates can all be readily rationalized by pathways involving episulfonium ion intermediates with nucleophilic attacks on the unsubstituted methylene carbon in every case. Some products considered in

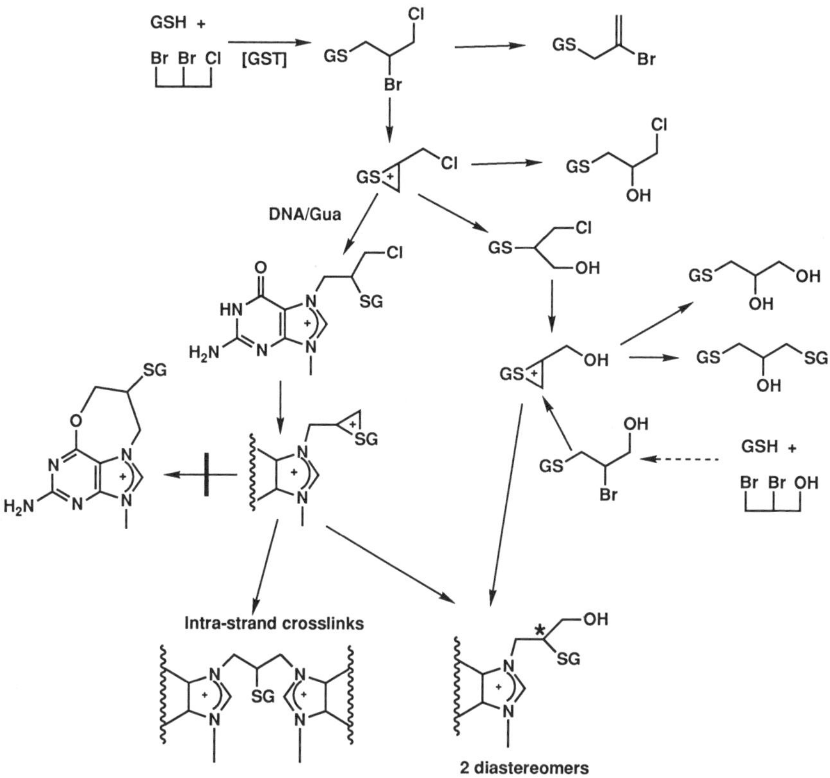

Fig. 4 Activation of 1,2-dibromo-3-chloropropane by GSH conjugation and structures of DNA adducts found in *in vitro* experiments (Humphreys *et al.*, 1991).

our earlier report are of interest. In one case *S*-(2-bromo-2-propenyl)glutathione is apparently formed in the reaction of GSH with the dehydrochlorinated product of DBCP (Fig. 4). A minor DNA adduct is highly fluorescent but could not be characterized in that study. We also considered the possibility that generation of the episulfonium ion at the guanyl adduct state could lead to a reaction at the guanyl O^6 atom, but no evidence for such a product could be obtained (Fig. 4) (Humphreys *et al.*, 1991).

However, the biological relevance of these DBCP adducts still remains unclear. Although the levels of the N^7-guanyl adducts formed *in vitro* are nearly as high as measured in the case of ethylene dibromide, only a trace of material migrating on HPLC with standard N^7-guanyl adducts could be detected in liver DNA isolated from rats treated with a high dose of DBCP, even when disulfiram was administered to block oxidative metabolism (Humphreys *et al.*, 1991). Also, when the GSH–DBCP conjugate (half-mustard) was added to *S. typhimurium* TA100, only a low level of N^7-guanyl adduct was found in bacterial DNA and no (base-pair) mutations were observed. The result could not be rationalized simply by competition of protein with DNA for reaction with the half-mustard. Finally, considerable evidence has been presented that administration of DBCP to rats leads to extensive DNA strand breaks (Søderlung *et al.*, 1988; Brunborg *et al.*, 1990). However, conversion of plasmid pBR322 DNA Form I to Form II (nicked) could not be detected with either *S*-(2-chloroethyl)glutathione or the DBCP–GSH half-mustard, but *S*-(2-chloroethyl)cysteine did produce nicks (Humphreys *et al.*, 1991). Whether any of the N^7-guanyl adducts formed *in vitro* has any connection with the *in vivo* toxicity of DBCP must still be considered an open question. The bis-guanyl DNA adduct (Fig. 4) could possibly be linked to toxicity, if traces of strand crosslinking occur. However, no evidence for its *in vivo* occurrence has been found.

B. 1,4-Diiodobutane

As pointed out earlier, increasing the number of methylenes of >2 in α,ω-dihaloalkanes has the effect of blocking the ability to form alkylated products because of the lack of anchimeric assistance (Buijs *et al.*, 1984). However, with 1,4-dihalobutanes the extended size of the chain allows the attack of the sulfur atom on the terminal methylene group and it is possible to form a 5-membered-ring thialonium ion (Marchand and Abdel-Monem, 1985) (Fig. 5). This product appears to be stable but is decomposed by alkalai to yield tetrahydrothiophene. There is currently no evidence that this conjugation product leads to any toxicity.

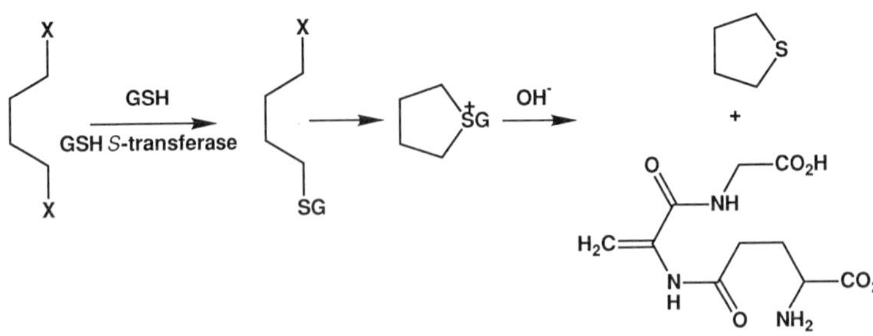

Fig. 5 Conjugation of 1,4-diiodobutane with GSH and formation of a stable episulfonium species (Marchand and Abdel-Monem, 1985).

C. Methylene Dihalides

The toxicology of methylene dihalides is of considerable interest because of the widespread use of methylene dichloride and the observation that it can produce mouse (but not rat) liver tumors (Andersen *et al.,* 1987). Methylene dihalides also produce base-pair mutations in *S. typhimurium* TA100 (van Bladeren *et al.,* 1980b; Jongen *et al.,* 1978; Green, 1983; Osterman-Golkar *et al.,* 1983). These compounds are metabolized via two initial routes, as demonstrated by Anders and his associates (Fig. 6) (Ahmed and Anders, 1976, 1978; Kubic and Anders, 1978). Oxidation leads to carbon monoxide via formyl chloride. Conjugation with GSH yields HCHO following decomposition of the intermediate.

The suggestion has been made that the GSH conjugation pathway is relevant to tumorigenesis since both phenomena show a lack of saturation with increasing dose, and pharmacokinetic models have been prepared (Andersen *et al.,* 1987; Reitz *et al.,* 1989). If this is the case, then humans would appear to be at decreased risk relative to mice because of lower enzyme activity (Reitz *et al.,* 1989). However, the role of GSH in the bacterial mutagenicity is not particularly clear (van Bladeren *et al.,* 1980b; Jongen *et al.,* 1978; Green, 1983; Osterman-Golkar *et al.,* 1983). Recently the conjugation of CH_2Cl_2 has been attributed to glutathione *S*-transferase enzymes in the so-called θ family (e.g., rat 5-5) (Meyer *et al.,* 1991; La Roche and Leisinger, 1990). Whether these reactions lead to DNA alkylation is still not proven. Heck and associates have suggested that the HCHO released in the reaction leads to the formation of DNA–protein crosslinks and that these are relevant to tumorigenesis (Heck *et al.,* 1990). However, HCHO is not very mutagenic, and the relevance of metabolism to tumorigenesis has remained unclear.

Recently a cDNA for rat glutathione *S*-transferase 5-5 has been expressed in *S. typhimurium* TA1535 and shown to yield revertants when

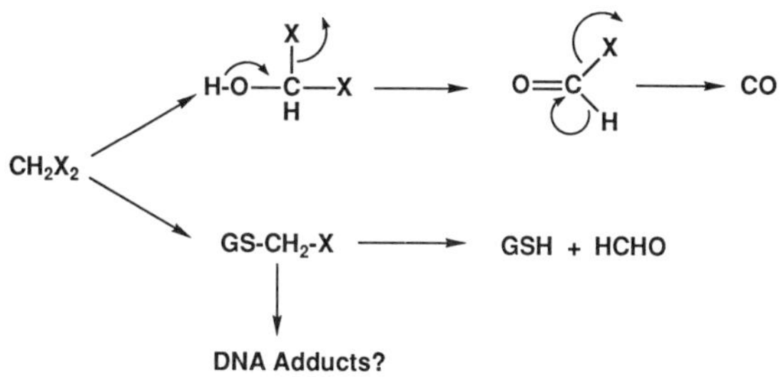

Fig. 6 Postulated mechanisms of oxidative and conjugative transformation of dihalothanes (Ahmed and Anders, 1978; Kubic and Anders, 1978).

the bacteria are exposed to CH_2BR_2 (Thier *et al.*, 1993). Mutations were not produced when the "reverse" construct was inserted or when the bacteria were treated with HCHO. These studies suggest that a GSH adduct is actually responsible for the mutagenicity.

IV. Basis of Mutagenicity of Ethylene Dibromide

Ethylene dibromide is a tumor initiator (Moslen *et al.*, 1985), and this biological activity is thought to be the result of mutations and possibly more complex genetic damage associated with DNA alkylation (Rannug, 1980). Only base-pair mutations are induced, either by ethylene dibromide or by ethylene dichloride activated by rodent cytosol or by *S*-(2-chloroethyl)glutathione (Rannug, 1980; Rannug and Beije, 1979; Rannug *et al.*, 1978). The working hypothesis is that the mutagenicity is the result of the major adduct *S*-[2-(N^7-guanyl)ethyl]glutathione although, as emphasized later, direct proof for this proposal has not yet been obtained.

It has been possible to synthesize defined deoxyribooligonucleotides with the adduct *S*-[2-(N^7-guanyl)ethyl]glutathione inserted at specific positions (Oida *et al.*, 1991; Min and Guengerich, 1993). The presence of the adduct has a considerable disruptive effect on hybridization of an oligonucleotide with its complement, e.g., a decrease of ~15°C in the T_m of a 7-mer or 9-mer (Oida *et al.*, 1991; Kim *et al.*, 1993). The structure of the modified 7-mer d(CATG*CCT)/d(AGGCATG), where G* = *S*-[2-(N^7-guanyl)ethyl]glutathione, has been examined in detail by NMR methods with two-dimensional correlated spectroscopy (COSY) to define atoms and two-dimensional nuclear Overhauser enhanced spectroscopy (NOESY) to probe spatial arrangements. The basic B-helix appears to be intact and no detectable interactions between the GSH moiety and the sugar or nucleic acid base hydrogens were observed (Oida *et al.*, 1991). Analysis of the hydrogen bonds and the H1′ deoxyribose resonance shifts indicated that disruption tends to be localized at the G* : C mismatch (Oida *et al.*, 1991).

Double-stranded bacteriophage M13mp18 DNA was treated with *S*-(2-chloroethyl)glutathione and allowed to replicate in *S. typhimurium;* analysis of the mutants in the *lac*Z α-complementation region revealed that essentially only base-pair mutations occurred and these were strongly dominated by GC to AT transitions (Cmarik *et al.*, 1992). In the same set of experiments, the positions of N^7-guanyl adducts were defined by sequence analysis following treatment of a segment of the DNA with hot piperidine. Both alkylations and mutations were clustered in runs of guanines, but outside of these there was no correspondence of alkylations and mutations (Cmarik *et al.*, 1992) (Fig. 7). These results have been interpreted to mean that there is considerable sequence selectivity in the production of mutations. The finding that GC to AT transitions are dominant (Table I) argues that guanyl adducts are the most mutagenic, since no cytosinyl adducts have been identified. Further, the results argue that apurinic sites are not the principal species responsible for mutations, since they would be expected to yield G to T transversions because of their noninstructional nature (Cmarik *et al.*, 1992; Sagher and Strauss, 1983).

These findings with *S*-(2-chloroethyl)glutathione are consistent with the observations of Foster *et al.* (1988) with ethylene dibromide at the *hisG46*

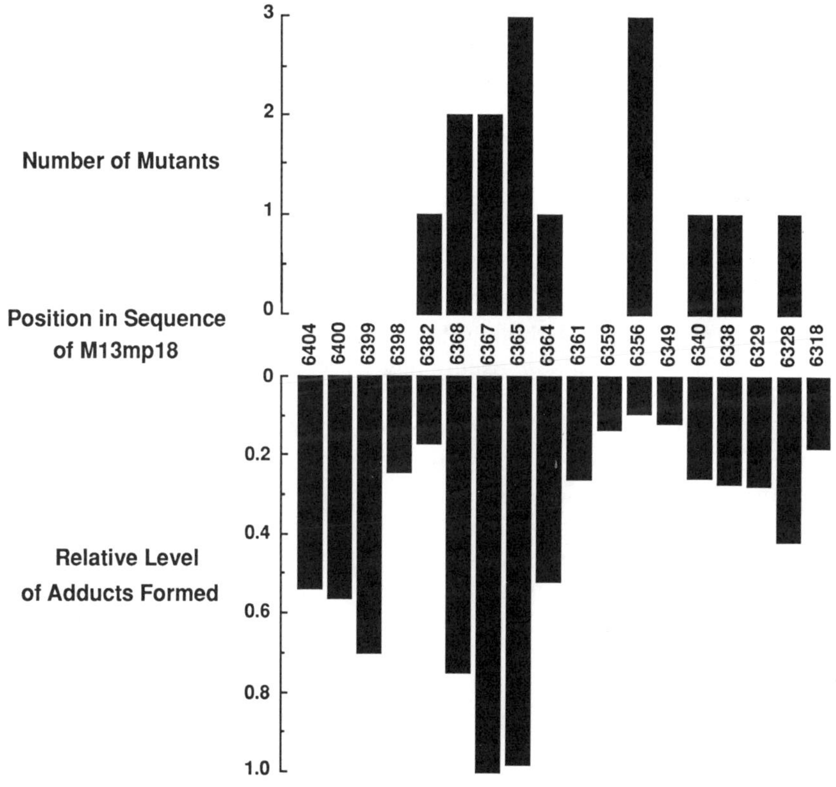

Fig. 7 Correlation of positions of mutations in bacteriophage M13mp18 DNA with levels of guanyl-N^7 adduct formation after treatment with *S*-(2-chloroethyl)glutathione (Cmarik *et al.*, 1992). Only the guanine positions at which phenotypic mutations are known to occur are presented.

Table I

Spontaneous and *S*-(2-Chloroethyl)glutathione-Induced Mutations in Bacteriophage M13mp18

Frameshifts	Spontaneous	$GSCH_2CH_2Cl$-induced
Small deletions, insertions (±1)	3 (23%)	1 (2%)
Large deletions, insertions (>2)	1 (8%)	2 (5%)
Base-pair substitutions		
G:C → A:T	3 (23%)	30 (70%)
A:T → G:C	2 (15%)	5 (12%)
G:C → T:A	3 (23%)	2 (5%)
A:T → C:G	1 (8%)	1 (2%)
G:C → C:G	0	0
A:T → T:A	0	2 (5%)

Note. M13mp18 RF DNA, untreated or treated to give an *S*-[2-(N^7-guanyl)ethyl]glutathione adduct level of 8 nmol [mg DNA]$^{-1}$, was used to transform competent *S. typhimurium* TA100. Mutants in the *lac* region were identified as deficient in β-galactosidase activity and sequenced. Values in parentheses indicate the percentage of the total number of mutations.

and *hisG428* loci in an *Escherichia coli* system. Their ethylene dibromide-induced revertants were also dominated by GC to AT and AT to GC transitions (cf. Table I) and the mutations were independent of SOS-mutagenic processing. Although the work indicated a high level of mutations (~50%) in *E. coli* supposedly devoid of GSH, the bacteria had not actually been assayed for GSH levels. (It has been our own experience that GSH-deficient mutants are unstable.) The authors also argue that apurinic sites are unlikely to account for the observed mutations.

In another study, various glutathionyl and cysteinyl half-mustards were used to treat *S. typhimurium* TA100 and both the level of adduct formation and the number of his revertants were analyzed in each case (Humphreys *et al.,* 1990) (Fig. 8). Both the extent of alkylation and the number of mutants varied considerably in the series of mustards used, partly due to transport into the cells. However, the ratio of revertants : N^7-guanyl adducts varied by two orders of magnitude. Since only two (adjacent) guanyl residues are known to be targets that give rise to the phenotypic reversion, the results are interpreted to mean that the nature of the side chain on the N^7-guanyl adduct is very important in influencing the mutations that result (Humphreys *et al.,* 1990). For example, the ratio of mutants : N^7-guanyl adducts was decreased an order of magnitude by methylation of the two carboxyl groups.

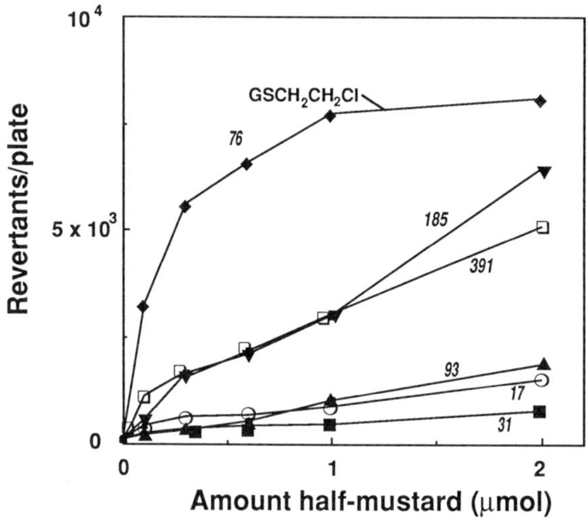

Fig. 8 His-independent *S. typhimurium* mutants formed by *S*-(2-haloethyl)thiol derivatives. The indicated amounts of these individual compounds were used: *S*-(2-chloroethyl)cysteine (■), *S*-(2-chloroethyl)cysteine methyl ester (○), *N*-acetyl-*S*-(2-chloroethyl)cysteine (▲), *N*-acetyl-*S*-(2-bromoethyl)cysteine methyl ester (▼), *S*-(2-chloroethyl)glutathione dimethyl ester (□) and *S*-(2-chloroethyl)glutathione (◆). DNA N^7-guanyl adduct levels are presented in numerals in each case, in pmol (mg DNA)$^{-1}$ [measured at the 1.0 μmol dose of half-mustard (Humphreys *et al.*, 1990)].

The adduct *S*-[2-(N^7-guanyl)ethyl]glutathione (= G*) was inserted into both the oligonucleotides d(CATG*CCG) and the d(TGCTG*CAAG), the latter of which corresponds to region where several M13mp18 mutations were seen in the mutation spectrum work with *S*-(2-chloroethyl)glutathione (Cmarik *et al.*, 1992). When binding to the complements was examined with UV spectroscopic techniques, the G* : C pairing was reduced relative to G : C pairing but the pairing of G* to C was still far stronger than to any of the other three bases, including the T expected from the mutant spectrum work. Therefore, the conclusion is reached that thermodynamic tendency to pair another base is not a major force in influencing mutation. The change in free energy of binding ($\Delta\Delta G^\circ$) of the oligomer d(TGCTG*CAAG) due to substitution of G* = *S*-[2-(N^7-guanyl)ethyl]glutathione for G is about +1 kcal mol^{-1}, which is not particularly large in comparison with the activation energies for several steps in polymerase reactions (Eger *et al.*, 1991).

The decreased free energy for hybridization of d(TGCTG*CAAG) with its complement upon substitution of G with G* = *S*-[2-(N^7-guanyl)ethyl]glutathione ($\Delta\Delta G^\circ$ = +1.3 kcal mol^{-1}) appears to be primarily due to an entropic component ($\Delta\Delta H^\circ = -14$ kcal mol^{-1}; $\Delta\Delta S^\circ = -49$ cal mol^{-1} deg^{-1}) (Kim *et al.*, 1993). This entropic contribution suggests that there is an interaction of the GSH moiety with the bases in the oligonucleotide, which was not seen in the NMR studies with d(CATG*CCT) and its complement (Oida *et al.*, 1991). The decreased pairing of G* = *S*-[2-(N^7-guanyl)ethyl]glutathione to C, relative to G, is also not simply due to introduction of a positive charge into the imidazole ring, since substitution of G* = N^7-methylguanine into d(CATG*CCT) did not affect the T_m for binding to the complement very much. Similar findings have been reported by Ezaz-Nikpay and Verdine (1992) for a palindromic dodecamer. It is our current conclusion that hydrogen bonds are critical in influencing the interaction of the GSH moiety of *S*-[2-(N^7-guanyl)ethyl]glutathione with DNA and with polymerases. The latter interactions are thought to be critical to the mutation process and are the subject of further investigation.

V. Conclusions

The most thoroughly studied of the dihaloalkanes is ethylene dibromide. Considerable evidence now exists to support the view that GSH conjugation is critical to mutagenicity and probably tumorigenesis. The major DNA adduct is *S*-[2-(N^7-guanyl)ethyl]glutathione and this adduct is most likely to be responsible for the mutations. Several indirect lines of evidence support this view but direct proof is still not available. Three minor guanyl and adenyl adducts have been shown to be present in DNA as well. The events described for ethylene bromide are probably also generally relevant to ethylene dichloride and any other 1,2-dihaloalkanes. DBCP is also activated by GSH conjugation involving similar chemical reactions (i.e., episulfonium ion formation) and N^7-guanyl DNA adducts have been characterized. However, the low levels of these formed *in vivo* make their relevance unclear at this time. Methylene dihalides react with GSH, forming HCHO, but the relevance of this reaction to the mutagenicity of the compounds and their weak carcinogenicity in mice is still unclear. Recent evidence suggests the reaction of a GSH conjugate with DNA to yield the base-pair mutations.

References

Agency for Toxic Substances and Disease Registry, U. S. Public Health Service, and U. S. Environmental Protection Agency (1989). "Toxicological Profile for 1,2-Dichloroethane." Public Health Service, Centers for Disease Control, Atlanta, GA.

Ahmed, A. E., and Anders, M. W. (1976). Metabolism of dihalomethanes to formaldehyde and inorganic halide. I. In vitro studies. *Drug Metab. Dispos.* **4**, 357–361.

Ahmed, A. E., and Anders, M. W. (1978). Metabolism of dihalomethanes to formaldehyde and inorganic halide. II. *Biochem. Pharmacol.* **27,** 2021–2025.

Ames, B., Infante, P., and Reitz, R., eds. (1980). "Banbury Report 5. Ethylene Dichloride: A Potential Health Risk?" Cold Spring Harbor Lab., Cold Spring Harbor, New York, 1980.

Anders, M. W., and Livesey, J. C. (1980). Metabolism of 1,2-dihaloethanes. *In* "Banbury Report 5. Ethylene Dichloride: A Potential Health Risk?" (B. Ames, P. Infante, and R. Reitz, eds.), pp. 331–341. Cold Spring Harbor Lab., Cold Spring Harbor, New York.

Andersen, M. E., Clewell, H. J., III, Gargas, M. L., Smith, F. A., and Reitz, R. H. (1987). Physiologically based pharmacokinetics and the risk assessment process for methylene chloride. *Toxicol. Appl. Pharmacol.* **87,** 185–205.

Anonymous (1983). Pesticide suit hits Dow: Six workers awarded $5 million. *Chem. Eng. News* **61,** 4.

Banerjee, S., and Van Duuren, B. L. (1979). Binding of carcinogenic halogenated hydrocarbons to cell macromolecules. *JNCI, J. Natl. Cancer Inst.* **63,** 707–711.

Banerjee, S., Van Duuren, B. L., and Oruambo, F. I. (1980). Microsome-mediated covalent binding of 1,2-dichloroethane to lung microsomal protein and salmon sperm DNA. *Cancer Res.* **40,** 2170–2173.

Brady, J. F., Xiao, F., Wang, M. H., Li, Y., Ning, S. M., Gapac, J. M., and Yang, C. S. (1991). Effects of disulfiram on hepatic P450IIE1, other microsomal enzymes, and hepatotoxicity in rats. *Toxicol. Appl. Pharmacol.* **108,** 366–373.

Brunborg, G., Holme, J., Søderlund, E. J., and Dybing, E. (1990). Organ-specific genotoxic effects of chemicals: The use of alkaline elution to detect DNA damage in various organs of *in vivo*-exposed animals. *In* "Mutation and the Environment" (M. L. Mendelsohn and R. J. Albertini, eds.), Part D, pp. 43–52. Wiley-Liss, New York.

Buijs, W., van der Gen, A., Mohn, G. R., and Breimer, D. D. (1984). The direct mutagenic activity of α,ω-dihalogenoalkanes in *Salmonella typhimurium*. *Mutat. Res.* **141,** 11–14.

Cheever, K. L., Cholakis, J. M., El-Hawari, A. M., Kovatch, R. M., and Weisburger, E. K. (1990). Ethylene dichloride: The influence of disulfiram or ethanol on oncogenicity, metabolism, and DNA covalent binding in rats. *Fundam. Appl. Toxicol.* **14,** 243–261.

Cmarik, J. L., Inskeep, P. B., Meyer, D. J., Meredith, M. J., Ketterer, B., and Guengerich, F. P. (1990). Activation of ethylene dibromide by glutathione conjugation: Selectivity of rat and human glutathione S-transferases and DNA binding and induction of unscheduled DNA synthesis in human hepatocytes. *Cancer Res.* **50,** 2747–2752.

Cmarik, J. L., Humphreys, W. G., Bruner, K. L., Lloyd, R. S., Tibbetts, C., and Guengerich, F. P. (1992). Mutation spectrum and sequence alkylation selectivity resulting from modification of bacteriophage M13mp18 with *S*-(2-chloroethyl)glutathione. Evidence for a role of *S*-[2-(N^7-guanyl)ethyl]glutathione as a mutagenic lesion formed from ethylene dibromide. *J. Biol. Chem.* **267,** 6672–6679.

Crespi, C. L., Seixas, G. M., Turner, T. R., Ryan, C. G., and Penman, B. W. (1985). Mutagenicity of 1,2-dichloroethane and 1,2-dibromoethane in two human lymphoblastoid cell lines. *Mutat. Res.* **142,** 133–140.

Dohn, D. R., and Casida, J. E. (1987). Thiiranium ion intermediates in the formation and reactions of S-(2-haloethyl)-L-cysteines. *Bioorg. Chem.* **15,** 115–124.

Eger, B. T., Kuchta, R. D., Carroll, S. S., Benkovic, P. A., Dahlberg, M. E., Joyce, C. M., and Benkovic, S. J. (1991). Mechanism of DNA replication fidelity for three mutants of DNA polymerase I: Klenow fragment KF(*exo*$^+$), KF(*pol*A5), and KF(*exo*$^-$). *Biochemistry* **30,** 1441–1448.

Ezaz-Nikpay, K., and Verdine, G. L. (1992). Aberrantly methylated DNA: site-specific introduction of N^7-methyl-2′-deoxyguanosine into the Dickerson/Drew dodecamer. *J. Am. Chem. Soc.* **114,** 6562–6563.

Foster, P. L., Wilkinson, W. G., Miller, J. K., Sullivan, A. D., and Barnes, W. M. (1988). An analysis of the mutagenicity of 1,2-dibromoethane to *Escherichia coli:* Influence of DNA repair activities and metabolic pathways. *Mutat. Res.* **194,** 171–181.

Green, T. (1983). The metabolic activation of dichloromethane and chlorofluoromethane in a bacterial mutation assay using *Salmonella typhimurium. Mutat. Res.* **118,** 277–288.

Guengerich, F. P., Crawford, W. M., Jr., Domoradzki, J. Y., Macdonald, T. L., and Watanabe, P. G. (1980). *In vitro* activation of 1,2-dichloroethane by microsomal and cytosolic enzymes. *Toxicol. Appl. Pharmacol.* **55,** 303–317.

Guengerich, F. P., Mason, P. S., Stott, W. T., Fox, T. R., and Watanabe, P. G. (1981). Roles of 2-haloethylene oxides and 2-haloacetaldehydes derived from vinyl bromide and vinyl chloride in irreversible binding to protein and DNA. *Cancer Res.* **41,** 4391–4398.

Guengerich, F. P., Koga, N., Inskeep, P. B., and Ozawa, N. (1986). Isotopic studies applied to the mechanism of alkylation of DNA by ethylene dibromide. *In* "Synthesis and Applications of Isotopically Labeled Compounds 1985" (R. R. Muccino, ed.), pp. 65–68. Elsevier, Amsterdam.

Guengerich, F. P., Kim, D.-H., and Iwasaki, M. (1991). Role of human cytochrome P-450 IIE1 in the oxidation of several low molecular weight cancer suspects. *Chem. Res. Toxicol.* **4,** 168–179.

Hales, D. B., Ho, B., and Thompson, J. A. (1987). Inter- and intramolecular deuterium isotope effects on the cytochrome P-450-catalyzed oxidative dehalogenation of 1,1,2,2-tetrachloroethane. *Biochem. Biophys. Res. Commun.* **149,** 319–325.

Heck, H. d'A., Casanova, M., and Starr, T. B. (1990). Formaldehyde toxicity—new understanding. *CRC Crit. Rev. Toxicol.* **20,** 397–426.

Hill, D. L., Shih, T. W., Johnston, T. P., and Struck, R. F. (1978). Macromolecular binding and metabolism of the carcinogen 1,2-dibromoethane. *Cancer Res.* **38,** 2438–2442.

Huff, J. E. (1983a). 1,2-Dibromo-3-chloropropane. *Environ. Health Perspect.* **47,** 365–369.

Huff, J. E. (1983b). 1,2-Dibromoethane (ethylene dibromide). *Environ. Health Perspect.* **47,** 359–363.

Humphreys, W. G., Kim, D.-H., Cmarik, J. L., Shimada, T., and Guengerich, F. P. (1990). Comparison of the DNA alkylating properties and mutagenic responses caused by a series of *S*-(2-haloethyl)-substituted cysteine and glutathione derivatives. *Biochemistry* **29,** 10342–10350.

Humphreys, W. G., Kim, D.-H., and Guengerich, F. P. (1991). Isolation and characterization of N^7-guanyl adducts derived from 1,2-dibromo-3-chloropropane. *Chem. Res. Toxicol.* **4,** 445–453.

Igwe, O. J., Que Hee, S. S., and Wagner, W. D. (1986). Inhalation pharmacokinetics of 1,2-dichloroethane after different dietary pretreatments of male Sprague-Dawley rats. *Arch. Toxicol.* **59,** 127–134.

Inskeep, P. B., and Guengerich, F. P. (1984). Glutathione-mediated binding of dibromoalkanes to DNA: Specificity of rat glutathione S-transferases and dibromoalkane structure. *Carcinogenesis (London)* **5,** 805–808.

Inskeep, P. B., Koga, N., Cmarik, J. L., and Guengerich, F. P. (1986). Covalent binding of 1,2-dihaloalkanes to DNA and stability of the major DNA adduct, *S*-[2-(N^7-guanyl)ethyl]glutathione. *Cancer Res.* **46,** 2839–2844.

Jean, P. A., and Reed, D. J. (1989). In vitro dipeptide, nucleoside, and glutathione alkylation by S-(2-chloroethyl)glutathione and *S*-(2-chloroethyl)-L-cysteine. *Chem. Res. Toxicol.* **2,** 455–460.

Jean, P. A., and Reed, D. J. (1992). Utilization of glutathione during 1,2-dihaloethane metabolism in rat hepatocytes. *Chem. Res. Toxicol.* **5,** 386–391.

Jongen, W. M. F., Alink, G. M., and Koeman, J. H. (1978). Mutagenic effect of dichloromethane on *Salmonella typhimurium. Mutat. Res.* **56,** 245–248.

Kerklaan, P. R. M., Zoetemelk, C. E. M., and Mohn, G. R. (1985). Mutagenic activity of various chemicals in *Salmonella* strain TA100 and glutathione-deficient derivatives: On the role of glutathione in the detoxification or activation of mutagens inside bacterial cells. *Biochem. Pharmacol.* **34,** 2151–2156.

Kim, D.-H., and Guengerich, F. P. (1989). Excretion of the mercapturic acid *S*-[2-(N^7-guanyl)ethyl]-*N*-acetylcysteine in urine following administration of ethylene dibromide to rats. *Cancer Res.* **49,** 5843–5851.

Kim, D.-H., and Guengerich, F. P. (1990). Formation of the DNA adduct *S*-[2-(N^7-guanyl)ethyl]glutathione from ethylene dibromide: Effects of modulation of glutathione and glutathione S-transferase levels and the lack of a role for sulfation. *Carcinogenesis (London)* **11,** 419–424.

Kim, D.-H., Humphreys, W. G., and Guengerich, F. P. (1990). Characterization of *S*-[2-(N^1-adenyl)ethyl]glutathione formed in DNA and RNA from 1,2-dibromoethane. *Chem. Res. Toxicol.* **3,** 587–594.

Kim, M.-S., Persmark, M., Min, K.-S., and Guengerich, F. P. (1993). Interaction of N^7-guanyl methyl- and thioether-substituted deoxyribooligonucleotides with complementary sequences: Preparation and spectroscopic and thermodynamic characterization. *Proc. Am. Assoc. Cancer Res.* **34,** 135.

Koga, N., Inskeep, P. B., Harris, T. M., and Guengerich, F. P. (1986). *S*-[2-(N^7-Guanyl)ethyl]glutathione, the major DNA adduct formed from 1,2-dibromoethane. *Biochemistry* **25,** 2192–2198.

Kubic, V. L., and Anders, M. W. (1978). Metabolism of dihalomethanes to carbon monoxide. III. *Biochem. Pharmacol.* **27,** 2349–2355.

La Roche, S. D., and Leisinger, T. (1990). Sequence analysis and expression of the bacterial dichloromethane dehalogenase structural gene, a member of the glutathione S-transferase supergene family. *J. Bacteriol.* **172,** 164–171.

Leonard, N. J. (1984). Etheno-substituted nucleotides and coenzymes: Fluorescence and biological activity. *CRC Crit. Rev. Biochem.* **15,** 125–199.

Letz, G. A., Pond, S. M., Osterloh, J. D., Wade, R. L., and Becker, C. E. (1984). Two fatalities after acute occupational exposure to ethylene dibromide. *J. Am. Med. Assoc.* **252,** 2428–2431.

Lin, E. L. C., Mattox, J. K., and Pereira, M. A. (1985). Glutathione plus cytosol- and microsome-mediated binding of 1,2-dichloroethane to polynucleotides. *Toxicol. Appl. Pharmacol.* **78,** 428–435.

Livesey, J. C., and Anders, M. W. (1979). *In vitro* metabolism of 1,2-dihaloethanes to ethylene. *Drug Metab. Dispos.* **7,** 199–203.

Livesey, J. C., Anders, M. W., Langvardt, P. W., Putzig, C. L., and Reitz, R. H. (1982). Stereochemistry of the glutathione-dependent biotransformation of *vicinal*-dihaloalkanes to alkenes. *Drug Metab. Dispos.* **10,** 201–204.

Ludlum, D. B., Kent, S., and Mehta, J. R. (1986). Formation of O^6-ethylthioethylguanine in DNA by reaction with the sulfur mustard, chloroethyl sulfide, and its apparent lack of repair by O^6-alkylguanine-DNA alkyltransferase. *Carcinogenesis (London)* **7,** 1203–1206.

Marchand, D. H., and Abdel-Monem, M. M. (1985). Glutathione S-transferase catalyzed conjugation of 1,4-disubstituted butanes with glutathione in vitro. *Biochem. Biophys. Res. Commun.* **128,** 360–367.

Marchand, D. H., and Reed, D. J. (1989). Identification of the reactive glutathione conjugate *S*-(2-chloroethyl)glutathione in the bile of 1-bromo-2-chloroethane-treated rats by high-pressure liquid chromatography and precolumn derivatization with o-phthalaldehyde. *Chem. Res. Toxicol.* **2,** 449–454.

Meyer, D. J., Coles, B., Pemble, S. E., Gilmore, K. S., Fraser, G. M., and Ketterer, B.

(1991). Theta, a new class of glutathione transferases purified from rat and man. *Biochem. J.* **274,** 409–414.

Milks, M. M., Wilt, S. R., Ali, I., Pereira, M. A., and Couri, D. (1982). Dibromoethane effects on the induction of γ-glutamyl-transpeptidase positive foci in rat liver. *Arch. Toxicol.* **51,** 27–35.

Min, K.-S., and Guengerich, F. P. (1993). Incorporation of N^7-guanyl adduct formed from ethylene dibromide into phage DNA and site-specific mutagenesis. *Proc. Am. Assoc. Cancer Res.* **34,** 171.

Moslen, M. T., Ahluwalia, M. B., and Farber, E. (1985). 1,2-Dibromoethane initiation of hepatic nodules in Sprague-Dawley rats selected with Solt-Farber system. *Arch. Toxicol.* **58,** 118–119.

Nachtomi, E. (1970). The metabolism of ethylene dibromide in the rat: The enzymic reaction with glutathione in vitro and in vivo. *Biochem. Pharmacol.* **19,** 2853–2860.

Nachtomi, E. (1981). Role of diethyldithiocarbamate in ethylene dibromide metabolism and covalent binding. *Toxicol. Appl. Pharmacol.* **57,** 247–253.

National Institute of Occupational Safety and Health (1981). "NIOSH Current Intelligence Bulletin 37: Ethylene Dibromide (EDB)," pp. 1–13. U.S. Government Printing Office, Washington, DC.

National Toxicology Program (1982). "NTP Technical Report on the Carcinogenesis Bioassay of 1,2-Dibromoethane in F344 Rats and B6C3F1 Mice (Inhalation Study)," pp. 1–163. U.S. Government Printing Office, Washington, DC.

Oida, T., Humphreys, W. G., and Guengerich, F. P. (1991). Preparation and characterization of oligonucleotides containing *S*-[2-(N^7-guanyl)ethyl]glutathione. *Biochemistry* **30,** 10513–10522.

Olson, W. A., Habermann, R. T., Weisburger, E. K., Ward, J. M., and Weisburger, J. H. (1973). Induction of stomach cancer in rats and mice by halogenated aliphatic fumigants. *J. Natl. Cancer Inst.* (*U.S.*) **51,** 1993–1995.

Omichinski, J. G., Søderlund, E. J., Dybing, E., Pearson, P. G., and Nelson, S. D. (1988). Detection and mechanism of formation of the potent direct-acting mutagen 2-bromoacrolein from 1,2-dibromo-3-chloropropane. *Toxicol. Appl. Pharmacol.* **92,** 286–294.

Osterman-Golkar, S., Hussain, S., Walles, S., Anderstam, B., and Sigvardsson, K. (1983). Chemical reactivity and mutagenicity of some dihalomethanes. *Chem.-Biol. Interact.* **46,** 121–130.

Ozawa, N., and Guengerich, F. P. (1983). Evidence for formation of an *S*-[2-(N^7-guanyl)ethyl]glutathione adduct in glutathione-mediated binding of 1,2-dibromoethane to DNA. *Proc. Natl. Acad. Sci. U.S.A.* **80,** 5266–5270.

Pearson, P. G., Søderlund, E. J., Dybing, E., and Nelson, S. D. (1990). Metabolic activation of 1,2-dibromo-3-chloropropane: Evidence for the formation of reactive episulfonium ion intermediates. *Biochemistry* **29,** 4971–4981.

Peterson, L. A., Harris, T. M., and Guengerich, F. P. (1988). Evidence for an episulfonium ion intermediate in the formation of *S*-[2-(N^7-guanyl)ethyl]glutathione in DNA. *J. Am. Chem. Soc.* **110,** 3284–3291.

Rannug, U. (1980). Genotoxic effects of 1,2-dibromoethane and 1,2-dichloroethane. *Mutat. Res.* **76,** 269–295.

Rannug, U., and Beije, B. (1979). The mutagenic effect of 1,2-dichloroethane on *Salmonella typhimurium*. II. Activation by the isolated perfused rat liver. *Chem.-Biol. Interact.* **24,** 265–285.

Rannug, U., Göthe, R., and Wachtmeister, C. A. (1976). The mutagenicity of chloroethylene oxide, chloroacetaldehyde, 2-chloroethanol and chloroacetic acid, conceivable metabolites of vinyl chloride. *Chem.-Biol. Interact.* **12,** 251–263.

Rannug, U., Sundvall, A., and Ramel, C. (1978). The mutagenic effect of 1,2-dichloroethane on *Salmonella typhimurium*. I. Activation through conjugation with glutathione in vitro. *Chem.-Biol. Interact.* **20,** 1–16.

Rao, K. S., Burek, J. D., Murray, F. J., John, J. A., Schwetz, B. A., Bell, T. J., Potts, W. J., and Parker, C. M. (1983). Toxicologic and reproductive effects of inhaled 1,2-dibromo-3-chloropropane in rats. *Fundam. Appl. Toxicol.* **3,** 104–110.

Reitz, R. H., Fox, T. R., Ramsey, J. C., Quast, J. F., Langvardt, P. W., and Watanabe, P. G. (1982). Pharmacokinetics and macromolecular interactions of ethylene dichloride in rats after inhalation or gavage. *Toxicol. Appl. Pharmacol.* **62,** 190–204.

Reitz, R. H., Mendrala, A., and Guengerich, F. P. (1989). *In vitro* metabolism of methylene chloride in human and animal tissues: Use in physiologically based pharmacokinetic models. *Toxicol. Appl. Pharmacol.* **97,** 230–246.

Sagher, D., and Strauss, B. (1983). Insertion of nucleotides opposite apurinic/apyrimidinic sites in deoxyribonucleic acid during *in vitro* synthesis: Uniqueness of adenine nucleotides. *Biochemistry* **22,** 4518–4526.

Schasteen, C. S., and Reed, D. J. (1983). The hydrolysis and alkylation activities of S-(2-haloethyl)-L-cysteine analogs—evidence for extended half-life. *Toxicol. Appl. Pharmacol.* **70,** 423–432.

Shih, T. W., and Hill, D. L. (1981). Metabolic activation of 1,2-dibromoethane by glutathione transferase and by microsomal mixed function oxidase: Further evidence for formation of two reactive metabolites. *Res. Commun. Chem. Pathol. Pharmacol.* **33,** 449–461.

Søderlung, E. J., Brunborg, G., Omichinski, J. G., Hohme, J., Dahl, J. E., Nelson, S. D., and Dybing, E. (1988). Testicular necrosis and DNA damage caused by selectively deuterated and methylated analogs of 1,2-dibromo-3-chloropropane in the rat. *Toxicol. Appl. Pharmacol.* **91,** 358–370.

Sun, M. (1984). EDB contamination kindles federal action. *Science* **223,** 464–466.

Thier, R., Pemble, S. E., Taylor, J. B., Humphreys, W. G., Persmark, M., Ketterer, B., and Guengerich, F. P. (1993). Expression of rat glutathione *S*-transferase 5-5 in *Salmonella typhimuriam* TA1535 leads to base-pair mutations upon exposure to dihalomethames. *Proc. Natl. Acad. Sci. U.S.A.* **90,** 8576–8580.

Tomasi, A., Albano, E., Dianzani, M. U., Slater, T. F., and Vannini, V. (1983). Metabolic activation of 1,2-dibromoethane to a free radical intermediate by rat liver microsomes and isolated hepatocytes. *FEBS Lett.* **160,** 191–194.

Vadi, H. V., Schasteen, C. S., and Reed, D. J. (1985). Interactions of S-(2-haloethyl)-mercapturic acid analogs with plasmid DNA. *Toxicol. Appl. Pharmacol.* **80,** 386–396.

van Bladeren, P. J., van der Gen, A., Breimer, D. D., and Mohn, G. R. (1979). Stereoselective activation of vicinal dihalogen compounds to mutagens by glutathione conjugation. *Biochem. Pharmacol.* **28,** 2521–2524.

van Bladeren, P. J., Breimer, D. D., Rotteveel-Smijs, G. M. T., de Jong, R. A. W., Buijs, W., van der Gen, A., and Mohn, G. R. (1980a). The role of glutathione conjugation in the mutagenicity of 1,2-dibromoethane. *Biochem. Pharmacol.* **29,** 2975–2982.

van Bladeren, P. J., Breimer, D. D., Rotteveel-Smijs, G. M. T., and Mohn, G. R. (1980b). Mutagenic activation of dibromomethane and diiodomethane by mammalian microsomes and glutathione S-transferases. *Mutat. Res.* **74,** 341–346.

van Bladeren, P. J., Breimer, D. D., Rotteveel-Smijs, G. M. T., de Knijff, P., Mohn, G. R., van Meeteren-Wälchi, B., and van der Gen, A. (1981a). The relation between the structure of vicinal dihalogen compounds and their mutagenic activation via conjugation to glutathione. *Carcinogenesis* (*London*) **2,** 499–505.

van Bladeren, P. J., Breimer, D. D., van Huijgevoort, J. A. T. C. M., Vermeulen, N. P. E., and van der Gen, A. (1981b). The metabolic formation of N-acetyl-S-2-hydroxyethyl-L-cysteine from tetradeutero-1,2-dibromoethane. Relative importance of oxidation and glutathione conjugation *in vivo*. *Biochem. Pharmacol.* **30,** 2499–2502.

van Bladeren, P. J., Hoogeterp, J. J., Breimer, D. D., and van der Gen, A. (1981c). The influence of disulfiram and other inhibitors of oxidative metabolism on the formation of 2-hydroxyethyl-mercapturic acid from 1,2-dibromoethane by the rat. *Biochem. Pharmacol.* **30,** 2983–2987.

Warren, D. L., and Reed, D. J. (1991). Modification of hepatic vitamin E stores *in vivo*. III. Vitamin E depletion by 1,2-dibromoethane may be related to initial conjugation with glutathione. *Arch. Biochem. Biophys.* **288,** 449–455.

Warren, D. L., Brown, M. K., and Reed, D. J. (1991). Modification of hepatic vitamin E stores *in vivo*. II. Alteration in plasma and liver vitamin E content by 1,2-dibromoethane. *Arch. Biochem. Biophys.* **288,** 440–448.

Webb, W. W., Elfarra, A. A., Webster, K. D., Thom, R. E., and Anders, M. W. (1987). Role for an episulfonium ion in *S*-(2-chloroethyl)-DL-cysteine-induced cytotoxicity and its reaction with glutathione. *Biochemistry* **26,** 3017–3023.

White, R. D., Sipes, I. G., Gandolfi, A. J., and Bowden, G. T. (1981). Characterization of the hepatic DNA damage caused by 1,2-dibromoethane using the alkaline elution technique. *Carcinogenesis (London)* **2,** 839–844.

White, R. D., Gandolfi, A. J., Bowden, G. T., and Sipes, I. G. (1983). Deuterium isotope effect on the metabolism and toxicity of 1,2-dibromoethane. *Toxicol. Appl. Pharmacol.* **69,** 170–178.

White, R. D., Petry, T. W., and Sipes, I. G. (1984). The bioactivation of 1,2-dibromoethane in rat hepatocytes: Deuterium isotope effect. *Chem.-Biol. Interact.* **49,** 225–233.

Wong, L. C. K., Winston, J. M., Hong, C. B., and Plotnick, H. (1982). Carcinogenicity and toxicity of 1,2-dibromoethane in the rat. *Toxicol. Appl. Pharmacol.* **63,** 155–165.

Working, P. K., Smith-Oliver, T., and Butterworth, B. E. (1986). Induction of DNA repair in rat spermatocytes and hepatocytes by 1,2-dibromoethane: The role of glutathione conjugation. *Carcinogenesis (London)* **7,** 467–472.

Bioactivation of Thiols by One-Electron Oxidation

Rex Munday
Ruakura Agricultural Research Centre
Hamilton, New Zealand

I. Introduction

To most biochemists and toxicologists, the term "thiol" immediately calls to mind the tripeptide glutathione, which exists at millimolar concentrations in mammalian cells and which has many functions in cellular metabolism. In particular, glutathione is comparatively easily oxidized, thereby protecting other cellular constituents against oxidation; this property of glutathione is recognized as an important factor in the cellular defenses against oxidative stress.

Oxidation of glutathione may be a two-electron process, as in its reaction with hydrogen peroxide, which occurs both spontaneously and through catalysis by glutathione peroxidase:

$$2GSH + H_2O_2 \rightarrow GSSG + 2H_2O \tag{1}$$

In converting the harmful hydrogen peroxide directly to the innocuous water and glutathione disulfide, this is certainly a detoxication process; the antioxidant defense of the cell can be maintained by rereduction of the disulfide through the reaction catalyzed by glutathione reductase.

Oxidation may also occur by a one-electron process, leading to formation of the glutathione thiyl radical, $GS^{\cdot}$:

$$GSH - e^- \rightarrow GS^{\cdot} + H^+ \tag{2}$$

Advances in Pharmacology, Volume 27

Again, this is an antioxidant reaction, and it is generally accepted that this is a second cellular defense mechanism involving glutathione, which leads to detoxication of intracellularly generated free radicals. Unlike the two-electron process, however, it is not intrinsically a detoxication reaction since a free radical is produced, rather than the unreactive GSSG. This process will be beneficial, therefore, only if $GS^{\cdot}$ is converted to products that do not lead to tissue damage. In some circumstances, this condition is fulfilled, and thiyl radical formation is associated with cellular protection. In others, however, one-electron oxidation of glutathione leads to detrimental effects.

Apart from glutathione and other endogenous thiols, such as cysteine and ergothioneine, there exists in our environment a range of thiols of different structural types, some of natural origin and others introduced through various aspects of human activity. These compounds similarly undergo one-electron oxidation to their respective thiyl radicals, although the ease with which they do so is strongly dependent upon their structure. It has been shown that the rate of oxidation of a number of xenobiotic thiols *in vitro* is directly proportional to the severity of their toxic effects *in vivo,* suggesting that for these compounds, one-electron oxidations are invariably deleterious processes.

In this chapter, the reactions of the thiyl radical are described, together with the ways in which this species may be generated and the structural features that govern its rate of formation. Lastly, the possible mechanisms by which one-electron oxidation of thiols leads to both beneficial and harmful effects in living systems are discussed.

II. Reactions of the Thiyl Radical

A. Formation of Disulfide via the Disulfide Radical Anion

$$RS^{\cdot} + RS^{-} \rightleftharpoons (RSSR)^{\cdot -} \quad (3)$$

$$(RSSR)^{\cdot -} + O_2 \rightarrow RSSR + O_2^{\cdot -} \quad (4)$$

This is the predominant pathway for thiyl radical degradation under physiological conditions (Wardman, 1988; Asmus, 1990; Koppenol, 1993) and can explain the characteristics of many of the reactions involving one-electron oxidation of thiols (Section III). Formation of the disulfide radical anion [Reaction (3)] is an equilibrium reaction involving RS^- and will, therefore, be favored by low thiol pK_a values or high concentrations of RSH, or both. The equilibrium will, therefore, be to the right with the highly ionized aromatic thiols (Section IV), and even with glutathione, with a comparatively high pK_a, appreciable levels of $(GSSG)^{\cdot -}$ are to be expected under equilibrium conditions at neutral pH in the presence of millimolar concentrations of thiol (Wardman, 1988). Furthermore, in the presence of oxygen, Reaction (3) will be driven to the right by the rapid, irreversible, oxidation of the radical anion [Reaction (4)].

Superoxide generated in the latter reaction will oxidize more thiol via Reaction (5), thereby establishing a radical chain reaction for thiol oxidation:

$$RSH + O_2^{\cdot -} + H^+ \rightarrow RS^{\cdot} + H_2O_2 \tag{5}$$

Hydrogen peroxide formed by reduction of superoxide will also oxidize the thiol according to Reaction (1).

The conversion of the thiyl radical to disulfide by this pathway is thus characterized by oxygen utilization and by formation not only of the disulfide radical anion but also of superoxide radical and hydrogen peroxide. Because of the known toxicity of the latter species and the possibility of hydroxyl radical formation from hydrogen peroxide, the operation of this pathway within cells must be considered potentially harmful.

B. Reaction with Oxygen

Formation of the thioperoxyl radical via Reaction (6) was first suggested by Packer (1963). Since that time, many experiments on the reaction between thiyl radicals and molecular oxygen have been conducted, although much remains unknown about the nature and significance of this process.

$$RS^{\cdot} + O_2 \rightarrow RSOO^{\cdot} \tag{6}$$

In early pulse radiolysis studies, the concentration of $(RSSR)^{\cdot -}$ in the pulsed solution was found to be markedly decreased in the presence of oxygen. This was interpreted as reflecting competition between RS^- and oxygen for $RS^{\cdot}$ and, on this basis, rates of Reaction (6) of 10^8–10^9 M^{-1} s^{-1} were calculated for a number of aliphatic and aminothiols (Barton and Packer, 1970; Quintiliani *et al.*, 1977; Schäfer *et al.*, 1978). Using a different competition method, however, Mönig *et al.* (1987) found rates 1–2 orders of magnitude lower and concluded that Reaction (6) was either much slower than previously supposed or reversible. To add to the confusion, recent studies in the gas phase, with a direct assay for the thiyl radical, have given values of $<1.5 \times 10^3$ M^{-1} s^{-1} for the rate of reaction of the methanethiyl and ethanethiyl radicals with oxygen (Black *et al.*, 1988; Tyndall and Ravishankara, 1989). It has also been argued that the fact that thioperoxyl radicals are not involved in the cooxidation of thiols and olefins indicates a low reactivity of thiyl radicals toward oxygen (Ito and Matsuda, 1988).

Irrespective of rate, the thioperoxyl radical does appear to be the primary product of the reaction between $RS^{\cdot}$ and oxygen, this species having been identified in irradiated frozen glasses of cysteine and glutathione (Sevilla *et al.*, 1990). $RSOO^{\cdot}$ is highly unstable, however, and subsequent reactions yield the sulfinyl radical ($RSO^{\cdot}$) and the sulfonyl radical ($RSO_2^{\cdot}$). It is possible that the last-named species are precursors of the sulfinic and sulfonic acids formed in some oxidation reactions (Tamba *et al.*, 1986; Sevilla *et al.*, 1990). This would be consistent with the observation that production of such species is favored at high oxygen tensions (Wefers and Sies, 1983) and at low pH and low thiol concentration, when disulfide radical anion formation is less likely (Lal, 1976). The importance of this pathway under physiological conditions is unclear, although the involvement of $RSOO^{\cdot}$ in certain aspects of radiation toxicity has been suggested (Section V,B).

C. Oxidation Reactions

The thiyl radical is capable of oxidizing many organic compounds (Table I) in reactions involving hydrogen donation or electron donation:

$$RS^{\cdot} + XH \rightleftharpoons RSH + X^{\cdot} \tag{7}$$

$$RS^{\cdot} + X^{-} \rightleftharpoons RS^{-} + X^{\cdot} \tag{8}$$

These reactions are reversible, and in the case of the glutathione thiyl radical and aminopyrine (Wilson *et al.*, 1986) and the cysteine thiyl radical and paracetamol (Bisby and Tabassum, 1988), the equilibrium lies to the right. It has been shown, however, that in both these cases the reaction proceeds from right to left, due, it is suggested, to displacement of the equilibrium through destruction of the thiyl radical via Reactions (3) and (4) (Wardman, 1988).

This will presumably be true also of the other oxidation reactions listed in Table I, so that under conditions in which formation and oxidation of the disulfide radical anion is feasible, oxidation by the thiyl radical is unlikely. It has been shown, however, that oxidation by $RS^{\cdot}$ is favored at low pH, when Reaction (3) is slow because of low concentrations of RS^{-} (Ahmad and Armstrong, 1984). The oxidizing ability of $RS^{\cdot}$ may also be expressed if it is removed to a lipophilic environment, inaccessible to RS^{-}. In this context, the report of Schöneich *et al.* (1992) that oxidation

Table I

Compounds Reported to be Oxidized by the Thiyl Radical

Compound	Source of thiyl radical[a]	Reference
Aromatic amines		
Aminopyrine	GSH	Wilson *et al* (1986)
N,*N*,*N*′,*N*′-Tetramethyl-p-phenylenediamine	Ethanethiol	Huston *et al.* (1992)
Paracetamol (*N*-[4-hydroxyphenyl] acetamide)	CYS	Bisby and Tabassum (1988)
Phenols		
α-Tocopherol	PEN	Schöneich *et al.* (1989a)
Phenothiazines		
Chlorpromazine	GSH, CYS, CAM, PEN	Forni *et al.* (1983)
	CYS, CAM, PEN, DTT, ME	Lal and Mahal (1990)
	GSH	Tamba and O'Neill (1991)
Promethazine	GSH, CYS, CAM, PEN	Forni *et al.* (1983)
	CYS, CAM, PEN, DTT, ME	Lal and Mahal (1990)
Metiazinic acid	LIP	Bahnemann *et al.* (1981)
Miscellaneous		
Aliphatic alcohols	GSH, CYS, PEN	Schöneich *et al.* (1989a); Schöneich and Asmus (1990)
Ascorbic acid	GSH, CYS, CAM, PEN	Forni *et al.* (1983)
	CYS, CAM, PEN, DTT, ME	Lal and Mahal (1990)
	GSH	D'Aquino *et al.* (1989); Tamba and O'Neill (1991)
NADH	GSH	Willson *et al.* (1985)
Ferrocytochrome *c*	GSH, CYS, PEN	Forni and Willson (1986)
Reduced flavins	ME, MPA	Ahmad and Armstrong (1982, 1984)
Vitamin A	GSH	D'Aquino *et al.* (1989)
Polyunsaturated fatty acids	GSH	D'Aquino *et al.* (1989)
	CYS	Schöneich *et al.* (1989b)
	GSH, CYS, PEN	Schöneich and Asmus (1990)
	GSH, CYS, PEN, ME, DTT	Schöneich *et al.* (1992)
Alkylbenzene derivatives	Cyclohexanethiol, benzenethiol	Pryor *et al.* (1978)
Diammonium 2,2′-azinobis-(3-ethylbenzthiazoline-6-sulfonate)	GSH, CYS, CAM	Wolfenden and Willson (1982)
	CYS, CAM, PEN, DTT, ME	Lal and Mahal (1990)
	CYS, ethanethiol	Huston *et al.* (1992)
2,5-Dimethyltetrahydrofuran	DTT	Akhlaq *et al.* (1987)
1,6-Diazabicyclo[4,4,4]dodecane radical cation	2-Methyl-2-propanethiol	Alder *et al.* (1986)

[a] Abbreviations used: GSH, glutathione; CYS, cysteine; CAM, cysteamine; PEN, penicillamine; DTT, dithiothreitol; ME, 2-mercaptoethanol; LIP, lipoic acid; MPA, 2-mercaptopropionic acid.

of polyunsaturated fatty acids increases with increasing lipophilicity of the thiyl radical is of particular interest.

D. Dimerization

In any oxidation system, the steady-state concentration of thiyl radicals will be low. Therefore, although the dimerization reaction is fast, it is unlikely to compete with alternative pathways for RS˙ decay (Wardman, 1988).

E. Addition to Unsaturated Compounds

Thiyl radicals, being electrophilic, readily add to olefins and acetylenes, forming carbon-centered radicals. In the presence of oxygen, peroxyl radicals are formed from the latter, which decay to a variety of products. Many thiyl radicals undergo this reaction, with those from aromatic thiols being particularly well studied (Ito and Matsuda, 1988). Glutathione thiyl radicals have been shown to add to styrene (Stock *et al.*, 1986; Ortiz de Montellano and Grab, 1986) and to dihydrodiols of polycyclic hydrocarbons (Foureman and Eling, 1989).

F. Rearrangement

Under anaerobic conditions, rearrangement of the thiyl radical of glutathione to a carbon-centered radical has been observed (Becker *et al.*, 1988). Rearrangement is favored at low concentrations of glutathione and at high pH (Grierson *et al.*, 1992), and this reaction is, therefore, unlikely to be important under physiological conditions.

III. Pathways for One-Electron Oxidation of Thiols

A. Metal-Catalyzed Oxidation of Thiols

Direct oxidation of thiols by molecular oxygen is very slow (Harrison, 1924; Barron *et al.*, 1947; Cullis *et al.*, 1968a). Oxidation is greatly accelerated, however, by certain transition metals, which act as electron carriers between the thiol and oxygen. The catalytic activity of a particular metal varies according to the structure of the thiol; salts of iron (Mathews and Walker, 1909; Gardner and Jursinic, 1981) and copper (Voegtlin *et al.*, 1931; Yee and Shipe, 1982) are generally very good catalysts of thiol oxidation, while compounds of cobalt, nickel, and manganese are also effective under certain conditions (Tsen and Tappel, 1958; Cullis *et al.*, 1968b).

In the case of simple metal salts, it is likely that the active catalyst is a thiol–metal complex (Schubert, 1932). Complexes of metals with ligands such as *o*-phenanthroline (Kobashi, 1968), diisopropyl salicylic acid (Khan and Sorenson, 1991), adriamycin (Zweier *et al.*, 1986; Muindi *et al.*, 1985), and bleomycin (Antholine *et al.*, 1991) are also effective, and complexes with porphyrin derivatives such as hemin, hematin, vitamin B_{12}, and synthetic phthalocyanines (Barron *et al.*, 1947; Cullis and Trimm, 1968; Skorobogaty and Smith, 1982; Shirai *et al.*, 1991) are particularly good promoters of thiol oxidation. Even metal bound to chelating agents, such as EDTA and DTPA, which are commonly added to media with the intent of inhibiting thiol oxidation, may be catalytically active; CuEDTA and MnEDTA are excellent catalysts for the oxidation of cysteine (Hanaki and Kamide, 1983) and mercaptoethanol (Paoletti *et al.*, 1990), respectively. The oxidation of some thiols is also promoted by protein-bound metals, as in transferrin, ceruloplasmin, methemoglobin, and cytochrome c (Holmquist and Vinograd, 1963; Anderson and Tomasi, 1977; Chidambaram *et al.*, 1984; Starkebaum and Harlan, 1986; Holler and Hopkins, 1990) and in the ferryl derivatives of peroxidases and myoglobin (Harman *et al.*, 1984, 1986; Mottley *et al.*, 1987; Romero *et al.*, 1992). The special case of thiol oxidation by hemoglobin is discussed below.

Studies of the stoichiometry of oxygen uptake during metal-catalyzed thiol oxidation (Meldrum and Dixon, 1930; Barron *et al.*, 1947; Zwart *et al.*, 1981) point to disulfide as the sole sulfur-containing oxidation product via the reaction

$$4RSH + O_2 \rightarrow 2RSSR + 2H_2O \qquad (9)$$

The oxidation is initiated by electron transfer from the thiolate anion to metal, with the catalytic cycle being maintained by autoxidation of the reduced metal:

$$M^{n+} + RS^- \rightarrow M^{(n-1)+} + RS^{\cdot} \qquad (10)$$

$$M^{(n-1)+} + O_2 \rightarrow M^{n+} + O_2^{\cdot -} \qquad (11)$$

Subsequent reactions of the thiyl radical and superoxide will be as described previously (Section II). The thiyl radical, superoxide radical, and hydrogen peroxide have been identified as intermediates in metal-catalyzed thiol oxidation, and under some conditions, hydroxyl radical has also been detected (Munday, 1989). The last-named species is believed to arise via the Fenton reaction:

$$Fe^{II} + H_2O_2 \rightarrow Fe^{III} + {}^{\cdot}OH + OH^- \qquad (12)$$

The cycle is maintained through reduction of ferric iron by thiol [Reaction (10)]; reduction of iron by the superoxide radical [which, in conjunction with Reaction (12), would constitute the metal-catalyzed Haber–Weiss reaction] does not play a significant role in thiol-induced hydroxyl radical formation (Searle and Tomasi, 1982; Florence, 1984).

B. Oxidation of Thiols by Hemoglobin

Oxyhemoglobin is oxidized to methemoglobin in the presence of thiols, with concomitant formation of hydrogen peroxide (Strömme, 1963; Eyer *et al.*, 1975; Munday, 1985a; Munday *et al.*, 1990b). This process can be understood in terms of the oxidase activity of hemoglobin as a ferric superoxide (Carrell *et al.*, 1977):

$$(Hb)Fe^{III}O_2^{\cdot -} + RS^- + 2H^+ \rightarrow (Hb)Fe^{III} + RS^{\cdot} + H_2O_2 \quad (13)$$

Superoxide radical is also generated during the interaction between thiols and hemoglobin (Munday, 1985a; Munday *et al.*, 1990b). This presumably arises during conversion of $RS^{\cdot}$ to disulfide via Reactions (3) and (4).

C. Oxidation of Thiols by Free Radicals

Thiols may reduce free radicals by hydrogen donation from the thiol or by electron donation from the thiolate anion—the reverse of Reactions (7) and (8):

$$RSH + X^{\cdot} \rightleftharpoons RS^{\cdot} + XH \quad (14)$$

$$RS^- + X^{\cdot} \rightleftharpoons RS^{\cdot} + X^- \quad (15)$$

These reactions have been extensively studied with regard to thiol-induced protection against radiation damage (Section V,B), but they are not confined to radiolytically derived radicals and have been demonstrated with a range of radicals generated by a variety of oxidative processes.

As mentioned previously, thiols are oxidized by superoxide [Reaction (5)] and this process has been observed with the radical being generated radiolytically (Al-Thannon *et al.*, 1974), chemically (Crank and Makin, 1984), or enzymatically (Asada and Kanematsu, 1976; Wefers and Sies, 1983; Ross *et al.*, 1985d; Ross and Moldéus, 1986). Among other inorganic radicals, hydroxyl is particularly active, reacting with thiols at close to diffusion-controlled rates (Packer, 1974). Halogen and pseudohalogen radicals are also powerful oxidants (Redpath, 1973), as are phosphite (Schäfer and Asmus, 1981) and azide radicals (Abedinzadeh *et al.*, 1991).

A large number of organic radicals have been shown to oxidize thiols (Table II). As mentioned previously (Section II,C), this reaction is reversible, but is driven toward thiyl radical formation by destruction of the latter via Reactions (3) and (4). In accord with this mechanism, oxygen is taken up during the reaction between thiols and free radicals (Barringer, 1955; Rosenwald, 1956; Løvstad 1974; Moldéus *et al.*, 1983; Ross *et al.*, 1985a,b,c; Subrahmanyam and O'Brien, 1985; Subrahmanyam *et al.*, 1987; Munday, 1987; Winterbourn, 1989; Winterbourn and Munday, 1989; Rao *et al.*, 1990; Munday *et al.*, 1990a; Thompson and Eling, 1991) and superoxide radical is produced (Subrahmanyam and O'Brien, 1985; Subrahmanyam *et al.*, 1987; Munday, 1987; Winterbourn, 1989; Winterbourn and Munday, 1989; Munday *et al.*, 1990a). In studies in which products have been quantitated, disulfide has been found to account for more than 85% of the thiol oxidized (Lal, 1976; Eyer and Lengfelder, 1984; Subrahmanyam and O'Brien, 1985; Munday, 1987; Winterbourn and Munday, 1989). Small amounts of glutathione sulfinic acid (Lal, 1976) or glutathione sulfonic acid (Wefers and Sies, 1983) are reported to be formed in some experiments, but no higher oxidation products of the thiol were detected in others (Subrahmanyam *et al.*, 1987; Ross *et al.*, 1985d).

D. Oxidation of Thiols by Quinones and by Alloxan

Gause *et al.* (1967) reported the formation of semiquinone radicals ($QH^{\cdot}$) in the reaction between cysteine and tetrachloro-*p*-benzoquinone or 2,3-dichloro-1,4-naphthoquinone, suggesting a one-electron redox process for their production:

$$Q + RSH \rightleftharpoons QH^{\cdot} + RS^{\cdot} \tag{16}$$

Later studies have similarly shown semiquinone radical formation from thiols and *p*-benzoquinone, 1,4-naphthoquinone, and 2-methyl- and 5-hydroxy-1,4-naphthoquinone (Gant *et al.*, 1986; Takahashi *et al.*, 1987; Miura *et al.*, 1992).

With quinones unsubstituted at the 2- or 3-positions, net one-electron reduction occurs at low glutathione concentrations (0.5 mM), consistent with Reaction (16), whereas at higher levels of thiol (2 mM), the quinone undergoes reductive conjugation to the hydroquinone thioether (Takahashi *et al.*, 1987). It has been suggested (Gant *et al.*, 1986; Takahashi *et al.*, 1987) that in this situation, semiquinone radicals could be generated through comproportionation of the quinols (QH_2) with quinone:

$$Q + QH_2 \rightleftharpoons 2QH^{\cdot} \tag{17}$$

Table II

Radicals Reported to Oxidize Thiols

Source of radical	Method of radical production[a]	Thiol(s) employed[b]	Reference
Aromatic amines			
p-Phenetidine (4-ethoxyanilline)	HRP, PGS	GSH	Ross *et al.* (1985a)
	HRP	GSH, CYS, ACYS	Ross *et al.* (1985b)
	HRP	GSH	Subrahmanyam *et al.* (1987)
2- and 4-Methoxyaniline	HRP	GSH	Thompson and Eling, (1991)
N,N-Dimethylaniline	HRP	GSH	Moldéus *et al.* (1983)
	HRP	GSH	Subrahmanyam *et al.* (1987)
4-Methyl-*N,N*-dimethylaniline	HRP	GSH	Subrahmanyam and O'Brien (1985)
N,N,N',N'-Tetramethyl-*p*-phenylenediamine	Chemical synthesis	GSH	Vinogradov *et al.* (1979); Störle and Eyer (1991)
2,3,5,6-Tetramethyl-*p*-phenylenediamine	Autoxidation	GSH	Munday *et al.* (1990a)
N,N'-di-sec-butyl-*p*-phenylenediamine	Autoxidation	Butanethiol	Barringer (1955); Rosenwald (1956)
1,2,4-Triaminobenzene	Autoxidation	GSH	Munday (1987)
N-Methyl-4-aminoazobenzene	HRP	GSH	Subrahmanyam and O'Brien (1985)
2-Naphthylamine	HRP	GSH	Subrahmanyam and O'Brien (1985)
N-Phenyl-2-naphthylamine	Chemical synthesis	Methanethiol	Bridger (1972)
2-Aminofluorene	HRP	GSH	Subrahmanyam and O'Brien (1985)
	HRP	GSH	Subrahmanyam *et al.* (1987)
Aminopyrine	HRP	GSH	Moldéus *et al.* (1983)
	HRP	GSH	Subrahmanyam *et al.* (1987)
	PGS	GSH	Eling *et al.* (1985)
Phenols			
Phenol	HRP	GSH	Subrahmanyam and O'Brien (1985)
	PGS	GSH	Schreiber *et al.* (1989)
p-Aminophenol	HRP	GSH	Subrahmanyam *et al.* (1987)
N,N-Dimethylaminophenol	HRP	GSH	Subrahmanyam and O'Brien (1985)
	Autoxidation	GSH	Eyer and Lengfelder (1984)

Paracetamol (*N*-[4-hydroxyphenyl] acetamide)	PGS	GSH	Moldéus *et al.* (1982)
	HRP, PGS	GSH, CYS, ACYS	Ross *et al.* (1984)
	HRP	GSH	Ross *et al.* (1985b)
	HRP	GSH	Subrahmanyam *et al.* (1987)
	HRP	GSH, CYS	Rao *et al.* (1990)
2,2′-Dihydroxybiphenyl	HRP	GSH	Subrahmanyam and O'Brien (1985)
1-Naphthol	HRP, pulse radiolysis	GSH	d'Arcy Doherty *et al.* (1986)
Diethylstilbestrol	HRP, PGS	GSH	Ross *et al.* (1985c)
Etopaside	HRP	GSH, CYS	Katki *et al.* (1987)
2,6-Di-*tert*-butyl-4-(3,5-di-*tert*-butyl-4-hydroxybenzylidine2,5-cyclohexadien-1-one	Chemical synthesis	GSH, CYS	Tsuchiya *et al.* (1985)
2,5-Bis-(1-aziridinyl)-3,6-bis(ethoxycarbonylamino)-1,4-benzenediol	Autoxidation	GSH	Ordoñez and Cadenas (1992)
α-Tocopherol	Chemical synthesis	GSH	Niki *et al.* (1982)
Phenothiazines			
Promazine	Autoxidation	GSH	Løvstad (1974)
Chlorpromazine	HRP	GSH	Ohnishi *et al.* (1969)
	HRP	GSH	Subrahmanyam *et al.* (1987)
	Autoxidation	GSH	Løvstad (1974)
Promethazine	HRP	GSH	Subrahmanyam *et al.* (1987)
Pyrimidines			
Dialuric acid	Autoxidation	GSH, CYS	Winterbourn and Munday (1989)
Divicine, isouramil	Autoxidation	GSH, CYS	Winterbourn (1989)
Miscellaneous compounds			
Phenylhydrazine	Autoxidation	GSH	Maples *et al.* (1990)
1,1-Diphenyl-2-(2,4,6-trinitrophenyl)hydrazine	Chemical synthesis	Butanethiol, benzenethiol	Brook *et al.* (1958)
	Chemical synthesis	GSH, CYS	Blois (1958)
	Chemical synthesis	Various aliphatic	Ewald (1959)
Triphenylmethane	Chemical synthesis	Benzenethiol	Colle *et al.* (1978)
		2,4,6-Trimethylbenzenethiol	Lewis and Butler (1976)

continues

Table II *continued*

Source of radical	Method of radical production[a]	Thiol(s) employed[b]	Reference
Aliphatic alcohols	Pulse radiolysis	GSH, CAM	Baker *et al*. (1982); Adams *et al*. (1968); Willson (1983)
Glucose and other sugars	Pulse radiolysis	GSH	Baker *et al*. (1982)
Nucleic acids	Pulse radiolysis	DTT, CAM	Held (1988)
Butanenitrile	Chemical synthesis	Various aliphatics, benzenethiol	Bruin *et al*. (1952)
Hexanenitrile	Chemical synthesis	Various aromatics	Schaafsma *et al*. (1957)
Dipotassium nitrosodisulfonate	Chemical synthesis	GSH, ovothiol A	Holler and Hopkins (1990)
4-Hydroxylaminoquinoline-1-oxide	Autoxidation	GSH, CYS	Hozumi *et al*. (1967)

[a] Abbreviations used: HRP, horseradish peroxidase; PGS, prostaglandin H synthase.
[b] Abbreviations used: GSH, glutathione; CYS, cysteine; ACYS, acetylcysteine; CAM, cysteamine; DTT, dithiothreitol.

Reaction (16) is, in fact, thermodynamically unfavorable (Wardman, 1990; Butler and Hoey, 1992). Again, it has been argued that the equilibrium may be displaced to the right by rapid decay of the radicals, particularly under aerobic conditions (Wardman, 1990). Disulfide radical anion, superoxide radical, and hydrogen peroxide have been detected in thiol–quinone interactions (Ross *et al.*, 1985e; Butler and Hoey, 1992), consistent with thiyl radical breakdown by Reactions (3) and (4).

Reaction of the pyrimidine derivative alloxan with glutathione or cysteine leads to the formation of the corresponding radical, and a one-electron process analogous to Reaction (16) has been proposed (Lagercrantz and Yhland, 1963). Since alloxan is readily reduced to its two-electron reduction product, dialuric acid, by thiols (Winterbourn and Munday, 1989), formation of alloxan radical by comproportionation is again a possibility.

Generation of superoxide radical, hydrogen peroxide, and, in the presence of iron salts, hydroxyl radical has been observed during the oxidation of thiols by alloxan (Winterbourn and Munday, 1989; Sakurai and Ogiso, 1991).

IV. Structural Effects on One-Electron Oxidation of Thiols

Although one-electron oxidation is a common property of thiols, the rate of this reaction is strongly dependent upon structure. For example, the rate of metal-catalyzed oxidation of aminothiols such as glutathione, cysteine, and cysteamine at neutral pH is greater than that of saturated aliphatic thiols, such as methanethiol (**I**) or its derivative, phenylmethanethiol (**II**). In contrast, the aromatic thiol benzenethiol (**III**) is oxidized much more rapidly than aminothiols under these conditions (Munday, 1989). The heterocyclic dithiol 3,6-dimercapto-1,4-dimethyl-2,5-piperazinedione (**IV**) is also rapidly oxidized in neutral solution (Eichner *et al.*, 1988), as is the related compound (**VI**), which is the reduced form of the mycotoxin sporidesmin (Munday, 1982). Similarly, the α,β-unsaturated thiol ovothiol A (**V**) reacts faster with free radicals than does glutathione (Holler and Hopkins, 1990), and aromatic thiols are more effective radical scavengers than aliphatic thiols (Simic, 1988; Bruin *et al.*, 1952).

Ring substitution in aromatic thiols greatly influences the rate of metal-catalyzed oxidation (Table III). Electron-donating groups (alkyl, amino, methoxy) in the 4-position markedly increase the rate of oxidation, although the effect of such substituents in the 2-position is much smaller. The rate of oxidation of 2-alkylbenzenethiols decreases with increasing

CH_3SH

(I)

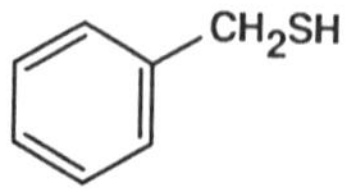

(II)

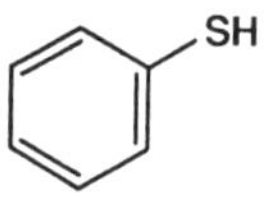

(III)

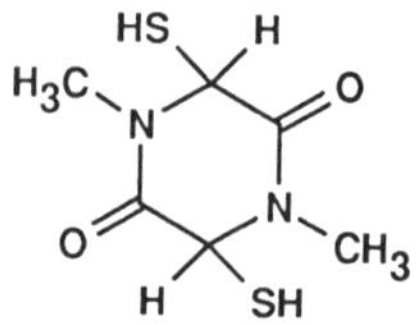

(IV)

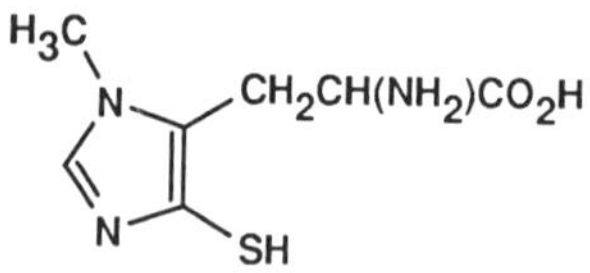

(V)

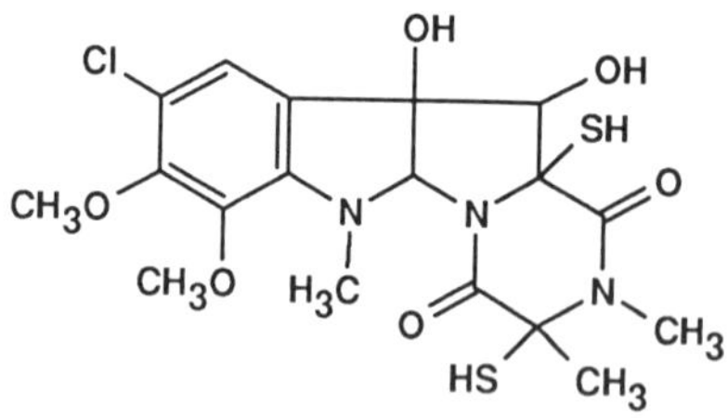

(VI)

size of the alkyl group, whereas electron-withdrawing substituents (nitro, carboxyl) decrease the oxidation rate, irrespective of their position in the ring.

Aromatic thiols also react much more rapidly with oxyhemoglobin than aliphatic thiols, and the same structure–activity relationships are observed with benzenethiol derivatives (Munday, 1985a).

These results can be rationalized in terms of degree of ionization of the thiol and the stability of the derived thiyl radical.

Electron-transfer reactions of thiols, as in oxidations mediated by metals and by hemoglobin and in some radical scavenging reactions, involve the thiolate anion in the rate-limiting step. The rate of the reaction will, therefore, increase with the concentration of RS^- in solution and hence will be related to the ionization constants of the thiols. The pK_a values of aminothiols are of the order of 8.5 in aqueous solution (Jocelyn, 1972), and these compounds will, therefore, be ionized at neutral pH to a greater extent than saturated aliphatic thiols ($pK_a \sim 10$). Ionization of aromatic

Table III

Effect of Ring Substitution on the Rate of Metal-Catalyzed Oxidation of Benzenethiol Derivatives

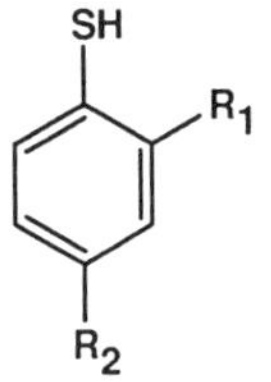

Compound	R_1	R_2	Relative rate of oxidation[a]
4-Aminobenzenethiol	H	NH_2	15.42
2-Aminobenzenethiol	NH_2	H	9.70
4-Methoxybenzenethiol	H	OCH_3	4.82
4-Methylbenzenethiol	H	CH_3	1.88
2-Methylbenzenethiol	CH_3	H	1.15
2-Ethylbenzenethiol	C_2H_5	H	1.08
2-Methoxybenzenethiol	OCH_3	H	1.07
Benzenethiol	H	H	1.00
2-Isopropylbenzenethiol	$^iC_3H_7$	H	0.92
4-Nitrobenzenethiol	H	NO_2	0.27
Benzenethiol-2-carboxylic acid	CO_2H	H	0.11

[a] Benzenethiol = 1; oxidation rate measured at pH 7 in the presence of hematin as catalyst (Munday, 1985a; Munday *et al.* 1990b).

thiols is facilitated by charge delocalization, and pK_a values for these compounds are generally less than 7 (Danehy and Parameswaran, 1968). Charge delocalization and hence low pK_a values are to be expected in any α,β-unsaturated thiol, such as Compound **V** (Holler and Hopkins, 1990), but when the thiol group is separated from the ring, as in Compound **II,** delocalization is not possible, and the pK_a of the latter substance is 9.4 (Danehy and Parameswaran, 1968). Compound **VI** is highly dissociated at neutral pH, ionization being promoted by hyperconjugation (Munday, 1982); the same may well be true of compound **IV.**

The rate of one-electron oxidation of thiols is increased by stabilization of the incipient radical. Thiyl radicals are stabilized by resonance, so they are more easily formed from aromatic and α,β-unsaturated thiols than from saturated thiols. In aromatic compounds, electron-donating groups increase the stability of the radical, whereas electron-withdrawing groups destabilize the radical; the different effects of substituents in the 2- and 4-positions can be explained in terms of steric hindrance at the reaction site when bulky groups are present at the position next to the thiol group (Munday, 1989).

V. Beneficial and Harmful Effects of Thiyl Radical Formation

A. Glutathione as a Radical Scavenger

If a radical $X^{\cdot}$ is scavenged by glutathione via Reaction (14), followed by destruction of the thiyl radical via Reactions (3) and (4), the overall stoichiometry of the process is

$$2GSH + X^{\cdot} + O_2 \rightarrow GSSG + XH + O_2^{\cdot -} + H^+ \qquad (18)$$

In this way, one radical ($X^{\cdot}$) is exchanged for another ($O_2^{\cdot -}$). Since, as discussed previously, superoxide radical can initiate a radical chain reaction for thiol oxidation, this process in itself is not beneficial. In the presence of superoxide dismutase, however, which destroys superoxide via Reaction (19), the radical chain is broken and radical scavenging by this concerted action of glutathione and superoxide dismutase has been proposed as an important cellular defense mechanism (Munday and Winterbourn, 1989). Furthermore, this reaction pathway would result in the conversion of all radicals into superoxide, so that a single enzyme, superoxide dismutase, could, in effect, destroy all the multitude of radicals that could conceivably be formed within cells (Winterbourn, 1993).

$$2O_2^{\cdot -} + 2H^+ \rightarrow H_2O_2 + O_2 \qquad (19)$$

This mechanism does require, however, the presence of adequate levels of superoxide dismutase to break the chain and of enough catalase or glutathione peroxidase to destroy the hydrogen peroxide formed in Reaction (19). In the absence of sufficient enzymatic protection, however, radical chain reactions leading to uncontrolled formation of "active oxygen" species would ensue, with severe consequences for cellular viability. The distribution of such enzymes among organs and tissues is by no means uniform, with, for example, endocrine pancreas and striated muscle containing relatively low levels (Grankvist *et al.*, 1981). It has been suggested that because of this, glutathione-mediated chain reactions may contribute to the diabetogenicity of alloxan (Winterbourn and Munday, 1989) and to the selective myotoxicity of *p*-phenylenediamine derivatives (Munday, 1992).

B. Radiation Protection by Thiols and the Oxygen Effect

Thiols have been shown to offer significant protection against ionizing radiation both *in vitro* and *in vivo*. Free-radical scavenging appears to play a major role in this effect, and two processes, "protection" and "repair," have been identified. In the first, thiols react with radiation-induced radicals before the latter are able to exert their deleterious effects upon a tissue component; in the second, damaged macromolecules (particularly DNA radicals) are repaired by one-electron reduction by a thiol (von Sonntag, 1987; Bump and Brown, 1990).

It has been known for many years that oxygen increases cellular sensitivity to ionizing radiation and diminishes the protective activity of thiols—the so-called "oxygen effect" (von Sonntag, 1987; Held, 1988). This has generally been accepted to reflect failure of the repair mechanism, with reaction of a damaged tissue component with oxygen yielding a peroxyl radical irreparable by thiols (Howard-Flanders, 1960). In both the protection and the repair processes, however, the thiyl radical will be produced, and more recent studies have focused upon the possible involvement of species derived from this.

Quintiliani (1986) proposed that the thiyl radical is activated by reaction with oxygen [Reaction (6)] with the thioperoxyl radical being responsible for the potentiating effect of oxygen. As discussed previously (Section II), however, formation of the disulfide radical anion is the predominant reaction of $RS^{\cdot}$ under physiological conditions, and this mechanism, therefore, seems unlikely. Indeed, it has been suggested (Prütz and Mönig, 1987; Prütz, 1989; Prütz *et al.*, 1989) that the disulfide radical anion itself is important in the repair process, reducing damaged macromolecules via Reaction (20):

$$X^{\cdot} + (RSSR)^{\cdot -} + H^{+} \rightarrow XH + RSSR \qquad (20)$$

With this mechanism, the oxygen effect would be explained by the rapid oxidation of $(RSSR)^{\cdot -}$ via Reaction (4), so that in the presence of oxygen, the concentration of the beneficial species would be decreased.

An alternative explanation is that "active oxygen" species, formed by oxidation of $(RSSR)^{\cdot -}$, are responsible for the oxygen effect. No experiments on this possibility have been reported, and, although it has been shown that intracellular generation of such species is not of major importance in determining radiation sensitivity in the absence of added thiols (Misra and Fridovich, 1976; Niwa *et al.*, 1978; Marklund *et al.*, 1984), studies on their significance in the presence of radioprotective thiols would be of interest.

C. Cytotoxicity of Aminothiols

Toxic effects have been identified in a number of *in vitro* cell systems after exposure to aminothiols, such as glutathione, cysteine, and cysteamine. Toxicity is increased by addition of copper or iron salts to the incubation medium and decreased by catalase, consonant with the involvement of metal-catalyzed thiol oxidation in the toxic process and with hydrogen peroxide, or a product derived from it, being the ultimate cytotoxic species (Munday, 1989).

In many instances, an unusual dose–response relationship is observed in aminothiol cytotoxicity (Munday, 1989). With increasing levels of the test material, the severity of the toxic response increases to a maximum, after which further increments in aminothiol concentration result in a progressive decrease in toxicity. As suggested by Takagi *et al.* (1974), this may reflect the fact that in this test system the aminothiol both generates and destroys "active oxygen" species. At low concentrations, the prooxidant activity of the thiol may predominate, but at high levels, even though the rate of production of oxidizing species increases, more thiol is available for their destruction and the antioxidant effect will become more important.

D. Mutagenicity of Thiols

The bacterial mutagenicity of aminothiols has been extensively studied. Yamaguchi and Yamashita (1981) found cysteine and penicillamine to be mutagenic in *Salmonella typhimurium* TA100 when activated by liver postmitochondrial supernatant or by high levels of enzymes, such as catalase, pepsin, or D-amino acid oxidase. The activation was not enzymatic, however, because the activating ability of both the supernatant and the enzymes was maintained after heat inactivation.

Later studies confirmed these results with cysteine and showed that tissue fractions or homogenates from a variety of sources were effective activators (Glatt *et al.*, 1983, 1990; Glatt, 1990). In contrast, glutathione was mutagenic only in the presence of postmitochondrial supernatant from kidney, and a role for γ-glutamyltranspeptidase was postulated (Glatt *et al.*, 1983). This was supported by the results of Stark *et al.* (1987), showing that purified γ-glutamyltranspeptidase activated glutathione to a mutagen, and cysteinylglycine was suggested as the agent responsible. The latter substance was subsequently shown to be a potent direct-acting mutagen in *S. typhimurium* TA102 (Stark *et al.*, 1989). Among other thiols tested in this strain, activity decreased in the order cysteine ethyl ester > penicillamine > cysteine > glutathione (Stark *et al.*, 1989).

Effects of thiols in mammalian cells have also been recorded. Sister-chromatid exchanges were observed with cysteamine (Speit *et al.*, 1980; Speit and Vogel, 1982) and with penicillamine (Speit and Haupter, 1987) in Chinese hamster V79 cells, although no effect was seen with cysteine (Speit *et al.*, 1980). Positive effects were observed with cysteine in CHO cells (MacRae and Stich, 1979), but neither cysteine nor cysteamine induced exchanges in lymphocytes (Speit and Vogel, 1982). Chromosomal aberrations were observed in CHO cells exposed to cysteine, cysteamine, and glutathione in the presence of copper (Stich *et al.,* 1978) and in V79 cells incubated with high concentrations of penicillamine (Speit and Haupter, 1987).

It is generally agreed that hydrogen peroxide, formed through metal-catalyzed oxidation of the thiols, is responsible for the mutagenic effects of these compounds in mammalian cells (Stich *et al.,* 1978; MacRae and Stich, 1979; Speit and Vogel, 1982; Speit and Haupter, 1987). There has been much debate about the proximate mutagen in bacterial systems, but recent data suggest that "active oxygen" species are again responsible. An association between the mutagenic activity of various aminothiols and their pK_a values has been recorded (Stark *et al.,* 1989), with the most highly ionized (and hence most readily oxidized) compounds showing the greatest activity. The mutagenicity of glutathione, activated by purified γ-glutamyltranspeptidase, was shown to be dependent upon the presence of oxygen; it was enhanced by iron and inhibited by metal-chelating agents (Stark *et al.,* 1988). Superoxide dismutase had no effect on mutagenic activity under these conditions, but partial protection was given by catalase. The failure of catalase completely to prevent mutagenicity was explained (Stark *et al.,* 1988) on the basis of the high K_m of this enzyme for hydrogen peroxide, but the fact that high levels of this enzyme potentiate mutagenicity (Yamaguchi and Yamashita, 1981) suggests that the response may be biphasic; a careful dose–response study is required. Added catalase and superoxide dismutase had no effect upon the mutagenicity of glutathione when activated by kidney postmitochondrial supernatant, and it has been argued that this precludes a role for "active oxygen" species in the mutagenic process (Glatt, 1989, 1990). However, the tissue fraction itself would contain significant amounts of both catalase and superoxide dismutase, and the mutagenicity measured under these conditions would be that fraction that is not inhibited by catalase; further addition of this enzyme would not, therefore, be expected to inhibit.

Overall, the data on the mutagenicity of aminothiols are consistent with the involvement of "active oxygen" species formed by oxidation, with the role of "activating" enzymes or tissue fractions (except in the case of glutathione) being simply to provide a source of catalytically active metal. The nature of such metal may be of crucial importance, however, since thiols oxidizing in the presence of copper caused little or no mutagenicity at nontoxic levels (Stich *et al.,* 1978). A detailed study of the role of metals, particularly iron, in aminothiol mutagenicity would be valuable.

Few studies on the mutagenicity of other thiols have been conducted. Dithiothreitol, in the presence of added iron, is mutagenic in *S. typhimurium* TA100 (Yamaguchi, 1981), and 4-aminobenzenethiol is also a bacterial mutagen (Topham, 1980). A study of structure–activity relationships among aliphatic and aromatic thiols would be useful for establishing the mechanism of mutagenicity of such compounds.

It should be noted that, although the aminothiols themselves are mutagenic under certain conditions, they are also capable of protecting against the mutagenicity of enzymatically generated "active oxygen" species (De Flora *et al.*, 1989). This again reflects the ability of these compounds to act as both pro- and anti-oxidants.

E. Lipid Peroxidation

In the presence of iron salts, thiols induce peroxidation of unsaturated lipids. The process is inhibited by iron chelators and by antioxidants, such as butylated hydroxytoluene and α-tocopherol (Searle and Willson, 1983; Kanner *et al.*, 1986), but not by superoxide dismutase or catalase or by scavengers of the hydroxyl radical (Tien *et al.*, 1982; Bucher *et al.*, 1983; Searle and Willson, 1983; Kanner *et al.*, 1986). With some thiols, a biphasic dose–response is observed, with promotion of peroxidation at low levels of thiol and inhibition at high concentrations (Tien *et al.*, 1982).

It has been suggested that peroxidation involves metal-catalyzed thiol oxidation, with the process being initiated by "site-specific" generation of hydroxyl radical, formed from hydrogen peroxide via Reaction (12). The failure of catalase and hydroxyl radical scavengers to prevent peroxidation is explained on the basis of $^{\cdot}OH$ formation close to the lipid target, at a site that is not accessible to the potential protectants. The more lipophilic antioxidants may scavenge peroxyl radicals, thereby preventing amplification of peroxidation through propagation reactions (Searle and Willson, 1983; Kanner *et al.*, 1986); high levels of thiols could presumably behave similarly.

More recently, it has been proposed that the thiyl radical itself may initiate lipid peroxidation (Schöneich *et al.*, 1992). In this model, scavengers of "active oxygen" species would not be expected to inhibit peroxidation, whereas antioxidants could compete with lipid for reaction with the thiyl radical. In the lipid environment, the oxidizing ability of the thiyl radical (Section II,C) may be more readily expressed than in an aqueous system since its destruction via reaction with RS^- is less likely; this process could, however, explain the protection observed at high concentrations of thiol.

F. Reactions of Glutathione Involving Peroxidases

Compounds such as 2-aminofluorene, diethylstilbestrol, paracetamol, *p*-phenetidine, and 2-methoxyaniline are carcinogenic. The target organs of these substances (kidney, urinary bladder, uterus, and Zymbal and Harderian glands) are particularly rich in peroxidases, and it has been suggested that enzymatic one-electron oxidation may be involved in their carcinogenic action (Subrahmanyam and O'Brien, 1985; Ross and Moldéus, 1985; Thompson and Eling, 1991). Since, as discussed earlier (Section III,C), the radicals derived from these compounds are reduced by glutathione, chain reactions involving RS˙ could also be involved in cancer induction if antioxidant enzyme levels in the target tissue were low. Although this appears unlikely in the case of the kidney, which is known to contain high levels of superoxide dismutase, catalase, and glutathione peroxidase (Grankvist *et al.*, 1981), study of the levels of such enzymes in the other target organs would be of interest.

Oxidation of glutathione itself by peroxidase, leading to conjugate formation through addition of the thiyl radical to aromatic compounds, has been suggested as a detoxication pathway (Stock *et al.*, 1986; Foureman and Eling, 1989). As yet, however, there is no evidence for the operation of this pathway *in vivo*.

G. Anti-Cancer Activity of Metal Complexes

It has been suggested (Petering, 1980) that thiosemicarbazonato–copper complexes may exert their anti-cancer effects through intracellular generation of "active oxygen" species by catalysis of the oxidation of endogenous thiols. A similar mechanism has been proposed for the iron–adriamycin complex (Muindi *et al.*, 1985; Zweier *et al.*, 1986).

Complex-mediated thiol oxidation is potentially of considerable therapeutic benefit, since tumor cells, which contain low levels of antioxidant enzymes (Powis, 1987), would be particularly liable to uncontrolled "active oxygen" production via chain reactions of RS˙.

H. Hemolytic Activity of Thiols and Disulfides

A number of aliphatic, aromatic, and heterocyclic thiols and disulfides have been shown to cause hemolytic anemia in animals and in humans (Munday and Manns, 1985; Munday, 1989; Munday *et al.*, 1990b). The hemolysis is of the oxidative type, being characterized by precipitation of oxidatively denatured hemoglobin (Heinz bodies) within erythrocytes. These compounds are selectively toxic to red blood cells, with no other harmful effects being recorded.

If has been suggested (Munday, 1989) that the hemolysis induced by these compounds is initiated by one-electron oxidation of the thiol by oxyhemoglobin [Reaction (13)]. Hydrogen peroxide formed in this reaction, together with more "active oxygen" species produced in reactions of the thiyl radical, would, if produced sufficiently rapidly to overwhelm the defenses of the cell, cause the oxidative damage responsible for erythrocyte destruction. This mechanism would explain the target-organ specificity of these compounds; it is supported by the fact that the severity of the hemolysis induced by the various compounds *in vivo* is directly proportional to their rate of reaction with oxyhemoglobin (Munday and Manns, 1985) and to their ability to generate hydrogen peroxide (Munday, 1985a) and to cause oxidative damage (Munday, 1985b; Amrolia *et al.*, 1989) in erythrocytes *in vitro*.

Disulfides of these types are readily reduced by thiol–disulfide exchange with glutathione [Reaction (21)] and both the *in vivo* toxicity of these substances and their effects on erythrocytes *in vitro* are the same as those of the corresponding thiols (Munday, 1989).

$$\text{RSSR} + 2\text{GSH} \rightarrow 2\text{RSH} + \text{GSSG} \tag{21}$$

Because of the reductive ability of glutathione, a redox cycle is established in which the xenobiotic thiol, in effect, catalyzes the one-electron oxidation of the endogenous thiol. Therefore, in a manner analogous to the *in vitro* effects of aminothiols (Section V,C), glutathione in erythrocytes would be responsible for both the generation and the destruction of "active oxygen" species, and modulation of cellular levels of GSH could have either beneficial or detrimental effects. High erythrocytic levels of glutathione are reported to protect against the hemolysis induced in sheep by feeding high dietary levels of kale, the hemolytic component of which is dimethyl disulfide (Smith, 1980). In contrast, dogs with an inherited high concentration of erythrocytic glutathione are more susceptible to the hemolytic action of 4,4′-diaminodiphenyl disulfide (Maede *et al.*, 1989) and to the hemolysis induced by feeding onions (Yamoto and Maede, 1992), the toxic constituents of which are prop-1-enyl disulfides (Munday and Manns, 1994).

I. Toxicity of Epidithiodioxopiperazine Derivatives

More than 50 mycotoxins containing the epidithiodioxopiperazine nucleus have been described; two examples are gliotoxin (**VII**) and sporidesmin (**VIII**). As mentioned previously (Section IV), the reduced (dithiol) form of sporidesmin readily undergoes one-electron oxidation in the presence of catalytic metals. Copper is by far the most effective catalyst, with

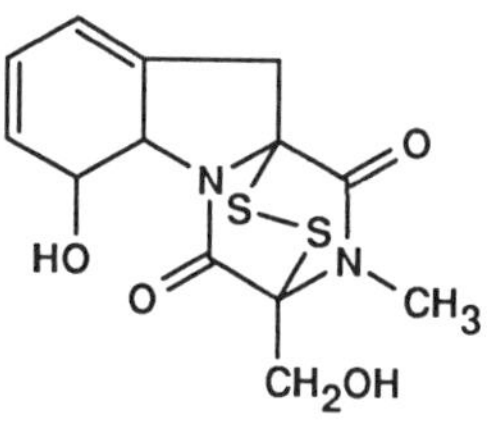

(VII)

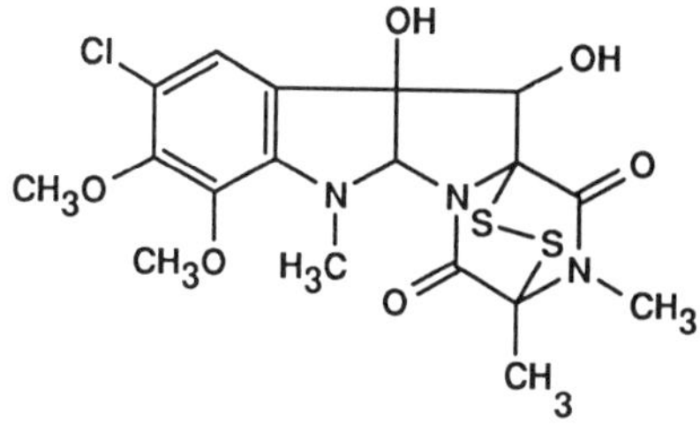

(VIII)

effects being recorded at submicromolar concentrations (Munday, 1982). Furthermore, both sporidesmin (Munday, 1982) and gliotoxin (Waring *et al.*, 1988) undergo redox cycling in the presence of glutathione and other thiols, generating "active oxygen" species.

Sporidesmin, produced by the saprophytic pastoral fungus *Pithomyces chartarum*, is responsible for the hepatogenous photosensitization disease "facial eczema" in grazing ruminants. The primary target site of the mycotoxin is the biliary epithelium, in which inflammatory changes and necrosis are rapidly produced. As discussed previously (Munday, 1989), the toxicity of sporidesmin can be explained on the basis of copper-catalyzed redox cycling of the mycotoxin, with concomitant generation of "active oxygen" species within the target tissue.

Gliotoxin is a secondary metabolite of several fungal species, which has been shown to possess antimicrobial and immunomodulating activity (Waring *et al.*, 1988); the latter property is under investigation for possible beneficial effects in organ transplantation (Waring and Müllbacher, 1992). In the presence of glutathione or dithiothreitol, gliotoxin causes oxidative damage to DNA and, again, the biological effects of this substance may be due to "active oxygen" species formed during redox cycling (Eichner *et al.*, 1988).

J. Formation of Cytotoxic Species through One-Electron Oxidation of Glutathione by Xenobiotics

The possible involvement of the oxidation of glutathione via Reaction (16) in the toxicity of simple quinones has not been explored. The observation (Harley *et al.*, 1982) that glutathione-deficient mutants of *Escherichia coli* are more resistant to the toxic effects of streptonigrin, a quinolinequinone derivative, suggests that such involvement may be important and more work in this area would be valuable.

It has been suggested (Winterbourn and Munday, 1989) that redox reactions between glutathione and alloxan may be significant in the diabetogenic action of the latter; chain reactions of the thiyl radical may be possible in the B-cells of the pancreas due to their low content of antioxidant enzymes (Section V,A).

VI. Conclusion

For many years, the thiyl radical was considered to be a rather unreactive species whose production within cells was either benign or beneficial. More recently, however, the true reactivity of this species has been recognized and the harmful, as well as beneficial, effects of intracellular thiyl radical production have been explored.

The major pathway for decay of the thiyl radical under physiological conditions involves formation of the corresponding disulfide with concomitant production of "active oxygen" species; there is evidence that the latter are responsible for many of the toxic effects associated with one-electron oxidation of thiols. Furthermore, the beneficial effects of thiyl radical formation, as, for example, in free-radical scavenging, will only be expressed in the presence of cellular systems for destruction of "active oxygen" species.

One-electron oxidation of thiols can be mediated in many ways. One of the most important routes involves reaction with molecular oxygen, catalyzed by transition metals. Copper and iron are the most important catalytic metals, the latter being of especial significance because of its ability to promote hydroxyl radical formation from the hydrogen peroxide formed in earlier stages of the oxidative process. Although intracellular levels of catalytically active metals are low, some thiols, such as those derived from epidithiodioxopiperazines, are sensitive to extremely low concentrations of metal, and this may account for their *in vivo* toxic effects. Others, such as aliphatic and aromatic thiols, are oxidized by protein-bound metal, specifically oxyhemoglobin. For this reason, such compounds are selectively toxic to erythrocytes, with their relative hemolytic activity in animals being governed by their rate of reaction with oxyhemoglobin.

The ease with which thiols undergo one-electron oxidation is critically dependent upon structure, with such factors as degree of ionization, stability of the thiyl radical and steric hindrance being important. Saturated aliphatic thiols are oxidized very slowly, whereas aromatic compounds undergo rapid reaction; aminothiols, such as glutathione, are oxidized at an intermediate rate. The facility with which glutathione is oxidized may well be optimal for its activity as a cellular protectant, being sufficiently easily oxidized to act as a radical scavenger but not reactive enough to be oxidized by the low levels of metals within cells or by protein-bound metal. Although this compound is oxidized *in vitro,* either by deliberately added metals or by the metals that invariably contaminate tissue culture media, leading to cytotoxic or mutagenic effects in cells or to lipid peroxidation, this is not a problem *in vivo* under normal conditions. An interesting possibility arises, however, in that oxidation of endogenous thiols by administration of particularly potent oxidation catalysts may be therapeutically beneficial, as in the anti-cancer iron and copper complexes.

Although much has been achieved with regard to understanding the consequences of one-electron oxidation of both endogenous and xenobiotic thiols, much remains to be done. In particular, a fuller understanding of the delicate balance that exists between the pro- and the anti-oxidant activities of these compounds would be valuable, not only in determining mechanisms of toxicity and protection but also, perhaps, in the design of cytotoxic drugs for therapeutic purposes.

References

Abedinzadeh, Z., Gardès-Albert, M., and Ferradini, C. (1991). One-electron oxidation of glutathione by azide radicals in neutral medium: A gamma and pulse radiolysis study. *Radiat. Phys. Chem.* **38,** 1–5.

Adams, G. E., McNaughton, G. S., and Michael, B. D. (1968). Pulse radiolysis of sulphur compounds. Part 2. Free radical "repair" by hydrogen transfer from sulphydryl compounds. *Trans. Faraday Soc.* **64,** 902–910.

Ahmad, R., and Armstrong, D. A. (1982). Reduction, oxidation, and addition reactions between free radicals and flavins. *Biochemistry* **21,** 5445–5450.

Ahmad, R., and Armstrong, D. A. (1984). The effect of pH and complexation on redox reactions between RS· radicals and flavins. *Can. J. Chem.* **62,** 171–177.

Akhlaq, M. S., Schuchmann, H.-P., and von Sonntag, C. (1987). The reverse of the 'repair' reaction of thiols: H-abstraction at carbon by thiyl radicals. *Int. J. Radiat. Biol.* **51,** 91–102.

Alder, R. W., Bonifačić, M., and Asmus, K.-D. (1986). Reaction of a stable N∴N bonded cation with free radicals generated by pulse radiolysis: exceedingly rapid hydrogen abstraction for C-H bonds. *J. Chem. Soc., Perkin Trans. 2,* pp. 277–284.

Al-Thannon, A. A., Barton, J. P., Packer, J. E., Sims, R. J., Trumbore, C. N., and Winchester, R. V. (1974). The radiolysis of aqueous solutions of cysteine in the presence of oxygen. *Int. J. Radiat. Phys. Chem.* **6,** 233–248.

Amrolia, P., Sullivan, S. G., Stern, A., and Munday, R. (1989). Toxicity of aromatic thiols in the human red blood cell. *J. Appl. Toxicol.* **9,** 113–118.

Anderson, W. L., and Tomasi, T. B. (1977). Stimulation of reduced lysozyme regeneration by transferrin and lactoferrin. *Arch. Biochem. Biophys.* **182,** 705–711.

Antholine, W. E., Kalyanaraman, B., Templin, J. A., Byrnes, R. W., and Petering, D. H. (1991). Spin-trapping studies of the oxidation-reduction reactions of iron bleomycin in the presence of thiols and buffer. *Free Radial Biol. Med.* **10,** 119–123.

Asada, K., and Kanematsu, S. (1976). Reactivity of thiols with superoxide radicals. *Agric. Biol. Chem.* **40,** 1891–1892.
Asmus, K. D. (1990). Sulfur-centered free radicals. *In* "Methods in Enzymology" (L. Packer and A. Glazer, eds.) Vol. 186, pp. 168–180.
Bahnemann, D., Asmus, K.-D., and Willson, R. L. (1981). Free radical reactions of the phenothiazine, metiazinic acid. *J. Chem. Soc., Perkin Trans. 2,* pp. 890–895.
Baker, M. Z., Badiello, R., Tamba, M., Quintiliani, M., and Gorin, G. (1982). Pulse radiolytic study of hydrogen transfer from glutathione to organic radicals. *Int. J. Radiat. Biol. Relat. Stud. Phys. Chem. Med.* **41,** 595–602.
Barringer, C. M. (1955). Antioxidant sweetening of gasolines. *Ind. Eng. Chem.* **47,** 1022–1027.
Barron, E. S. G., Miller, Z. B., and Kalnitsky, G. (1947). The oxidation of dithiols. *Biochem. J.* **41,** 62–68.
Barton, J. P., and Packer, J. E. (1970). The radiolysis of oxygenated cysteine solutions at neutral pH. The role of $RS\bar{S}R$ and O_2^-. *Int. J. Radiat. Phys. Chem.* **2,** 159–166.
Becker, D., Swarts, S., Champagne, M., and Sevilla, M. D. (1988). An ESR investigation of the reactions of glutathione, cysteine and penicillamine thiyl radicals: Competitive formation of RSO·, R·, RSSR· –, and RSS·. *Int. J. Radiat. Biol.* **53,** 767–786.
Bisby, R. H., and Tabassum, N. (1988). Properties of the radicals formed by one-electron oxidation of acetaminophen—a pulse radiolysis study. *Biochem. Pharmacol.* **37,** 2731–2738.
Black, G., Jusinski, L. E., and Patrick, R. (1988). Kinetics of the reactions of C_2H_5S with NO_2, NO, and O_2 at 296K. *J. Phys. Chem.* **92,** 5972–5977.
Blois, M. S. (1958). Antioxidant determinations by the use of a stable free radical. *Nature (London)* **181,** 1199–1200.
Bridger, R. F. (1972). Reactions of N-phenyl-2-naphthylamino radicals. *J. Am. Chem. Soc.* **94,** 3124–3132.
Brook, A. G., Anderson, R. J., and Tissot van Patot, J. (1958). Hydrogen abstraction reactions of diphenylpicrylhydrazyl. *Can. J. Chem.* **36,** 159–166.
Bruin, P., Bickel, A. F., and Kooyman, E. C. (1952). Free-radical reactions involving sulphur compounds. I. The interaction of 2-cyano-2-propyl radicals and thiols. *Recl. Trav. Chim. Pays-Bas.* **71,** 1115–1123.
Bucher, J. R., Tien, M., Morehouse, L. A., and Aust, S. D. (1983). Redox cycling and lipid peroxidation: the central role of iron chelates. *Fundam. Appl. Toxicol.* **3,** 222–226.
Bump, E. A., and Brown, J. M. (1990). Role of glutathione in the radiation response of mammalian cells *in vitro* and *in vivo*. *Pharmacol. Ther.* **47,** 117–136.
Butler, J., and Hoey, B. M. (1992). Reactions of glutathione and glutathione radicals with benzoquinones. *Free Radical Biol. Med.* **12,** 337–345.
Carrell, R. W., Winterbourn, C. C., and French, J. K. (1977). Haemoglobin—A frustrated oxidase? Implications for red cell metabolism. *Hemoglobin* **1,** 815–827.
Chidambaram, M. V., Zgirski, A., and Frieden, E. (1984). The reaction of cysteine with ceruloplasmin copper. *J. Inorg. Biochem.* **21,** 227–239.
Colle, T. H., Glaspie, P. S., and Lewis, E. S. (1978). The triphenylmethyl radical: Equilibrium measurements and the reaction with thiophenol. *J. Org. Chem.* **43,** 2722–2725.
Crank, G., and Makin, M. I. H. (1984). Oxidation of thiols by superoxide ion. *Aust. J. Chem.* **37,** 2331–2337.
Cullis, C. F., and Trimm, D. L. (1968). Homogenous catalysis of the oxidation of thiols by metal ions. *Discuss. Faraday Soc.* **46,** 144–149.
Cullis, C. F., Hopton, J. D., and Trimm, D. L. (1968a). Oxidation of thiols in gas-liquid systems. I. Reaction in the absence of added metal catalysts. *J. Appl. Chem.* **18,** 330–335.

Cullis, C. F., Hopton, J. D., Swan, C. J., and Trimm, D. L. (1968b). Oxidation of thiols in gas-liquid systems. II. Reaction in the presence of added metal catalysts. *J. Appl. Chem.* **18,** 335–344.

Danehy, J. P., and Parameswaran, K. N. (1968). Acidic dissociation constants of thiols. *J. Chem. Eng. Data* **13,** 386–389,

D'Aquino, M., Dunster, C., and Willson, R. L. (1989). Vitamin A and glutathione-mediated free radical damage: competing reactions with polyunsaturated fatty acids and vitamin C. *Biochem. Biophys. Res. Commun.* **161,** 1199–1203.

d'Arcy Doherty, M., Wilson, I., Wardman, P., Basra, J., Patterson, L. H., and Cohen, G. M. (1986). Peroxidase activation of 1-naphthol to naphthoxy or naphthoxy-derived radicals and their reaction with glutathione. *Chem.-Biol. Interact.* **58,** 199–215.

De Flora, S., Bennicelli, C., Zanacchi, P., D'Agostini, F., and Camoirano, A. (1989). Mutagenicity of active oxygen species in bacteria and its enzymatic or chemical inhibition. *Mutat. Res.* **214,** 153–158.

Eichner, R. D., Waring, P., Geue, A. M., Braithwaite, A. M., and Müllbacher, A. (1988). Gliotoxin causes oxidative damage to plasmid and cellular DNA. *J. Biol. Chem.* **263,** 3772–3777.

Eling, T. E., Mason, R. P., and Sivarajah, K. (1985). The formation of aminopyrine cation radical by the peroxidase activity of prostaglandin H synthase and subsequent reactions of the radical. *J. Biol. Chem.* **260,** 1601–1607.

Ewald, A. H. (1959). Hydrogen-atom transfer reactions at high pressures. The reaction between mercaptans and diphenylpicrylhydrazyl. *Trans. Faraday Soc.* **55,** 792–797.

Eyer, P., and Lengfelder, E. (1984). Radical formation during autoxidation of 4-dimethylaminophenol and some properties of the reaction products. *Biochem. Pharmacol.* **33,** 1005–1013.

Eyer, P., Hertle, H., Kiese, M., and Klein, G. (1975). Kinetics of ferrihemoglobin formation by some reducing agents and the role of hydrogen peroxide. *Mol. Pharmacol.* **11,** 326–334.

Florence, T. M. (1984). The production of hydroxyl radical from hydrogen peroxide. *J. Inorg. Biochem.* **22,** 221–230.

Forni, L. G., and Willson, R. L. (1986). Thiyl free radicals and the oxidation of ferrocytochrome c. Direct observation of coupled hydrogen-atom- and electron-transfer reactions. *Biochem. J.* **240,** 905–907.

Forni, L. G., Mönig, J., Mora-Arellano, V. O., and Willson, R. L. (1983). Thiyl free radicals: Direct observations of electron transfer reactions with phenothiazines and ascorbate. *J. Chem. Soc., Perkin Trans. 2,* pp. 961–965.

Foureman, G. L., and Eling, T. E. (1989). Peroxidase-mediated formation of glutathione conjugates from polycyclic aromatic dihydrodiols and insecticides. *Arch. Biochem. Biophys.* **269,** 55–68.

Gant, T. W., d'Arcy Doherty, M., Odowole, D., Sales, K. D., and Cohen, G. M. (1986). Semiquinone anion radicals formed by the reaction of quinones with glutathione or amino acids. *FEBS Lett.* **201,** 296–300.

Gardner, H. W., and Jursinic, P. A. (1981). Degradation of linoleic acid hydroperoxides by a cysteine·$FeCl_3$ catalyst as a model for similar biochemical reactions. I. Study of oxygen requirement, catalyst and effect of pH. *Biochim. Biophys. Acta* **665,** 100–112.

Gause, E. M., Montalvo, D. A., and Rowlands, J. R. (1967). Electron spin resonance studies of the chloranil-cysteine, dichlone-cysteine and dichlone-glutathione reactions. *Biochim. Biophys. Acta* **141,** 217–219.

Glatt, H. (1989). Mutagenicity spectra in *Salmonella typhimurium* strains of glutathione, L-cysteine and active oxygen species. *Mutagenesis* **4,** 221–227.

Glatt, H. (1990). Endogenous mutagens derived from amino acids. *Mutat. Res.* **238,** 235–243.

Glatt, H., Protić-Sabljić, M., and Oesch, F. (1983). Mutagenicity of glutathione and cysteine in the Ames test. *Science* **220,** 961–963.
Glatt, H., Utesch, D., Herbst, M., and Oesch, F. (1990). Mutagenicity experiments on L-cysteine and D-penicillamine using V79 cells as indicators and for metabolic activation. *Mutat. Res.* **243,** 187–193.
Grankvist, K., Marklund, S. L., and Täljedal, I.-B. (1981). CuZn-superoxide dismutase, Mn-superoxide dismutase, catalase and glutathione peroxidase in pancreatic islets and other tissues in the mouse. *Biochem. J.* **199,** 393—398.
Grierson, L., Hildenbrand, K., and Bothe, E. (1992). Intramolecular transformation reaction of the glutathione thiyl radical into a non-sulphur-centered radical: A pulse radiolysis and EPR study. *Int. J. Radiat. Biol.* **62,** 265–277.
Hanaki, A., and Kamide, H. (1983). The copper-catalyzed autoxidation of cysteine. The amount of hydrogen peroxide produced under various conditions and the stoichiometry of the reaction. *Bull. Chem. Soc. Jpn.* **56,** 2065–2068.
Harley, J. B., Fetterolf, C. J., Bello, C. A., and Flaks, J. G. (1982). Streptonigrin toxicity in *Escherichia coli:* Oxygen dependence and the role of the intracellular oxidation-reduction state. *Can. J. Microbiol.* **28,** 545–552.
Harman, L. S., Mottley, C., and Mason, R. P. (1984). Free radical metabolites of L-cysteine oxidation. *J. Biol. Chem.* **259,** 5606–5611.
Harman, L. S., Carver, D. K., Schreiber, J., and Mason, R. P. (1986). One- and two-electron oxidation of reduced glutathione by peroxidases. *J. Biol. Chem.* **261,** 1642–1648.
Harrison, D. C. (1924). The catalytic action of traces of iron on the oxidation of cysteine and glutathione. *Biochem. J.* **18,** 1009–1022.
Held, K. D. (1988). Models for thiol protection of DNA in cells. *Pharmacol. Ther.* **39,** 123–131.
Holler, T. P., and Hopkins, P. B. (1990). Ovothiols as free-radical scavengers and the mechanism of ovothiol-promoted NAD(P)H-O_2 oxidoreductase activity. *Biochemistry* **29,** 1953–1961.
Holmquist, W. R., and Vinograd, J. R. (1963). Reduction of human ferrihemoglobin A in the presence of cysteine and the effect of carbon monoxide. *Biochim. Biophys. Acta* **69,** 337–354.
Howard-Flanders, P. (1960). Effect of oxygen on the radiosensitivity of bacteriophage in the presence of sulphydryl compounds. *Nature* (*London*) **186,** 485–487.
Hozumi, M., Inuzuka, S., and Sugimura, T. (1967). Oxidation of sulfhydryl compounds *in vitro* by 4-hydroxyaminoquinoline-1-oxide, a carcinogenic metabolite of 4-nitroquinoline-1-oxide. *Cancer Res.* **27,** 1378–1383.
Huston, P., Espenson, J. H., and Bakac, A. (1992). Kinetics of formation and reactions of thiyl radicals in aqueous solution. *Inorg. Chem.* **31,** 720–722.
Ito, O., and Matsuda, M. (1988). Chemical kinetics of reactions of sulfur-containing organoradicals. *In* "Chemical Kinetics of Small Organic Radicals" (Z. B. Alfassi, ed.), Vol. 3, pp. 133–164. CRC Press, Boca Raton, FL.
Jocelyn, P. C. (1972). "Biochemistry of the SH Group." Academic Press, London.
Kanner, J., Harel, S., and Hazan, B. (1986). Muscle membranal lipid peroxidation by an "iron redox cycle" system: Initiation by oxy radicals and site-specific mechanism. *J. Agric. Food Chem.* **34,** 506–510.
Katki, A. G., Kalyanaraman, B., and Sinha, B. K. (1987). Interactions of the antitumor drug, etoposide, with reduced thiols in vitro and in vivo. *Chem.-Biol. Interact.* **62,** 237–247.
Khan, M. F., and Sorenson, J. R. J. (1991). Copper(II)(3,5-diisopropylsalicylate)$_2$ oxidizes thiols to symmetrical disulfides and oxidatively converts mixtures of 5-thio-2-nitrobenzoic acid and nonsymmetrical 5-thiol-2-nitrobenzoic acid disulfides to symmetrical disulfides. *J. Inorg. Biochem.* **41,** 221–234.

Kobashi, K. (1968). Catalytic oxidation of sulfhydryl groups by o-phenanthroline copper complex. *Biochim. Biophys. Acta* **158,** 239–245.

Koppenol, W. H. (1993). A thermodynamic appraisal of the radical sink hypothesis. *Free Radical Biol. Med.* **14,** 91–94.

Lagercrantz, C., and Yhland, M. (1963). Free radicals in the reaction of alloxan with glutathione and ascorbic acid. *Acta Chem. Scand.* **17,** 1677–1682.

Lal, M. (1976). ^{60}Co γ radiolysis of reduced glutathione in aerated solutions at pH values between 1-7.0. *Can. J. Chem.* **54,** 1092–1097.

Lal, M., and Mahal, H. S. (1990). Reactions of thiyl radicals with phenothiazines and other antioxidants in aqueous solutions. A pulse radiolysis study. *Can. J. Chem.* **68,** 1376–1382.

Lewis, E. S., and Butler, M. M. (1976). Isotope effects of hydrogen atom transfers. V. The reactions of olefins with mercaptans. *J. Am. Chem. Soc.* **98,** 2257–2260.

Løvstad, R. A. (1974). Interaction of phenothiazine derivatives with human ceruloplasmin. *Biochem. Pharmacol.* **23,** 1045–1052.

MacRae, W. D., and Stich, H. F. (1979). Induction of sister-chromatid exchanges in Chinese hamster ovary cells by thiol and hydrazine compounds. *Mutat. Res.* **68,** 351–365.

Maede, Y., Kuwabara, M., Sasaki, A., Inaba, M., and Hiraoka, W. (1989). Elevated glutathione accelerates oxidative damage to erythrocytes produced by aromatic disulfide. *Blood* **73,** 312–317.

Maples, K. R., Eyer, P., and Mason, R. P. (1990). Aniline-, phenylhydroxylamine-, nitrosobenzene-, and nitrobenzene-induced hemoglobin thiyl free radical formation *in vivo* and *in vitro*. *Mol. Pharmacol.* **37,** 311–318.

Marklund, S. L. Westman, N. G., Roos, G., and Carlsson, J. (1984). Radiation resistance and the CuZn superoxide dismutase, Mn superoxide dismutase, catalase, and glutathione peroxidase activities of seven human cell lines. *Radiat. Res.* **100,** 115–123.

Mathews, A. P., and Walker, S. (1909). The action of metals and strong salt solutions on the spontaneous oxidation of cystein. *J. Biol. Chem.* **6,** 299–312.

Meldrum, N. U., and Dixon, M. (1930). The properties of pure glutathione. *Biochem. J.* **24,** 472–496.

Misra, H. P., and Fridovich, I. (1976). Superoxide dismutase and the oxygen enhancement of radiation lethality. *Arch. Biochem. Biophys.* **176,** 577–581.

Miura, T., Muraoka, S., and Ogiso, T. (1992). Generation of semiquinone and oxygen radicals by the reaction of menadione with reduced glutathione at various pH. *Chem. Pharm. Bull.* **40,** 709–712.

Moldéus, P., Andersson, B., Rahimtula, A., and Berggren, M. (1982). Prostaglandin synthetase catalyzed activation of paracetamol. *Biochem. Pharmacol.* **31,** 1363–1368.

Moldéus, P., O'Brien, P. J., Thor, H., Berggren, M., and Orrenius, S. (1983). Oxidation of glutathione by free radical intermediates formed during peroxidase-catalyzed N-demethylation reactions. *FEBS Lett.* **162,** 411–415.

Mönig, J., Asmus, K.-D., Forni, L. G. and Willson, R. L. (1987). On the reaction of molecular oxygen with thiyl radicals: a re-examination. *Int. J. Radiat. Biol.* **52,** 589–602.

Mottley, C., Toy, K., and Mason, R. P. (1987). Oxidation of thiol drugs and biochemicals by the lactoperoxidase/hydrogen peroxide system. *Mol. Pharmacol.* **31,** 417–421.

Muindi, J., Sinha, B. K., Gianni, L., and Myers, C. (1985). Thiol-dependent DNA damage produced by anthracycline-iron complexes. The structure-activity relationships and molecular mechanisms. *Mol. Pharmacol.* **27,** 356–365.

Munday, R. (1982). Studies on the mechanism of toxicity of the mycotoxin sporidesmin. I. Generation of superoxide radical by sporidesmin. *Chem.-Biol. Interact.* **41,** 361–374.

Munday, R. (1985a). Toxicity of aromatic disulphides. I. Generation of superoxide radical and hydrogen peroxide by aromatic disulphides in vitro. *J. Appl. Toxicol.* **5,** 402–407.

Munday, R. (1985b). Toxicity of aromatic disulphides. II. Intra-erythrocytic hydrogen peroxide formation and oxidative damage by aromatic disulphides. *J. Appl. Toxicol.* **5,** 409–413.

Munday, R. (1987). Oxidation of glutathione and reduced pyridine nucleotides by the myotoxic and mutagenic aromatic amine 1,2,4-triaminobenzene. *Chem.-Biol. Interact.* **62,** 131–141.

Munday, R. (1989). Toxicity of thiols and disulphides: involvement of free radical species. *Free Radical Biol. Med.* **7,** 659–673.

Munday, R. (1992). Mitochondrial oxidation of p-phenylenediamine derivatives in vitro: Structure-activity relationships and correlation with myotoxic activity in vivo. *Chem.-Biol. Interact.* **82,** 165–179.

Munday, R., and Manns, E. (1985). Toxicity of aromatic disulphides. III. *In vivo* haemolytic activity of aromatic disulphides. *J. Appl. Toxicol.* **5,** 414–417.

Munday, R., and Manns, E. (1994). Comparative toxicity of prop(en)yl disulfides derived from *Alliaceae:* Possible involvement of 1-propenyl disulfides in onion-induced hemolytic anemia. *J. Agric. Food Chem.* (in press).

Munday, R., and Winterbourn, C. C. (1989). Reduced glutathione in combination with superoxide dismutase as an important biological antioxidant defence mechanism. *Biochem. Pharmacol.* **38,** 4349–4352.

Munday, R., Manns, E., Fowke, E. A., and Hoggard, G. K. (1990a). Structure-activity relationships in the myotoxicity of ring-methylated p-phenylenediamines in rats and correlation with autoxidation rates in vitro. *Chem.-Biol. Interact.* **76,** 31–45.

Munday, R., Manns, E., and Fowke, E. A. (1990b). Steric effects on the haemolytic activity of aromatic disulphides in rats. *Food Chem. Toxicol.* **28,** 561–566.

Niki, E., Tsuchiya, J., Tanimura, R., and Kamiya, Y. (1982). Regeneration of vitamin E from α-chromanoxyl radical by glutathione and vitamin C. *Chem. Lett.*, pp. 789–792.

Niwa, T., Yamaguchi, H., and Yano, K. (1978). Oxygen-dependent radiosensitivity of *Escherichia coli* and mitigation in lethality by superoxide dismutase. *Agric. Biol. Chem.* **42,** 689–695.

Ohnishi, T., Yamazaki, H., Iyanagi, T., Nakamura, T., and Yamazaki, I. (1969). One-electron-transfer reactions in biochemical systems. II. The reaction of free radicals formed in the enzymic oxidation. *Biochim. Biophys. Acta* **172,** 357–369.

Ordoñez, I. D., and Cadenas, E. (1992). Thiol oxidation coupled to DT-diaphorase-catalysed reduction of diaziquone. Reductive and oxidative pathways of diaziquone semiquinone modulated by glutathione and superoxide dismutase. *Biochem. J.* **286,** 481–490.

Ortiz de Montellano, P. R. and Grab, L. A. (1986). Cooxidation of styrene by horseradish peroxidase and glutathione. *Mol. Pharmacol.* **30,** 666–669.

Packer, J. E. (1963). The action of ^{60}Co-gamma-rays on aqueous solutions of hydrogen sulphide and of cysteine hydrochloride. *J. Chem. Soc.* pp. 2320–2325.

Packer, J. E. (1974). The radiation chemistry of thiols. *In* "The Chemistry of the Thiol Group" (S. Patai, ed.), pp. 481–517. Wiley, London.

Paoletti, F., Mocali, A., and Aldinucci, D. (1990). Superoxide-driven NAD(P)H oxidation induced by EDTA-manganese complex and mercaptoethanol. *Chem.-Biol. Interact.* **76,** 3–18.

Petering, D. H. (1980). Carcinostatic copper complexes. *Met. Ions Biol. Sys.* **11,** 197–229.

Powis, G. (1987). Metabolism and reactions of quinoid anticancer agents. *Pharmacol. Ther.* **35,** 57–162.

Prütz, W. A. (1989). 'Chemical repair' in irradiated DNA solutions containing thiols and/or disulphides. Further evidence for disulphide radical anions acting as electron donors. *Int. J. Radiat. Biol.* **56,** 21–33.

Prütz, W. A., and Mönig, H. (1987). On the effect of oxygen or copper(II) in radiation-induced degradation of DNA in the presence of thiols. *Int. J. Radiat. Biol.* **52,** 677–682.

Prütz, W. A., Butler, J., Land, E. J., and Swallow, A. J. (1989). The role of sulphur peptide functions in free radical transfer: a pulse radiolysis study. *Int. J. Radiat. Biol.* **55,** 539–556.

Pryor, W. A., Gojon, G., and Church, D. F. (1978). Relative rate constants for hydrogen atom abstraction by the cyclohexanethiyl and benzenethiyl radicals. *J. Org. Chem.* **43,** 793–800.

Quintiliani, M. (1986). The oxygen effect in radiation inactivation of DNA and enzymes. *Int. J. Radiat. Biol.* **50,** 573–594.

Quintiliani, M., Badiello, R., Tamba, M., Esfandi, A., and Gorin, G. (1977). Radiolysis of glutathione in oxygen-containing solutions of pH 7. *Int. J. Radiat. Biol.* **32,** 195–202

Rao, D. N. R., Fischer, V., and Mason, R. P. (1990). Glutathione and ascorbate reduction of the acetaminophen radical formed by peroxidase. Detection of the glutathione disulfide radical anion and the ascorbyl radical. *J. Biol. Chem.* **265,** 844–847.

Redpath, J. L. (1973). Pulse radiolysis of dithiothreitol. *Radiat. Res.* **54,** 364–374.

Romero, F. J., Ordoñez, I., Arduini, A., and Cadenas, E. (1992). The reactivity of thiols and disulfides with different redox states of myoglobin. Redox and addition reactions and formation of thiyl radical intermediates. *J. Biol. Chem.* **267,** 1680–1688.

Rosenwald, R. H. (1956). Chemistry of inhibitor sweetening. Effects of variations in olefins and caustic. *Pet. Process.* **11,** 91–95.

Ross, D., and Moldéus, P. (1985). Generation of reactive species and fate of thiols during peroxidase-catalyzed metabolic activation of aromatic amines and phenols. *Environ. Health Perspect.* **64,** 253–257.

Ross, D., and Moldéus, P. (1986). Thiyl radicals—their generation and further reactions. *Adv. Exp. Med. Biol.* **197,** 329–335.

Ross, D., Albano, E., Nilsson, U., and Moldéus, P. (1984). Thiyl radicals—formation during peroxidase-catalyzed metabolism of acetaminophen in the presence of thiols. *Biochem. Biophys. Res. Commun.* **125,** 109–115.

Ross, D., Larsson, R., Andersson, B., Nilsson, U., Lundquist, T., Lindeke, B., and Moldéus, P. (1985a). The oxidation of p-phenetidine by horseradish peroxidase and prostaglandin synthase and the fate of glutathione during such oxidations. *Biochem. Pharmacol.* **34,** 343–351.

Ross, D., Norbeck, K., and Moldéus, P. (1985b). The generation and subsequent fate of glutathionyl radicals in biological systems. *J. Biol. Chem.* **260,** 15028–15032.

Ross, D., Mehlhorn, R. J., Moldéus, P., and Smith, M. T. (1985c). Metabolism of diethylstilbestrol by horseradish peroxidase and prostaglandin-H synthase. Generation of a free radical intermediate and its interaction with glutathione. *J. Biol. Chem.* **260,** 16210–16214.

Ross, D., Cotgreave, I., and Moldéus, P. (1985d). The interaction of reduced glutathione with active oxygen species generated by xanthine-oxidase-catalyzed metabolism of xanthine. *Biochim. Biophys. Acta* **841,** 278–282.

Ross, D., Thor, H., Orrenius, S., and Moldeus, P. (1985e). Interaction of menadione (2-methyl-1,4-naphthoquinone) with glutathione. *Chem.-Biol. Interact.* **55,** 177–184.

Sakurai, K., and Ogiso, T. (1991). Inhibitory effect of glutathione on the generation of hydroxyl radicals in the reaction system of glutathione-alloxan. *Chem. Pharm. Bull.* **39,** 737–742.

Schaafsma, Y., Bickel, A. F., and Kooyman, E. C. (1957). Free radical reactions involving sulphur compounds. II. The relative reactivities of substituted thiophenols with respect to hydrogen abstraction by a free trivalent carbon radical. *Recl. Trav. Chim. Pays-Bas* **76,** 180–186.

Schäfer, K., and Asmus, K.-D. (1981). Reaction of thiols and disulfides with phosphite radicals. A chain mechanism and $RS\cdot/PO_3^{2-}\cdot$ equilibrium. *J. Phys. Chem.* **85,** 852–855.

Schäfer, K., Bonifačić, M., Bahnemann, D., and Asmus, K.-D. (1978). Addition of oxygen to organic sulfur radicals. *J. Phys. Chem.* **82,** 2777–2780.

Schöneich, C., and Asmus, K.-D. (1990). Reaction of thiyl radicals with alcohols, ethers and polyunsaturated fatty acids: A possible role of thiyl free radicals in thiol mutagenesis? *Radiat. Environ. Biophys.* **29,** 263–271.

Schöneich, C., Bonifačić, M., and Asmus, K.-D. (1989a). Reversible H-atom abstraction from alcohols by thiyl radicals: Determination of absolute rate constants by pulse radiolysis. *Free Radical Res. Commun.* **6,** 393–405.

Schöneich, C., Asmus, K.-D., Dillinger, U., and von Bruchhausen, F. (1989b). Thiyl radical attack on polyunsaturated fatty acids: A possible route to lipid peroxidation. *Biochem. Biophys, Res. Commun.* **161,** 113–120.

Schöneich, C., Dillinger, U., von Bruchhausen, F., and Asmus, K.-D. (1992). Oxidation of polyunsaturated fatty acids and lipids through thiyl and sulfonyl radicals: Reaction kinetics, and influence of oxygen and structure of thiyl radicals. *Arch. Biochem. Biophys.* **292,** 456–467.

Schreiber, J., Foureman, G. L., Hughes, M. F., Mason, R. P., and Eling, T. E. (1989). Detection of glutathione thiyl free radical catalyzed by prostaglandin H synthase present in keratinocytes. *J. Biol. Chem.* **264,** 7936–7943.

Schubert, M. (1932). Complex types involved in the catalytic oxidation of thiol acids. *J. Am. Chem. Soc.* **54,** 4077–4085.

Searle, A. J. F., and Tomasi, A. (1982). Hydroxyl free radical production in iron-cysteine solutions and protection by zinc. *J. Inorg. Biochem.* **17,** 161–166.

Searle, A. J. F., and Willson, R. L. (1983). Stimulation of microsomal lipid peroxidation by iron and cysteine. Characterization and the role of free radicals. *Biochem. J.* **212,** 549–554.

Sevilla, M. D., Becker, D., and Yan, M. (1990). The formation and structure of the sulfoxyl radicals RSO·, RSOO·, RSO_2·, and RSO_2OO· from the reaction of cysteine, glutathione and penicillamine thiyl radicals with molecular oxygen. *Int. J. Radiat. Biol.* **57,** 65–81.

Shirai, H., Tsuiki, H., Masuda, E., Koyama, K., Hanabusa, K., and Kobayashi, N. (1991). Functional metallomacrocycles and their polymers. 25. Kinetics and mechanism of the biomimetic oxidation of thiol by oxygen catalyzed by homogeneous (polycarboxyphthalocyaninato) metals. *J. Phys. Chem.* **95,** 417–423.

Simic, M. G. (1988). Mechanisms of inhibition of free-radical processes in mutagenesis and carcinogenesis. *Mutat. Res.* **202,** 377–386.

Skorobogaty, A., and Smith, T. D. (1982). Autoxidation of thiols catalyzed by cobalt(II)tetrasulphophthalocyanine absorbed on an ion exchange resin. *J. Mol. Catal.* **16,** 131–147.

Smith, R. H. (1980). Kale poisoning: The brassica anaemia factor. *Vet. Rec.* **107,** 12–15.

Speit, G., and Haupter, S. (1987). Cytogenetic effects on penicillamine. *Mutat. Res.* **190,** 197–203.

Speit, G., and Vogel, W. (1982). The effect of sulfhydryl compounds on sister-chromatid exchanges. II. The question of cell specificity and the role of H_2O_2. *Mutat. Res.* **93,** 77–85.

Speit, G.,Wolf, M., and Vogel, W. (1980). The effect of sulfhydryl compounds on sister-chromatid exchanges. *Mutat. Res.* **78,** 267–272.

Stark, A.-A., Zeiger, E., and Pagano, D. A. (1987). Glutathione mutagenesis in *Salmonella typhimurium* TA100: Dependence on a single enzyme, γ-glutamyltranspeptidase. *Mutat. Res.* **177,** 45–52.

Stark, A.-A., Zeiger, E., and Pagano, D. A. (1988). Glutathione mutagenesis in *Salmonella typhimurium* is a γ-glutamyltranspeptidase-enhanced process involving active oxygen species. *Carcinogenesis (London)* **9,** 771–777.

Stark, A.-A., Arad, A., Siskindovich, S., Pagano, D. A., and Zeiger, E. (1989). Effect of pH on mutagenesis by thiols in *Salmonella typhimurium* TA102. *Mutat. Res.* **224,** 89–94.

Starkebaum, G., and Harlan, J. M. (1986). Endothelial cell injury due to copper-catalyzed hydrogen peroxide generation from homocysteine. *J. Clin. Invest.* **77,** 1370–1376.

Stich, H. F., Wei, L., and Lam, P. (1978). The need for a mammalian test system for mutagens: action of some reducing agents. *Cancer Lett.* (*Shannon. Irel.*) **5,** 199–204.

Stock, B. H., Schreiber, J., Guenat, C., Mason, R. P., Bend, J. R., and Eling, T. E. (1986). Evidence for a free radical mechanism of styrene-glutathione conjugate formation catalyzed by prostaglandin H synthase and horseradish peroxidase. *J. Biol. Chem.* **261,** 15915–15922.

Störle, C., and Eyer, P. (1991). Reactions of the Wurster's blue radical cation with thiols, and some properties of the reaction products. *Chem.-Biol. Interact.* **78,** 333–346.

Strömme, J. H. (1963). Methaemoglobin formation induced by thiols. *Biochem. Pharmacol.* **12,** 937–948.

Subrahmanyam, V. V., and O'Brien, P. J. (1985). Peroxidase catalysed oxygen activation by arylamine carcinogens and phenol. *Chem.-Biol. Interact.* **56,** 185–199.

Subrahmanyam, V. V., McGirr, L. G., and O'Brien, P. J. (1987). Glutathione oxidation during peroxidase catalysed drug metabolism. *Chem.-Biol. Interact.* **61,** 45–59.

Takagi, Y., Shikita, M., Terasima, T., and Akaboshi, S. (1974). Specificity of radioprotective and cytotoxic effects of cysteamine in HeLa S_3 cells: Generation of peroxide as the mechanism of paradoxical toxicity. *Radiat. Res.* **60,** 292–301.

Takahashi, N., Schreiber, J., Fischer, V., and Mason, R. P. (1987). Formation of glutathione-conjugated semiquinones by the reaction of quinones with glutathione: An ESR study. *Arch. Biochem. Biophys.* **252,** 41–48.

Tamba, M., and O'Neill, P. (1991). Redox reactions of thiol free radicals with the antioxidants ascorbate and chlorpromazine: Role in radioprotection. *J. Chem. Soc., Perkin Trans. 2,* pp. 1681–1685.

Tamba, M., Simone, G., and Quintiliani, M. (1986). Interactions of thiyl free radicals with oxygen: A pulse radiolysis study. *Int. J. Radiat. Biol.* **50,** 595–600.

Thompson, D. C., and Eling, T. E. (1991). Reactive intermediates formed during the peroxidative oxidation of anisidine isomers. *Chem. Res. Toxicol.* **4,** 474–481.

Tien, M., Bucher, J. R., and Aust, S. D. (1982). Thiol-dependent lipid peroxidation. *Biochem. Biophys. Res. Commun.* **107,** 279–285.

Topham, J. C. (1980). The detection of carcinogen-induced sperm head abnormalities in mice. *Mutat. Res.* **69,** 149–155.

Tsen, C. C., and Tappel, A. L. (1958). Catalytic oxidation of glutathione and other sulfhydryl compounds by selenite. *J. Biol. Chem.* **233,** 1230–1232.

Tsuchiya, J., Yamada, T., Niki, E., and Kamiya, Y. (1985). Interaction of galvinoxyl radical with ascorbic acid, cysteine, and glutathione in homogeneous solution and in aqueous dispersions. *Bull. Chem. Soc. Jpn.* **58,** 326–330.

Tyndall, G. S., and Ravishankara, A. R. (1989). Kinetics and mechanism of the reactions of CH_3S with O_2 and NO_2 at 298K. *J. Phys. Chem.* **93,** 2426–2435.

Vinogradov, A. D., Grivennikova, V. G., and Gavrikova, E. V. (1979). Studies of the succinate dehydrogenating system. I. Kinetics of the succinate dehydrogenase interaction with a semiquindiimine radical of N,N,N′,N′-tetramethyl-p-phenylenediamine. *Biochim. Biophys. Acta* **545,** 141–154.

Voegtlin, C., Johnson, J. M., and Rosenthal, S. M. (1931). The oxidation catalysis of crystalline glutathione with particular reference to copper. *J. Biol. Chem.* **93,** 435–453.

von Sonntag, C. (1987). "The Chemical Basis of Radiation Biology." Taylor & Francis, London.

Wardman, P. (1988). Conjugation and oxidation of glutathione via thiyl free radicals. *In*

"Glutathione Conjugation: Mechanisms and Biological Significance" (H. Sies and B. Ketterer, eds.), pp. 42–72. Academic Press, San Diego, CA.

Wardman, P. (1990). Bioreductive activation of quinones: Redox properties and thiol reactivity. *Free Radical Res. Commun.* **8,** 219–229.

Waring, P., and Müllbacher, A. (1992). The possible role of gliotoxin in health and disease. *Endeavour* **16,** 14–16.

Waring, P., Eichner, R. D., and Müllbacher, A. (1988). The chemistry and biology of the immunomodulating agent gliotoxin and related epipolythiodioxopiperazines. *Med. Res. Rev.* **8,** 499–524.

Wefers, H., and Sies, H. (1983). Oxidation of glutathione by the superoxide radical to the disulfide and the sulfonate yielding singlet oxygen. *Eur. J. Biochem.* **137,** 29–36.

Willson, R. L. (1983). Free radical repair mechanisms and the interactions of glutathione and vitamins C and E. *In* "Radioprotectors and Anticarcinogens" (O. F. Nygaard and M. G. Simic, eds.), pp. 1–22. Academic Press, New York.

Willson, R. L., Dunster, C. A., Forni, L. G., Gee, C. A., and Kittridge, K. J. (1985). Organic free radicals and proteins in biochemical injury: Electron- or hydrogen-transfer reactions? *Philos. Trans. R. Soc. London, Ser. B* **311,** 545–563.

Wilson, I., Wardman, P., Cohen, G. M., and d'Arcy Doherty, M. (1986). Reductive role of glutathione in the redox cycling of oxidizable drugs. *Biochem. Pharmacol.* **35,** 21–22.

Winterbourn, C. C. (1989). Inhibition of autoxidation of divicine and isouramil by the combination of superoxide dismutase and reduced glutathione. *Arch. Biochem. Biophys.* **271,** 447–455.

Winterbourn, C. C. (1993). Superoxide as an intracellular radical sink. *Free Radical Biol. Med.* **14,** 85–90.

Winterbourn, C. C., and Munday, R. (1989). Glutathione-mediated redox cycling of alloxan. Mechanisms of superoxide dismutase inhibition and of metal-catalyzed $OH^{\cdot}$ formation. *Biochem. Pharmacol.* **38,** 271–277.

Wolfenden, B. S., and Willson, R. L. (1982). Radical-cations as reference chromogens in kinetic studies of one-electron transfer reactions: Pulse radiolysis studies of 2,2′-azinobis-(3-ethylbenzthiazoline-6-sulphonate). *J. Chem. Soc., Perkin Trans. 2* pp. 805–812.

Yamaguchi, T. (1981). Mutagenicity of low molecular substances in various superoxide generating systems. *Agric. Biol. Chem.* **45,** 327–330.

Yamaguchi, T., and Yamashita, Y. (1981). Activating effect of several enzyme-proteins on the mutagenicity of amino-thiol compounds. *Agric. Biol. Chem.* **45,** 2157–2160.

Yamoto, O., and Maede, Y. (1992). Susceptibility to onion-induced hemolysis in dogs with hereditary high erythrocyte reduced glutathione and potassium concentrations. *Am. J. Vet. Res.* **53,** 134–137.

Yee, J. J., and Shipe, W. F. (1982). Effects of sulfhydryl compounds on lipid oxidations catalyzed by copper and heme. *J. Dairy Sci.* **65,** 1414–1420.

Zwart, J., van Wolput, J. H. M. C., and Koningsberger, D. C. (1981). The mechanism of the copper ion catalyzed autoxidation of cysteine in alkaline medium. *J. Mol. Catal.* **12,** 85–101.

Zweier, J. L., Gianni, L., Muindi, J., and Myers, C. E. (1986). Differences in O_2 reduction by the iron complexes of adriamycin and daunomycin: The importance of the sidechain hydroxyl group. *Biochim. Biophys. Acta* **884,** 326–336.

Glutathione Mercaptides as Transport Forms of Metals

N. Ballatori

Department of Environmental Medicine
University of Rochester School of Medicine
Rochester, New York 14642

I. Introduction

Identification of the mechanisms for the selective cellular accumulation and toxicity of metals remains one of the greatest challenges for the toxicologist. This information is critical for the evaluation of the potential adverse effects of metal exposure and for designing therapeutic strategies.

In contrast to the majority of organic chemicals that undergo metabolic degradation, metal elements are indestructible in biological tissues. Once incorporated into an organism, their physiological and toxicological effects are regulated by two general mechanisms: binding to specific ligands and excretion. Binding to amino acids, peptides, proteins, phospholipids, and other tissue constituents modulates both their reactivity (toxicity) and biological effects. Because most heavy metals are present in biological tissues as complexes with specific ligands rather than as the free cations, these ligands play a critical role in metal homeostasis.

Among the metal binding ligands, glutathione (GSH) is one of the most versatile and pervasive. GSH is the most abundant nonprotein sulfhydryl-containing compound within cells at concentrations of 1–10 m*M*. The sulfhydryl group of the cysteine moiety has a high affinity for metals, forming thermodynamically stable but kinetically labile mercaptides with a number of metals, including mercury, silver, cadmium, arsenic, lead, gold, zinc, and copper. Recent studies demonstrate that the abundance

Advances in Pharmacology, Volume 27

of GSH and the high turnover rate for this tripeptide and its adducts via both intra- and interorgan cycles play key roles in metal transport and metabolism.

GSH modulates the disposition and toxicity of metals by at least four mechanisms: (*a*) it functions in the mobilization and delivery of metals between ligands; (*b*) it functions to transport metals across cell membranes as GSH complexes; (*c*) it serves as a source of cysteine, an amino acid that plays a central role in metal homeostasis; and (*d*) it serves as a cofactor for redox reactions, yielding metal compounds with different speciation or biochemical forms. The present discussion reviews studies that examine these roles of GSH, focusing on tissues critical to the disposition of metals: kidney, liver, gastrointestinal tract, and blood-brain barrier.

II. Formation of Glutathione Mercaptides

GSH contains six potential coordination sites for metal binding: the cysteinyl thiolate, the glutamyl amino, and the glycyl and glutamyl carboxyl groups, and the peptide linkages. Of these, the thiolate residue exhibits the highest affinity for a number of metals, including Hg, Cd, Cu, Zn, Ag, As, and Pb. Mercury binds almost exclusively to the cysteinyl thiolate group (Fuhr and Rabenstein, 1973), whereas other heavy metals also exhibit an affinity for the amino and carboxyl groups and the peptide linkages. For example, zinc and cadmium interact strongly with the thiolate group, but also exhibit a weak interaction with the carboxyl and amino groups (Fuhr and Rabenstein, 1973). In contrast, Ni(II) binds GSH primarily through the glycyl carboxyl and glutamyl carboxyl and amino groups (Jezowska-Trzebiatowska *et al.*, 1976). A metal bound to the thiolate group of GSH can be stabilized by coordination with one of the other potential binding sites within the tripeptide. In general, however, more stable structures are obtained when these divalent metals form a 1 : 2 (M^{2+} : S^-) complex to generate either GS—M—SG, GS—M—SR or RS—M—SR, where RS— is an additional ligand [Eq. (1) and (2)]. The chemistry of monovalent heavy-metal compounds (M^+) is more straightforward, forming 1 : 1 complexes (Martell and Calvin, 1952; Lenz and Martell, 1964).

$$RS^- + M^{2+} \underset{k_{-1}}{\overset{k_1}{\rightleftharpoons}} RS{-}M^+ \qquad (1)$$

$$RS^- + RS\text{—}M^+ \underset{k_{-2}}{\overset{k_2}{\rightleftharpoons}} RS\text{—}M\text{—}SR \qquad (2)$$

Formation of GSH mercaptides is highly favored thermodynamically and occurs spontaneously (nonenzymatically) under physiological conditions. The reaction is reversible, but the equilibrium is shifted toward the formation of the weakly dissociating mercaptides. The formation constants of selected mercaptides are listed in Table I. As illustrated in this table, the formation constants are high, but vary considerably among metals. The alkaline earth metals, magnesium, calcium, and strontium (group IIA metals), have no appreciable affinity for sulfhydryl groups (Table I). On the other hand, the formation constants of cysteine and glutathione mercaptides of mercury are exceptionally high, log k_1 = 14–16 (Table I). These formation constants are 8 orders of magnitude larger than those for mercury binding to amino groups (Carty and Malone, 1979). Table I also illustrates that for the zinc–glutathione complex, log k_2 is much smaller

Table I

Formation Constants of Cysteine and GSH Mercaptides

Metal	Ligand	log k (mol/liter) k_1	k_2	Reference
CH_3Hg^+	Cys	15.7		Simpson (1961)
	GSH	15.9		Simpson (1961)
Hg^{2+}	Cys	14.2		Lenz and Martell (1964); Stricks and Kolthoff (1953)
	GSH	14		Kapoor *et al.* (1965); Stricks and Kolthoff (1953)
Pb^{2+}	Cys	12.2		Li and Manning (1955)
	GSH	10.6		Li and Manning (1955)
Cd^{2+}	GSH	10.5		Li and Manning (1955)
Zn^{2+}	Cys	9.8	8.8	Li and Manning (1955)
	GSH	8.3	4	Li *et al.* (1954); Perrin and Watt (1971)
Cu^+	Cys	19.4		Stricks and Kolthoff (1951)
Co^{2+}	Cys	9.3	7.6	Albert (1952); Lenz and Martell (1964)
Ni^{2+}	Cys	9.6	9.4	Lenz and Martell (1964); Albert (1952)
Mn^{2+}	Cys	4.6		Lenz and Martell (1964)
Mg^{2+}	Cys	—[a]		Lenz and Martell (1964)
Ca^{2+}	Cys	—[a]		Lenz and Martell (1964)

[a] No evidence of chelate formation.

than log k_1 (4 versus 8.3), indicating that the second GSH molecule is bonded to the zinc only through the γ-glutamyl carboxyl and amino groups, rather than through the sulfhydryl group (Perrin and Watt, 1971). Martin and Edsall (1959) estimated that for divalent Zn, Ni, Co, and Mn, approximately 80, 10, 30, and 15% of the bound metal is attached to the sulfur moiety, respectively.

Despite the thermodynamic stability of these metal–sulfhydryl complexes, they are generally kinetically labile. That is, the metal rapidly exchanges between available sulfhydryl ligands. Perhaps the fastest exchange rates are observed between mercury and sulfhydryl-containing molecules, with rates approaching those of diffusion-controlled bimolecular reactions:

$$CH_3Hg\text{—}SG + {}^*GS^- \leftrightarrow CH_3Hg\text{—}SG^* + GS^-,$$
$$k = 5.8 \times 10^8\ M^{-1} \cdot s^{-1}$$ (Rabenstein and Fairhurst, 1975).

These fast rates of exchange are due to the facile nucleophilic displacement of complexed glutathione by GSH thiolate on the linear two-coordinate mercury. This property is the key to the rapid mobility of methylmercury in biological systems (Clarkson, 1972, 1993).

III. Roles of Glutathione in the Disposition and Toxicity of Metals

A. Mobilization and Delivery of Metals between Ligands

Because metal toxicity is usually attributed to the reactivity of the "free" metal, highly reactive essential metals must be sequestered and not allowed to bind prematurely or at inappropriate sites during metalloprotein synthesis. At the same time, the cell must employ a mechanism to rapidly mobilize the metal and make it available for delivery and incorporation into apometalloproteins. The concentration of free, nonessential (toxic) metals must also be minimized if toxicity is to be avoided, but the metal should be maintained in a chemical form that does not preclude its interaction with specific membrane transport sites and its excretion from the cell.

There is increasing evidence that the normal transport, storage, and metabolism of metals are inextricably linked to systems that protect against their toxicity and, in particular, to GSH. First, because cellular GSH concentrations are in the millimolar range, whereas essential heavy metals and toxic metals are present at low micromolar concentrations, GSH efficiently chelates metals, limiting the concentration of the reactive species. The observation that changes in intracellular GSH concentration modulate metal toxicity *in vivo* (Fukino *et al.*, 1986; Sager and Syversen,

1986; Singhal *et al.*, 1987; Kromidas *et al.*, 1990) and in cultured cells (Hsu, 1981; Stacey, 1986; Siegers *et al.*, 1986; Andrews *et al.*, 1987; Kang and Enger, 1988; Ochi *et al.*, 1988; Naganuma *et al.*, 1990; Kuchan and Milner, 1992; Chan and Cherian, 1992; Houser *et al.*, 1992) supports the hypothesis that GSH is capable of chelating and detoxifying metals soon after they enter the cell. Likewise, resistance to some metals is associated with an increase in intracellular GSH in many cell types (Seagrave *et al.*, 1983; Andrews *et al.*, 1987; Bannai *et al.*, 1991; Li and Chou, 1992; Zinkewich-Peotti and Andrews, 1992; Miura and Clarkson, 1993) Second, GSH–metal complexes readily diffuse throughout the cytosol, facilitating delivery of the metal to all cellular compartments. GSH in mitochondria, nuclei, and endoplasmic recticulum may function similarly within these organelles. Third, because of the kinetic lability of most mercaptide bonds, the metal can be mobilized by other ligands and, in particular, apometalloproteins. Some examples are given below.

1. Copper

Recent studies demonstrate that GSH is responsible for the mobilization and delivery of copper during the biosynthesis of copper proteins. GSH maintains the intracellular pool of free copper in the Cu(I) oxidation state by providing reducing equivalents for the conversion of Cu(II) to Cu(I) (Freedman *et al.*, 1989). In the Cu(I) oxidation state, copper forms stable GSH complexes (Ciriolo *et al.*, 1990), and these complexes serve as a source of Cu(I) for incorporation into apometallothionein (Freedman *et al.*, 1989), aposuperoxide dismutase (Steinkuhler *et al.*, 1991), and apohemocyanin (Brouwer and Brouwer-Hoexum, 1992).

Using a hepatoma cell line, Freedman *et al.* (1989) demonstrated that more than 60% of the copper in the cytosol of the wild-type and a copper-resistant cell line was bound to GSH. The resistant cells had nearly four times the concentration of GSH than the wild type and had highly elevated levels of metallothionein. Resistance to the toxic effects of copper was correlated with intracellular GSH concentrations (Freedman *et al.*, 1989). These investigators demonstrated that intracellular Cu(I) was transferred from GSH to metallothionein in a reversible reaction. Formation of the Cu(I)-metallothionein complex required the presence of the Cu(I)–glutathione complex and was not readily achieved in its absence. Transfer of copper from GSH to metallothionein was suggested to occur via a GS–Cu–metallothionein intermediate (Freedman *et al.*, 1989). Similarly, Ciriolo *et al.* (1990) demonstrated that the Cu(I)–glutathione complex serves as the substrate for introducing copper into CuZn–superoxide dismutase. GSH complexes of Cu(I) but not Cu(II) were able to reconstitute enzyme activity. EPR signals of the reconstituted enzyme were indistinguishable from the native enzyme, indicating the correct localization and binding of copper to the protein.

Moreover, Brouwer and Brouwer-Hoexum (1992) recently demonstrated that the Cu(I)–glutathione complex, but not the Cu(I)–metallothionein complex, is able to serve as a source of Cu(I) in the *in vitro* reconstitution of lobster apohemocyanin. Although metallothioneins have long been implicated in the delivery of copper for the biosynthesis of metalloproteins (see Harris, 1991, for review), the results of Brouwer and Brouwer-Hoexum (1992) indicate that the metallothionein complex is not able to accomplish this task for apohemocyanin. Thus, GSH not only functions to deliver copper to metallothionein during its biosynthesis (Freedman *et al.*, 1989), but when the cellular supply of copper is limited, GSH can mobilize the copper from metallothionein for delivery and incorporation into apometalloproteins. In this case, metallothionein apparently serves as a temporary storage for cytoplasmic copper, whereas GSH serves both as a vehicle for delivery of copper to metallothionein and as a vehicle for removing the metal from metallothionein in times of need. In contrast, metallothionein plays a more important role in the long-term detoxification of high concentrations of copper (Harris, 1991). Metallothionein levels directly reflect intracellular copper concentrations (Sone *et al.*, 1987).

Binding of copper to albumin, a major copper-binding protein in blood plasma, also involves GSH. Suzuki *et al.* (1989) demonstrated that cupric ions injected intravenously into rats were gradually incorporated into an albumin–Cu–cysteine or albumin–Cu–glutathione complex. These albumin complexes constitute the largest labile pool of plasma copper and probably play a key role in the distribution and cellular uptake of this metal.

2. Cadmium

Another interesting role of GSH in the mobilization of metals was recently described by Dameron *et al.* (1989). In the yeast *Candida glabrata* cultured in the presence of cadmium salts, intracellular cadmium-sulfide crystallites form that function to sequester and detoxify cadmium. These investigators noted that these particles are coated with peptides consisting of a mixture of GSH and γ-glutamylcysteine, a dipeptide constituent of GSH. Although the role of GSH in the crystallization process is not clear, the GSH–cadmium complexes apparently stabilize the particles and effectively sequester this toxic metal. These $(\gamma\text{-GluCys})_n$Gly peptides are present in most plants and fungi (Grill *et al.*, 1985).

B. Transport of Metals across Cell Membranes as Glutathione Complexes

The fundamental mechanisms by which transition metals are transported across mammalian cell membranes remain poorly defined (Ballatori and Clarkson, 1985b; Ballatori, 1991a,b; Clarkson, 1993). Putative membrane carriers have been suggested for a number of metals based largely on kinetic data obtained *in vivo* or in isolated cell systems. However, none of these putative transport proteins have been isolated or studied at the subcellular or molecular level. Moreover, the specific chemical form of the transported metal has not yet been elucidated for any of these putative carriers.

There is now increasing evidence that membrane carriers for GSH may be responsible for the transport of some metals as GSH mercaptides. This evidence is discussed below. Potential mechanisms for cellular efflux and uptake of GSH-containing compounds are considered separately.

1. Efflux of GSH and GSH Conjugates from Cells

GSH is synthesized within all mammalian cells and is released by transport across the cell membrane (Fig. 1; Meister and Anderson, 1983). Transport of GSH out of the cell is the initial and perhaps rate-limiting step in GSH degradation. Once outside the cell, GSH and GSH adducts become substrates for γ-glutamyltransferase, the only enzyme known to initiate GSH catabolism (Fig. 1). This enzyme is membrane-bound, with its active site on the extracellular surface.

At least three different transport systems are thought to be involved in the efflux of GSH, glutathione disulfide (GSSG), and glutathione *S*-conjugates (1–3 in Fig. 1). Most studies of GSH and GSH-conjugate efflux have been performed in liver, kidney, intestines, heart, and red blood cells (Board, 1981; Inoue and Morino, 1985; LaBelle *et al.,* 1986; Deleve and Kaplowitz, 1990; Vincenzini *et al.,* 1992; Garcia-Ruiz *et al.,* 1992). In each of these tissues the driving force for GSH efflux is the membrane potential and the large outwardly directed chemical gradient (m*M* inside, but only μM in most extracellular fluids). In contrast, efflux of GSSG and GSH *S*-conjugates is mediated by ATP-dependent transport processes. In human red blood cells, distinct transport mechanisms appear to mediate efflux of GSSG and GSH conjugates (LaBelle *et al.,* 1986).

Evidence that GSH mercaptides may also be substrates for these GSH efflux transporters (i.e., 1–3 in Fig. 1), or possibly transport systems more selective for mercaptides (4, Fig. 1), is discussed below.

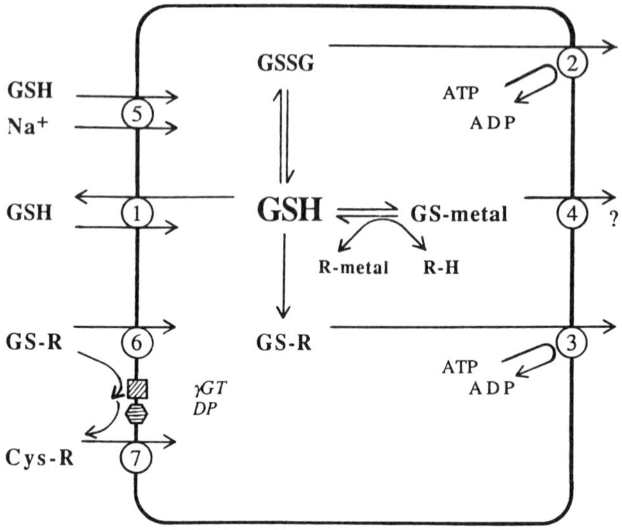

Fig. 1 Cellular uptake and efflux mechanisms for GSH and its *S*-conjugates and mercaptides. This diagram depicts a working model of the interrelation between the putative plasma membrane transport systems for GSH and GSH adducts in a nonpolarized cell. Note, however, that the identity, number, cellular distribution, substrate specificity, and driving forces for the transport proteins that actually carry out these functions are not well described for any tissue. γGT, γ-glutamyltransferase; DP, dipeptidase; GS-R, glutathione *S*-conjugate.

2. Transport of GSH Mercaptides out of Hepatocytes

Perhaps the first indication that GSH could mediate cellular efflux of metals came from studies on the biliary excretion of methylmercury by Norseth and Clarkson (1971) and Norseth (1973a,b,c; Refsvik and Norseth, 1975). Most of the methylmercury in bile is complexed with GSH, cysteinylglycine, and cysteine. The latter two breakdown products of GSH are formed within biliary spaces from GSH by the ectoproteins γ-glutamyltransferase and dipeptidases present on the luminal surface of biliary epithelia (Hirata and Takahashi, 1981; Ballatori *et al.*, 1986b, 1988, 1989; Gregus *et al.*, 1987; Stein *et al.*, 1988; Urano *et al.*, 1988). Because biliary γ-glutamyltransferase activity differs among species, the relative amounts of the GSH hydrolysis products in bile also differ among species (Ballatori *et al.*, 1988; Stein *et al.*, 1988).

Although most of the methylmercury in bile is bound to GSH, this in itself is not sufficient evidence to conclude that methylmercury is transported across the canalicular membrane into bile as a GSH complex. Methylmercury could have been transported as a chloride salt or as a complex with some other ligand, and the GSH complex formed within biliary spaces. Indeed, a GSH–mercury complex is readily formed *in vitro* when CH_3HgCl is mixed with rat bile, suggesting that the complex could easily have been formed within the biliary tree subsequent to the secretion of the individual components. Alternatively, methylmercury could also have entered bile in response to a compartmentalized sulfhydryl gradient.

To resolve some of these issues, Refsvik (1978) examined the effects of agents that deplete hepatic and biliary GSH levels on methylmercury transport into rat bile. These agents produced a parallel decrease in biliary excretion of GSH and methylmercury, suggesting a close coupling between efflux of these two substances. However, because hepatic mercury levels were not measured in this study, it is impossible to distinguish whether inhibition of biliary excretion was causally related to the decreased GSH in bile, to a decreased hepatic level of mercury, or possibly to nonspecific toxic effects of the GSH depletors on intracellular transport or metabolism of mercury.

Subsequent studies in the author's laboratory provided a more detailed description of the mechanism for the biliary transport of methylmercury and inorganic mercury (Ballatori and Clarkson, 1982, 1983, 1984a,b, 1985a; Dutczak and Ballatori, 1992) and, more recently, have identified a specific carrier-mediated mechanism for canalicular transport of methylmercury as the GSH complex (Dutczak *et al.*, 1993). These *in vivo* studies suggested the existence of a mediated GSH transport system on the canalicular membrane and that this transport system was the primary determinant of biliary methylmercury excretion. The inability of the suckling rat to excrete methylmercury into bile was related to its inability to excrete GSH into bile (Ballatori and Clarkson, 1982). Development of the ability to excrete methylmercury paralleled the ability to excrete GSH. Sex and individual differences in methylmercury excretion were correlated with differing rates of GSH excretion. In addition, changes in the rates of methyl-mercury excretion induced by sulfobromophthalein (BSP), phenol-3,6-dibromphthalein disulfonate (DBSP), indocyanine green (ICG), and phenobarbital were associated with concomitant changes in the rates of GSH excretion. Indeed, at high doses of BSP and DBSP, biliary excretion of both GSH and methylmercury was nearly completely inhibited (Ballatori and Clarkson, 1983, 1985a). These changes were largely independent of hepatic mercury and GSH concentrations. Furthermore, the rate of methylmercury transport into bile was independent of methylmercury concentrations in bile or the sulfhydryl gradient between liver and bile, but was directly related to the rate of GSH excretion into bile (Ballatori and Clarkson, 1982, 1983, 1985a; Refsvik, 1982). Taken together, these observations indicate that the GSH–methylmercury complex is transported across the canalicular membrane into bile. Because transport of this complex behaves in a manner similar to that of GSH itself, the data also suggest that the complex may in fact be a substrate for the canalicular GSH transport proteins (Ballatori and Clarkson, 1985b).

To test this possibility, recent studies have examined the mechanism of methylmercury transport in canalicular membrane vesicles (Dutczak *et al.*, 1993). The findings demonstrate that transport of the GSH–[^{203}Hg]methylmercury complex is saturable, temperature sensitive, electrogenic, and unaffected by an inwardly directed Na^+ gradient or the presence of ATP, but is inhibited by GSH. These data establish for the first time the existence of a mediated transport mechanism for a GSH mercaptide. Furthermore, they provide strong evidence that this mercaptide serves as a substrate for a canalicular GSH transport protein, but not for the canalicular GSSG or GS-conjugate transport proteins. Because GSH transport proteins are present on all cells, these findings have broad implications for the detoxification and excretion of methylmercury and other metals that form GSH mercaptides.

Indeed, a number of other metals also appear to be transported out of hepatocytes as GSH mercaptides; however, direct evidence for intracellular formation and membrane transport of these other GSH complexes has not yet been obtained. A role for GSH in the efflux of copper and zinc (Alexander and Aaseth, 1980; Alexander *et al.*, 1981; Nederbragt, 1989; Houwen *et al.*, 1990), and silver, lead, arsenic and chromium (Alexander and Aaseth, 1981; Anundi *et al.*, 1982; Norseth *et al.*, 1982; Alexander *et al.*, 1986; Gyurasics *et al.*, 1991a,b, 1992) from liver cells into bile has been proposed based on the observation that hepatic GSH depletion leads to an inhibition of biliary transport of these metals. Mutant Wistar rats (GY) that are deficient in their ability to secrete GSH into bile also have diminished ability to transport zinc and excess copper into bile (Houwen *et al.*, 1990). In contrast, basal excretion of endogenous copper is unaffected in the mutant Wistar rats (Houwen *et al.*, 1990). Gyurasics *et al.* (1991b, 1992) reported that biliary transport of arsenic, antimony, and bismuth is also inhibited after hepatic GSH depletion. Biliary excretion of arsenic after either arsenite or arsenate administration is inhibited by the GSH transport inhibitors BSP, ICG, and diethylmaleate (DEM). The inhibition of arsenic transport was directly proportional to the inhibition of GSH transport, suggesting a tight coupling between these processes. Cadmium has also been reported to be complexed with GSH in rat bile (Cherian and Vostal, 1977; Gregus and Varga, 1985). Cherian and Vostal (1977) described an inhibition of biliary cadmium excretion by BSP administration that was probably related to the ability of BSP to inhibit biliary GSH excretion (Ballatori and Clarkson, 1983, 1985a). As with mercury, the underlying mechanism presumably involves complexation of metals with GSH within the cell and the subsequent transport of the GSH complexes across the canalicular membrane into bile.

Once secreted across the canalicular membrane into bile, GSH and GSH adducts are degraded by the γ-glutamyltransferase and dipeptidase activities on the canalicular membrane of hepatocytes and the luminal surface of bile ductular epithelia (Ballatori *et al.*, 1986b; Hinchman *et al.*, 1991; Simmons *et al.*, 1991), and the resulting products are partially reabsorbed from bile back to liver (Ballatori *et al.*, 1986a, 1988; Moseley *et al.*, 1988; Hinchman *et al.*, 1991; Simmons *et al.*, 1992). Species differences in biliary glutathione degradation result from differences among species in γ-glutamyltransferase activity and localization within the liver (Ballatori *et al.*, 1988; Stein *et al.*, 1988; Hinchman and Ballatori, 1990).

Recent studies demonstrate that the methylmercury–glutathione complex is metabolized within the biliary tree by a similar mechanism, and the resulting methylmercury–cysteine complex is reabsorbed back into liver (Dutczak and Ballatori, 1992). Reabsorption diminishes the amount excreted in bile, the main route of methylmercury elimination, and may account for the relatively long biological half-life and toxicity of this metal.

3. Kidney and Rat Pheochromocytoma (PC12) Cells

A similar GSH-dependent mechanism may be responsible for methylmercury efflux from kidney cells (Yasutake *et al.*, 1989) and PC12 cells (Miura and Clarkson, 1993). The mechanism for the urinary excretion of methylmercury was examined in male C57BL/6N mice by Yasutake *et al.* (1989). These investigators demonstrated that methylmercury rapidly accumulated in the kidney after oral administration and that some of the methylmercury was excreted in urine predominantly as the cysteine complex. However, within kidney tissue, 50% of the methylmercury was present as the GSH complex and the rest was bound to proteins. In contrast, nearly all of the methylmercury in plasma was bound to proteins, indicating that renal uptake was most likely due to a nonfiltrating peritubular mechanism. The chemical form of the protein-bound methylmercury was not identified; however, their preliminary results indicate that not only CH_3Hg^+, but also its GSH, cysteinylglycine, and cysteine complexes can also bind tightly to both mercapt- and nonmercaptalbumin (Yasutake *et al.*, 1989). Takahashi *et al.* (1988) also reported the presence of the cysteine and GSH complexes of methylmercury in blood plasma. Yasutake *et al.* (1989) noted a large increase in urinary excretion of methylmercury in mice treated with acivicin, an inhibitor of γ-glutamyltransferase activity, and that most of the methylmercury was present in urine as the GSH complex. These data indicate that the methylmercury–glutathione complex within renal tubular cells is transported into the luminal space and hydrolyzed to the cysteine complex. Although most of the methylmercury is reabsorbed within the tubular lumen, some escapes into urine (Yasutake *et al.*, 1989).

Miura and Clarkson (1993) isolated a methylmercury-resistant PC12 cell line that accumulates smaller amounts of methylmercury and contains four times the GSH concentration of parental PC12 cells. The reduced accumulation in the resistant cells was associated with a more rapid efflux of methylmercury. Methylmercury efflux was inhibited by DBSP, an inhibitor of GSH transport in the liver (Ballatori and Clarkson, 1983, 1985a), suggesting that methylmercury is released from PC12 cells as a GSH complex.

4. Erythrocytes

Gold(I) efflux from red blood cells incubated with auranofin, an organogold compound, also involves GSH (Shaw *et al.*, 1990). Release of gold(I) from preloaded erythrocytes paralleled the release of [^{14}C]GSH, and a [^{14}C]GSH–gold–albumin complex was identified in the extracellular media. Because of the structural similarity among the bis(glutathione)–gold(I) complex, $Au(SG)_2$, and GSH conjugates of organic electrophiles, these investigators suggested that this mercaptide may be a substrate for the glutathione conjugate transport system on the erythrocyte membrane (Board, 1981). However, their data cannot distinguish whether gold(I) was actually transported as a GSH complex or whether these were released separately and the mercaptide formed extracellularly.

5. Yeast

Glaeser and co-workers (1991) isolated a number of GSH-deficient mutants (*gsh*$^-$) of the fission yeast *Schizosaccharomyces pombe* and examined the ability of the mutants to transport and detoxify cadmium and other heavy metals. The *gsh*$^-$ mutants were sensitive to cadmium, as well as to Hg, Ag, Pb, Bi, and Sb. Sensitivity to cadmium was inversely proportional to the residual GSH content. Moreover, although the wild-type strains were able to excrete cadmium (presumably as a mercaptide), the *gsh*$^-$ mutants were unable to do so, and the metal accumulated intracellularly.

C. Cellular Uptake of Glutathione and Glutathione Adducts: Inter- and Intraorgan Metabolism of Glutathione–Metal Complexes

Three distinct mechanisms for cellular uptake of intact GSH and GSH adducts have been described (1,5,6 in Fig. 1). However, these are not as well characterized as the efflux systems, and their physiological significance remains controversial.

The first mechanism for uptake of extracellular GSH (or possibly GSH adducts) is by exchange with intracellular GSH (1 in Fig. 1). Studies in membrane vesicles isolated from different tissues demonstrate that the GSH transport proteins exhibit the phenomenon of *trans*-stimulation or countertransport, suggesting that these proteins should function as exchangers. However, studies in hepatocytes (Hahn *et al.,* 1978) and other tissues demonstrate that most cells are unable to take up GSH when it is present at physiological concentrations. An explanation for this apparent discrepancy between vesicle studies and observations in intact cells is provided by the recent work of Garcia-Ruiz *et al.* (1992). These investigators demonstrated that the hepatic sinusoidal GSH efflux carrier functions as an exchanger, but that it exhibits low-affinity kinetics with sigmoidal features for both GSH uptake and efflux. Thus, at physiological concentrations the transporter operates largely in the unidirectional efflux of GSH from the cell into plasma, but at high extracellular concentrations it can exchange intracellular for extracellular GSH. Because the apparent affinity for GSH in its external orientation is more than 10-fold higher than its internal orientation (0.34 versus 3–7 m*M*), the transporter may affect the net uptake of GSH when external concentrations are above 0.5–1 m*M* (Garcia-Ruiz *et al.,* 1992). These investigators also demonstrated that release of [^{35}S]GSH from hepatocytes was stimulated by GSH, by high extracellular concentrations (5 m*M*) of GSSG, by ophthalmic acid (a GSH analogue), and by the GSH conjugate of sulfobromophthalein (BSP-SG). In sinusoidal membrane vesicles, BSP-SG, γ-glutamylglutamate, γ-glutamylalanine and γ-glutamylcysteine *trans*-stimulated GSH uptake, whereas aspartate, succinate, glutamate, malate, oxaloacetate, citrate, and isocitrate had no effect (Garcia-Ruiz *et al.,* 1992). These findings raise the interesting possibility that a mechanism for the hepatic uptake of γ-glutamyl compounds and GSH analogues is via exchange or countertransport with intracellular GSH. Thus, transport of GSH out of hepatocytes may serve as a driving force for the uptake of these and perhaps other as yet unidentified organic anions. Clearly, GSH mercaptides may be substrates for this exchanger; however, there is no experimental evidence for this.

Two additional mechanisms for uptake of intact GSH and GSH adducts have been suggested based on studies in enterocytes, kidney cells, hepatocytes, and brain capillary endothelial cells, and are denoted as steps 5 and 6 in Fig. 1 (Hinchman *et al.,* 1993; Kannan *et al.,* 1992; Kramer *et al.,* 1987; Lash and Jones, 1984; Vincenzini *et al.,* 1992). Although there is considerable uncertainty about the identity, substrate specificity, and physiological significance of these transport mechanisms, they may also be involved in uptake of GSH complexes.

In addition, there is increasing evidence for a key role of membrane-bound γ-glutamyltransferase and dipeptidase activities in facilitating uptake of metals complexed with GSH. These enzymes promote the stepwise conversion of GSH complexes to the cysteinylglycine and cysteine complexes, species that are more readily taken up into cells. Cysteine–metal complexes can be formed from GSH by two mechanisms: the metals can bind to the cysteine liberated upon GSH hydrolysis (ligand exchange), or they can be formed from the direct degradation of GSH–metal complexes by γ-glutamyltransferase and dipeptidase activities. Naganuma *et al.* (1988) have provided evidence that the CH_3Hg–SG complex is a substrate for γ-glutamyltransferase.

Figure 1 illustrates the γ-glutamyltransferase- and dipeptidase-catalyzed breakdown of a GSH adduct to the cysteine adduct and reabsorption of the amino acid constituents. The cysteine conjugates or complexes are most likely transported by amino acid carriers (7 in Fig. 1), as discussed below.

1. Role of GSH in the Renal Uptake and Accumulation of Metals

The kidney is the principal site of accumulation and toxicity of heavy metals. Renal incorporation of inorganic mercury, for example, occurs rapidly, with as much as 50% of the dose deposited in the kidneys in a few hours (Tanaka *et al.*, 1990). Mercury enters renal tubular cells both across the brush-border membranes after glomerular filtration and across the basolateral membrane (for review, see Clarkson, 1972). However, the mechanisms involved in this accumulation remain poorly defined.

GSH and GSH conjugates are catabolized by the highly active renal γ-glutamyltransferase and dipeptidase activities, and the resulting amino acids and amino acid conjugates are reabsorbed. Presumably, the GSH–metal complexes are also metabolized by these enzymes, and the resulting cysteine–metal complexes could then be reabsorbed by the kidney.

Richardson and Murphy (1975) provided the first evidence that GSH is a determinant of renal methylmercury uptake. They demonstrated that the deposition of methylmercury in kidneys is decreased in rats treated with diethylmaleate and other compounds capable of depleting cellular GSH concentrations. Alexander and Aaseth (1982) confirmed the effects of diethylmaleate and further demonstrated that renal uptake of methylmercury was enhanced when the metal was administered intravenously as either the GSH or the cysteine complex, when compared to CH_3HgCl. Renal accumulation toxicity of inorganic mercury were also diminished in rats treated with diethylmaleate (Johnson, 1982) or buthionine sulfoximine (Berndt *et al.*, 1985), an inhibitor of GSH biosynthesis. Although diethylmaleate treatment decreased renal methylmercury accumulation when given as CH_3HgCl, it had no effect on methylmercury uptake when administered as the GSH complex. Because the CH_3Hg–SG complex is extracted by rat kidney as rapidly as GSH itself, Alexander and Aaseth (1982) suggested that a similar mechanism is involved in its removal.

Subsequent studies have provided data that support a role for γ-glutamyltransferase in facilitating renal uptake of both methyl- and inorganic mercury. Inhibition of renal γ-glutamyltransferase results in a marked increase in urinary excretion of both GSH and mercury and a decrease in renal mercury uptake (Mulder and Kostyniaki, 1985a; Berndt *et al.*, 1985; Gregus *et al.*, 1987; Naganuma *et al.*, 1988; Yasutake *et al.*, 1989; de Ceaurriz and Ban, 1990; Di Simplicio *et al.*, 1990; Tanaka *et al.*, 1990, 1991, 1992). Strain differences in renal methylmercury excretion are related to differences in urinary GSH excretion and γ-glutamyltransferase activity (Mulder and Kostyniak, 1985b; Tanaka *et al.*, 1991). Furthermore, age, sex, and species differences in renal methylmercury accumulation correlate with differences in renal γ-glutamyltransferase activity (Hirayama *et al.*, 1987; Stein *et al.*, 1988; Tanaka *et al.*, 1991, 1992).

A role for hepatic GSH in the transport of methylmercury (Naganuma *et al.*, 1988) and inorganic mercury (Tanaka *et al.*, 1990) into mouse kidney has been identified. Treatment of animals with 1,2-dichloro-4-nitrobenzene at a dose that depletes hepatic GSH concentrations without affecting renal GSH levels significantly decreased renal accumulation of mercury. Furthermore, renal methylmercury accumulation in mice given GSH mercaptides intravenously was higher than that in mice given chloride salts, suggesting that hepatic GSH released into the circulation plays a key role in renal methylmercury accumulation (Naganuma *et al.*, 1988; Tanaka *et al.*, 1990). Treatment with acivicin significantly inhibited renal methylmercury uptake. Thus, the GSH–methylmercury complexes that are delivered to the kidney appear to be substrates for γ-glutamyltransferase and dipeptidases, and the resulting metabolic products are readily taken up by the kidney cells.

The L-cysteine complex of cadmium is also efficiently taken up by the kidney and is concentrated in the straight segments of the proximal tubule (Murakami and Webb, 1981; Murakami *et al.*, 1987). This rapid cadmium uptake is associated with extensive renal damage.

GSH administration to patients undergoing treatment with cisplatin affords protection against nephrotoxicity without reducing antitumor activity, presumably by changing the pharmacokinetics of the metal compound (Leone *et al.*, 1992).

2. Intestinal Uptake

Anundi *et al.* (1984) reported that both extracellular and intracellular GSH stimulate uptake of selenium from selenite in isolated enterocytes. Similarly, intestinal absorption of selenium as selenite is stimulated by GSH and cysteine *in situ* (luminally perfused distal jejunum; Senn *et al.*, 1992) and in ligated intestinal loops (Vendeland *et al.*, 1992). Selenium uptake does not appear to be due to transport of the intact GSH complexes, as its uptake by isolated jejunal segments was actually inhibited by GSH, but stimulated by cysteine, after short-term (3 min) exposure (Senn *et al.*, 1992). Furthermore, Anundi *et al.* (1984) observed a decrease in selenium uptake in intestinal cells treated with serine-borate, a γ-glutamyltransferase inhibitor. Taken together, these data indicate that one mechanism by which GSH facilitates selenium uptake involves the degradation of the GSH complexes by luminal γ-glutamyltransferase and dipeptidase activities to form the L-cysteine mercaptides. Selenium–cysteine complexes are readily transported across the intestinal brush-border membrane by a process that is competitively inhibited by nonthiol L-amino acids (Wurmli *et al.*, 1989), suggesting transport of the cysteine complex on amino acid carriers.

Intestinal absorption of methylmercury complexed with cysteine, cysteinylglycine, or GSH was studied in ligated rat intestinal segments by Urano *et al.* (1990). Absorption of methylmercury–cysteine and methylmercury–cysteinylglycine was similar and was 1.5 times greater than that of the GSH complex. Absorption of the GSH complex was reduced by 50% in tissue treated with acivicin, indicating a role for γ-glutamyltransferase. Treatment with both acivicin and probenecid nearly completely blocked uptake of methylmercury when given as the GSH complex, suggesting a second, and presumably γ-glutamyltransferase-independent, mechanism of uptake. Transport of GSH itself or glutathione *S*-conjugates by renal cells is also inhibited by both probenecid and acivicin. Urano *et al.* (1990) concluded that intestinal uptake of methylmercury, as for the renal uptake, is due largely to the γ-glutamyltransferase-mediated catabolism of the GSH complex.

The absorption and distribution of cadmium in rats after oral subchronic low-level administration of the GSH, metallothionein, or acetate forms of the metal were examined by Muller *et al.* (1986). Administration of the cadmium–glutathione complex resulted in higher intestinal absorption and retention of cadmium, but the mechanism involved was not examined further.

In Caco-2 cells, which possess many morphological and functional characteristics of mature enterocytes, histidine, cysteine, proline, and GSH did not have any effect on zinc uptake (Raffaniello *et al.*, 1992).

3. Transport across the Blood–Brain Barrier

Recent studies from our laboratory demonstrate that GSH facilitates transport of methylmercury across the blood–brain barrier (Kerper *et al.*, 1992a,b), but it is not clear whether the intact GSH mercaptide is transported or whether it is first degraded to the cysteine mercaptide by γ-glutamyltransferase and dipeptidase activities present on the luminal surface of brain capillaries. The cysteine–methylmercury complex resulting from such degradation could then be taken up by the neutral amino acid transport system on the lumenal surface of brain capillaries (Kerper *et al.*, 1992a).

Similarly, uptake of methylmercury into brain after a single lateral ventricular injection was increased by coadministration of GSH (Watanabe *et al.*, 1988), presumably by a mechanism that involved the γ-glutamyltransferase-mediated conversion to the cysteine complex.

4. Hepatic and Gallbladder Uptake of Metals

Accumulation of cadmium by isolated hepatocytes appears to be mediated by membrane sulfhydryl groups (Failla *et al.*, 1979; Gerson and Shaikh, 1984). Uptake of cadmium, but not of inorganic mercury, is inhibited by *N*-ethylmaleimide and by *p*-chloromercuribenzenesulfonate (Gerson and Shaikh, 1984). Uptake by isolated hepatocytes is inhibited by 1–10 M excesses of cysteine or penicillamine, but is stimulated by 2,3-dimercaptopropanol or dithiothreitol. Similarly, in the perfused liver, Graf and Sies (1984) reported that cadmium uptake is increased by dithiothreitol, a membrane-permeable thiol, but is decreased by GSH and albumin.

The initial rate of methylmercury uptake by the liver is also enhanced by coadministration or by subsequent administration of cysteine or GSH (Thomas and Smith, 1982; Hirayama, 1975; Ballatori and Clarkson, 1983). In contrast, the initial rates of inorganic mercury uptake are slightly diminished by cysteine or GSH administration (Thomas and O'Tuama, 1979; Ballatori and Clarkson, 1984b).

Recent studies demonstrate that methylmercury secreted into bile is extensively reabsorbed from the gallbladder (Dutczak *et al.*, 1991) and from intrahepatic biliary spaces (Dutczak and Ballatori, 1992). Methylmercury reabsorption from guinea pig biliary spaces back into the liver was facilitated by the γ-glutamyltransferase-mediated conversion of the GSH complex to the cysteine complex. Guinea pig liver has relatively high γ-glutamyltransferase activity (Hinchman and Ballatori, 1990), such that nearly all of the GSH (Ballatori *et al.*, 1988) and GSH conjugates (Hinchman *et al.*, 1991) secreted into bile are catabolized within the biliary tree. Dutczak and Ballatori (1992) demonstrated that γ-glutamyltransferase also catalyzes the initial step in the conversion of the biliary methylmercury–glutathione complex to the cysteine complex and that in the latter form methylmercury is readily reabsorbed from the biliary tree.

Similarly, methylmercury reabsorption from guinea pig gallbladder was more extensive when the metal was given as a cysteine or GSH complex, than as an albumin complex (82, 64, and 50% of the methylmercury reabsorbed in 2 h, respectively; Dutczak *et al.*, 1991). Because the gallbladder also has γ-glutamyltransferase activity (Hinchman and Ballatori, 1990), a similar mechanism may be involved; however, there is no evidence for this hypothesis.

D. Glutathione as Reductant or Cofactor for Redox Reactions of Metal Compounds

1. Copper

As discussed earlier, GSH functions in the reduction of Cu(II) to Cu(I), the form incorporated into apometalloproteins, including metallothionein, superoxide dismutase, and lobster hemocyanin (Freedman *et al.*, 1989; Steinkuhler *et al.*, 1991; Brouwer and Brouwer-Hoexum, 1992). Cu(I) forms a relatively stable diglutathione complex and as such is more energetically favored to exchange with and bind to internal metal-binding sites of metalloproteins.

2. Selenium

GSH is involved in the reductive detoxification of selenite (Ganther, 1966), but may also promote the formation of toxic seleno compounds (Vernie *et al.*, 1979; Frenkel and Falvey, 1989). The selenite-mediated inhibition of protein and nucleic acid synthesis is potentiated by the addition of GSH, but the mechanism is not well defined (Vernie *et al.*, 1979; Frenkel and Falvey, 1989). The reaction of GSH with selenite (SeO_3^{-2}) produces selenodiglutathione (GSSeSG), glutathione selenopersulfide (GSSeH), elemental selenium, and selenide.

Selenium decreases the toxicity of methylmercury in experimental animals by promoting the conversion of methylmercury to a biologically less active form, bis(methylmercuric)selenide. GSH facilitates the reduction of selenium to a chemical form (probably H_2Se) that can react with methylmercury to yield bis(methylmercuric)selenide (Iwata *et al.*, 1981).

3. Vanadium

Vanadium, which most likely enters cells through phosphate or other anion transport systems as vanadate (Cantley *et al.*, 1978; Dingley *et al.*, 1981), is also reduced intracellularly by GSH. Studies in erythrocytes (Macara *et al.*, 1980) and adipocytes (Degani *et al.*, 1981) demonstrate that vanadate is slowly but quantitatively reduced to V(IV) nonenzymatically by GSH. The vanadyl ion can then compete with other transition metals for binding sites on metalloproteins or other ligands. In contrast to vanadate, the vanadyl ion (V(IV)) is not a good inhibitor of NaK-ATPase (Cantley and Aisen, 1979), and its physiological effects are not well characterized.

4. Arsenic

As with vanadium, the arsenate oxyanion is structurally similar to phosphate (Wetterhahn-Jennette, 1981). Arsenate enters cells and exerts its toxicity by substituting for phosphate. Reduction of arsenate to arsenite, the thiol-reactive form, also involves GSH. As(V), the dominant species in drinking water, reacts with GSH to produce GSSG and As(III), which can then react with the sulfhydryl groups of additional GSH molecules to form the $(GS)_3As(III)$ complex (Delnomdedieu *et al.*, 1993).

5. Chromium

Tetrahedral chromium(VI) oxyanions in the form of chromate readily enter cells on the sulfate or anion carriers, whereas octahedral chromium(III) compounds are not readily carried across cell membranes (Wetterhahn-Jennette, 1981). Within the cell, GSH serves as a reductant in the conversion of Cr(VI) compounds to Cr(III) compounds (Wiegand *et al.*, 1984). This conversion occurs at a significant rate at physiological pH *in vitro* and also appears to occur *in vivo* (Standeven and Wetterhahn, 1991; Norseth *et al.*, 1982). Cr(VI) and Cr(III) form stable complexes with GSH (Brauer and Wetterhahn, 1991); however, during the reduction to the trivalent form other intermediate oxidation states are formed (V and IV) and toxic free radicals are produced. In particular, the reaction of GSH with Cr(VI) to produce the Cr(V)–glutathione complex also leads to the release of thiyl radical (GS·), H_2O_2, and hydroxyl radicals (·OH), but the mechanisms involved are not well defined (for review, see Sugiyama, 1992). Elevation of GSH levels in cultured hepatocytes produces increased DNA damage by Cr(VI) (Cupo and Wetterhahn, 1985). More recently, Borges and Wetterhahn (1989) demonstrated that the intermediates formed by reaction of Cr(VI) with GSH or cysteine are capable of interacting with DNA, with preferential binding of the Cr-mercaptides to guanine-containing polynucleotides. Formation of these GSH–Cr–DNA adducts may be responsible for the genotoxicity of chromium compounds (Borges and Wetterhahn, 1989).

The role of GSH in the nephrotoxicity of Cr(VI) is controversial. Appenroth and Kersten (1990) suggest that cellular GSH enhances Cr(VI) nephrotoxicity, whereas Standeven and Wetterhahn (1991) and Na *et al.* (1992) describe a protective role for GSH.

IV. Summary

Among the many cellular functions of GSH, the roles of this tripeptide in metal transport, storage, and metabolism have recently received considerable attention. Although these roles had often been overlooked, they are critical for normal cellular metabolism and for protection from xenobiotics. Indeed, a number of the protective and regulatory functions of GSH are related to its ability to chelate reactive metals. GSH functions in the mobilization and delivery of metals between ligands, in the transport of metals across cell membranes, as a source of cysteine for metal binding, and as a reductant or cofactor in redox reactions involving metals. However, the interaction between GSH and metals can also produce or exacerbate cell injury. For example, GSH appears to be involved in the renal accumulation and toxicity of a number of metals, and in the carcinogenicity of chromium. Additional work is clearly needed to identify the mechanisms involved, and to better define the roles of GSH in metal homeostasis.

Acknowledgments

The author thanks T. W. Clarkson, W. J. Dutczak, C. A. Hinchman, L. E. Kerper, E. M. Mokrzan, and A. T. Truong for their comments on the manuscript. Preparation of this overview was supported in part by National Institutes of Health Grant DK39165, Program Project Grant ES05197, and by NIEHS Center Grant ES01247.

References

Albert, A. (1952). Quantitative studies of the avidity of naturally occurring substances for trace metals. *Biochem. J.* **50,** 690–697.

Alexander, J., and Aaseth, J. (1980). Biliary excretion of copper and zinc in the rat as influenced by diethylmaleate, selenite and diethyldithiocarbamate. *Biochem. Pharmacol.* **29,** 2129–2133.

Alexander, J., and Aaseth, J. (1981). Hepatobiliary transport and organ distribution of silver in the rat as influenced by selenite. *Toxicology* **21,** 179–186.

Alexander, J., and Aaseth, J. (1982). Organ distribution and cellular uptake of methyl mercury in the rat as influenced by the intra- and extracellular glutathione concentration. *Biochem. Pharmacol.* **31,** 685–690.

Alexander, J., Aaseth, J., and Refsvik, T. (1981). Excretion of zinc in rat bile. A role of glutathione. *Acta Pharmacol. Toxicol.* **49,** 190–194.

Alexander, J., Aaseth, J., and Mikalsen, A. (1986). Excretion of lead in rat bile—the role of glutathione. *Acta Pharmacol. Toxicol.* **59,** 486–489.

Andrews, P. A., Murphy, M. P., and Howell, S. B. (1987). Metallothionein-mediated cisplatin resistance in human ovarian carcinoma cells. *Cancer Chemother. Pharmacol.* **19,** 149–154.

Anundi, I., Hogberg, J., and Vahter, M. (1982). GSH release in bile as influenced by arsenite. *FEBS Lett.* **145,** 285–288.

Anundi, I., Hogberg, J., and Stahl, A. (1984). Absorption of selenite in the rat small intestine: Interactions with glutathione. *Acta Pharmacol. Toxicol.* **54,** 272–277.

Appenroth, D., and Kersten, L. (1990). The activity of chromate reduction in renal tissue corresponds to chromate nephrotoxicity-developmental aspects. *Toxicol. Lett.* **53,** 157–159.

Ballatori, N. (1991a). Mechanisms of metal transport across liver cell plasma membranes. *Drug Metab. Rev.* **23,** 83–132.

Ballatori, N. (1991b). Biliary excretion of metals. *In* "Biliary Excretion of Drugs and Other Chemicals" (C.-P. Siegers and J. B. Watkins, III, eds.), Chapter 3.7, pp. 283–306. Fischer, Stuttgart and New York.

Ballatori, N., and Clarkson, T. W. (1982). Developmental changes in the biliary excretion of methylmercury and glutathione. *Science* **216,** 61–63.

Ballatori, N., and Clarkson, T. W. (1983). Biliary transport of glutathione and methylmercury. *Am. J. Physiol.* **244,** G435–G441.

Ballatori, N., and Clarkson, T. W. (1984a). Inorganic mercury secretion into bile as a low molecular weight complex. *Biochem. Pharmacol.* **33,** 1087–1092.

Ballatori, N., and Clarkson, T. W. (1984b). Dependence of biliary secretion of inorganic mercury on the biliary transport of glutathione. *Biochem. Pharmacol.* **33,** 1093–1098.

Ballatori, N., and Clarkson, T. W. (1985a). Sulfobromophthalein inhibition of glutathione and methylmercury secretion into bile. *Am. J. Physiol.* **248,** G238–G245.

Ballatori, N., and Clarkson, T. W. (1985b). Biliary secretion of glutathione and of glutathione-metal complexes. *Fundam. Appl. Toxicol.* **5,** 816–831.

Ballatori, N., Moseley, R. H., and Boyer, J. L. (1986a). Sodium gradient-dependent L-glutamate transport is localized to the canalicular domain of liver plasma membranes. Studies in rat liver sinusoidal and canalicular membrane vesicles. *J. Biol. Chem.* **261,** 6216–6221.

Ballatori, N., Jacob, R., and Boyer, J. L. (1986b). Intrabiliary glutathione hydrolysis—a source of glutamate in bile. *J. Biol. Chem.* **261,** 7860–7865.

Ballatori, N., Jacob, R., Barrett, C., and Boyer, J. L. (1988). Biliary catabolism of glutathione and differential reabsorption of its amino acid constituents. *Am. J. Physiol.* **254,** G1–G7.

Ballatori, N., Truong, A. T., Ma, A. K., and Boyer, J. L. (1989). Determinants of glutathione efflux and biliary GSH/GSSG ratio in perfused rat liver. *Am. J. Physiol.* **256,** G482–G490.

Bannai, S., Sato, H., Ishii, T., and Taketani, S. (1991). Enhancement of glutathione levels in mouse peritoneal macrophages by sodium arsenite, cadmium chloride and glucose/glucose oxidase. *Biochim. Biophys. Acta* **1092,** 175–179.

Berndt, W. O., Baggett, J.McC., Blacker, A., and Houser, M. (1985). Renal glutathione and mercury uptake by kidney. *Fundam. Appl. Toxicol.* **5,** 832–839.

Board, P. G. (1981). Transport of glutathione S-conjugate from human erythrocytes. *FEBS Lett.* **124,** 163–165.

Borges, K., and Wetterhahn, K. E. (1989). Chromium cross-links glutathione and cysteine to DNA. *Carcinogenesis (London)* **10,** 2165–2168.

Brauer, S. L., and Wetterhahn, K. E. (1991). Chromium(VI) forms a thiolate complex with glutathione. *J. Am. Chem. Soc.* **113,** 3001–3007.

Brouwer, M., and Brouwer-Hoexum, T. (1992). Glutathione-mediated transfer of copper(I) into American lobster apohemocyanin. *Biochemistry* **31,** 4096–4102.

Cantley, L. C., Jr., and Aisen, P. (1979). The fate of cytoplasmic vanadium. Implications on (Na,K)-ATPase inhibition. *J. Biol. Chem.* **254,** 1781–1784.

Cantley, L. C., Jr., Resh, M. D., and Guidotti, G. (1978). Vanadate inhibits the red cell (Na^+,K^+)ATPase from the cytoplasmic side. *Nature (London)* **272,** 552–554.

Carty, A. J., and Malone, S. F. (1979). The chemistry of mercury in biological systems. *In* "The Biogeochemistry of Mercury in the Environment" (J. O. Nriagu, ed.), Vol. 3, Chapter 17, pp. 433–479. Elsevier/North-Holland Biomedical Press, Amsterdam.

Chan, H. M., and Cherian, M. G. (1992). Protective roles of metallothionein and glutathione in hepatotoxicity of cadmium. *Toxicology* **72,** 281–290.

Cherian, M. G., and Vostal, J. J. (1977). Biliary excretion of cadmium in rat. I. Dose-dependent biliary excretion and form of cadmium in the bile. *J. Toxicol. Environ. Health* **2,** 945–954.

Ciriolo, M. R., Desideri, A., Paci, M., and Rotilio, G. (1990). Reconstitution of Cu, Zn-superoxide dismutase by the Cu(I)·glutathione complex. *J. Biol. Chem.* **265,** 11030–11034.

Clarkson, T. W. (1972). The pharmacology of mercury compounds. *Annu. Rev. Pharmacol.* **22,** 375–406.

Clarkson, T. W. (1993). Molecular and ionic mimicry of toxic metals. *Annu. Rev. Pharmacol. Toxicol.* **32,** 545–571.

Cupo, D. Y., and Wetterhahn, K. E. (1985). Modification of chromium(IV)-induced DNA damage by glutathione and cytochromes P-450 in chicken embryo hepatocytes. *Proc. Natl. Acad. Sci. U.S.A.* **82,** 6755–6759.

Dameron, C. T., Smith, B. R., and Winge, D. R. (1989). Glutathione-coated cadmium-sulfide crystallites in *Candida glabrata. J. Biol. Chem.* **264,** 17355–17360.

deCeaurriz, J., and Ban, M. (1990). Role of gamma-glutamyltranspeptidase and beta-lyase in the nephrotoxicity of hexachloro-1,3-butadiene and methyl mercury in mice. *Toxicol. Lett.* **50,** 249–256.

Degani, H., Gochin, M., Karlish, S. J. D., and Shechter, Y. (1981). Electron paramagnetic resonance studies and insulin-like effects of vanadium in rat adipocytes. *Biochemistry* **20,** 5795–5799.

Deleve, L. D. and Kaplowitz, N. (1990). Importance and regulation of hepatic glutathione. *Semin. Liver Dis.* **10,** 251–266.

Delnomdedieu, M., Basti, M. M., Otvos, J., and Thomas, D. J. (1993). Nuclear magnetic resonance studies of the solution chemistry of arsenic-glutathione complexes. *Toxicologist* **13,** 159.

Dingley, A. L., Kustin, K., Macara, I. G., and McLeod, G. C. (1981). Accumulation of vanadium by tunicate blood cells occurs via a specific anion transport system. *Biochim. Biophys. Acta* **649,** 493–502.

Di Simplicio, P., Gorelli, M., Ciuffreda, P., and Leonzio, C. (1990). The relationship between γ-glutamyltranspeptidase and Hg levels in Se/Hg antagonism in mouse liver and kidney. *Pharmacol. Res.* **22,** 515–526.

Dutczak, W. J., and Ballatori, N. (1992). γ-Glutamyltransferase-dependent biliary-hepatic recycling of methyl mercury in the guinea pig. *J. Pharmacol. Exp. Ther.* **262,** 619–623.

Dutczak, W. J., Clarkson, T. W., and Ballatori, N. (1991). Biliary-hepatic recycling of a xenobiotic. Gallbladder absorption of methylmercury. *Am. J. Physiol.* **260,** G873–G880.

Dutczak, W. J., Truong, A. T., and Ballatori, N. (1993). Transport of the glutathione-methylmercury complex in rat liver canalicular membrane vesicles. *Toxicologist* **13,** 191.

Failla, M. L., Cousins, R. J., and Mascenik, M. J. (1979). Cadmium accumulation and metabolism by rat liver parenchymal cells in primary monolayer culture. *Biochim. Biophys. Acta* **583,** 63–72.

Freedman, J. H., Ciriolo, M. R., and Peisach, J. (1989). The role of glutathione in copper metabolism and toxicity. *J. Biol. Chem.* **264,** 5598–5605.

Frenkel, G. D., and Falvey, D. (1989). Involvement of cellular sulfhydryl compounds in the inhibition of RNA synthesis by selenite. *Biochem. Pharmacol.* **38,** 2849–2852.

Fuhr, B. J., and Rabenstein, D. L. (1973). Nuclear magnetic resonance studies of the solution chemistry of metal complexes. IX. The binding of cadmium, zinc, lead and mercury by glutathione. *J. Am. Chem. Soc.* **95,** 6944–6950.

Fukino, H., Hirai, M., Hsueh, Y. M., Moriyasu, S., and Yamane, Y. (1986). Mechanism of protection by zinc against mercuric chloride toxicity in rats: Effects of zinc and mercury on glutathione metabolism. *J. Toxicol. Environ. Health* **19,** 75–89.

Ganther, H. E. (1966). Enzymatic synthesis of dimethyl selenide from selenite in mouse liver extracts. *Biochemisty* **5,** 1089–1098.

Garcia-Ruiz, C., Fernandez-Checa, J. C., and Kaplowitz, N. (1992). Bidirectional mechanism of plasma membrane transport of reduced glutathione in intact rat hepatocytes and membrane vesicles. *J. Biol. Chem.* **267,** 22256–22264.

Gerson, R. J., and Shaikh, Z. A. (1984). Differences in the hepatic uptake of cadmium and mercury by rat hepatocyte primary cultures: role of sulfhydryl carrier. *Biochem. Pharmacol.* **33,** 199–203.

Glaeser, H., Coblenz, A., Kruczek, R., Ruttke, I., Ebert-Jung, A., and Wolf, K. (1991). Glutathione metabolism and heavy metal detoxification in *Schizosaccharomyces pombe*. *Curr. Genet.* **19,** 207–213.

Graf, P., and Sies, H. (1984). Hepatic uptake of cadmium and its biliary release as affected by dithioerythritol and glutathione. *Biochem. Pharmacol.* **33,** 639–643.

Gregus, Z., and Varga, F. (1985). Role of glutathione and hepatic glutathione S-transferase in the biliary excretion of methyl mercury, cadmium and zinc: A study with enzyme inducers and glutathione depletors. *Acta Pharmacol. Toxicol.* **56,** 398–403.

Gregus, Z., Stein, A. F., and Klaassen, C. D. (1987). Effect of inhibition of gamma-glutamyltranspeptidase on biliary and urinary excretion of glutathione-derived thiols and methylmercury. *J. Pharmacol. Exp. Ther.* **242,** 27–32.

Grill, E., Winnacker, E.-L., and Zenk, M. H. (1985). Phytochelatins: The principal heavy-metal complexing peptides of higher plants. *Science* **230,** 674–676.

Gyurasics, A., Varga, F., and Gregus, Z. (1991a). Effect of arsenicals on biliary excretion of endogenous glutathione and xenobiotics with glutathione-dependent hepatobiliary transport. *Biochem. Pharmacol.* **41,** 937–944.

Gyurasics, A., Varga, F., and Gregus, Z. (1991b). Glutathione-dependent biliary excretion of arsenic. *Biochem. Pharmacol.* **42,** 465–468.

Gyurasics, A., Koszorus, L., Varga, F., and Gregus, Z. (1992). Increased biliary excretion of glutathione is generated by the glutathione-dependent hepatobiliary transport of antimony and bismuth. *Biochem. Pharmacol.* **44,** 1275–1281.

Hahn, R., Wendel, A., and Flohé, L. (1978). The fate of extracellular glutathione in the rat. *Biochim. Biophys. Acta* **539,** 324–337.

Harris, E. D. (1991). Copper transport: an overview. *Proc. Soc. Exp. Biol. Med.* **196,** 130–140.

Hinchman, C. A., and Ballatori, N. (1990). Glutathione-degrading capacities of liver and kidney in different species. *Biochem. Pharmacol.* **40,** 1131–1135.

Hinchman, C. A., Matsumoto, H., Simmons, T. W., and Ballatori, N. (1991). Intrahepatic conversion of a glutathione conjugate to its mercapturic acid. Metabolism of 1-chloro-

2,4-dinitrobenzene in isolated perfused rat and giunea pig livers. *J. Biol. Chem.* **266,** 22179–22185.

Hinchman, C. A., Truong, A. T., and Ballatori, N. (1993). Hepatic uptake of intact glutathione *S*-conjugate, inhibition by organic anions, and sinusoidal catabolism. *Am. J. Physiol.* **265,** G547–G554.

Hirata, E., and Takahashi, H. (1981). Degradation of methyl mercury glutathione by the pancreatic enzymes in bile. *Toxicol. Appl. Pharmacol.* **58,** 483–491.

Hirayama, K. (1975). Transport mechanism of methyl mercury: Intestinal absorption, biliary excretion and distribution of methyl mercury. *Kumamoto Med. J.* **28,** 151–163.

Hirayama, K., Yasutake, A., and Inoue, M. (1987). Effect of sex hormones on the fate of methylmercury and on glutathione metabolism in mice. *Biochem. Pharmacol.* **36,** 1919–1924.

Houser, M. T., Milner, L. S., Kolbeck, P. C., Wei, S. H., and Stohs, S. J. (1992). Glutathione monoethyl ester moderates mercuric chloride-induced acute renal failure. *Nephron* **61,** 449–455.

Houwen, R., Dijkstra, M., Kuipers, F., Smit, E. P., Havinga, R., and Vonk, R. J. (1990). Two pathways for biliary copper excretion in the rat. The role of glutathione. *Biochem. Pharmacol.* **39,** 1039–1044.

Hsu, J. M. (1981). Lead toxicity as related to glutathione metabolism. *J. Nutr.* **111,** 26–33.

Inoue, M., and Morino, Y. (1985). Direct evidence for the role of the membrane potential in glutathione transport by renal brush-border membranes. *J. Biol. chem.* **260,** 326–331.

Iwata, H., Masukawa, T., Kito, H., and Hayashi, M. (1981). Involvement of tissue sulfhydryls in the formation of a complex of methylmercury with selenium. *Biochem. Pharmacol.* **30,** 3159–3163.

Jezowska-Trzebiatowska, B., Formicka-Kozlowska, G., and Kozlowski, H. (1976). Metal-glutathione interaction in water solution. II. NMR and electron spectroscopy study of Ni(II)-glutathione complexes in aqueous solution. *Chem. Phys. Lett.* **42,** 242–245.

Johnson, D. R. (1982). Role of renal cortical sulfhydryl groups in development of mercury-induced renal toxicity. *J. Toxicol. Environ. Health* **9,** 119–126.

Kang, Y.-J., and Enger, D. (1988). Glutathione is involved in the early cadmium cytotoxic response in human lung carcinoma cells. *Toxicology* **48,** 93–101.

Kannan, R., Kuhlenkamp, J. F., Ookhtens, M., and Kaplowitz, N. (1992). Transport of glutathione at blood-brain barrier of the rat: Inhibition by glutathione analogs and age-dependence. *J. Pharmacol. Exp. Ther.* **263,** 964–970.

Kapoor, R. C., Doughty, G., and Gorin, G. (1965). The reaction and assay of glutathione with Hg^{2+} and alkali. *Biochim. Biophys. Acta* **100,** 376–383.

Kerper, L. E., Ballatori, N., and Clarkson, T. W. (1992a). Methylmercury transport across the blood-brain barrier by an amino acid carrier. *Am. J. Physiol.* **262,** R761–R765.

Kerper, L. E., Ballatori, N., and Clarkson, T. W. (1992b). A role for glutathione in the transport of methylmercury across the blood-brain barrier. *Toxicologist* **12,** 307.

Kramer, R. A., Foureman, G., Greene, K, E., and Reed, D. J. (1987). Nephrotoxicity of *S*-(2-chloroethyl)glutathione in the Fischer rat: Evidence for gamma-glutamyltranspeptidase-independent uptake by the kidney. *J. Pharmacol. Exp. Ther.* **242,** 741–748.

Kromidas, L., Trombetta, L. D., and Jamall, I. S. (1990). The protective effects of glutathione against methylmercury cytotoxicity. *Toxicol. Lett.* **51,** 67–80.

Kuchan, M. J., and Milner, J. A. (1992). Influence of intracellular glutathione on selenite-mediated growth inhibition of canine mammary tumor cells. *Cancer Res.* **52,** 1091–1095.

LaBelle, E. F., Singh, S. V., Srivastava, S. K., and Awasthi, Y. C. (1986). Evidence for different transport systems for oxidized glutathione and *S*-dinitrophenyl glutathione in human erythrocytes. *Biochem. Biophys. Res. Commun.* **139,** 538–544.

Lash, L. H., and Jones, D. P. (1984). Renal glutathione transport: Characteristics of the sodium-dependent system in the basal-lateral membrane. *J. Biol. Chem.* **259,** 14508–14514.

Lenz, G. R., and Martell, A. E. (1964). Metal chelates of some sulfur-containing amino acids. *Biochemistry* **3,** 745–750.

Leone, R., Fracasso, M. E., Soresi, E., Cimino, G., Tedeschi, M., Castoldi, D., Monzani, V., Colimbi, L., Usari, T., and Bernareggi, A. (1992). Influence of glutathione administration on the disposition of free and total platinum in patients after administration of cisplatin. *Cancer Chemother. Pharmacol.* **29,** 385–390.

Li, N. C., and Manning, R. A. (1955). Some metal complexes of sulfur-containing amino acids. *J. Am. Chem. Soc.* **77,** 5225–5228.

Li, N. C., Gawron, O., and Bascuas, G. (1954). Stability of zinc complexes with glutathione and oxidized glutathione. *J. Am. Chem. Soc.* **76,** 225–229.

Li, W., and Chou, I.-N. (1992). Effects of sodium arsenite on the cytoskeleton and cellular glutathione levels in cultured cells. *Toxicol. Appl. Pharmacol.* **114,** 132–139.

Macara, I. G., Kustin, K., and Cantley, L. C., Jr. (1980). Glutathione reduces cytoplasmic vanadate. Mechanism and physiological implications. *Biochim. Biophys. Acta* **629,** 95–106.

Martell, A. E., and Calvin, M. (1952). "Chemistry of the Metal Chelate Compounds," Prentice-Hall, New York.

Martin, R. B., and Edsall, J. T. (1959). The association of divalent cations with glutathione. *J. Am. Chem. Soc.* **81,** 4044–4047.

Meister, A., and Anderson, M. E. (1983). Glutathione. *Annu. Rev. Biochem.* **52,** 711–760.

Miura, K., and Clarkson, T. W. (1993). Reduced methylmercury accumulation in a methylmercury-resistant rat pheochromocytoma PC12 cell line. *Toxicol. Appl. Pharmacol.* **118,** 39–45.

Moseley, R. H., Ballatori, N., and Murphy, S. M. (1988). Na^+-Glycine cotransport in canalicular liver plasma membrane vesicles. *Am. J. Physiol.* **255,** G253–G259.

Mulder, K. M., and Kostyniak, P. J. (1985a). Effect of L-(αS, 5S)-α-amino-3-chloro-4,5-dihydro-5-isoxazoleacetic acid on urinary exretion of methylmercury in the mouse. *J. Pharmacol. Exp. Ther.* **234,** 156–160.

Mulder, K. M., and Kostyniak, P. J. (1985b). Involvement of glutathione in the enhanced renal excretion of methyl mercury in CFW Swiss mice. *Toxicol. Appl. Pharmacol.* **78,** 451–457.

Muller, L., Abel, J., and Ohnesorge, F. K. (1986). Absorption and distribution of cadmium (Cd), copper and zinc following oral subchronic low level administration to rats of different binding forms of cadmium (Cd-acetate, Cd-metallothionein, Cd-glutathione). *Toxicology* **39,** 187–195.

Murakami, M., and Webb, M. (1981). A morphological and biochemical study of the effects of L-cysteine on the renal uptake and nephrotoxicity of cadmium. *Br. J. Exp. Pathol.* **62,** 115–130.

Murakami, M., Sano, K., and Webb, M. (1987). The effect of L-cysteine on the portion-selective uptake of cadmium in the renal proximal tubule. *Arch. Toxicol.* **60,** 365–369.

Na, K. J., Jeong, S. Y., and Lim, C. H. (1992). The role of glutathione in the acute nephrotoxicity of sodium dichromate. *Arch. Toxicol.* **66,** 646–651.

Naganuma, A., Oda-Urano, N., Tanaka, T., and Imura, N. (1988). Possible role of hepatic glutathione in transport of methylmercury into mouse kidney. *Biochem. Pharmacol.* **37,** 291–296.

Naganuma, A., Anderson, M. E., and Meister, A. (1990). Cellular glutathione as a determinant of sensitivity to mercuric chloride toxicity. Prevention of toxicity by giving glutathione monoester. *Biochem. Pharmacol.* **40,** 693–697.

Nederbragt, H. (1989). Effect of the glutathione-depleting agents diethylmaleate, phorone and buthionine sulfoximine on biliary copper excretion in rats. *Biochem. Pharmacol.* **38,** 3399–3406.

Norseth, T. (1973a). The effect of chelating agents on biliary excretion of methyl mercuric salts in the rat. *Acta Pharnacol. Toxicol.* **32,** 1–10.

Norseth, T. (1973b). Biliary complexes of methylmercury: A possible role in organ distribution. *In* "Mercury, Mercurials and Mercaptans" (M. W. Miller and T. W. Clarkson, eds.), pp. 264–276. Thomas, Springfield, IL.

Norseth, T. (1973c). Biliary excretion and intestinal reabsorption of mercury in the rat after injection of methyl mercuric chloride. *Acta Pharmacol. Toxicol.* **33,** 280–288.

Norseth, T., and Clarkson, T. W. (1971). Intestinal transport of ^{203}Hg-labeled methyl mercury chloride. *Arch. Environ. Health* **22,** 568–577.

Norseth, T., Alexander, J., Aaseth, J., and Langard, S. (1982). Biliary excretion of chromium in the rat: A role of glutathione. *Acta Pharmacol. Toxicol.* **51,** 450–455.

Ochi, T., Otsuka, F., Takahashi, K., and Ohsawa, M. (1988). Glutathione and metallothioneins as cellular defense against cadmium toxicity in cultured Chinese hamster cells. *Chem.-Biol. Interact.* **65,** 1–14.

Perrin, D. D., and Watt, A. E. (1971). Complex formation of zinc and cadmium with glutathione. *Biochim. Biophys. Acta* **230,** 96–104.

Rabenstein, D. L., and Fairhurst, M. T. (1975). Nuclear magnetic resonance studies of the solution chemistry of metal complexes. XI. The binding of methylmercury by sulfhydryl-containing amino acids and by glutathione. *J. Am. Chem. Soc.* **97,** 2086–2092.

Raffaniello, R. D., Lee, S.-Y., Teichberg, S., and Wapnir, R. A. (1992). Distinct mechanisms of zinc uptake at the apical and basolateral membranes of Caco-2 cells. *J. Cell. Physiol.* **152,** 356–361.

Refsvik, T. (1978). Excretion of methyl mercury in rat bile: The effect of diethylmaleate, cyclohexene oxide and acrylamide. *Acta Pharmacol. Toxicol.* **42,** 135–141.

Refsvik, T. (1982). Excretion of methyl mercury in rat bile: The effect of thioctic acid, thionalide, hexadecyl- and octadecylmercaptoacetate. *Acta Pharmacol. Toxicol.* **50,** 196–205.

Refsvik, T., and Norseth, T. (1975). Methyl mercuric compounds in rat bile. *Acta Pharmacol. Toxicol.* **36,** 67–78.

Richardson, R. J., and Murphy, S. D. (1975). Effect of glutathione depletion on tissue deposition of methylmercury in rats. *Toxicol. Appl. Pharmacol.* **31,** 505–519.

Sager, P. R., and Syversen, T. L. M. (1986). Disruption of microtubules by methylmercury. *In* "The Cytoskeleton: A Target for Toxic Agents" (T. W. Clarkson, P. R. Sager, and T. L. M. Syversen, eds.), pp. 97–116. Plenum, New York.

Seagrave, J. C., Tobey, R. A., and Hildebrand, C. E. (1983). Zinc effects on glutathione metabolism: Relationship to zinc-induced protection from alkylating agents. *Biochem. Pharmacol.* **32,** 3017–3021.

Senn, E., Scharrer, E., and Wolffram, S. (1992). Effects of glutathione and of cysteine on intestinal absorption of selenium from selenite. *Biol. Trace Elem. Res.* **33,** 103–108.

Shaw, C. F., III, Isab, A. A., Coffer, M. T., and Mirabelli, C. K. (1990). Gold(I) efflux from auranofin-treated red blood cells. Evidence for a glutathione-gold-albumin metabolite. *Biochem. Pharmacol.* **40,** 1227–1234.

Siegers, C.-P., Sharma, S. C., and Younes, M. (1986). Hepatotoxicity of metals in glutathione-depleted mice. *Toxicol. Lett.* **34,** 185–191.

Simmons, T. W., Hinchman, C. A., and Ballatori, N. (1991). Polarity of hepatic glutathione and glutathione S-conjugate efflux, and intraorgan mercapturic acid formation in the skate. *Biochem. Pharmacol.* **42,** 2221–2228.

Simmons, T. W., Anders, M. W., and Ballatori, N. (1992). Cysteine and S-(1,2-dicholorvinyl)-L-cysteine transport in rat liver canalicular membrane vesicles: Potential reabsoprtion mechanisms for biliary metabolites of glutathione and its S-conjugates. *J. Pharmacol. Exp. Ther.* **262,** 1182–1188.

Simpson, R. B. (1961). Association constants of methylmercury with sulfhydryls and other bases. *J. Am. Chem. Soc.* **83,** 4711–4717.

Singhal, R. K., Anderson, M. E., and Meister, A. (1987). Glutathione, a first line of defense against cadmium toxicity. *FASEB J.* **1,** 220–223.

Sone, T., Yamaoka, K., Minami, Y., and Tsunoo, H. (1987). Induction of metallothionein synthesis in menkes' and normal lymphoblastoid cells is controlled by the level of intracellular copper. *J. Biol. Chem.* **262,** 5878–5885.

Stacey, N. H. (1986). The amelioration of cadmium-induced injury in isolated hepatocytes by reduced glutathione. *Toxicology* **42,** 85–93.

Standeven, A. M., and Wetterhahn, K. E. (1991). Possible role of glutathione in chromium (VI) metabolism and toxicity in rats. *Pharmacol. Toxicol.* **68,** 469–476.

Stein, A. F., Gregus, Z., and Klaassen, C. D. (1988). Species variations in biliary excretion of glutathione-related thiols and methylmercury. *Toxicol. Appl. Pharmacol.* **93,** 351–359.

Steinkuhler, C., Sapora, O., Carri, M. T., Nagel, W., Marcocci, L., Ciriolo, M. R., Weser, U., and Rotilio, G. (1991). Increase of Cu, Zn-superoxide dismutase activity during differentiation of human K562 cells involves activation by copper of a constantly expressed copper-deficient protein. *J. Biol. Chem.* **266,** 24580–24587.

Stricks, W., and Kolthoff, I. M. (1951). Polarographic investigations of reactions in aqueous solutions containing copper and cysteine (cystine). I. Cuprous copper and cysteine in ammoniacal medium. The dissociation constant of cuprous cysteinate. *J. Am. Chem. Soc.* **73,** 1723–1727.

Stricks, W., and Kolthoff, I. M. (1953). Reactions between mercuric mercury and cysteine and glutathione. Apparent dissociation constants, heats and entropies of formation of various forms of mercuric mercapto-cysteine and -glutathione. *J. Am. Chem. Soc.* **75,** 5673–5681.

Sugiyama, M. (1992). Role of physiological antioxidants in chromium(VI)-induced cellular injury. *Free Radical Biol. Med.* **12,** 397–407.

Suzuki, K. T., Karasawa, A., and Yamanaka, K. (1989). Binding of copper to albumin and participation of cysteine *in vivo* and *in vitro*. *Arch. Biochem. Biophys.* **273,** 572–577.

Takahashi, H., Wasa, S., Suda, I., Totoki, S., and Ohota, J. (1988). Toxicological and metabolic aspects of organic mercury poisoning. *J. UOEH, Suppl.* **10,** 117–126.

Tanaka, T., Naganuma, A., and Imura, N. (1990). Role of γ-glutamyltranspeptidase in renal uptake and toxicity of inorganic mercury in mice. *Toxicology* **60,** 187–198.

Tanaka, T., Naganuma, A., Kobayashi, K., and Imura, N. (1991). An explanation for strain and sex differences in renal uptake of methylmercury in mice. *Toxicology* **69,** 317–329.

Tanaka, T., Naganuma, A., Miura, N., and Imura, N. (1992). Role of testosterone in γ-glutamyltranspeptidase-dependent renal methylmercury uptake in mice. *Toxicol. Appl. Pharmacol.* **112,** 58–63.

Thomas, D. J., and O'Tuama, L. A. (1979). Effects of L- or D-cysteine coadministration on short-term distribution of inorganic mercury in the rat. *Toxicol. Appl. Pharmacol.* **50,** 219–227.

Thomas, D. J., and Smith, J. C. (1982). Effect of coadministered low-molecular-weight thiol compounds on short-term distribution of methylmercury in the rat. *Toxicol. Appl. Pharmacol.* **62,** 104–110.

Urano, T., Naganuma, A., and Imura, N. (1988). Species differences in biliary excretion of methylmercury. Role of non-protein sulfhydryls in bile. *Res. Commun. Chem. Pathol. Pharmacol.* **62,** 339–351.

Urano, T., Iwasaki, A., and Himeno, S. (1990). Absorption of methylmercury compounds from rat intestine. *Toxicol. Lett.* **50,** 159–164.

Vendeland, S. C., Deagen, J. T., and Whanger, P. D. (1992). Uptake of selenotrisulfides of glutathione and cysteine by brush border membranes from rat intestines. *J. Inorg. Biochem.* **47,** 131–140.

Vernie, L. N., Collard, J. G., Ekker, A. P., DeWildt, A., and Wilders, I. T. (1979). Studies on the inhibition of protein synthesis by selenodiglutathione. *Biochem. J.* **180,** 213–218.

Vincenzini, M. T., Favilli, F., and Iantomasi, T. (1992). Intestinal uptake and transmembrane transport systems of intact GSH; Characteristics and possible biological role. *Biochim. Biophys. Acta* **1113,** 13–23.

Watanabe, H., Shimojo, N., Sano, K., and Yamaguchi, S. (1988). The distribution of total mercury in the brain after the lateral ventricular single injection of methylmercury and glutathione. *Res. Commun. Chem. Pathol. Pharmacol.* **60,** 57–69.

Wetterhahn-Jennette, K. (1981). The role of metals in carcinogenesis: Biochemistry and metabolism. *Environ. Health Perspect.* **40,** 233–252.

Wiegand, H. J., Ottenwalder, H., and Bolt, H. M. (1984). The reduction of chromium (VI) to chromium (III) by glutathione: An intracellular redox pathway in the metabolism of the carcinogen chromate. *Toxicology* **33,** 341–348.

Wurmli, R., Wolffram, S., Stingelin, Y., and Scharrer, E. (1989). Stimulation of mucosal uptake of selenium from selenite by L-cysteine in sheep small intestine. *Biol. Trace Elem. Res.* **20,** 75–85.

Yasutake, A., Hirayama, K., and Inoue, M. (1989). Mechanism of urinary excretion of methylmercury in mice. *Arch. Toxicol.* **63,** 479–483.

Zinkewich-Peotti, K., and Andrews, P. A. (1992). Loss of *cis*-Diamminedichloroplatinum (II) resistance in human ovarian carcinoma cells selected for Rhodamine 123 resistance. *Cancer Res.* **52,** 1902–1906.

Part II

Sulfate Conjugate-Dependent Toxicity

Biochemistry of Cytosolic Sulfotransferases Involved in Bioactivation

Charles N. Falany and Teresa W. Wilborn
Department of Pharmacology
University of Alabama at Birmingham
Birmingham, Alabama 35294

I. Introduction

The formation of sulfate conjugates of xenobiotic compounds was first reported by Baumann in 1876, when he detected the presence of phenyl sulfate in the urine of a patient who had been adminstered phenol. Sulfation is now recognized as one of the major Phase-II conjugation reactions in the biotransformation of a range of drugs and xenobiotic compounds as well as many endogenous compounds.

Although the widespread occurrence of sulfate conjugates was readily recognized from *in vivo* studies, insight into the mechanism of sulfation was not obtained until 1956 when Robbins and Lipmann determined the structure of the sulfate donor compound, "active sulfate," as 3′-phosphoadenosine 5′-phosphosulfate (PAPS). The transfer of the sulfonate moiety of PAPS to an acceptor compound to form either a sulfate ester or a sulfamate is catalyzed by the family of enzymes termed sulfotransferases (STs). Examples of the formation of sulfate esters and sulfamates are shown in Fig. 1.

PAPS is probably synthesized in the cytoplasm of all animal cells by a two-step reaction utilizing ATP and inorganic sulfate (Fig. 2). The first step involves the formation of adenosine 5′-phosphosulfate (APS) by ATP-sulfurylase. Inorganic sulfate used in the synthesis of PAPS may come from the diet or from the catabolism of the amino acids cysteine and

Advances in Pharmacology, Volume 27

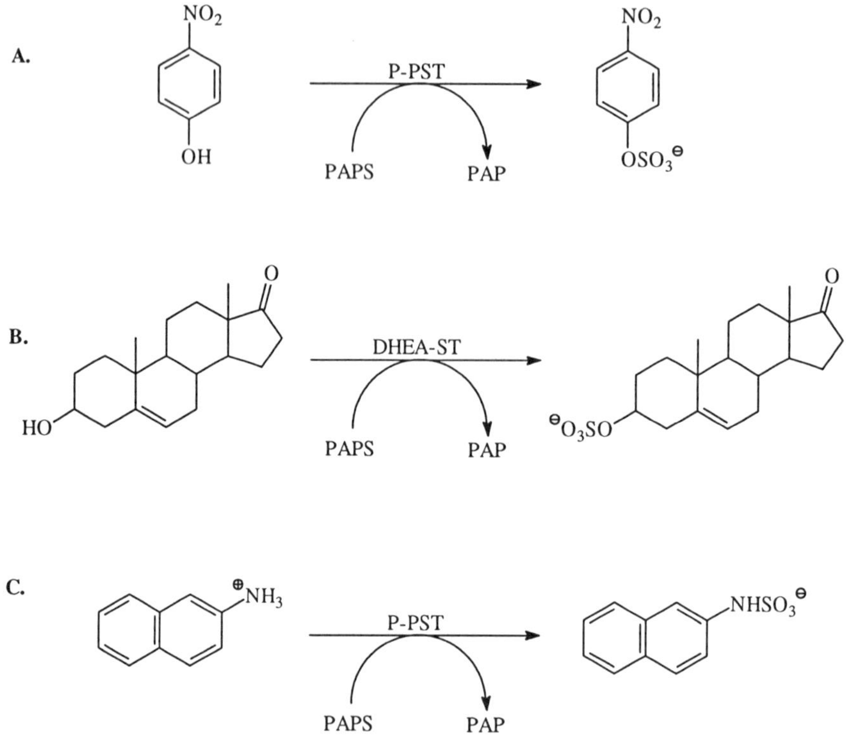

Fig. 1 Examples of reactions catalyzed by human cytosolic sulfotransferases. Reaction (A) is the sulfation of 4-nitrophenol by the phenol sulfating form of phenol sulfotransferase (P-PST). Reaction (B) is the conjugation of dehydroepiandrosterone by human dehydroepiandrosterone sulfotransferase. Reaction (C) shows the formation of 2-naphthylamine sulfamate by P-PST.

methionine. The second step results in the formation of PAPS from APS by APS-kinase. The equilibrium of APS formation is thermodynamically unfavorable, and the steady-state levels of PAPS in a cell may be relatively low. However, normally the two enzymes involved in the synthesis of PAPS are tightly coupled and conversion of APS and ATP to PAPS proceeds rapidly. Also, the pyrophosphate that is formed in the reaction catalyzed by ATP-sulfurylase is hydrolyzed by pyrophosphatase preventing the reversibility of the reaction.

Although PAPS synthesis may be very rapid, the sulfation of drugs in humans may be limited by PAPS availability in certain instances. PAPS synthesis may also be limited by depletion of inorganic sulfate in plasma by high doses of substrates for sulfation or by genetic defects in the synthesis of PAPS (Pang, 1990). The decrease in sulfation due to the

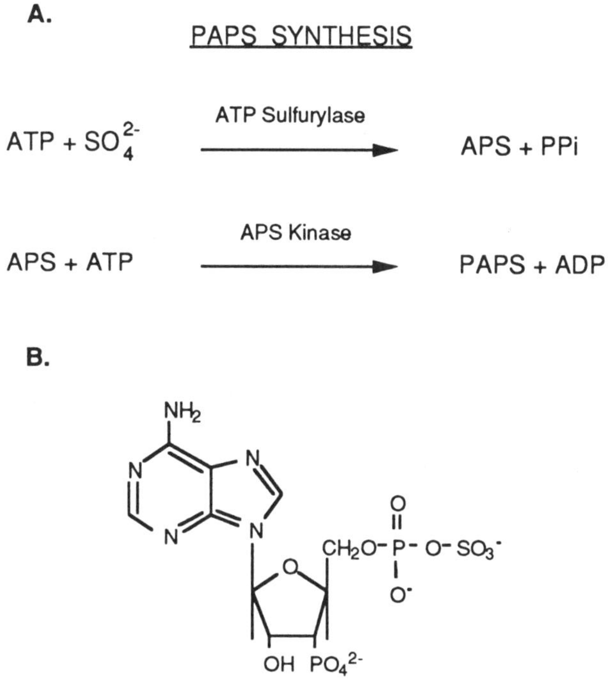

Fig. 2 The synthesis and structure of PAPS. (A) The two-step synthesis of 3′-phosphoadenosine 5′-phosphosulfate (PAPS) from inorganic sulfate and ATP. ATP, adenosine triphosphate; ADP, adenosine diphosphate; APS, adenosine 5′-phosphosulfate. (B) The structure of PAPS.

depletion of PAPS may have a role in the toxicity of some drugs such as acetaminophen.

A. Importance of Sulfation

Although sulfation was initially recognized as a major reaction involved in the biotransformation and excretion of many drug and xenobiotics, recent studies have greatly increased the range of compounds recognized as substrates for conjugation. One reason for the large number of different compounds that are sulfated is that many drugs and xenobiotics are first metabolized by the Phase-1 or oxidative enzyme systems to phenols and alcohols, which may be conjugated with sulfate. Many other compounds already possess the appropriate chemical structures for direct conjugation with sulfate.

The diversity of xenobiotics and drugs that can be conjugated with sulfate has been well established (Mulder, 1981). Sulfation of endogenous compounds is now also recognized as an important process. Examples of endogenous substrates for sulfation include steroids, glycosaminoglycans, tyrosines in proteins and peptides, thyroid hormones, monoamine neurotransmitters, and bile acids.

The major physiologic result of conjugation of small xenobiotic compounds with the charged sulfonate moiety is a decrease in their biologic activity and an increase in their hydrophilicity. The increased water solubility of the sulfate conjugates almost always leads to an increase in their rate of excretion into the urine or bile. These properties represent the major function of sulfation in xenobiotic metabolism. However, the STs are also involved in the less frequent but very important process of bioactivation of certain drugs and xenobiotics. This includes the conversion of certain procarcinogens to potent electrophilic mutagens and carcinogens via sulfate esterification (see Michejda and Kroeger-Koepke, this volume).

B. Bioactivation by Sulfation

A small but growing list of compounds, including *N*-hydroxy-2-acetylaminofluorene (NOHAAF), 1′-hydroxysafrole, *N*-hydroxy-4-aminobiphenyl, and 7-hydroxymethyl-12-methylbenz[*a*]anthracene (HMBA), are reported to be activated to reactive electrophilic metabolites via sulfation (Watabe *et al.*, 1982; Miller *et al.*, 1985; Mulder and Jakoby, 1990; Surh *et al.*, 1990). The mechanism of the activation of these compounds involves the formation of unstable sulfate esters, which spontaneously rearrange to release sulfate and generate electrophilic nitrenium or carbonium ions. These electrophilic metabolites react with cellular nucleophiles to elicit mutagenic or cytotoxic responses, or both. Endogenous nucleophiles such as proteins, DNA, and RNA are targets for the procarcinogens that undergo bioactivation to reactive electrophiles via sulfation. The role of sulfation in the bioactivation of NOHAAF to an electrophilic form capable of binding DNA is shown in Fig. 3.

In addition to the bioactivation of certain carcinogens, sulfation is involved in forming the active species of a widely used therapeutic agent, minoxidil. The *N,O*-sulfate ester of minoxidil is responsible for both the therapeutic antihypertensive (DuCharme *et al.*, 1973) and hair-growth-stimulating properties (Buhl *et al.*, 1990) of minoxidil. This metabolite is very active in terms of its ability to donate the sulfate group nonenzymatically to proteins. In an *in vitro* study, over 50 peptides were screened in order to characterize sulfation by minoxidil sulfate (Groppi *et al.*, 1990). This study revealed that sulfation of peptides preferentially occurred at N-termini and on histidine residues. The sulfated peptides were stable in exponentially growing cells with a half-life of approximately 12 h. It has been suggested that the molecular mechanism by which minoxidil sulfate produces its vascular effects may be due to its intrinsic ability to donate

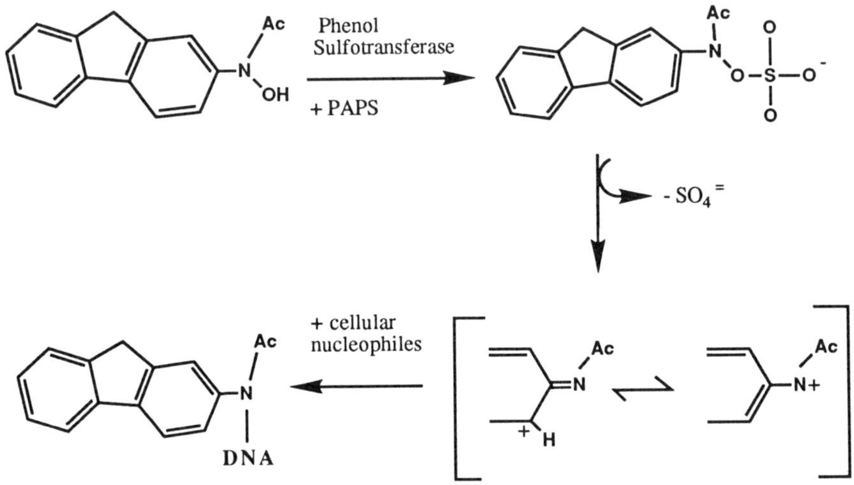

Fig. 3 The role of sulfation in the bioactivation of *N*-hydroxy-2-acetylaminofluorene (NOHAAF) by rat phenol sulfotransferase.

the sulfate group to smooth muscle membrane proteins including a class of ATP-sensitive K^+ channels (Meisheri *et al.*, 1988, 1993). Another interesting characteristic of minoxidil sulfate is related to the increased lipophilicity of the metabolite, compared to the parent compound, due to the formation of an "inner salt" (DuCharme *et al.*, 1973). This finding is in contrast to the general premise that sulfation increases water solubility.

C. Classification of Sulfotransferases

A major factor in the wide diversity of the substrates for sulfation is the multiplicity of STs catalyzing this reaction. The family of STs can be divided into two broad classes: the membrane-bound enzymes located in the Golgi apparatus, which are involved in the sulfation of glucosaminoglycans and tyrosines in proteins, and the cytosolic STs, which are involved with the conjugation of drugs, xenobiotics, and small endogenous compounds. The sulfation of proteins, peptides, and glycosaminoglycans by the membrane-associated Golgi STs has been shown to have important effects on the biologic activity of these compounds, their transit time through the Golgi apparatus, and their rates of secretion from the cell (Roth, 1986; Huttner, 1988). However, a role for the Golgi STs in the metabolism of small xenobiotic or endogenous compounds has not been described, and these enzymes are most likely not involved in the bioactivation of mutagens and carcinogens by sulfation. This chapter will, therefore, be limited to a discussion to the biochemistry of the cytosolic STs that appear to be involved in the conjugation and activation of small xenobiotic substrates to reactive electrophiles.

II. Classes of Cytosolic Sulfotransferase Involved in the Bioactivation of Xenobiotics

The involvement of sulfation in the bioactivation of several small xenobiotic compounds was initially well established by the laboratory of James and Elizabeth Miller (Miller *et al.*, 1985). These studies demonstrated that sulfation by cytosolic PAPS-dependent STs was involved in the bioactivation of several carcinogenic and mutagenic compounds, including NOHAAF and 1′-hydroxysafrole, but the identity of the specific enzymes involved in the sulfation of the different substrates is still being established.

The cytosolic STs known to be involved in the bioactivation of procarcinogens may be divided into two classes, the hydroxysteroid or alcohol STs and the phenol STs, based on kinetic, immunologic, and structural similarities including nucleotide and amino acid sequences. Both classes of enzymes contain multiple isoforms of related enzymes which may or may not have similar properties or overlapping substrate reactivities. An important direction for the investigation of the STs is to determine and characterize the isoforms of these enzymes involved in the sulfation and generation of reactive electrophiles from procarcinogens identified in the animal studies. A major factor in this analysis has been the investigation of the heterogeneity of the STs in rodent tissues. Purification and characterization studies have indicated that both the hydroxysteroid and the phenol ST families in rat liver consist of multiple closely related isoforms. Within these families the isoforms may be very closely related at both the amino acid and the nucleotide sequence levels.

As shown in Fig. 4, analysis of the nucleotide sequences of the published cytosolic ST cDNA sequences shows the delineation between the two families. The available amino acid sequences were compared by sequential pairwise alignment with the University of Wisconsin Genetic Computer Groups (UWGCG) program, Pileup (Devereux *et al.*, 1984). The cDNAs encoding members of the hydroxysteroid ST subfamily include rSTa (Ogura *et al.*, 1990a), rST20 (Ogura *et al.*, 1989), rSMP-2 (Chatterjee *et al.*, 1987), and hDHEA-ST (Otterness *et al.*, 1992; Comer *et al.*, 1993). Interspecies conservation of amino acid sequence of members of this subfamily is demonstrated by the considerable homology (60–63%) found to exist between hDHEA-ST and the rat enzymes. The cDNAs encoding members of the phenol ST subfamily include rat estrogen ST (rEST) (Demyan *et al.*, 1992), bovine estrogen ST (bEST) (Nash *et al.*, 1988), rat

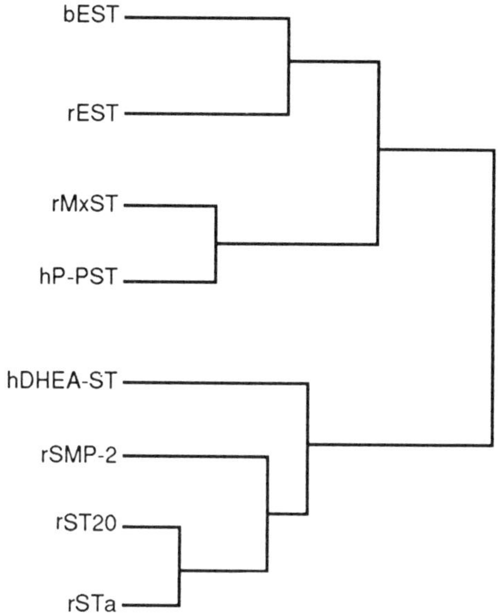

Fig. 4 Dendogram generated from sequence alignment of sulfotransferase cDNAs. The UWGCG program Pileup, creates a multiple sequence alignment beginning with pairwise alignment of the two most similar sequences, producing a cluster of two aligned sequences (Devereux *et al.*, 1984). This cluster is then aligned to the next most related sequence or cluster of aligned sequences. The final alignment is made by a series of progressive, pairwise alignments that are included in the final alignment. Although this is not a phylogenetic tree, the distance along the vertical axis is proportional to the difference between sequences.

minoxidial ST (rMx-ST) (Hirshey *et al.*, 1992), also known as PST-1 (Ozawa *et al.*, 1990), and the human phenol-sulfating form of phenol ST (hP-PST-1) (Wilborn *et al.*, 1993). Although the estrogen STs are more closely related to the phenol STs than to the hydroxysteroid STs and are included in this subfamily, it is not yet known whether the estrogen STs are capable of conjugating procarcinogens believed to be activated by sulfation.

A. Cytosolic Hydroxysteroid Sulfotransferases

1. Rodent Hydroxysteroid Sulfotransferases

Polyaromatic hydrocarbons (PAH), such as dimethylbenz[*a*]anthracene (DMBA) and benzo[*a*]pyrene, are potent mutagens and carcinogens that may be bioactivated to reactive electrophiles by several pathways. One pathway for the activation of a hydroxymethyl metabolite of DMBA is shown in Fig. 5. DMBA is metabolized via the mixed-function-oxidase

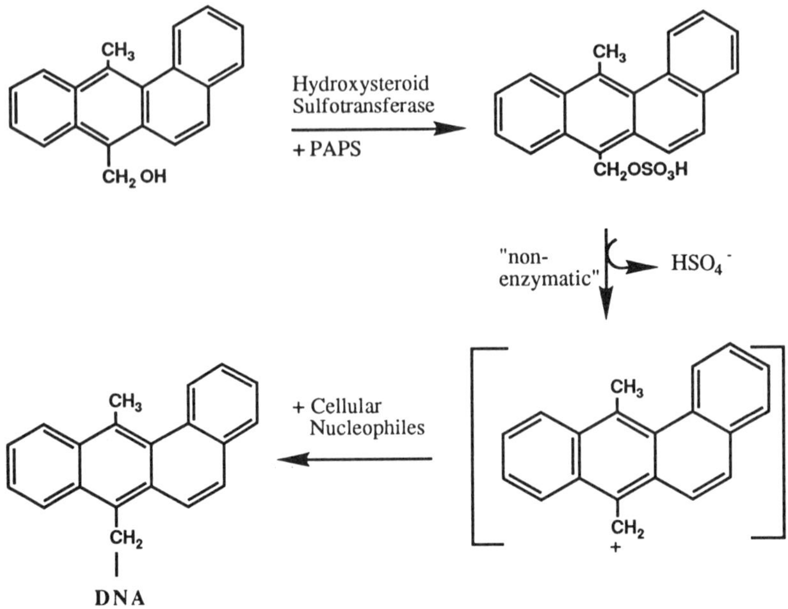

Fig. 5 The role of sulfation in the bioactivation of 7-hydroxymethyl-12-methylbenz-[*a*]anthracene by rat hydroxysteroid sulfotransferase activity.

system to the 7-hydroxymethyl metabolite, which is subsequently esterified with sulfate, acetate, or phosphate (Flesher and Sydnor, 1971, 1973). Several reports from the laboratories of Watabe (Watabe *et al.*, 1982, 1985) and the Millers (Surh *et al.*, 1987, 1989) have established the role of sulfation of hydroxymethyl PAHS in the *in vivo* formation of DNA adducts.

In these studies the characteristics of the ST activity involved in the formation of the DNA adducts from the hydroxymethyl PAHs were similar to the properties of the hydroxysteroid STs. The rat liver ST activity responsible for the sulfation of hydroxymethyl PAHs and the *in vivo* formation of DNA adducts was relatively insensitive to inhibition by pentachlorophenol (PCP), a potent inhibitor of rat phenol ST activity (Mulder and Scholtens, 1977). Also, the bioactivation of 7-hydroxymethyl, 12-methylbenz[a]anthracene (HMBA) was more readily inhibited by dehydroepiandrosterone (DHEA), an important substrate and inhibitor of hydroxysteroid ST activity (Surh *et al.*, 1987, 1989).

The observation that the ST activity responsible for the sulfation of HMBA might also be responsible for the sulfation of hydroxysteroids was supported by the age- and sex-related changes observed in these activities. Hydroxymethyl PAH sulfation and hydroxysteroid ST activity are approximately equally expressed in the livers of young rats and mice; however, after puberty both ST activities are expressed at much higher levels in females than in males (Miller *et al.*, 1991). The expression of the STs in rodents is regulated in part by steroids. In male and female rats treatment with estrogens significantly induced hydroxymethyl PAH sulfation activity and hydroxysteroid ST activity. Treatment of animals with testosterone markedly lowers these activities (Surh *et al.*, 1991).

The role of hydroxysteroid STs in the bioactivation of hydroxymethyl PAHs has been confirmed by the demonstration that purified forms of rat liver hydroxysteroid ST are capable of catalyzing the reaction *in vitro*. Two forms of purified rat liver hydroxysteroid ST have been shown to sulfate and activate hydroxymethyl PAHs to reactive electrophiles. Ogura *et al.* (1990b) purified rSTa from female rat liver cytosol and demonstrated that the enzyme was capable of sulfating several hydroxymethyl PAHs, as well as DHEA. During the purification procedure, rSTa was resolved from other peaks of hydroxysteroid ST activity by anion-exchange chromatography. Falany *et al.* (1992), using purified bile acid sulfotransferase 1 (rBAST 1) from female rat liver cytosol, showed that rBAST 1 was able to sulfate and bioactivate HMBA. Rat BAST 1 was originally purified by Barnes *et al.* (1989) because of its ability to sulfate bile acids such as lithocholic acid. The pure enzyme was also shown to sulfate steroids, such as DHEA and cortisol. The levels of rBAST 1 activity and immunoreactivity were also shown to be 25- to 30-fold greater in female rat liver cytosol than in male rat liver cytosol.

a. Molecular Characterization of Rat Hydroxysteroid Sulfotransferases Comparison of the amino terminal sequences of rSTa and rBAST 1 indicates there is a great deal of homology between these proteins. The N-terminus of pure rBAST 1 has been reported (Barnes *et al.*, 1989), whereas the full-length cDNA for rSTa has been isolated and sequenced (Ogura *et al.*, 1990a). Comparison of the amino-terminal sequences shows that these two hydroxysteroid STs are approximately 90% similar. Another rat liver cDNA, the senescence marker protein-2, rSMP-2 (Chatterjee *et al.*, 1987), has been isolated and is closely related to rSTa and rBAST 1 based on sequence similarities. The deduced amino acid sequence of rSTa is 74% homologous with the translation product of rSMP-2. The biologic or kinetic properties of rSMP-2 have not been reported; however, the purified protein demonstrates hormonal regulation similar to rBAST 1 and rSTa. Based on sequence analysis and comparison to rSTa, SMP-2 has been suggested to be another form of rat hydroxysteroid ST (Ogura *et al.*, 1990a).

Rat SMP-2 was isolated in an investigation of age-related changes in hepatic protein expression associated with androgen sensitivity (Chatterjee *et al.*, 1987). Analysis of approximately 2.2 kb of the 5′ flanking sequence of several rSMP-2-related genes has provided initial insight into the molecular mechanism involved in the expression of this protein (Song *et al.*, 1989). The cis-acting regulatory elements found in the promoter region included the palindromic glucocorticoid-response element (GRE), a half-palindromic estrogen-response element (ERE), and several motifs for a half-palindromic androgen-response element (ARE). Although the functional significance of the GRE and ERE are not known, the presence of half-palindromic AREs is consistent with the report that rSMP-2 is predominantly under androgenic repression (Chatterjee *et al.*, 1987).

The isolation of the closely related hydroxysteroid ST cDNAs and proteins has led to the concept that a family of closely distinct genes encoding these enzymes is present in rats. Whereas allelic forms of the individual genes may be present, the different closely related hydroxysteroid STs probably also represent microheterogeneity at the genome level. To investigate this possibility, restriction-digested DNA isolated from Sprague–Dawley rat liver was probed with a ^{32}P-labeled rat hydroxysteroid ST cDNA, rST2-4. ST2-4 was isolated from a male Sprague–Dawley rat liver λgt11 cDNA library and is 90 and 87% similar to STa and SMP-2, respectively, in amino acid sequence. Figure 6 shows that a relatively large number of DNA fragments are detected during Southern blotting of restriction enzyme-digested rat liver DNA. In digests including restriction enzymes with 6-base recognition sites, 8–10 DNA fragments hybridize with the hydroxysteroid ST cDNA. This result supports the idea that there are several genes encoding closely related hydroxysteroid STs in rats. The relationship between these genes and the protein products have only begun to be described; however, to date only rSTa and rBAST 1 have been shown to sulfate and activate hydroxymethyl PAHs. However, it is likely that other forms of the closely related hydroxysteroid STs are also involved in this process in rats.

2. Human Cytosolic Hydroxysteroid Sulfotransferase

Although multiple forms of hydroxysteroid ST have been identified and purified from rat liver, in human tissues only a single form of the enzyme has been purified and characterized. Initial reports by Adams and McDonald (1979, 1980) characterized hydroxysteroid ST activity in human fetal and adult adrenal tissue. This enzyme activity displayed high activity with DHEA, but was also capable of conjugating several other hydroxysteroids. In contrast to rodents, DHEA sulfate is a major secretory product of both human fetal and adult adrenals. DHEA sulfate in the circulation can be converted into both androgens and estrogens in peripheral tissues (Baxter and Tyrrell, 1987). The high levels of estrogen sulfates detected in the plasma during pregnancy are derived from DHEA sulfate secreted by the fetal adrenal.

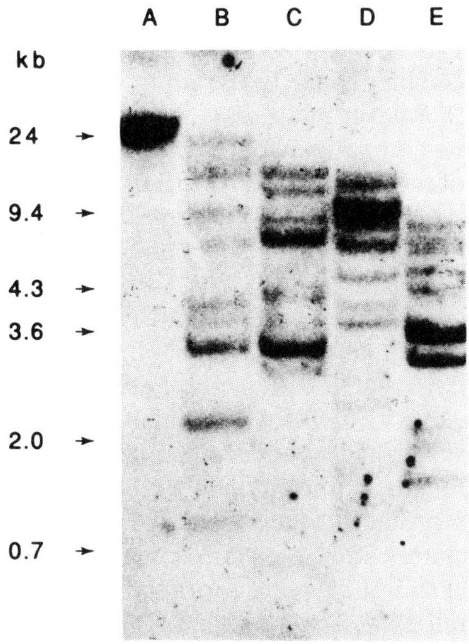

Fig. 6 Southern blot analysis of rat liver DNA probed with hydroxysteroid sulfotransferase cDNA. High-molecular-weight genomic DNA obtained from rat liver (lane A) was resolved in a 0.8% agarose gel adjacent to DNA digested with an excess of *Bam*HI (lane B), *Eco*RI (lane C), *Hin*dIII (lane D), and *Taq*I (lane E). The DNA was transferred to a nylon membrane, hybridized with 1.5×10^6 dpm/ml of [^{32}P] nick-translated rST2-4, washed under high stringency conditions, and exposed to autoradiograph film for 42 h.

Hydroxysteroid sulfation is also a prominent process in human liver. The liver form of hDHEA-ST has been purified (Falany *et al.*, 1989) and shown to sulfate a number of hydroxysteroids, including androsterone, pregnenolone, β-estradiol, estrone, and testosterone. Liver hDHEA-ST activity displayed substrate inhibition with increasing concentrations of DHEA and showed maximal activity at approximately 3 μM. Radominska *et al.* (1990) have reported that hDHEA-ST is responsible for most, if not all, of the bile acid ST activity in human liver. This is consistent with the similarity in the structure of hydroxysteroids and bile acids and with the observation that the rat hydroxysteroid STs conjugate bile acids (Barnes *et al.*, 1989). Although differences have been reported in the kinetic properties of the adrenal and liver forms of hDHEA-ST, Comer and Falany (1992) have shown that the adrenal enzyme is immunologically, structurally, and kinetically very similar to the liver form of hDHEA-ST.

Molecular cloning of DHEA-ST has been reported (Otterness *et al.*, 1992; Comer *et al.*, 1993) and the active enzyme has been expressed in COS Green monkey kidney cells. The enzyme expressed in COS cells reacts with rabbit anti-DHEA-ST antibodies and migrates during SDS–PAGE with the same molecular mass as the liver enzyme (Comer and Falany, 1992; Comer *et al.*, 1993). Southern blot analysis of restriction enzyme-digested human liver DNA with the DHEA-ST cDNA detected very few fragments compared with the multiple bands observed in Southern blot analysis of rat genomic DNA with a rat hydroxysteroid ST cDNA (Fig. 6) (Comer *et al.*, 1993). Cloning and expression studies of hDHEA-ST have not suggested the presence of more than one hydroxysteroid ST in human tissues.

Rat and human hydroxysteroid STs share many kinetic and immunological properties. Hydroxysteroid STs from both species are capable of conjugating hydroxysteroids and bile acids, and rabbit antibodies raised to rBAST 1 cross-react with hDHEA-ST (Falany *et al.*, 1992). This similarity is supported by amino acid sequence comparisons of the cloned rat and human enzymes. Figure 7 shows the alignment of the protein sequences of rSTa with the protein sequence of hDHEA-ST. Human DHEA-ST displays approximately 78% sequence similarity and 63% sequence identity with rSTa and rST-20, an allelic form of rSTa (Ogura *et al.*, 1989). In addition, an overall 79% sequence similarity and 61% sequence identity was observed between rSMP-2 and hDHEA-ST.

Although hDHEA-ST is very similar to the rat hydroxysteroid STs, no reports have appeared describing the ability of human enzyme to sulfate

```
                 .         .         .         .         .
rSTa           MPDYTWFEGIPFHAFGISKETLQNVCNKFVVKDEDLILLAYPKSGTNWL  49
               .|:  |||||:|..:|:..|||..|  :.||::|||:|:|.||||||||||
hDHEA-ST      MSDDFLWFEGIAFPTMGFRSETLRKVRDEFVIRDEDVIILTYPKSGTNWL  50

                 .         .         .         .         .
              IEIVCLIQTKGDPKWIQSVTIWDRSPWIETDVGYDILIKKKGPRLMTSHL  99
               ||:||::.|||:||||||.||:||||:|.::||. |  ...:||||:.|||
              AEILCLMHSKGDAKWIQSVPIWERSPWVESEIGYTALSESESPRLFSSHL  100

                 .         .         .         .         .
              PMHLFSKSLFSSKAKVIYLVRNPRDVLVSGYYFWGNSTLAKKPDSLGTYV  149
              |::||.||:|||||||||||:|||||||||||||:||  |  .:  |||.|::.|.
              PIQLFPKSFFSSKAKVIYLMRNPRDVLVSGYFFWKNMKFIKKPKSWEEYF  150

                 .         .         .         .         .
              EWFLKGNVLYGSWFEHIRAWLSMQEWDNFLLLYYEDMKKDTMGTIKKICD  199
              |||  .|.||||||||:||::|:.|.|  .|||||  ||::|.||   ||.|||:
              EWFCQGTVLYGSWFDHIHGWMPMREEKNFLLLSYEELKQDTGRTIEKICQ  200

                 .         .         .         .         .
              FLGKKLEPDELDLVLKYSSFQVMKENDMSNYSLLMKKSIFTGTGLMRKGT  249
              ||||.|||:||:|:||  ||||  ||||.||||||||   .  :..  ..|:|||.
              FLGKTLEPEELNLILKNSSFQSMKENKMSNYSLLSVDYVVDKAQLLRKGV  250

                 .         .         .         .
              VGDWKNHFTVSQAEAFDKVFQEKMAGFPPGMFPWE                 299
               |||||||||.|||.|||:||||||::|.::||||
              SGDWKNHFTVAQAEDFDKLFQEKMADLPRELFPWE                 286
```

Fig. 7 Amino acid sequence comparison of human liver DHEA-ST with rat liver hydroxysteroid sulfotransferase, STa. The GAP program (Devereux *et al.*, 1984) was used to optimize the sequence alignment of the deduced amino acid sequence of h-DHEA-ST (Comer *et al.*, 1993) and rSTa (Ogura *et al.*, 1989). Vertical lines indicate seqence identity, two dots indicate highly conservative changes, and one dot indicates conservative changes.

and bioactivate hydroxymethyl PAHs to reactive electrophiles. An in-depth investigation of this process has not been reported. One difference between the kinetic properties of the rat hydroxysteroid STs and hDHEA-ST is the reported inability of the human enzyme to sulfate cortisol (11β,17α,21-trihydroxypreg-4-nene-3,20-dione) (Comer and Falany, 1992). Cortisol is sulfated at the 21-position by the rat hydroxysteroid STs (Hobkirk, 1985; Barnes *et al.,* 1989). This result suggests that hDHEA-ST may have an altered substrate reactivity compared with the rat hydroxysteroid STs. The recent cloning and expression of hDHEA-ST, however, provide the tools for the detailed analysis and characterization of the properties of this enzyme, including the determination of its possible role in the bioactivation of hydroxymethyl PAH carcinogens.

An additional difference between rat and human hydroxysteroid ST activities is related to the regulation of their expression. Hydroxysteroid ST activity is expressed at much higher levels in female rats compared to males (Hobkirk, 1985; Miller *et al.,* 1991). Human DHEA-ST does not appear to be subject to the same degree of regulation by hormonal influences as do the ST activities in rodents (Falany, 1991), since in humans, DHEA-ST activity is equally expressed in both sexes (Otterness *et al.,* 1992). The high levels of DHEA-ST activity in human adrenals and the consistent expression of DHEA-ST activity in liver suggest that the hydroxysteroid STs differ, at least in part, in the functions that they perform in rats and humans.

B. Cytosolic Phenol/Hydroxyarylamine Sulfotransferases

The second major family of cytosolic STs that are involved in the bioactivation of procarcinogens is the phenol STs. This family of enzymes consists of two groups of STs, termed phenol STs and the estrogen STs. Much less is known about the estrogen STs, although the cloning of the bovine and rat enzymes will provide important probes for their investigation. The estrogen STs have been included in the larger family of phenol STs based on the ability of both groups of STs to sulfate the 3-phenolic position of estrogens and their sequence similarity to the phenol STs (Fig. 4). No evidence that the estrogen STs are involved in the bioactivation of carcinogens has been reported. It is of interest, however, that the phenol STs, such as hP-PST, are also capable of conjugating estrogens, although usually with K_m values in the micromolar range (Hernandez *et al.,* 1992).

The phenol STs as a family have been demonstrated to conjugate a variety of phenolic compounds as well as a number of different types of compounds. The substrate classes for these enzymes include simple phenols, tyrosines in peptides, *N*-oxides, aromatic amines, benzylic alcohols, arylhydroxylamines, and thyroid hormones. The substrates for the phenol STs include compounds that have been identified as procarcinogens, such as the alkenylbenzenes, arylhydroxylamines, and arylhydroxylamides (Miller *et al.*, 1985; Abu-Zeid *et al.*, 1992). However, the variety of different types of compounds conjugated by the phenol STs has hampered the identification and purification of these enzymes from different species. Most of the initial studies identifying the role of phenol ST activity in the bioactivation of procarcinogens were performed in rodents that possess multiple members of the family with different but overlapping substrate specificities (Jakoby *et al.*, 1980).

1. Rodent Phenol Sulfotransferases One of the major areas of investigation in sulfation has been the bioactivation of arylhydroxylamines and arylhydroxylamides by sulfate esterification. The target of much of this effort has been the investigation of the bioactivation of NOHAAF as a model compound for the role of sulfation in mutagenesis and carcinogenesis. The involvement of the phenol STs in the bioactivation of arylhydroxylamines and alkenylbenzenes has been investigated primarily in rats and mice. As will be discussed, NOHAAF sulfation activity has been used for the identification and purification of a number of rat phenol STs. The metabolic activation of NOHAAF to a mutagen involves several enzymatic reactions. Formation of *N*-sulfooxy-2-acetylaminofluorene by phenol ST activity is only one of these mechanisms and its importance in the generation of DNA adducts and tumorigenesis apparently varies with the age, sex, and species involved.

In rats and mice, the levels of phenol ST activity in liver cytosol display differences with respect to the age and sex of the animals being investigated. The overall rates of phenol ST and NOHAAF ST activity are higher in male rats than female rats (Wu and Straub, 1976; Gong *et al.*, 1991). Also, the susceptibility of rats to NOHAAF-induced hepatocarcinogenesis correlates with the level of NOHAAF ST activity (Wu and Straub, 1976). The association between phenol ST and NOHAAF activities in the activation of NOHAAF has been supported by the observations that potent inhibitors of phenol ST activity, such as PCP, decreased the bioactivation of NOHAAF both *in vivo* and *in vitro* (Miller *et al.*, 1985). The association between phenol ST and NOHAAF ST activities was subsequently established in purification studies with these substrates as markers for enzyme activity (Wu and Straub, 1976; Sekura and Jakoby, 1981; Gong *et al.*, 1991).

The initial investigations into the ability of rats to bioactivate NOHAAF demonstrated a distinct sex-related difference in that male rats displayed higher activity than females (Wu and Straub, 1976; Sekura and Jakoby, 1979). The differences observed between the sexes were also observed in the sulfation of small phenols such as 1-naphthol and *p*-nitrophenol. The high levels of phenol ST activity in male rats as compared with females is in contrast to the higher levels of hydroxysteroid ST activity in females than males. Many of the studies describing the purification of phenol ST activities have used male rat liver as a source of the enzyme activity due to the high levels of NOHAAF ST activity. However, it should be noted that with an increased understanding of the properties of the individual rat phenol STs, it is apparent that the levels of the expression of the individual isoforms of phenol ST may vary greatly (Sekura and Jakoby, 1979; Hirshey and Falany, 1990; Gong *et al.*, 1991). The initial studies indicated that the expression of NOHAAF ST activity was regulated in part by steroid hormones. DeBaun *et al.* (1970) reported that female rats were relatively resistant to the hepatocarcinogenicity of NOHAAF compared to male rats. Testosterone propionate administration resulted in 50–100% increases in NOHAAF ST activity in female ovariectomized rats, but had little effect on enzyme activity in castrated male rats. Larger effects were observed when normal and gonadectomized rats were treated with β-estradiol; NOHAAF ST activity was decreased to 20–50% of controls in both normal and gonadectomized rats.

A role for growth hormone in the regulation of phenol ST and NOHAAF ST activities in rats has been reported (Gong *et al.*, 1991). Yamazoe and co-workers (1987) have shown that steroid hormones may not directly regulate ST expression but may act via effects on the release of growth hormone (GH). The pattern of secretion of GH apparently has an important role in regulating the expression of phenol ST activities in rat liver. In male rats, GH is secreted in a pulsatile pattern, whereas in females the secretion is more constant. In hypophysectomized rats, the intermittent injection of GH was very effective at restoring NOHAAF ST activity. The ability of testosterone to regulate NOHAAF ST activities has been proposed to have an effect by influencing the pituitary to secrete GH in a male pulsatile pattern (Gong *et al.*, 1991). The analysis of the regulation of NOHAAF and phenol ST activities, however, is complicated by the presence of multiple forms of these enzymes in rats that apparently respond differently to regulation by steroid and pituitary hormones.

The involvement of sulfation and phenol ST activity in the bioactivation of *N*-hydroxy aromatic amines and alkenylbenzenes was established by two processes. One involved animal studies investigating the correlations between age, sex, inhibition of ST activities, and decreases in procarcinogen activation in brachymorphic mice that are defective in PAPS synthesis. The second approach involved the purification and, more recently, the cloning of the STs involved in catalyzing these reactions.

a. Characterization of Rat Phenol Sulfotransferases Involved in Bioactivation Wu and Straub (1976) purified a form of cytosolic ST from male rats that was capable of sulfating NOHAAF and 4-nitrophenol (PNP). Only trace amounts of this enzyme were detected in female rat liver cytosol. The purified enzyme had a subunit molecular mass of 38,250 Da and the active form of the enzyme existed as a dimer. Sekura and Jacoby (1979, 1981) subsequently isolated four forms of phenol ST from male rat liver cytosol. These isoforms of phenol ST were named aryl sulfotransferase (AST) I, II, III, and IV based on their elution during anion-exchange chromatography. The four ASTs were divided into two groups. AST I and II represent one group since these activities displayed similar pH optima for activity and similar substrate specificity for phenols. AST I and II were capable of sulfating phenol but with $K_m$values in range 1.8–2.5 m*M*. Antibodies raised to either AST I or II react with the other enzyme. AST I and II were also reported not to sulfate catecholamines and hydroxylamines including NOHAAF, and their roles in bioactivation require further investigation.

AST III and IV were considered to be a separate group of enzymes based on several kinetic and immunologic properties. AST III, however, has not been well characterized because of its instability during purification. AST III and IV displayed more acidic pH maxima for 2-naphthol sulfation (pH 5.5 vs 6.5) compared with AST I and II. Antibodies raised against AST IV do not cross-react with either AST I or II. AST IV is capable of sulfating amino-terminal tyrosines in small peptides and aromatic hydroxylamines (Sekura and Jacoby, 1981). The enzyme purified by Wu and Straub (1976) may represent either AST III or IV or a mixture of these two enzymes. Gilissen *et al.* (1992) have reported that pure AST IV is capable of conjugating several different hydroxamic acids, including NOHAAF, but the pure enzyme was not capable of sulfating all the hydroxylamines tested as substrates. AST IV did conjugate *N*-hydroxy-2-amino-5-phenyl-pyridine and *N*-hydroxy-2-amino-3-methyl-5-phenyl-pyridine. Gilissen *et al.* (1992) suggest that comparison of the activity of pure AST IV with male and female rat liver cytosol with a variety of *N*-hydroxylamine and *N*-hydroxaminic acids indicates that multiple STs in rat liver are involved in their bioactivation.

Further studies of the activity and immunolocalization of AST IV by Duffel and co-workers (1991; Binder and Duffel, 1988) have shown that the enzyme can also catalyze the sulfation of benzylic alcohols, such as 1′-hydroxysafrole and 1-naphthalenemethanol. Immunologic characterization of AST IV in rat liver has shown that the enzyme is more abundant in centrilobular cells than in midzonal or periportal cells (Duffel *et al.*, 1991). In rats fed 2-acetylaminofluorene (2-AAF), more cellular necrosis is observed in the centrilobular cells than in other liver areas upon histologic examination (Groothius *et al.,* 1983). These results are consistant with AST IV being a major form of phenol ST involved in the bioactivation of *N*-hydroxylated arylamines in rats.

A phenol ST was subsequently purified and characterized from male rat liver cytosol by Borchardt and Schasteen (1982). This phenol ST shows some similarities to AST IV in kinetic properties; both enzymes conjugate small phenols, such as PNP and 2-naphthol, and catechloamines, such as dopamine and epinephrine; however, the ability of this phenol ST to sulfate NOHAAF was not reported. A number of differences were also noted between this PST and AST IV including subunit molecular mass, the K_m for PAPS, the inhibition constant for PAP, and inhibition by NaCl. This phenol ST and AST IV are apparently closely related but the role of the PST in the study by Borchardt and Schasteen (1982) in bioactivation is not known.

The purification of a phenol ST from male Sprague–Dawley rats has been described by Hirshey and Falany (1990). This phenol ST was purified on the basis of its ability to sulfate minoxidil to form an *N,O*-sulfate ester and has been named minoxidil-ST (Mx-ST). This ST is apparently responsible for the majority of minoxidil sulfation in rat liver cytosol. Pure Mx-ST is capable of sulfating PNP but maximal activity was observed with relatively low PNP concentrations (1.2 μM); above this concentration, potent substrate inhibition was observed. Mx-ST possessed a subunit molecular mass of approximately 35,000 Da, as determined by SDS–PAGE, and an isoelectric point of 4.8. This isoelectric point is slightly more acidic than the values reported for AST I–IV (Sekura and Jakoby, 1979, 1981). The ability of pure Mx-ST to sulfate NOHAAF has not been tested; however, Mx-ST activity in rat liver cytosol is resolved from the majority of NOHAAF sulfation during anion-exchange chromatography (Fig. 8).

Rabbit polyclonal antibodies raised against Mx-ST detect a single band during Western blot analysis of male and female rat liver cytosol; immunoreactive Mx-ST is slightly more abundant in males than in females. This slight difference in immunoreactive Mx-ST between males and females is similar to the two-fold greater level of minoxidil sulfation in male rat liver cytosol compared with female rat liver cytosol (Hirshey and Falany, 1990). No immunoreactive proteins were observed in cytosol prepared from male rat kidney, intestine, lung, or brain. These tissues did not possess Mx-ST activity but did possess PST activity. Immunohistochemical localization studies detected Mx-ST in hepatocytes and in the outer root sheath cells of rat and mouse hair follicles (Dooley *et al.,* 1991). Mx-ST activity has also been detected in hair follicles (Hamamoto and Mori, 1989), and the

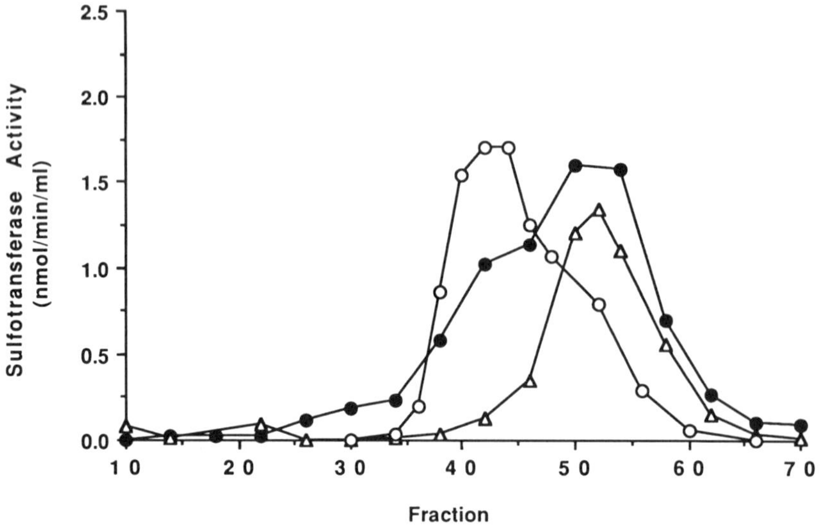

Fig. 8 Resolution of rat liver minoxidil, *p*-nitrophenol, and *N*-hydroxy-2-acetylaminofluorene sulfotransferase activities during DEAE-Sepharose Cl-6B chromatography. Cytosol prepared from male rat liver (Hirshey and Falany, 1990) was applied to a DEAE-Sepharose Cl-6B column and ST activities were eluted with a gradient of 50–200 m*M* NaCl in 10 m*M* Tris–HCl, pH 7.4, 0.25 *M* sucrose, 1 m*M* DTT, and 10% glycerol. Assays were performed as described previously (Hirshey and Falany, 1990; Falany *et al.*, 1992) with 1.2 μM PNP, 1 m*M* minoxidil, and 7.5 μM NOHAAF. Solid circles indicate PNP sulfation activity, open triangles indicate minoxidil sulfation activity, and open circles indicate NOHAAF sulfation activity ($\times 10$).

presence of Mx-ST activity in hair follicles may be important in the hair growth-stimulating action of minoxidil (Meisheri *et al.*, 1993).

Coughtrie and Sharp (1990) have isolated a PST from male rat liver cytosol by following paracetamol sulfation activity. The pure enzyme had a subunit molecular mass of 35,000 Da and was capable of conjugating 1-naphthol and 1-naphthalenemethanol. Antibodies raised against the pure protein recognized only a single polypeptide during Western blot analysis of rat liver cytosol. The tissue distribution of paracetamol ST is similar to that observed for Mx-ST, including the approximately two-fold difference between enzyme levels in male and female rats. These similarities suggest that paracetamol ST and Mx-ST may represent the same or very similar enzymes (Coughtrie and Sharp, 1990; Hirshey and Falany, 1990).

Two STs capable of sulfating NOHAAF have recently been purified from male rat liver cytosol by Gong *et al.* (1991). These STs have been termed hydroxyamine ST I and II (HAST). A separate phenol ST, named

PST-1, capable of conjugating NOHAAF but at a relatively low rate compared with the HASTs, was also isolated during these studies. All three enzymes comigrated during SDS–PAGE with an apparent molecular mass of 33,000 Da. Antibodies raised to PST-1 displayed little cross-reactivity with the HASTs, indicating some degree of dissimilarity between these enzymes. The sex-related expression of these enzymes in rats was also investigated. The expression HAST I was male-dominant and HAST II was male-specific; HAST I is present in higher levels in males than in females, whereas very little HAST II is detectable in female rats compared with males. The expression of both enzymes was increased in hypophysectomized male rats by the male-pattern intermittent injection of GH. PST I, in contrast to the HASTs, showed no clear sex-related difference in hepatic content and its expression was not affected by GH treatment (Gong *et al.*, 1991).

Based on a comparison of kinetic and physical properties, AST IV apparently represents one of the two forms of HAST. The other form of HAST could correspond to the poorly characterized AST III. Based on elution patterns during anion-exhange chromatography it is likely that HAST II is similar to AST IV (Jakoby *et al.*, 1980; Gong *et al.*, 1991). The PST I purifed by Gong *et al.* (1991) may correspond to Mx-ST or paracetamol ST. Both PST I and Mx-ST elute during anion-exchange chromatography after the majority of NOHAAF sulfation activity and are relatively well expressed in male and female rats; antibodies raised to Mx-ST, paracetamol-ST, or PST-I do not readily cross-react with other rat STs (Coughtrie and Sharp, 1990; Hirshey and Falany, 1990; Gong *et al.*, 1991). This interpretation would allow for five to six identified members of the PST family in rats. Three members of this family would have roles in bioactivation of drugs and carcinogens, HAST I and II (AST III and IV), and Mx-ST/PST-I/paracetamol-ST, although the involvement of the individual enzymes in the bioactivation of the different procarcinogens remains to be established.

b. Molecular Cloning of Rat Phenol STs The first molecular cloning of a rat PST was reported by Ozawa *et al.* (1990); however, little supporting evidence was presented with the cDNA sequence to identify the activity or characteristics of the translated protein. The research group that reported the PST-I cDNA sequence subsequently reported the purification of the PST I protein (Ozawa *et al.*, 1990). The PST I cDNA was subsequently isolated by Chen *et al.* (1992) with specific oligonucleotide primers and PCR, and the resulting amplimers were subcloned into a bacterial expression vector and expressed in *E. coli* DH5α cells. The expressed PST I cDNA generated two recombinant products. Both recombinant proteins possessed approximately 33,500-Da subunits, reacted witt rabbit anti-AST

IV antibodies, and conjugated 2-naphthol, 1-naphthalenemethanol, and tyrosine methyl ester; however, the recombinant proteins showed significantly different kinetic parameters when directly compared with pure AST IV. Prior to the report of Chen *et al.* (1992), a report describing the cloning of AST IV (Yerokun *et al.*, 1992) and a report of the cloning and expression of Mx-ST (Hirshey *et al.*, 1992) were published. Both of the cDNAs in these reports also have open reading frames identical to that of PST I.

The cDNA for Mx-ST was isolated from a female rat liver cDNA library to decrease the likelihood of isolating a cDNA for AST III or IV (Hirshey *et al.*, 1992). Two cDNAs were isolated, and the larger cDNA possessed an open-reading frame encoding a 33,909-Da polypeptide identical to the translation of PST I. Expression of the cDNA in COS-1 cells gave rise to a protein that comigrated with Mx-ST during SDS–PAGE, reacted with rabbit anti-Mx-ST antibodies, and was capable of conjugating PNP and minoxidil. Northern blot analysis of poly(A) RNA from male and female Sprague–Dawley rat livers showed about twofold more message in males than females. Immunoreactive Mx-ST is also approximately two-fold more abundant in male rat liver compared with female liver (Hirshey and Falany, 1990).

Yerokun *et al.* (1992) have reported the isolation and expression of a single partial cDNA from a male Sprague–Dawley rat liver λgt11 cDNA library using a monoclonal antibody raised to AST IV. This cDNA is reported to encode AST IV and is essentially identical to PST I, but the cDNA lacked 18 nucleotides at the 5′-end of the open reading frame compared with PST I. Because the cDNA was slightly truncated the cDNA was expressed as a fusion protein *E. coli* XL-1 Blue cells. A protein was generated that reacted with the anti-AST IV monoclonal antibody and also displayed a low level of NOHAAF sulfation activity. No other compounds were tested as substrates for the expressed enzyme. Equivalent levels of mRNA were detected in male and female Sprague–Dawley rat liver by Northern blot analysis but male rats expressed five-fold higher levels of NOHAAF-ST activity (Ringer *et al.*, 1990).

Sequence analysis of the cloned rat cDNAs indicates that they are identical, but the cDNA is proposed to encode AST IV, PST I, and Mx-ST. Amino acid sequence obtained from pure AST IV (Yerokun *et al.*, 1992) and Mx-ST (Hirshey *et al.*, 1992) is used to support each of these interpretations. It is unlikely that the PST I sequence encodes all these proteins because of the disparity in the kinetic and immunologic properties of the different purified proteins. Southern blot analysis of total rat DNA also indicates that the PST I/Mx-ST sequence hybridizes to a limited number of fragments in restricted genomic DNA (Hirshey *et al.*, 1992), indicating that the PST I sequence is probably not a member of a closely related gene family as seen with the rat OHSTs (Fig. 4). The PST I sequence is expressed at relatively high levels in both male and female rats and multiple copies of the cDNA have been isolated from a female rat liver cDNA library (Hirshey *et al.*, 1992). This makes it unlikely that the PST I cDNA encodes the male-specific HAST II protein. Whether the PST-I cDNA translates a protein with the properties of HAST I has not been reported. It is significant that the group that purified HAST I, HAST II, and PST I also isolated the PST-I cDNA (Ozawa *et al.*, 1990; Gong *et al.*, 1991).

2. Human Phenol Sulfotransferases

To date only two forms of PST have been isolated and well characterized from human tissues (Weinshilboum, 1986; Falany, 1991; Falany and Roth, 1993). These are the phenol-sulfating form of PST (P-PST) and the monoamine-sulfating form of PST (M-PST). The names refer to the observation that at relatively low substrate concentrations, M-PST specifically catalyzes the sulfation of monoamines such as dopamine and epinephrine, whereas P-PST sulfates neutral phenols such as PNP and 2-naphthol. P-PST is also capable of conjugating the *N*-oxide moiety of minoxidil, estrogenic steroids, and amino-terminal and internal tyrosine residues in peptides, and aromatic amines, such as 2-naphthylamine. Both of the human PSTs are apparently capable of sulfating triiodothyronine and acetaminophen. P-PST and M-PST are also frequently referred to as the thermostable and thermolabile forms of PST, respectively, based on their sensitivities to thermal inactivation (Weinshilboum, 1986).

The observation that human blood platelets contained readily measurable levels of PST activity opened the way for the characterization of these activities in human tissues (Hart *et al.*, 1979). Weinshilboum and co-workers (Price *et al.*, 1988; Weinshilboum, 1990) have analyzed the variability and pharmacogenetic properties of PST activity in human platelets and liver. PST activities in platelets show a greater than 15-fold range in specific activity. The variation detected in the PST activities shows a high degree of heritability and for P-PST is apparently due to polymorphism at a single genetic locus (Weinshilboum, 1990). Also, in contrast to the sex differences observed in PST activities in rats, neither P-PST or M-PST in platelets or liver show a difference in activity between males and females (Weinshilboum, 1990; Falany, 1991).

M-PST and P-PST have been purified to homogeneity from human blood platelets and liver, respectively (Heroux and Roth, 1988; Falany *et al.*, 1990). The molecular masses of M-PST and P-PST, as determined by SDS–PAGE, are 34,000 and 32,000, respectively. Rabbit polyclonal antibodies raised against M-PST purified from platelets cross-react strongly with P-PST, indicating that the proteins are structurally related (Heroux *et al.*, 1989). The two PST activities have been identified enzymatically and immunologically in various human tissues, including adrenals, intestine, lung, brain, and pituitary (Heroux *et al.*, 1989; Zou *et al.*, 1990; Comer and Falany, 1992; Falany and Roth, 1993).

a. Molecular Cloning of Human PSTs The molecular cloning of the liver form of P-PST has been recently reported (Wilborn *et al.*, 1993). The cDNA was isolated from a human liver λgt11 cDNA library using the rat Mx-ST cDNA as a probe (Hirshey *et al.*, 1992). Multiple copies of the cDNA were isolated, and the longest clone, P-PST-1, possessed an open-reading frame encoding a 295 amino acid protein with a subunit mass of 34,097 Da. High levels of enzymatically active P-PST have been expressed in COS-7 cells with the pSV-SPORT-1 expression vector. The expressed enzyme comigrates with liver P-PST during SDS–PAGE, reacts strongly with the rabbit anti-human PST antibody, and is capable of high levels of PNP and minoxidil sulfation. P-PST expressed in COS-7 cells does not conjugate dopamine at low concentrations but is capable of sulfating β-estradiol and estrone. The expressed enzyme does not conjugate DHEA or pregnenolone (C. N. Falany *et al.*, 1994). Although high levels of hP-PST activity can be expressed in cultured cells, it has not been reported whether the expressed enzyme can conjugate NOHAAF or other similar procarcinogens (Wilborn *et al.*, 1993).

The similarity between rMx-ST and hP-PST includes comparable kinetic properties and immunologic cross-reactivity (Hirshey and Falany, 1990; Falany and Kerl, 1990). These similarities are also observed in the amino acid sequences of these proteins. Figure 9 shows a comparison of the amino acid sequences of rat Mx-ST and human P-PST. The amino acid sequences are 80% identical and 89% similar. The nucleotide sequences of the two cDNAs are also approximately 80% identical.

b. Involvement of Human PST Activity in Bioactivation The investigation of the role of the human STs in the bioactivation of procarcinogens and drugs has not been extensively studied. The bioactivation of minoxidil by P-PST activity has been demonstrated with both pure and expressed enzymes; however, the functional response to minoxidil sulfation has been elucidated in non-human species. The investigation of the bioactivation of *N*-hydroxylarylamines has been studied with human liver cytosol and hepatocytes. In a preliminary study, Chou *et al.* (1991) reported that human liver cytosol did not sulfate NOHAAF but was capable of conjugating *N*-hydroxy-2-aminofluorene (NOHAF) and *N*-hydroxy-4-aminobiphenyl. Subsequently, Abu-Zeid *et al.* (1992) have detected sulfation activity in human liver cytosol toward NOHAAF, NOHAF, and 2-hydroxyamino-

```
hP-PST   MELIQDTSRPPLEYVKGVPLIKYFAEALGPLQSFQARPDDLLISTYPKSG 50
         ||:    ||||| .|||:||||||||.:||||.| |:|||||||||||||
rMxST    MEF....SRPPLVHVKGIPLIKYFAETIGPLQNFTAWPDDLLISTYPKSG 46
                  .         .         .         .         .
         TTWVSQILDMIYQGGDLEKCHRAPIFMRVPFLEFKAPGIPSGMETLKDTP 100
         |||:|:|||||||||.|||| ||||: ||||||||.||:|||:|||.:||
         TTWMSEILDMIYQGGKLEKCGRAPIYARVPFLEFKCPGVPSGLETLEETP 96
                  .         .         .         .         .
         APRLLKTHLPLALLPQTLLDQKVKVVYVARNAKDVAVSYYHFYHMAKVHP 150
         |||||||||||.||||.||||||||:|:|||||||.||||:||:|||:||
         APRLLKTHLPLSLLPQSLLDQKVKVIYIARNAKDVVVSYYNFYNMAKLHP 146
                  .         .         .         .         .
         EPGTWDSFLEKFMVGEVSYGSWYQHVQEWWELSRTHPVLYLFYEDMKENP 200
         :|||||||||.|| ||||||||||||.|||||.:|||||||||||:||||
         DPGTWDSFLENFMDGEVSYGSWYQHVKEWWELRHTHPVLYLFYEDIKENP 196
                  .         .         .         .         .
         KREIQKILEFVGRSLPEETVDFMVQHTSFKEMKKNPMTNYTTVPQEFMDH 250
         ||||.|||||:|||||||||| :|:|||||.||.|.||||||:| |:|||
         KREIKKILEFLGRSLPEETVDSIVHHTSFKKMKENCMTNYTTIPTEIMDH 246
                  .         .         .         .
         SISPFMRKGMAGDWKTTFTVAQNERFDADYAKKMAGCSLTFRSEL      295
         .:|||||||  .||||.||||||||||||.|||.|.:|.:.||:||
         NVSPFMRKGTTGDWKNTFTVAQNERFDAHYAKTMTDCDFKFRCEL      291
```

Fig. 9 Amino-acid sequence comparison of human liver in P-PST with rat liver minoxidil sulfotransferase. The GAP program (Devereux *et al.*, 1984) was used to optimize the sequence alignment of the deduced amino acid sequence of hP-PST-1 (Wilborn *et al.*, 1993) and rMxST (Hirshey *et al.*, 1992). Vertical lines indicate sequence identity, two dots indicate highly conservative changes, and one dot indicates conservative changes.

6-methyldipyrido[1,2-*a* : 3′,2′-*d*]imidizole (NOH-Glu-P-1). The sulfation of NOHAF and PNP was also monitored in fractions of human liver cytosol obtained after DEAE-HPLC. PNP sulfation activity eluted in two peaks, and the sulfation of NOHAAF was associated with one of the fractions containing PNP sulfation activity. The identity of the ST catalyzing the PNP sulfation is not known, but the fraction with the NOHAF sulfation activity possessed a protein that reacted with an anti-rat HAST antibody; however, the levels of the immunoreactive protein did not correlate with the bioactiviation of the *N*-hydroxyarylamines. Possibly a previously unidentified PST is present in human liver cytosol capable of HAST activity.

The sulfation of NOHAF and NOHAAF by human liver cytosol was strongly inhibited by PCP, an inhibitor of PST activity (Abu-Zeid *et al.*, 1992). PCP is an inhibitor of rat PST activity and has been shown to decrease the bioactivation of AAF in isolated rat hepatocytes (Mangold *et al.*, 1990). In isolated human hepatocytes treated with AAF, PCP treatment increased rather than decreased the levels of NDA adduct formation (Monteith, 1992). The major adduct formed in human hepatocytes treated with AAF is dG-C8-AF. There is a lack of formation of dG-C8-AAF and dG-N2-AAF adducts expected to be formed from the sulfation of NOHAAF (Monteith and Gupta, 1991). These results indicate that the elucidation of the role of sulfation in bioactivation in human tissues requires a further understanding of the heterogeneity of the human STs and of the characterization of the enzymatic properties.

III. Summary

Numerous studies have indicated that two classes of cytosolic STs are involved in the bioactivation of procarcinogens and drugs to reactive electrophiles, especially in rodent tissues. These two classes of STs are the hydroxysteroid STs, which are involved in the conjugation of hydroxymethyl PAHs, and the phenol STs involved in the sulfation of alkenylbenzenes and *N*-hydroxyarylamines. Purification studies of rat liver STs have clearly indicated that specific isoforms of hydroxysteroid and phenol STs are capable of sulfating procarcinogens *in vitro*. Rat liver STa and BAST I are structurally similar hydroxysteroid STs, which have been shown to sulfate and bioactive HMBA. Molecular cloning studies of the rat hydroxysteroid STs indicate that these enzymes are probably part of a family of closely related genes. The single human hydroxysteroid ST that has been characterized is very similar to the rat enzymes, but its role in the bioactivation of hydroxymethyl PAHs has not been established.

Phenol STs have been demonstrated to have an important role in the bioactivation of alkenylbenzenes and *N*-hydroxyarylamines. Purification of rat phenol STs has identified several different forms, but only some appear to be involved in bioactivation of procarcinogens. Four isoforms (HAST I and II, AST III and IV) are apparently responsible for the majority of *N*-hydroxyarylamine sulfation. The relationship between these enzymes has not been established but they may represent similar enzymes. Different isoforms of rat phenol ST are also involved in the bioactivation of procarcinogens and drugs. However, the role of these phenol STs, PST-1, Mx-ST, and paracetamol ST, in carcinogenesis requires further study. In human tissues, only two phenol STs, P-PST and M-PST, have been identified. The role of these enzymes or unidentified STs in the sulfation of *N*-hydroxyarylamine procarcinogens has not yet been established. Initial reports of the molecular cloning and expression of the rat and human phenol ST genes will provide a valuable mechanism for the characterization of roles of the individual enzymes in bioactivation.

Acknowledgments

The authors thank Dr. Sharon Hirshey for her data on the characterization of rat minoxidil ST shown in Fig. 8. This work was supported in part by NIH Grants GM38953 and GM40440 to CNF.

References

Abu-Zeid, M., Yamazoe, Y., and Kato, R. (1992). Sulfotransferase-mediated DNA binding of N-hydroxyarylamines(amide) in liver cytosols from human and experimental animals. *Carcinogenesis (London)* **13,** 1307–1314.

Adams, J. B., and McDonald, D. (1979). Enzymatic synthesis of steroid sulfates. XII. Isolation of dehydroepiandrosterone sulfotransferase from human adrenals by affinity chromatography. *Biochim. Biophys. Acta* **567,** 144–153.

Adams, J. B., and McDonald, D. (1980). Enzymatic synthesis of steroid sulfates. XIII. Isolation and properties of dehydroepiandrosterone sulfotransferase from human foetal adrenals. *Biochim. Biophys. Acta* **615,** 275–278.

Barnes, S., Buchina, E. S., King, R. J., McBurnett, T., and Taylor, K. B. (1989). Bile acid sulfotransferase I from rat liver sulfates bile acids and 3-hydroxy steroids: Purification, N-terminal amino acid sequence, and kinetic properties. *J. Lipid Res.* **30,** 529–540.

Baumann, E. (1876). Ueber Sulfosauren im Harn. *In* "Ber. Deutsch Chem. Ges.", pp. 54–58.

Baxter, J. D., and Tyrrell, J. B. (1987). The adrenal cortex. *In* "Endocrinology and Metabolism" (P. Felig, J. D. Baxter, A. E. Broadus, and L. A. Frohman, eds.), pp. 511–632. McGraw-Hill, New York.

Binder, T. P., and Duffel, M. W. (1988). Sulfation of benzylic alcohols catalyzed by aryl sulfotransferase. IV. *Mol. Pharmacol.* **33,** 477–479.

Borchardt, R. T., and Schasteen, C. S. (1982). Phenol sulfotransferase. 1. Purification of a rat liver enzyme by affinity chromatography. *Biochim. Biophys. Acta* **708,** 272–279.

Buhl, A. E., Waldon, D. J., Baker, C. A., and Johnson, G. A. (1990). Minoxidil sulfate is the active metabolite that stimulates hair follicles. *J. Invest. Dermatol.* **95,** 553–557.

Chatterjee, B., Majumdar, D., Ozbilen, O., Ramana Murty, C. V., and Roy, A. K. (1987). Molecular cloning and characterization of cDNA for androgen-repressible rat liver protein, SMP-2. *J. Biol. Chem.* **262,** 822–825.

Chen, X., Yang, Y., Zheng, Y., Martin, B. M., Duffel, M. W., and Jakoby, W. B. (1992). Tyrosin-ester sulfotransferase from rat liver: Bacterial expression and identification. *Protein Express. Purif.* **3,** 1–6.

Chou, H. C., Lang, N. P., and Kadlubar, F. F. (1991). Metabolic activation of N-hydroxy arylamines and N-hydroxy heterocyclic amines by human liver sulfotransferase. *Proc. Am. Assoc. Cancer Res.* **32,** 120.

Comer, K. A., and Falany, C. N. (1992). Immunological characterization of dehydroepiandrosterone sulfotransferase from human liver and adrenal. *Mol. Pharmacol.* **41,** 645–651.

Comer, K. A., Falany, J. L., and Falany, C. N. (1993). Cloning and expression of human liver dehydroepiandrosterone sulphotransferase. *Biochem. J.* **289,** 233–240.

Coughtrie, M. W. H., and Sharp, S. (1990). Purification and immunochemical characterization of a rat liver sulfotransferase conjugating paracetamol. *Biochem. Pharmacol.* **40,** 2305–2313.

DeBaun, J. R., Miller, E. C., and Miller, J. A. (1970). N-Hydroxy-2-acetylaminofluorene sulfotransferase: Its probable role in carcinogenesis and in protein-(methion-S-yl) binding in rat liver. *Cancer Res.* **30,** 577–595.

Demyan, W. F., Song, C. S., Kim, D. S., Her, S., Gallwitz, W., Rao, T. R., Slomczynska, M., Chatterjee, and Roy, A. K. (1992). Estrogen sulfotransferase of the rat liver: Complementary DNA cloning and age- and sex-specific regulation of messenger RNA. *Mol. Endocrinol.* **6,** 589–597.

Devereux, J., Haeberli, P., and Smithies, O. (1984). A comprehensive set of sequence analysis programs for the VAX. *Nucleic Acids Res.* **12,** 387–395.

Dooley, T. P., Walker, C. J., Hirshey, S. J., Falany, C. N., and Diani, A. R. (1991). Localization of minoxidil sulfotransferase in the liver and outer root sheath of rat anagen pelage and vibrissa follicles. *J. Invest. Dermatol.* **96,** 65–70.
DuCharme, D. W., Freyburger, W. A., Graham, B. E., and Carlson, R. G. (1973). Pharmacologic properties of minoxidil: A new hypotensive agent. *J. Pharmacol. Exp. Ther.* **184,** 662–670.
Duffel, M. W., Binder, T. P., Hosie, L., Baden, H. A., Snaders, J. A., Knapp, S. A., and Baron, J. (1991). Purification, Immunochemical characterization and immunohistochemical localization of rat hepatic aryl sulfotransferase IV. *Mol. Pharmacol.* **40,** 36–44.
Falany, C. N. (1991). Molecular enzymology of human cytosolic sulfotransferases. *Trends Pharmacol. Sci.* **12,** 255–259.
Falany, C. N., and Kerl, E. (1990). Sulfation of minoxidil by human liver phenol sulfotransferase. *Biochem. Pharmacol.* **40,** 1027–1032.
Falany, C. N., and Roth, J. A. (1993). Properties of human cytosolic sulfotransferases involved in human drug metabolism. *In* "Human Drug Metabolism: From Molecular Biology to Man" (E. H. Jeffery, ed.), pp. 101–115. CRC Press, Boca Raton, FL.
Falany, C. N., Vazquez, M. E., and Kalb, J. M. (1989). Purification and characterization of human liver dehydroepiandrosterone sulfotransferase. *Biochem. J.* **260,** 641–646.
Falany, C. N., Vazquez, M. E., Heroux, J. A., and Roth, J. A. (1990). Purification and characterization of human liver phenol sulfating phenol sulfotransferase. *Arch. Biochem. Biophys.* **178,** 312–318.
Falany, C. N., Wheeler, J., Coward, L., Keehan, D., Falany, J. L., and Barnes, S. (1992). Bioactivation of 7-hydroxymethyl-12-methylbenz[*a*]anthracene by rat liver bile acid sulfotransferase I. *J. Biochem. Toxicol.* **7,** 241–248.
Falany, C. N., Wheeler, J., Oh, T. S., and Falany, J. (1994). Steroid sulfation by expressed human cytosolic sulfotransferases. *J. Steroid Biochem. Mol. Biol.* (in press).
Flesher, J. W., and Sydnor, K. L. (1971). Carcinogenicity of derivatives of 7,12-dimethylbenz-[*a*]anthracene. *Cancer Res.* **31,** 1951–1954.
Flesher, J. W., and Sydnor, K. L. (1973). Possible role of 6-hydroxymethylbenzo[α]pyrene as a proximate carcinogen of benzo[*a*]pyrene and 6-methylbenzo[*a*]pyrene. *Int. J. Cancer* **11,** 433–437.
Gilissen, R. A. H. J., Ringer, D. P., Stavenuiter, H. J. F. C., Mulder, G. J., and Meerman, J. H. N. (1992). Sulfation of hydroxylamines and hydroxamic acids in liver cytosol from male and female rats and purified aryl sulfotransferase IV. Carcinogenesis (*London*) **13,** 1699–1703.
Gong, D.-W., Ozawa, S., Yamazoe, Y., and Kato, R. (1991). Purification of hepatic N-hydroxyarylamine sulfotransferases and their regulation by growth hormone and thyroid hormone in rats. *J. Biochem.* (*Tokyo*) **110,** 226–231.
Groothius, G. M., Meyer, D. K. F., and Hardonk, M. J. (1983). Methodological studies on acinar damage by N-hydroxy-2-acetylaminofluorene and carbon tetrachloride. *Naunyn-Schmeideberg's Arch. Pharmacol.* **322,** 298–308.
Groppi, V. E., Burnett, B. A., Maggoria, L., Zucher-Neeley, H. A., Heinrikson, R. L., Schostarez, H. J., and Bienkowski, M. J. (1990). Protein sulfation: A unique mechanism of action of minoxidil sulfate. *J. Invest. Dermatol.* **94,** 532.
Hamamot, T., and Mori, Y. (1989). Sulfation of minoxidil in keratinocytes and hair follicles. *Res. Commun. Chem. Pathol. Pharmacol.* **66,** 33–44.
Hart, R. F., Renskers, K. J., Nelson, E. B., and Roth, J. A. (1979). Localization and characterization of phenol sulfotransferase in human platelets. *Life Sci.* **24,** 125–130.
Hernandez, J. S., Watson, R. W. G., Wood, T. C., and Weinshilboum, R. M. (1992). Sulfation of estrone and 17β-estradiol in human liver: Catalysis by thermostable phenol

sulfotransferase and by dehydroepiandrosterone sulfotransferase. *Drug Metab. Dispos.* **20,** 413–422.

Heroux, J. A., and Roth, J. A. (1988). Physical characterization of a monoamine-sulfating form of phenol sulfotransferase from human platelets. *Mol. Pharmacol.* **34,** 194–199.

Heroux, J. A., Falany, C. N., and Roth, J. A. (1989). Immunological characterization of human phenol sulfotransferase. *Mol. Pharmacol.* **36,** 29–33.

Hirshey, S. J., and Falany, C. N. (1990). Purification and characterization of rat liver minoxidil sulfotransferase. *Biochem. Pharmacol.* **40,** 1027–1032.

Hirshey, S. J., Dooley, T. P., Reardon, I. M., Heinrikson, R. L., and Falany, C. N. (1992). Sequence analysis, *in vitro* translation and expression of the cDNA for rat liver minoxidil sulfotransferase. *Mol. Pharmacol.* **42,** 257–264.

Hobkirk, R. (1985). Steroid sulfotransferases and steroid sulfate sulfatases: Characteristics and biological roles. *Can. J. Biochem. Cell Biol.* **63,** 1127–1144.

Huttner, W. B. (1988). Tyrosine sulfation and the secretory pathway. *Annu. Rev. Physiol.* **50,** 161–176.

Jakoby, W. B., Sekura, R. D., Lyon, E. S., Marcus, C. J., and Wang, J.-L. (1980). Sulfotransferases. *In* "Enzymatic Basis of Detoxication" (W. B. Jakoby, ed.), Vol. 2, pp. 199–227. Academic Press, New York.

Mangold, B. L., Erickson, J., Lohr, C., McCan, D. J., and Mangold, J. B. (1990). Self-catalyzed irreversible inactivation of rat hepatic aryl sulfotransferase IV by N-hydroxy-2-acetylaminofluorene. *Carcinogenesis (London)* **11,** 1563–1567.

Meisheri, K. D., Cipkus, L. A., and Taylor, C. J. (1988). Mechanism of action of minoxidil sulfate-induced vasodilation: A role for increasing K^+ permeability. *J. Pharmacol. Exp. Ther.* **245,** 751–760.

Meisheri, K. D., Johnson, G. A., and Puddington, L. (1993). Enzymatic and non-enzymatic sulfation mechanisms in the biological actions of minoxidil. *Biochem. Pharmacol.* **45,** 271–279.

Miller, E. C., Miller, J. A., Boberg, E. W., Delclos, K. B., Lai, C. C., Fennell, T. R., Wiseman, R. W., and Liem, A. (1985). Sulfuric acid esters as ultimate electrophilic and carcinogenic metabolites of some alkenylbenzenes and aromatic amines in mouse liver. *Carcinogenesis (London)* **11,** 93–107.

Miller, J. A., Surh, Y.-J., Liem, A., and Miller, E. C. (1991). Electrophilic sulfuric acid ester metabolites of hydroxy-methyl aromatic hydrocarbons as precursors of hepatic benzylic DNA adducts in vivo. *In* "Biological Reactive Intermediates. IV" (C. M. Witmer, R R. Snyder, D. J. Jolow, G. F. Kalf, J. J. Kocsis, and J. G. Sipes, eds.), pp. 555–567. Plenum, New York.

Monteith, D. K. (1992). Inhibition of sulfotransferase in primary cultures of human hepatocytes affecting metabolism and binding of 2-acetylaminofluorene. *Cancer Lett.* **64,** 109–115.

Monteith, D. K., and Gupta, R. C. (1991). Carcinogen-DNA adducts in cultures of rat and human hepatocytes. *Cancer Lett.* **62,** 87–93.

Mulder, G. J., ed. (1981). "Sulfation of Drugs and Other Compounds." CRC Press, Boca Raton, FL.

Mulder, G. J., and Jakoby, W. B. (1990). Sulfation. *In* "Conjugation Reactions in Drug Metabolism: An Integrated Aproach" (G. J. Mulder, ed.), pp. 107–161. Taylor & Francis, London.

Mulder, G. J., and Scholtens, E. (1977). Phenol sulfotransferase and uridine diphosphate glucuronyltransferase from rat liver *in vivo* and *in vitro*. 2,6-Dichloro-4-nitrophenol as a selective inhibitor of sulphation. *Biochem. J.,* **165,** 553–559.

Nash, A. R., Glenn, W. K., Moore, S. S., Kerr, J., Thompson, A. R., and Thompson, E. O. P. (1988). Oestrogen sulfotransferase: Molecular cloning and sequencing of cDNA for the bovine placental enzyme. *Aust. J. Biol. Sci.* **41,** 507–516.

Ogura, K., Kajita, J., Narihata, H., Watabe, T., Ozawa, S., Nagata, K., Yamazoe, Y., and Kato, R. (1989). Cloning and sequence analysis of a rat liver cDNA encoding hydroxysteroid sulfotransferase. *Biochem. Biophys. Res. Commun.* **165,** 168–174.

Ogura, K., Kajita, J., Narihata, H., Watabe, T., Ozawa, S., Nagata, K., Yamazoe, Y., and Kato, R. (1990a). cDNA cloning of the hydroxysteroid sulfotransferase STa sharing a strong homology in amino acid sequence with the senescence marker protein SMP-2 in rat livers. *Biochem. Biophys. Res. Commun.* **166,** 1494–1500.

Ogura, K., Kajita, J., Narihata, H., Watabe, T., Ozawa, S.., Nagata, K., Yamazoe, Y., and Kato, R. (1990b). Rat liver cytosolic hydroxysteroid sulfotransferase (Sulfotransferase a) catalyzes the formation of reactive sulfate esters from carcinogenic polycyclic hydroxymethylarenes. *Mol. Pharmacol.* **37,** 848–854.

Otterness, D. M., Wieben, E. D., Wood, T. C., Watson, R. W. G., Madden, B. J., McCormick, D. J., and Weinshilboum, R. M. (1992). Human liver dehydroepiandrosterone sulfotransferase: Molecular cloning and expression of cDNA. *Mol. Pharmacol.* **41,** 865–872.

Ozawa, S., Nagata, K., Gong, D., Yamazoe, Y., and Kato, R. (1990). Nucleotide sequence of a full-length cDNA (PST-1) for aryl sulfotransferase from rat liver. *Nucleic Acids Res.* **18,** 4001.

Pang, K. S. (1990). Kinetics of conjugation reactions in eliminating organs. *In* "Conjugation Reactions in Drug Metabolism" (G. J. Mulder, ed.), pp. 5–40. Taylor & Francis, London.

Price, R. A., Cox, N. J., Spielman, R. S., Van Loon, J. Maidak, B. L., and Weinshilboum, R. M. (1988). Inheritance of human platelet thermolabile phenol sulfotransferase (TL PST) activity. *Genet. Epidemiol.* **16,** 1–15.

Radominska, A., Comer, K. A., Zimniak, P., Falany, J., Iscan, M., and Falany, C. N. (1990). Human liver steroid sulfotransferase sulphates bile acids. *Biochem. J.* **272,** 597–604.

Ringer, D. P., Norton, T. R., and Howell, B. A. (1990). 2-Acetylaminofluorene-mediated alteration in the level of liver arylsulfotransferase IV during rat hepatocarcinogenesis. *Cancer Res.* **50,** 5301–5307.

Robbins, P. W., and Lipmann, F. (1956). Identification of enzymatically active sulfate as adenosine-3′-phosphate-5′-phosphosulfate. *J. Amer. Chem. Soc.* **78,** 2652–2657.

Roth, J. A. (1986). Sulfoconjugation: Role in neurotransmitter and secretory protein activity. *Trends Pharmacol. Sci.* **7,** 404–407.

Sekura, R. D., and Jakoby, W. B. (1979). Phenol sulfotransferases. *J. Biol. Chem.* **254,** 5658–5663.

Sekura, R. D., and Jakoby, W. B. (1981). Aryl sulfotransferase IV from rat liver. *Arch. Biochem. Biophys.* **211,** 352–359.

Song, C., Kim, J. M., Roy, A. K., and Chatterjee, B. (1989). Structure and regulation of the senescence marker protein 2 gene promoter. *Biochemistry* **29,** 542–551.

Surh, Y.-J., Lai, C.-C., Miller, J. A., and Miller, E. C. (1987). Hepatic DNA and RNA adduct formation from the carcinogen 7-hydroxymethyl-12-methylbenz[a]anthracene and its electrophilic sulfuric acid ester metabolite in preweanling rats and mice. *Biochem. Biophys. Res. Commun.* **144,** 576–582.

Surh, Y.-J., Liem, A., Miller, E. C., and Miller, J. A. (1989). Metabolic activation of the carcinogen 6-hydroxymethylbenzo[*a*]pyrene: Formation of an electrophilic sulfuric acid ester and benzylic DNA adducts in rat liver *in vivo* and in reactions *in vitro*. *Carcinogenesis (London)* **10,** 1519–1528.

Surh, Y.-J., Blomquist, J. C., Liem, A., and Miller, J. A. (1990). Metabolic activation of 9-hydroxymethyl-10-methylanthracene and 1-hydroxymethylpyrene to electrophilic, mutagenic and tumorigenic sulfuric acid esters by rat sulfotransferase activity. *Carcinogenesis (London)* **11,** 1451–1460.

Surh, Y.-J., Liem, A., Miller, E. C., and Miller, J. A. (1991). Age- and sex-related differences in activation of the carcinogen 7-hydroxymethyl-12-methylbenz[*a*]anthracene to an electrophilic sulfuric acid ester metabolite in rats: Possible involvement in hydroxysteroid sulfotransferase activity. *Biochem. Pharmacol.* **41,** 213–221.

Watabe, T., Ishizuka, T., Isobe, M., and Ozawa, N. (1982). A 7-hydroxymethyl sulfate ester as an active metabolite of 7,12-dimethylbenz[*a*]anthracene. *Science* **215,** 403–405.

Watabe, T., Hiratsuka, A., Ogura, K., and Endoh, K. (1985). A reactive hydroxymethyl sulfate ester formed regioselectively from the carcinogen, 7,12-dihydroxymethylbenz-[*a*]anthracene by rat liver sulfotransferase. *Biochem. Biophys. Res. Commun.* **131,** 694–699.

Weinshilboum, R. M. (1986). Phenol sulfotransferase in humans: Properties, regulation, and function. *Fed. Proc., Fed. Am. Soc. Exp. Biol.* **45,** 2223–2228.

Weinshilboum, R. M. (1990). Sulfotransferase pharmacogenetics. *Pharmacol. Ther.* **45,** 93.

Wilborn, T. W., Comer, K. A., Dooley, T. P., Reardon, I. M., Heinrikson, R. L., and Falany, C. N. (1993). Sequence analysis and expression of the cDNA for the phenol-sulfating form of human liver phenol sulfotransferase. *Mol. Pharmacol.* **43,** 70–77.

Wu, S. G., and Straub, K. D. (1976). Purification and characterization of N-hydroxy-2-acetylaminofluorene sulfotransferase from rat liver. *J. Biol. Chem.* **251,** 6529–6536.

Yamazoe, Y., Manabe, S., Murayama, S., and Kato, R. (1987). Regulation of hepatic sulfotransferase catalyzing the activation of N-hydroxyarylamide and N-hydroxyarylamine by growth hormone. *Mol. Pharmacol.* **32,** 536–541.

Yerokun, T., Etheredge, J. L., Norton, T. R., Carter, H. A., Chung, K. H., Birckbichler, P. J., and Ringer, D. P. (1992). Characterization of a complementary DNA for rat liver aryl sulfotransferase IV and use in evaluating the hepatic gene transcript levels of rats at various stages of 2-acetylaminofluorene-induced hepatocarcinogenesis. *Cancer Res.* **52,** 4779–4786.

Zou, J., Pentney, R., and Roth, J. A. (1990). Immunohistochemical detection of phenol sulfotransferase containing neurons in human brain. *J. Neurochem.* **55,** 1154–1158.

Carcinogen Activation by Sulfate Conjugate Formation

Christopher J. Michejda and Marilyn B. Kroeger Koepke
Molecular Aspects of Drug Design Section
Macromolecular Structure Laboratory
ABL-Basic Research Program
*NCI–Frederick Cancer Research and Development Center**
Frederick, Maryland 21702

I. Introduction

Activation of xenobiotics to toxic, mutagenic, or carcinogenic agents most often involves either oxidative or reductive metabolism. However, it has been recognized for some time that Phase-II metabolism can also result in activation. Phase-I metabolites that are not sufficiently hydrophilic to be rapidly excreted in the urine are conjugated by endogenous reactants to render them more polar. The most common conjugation reactions are glucuronidation, acetylation, methylation, and conjugation with glutathione, sulfate, or glycine. This chapter will focus on sulfation of hydroxylic substrates as an activation pathway.

Sulfation is the formation of sulfate esters of hydroxyl groups of alcohol, phenols, hydroxylamines, or amines (Fig. 1). The reactions are catalyzed by one of several sulfotransferase enzymes (see Falany and Wilborn, this volume). Inorganic sulfate is activated by conversion to 3′-phosphoadenosine 5′-phosphosulfate (PAPS), which is then utilized as a cofactor by all of the sulfotransferases (ST). The sulfuric acid monoesters are relatively

* The U.S. Government's right to retain a nonexclusive, royalty-free license in and to the copyright covering this paper, for governmental purposes, is acknowledged.

Advances in Pharmacology, Volume 27

$$ArOH + PAPS \xrightarrow{ST} ArOSO_3H$$

$$RR'CHOH + PAPS \xrightarrow{ST} RR'CHOSO_3H$$

$$RR'NOH + PAPS \xrightarrow{ST} RR'NOSO_3H$$

Fig. 1 General reactions catalyzed by sulfotransferases.

strong acids and are usually fully ionized at physiological pH. Thus, the resulting sulfate conjugate is usually more hydrophilic than the precursor hydroxy compound and is more readily eliminated. Conjugation by sulfation was reviewed recently (Mulder and Jakoby, 1990).

Formation of sulfate esters can result in activation rather than detoxification when the resultant conjugate is susceptible to further reaction, which leads to an electrophilic intermediate. This can occur by one of two possible mechanisms, both of which rely on the fact that the sulfate group is a good leaving group in both S_N1 and S_N2 types of nucleophilic displacement reactions. The first mechanism operates if the positive ion formed by the departure of sulfate anion is relatively stable (Fig. 2). Several examples of this type of reaction, where the positive charge is located on either a carbon or a nitrogen, will be illustrated in this chapter. The second mechanism operates when the substrate molecule contains other functional groups in a correct orientation to participate in an intramolecular reaction, which displaces the sulfate anion and generates a new reactive species. This type of a mechanism can be illustrated by the sulfate conjugation of a β-aminoalcohol (Fig. 3). The formation of the aziridinium ion creates a highly reactive alkylating agent, which has the capacity of modifying a variety of cellular nucleophiles, including proteins and nucleic acids. Several examples of this type of activation, which can be considered to be an intramolecular S_N2 reaction, will be presented.

The sulfated intermediates of xenobiotics that are activated by sulfation are most frequently very unstable substances. Thus their isolation from biological fluids is rarely possible. For example, the *N-O*-sulfate of *N*-hydroxy-2-acetamido fluorene (*N*-OH-2-AAF) has been prepared by chemical synthesis (Beland *et al.*, 1983). This material was found to be so unstable that it could not be purified without decomposition. This chemical instability and the high reactivity of many of these sulfate conju-

$$ROSO_3^- \longrightarrow R^+ + SO_4^-$$

Fig. 2 Solvolysis of sulfate esters results in formation of cationic intermediates.

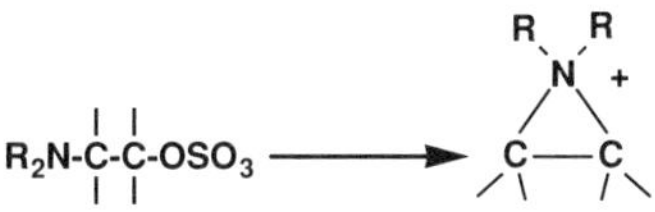

Fig. 3 Solvolysis of sulfate esters by neighboring group participation, illustrated by β-aminoethylsulfates.

gates require indirect approaches to the study of activation of xenobiotics by sulfation. Since all sulfotransferases require PAPS, a traditional method of determining the rate of sulfation involves the measurement of dependence of substrate disappearance on PAPS concentration. Obviously, this technique can be applied only *in vitro*. Another way of measuring the extent of sulfate involvement in activation is to trap the reactive sulfate conjugate by added nucleophiles. For example, methionine is an effective nucleophile for the scavenging of the sulfate ester of *N*-OH-2-AAF (DeBaun *et al.*, 1970a). This indirect method suffers from two major deficiencies. Although some sulfate conjugates are effectively trapped by added nucleophiles, others are not. Thus one must have some idea about the reactivity of a given nucleophile with a specific sulfate ester before data from trapping experiments can be used in a quantitative fashion. Another problem is that unless the involvement of a sulfate conjugate had been established in the formation of adducts to nucleophiles, there is always a possibility that a given adduct was formed from a different electrophile. For example, the adducts formed from the reaction of the sulfate ester of *N*-OH-2-AAF and deoxyguanosine are the same as the adducts formed from *N*-acetoxy-2-AAF and deoxyguanosine (Beland *et al.*, 1983). Although there is no evidence for an enzymatic formation of the *N*-acetoxy intermediate, this example illustrates the point well. However, the formation of adducts, taken together with other lines of evidence, can provide valuable data on sulfate activation. Many workers have utilized this approach, especially when the monitoring of DNA adducts was convenient. It was then necessary to provide additional independent evidence of sulfate ester involvement. This was done frequently by demonstrating that adduct formation depended on sulfate or PAPS.

The enzymology of the sulfotransferases is discussed elsewhere in this volume (see Falany and Wilborn). However, various enzymes of the sulfotransferase family will be mentioned in this chapter. It is useful for the purpose of the present discussion to distinguish between two classes of sulfotransferases: those that prefer substrates where the hydroxyl group is attached to an aromatic residue, the phenol sulfotransferases, and those that prefer to form sulfate esters of aliphatic hydroxyls. The latter group includes the sterol sulfotransferases, which catalyze the sulfation of various steroids, bile acids, and presumably various alcohols, including those derived from polycyclic aromatic hydrocarbons. Historically, these two broad classes were distinguished by their response to various inhibitors. Thus, the phenol sulfotransferases can be distinguished from other sulfotransferases because they are readily inhibited by various phenolic inhibitors, such as pentachlorophenol and 2,6-dichloro-4-nitrophenol. Conversely, the alcohol sulfotransferases can be competitively inhibited by various alcohols or by steroids, such as dehydroepiandrosterone (DHEA). Inhibition of enzymatic activity, as measured by secondary effects such as nucleic acid alkylation, is frequently difficult to interpret because of the multiple modes of action of many inhibitors. The recent isolation and cloning of the various rodent and human sulfotransferases (see Falany and Wilborn, this volume) have now made it possible to identify some of the relevant enzymes by much more direct techniques.

Sulfotransferases are a heterogeneous group of enzymes, which display a wide variation in the levels of various members of the family with respect to species, sex, and ontogeny. Only three sulfotransferases have been identified in the human liver. Two phenol sulfotransferases, one (M-PST) which favors monoamine substrates and the other (P-PST) which favors phenolic substrates, have been isolated from human platelets (Heroux and Roth, 1988) and human liver (Falany *et al.*, 1990), respectively. Only one sterol sulfotransferase (DHEA-ST) has been characterized in the human liver (Comer *et al.*, 1993). In contrast, there are at least six phenol sulfotransferases and seven steroid/bile acid sulfotransferases in the rat (Mulder and Jakoby, 1990). The enzyme (aryl sulfotransferase IV) that transforms *N*-OH-2-AAF to the active sulfate conjugate is found only in male rats, with only a trace in females (Wu and Straub, 1976). However, sulfotransferase activity toward tiaramide, a β-aminoethyl alcohol, was considerably higher in female rats than in males (Iwasaki *et al.*, 1983). There are still considerable gaps in our understanding of the sex differences in the sulfotransferase activities in various species, but what is known about this and other conjugation reactions has been reviewed by Mulder (1986). The age-related changes in the levels of various sulfotransferases are also imperfectly understood, but some of the enzymes appear to vary considerably with age, with particularly large changes occurring at sexual maturity (Balistreri *et al.*, 1984).

Sulfate conjugation is usually in competition with other Phase-II conjugation reactions, especially glucuronidation (Mulder, 1986). In fact, there is an inverse relationship between the two pathways. Where there is competition, several factors need to be considered. First, sulfotransferases are cytosolic enzymes, while UDP-glucuronosyl transferases, the enzymes responsible for glucuronidation, are microsomal. Thus one might expect that lipophilic substrates might be favored in the glucuronidation reaction, other factors being equal. Second, the specific kinetic parameters of the two types of transferases with respect to a given substrate, together with cofactor availability and depletion, will influence the balance of the two pathways. Usually, but not always, sulfotransferases exhibit lower K_m values than the glucuronosyl transferases. One might expect then that sulfotransferase activity would be saturated at a lower substrate concentration. This suggests that sulfotransferase activation might be most effective for low levels of xenobiotics. As is usual for generalizations of this sort, many exceptions probably exist, particularly for those substrates where structural characteristics impart similar kinetic properties with respect to the two types of enzymes.

Activation of xenobiotics by sulfate conjugation has been discussed in several earlier reviews, most notably those of Mulder (1981). This chapter will draw on some of the older material, but will focus mainly on the newer data. It is also our intent to discuss the mechanistic aspects of activation by sulfate conjugation in order to enhance the predictive value of this review. We will follow the traditional organization of considering the various substrate classes separately since that provides a logical framework for the discussion of both the chemistry and the toxicology.

II. Aromatic Hydroxylamines and Amides

A. Aminofluorenes

The seminal papers by King and Phillips (1968) and DeBaun *et al.* (1968) suggested strongly that sulfotransferase-catalyzed sulfation of *N*-hydroxy-2-AAF was an activation pathway for this carcinogen. These observations were the first indications that conjugation reactions, whose purpose is to enhance the elimination of xenobiotics, may increase the toxic potential of some chemicals. A considerable amount of work has been carried out on the metabolism, DNA adduct formation, and carcinogenicity of 2-AAF. This discussion will focus on the sulfation pathway in some detail, not only because 2-AAF is the prototype chemical but also because almost all of the techniques and ideas relative to activation by sulfation have been developed during the course of these investigations (Fig. 4).

The chemical synthesis of the sulfate ester of *N*-hydroxy-2-AAF has been reported (Beland *et al.,* 1983). This highly unstable compound was found to react with deoxyguanosine to form *N*-(deoxyguanosin-8-yl)-

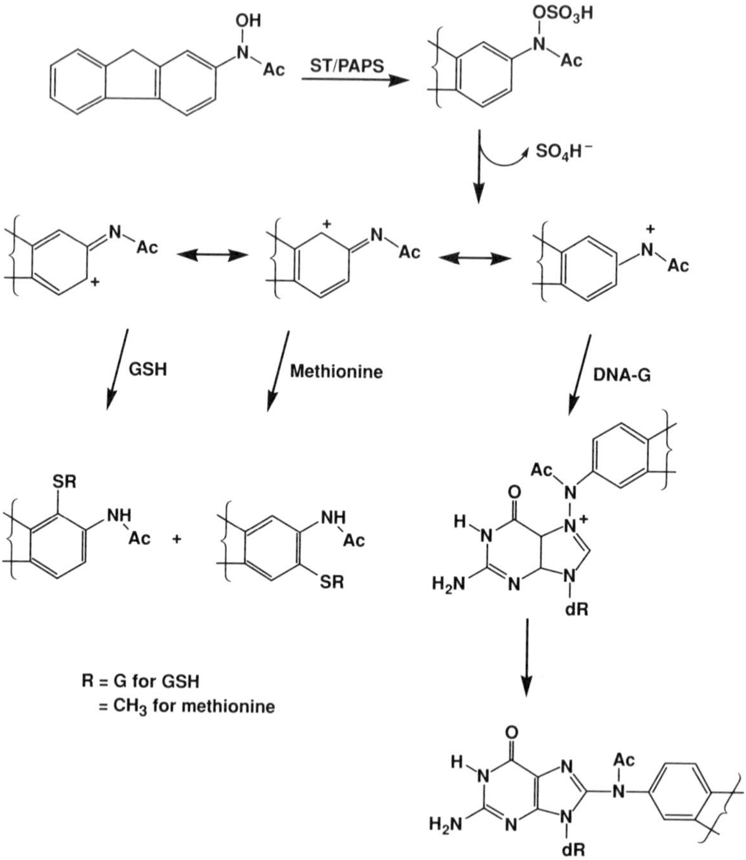

Fig. 4 Adduct formation from 2-AAF sulfate, (GSH, glutathione; ST, sulfotransferase, aryl sulfotransferase IV in this case; DNA-G, guanine in DNA).

2AAF, the same adduct that was obtained with the model carcinogen *N*-acetoxy-2-AAF. Moreover, the sulfate ester reacted with glutathione to produce four isomeric adducts, 1-, 3-, 4-, and 7-(glutathion-*S*-yl)-2-AAF. This simple experiment illustrated several important chemical problems associated with the reactions of the active sulfate ester of 2-AAF. First, the sole product that was formed with guanosine, and as shall be seen below is the major adduct in DNA both *in vivo* and *in vitro,* was the adduct in position 8. That position is usually associated with radical-derived adducts on guanine. As was mentioned above, sulfate esters produce adducts by virtue of their ability to form electrophilic agents; in the case of 2-AAF sulfate the electrophile is expected to be the nitrenium ion formed by the S_N1 dissociation of the ester. Cationic electophiles attack the exocyclic atoms of guanine or at the 7-position. It was recently shown (Humphreys *et al.*, 1992) that in the case of the acetyl ester of 2-aminofluorene (*N*-OAc-2-AF), the initial site of attack is also the 7-position, but, because of the thermodynamically unstable nature of the 7-adduct, a 1,2 sigmatropic shift occurs to yield the observed adduct in position 8. This is likely to be the mechanism of formation of all of the 8-guanine adducts derived from nitrenium ions.

The reaction of glutathione with the 2-AAF sulfate also merits comment. The nitrenium ion produced by the dissociation of the sulfate ester (or of some other leaving group) is capable of resonance delocalization with the fluorenyl ring system. On the basis of this, one would predict that the two sites, which are *ortho* to the acetylamino group, namely positions 1 and 3, should be particularly electrophilic. One should also expect that other sites would also be attacked to a somewhat lesser degree. The glutathione adducts obtained from the bile of rats treated with *N*-OH-2-AAF were shown to be the 1- and 3-(glutathion-*S*-yl)-2-AAF isomers, whereas the chemical reaction of *N*-OAc-2-AAF with glutathione yielded additionally smaller amounts of the 4- and 7-isomers. The 4-isomer was formed presumably by a 1,2-shift of the initially formed adduct at position 4a (Meerman *et al.*, 1982). The metabolism of *N*-OH-2-AAF also results in methionine adducts. These are formed by the reaction of the putative nitrenium ion at the sulfur center of the molecule. The reaction involves the transfer of the methylthio group to the 1- and 3-positions of the fluorene (Kadlubar *et al.*, 1976).

The initial hypothesis and preliminary evidence for activation of 2-AAF by sulfate conjugation were followed by a slow development of a concrete proof of the mechanism. The general scheme for the metabolism and activation of 2-AAF is shown in Fig. 5. It was demonstrated in 1960 by Miller *et al.* that 2-AAF was activated by the formation of *N*-OH-2-AAF. This reaction, however, was a necessary but not sufficient condition for DNA adduct formation (Gutman *et al.*, 1972) The aforementioned papers of DeBaun *et al.* (1968) and of King and Phillips (1968) only provided indirect evidence for the involvement of a subsequent sulfation step. Thus, King and Phillips showed that rat liver cytosol preparations transformed *N*-OH-2-AAF into an intermediate, presumably sulfate, which reacted with RNA and proteins. DeBaun *et al.* (1968) found that the macromolecular binding was a function of PAPS concentration and that the formation of the methylthio derivatives of 2-AAF was also PAPS-dependent. The ability to activate 2-AAF was found to be dependent on species, strain, and sex of the animal (King and Olive, 1975) and that in turn correlated with acetyltransferase and sulfotransferase levels in those animals. Male rat liver was much more active than the liver from female rats. Attempts to modulate *in vivo* inorganic sulfate levels gave somewhat equivocal

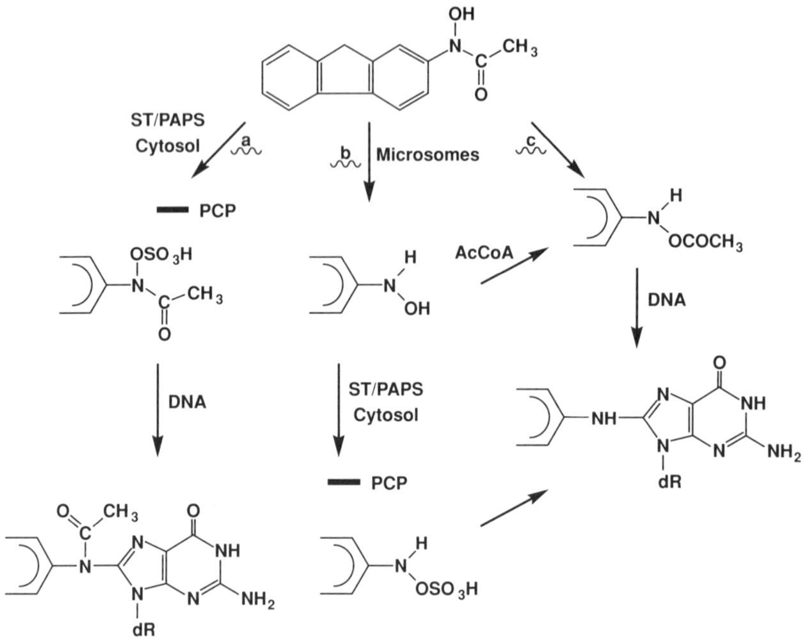

Fig. 5 Simplified scheme for the metabolism and DNA adduct formation from *N*-OH-2-AAF. The solid bar with PCP denotes inhibition of that step by PCP.

results, although consistent with the sulfation hypothesis (DeBaun *et al.*, 1970b). Strong evidence, but not proof, for activation by sulfation was provided by the inhibition of *N*-OH-2-AAF sulfotransferase activity *in vivo* in rats by treatment with pentachlorophenol (PCP), a specific inhibitor of aryl sulfotransferases (Meerman *et al.*, 1981). It was interesting to note that not all of the binding in that study was inhibited by PCP. The foregoing scheme suggests that there are in fact three possible activation reactions that transform *N*-OH-2-AAF to an electrophile capable of modifying DNA-guanine. Path **a** is the sulfation of the unmodified *N*-OH-2-AAF. Path **b** involves the microsomal *N*-deacetylation (Irving, 1966), followed by PAPS-dependent sulfate conjugation. Path **c,** however, does not depend on sulfation since the activation is accomplished by a cytosolic enzyme-catalyzed *N,O*-transacetylation reaction (Bartsch *et al.*, 1972, 1973; King and Allaben, 1980). The adducts due to path **a** retain the acetyl group of 2-AAF, which is lost in adducts from paths **b** and **c.**

Lai *et al.* (1985) studied the metabolism and DNA binding of *N*-OH-2-AAF in B6C3F$_1$ mice and in brachymorphic mice derived from that cross. The latter are constitutively impaired in their ability to synthesize PAPS (Sugahara and Schwartz, 1982), the active sulfate cofactor of all sulfotransferases. They found that acetylated adducts accounted for <10% of the adducts in the phenotypically normal mice, with the rest being the unacetylated adducts. All of the adducts were decreased by 90% by pretreatment with PCP, as was the induction of hepatomas in the mice produced by a dose of the carcinogen equivalent to that used in the adduct studies. The male brachymorphic mice formed only one quarter of the amount of adducts observed in the normal mice, and were also much less susceptible to *N*-OH-2-AAF carcinogenesis. The cytosol from the livers of the normal mice contained both *N*-OH-2-AAF and *N*-OH-2-AF activities, which was effectively inhibited by PCP. The other enzymatic activities, such the microsomal deacetylase and the *N,O*-acetyltransferases, were not affected by PCP. These data provided very strong evidence for the critical involvement of sulfate conjugation in the activation of 2-AAF and also indicated that path **b** in Fig. 5 was the predominant pathway for adduct formation and for the initiation of carcinogenesis by 2-AAF. In a subsequent study (Lai *et al.*, 1987), it was shown that *N*-sulfooxy-2-aminofluorene (*N*-2-AF sulfate) was the most important initiator of carcinogenesis and that the only adduct produced by it in the livers of $B6C3F_1$ mice, namely *N*-(deoxyguanosin-8-yl)-2-aminofluorene, was the adduct responsible for tumor initiation.

The sulfotransferase responsible for the conjugation of *N*-OH-2-AAF and *N*-OH-2-AF in the rat liver is aryl sulfotransferase IV (Ringer *et al.*, 1990) The cDNA for this enzyme has been cloned recently (Yerokun *et al.*, 1992). Interestingly, Falany was unable to find any activity in human liver cytosol that converted *N*-OH-2-AAF to DNA adducts (Falany, 1991), and Montieth (1992) was unable to inhibit AAF adduct formation in human hepatocytes by PCP; his study supports the hypothesis that *N,O*-acyltransferase is the critical enzyme in the activation of AAF in the human liver. Strong evidence was provided for irreversible inhibition of aryl sulfotransferase IV by 2-AAF sulfate, presumably by adduct formation between the enzyme and the activated product (Ringer *et al.*, 1992). This reaction undoubtedly limits the bioavailability of the activated carcinogen. Another reaction that may limit the availability of 2-AAF sulfate is the apparent serum albumin-catalyzed ortho-rearrangement of the *N*-sulfate to the inactive 1- and 3-(sulfooxy)-2-AAF (Kolanczyk *et al.*, 1991) (Fig. 6). It was shown that bovine serum albumin markedly enhanced the rearrangement *in vitro* and effectively blocked adduct formation.

B. Phenacetin and Its Metabolites

Phenacetin is an analgesic found in a number of nonprescription compounds in the United States (Vaught *et al.*, 1981). Several studies have shown it to induce tumor formation in a number of tissues in the rat,

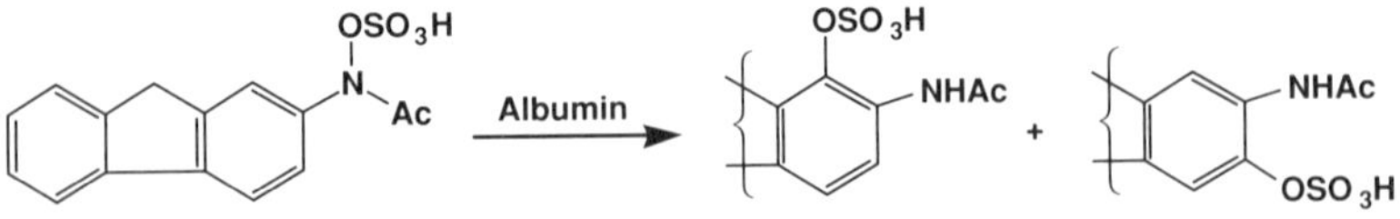

Fig. 6 Inactivation of 2-AAF sulfate by intramolecular rearrangement (Kolanczyk *et al.*, 1991).

including the urinary bladder (Isaka *et al.*, 1979; Johansson and Angervall, 1976). In animals, phenacetin has been to be converted to acetaminophen (Mulder *et al.*, 1978). The first step in the metabolism of phenacetin appears to be *N*-hydroxylation (Hinson and Mitchell, 1976). The *N*-hydroxy metabolite (*N*-OH-P) is mutagenic when activated by S9 rat liver supernatant (Shudo *et al.*, 1978) or microsomes (Wirth *et al.*, 1980).

Phenacetin has been shown to be activated to macromolecule-bound products by a phenol sulfotransferase. An early study (Nery, 1971) demonstrated that microsomes activated phenacetin to products that bind to nucleic acids as well as to proteins. Rat and hamster liver microsomes have been shown to activate phenacetin by several different pathways to bind to protein or react with glutathione. The formation of *N*-OH-P, which can be further metabolized to reactive intermediates by sulfation or glucuronidation, has been demonstrated (Mulder *et al.*, 1977). An additional report showed that sulfate conjugation of *N*-[ring-^{3}H]OH-P could lead to binding to transfer RNA (Vaught *et al.*, 1981). These authors also implicated acyl transfer and deacylation as activation pathways for the binding of *N*-OH-P to nucleic acids. Other *N*-hydroxy-arylacetamides can form sulfate derivatives. However, sulfation does not always result in appreciable covalent binding, as can be seen in the case of both *N*-hydroxyacetanilide and *N*-hydroxy-*p*-chloroacetanilide after addition of the PAPS precursor PAP and *p*-nitrophenol, but not by methionine. Since methionine is an inhibitor of *N*-hydroxy-2-AAF sulfotransferase-activated binding to protein, the mechanism of covalent binding of *N*-OH-P to protein may not involve sulfation.

Although the exact details of a mechanism of covalent protein binding for *N*-OH-P are not fully delineated, it has been proposed (Mulder *et al.*, 1977, 1978) that the *N*,*O*–sulfate conjugates eliminate to give a *N*-acetylbenzoquinoneimide intermediate (Fig. 7). The *N*,*O*–sulfate conjugates themselves are too unstable to isolate, but it has been shown that the final reactive metabolite lacks the ethyl group of phenacetin, consistent with the mechanism in the figure.

In addition to the sulfate conjugate, the *N*,*O*–glucuronide of *N*-*O*-P has also been identified as a second major product (Mulder *et al.*, 1978). It is

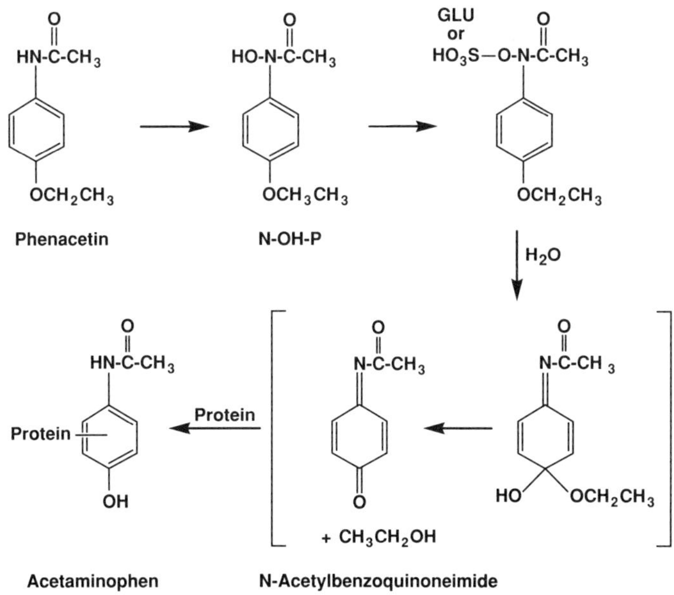

Fig. 7 Metabolism of phenacetin via a sulfate ester intermediate.

also bound to protein at pH 7.4, but at a slower rate that the *N,O*–sulfate conjugate (Mulder *et al.*, 1977). In fact, the rate of covalent binding of protein in the liver following glucuronidation showed a lag phase compared with the binding of the *N,O*–sulfate conjugate, which seemed to bind almost as rapidly as it was formed. In contrast to the *N,O*–sulfate conjugate, the *N,O*–glucuronide has been determined to have a half-life of about 8.7 h at pH 7.2 in Tris buffer (Hinson *et al.*, 1979). Although it seems that glucuronidation is less important for covalent binding in the liver than is sulfation, it may play a role in nephrotoxicity since the kidney is the collection site for glucuronides. Acetaminophen has been identified as a breakdown product from both the *N,O*–sulfate and *N,O*–glucuronide conjugates, suggesting that they are metabolized through the same *N*-acetylbenzoquinoneimide intermediate (Mulder *et al.*, 1978; Vaught *et al.*, 1981). Although approximately half of the *N-O*-P glucuronide breaks down to acetaminophen or its derivatives, the rest is converted directly to phenacetin or the 2-hydroxy glucuronide of phenacetin (Hinson *et al.*, 1979). Thus, although the relative importance of glucuronidation compared with sulfation is unknown, in part due to the unknown degree of latency of the UDP glucuronyl-transferase *in vivo* (Dutton, 1975), it is clear that more work needs to be carried out to assess its possible role as an activating process. Other pathways, such as deacetylation and acyltransfer, have also been identified for the metabolism of *N-O*-P (Vaught *et al.*, 1981; Wirth *et al.*, 1980). However, the relative importance of these pathways in the bioactivation of *N-O*-P to a carcinogenic agent remains open.

Acetaminophen itself is sulfated by hepatic arylsulfotransferase in isolated rat hepatocytes (Sweeny and Reinke, 1988) and also by the liver $105{,}000 \times g$ cytosolic fraction. There appears to be a sex difference in this sulfation activity (Raheja *et al.*, 1983; Green and Fisher, 1981); male rats eliminated acetaminophen more as sulfate conjugates and less as glucuronide conjugates than did females. This effect was concluded to be due to the influence of gonadal hormones on one of the two sulfotransferases identified in the hepatic cytoplasm (Kane and Chen, 1987). Acetaminophen is also a substrate for the thermostable and thermolabile forms of human platelet phenol transferase (Reiter and Weinshilboum, 1982). The level of sulfate pool is thought to be an important factor in the ability of the body to conjugate acetaminophen following ingestion of toxic doses of the drug. Thus, it appears that in the case of acetaminophen, sulfation results in detoxification rather than activation.

More recent studies have been carried out on the mechanism of action of acetaminophen, which is also called paracetamol. The removal of the ethyl group of phenacetin has been reported to be induced by cigarette smoke (Kahn *et al.*, 1985). Like phenacetin, the reactive metabolite of acetaminophen is thought to be the *N*-acetyl-*p*-benzoquinone imine (Coughtrie and Sharp, 1990). These authors have purified and characterized the aryl (or phenol) sulfotransferase responsible for the conjugation reaction. It is a heretofore unknown isozyme of rat liver sulfotransferase. There was a profound sex difference in paracetamol sulfotransferase ability, with males having approximately twice the levels observed in females. It was determined by immunoblot analysis that this difference was probably the result of reduced enzyme protein levels in the females. Humans have also been reported to *N*-hydroxylate phenacetin and acetaminophen, with about 30% of the metabolized compounds being detected as the *N,O*–sulfates (Veronese and McLean, 1991).

C. Aminoazo Dyes and Benzidines

The aminoazo dyes represent one of the oldest known human carcinogen groups (Rehn, 1895). Both *N,N*-dimethyl-4-aminoazobenzene (DAB) and *N*-methyl-4-aminoazobenzene (MAB) are potent liver carcinogens in Sprague–Dawley rats when administered in the diet (Miller and Baumann, 1945; Miller and Miller, 1948). The unmethylated derivative 4-aminoazobenzene (AB) was found to be a somewhat less potent carcinogen (Miller and Miller, 1953). AB, however, was found to be as potent a carcinogen as DAB and MAB after a single ip dose was administered to male 12-day-old B6C3F$_1$ mice (Delclos *et al.,* 1984). The principal persistent DNA adduct from the azobenzenes was established by these workers to be the *N*-(deoxyguanosin-8-yl)-4-aminoazobenzene. It was also shown that the apparent proximate carcinogens derived from AB and from MAB were the corresponding *N*-hydroxy derivatives (Kadlubar *et al.,* 1976). The *N*-demethylation of DAB and MAB, presumably by a cytochrome P450 monooxygenase, was demonstrated *in vivo* (Miller and Miller, 1953) and *in vitro* with rat hepatic microsomes (Mueller and Miller, 1953). The activation of the *N*-hydroxyaminoazo dyes by sulfation is a little more problematic than the total evidence for sulfation of AAF and its derivatives. Part of the reason is that the sulfate esters are much less stable, and indeed, the hydroxylamines themselves are difficult to isolate. Nevertheless, the evidence seems to indicate that sulfation is important, although activation by acetylation or possibly other conjugation reactions may also be involved (Fig. 8). It was shown that variation of dietary sulfate modulated the carcinogenicity of 3′-methyl-4-dimethylaminoazobenzene (3-Me-DAB). Rats treated with 0.84% sodium sulfate in their diet along with the carcinogen developed liver tumors with a shorter latency period than rats given a control diet (Blunck and Crowther, 1975). Experiments such as this are very difficult to interpret; whereas the results may support activation by sulfation, other explanations may be equally valid. Stronger evidence for the sulfation hypothesis was provided by Kadlubar *et al.* (1976). These workers showed that synthetic, but highly labile, *N*-hydroxy MAB was converted to a sulfate ester, which could be trapped by methionine (rather poorly) and guanosine. Male rat liver had a higher capacity to affect this transformation than female rat liver, a result consistent with what had been previously observed in the activation of *N*-hydroxy-2-AAF.

The critical experiment that established activation by sulfate ester formation in the case of the azo dyes was the finding that the binding of tritiated AB and DAB to the hepatic DNA of 12-day-old brachymorphic B6C3F$_2$ mice was 15 and 20% of that found in their phenotypically normal litter mates (Delclos *et al.,* 1986). As was mentioned earlier, the brachymorphic mice are deficient in their ability to synthesize PAPS. The incidence of hepatomas induced in these mice by AB and MAB was 11 and 29%, respectively, whereas the incidence of these tumors in the normal litter mates, treated under identical conditions, was 77 and 86%, respectively. Thus, there appears to be a lot of parallelism in the activation of the azo dyes and 2-AAF.

Another group of carcinogenic chemicals, important in the dye industry,

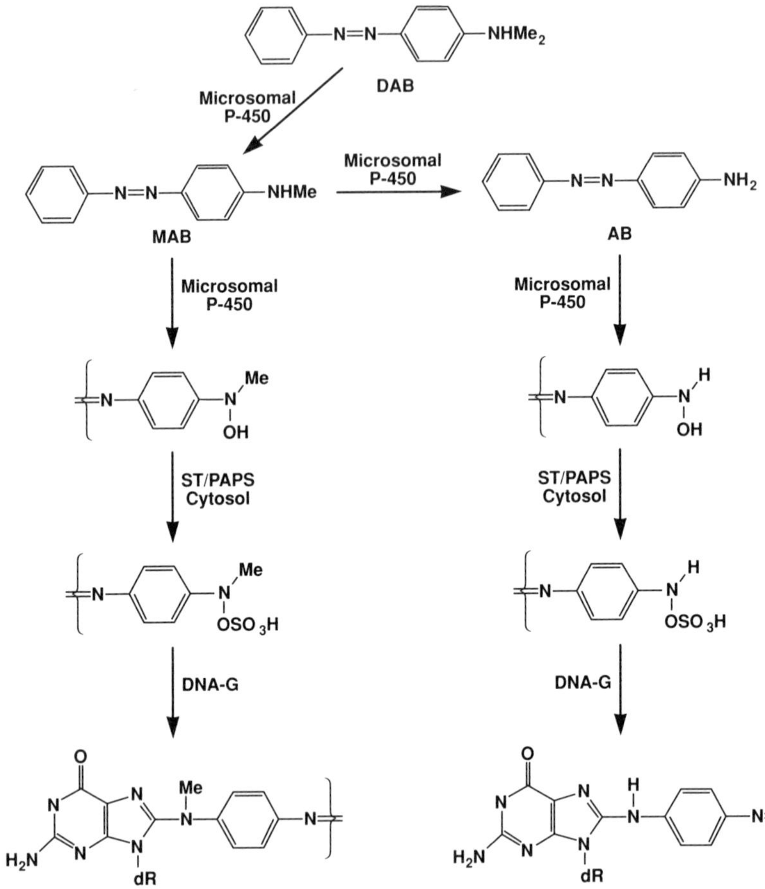

Fig. 8 Metabolism of aminoazobenzenes and DNA adduct formation from *N*-sulfate esters.

are the benzidines. The best evidence supports the hypothesis that these too are activated by sulfation. Benzidine itself is a potent human bladder carcinogen (Haley, 1975). In rodents, however, the liver is the major site of carcinogenic activity (Morton *et al.*, 1980). These workers studied the metabolic activation of *N*-hydroxy-*N,N*′-diacetylbenzidine, which itself is an *in vitro* metabolite of benzidine produced by rodent liver homogenates (Morton *et al.*, 1979). The metabolism of the chemical in rat and mouse liver postmicrosomal cytosol was dependent on the presence of PAPS, as was the reaction with methionine, which produced the methylmercapto adduct. The metabolism of the model liver carcinogen *N*-hydroxy-4′-fluoro-4-acetylaminobiphenyl (*N*-OH-FAAB) was studied with respect to sulfation in the male rat (van de Poll *et al.*, 1989). The chemical was a substrate for sulfotransferases *in vitro,* and its metabolism was inhibited by the phenolic sulfotransferase inhibitors PCP and DCNP. At least 21% of the dose was converted to the sulfate conjugate in the intact animal, and PCP and DCNP inhibited macromolecular binding *in vivo.* Deletion of inorganic sulfate from the perfusate of the isolated liver reduced DNA binding by 70–80%. Interestingly, both the acetylated and the deacetylated 8-deoxyguanosine adducts were formed, which suggests that there are two sulfotransferase-dependent pathways in the metabolism of *N*-OH-FAAB. The activation of the related compound *N*-hydroxy-4-acetylaminobiphenyl by cultured human uroepithelial cells appears to involve the *N,O*-acetyltransferase rather than a sulfotransferase (Swaminathan and Reznikoff, 1992). This result illustrates the danger of generalizing about the mechanisms of activation, even when the observed DNA adducts are the same.

D. Heteroaromatic Amines

One of the more important current problems regarding activation by sulfation concerns the heteroaromatic amines produced by the pyrolysis of meat and other foods (Sugimura and Sato, 1983). Compounds such as 2-amino-3-methylimidazo[4,5-*f*]quinoline (IQ) and 2-amino-6-methyldipyrido[1,2-a:3′,2′-*d*]imidazole (Glu-P-1) are converted to the *N*-2-hydroxyamino derivatives by cytochrome P4501A2 (Shimada *et al.,* 1989). The *N*-hydroxy derivatives are the presumed proximate carcinogens and are converted to the ultimate carcinogenic forms either by *o*-acetylation or by the formation of sulfate esters. A recent report on the *in vitro* conjugation by sulfate in human liver cytosol as well as by the cytosols from several animal species revealed interesting data (Abu-Zeid *et al.,* 1992). Thus, *N*-hydroxy-Glu-P-1 was readily converted to a DNA binding form by human hepatic cytosol in the presence of PAPS. This was also true for hepatic cytosols from the mouse, rat, guinea pig, hamster, rabbit, dog, and monkey. These cytosols also converted *N*-hydroxy-AF and *N*-hydroxy-AAF to the active sulfate esters. However, the cytosols (except that from the rat) failed to activate *N*-hydroxy-IQ to the sulfate ester, which is probably converted to the ultimate carcinogen by the acetylation pathway (Fig. 9).

It is important to realize, and the available data clearly support this, that sulfation is rarely the exclusive pathway of activation. Although at times it may be the dominant path, acetylation and possible other forms of activation may be more important for other substances, and conditions. It is also reasonably clear that sulfation works best at low substrate concentrations. Other pathways take over readily when conditions for sulfation are inappropriate.

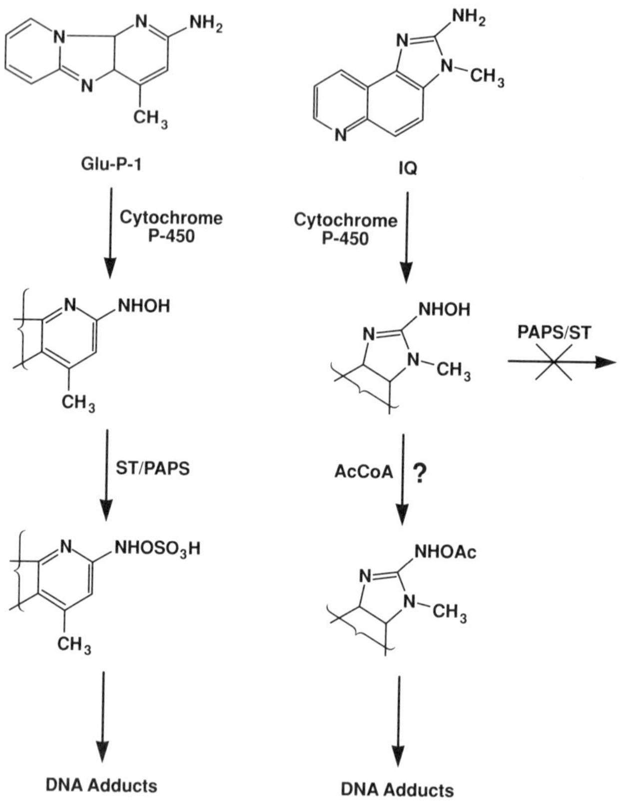

Fig. 9 Metabolism and activation of food-derived heteroaromatic amines Glu-P-1 and IQ. The former is activated by sulfation, whereas the latter is not.

III. Hydroxymethyl Aromatics and Alcohols

A. Hydroxymethyl Aromatics

The most widely accepted hypothesis for the activation of most polycyclic aromatic hydrocarbons (PAHs) involves three enzyme-catalyzed transformations. Oxidation of an aromatic ring by a cytochrome P450 monooxygenase to form an epoxide is followed by an epoxide hydrolase-catalyzed hydrolysis to form a dihydrodiol, whose isolated double bond is oxidized to form the bay region diol epoxide (Dipple *et al.*, 1984). However, various other metabolic reactions have been proposed for the activation of PAHs, including activation by sulfate formation. Thus, Watabe *et al.* (1982) proposed that the sulfate ester of 7-hydroxymethyl-12-

methylbenz[*a*]anthracene (7-OH-DMBA) was the active metabolite of 7,12-dimethylbenz[*a*]anthracene (DMBA). In subsequent papers, Watabe and co-workers demonstrated that 7-OH-DMBA was a substrate for cytosolic sulfotransferases from the rat liver and that the sulfate ester was able to bind covalently to the exocyclic amino groups of DNA adenine and guanine (Watabe *et al.*, 1985a) and to cytosolic proteins (Watabe *et al.*, 1983). These reactions are shown in Fig. 10.

Suhr *et al.* (1987) demonstrated that the hepatic DNA adducts formed

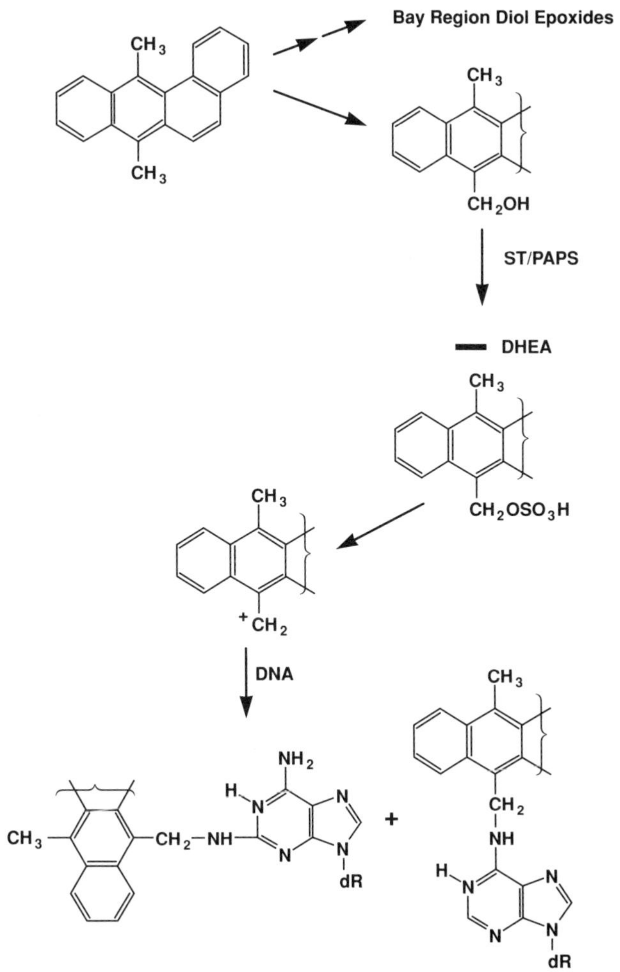

Fig. 10 Activation of 7-OH-DMBA by sulfotransferase and formation of DNA adducts: dihydroepiandrosterone (DHEA) blocks the sulfation step.

from HMBA and its sulfate ester were the same when the chemicals were administered ip to preweanling rats, and mice, although there were considerably more adducts from the ester than from the hydroxy compound. The adduct formation from HMBA was inhibited by prior administration of DHEA. These results strongly supported the hypothesis that HMBA was activated by sulfation. Subsequently, Suhr *et al.* (1991a) showed that while DHEA strongly inhibited binding in rat liver, the phenolic sulfotransferase inhibitor pentachlorophenol was not inhibitory. Thus, HMBA appears to be a good substrate for hydroxysteroid sulfotransferase. These workers also showed that while no sex differences were noted in preweanling rats, adult female rats had a much higher enzymatic activity, consistent with the known sex hormone regulation of this sulfotransferase.

Although the activation of HMBA by sulfation appears to be secure, the involvement of this pathway in the activation of the parent DMBA is much more problematical. In fact, Suhr *et al.* (1991b) demonstrated convincingly that neither DMBA or HMBA were activated to ultimate carcinogens via sulfation. The argument against sulfation rested on a number of experimental observations, but most importantly that there was no difference in the carcinogenicity of DMBA, HMBA, and the sulfate ester of HMBA and that the carcinogenicity of HMBA was not inhibited by DHEA, although the formation of adducts was inhibited. Thus it appears that although HMBA is metabolically transformed into a strongly electrophilic and mutagenic sulfate ester, this reaction is not very important in carcinogenesis by DMBA.

The activation of other hydroxymethylaromatics to electrophilic metabolites by sulfation appears to be relatively well established. This, however, is not true for the involvement of this process in the carcinogenesis by the parent hydrocarbons. Watabe *et al.* (1985b) showed that the minor metabolite of DMBA, 7,12-dihydroxymethylbenz[*a*]anthracene (DHBA), was regioselectively transformed to the 7-sulfate by rat liver cytosol in the presence of PAPS and that the sulfate was a strong bacterial mutagen. These workers (Watabe *et al.*, 1987) also showed that DHBA binding to DNA and the formation of DHBA 7-sulfate were strongly inhibited by DHEA, but not by the phenolic sulfotransferase inhibitors PCP and DCNP. The binding of DHBA *in vitro* to calf thymus DNA was strongly inhibited by glutathione transferases. This may, in fact, be a common detoxification mechanism for these metabolites.

The activation of 5-hydroxymethylchrysene by sulfation in hapatic cytosols from male and female rats, mice, guinea pigs, and hamsters was described by Okuda *et al.* (1989). This transformation was inhibited by DHEA, but not by PCP or DCNP, consistent with the involvement of a hydroxysteroid/alcohol sulfotransferase.

The important role of sulfation in the activation of the carcinogen 6-hydroxymethylbenzo[*a*]pyrene (HMBP) in the covalent binding to rat liver DNA has been demonstrated by Suhr *et al.* (1989). However, this study did not deal directly with the carcinogenicity of the compound; it would be interesting to compare HMBP with HMBA in this regard. The sulfation of 9-hydroxymethyl-10-methylanthracene (HMA) and 1-hydroxymethylpyrene (HMP) was also reported (Suhr *et al.*, 1990). These compounds were transformed to the sulfate esters *in vitro* by rat liver cytosol supplemented with PAPS, and it was shown that the adducts formed with calf thymus DNA were the same as those obtained from the synthetic sulfate esters. Cytosolic activation of HMA and HMP induced bacterial mutations, and it was shown that the sulfate esters were weak skin tumor initiators in the mouse, but stronger than the parent compounds. Interestingly, the mutagenicity of the sulfate ester of HMA was enhanced by chloride anion. Apparently the chloride reacted with the sulfate to produce the more mutagenic chloromethyl compound, as had been previously observed by Henschler and co-workers (1989). The chloride enhancement of the mutagenicity of HMP and of 4*H*-cyclopenta[*def*]chrysen-4-ol was defined further by the work of Glatt *et al.* (1993). This chloride effect was also observed in mutagenticity studies of the chemically synthesized sulfate ester of the important food-borne chemical 5-hydroxymethyfurfural (Suhr *et al.*, 1993). The mutagenicity of the sulfate was enhanced by addition of chloride. These types of data suggest that in some cases sulfate conjugation may provide an intermediate step in the ultimate activation of some xenobiotics.

B. Terpenes

Safrole (4-allyl-1,2-methylenedioxybenzene) is the best known terpene to be activated by sulfation. Safrole, a widely used spice constituent, is a hepatocarcinogen for both animals and humans (Wislocki *et al.*, 1976). The proximate carcinogenic metabolite of safrole has been determined to be 1′-hydroxysafrole, which has greater carcinogenic activity than safrole itself (Borchert *et al.*, 1973). 1′-Hydroxysafrole was found to bind to added RNA in hepatic cytosols from rats and mice in a reaction requiring the sulfate donor PAPS (Wislocki *et al.*, 1976). This implicated the sulfuric acid ester of 1′-hydroxysafrole (1′-sulfooxysafrole) as the reactive metabolite. This hypothesis was tested (Boberg *et al.*, 1983) by monitoring DNA and RNA adduct formation both *in vitro* with the specific phenol sulfotransferase inhibitor PCP as well as *in vivo* with brachymorphic mice. Adduct formation was significantly inhibited by PCP. Pretreatment of mice with PCP in the diet reduced the levels of hepatic DNA and RNA adducts from 1′-hydroxysafrole by 85% compared with those that did not receive PCP. Administration of 1′-hydroxysafrole to either adult or 12-day-old brachymorphic mice led to levels of hepatic DNA and RNA adducts that were 7- to 12-fold lower than those obtained from phenotypically normal mice of the same age. In addition, brachymorphic mice were much less responsive than their heterozygous littermates to the induction of liver tumors by 1′-hydroxysafrole. Taken together, these findings provide compelling evidence that 1′-sulfooxysafrole is the ultimate electrophilic and tumor-initiating metabolite of safrole.

The rearranged metabolite of safrole, 3′-hydroxy-isosafrole, has also been implicated in adduct formation in vitro (Boberg *et al.*, 1986). However, it was not nearly as potent in inducing hepatomas in mice as was 1′-hydroxysafrole, nor was it as efficient at alkylating hepatic nucleic acids and proteins (only 2–4 and 8–14% of the levels obtained after administration of an equimolar dose of 1′-[^{3}H]hydroxyisosafrole). The latter metabolite underwent rapid side-chain oxidation to yield 3,4-methyldioxycinnamic acid and 3,4-methylenedioxybenzoyl glycine. No glucuronide was detected from the 3′-hydroxyisosafrole. In contrast, urinary excretion of the glucuronide of 1′-hydroxysafrole acounted for 40% of the administered dose. A summary of the metabolism of both 1′-hydroxysafrole and the 3′-hydroxyisosafrole is shown in Fig. 11.

Estragole, the 4-methoxy analog of safrole, is also a hepatocarcinogen (Miller *et al.*, 1983; Drinkwater *et al.*, 1976). It appears to be metabolically activated to its 1′-hydroxy metabolite in a manner similar to that of safrole, which is then subsequently conjugated to form a reactive ester (Drinkwater *et al.*, 1976; Phillips *et al.*, 1981a). This reactive ester is more than likely a sulfuric acid ester, since the hepatocarcinogenic activities of estragole and 1′-hydroxyestragole in infant brachymorphic mice are strongly inhibitied by pretreatment with PCP (R. W. Wiseman, J. A. Miller, E. C. Miller,, and A. Liem unpublished data). The acetylenic analog of 1′-hydroxyestragole, 1′-hydroxy-2′,3′-dehydroestragole, was also found to be a potent hepatocarcinogen in mice (Fennell *et al.*, 1985). These authors demonstrated that estragole was subject to cytosolic sulfotransferase conjugation in mouse liver and found that 95% of this activity could be inhibited by PCP. The administration of PCP prior to dosing with 1′-hydroxy-2′,3′-dehydroestragole greatly inhibited (87–97%) both the covalent binding to macromolecules and the formation of hepatomas in mice. These data strongly suggest that 1′-sulfooxy-2′,3′dehydroestragole is the major electrophilic and carcinogenic metabolite of 1′-hydroxy-2′,3′-dehydroestragole in the mouse. It is curious that PCP is such a good inhibitor of sulfation of 1′-hydroxysafrole and its cogeners. These compounds are benzylic alcohols, related to the hydroxy methylaromatics discussed

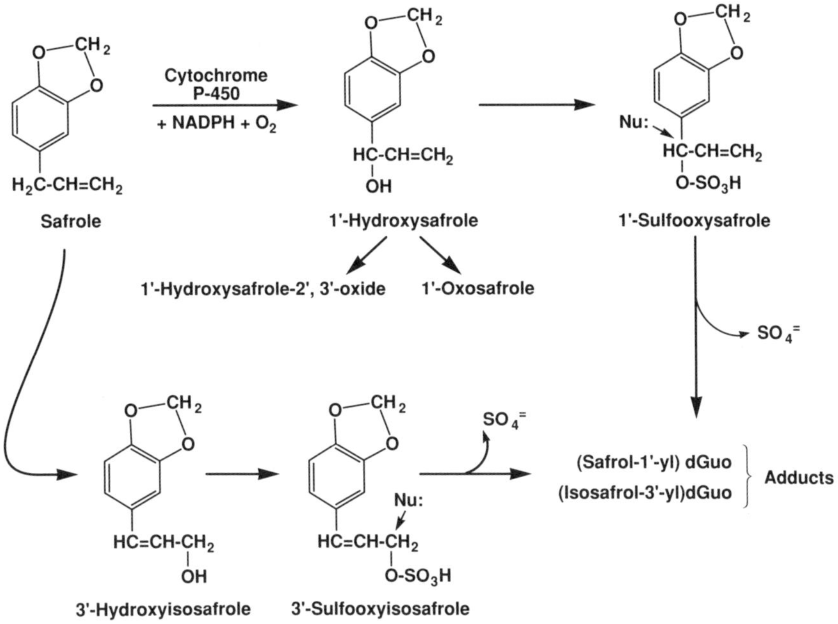

Fig. 11 Metabolic activation of the terpene safrole.

above. The sulfation of the latter was not inhibited by PCP or the related inhibitor DCNP. Based on these experiments, the hydroxy terpenes appear to be activated by the phenol sulfotransferases rather than by the hydroxysteroid alcohol enzymes.

Several pathways compete with sulfotransferases for activation of 1′-hydroxysafrole and it is likely that analogous processes occur for 1′-hydroxy-2′,3′-dehydroestragole. It is unclear, however, what level of competition exists between sulfation or glucuronidation (Stillwell *et al.*, 1974; Fennell *et al.*, 1984), oxidation to an α,β-unsaturated ketone (Oswald *et al.*, 1971), and epoxidation of the double bond (Swanson *et al.*, 1981). Both the 1′-oxo-safrole and the 1′-hydroxy-safrole-2′,3′-oxide have been shown to be formed *in vivo* and were found to be electrophilic, mutagenic, and possible carcinogenic (Wislocki *et al.*, 1977). However, the identification of the DNA adducts following *in vivo* exposure to mice to estragole and 1′-hydroxysafrole sheds some light on which intermediate may be involved in the activation of these compounds to ultimate carcinogens. In each case, 3 dGuo and 1 dAdo adducts were detected (Phillips *et al.*, 1981b) and all of these involved the covalent attachment of the carcinogen moiety to the exocyclic amino groups of the guanine or adenine residues.

The major product formed from each compound contained the 2-amino group of guanine attached to the 3′ position of *trans*-isoestragole or *trans*-isosafrole. The other two guanine adducts identified from estragole were N^2-(estragol-1′-yl)deoxyguanosine and N^2-(*cis*-isoestragol-3′-yl)deoxyguanosine. The two other guanine adducts detected from 1′-hydroxysafrole have been tentively identified as similar in structure to those of estragole. A minor product formed from both compounds was determined to involve the conjugation of the 6-amino group of adenine residues to the 3′-position of *trans*-isosafrole or *trans*-isoestragole. All of these DNA adducts are thought to be the products of active esters formed by S_N1, S_N2, or S_N2' reaction with the purine bases.

C. β-Aminoethyl Alcohols and Derivatives

The generation of new reactive metabolites by displacement of the sulfate ion by an intramolecular S_N2 reaction can be illustrated with compounds such as β-aminoalcohols. The similarity of this class to several biogenic amines makes the study of these substances especially relevant. The presence of PAPS and a sulfotransferase enzyme transforms a β-aminoalcohol such as pronethalol (Howe, 1965) to its active aziridine form. Some aziridines are known to be potent carcinogens (Lawley, 1984) in both rats and mice. The formation of 1,2-dimethyl-2-phenylaziridine (Fig. 12) from its β-aminoalcohol precursor, an isomer of ephedrine, has also been found to involve sulfate conjugation (Bicker and Fischer, 1974).

A special case of sulfate ester formation followed by the generation of a reactive intermediate by anchimeric displacement is that of β-

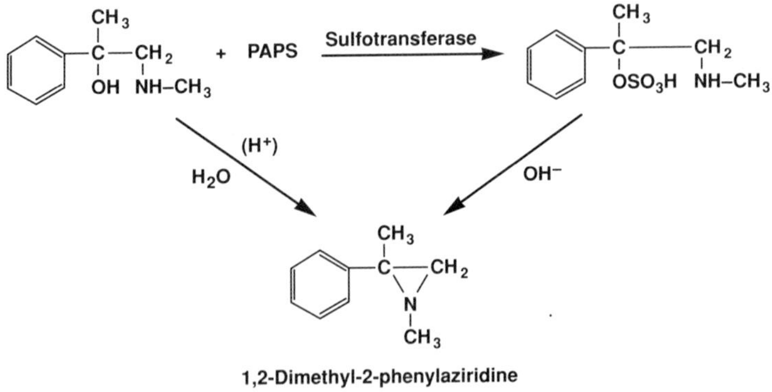

Fig. 12 Aziridine formation from 1-methyl-1-phenol-2-(methylamino)ethanol catalyzed by an alcohol sulfotransferase. This reaction is a prototype for a number of biologically active β-amino alcohols.

hydroxyalkylnitrosamines. Several members of this group [*N*-nitrosobis(2-oxopropyl)amine, *N*-nitroso(2-hydroxypropyl)(2-oxopropyl)amine, *N*-nitrobis(2-hydroxypropyl)amine, or *N*-nitroso-2,6-dimethylmorpholine)] are potent pancreatic carcinogens (Kokkinakis *et al.*, 1987), whereas others [*N*-nitrosomethyl(2-hydroxyethyl)amine (NMHEA) and *N*-nitrosodiethanolamine (NDELA)] are strong liver carcinogens (Koepke *et al.*, 1988a; Lijinsky *et al.*, 1980).

It was suggested that some β-hydroxyalkylnitrosamines could be activated by conjugation with a good leaving group such as sulfate (Michejda and Koepke, 1978). This hypothesis was based on purely chemical evidence. The *p*-toluenesulfonate (tosylate) ester of *N*-nitrosomethyl(2-hydroxyethyl)amine was shown to solvolyze very rapidly via the intermediacy of acyclic intermediate, the 3-methyloxadiazolinium ion. Both the tosylate ester and the oxadiazolinium ion were found to be potent bacterial mutagens, in contrast to NMHEA itself (Michejda *et al.*, 1979).

The alkylation of genomic DNA by NMHEA was demonstrated *in vivo* in F344 rats (Koepke *et al.*, 1988b). NMHEA gave rise to four alkylated guanine adducts when administered by gavage or in drinking water. The adducts were 7-methyl- and 7-hydroxyethylguanine and O^6-methyl- and O^6-hydroxyethylguanine. It was found that female rats had higher levels of alkylation in the liver than male rats. The alkylation of NMHEA *in vivo* was found to be almost totally blocked by the administration of propylene glycol, a competitive inhibitor of alcohol sulfotransferase (Spencer, 1960), prior to the administration of NMHEA (Kroeger-Koepke *et al.*, 1992). The only adduct that was not affected by the alcohol was 7-hydroxyethylguanine, whose formation was found to be inhibited by DCNP, a phenol sulfotransferase inhibitor (Mulder and Scholtens, 1977). This result indicated that 7-hydroxyethylguanine was formed by a different mechanism from the other adducts. These data illustrate the confusion that can arise from *in vivo* inhibition experiments. One might be led to a conclusion that DCNP inhibition of one adduct may be the result of the inhibition of a specific sulfation pathway. DCNP, and indeed other inhibitors, probably have several modes of action *in vivo*. It is highly unlikely in this case that the DCNP inhibition of adduct formation was due to the blockage of an aryl sulfotransferase.

The alkylation of DNA in primary rat liver hepatocytes was also determined (Kroeger-Koepke *et al.*, 1992) as a function of varying concentrations of added inorganic sulfate. The data revealed that methylation of DNA was sulfate-dependent up to concentrations of 3 m*M*; additional sulfate decreased the methylation. The alkylation of genomic DNA was also measured in cultured rat liver cells (Clone 9, ATCC) as a function both of added sulfate and of NMHEA concentrations. The results showed a dependence of methylation on sulfate and on NMHEA. Taken together, these experiments strongly suggest the involvement of a sulfotransferase in the activation of NMHEA to a genotoxic agent (Fig. 13). The data on NMHEA activation by sulfation illustrate an important point. It became clear from those studies that detoxification machinery in mammals is highly flexible. Pressure exerted on one pathway can cause compensatory activity increases in other pathways. It is reasonable to assume that this metabolic switching is a rather common phenomenon.

Evidence for activation of other β-hydroxynitrosamines to genotoxic agents by sulfation has also been reported. Sterzel and Eisenbrand (1986) showed that single-strand breaks in rat liver DNA induced by NDELA and *N*-nitroso-3-hydroxymorpholine could be abolished with DCNP, which was administered ip as a solution in propylene glycol. The single-strand breaks induced by NMHEA and *N*-nitrosoethyl(2-hydroxyethyl)amine could be partially abolished by pretreatment with the DCNP in propylene glycol, whereas no effect was observed when the single-strand breaks were induced by *N*-nitrosodiethylamine. It was

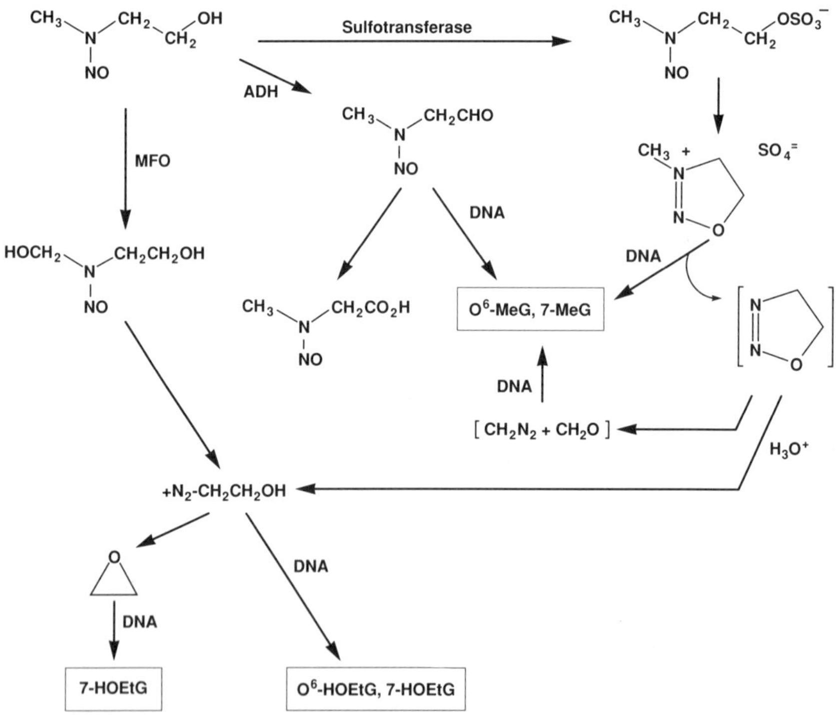

Fig. 13 The complex activation scheme of *N*-nitroso-methyl-(2-hydroxyethyl)amine. Activation by sulfotransferase is only one of several possible reactions.

later shown by these workers (personal communication) and by Kroeger Koepke *et al.* (1992) that this inhibition was due to propylene glycol rather than DCNP.

The sulfate ester of *N*-nitroso-(2-hydroxypropyl)(2-oxopropyl)amine (HPOP) has been detected in significant quantity in the urine of hamsters treated with the chemical, although the sulfate ester was not detected in the urine of similarly treated rats (Kokkinakis *et al.*, 1985). This sulfate ester resists closing the corresponding oxadiazolinium ion because that ring closure requires more energy than when the sulfate ester is on a primary carbon (Koepke *et al.*, 1979). Thus, it is likely that the formation of the sulfate ester in the case of HPOP probably leads to enhanced elimination rather than activation.

In addition, it has been suggested (Loeppky *et al.*, 1984) that β-hydroxyethylnitrosamines could be activated to alkylating agents by oxidation of the hydroxyl group to an aldehyde. They found that *N*-nitroso-(4-chlorophenyl)ethanalamine decomposed spontaneously to 4-chlorophenyldiazonium ion. In addition, it was found that NDELA, which is not mutagenic in the *Salmonella typhimurium* bacterial mutagenesis assay, was transformed to a potent mutagen by alcohol dehydrogenase (Eisenbrand *et al.*, 1984). Indirect evidence was found for the formation of the NDELA aldehyde *in vivo* by the isolation of the corresponding carboxylic acid, *N*-nitroso-*N*-hydroxyethylglycine, from the urine of NDELA-treated rats (Airoldi *et al.*, 1983).

D. Other Compounds

The activation of 2-nitropropane (2-NP) by a phenol sulfotransferase has recently been reported (Sodum *et al.*, 1993). This unlikely substrate for a sulfotransferase illustrates the fact that there may be other xenobiotics whose structures do not immediately suggest themselves as substrates for this activation pathway. 2-NP caused the formation of the novel lesion 8-aminoguanine (8-AG) in DNA. The formation of this modified base was inhibited in rats by PCP and DCNP but not by the alcohol sulfotransferase inhibitor propylene glycol. Partially purified sulfotransferase from rat liver catalyzed the formation of 8-AG from the nitronate form of 2-NP, and this reaction was inhibited by N-OH-2-AAF. These experiments suggest aryl sulfotransferase IV as the enzyme responsible for the activation. Although the details of the exact mechanism for this remarkable transformation have not been worked out, it is likely that the aminating agent is hydroxylamine-*O*-sulfate. Other secondary nitroalkanes, but not primary nitroalkanes, cause amination of DNA guanine.

Finally, some sulfate conjugation causes beneficial activation. The case of minoxidil is case in point. The parent compound (Fig. 14) is an *N*-oxide

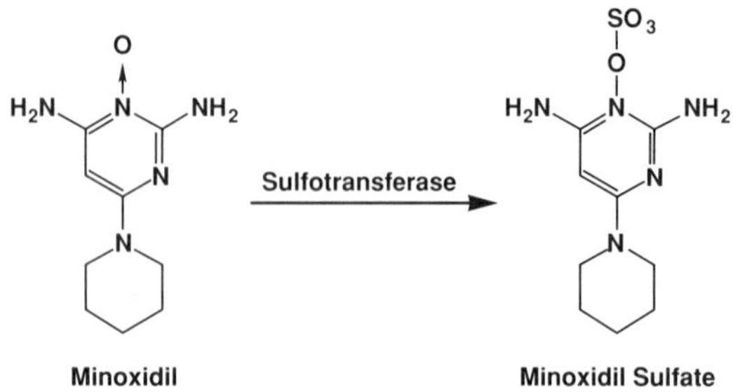

Fig. 14 Sulfotransferase activation of minoxodil.

that is a substrate for a phenol sulfotransferase (Meisheri *et al.*, 1993; Wong *et al.*, 1993). Minoxidil is an antihypertensive drug and is also used to combat male baldness. Both of these activities require the sulfate conjugate (Leblanc *et al.*, 1989; Buhl *et al.*, 1990).

IV. Summary

The foregoing pages presented a substantial body of data that established that sulfotransferase conjugation can transform many xenobiotics into agents that can modify cellular macromolecules. However, activation by sulfation is rarely the only metabolic pathway that is open to these compounds; other pathways can become more important in response to a variety of factors. This metabolic switching can be produced by substrate concentration, cofactor availability, kinetic factors that dictate the velocity of the various possible conjugation reactions, and, in some cases, competition between Phase-I and Phase-II metabolism. Also, it is important to realize that demonstration of activation by sulfate ester formation *in vitro* does not necessarily mean that a similar activation process will occur *in vivo*. Experience also teaches that argument by analogy can be very misleading in the case of sulfate activation. Small structural differences can upset the delicate balance between sulfate activation and the various other competing pathways. Nevertheless, sulfation is an important mechanism by which a number of chemicals are transformed to their activated forms.

Acknowledgment

Research sponsored by the National Cancer Institute, DHHS, under Contract NO1-CO-74101 with ABL. The contents of this publication do not necessarily reflect the views or policies of the Department of Health and Human Services, nor does mention of trade names, commerical products, or organizations imply endorsement by the United States Government.

References

Abu-Zeid, M., Yamazoe, Y., and Kato, R. (1992). Sulfotransferase-mediated DNA binding of N-hydroxyarylamines(amides) in liver cytosols from human and experimental animals. *Carcinogenesis* (*London*) **13,** 1307–1314.

Airoldi, L., Bonfanti, M., Benfenati, E., Tavecchia, P., and Fanelli, R. (1983). Identification of an acidic metabolite of N-nitrosodiethanoline isolated from rat urine. *Biomed. Mass Spectrom.* **10,** 334–337.

Balistreri, W. F., Zimmer, L., Suchy, F. J., and Bove, K. E. (1984). Bile salt sulfotransferase: Alterations during maturation and non-inducibility during substrate ingestion. *J. Lipid Res.* **25,** 228–235.

Bartsch, H., Dworkin, M., Miller, J. A., and Miller, E. C. (1972). Electrophilic N-acetoxyaminoarenes derived from carcinogenic N-hydroxyl-2-acetylaminoarenes by enzymatic deacetylation and transacetylation in liver. *Biochim. Biophys Acta* **286,** 272–298.

Bartsch, H., Dworkin, C., Miller, E. C., and Miller, J. A. (1973). Formation of electrophilic N-acetoxyarylamines in cytosols from rat mammary gland and other tissues by transacetylation from the carcinogen N-hydroxy-4-acetylaminobiphenyl. *Biochim. Biophys Acta* **304,** 42–55.

Beland, F. A., Miller, D. W., and Mitchum, R. K. (1983). Synthesis of the ultimate hepatocarcinogen, 2-acetylaminofluorene-N-sulphate. *J. Chem. Soc. Chem. Commun.* pp. 30–31.

Bicker, U., and Fischer, W. (1974). Enzymatic aziridine synthesis from β aminoalcohols - A new example of endogenous carcinogen formation. *Nature* (*London*) **249,** 344–345.

Blunck, J. M., and Crowther, C. E. (1975). Enhancement of azo dye carcinogenesis by dietery sodium sulfate. *Eur. J. Cancer* ***11,*** 23–31.

Boberg, E. W., Miller, E. C., Miller, J. A., Poland, A., and Liem, A. (1983). Strong evidence from studies with brachymorphic mice and pentachlorophenol that 1′-sulfooxysafrole is the major ultimate electrophilic and carcinogenic metabolite of 1′-hydroxysafrole in mouse liver. *Cancer Res.* **43,** 5163–5173.

Boberg, E., W., Miller, E. C., and Miller, J. A. (1986). The metabolic sulfonation and side-chain oxidation of 3′-hydroxyisosafrole in the mouse and its activity as a hepatocarcinogen relative to 1′-hydroxysafrole. *Chem.-Biol. Interact.* **59,** 73–97.

Borchert, P., Miller, J. A., Miller, E. C., and Shires, T. K. (1973). 1′-Hydroxysafrole, a proximate carcinogenic metabolite of safrole in the rat and mouse. *Cancer Res.* **33,** 590–600.

Buhl, A. E., Waldon, D. J., Baker, C. A., and Johnson, G. A. (1990). Minoxidil sulfate is the active metabolite that stimulates hair follicles. *J. Invest. Dermatol.* **95,** 553–557.

Comer, K. A., Falany, J. L., and Falany, C. N. (1993). Cloning and expression of human liver dehydroepiandrosterone sulphotransferase. *Biochem. J.* **289,** 233–240.

Coughtrie, M. W., and Sharp, S. (1990). Purification and immunochemical characterization of a rat liver sulphotransferase conjugating paracetamol. *Biochem. Pharmacol.* **40,** 2305–2313.

DeBaun, J. R., Rowley, J. Y., Miller, E. C., and Miller, J. A. (1968). Sulfotransferase activation of N-hydroxy-2-acetylaminofluorene in rodent livers susceptible and resistent to this carcinogen. *Proc. Soc. Exp. Biol. Med.* **129,** 268.

DeBaun, J. R., Miller, E. C., and Miller, J. A. (1970a). N-Hydroxy-2-acetylaminofluorene sulfotransferase: Its probable role in the carcinogenesis and in protein-(methion-S-yl) binding in rat liver. *Cancer Res.* **35,** 906.

DeBaun, J. R., Smith, E. Y. R., Miller, E. C., and Miller, J. A. (1970b). Reactivity *in vivo* of the carcinogen N-hydroxy-2-acetylaminofluorene: Increase by sulfate ion. *Science* **167,** 184–186.

Delclos, K. B., Tarplay, W. G., Miller, E. C., and Miller, J. A. (1984). 4-Aminoazobenzene and N,N-dimethyl-4-aminoazobenzene as equipotent carcinogens in male C57BL/6 × C3H/He F_1 mice and characterization of N-(deoxyguanosin-8-yl)-4-aminoazobenzene as the major persistent hepatic DNA-bound dye in these mice. *Cancer Res.* **44,** 2540–2550.

Delclos, K. B., Miller, E. C., Miller, J. A., and Liem, A. (1986). Sulfuric acid esters as major ultimate electrophilic and hepatocarcinogenic metabolites of 4-aminoazobenzene and its N-methyl derivatives in infant male C57BL/6J × C3H/HeJ F_1 (B6C3F_1) mice. *Carcinogenesis (London)* **7,** 277–287.

Dipple, A., Moschel, R. C., and Bigger, C. A. H. (1984). Polynuclear aromatic hydrocarbons. *ACS Monogr.* **182,** 41–174.

Drinkwater, N. R., Miller, E. C., Miller, J. A., and Pitot, H. C. (1976). The hepatocarcinogenicity of estragole(1-allyl-4-methoxybenzene) and 1′-hydroxyestragole in the mouse and the mutagenicity of 1′-acetoxyestragole in bacteria. *J. Natl. Cancer Inst. (U. S.)* **57,** 1323–1331.

Dutton, G. J. (1975). Control of UDP-glucuronyltransferase activity. *Biochem. Pharmacol.* **24,** 1835–1841.

Eisenbrand, G., Denkel, E., and Pool, B. (1984). Alcohol dehydrogenase as an activating enzyme for N-nitrosodiethanolamine (NDELA): *In vitro* activation of NDELA to a potent mutagen in *Salmonella typhimurium*. *J. Cancer Res. Clin. Oncol.* **108,** 76–80.

Falany, C. N. (1991). Molecular enzymology of human liver cytosolic sulfotransferases. *Trends Pharmacol. Sci.* **12,** 255–259.

Falany, C. N., Vazquez, M. E., Heroux, J. A., and Roth, J. A. (1990). Purification and characterization of human liver phenol-sulfating phenol sulfotransferase. *Arch. Biochem. Biophys.* **2,** 312–318.

Fennell, T. R., Miller, J. A., and Miller, E. C. (1984). Characterization of biliary and urinary glutathione and N-acetylcysteine metabolites of the hepatic carcinogen 1′-hydroxy-safrole and its 1′-oxo metabolite in rats and mice. *Cancer Res.* **44,** 3231–3240.

Fennell, T. R., Wiseman, R. W., Miller, J. A., and Miller, E. C. (1985). Major role of hepatic sulfotransferase activity in the metabolic activation, DNA adduct formation, and carcinogenicity of 1′-hydroxy-2′,3′-dehydroestragole in infant male C57BL/6J × C3H/HeJF_1 mice. *Cancer Res.* **45,** 5310–5320.

Glatt, H., Henschler, R., Frank, H., Seidel, A., Yang, C., Abu-Shqaru, E., and Harvey, R. G. (1993). Sulfotransferase-mediated mutagenicity of 1-hydroxymethylpyrene and 4H-cyclopenta[def]chrysen-4ol and its enhancement by chloride anions. *Carcinogenesis (London)* **14,** 599–602.

Green, M. D., and Fisher, L. J. (1981). Age and sex-related differences in acetaminophen metabolism in the rat. *Life Sci.* **29,** 2421–2428.

Gutmann, J. R., Malejka-Giganti, D., Barry, E. J., and Rydell, R. E. (1972). On the correlation between hepatocarcinogenicity of the carcinogen N-2-fluorenylacetamide and its metabolic activation by the rat. *Cancer Res.* **32,** 1554–1561.

Haley, T. J. (1975). Benzidine revisited: A review of the literature and problems associated with use of benzidine and its cogeners. *Clin. Toxicol.* **8,** 13–42.

Henschler, R., Pauly, K., Godtel, U., Seidel, A., Oesch, F., and Glatt, H. R. (1989). The mutagenicity of sulfate esters in *Salmonella typhimurium* is strongly affected by the ions present in the exposure medium. *Mutagenesis* **4,** 309.

Heroux, J. A., and Roth, J. A. (1988). Physical characterization of a monoamine sulfating form of phenol sulfotransferase from human platelets. *Mol. Pharmacol.* **34,** 194–199.

Hinson, J. A., and Mitchell, J. R. (1976). N-Hydroxylation of phenacetin by hamster liver microsomes *Drug Metab. Dispos.* **4,** 430–435.

Hinson, J. A., Nelson, S. D., and Gillette, J. R. (1979). Metabolism of [p-^{18}O]-phenacetin:

The mechanism of activation of phenacetin to reactive metabolites in hamsters. *Biochem. Pharmacol.* **15,** 419–427.

Howe, R. (1965). Carcinogenicity of "alderlin" (pronethalol) in mice. *Nature (London)* **207,** 594–595.

Humphreys, W. G., Kadlubar, F. F., and Guengerich, F. P. (1992). Mechanism of C^8 alkylation of guanine residues by activated arylamines: Evidence for initial adduct formation at the N^7 position. *Proc. Natl. Acad. Sci. U.S.A.* **89,** 8278–8282.

Irving, C. C. (1966). Enzymatic deacetylation of N-hydroxy-2-acetylaminofluorene by liver microsomes. *Cancer Res.* **26,** 1390–1396.

Isaka, H., Yoshii, H., Otsuji, A., Koike, M., Nagai, Y., Koura, M., Sugiyasu, K., and Kanabayashi, T. (1979). Tumors of Sprague-Dawley rats induced by long-term feeding of phenacetin. *Gann* **70,** 29–36.

Iwasaki, K., Shiraga, T., Noda, K., Tada, K., and Noguchi, N. (1983). Metabolism of tiaramide *in vitro*. III. Sulphoconjugation of alicyclic amino and hydroxyl group by hepatic 105,000 g supernatants. *Xenobiotica* **13,** 565–574.

Johansson, S., and Angervall, L. (1976). Urothelial changes of the renal papillae in Sprague-Dawley rats induced by long-term feeding of phenacetin. *Acta Pathol. Microbiol. Scand., Sect. A.* 84A, 375–383.

Kadlubar, F. F., Miller, J. A., and Miller, E. C. (1976). Hepatic metabolism of N-hydroxy-N-methyl-4-aminoazobenzene and other N-hydroxy arylamines to reactive sulfunic acid esters. *Cancer Res.* **36,** 2350–2359.

Kahn, G. C., Boobis, A. R., Brodie, M. J., Toverrud, E., Murry, S., and Davies, D. S. (1985). Phenacetin O-deethylase: An activity of a cytochrome P-450 showing genetic linkage with that catalysing the 4-hydroxylation of debrisoqine? *Br. J. Clin. Pharmacol.* **20,** 67–76.

Kane, R. E, and Chen, L. J. (1987). Influence of gonadal hormones upon rat hepatic acetaminophen sulfotransferases. *Drug Metab. Dispos.* **15,** 725–728.

King, C. M., and Allaben, W. T. (1980). Arylhydroxamic acid acyltransferase. *In* "Enzymatic Basis of Detoxificaiton" (W. B. Jacoby, ed.), Vol. 2, pp. 187–197. Academic Press, New York.

King, C. M., and Olive, C. W. (1975). Comparative effects of strain, species and sex on the acyltransferase and sulfotransferase-catalyzed activations of N-hydroxy-N-2-fluorenylacetamide. *Cancer Res.* **35,** 906–912.

King, C. M., and Phillips, B. (1968). Enzyme-catalyzed reactions of the carcinogen N-hydroxyl-2-fluorenylacetamide with nucleis acid. *Science* **159,** 1351.

Koepke, S. R., Kupper, R., and Michejda, C. J. (1979). Unusually facile solvolysis of primary tosylates. A case for participation by the N-nitroso group. *J. Org. Chem.* **44,** 2718–2722.

Koepke, S. R., Cresia, D. R., Knutsen, G. L., and Michejda, C. J. (1988a). Carcinogenicity of hydroxyalkylnitrosamines in F344 rats: Contrasting behavior of β- and α-hydroxylated nitrosamines. *Cancer Res.* **48,** 1533–1537.

Koepke, S. R., Kroeger-Koepke, M. B., Bosan, W., Thomas, B. J., Alvord, W. G., and Michejda, C. J. (1988b). Alkylation of DNA in rats by N-nitroso-(2-hydroxyethyl)amine: Dose response and persistence of the alkylated lesions *in vivo*. Cancer Res. **48,** 1537–1542.

Kokkinakis, D. M., Hollenberg, P. F., and Scarpelli, D. G. (1985). Major urinary metabolites in hamsters and rats treated with N-nitroso(2-hydroxypropyl)(2-oxopropyl)amine. *Cancer Res.* **45,** 3586–3592.

Kokkinakis, D. M., Scarpelli, D. G., Subbarao, V., and Hollenberg, P. F. (1987). Species differences in the metabolism of N-nitroso(2-hydroxypropyl)(2-oxopropyl)amine. *Carcinogenesis (London)* **8,** 295–303.

Kolanczyk, R. C., Rutks, I. R., and Gutmann, H. R. (1991). The catalytic effect of bovine serum albumin on the ortho rearrangement of the potential ultimate carcinogen, N-(sulfooxy)-2-(acetylamino)fluroene, generated enzymatically from N-hydroxy-2-(acetylamino)fluorene and evidence for substrate specificity of the enzymatic sulfonation of arylhydroxamic acids. *Chem. Res. Toxicol.* **4,** 187–194.

Kroeger Koepke, M. B., Koepke, S. R., Hernandez, L., and Michejda, C. J. (1992). Activation of β-hydroxyalkylnitrosamine to an alkylating agent: Evidence for the involvement of a sulfotransferase. *Cancer Res.* **52,** 3300–3305.

Lai, C.-C., Miller, J. A., Miller, E. C., and Liem, A. (1985). N-Sulfooxy-2-aminofluorene is the major ultimate electrophilic and carcinogenic metabolite of N-hydroxy-2-acetylaminofluorene in the liver of infant male C57BL/6J × C3H/HeJ F_1 (B6C3F_1) mice. *Carcinogenesis (London)* **6,** 1037–1045.

Lai, C.-C., Miller, E. C., Miller, J. A., and Liem, A. (1987). Initiation of hepatocarcinogenesis in infant male B6C3F_1 mice by N-hydroxy-2-aminofluorene or N-hydroxy-2-acetylaminofluorene depends primarily on metabolism to N-sulfooxy-2-aminofluorene and formation of DNA-(deoxyguanosin-8-yl)-2-aminofluorene adducts. *Carcinogenesis (London)* **8,** 471–478.

Lawley, P. D. (1984). Carcinogenesis by alkylating agents. *ACS Monogr.* **182,** 325–484.

Leblanc, N., Wilde, D. W., Keef, K. D., and Hume, J. R. (1989). Electrophysiological mechanisms of minoxidil sulfate induce vasodilation of rabbit portal vein. *Circ. Res.* **65,** 1102–1111.

Lijinsky, W., Reuber, M. D., and Manning, W. B. (1980). Potent carcinogenicity of N-nitrosodiethanolamine in rats. *Nature (London)* **288,** 589–590.

Loeppky, R. N., Tomasik, W., Kovacs, D. A., Outram, J. R., and Byington, K. H. (1984). Alternative bioactivation routes for β-hydroxynitrosamines. Biochemical and chemical model studies. *IARC Sci. Publ.* **57,** 429–436.

Meerman, J. H. N., Beland, F. A., and Mulder, G. J. (1981). Role of sulfation in the formation of DNA adducts from N-hydroxy-2-acetylaminofluorene in rat liver *in vivo*. Inhibition of N-acetylated aminofluorene adduct formation by pentachlorophenol. *Carcinogenesis (London)* **2,** 413–416.

Meerman, J. H. N., Beland, F. A., Ketterer, B., Srai, S. K. S., Bruins, A. P., and Mulder, G. J. (1982). Identification of glutathione conjugates formed from N-hydroxy-2-acetylaminofluorene in the rat. *Chem.-Biol. Interact.* **39,** 149–168.

Meisheri, K. D., Johnson, G. A., and Puddington, L. (1993). Enzymatic and non-enzymatic sulfation mechanisms in the biological actions of minoxidil. *Biochem. Pharmacol.* **45,** 271–279.

Michejda, C. J., and Koepke, S. R. (1978). Powerful anchimeric effect of the N-nitroso group. *J. Am. Chem. Soc.* **100,** 1959–1960.

Michejda, C. J., Andrews, A. W., and Koepke, S. R. (1979). Derivatives of side-chain hydroxylated nitrosamines: Direct acting mutagens in *Salmonella typhimurium*. *Mutat. Res.* **67,** 301–308.

Miller, E. C., Swanson, A. B., Phillips, D. H., Fletcher, T. L., Liem, A., and Miller, J. A. (1983). Structure-activity studies of the carcinogenicities in the mouse and rat of some naturally occurring and synthetic alkenylbenzene derivatives related to safrole and estragole. *Cancer Res.* **43,** 1124–1134.

Miller, J. A., and Baumann, C. A. (1945). The carcinogenicity of certain azo dyes related to p-dimethylaminoazobenzene. *Cancer Res.* **5,** 227–234.

Miller, J. A., and Miller, E. C. (1948). The carcinogenicity of certain derivatives of *p*-dimethylaminoazobenzene in the rat. *J. Exp. Med.* **87,** 139–156.

Miller, J. A., and Miller, E. C. (1953). The carcinogenic aminoazo dyes. *Adv. Cancer Res.* **1,** 339–396.

Miller, J. A., Crames, J. W., and Miller, E. C. (1960). The N- and ring-hydroxylation of 2-acetylaminofluorene during carcinogenesis in the rat. *Cancer Res.* **20,** 950–962.

Montieth, D. K. (1992). Inhibition of sulfotransferase in primary cultures of human hepatocytes affecting metabolism and binding of 2-acetylaminofluorene. *Cancer Lett.* **64,** 109–115.

Morton, K. C., King, C. M., and Baetcke, K. P. (1979). Metabolism of benzidine to N-hydroxy-N,N′-diacetylbenzidine and subsequent nucleic acid binding and mutagenicity. *Cancer Res.* **39,** 3107–3113.

Morton, K. C., Beland, F. A., Evans, F. E., Fullerton, N. F., and Kadlubar, F. F. (1980). Metabolic activation of N-hydroxy-N,N′-diacetylbenzidine by hepatic sulfotransferase. *Cancer Res.* **40,** 751–757.

Muller, G. C., and Miller, J. A. (1953). The metabolism of methylated aminoazo dyes. II. Oxidative demethylation by rat liver homogenates. *J. Biol. Chem.* **202,** 579–587.

Mulder, G. J. (1981). Generation of reactive intermediates from xenobiotics by sulfate conjugation: Potential role in chemical carcinogenesis. *In* "Sulfation of Drugs and Other Compounds" (G. J. Mulder, ed.), pp. 213–226. CRC Press, Boca Raton, FL.

Mulder, G. J. (1986). Sex differences in drug conjugation and their consequences for drug toxicity. Sulfation, glucuronidation, and glutathione conjugation. *Chem.-Biol. Interact.* **57,** 1–15.

Mulder, G. J., and Jakoby, W. B. (1990). *In* "Conjugation Reactions in Drug Metabolism: An Integrated Approach" (G. J. Mulder, ed.), pp. 107–161. Taylor & Francis, London.

Mulder, G. L., and Scholtens, E. (1977). Phenol sulphotransferase and uridine diphosphate glucuronyl transferase from rat liver *in vivo* and *in vitro*. *Biochem. J.* **165,** 533–559.

Mulder, G. J., Hinson, J. A., and Gillette, J. R. (1977). Generation of reactive metabolites of N-hydroxyphenacetin by glucuronidation and sulfation. *Biochem. Pharmacol.* **26,** 189–196.

Mulder, G. J., Hinson, J. A., and Gillette, J. R. (1978). Conversion of the N-O-glucuronide and N-O-sulfate conjugates of N-hyroxyphenacetin to reactive intermediates. *Biochem. Pharmacol.* **27,** 1641–1649.

Nery, R. (1971). The binding of radioactive label from labeled phenacetin and related compounds to rat tissues *in vivo* and to nucleic acids and bovine plasma albumin *in vitro*. *Biochem. J.* **122,** 311–315.

Okuda, H., Nojima, H., Watanabe, N., and Watabe, T. (1989). Sulphotransferase-mediated activation of the carcinogen 5-hydroxymethyl-chrysene. *Biochem. Pharmacol.* **38,** 3003–3009.

Oswald, E. O., Fishbein, L., Corbett, B. J., and Walker, M. P. (1971). Identification of tertiary aminomethylenedioxypropiophenones as urinary metabolites of safrole in the rat and guinea pig. *Biochim. Biophys. Acta* **230,** 237–247.

Phillips, D. H., Miller, J. A., Miller, E. C., and Adams, B. (1981a). Structures of the DNA adducts formed in mouse liver after administration of the proximate hepatocarcinogen 1′-hydroxyestragole. *Cancer Res.* **41,** 176–186.

Phillips, D. H., Miller, J. A., Miller, E. C., and Adams, B. (1981b). N^2 Atom of guanine and N^6 atom of adenine residues as sites for covalent binding of metabolically activated 1′-hydroxysafrole to mouse liver DNA *in vivo*. *Cancer Res.* **41,** 2664–2671.

Raheja, K. L., Linscheer, W. G., and Cho, C. (1983). Hepatotoxicity and metabolism of acetaminophen in male and female rats. *J. Toxicol. Environ. Health* **12,** 143–147.

Rehn, L. (1895). Blasengeschwülste bei Fuchsin-Arbeiten. *Arch. Klin. Chir.* **50,** 588–600.

Reiter, C., and Weinshilboum, R. (1982). Platelet phenol sulfotransferase activity: Correlation with sulfate conjugation of acetaminophen. *Clin. Pharmacol. Ther.* **32,** 612–621.

Ringer, D. P., Norton, T. R., and Howell, B. A. (1990). 2-Acetylaminofluorene-mediated alteration in the level of liver arylsulfotransferase IV during rat hepatocarcinogenesis. *Cancer Res.* **50,** 5301–5307.

Ringer, D. P., Norton, T. R., and Self, R. R. (1992). Reaction product inactivation of aryl sulfotransferase IV following electrophilic substitution by the sulfuric acid ester of N-hyroxy-2-acetylaminofluorene. *Carcinogenesis* (*London*) **13,** 107–112.

Shimada, T., Iwasaki, M., Martin, M. V., and Guengerich, F. P. (1989). Human liver microsomal cytochrome P-450 enzymes involved in the bioactivation of procarcinogens detected by umu gene response in *Salmonella typhimurium* TA 1535/pSK1002. *Cancer Res.* **49,** 3218–3228.

Shudo, K., Ohta, T., Orihara, Y., Okamoto, T., Nagao, M., Takahashi, Y., and Sugimura, T. (1978). Mutagenicities of phenacetin and its metabolites. *Mutat. Res.* **58,** 367–370.

Sodum, R. S., Sohn, O. S., Nie, G., and Fiala, E. S. (1993). Aryl sulfotransferase activation of 2-nitropropane (2-NP) and related secondary nitroalkanes. *Proc. Am. Assoc. Cancer Res.* **34,** Abstr. 933, 157.

Spencer, B. (1960). Endogenous sulfate acceptors in rat liver. *Biochem. J.* **77,** 294–304.

Sterzel, W., and Eisenbrand, G. (1986). N-Nitrosodiethanolamine is activated in the rat to an ultimate genotoxic metabolite by sulfotransferase. *J. Cancer Res. Clin. Oncol.* **111,** 20–24.

Stillwell, W. G., Carman, M. J., Bell, L. and Horning, M. G. (1974). The metabolism of safrole and 2′-3′-epoxysafrole in the rat and guinea pig. *Drug Metab. Dispos.* **2,** 489–498.

Sugahara, K., and Schwartz, N. B. (1982). Defect in 3′-phosphoadenosine-5′-phosphosulfate synthesis in brachymorphic mice. I. Characterization of the defect. *Arch. Biochem. Biophys.* **214,** 589–601.

Sugimura, T., and Sato, S. (1983). Mutagens-carcinogens in foods. *Cancer Res.* **43,** Suppl., 2415–2421.

Suhr, Y.-J., Lai, C.-C., Miller, J. A., and Miller, E. C. (1987). Hepatic DNA and RNA adduct formation from the carcinogen 7-hydroxy-12-methylbenz[*a*]anthracene and its electrophilic sulfuric acid ester metabolite in preweanling rats and mice. *Biochem. Biophys. Res. Commun.* **144,** 576–582.

Suhr, Y.-J., Leim, A., Miller, E. C., and Miller, J. A. (1989). Metabolic activation of the carcinogen 6-hydroxymethylbenzo[*a*]pyrene: Formation of an electrophilic sulfuric acid ester and benzylic DNA adducts in rat liver *in vivo* and in reactions *in vitro*. *Carcinogenesis* (*London*) **10,** 1519–1528.

Suhr, Y.-J., Blomquist, J. C., Liem, A., and Miller, J. A. (1990). Metabolic activation of 9-hydroxymethyl-10-methylanthracene and 1-hydroxymethylpyrene to electrophilic, mutagenic and tumorigenic sulfuric acid esters by rat hepatic sulfotransferase activity. *Carcinogenesis* (*London*) **11,** 1451–1460.

Suhr, Y.-J., Liem, A., Miller, E. C., and Miller, J. A. (1991a). Age- and sex-related differences in activation of the carcinogen 7-hydroxymethyl-12-methylbenz[*a*]anthracene to an electrophilic sulfuric acid ester metabolite in rats. *Biochem. Pharmacol.* **41,** 213–221.

Suhr, Y.-J., Liem, A., Miller, E. C., and Miller, J. A. (1991b). 7-Sulfooxymethyl-12-methylbenz[*a*]anthracene is an electrophilic mutagen, but does not appear to play a role in carcinogenesis by 7,12-dimethylbenz[*a*]anthracene or 7-hydroxymethyl-12-methylbenz[*a*]anthracene. *Carcinogenesis* (*London*) **12,** 339–347.

Suhr, Y.-J., Liem, A., Miller, J. A., and Tannenbaum, S. R. (1993). High mutagenicity and tumorigenicity of sulfooxymethyl and chloromethyl derivatives of 5-hydroxymethylfurfural. *Proc. Am. Assoc. Cancer Res.* **34,** Abstr. 710, 119.

Swaminathan, S., and Reznikoff, C. A. (1992). Metabolism and DNA binding of N-hyroxyl-4-acetylaminobiphenyl and N-acetox-4-acetylaminobiphenyl by cultured human uroepithelial cells. *Cancer Res.* 52, 3286–3294.

Swanson, A. B., Miller, E. C., and Miller, J. A. (1981). The side-chain epoxidation and hyroxylation of the hepatocarcinogens safrole and estragole and some related compounds by rat and mouse liver microsomes. *Biochim. Biophys. Acta* **673,** 504–516.

Sweeny, D. J., and Reinke, L. A. (1988). Sulfation of acetaminophen in isolated rat hepatocytes. *Drug Metab. Dispos.* **16,** 712–715.

van de Poll, M. L. M., Tijdens, R. B., Vondracek, P., Bruins, A. P., Meijer, D. K. F., and Meerman, J. H. N. (1989). The role of sulfation in the metabolic activation of N-hydroxy-4′-fluoro-4-acetylaminobiphenyl. *Carcinogenesis (London)* **10,** 2285–2291.

Vaught, J. B., McGarvey, P. B., Lee, M.-S., Garner, C. D., Wang, C. Y., Linsmaier-Bednar, E. M., and King, C. M. (1981). Activation of N-hydroxyphenacetin to mutagenic and nucleic acid-binding metabolites by acyltransfer, deacylation, and sulfate conjugation. *Cancer Res.* **41,** 3424–3429.

Veronese, M. E., and McLean, S. (1991). Metabolism of paracetam and phenacetin in relation to debrisoquine oxidation phenotype. *Eur. J. Clin. Pharmacol.* **40,** 547–552.

Watabe, T., Ishizuka, T., Isobe, M., and Ozawa, N. (1982). A 7-hydroxymethyl sulphate ester as an active metabolite of 7,12-dimethylbenz[*a*]anthracene. *Science* **215,** 403–405.

Watabe, T., Ishizuka, T., Hakamata, Y., Aizawa, T., and Isobe, M. (1983). Covalent binding of the proximate carcinogen 7-hydroxymethyl-12-methylbenz[*a*]anthracene (HMBA) to rat liver cytosolic protein via HMBA sulfate. *Biochem. Pharmacol.* **32,** 2120–2122.

Watabe, T., Fujieda, T., Hiratsuka, T., Ishizuka, T., Hakamata, Y., and Ogura, K. (1985a). The carcinogen, 7-hydroxymethyl-12-methylbenz[*a*]anthracene, is activated and binds covalently to DNA via a sulfate ester. *Biochem. Pharmacol.* **34,** 3002–3005.

Watabe, T., Hiratsuka, A., Ogura, K., and Endoh, K. (1985b). A reactive hydroxymethyl sulfate ester formed regioselectively from the carcinogen, 7,12-dihydroxymethylbenz[*a*]anthracene, by rat liver sulfotransferase. *Biochem. Biophys. Res. Commun.* **131,** 694–699.

Watabe, T., Hiratsuka, A., and Ogura, K. (1987). Sulphotransferase-mediated covalent binding of the carcinogen 7,12-dihydroxymethylbenz[*a*]anthracene to calf thymus DNA and its inhibition by glutathione transferase. *Carcinogenesis (London)* **8,** 445–453.

Wirth, P. J., Dybing, E., von Bahr, C., and Thorgeirsson, S. S. (1980). Mechanisms of N-hydroxyacetylarylamine mutagenicity in the Salmonella test system: Metabolic activation of N-hydroxyphenacetin by liver and kidney fraction from rat, mouse, hamster, and man. *Mol. Pharmacol.* **18,** 117–127.

Wislocki, P. G., Borchert, P., Miller, J. A., and Miller, E. (1976). The metabolic activation of the carcinogen 1′-hydroxysafrole *in vivo* and *in vitro* and the electrophilic reactivities of possible ultimate carcinogens. *Cancer Res.* **36,** 1686–1695.

Wislocki, P. G., Miller, E. C., Miller, J. A., McCoy, E. C., and Rosenkranz, H. S. (1977). Carcinogenic and mutagenic activities of safrole, 1′-hydroxysafrole, and some known or possible metabolites. *Cancer Res.* **37,** 1883–1891.

Wong, K. O., Tan, A. Y. H., Lim, B. G., and Wong, K. P. (1993). Sulphate conjugation of minoxidil in rat skin. *Biochem. Pharmacol.* **45,** 1180–1182.

Wu, S. G., and Straub, K. D. (1976). Purification and characterization of N-hydroxy-2-acetylaminofluorene sulfotransferase from rat liver. *J. Biol. Chem.* **251,** 5629–6536.

Yerokun, T., Etheredge, J. L., Norton, T. R., Carter, H. A., Chung, K. H., and Birckbichler, P. J. (1992). Characterization of a complementary DNA for rat liver aryl sulfotransferase IV and use in evaluating the hepatic gene transcript levels of rats at various stages of 2-acetylaminofluorene-induced hepatocarcinogenesis. *Cancer Res.* **52,** 4779–4786.

Part III

Glucuronide Conjugate-Dependent Toxicity

UDP-Glucuronosyltransferases and Their Role in Metabolism and Disposition of Carcinogens

Karl Walter Bock
Institute of Toxicology
University of Tübingen
72074 Tübingen, Germany

I. Introduction

Glucuronidation represents a major drug metabolizing reaction catalyzed by a superfamily of UDP-glucuronosyltransferases (UGT; EC 2.4.1.17). UGT isozymes are located in the endoplasmic reticulum and nuclear membrane of hepatocytes and in various extrahepatic tissues such as intestine, kidney, and olfactory epithelium. In general, they convert hydrophobic endo- and xenobiotics into less pharmacologically active, water-soluble products that can be excreted (with notable exceptions such as morphine 6-glucuronide, which appears to be more potent than the parent drug). Endogenous substrates include bilirubin, steroid hormones, and fat-soluble vitamins. Xenobiotic substrates include many drugs, plant constituents, environmental pollutants, and carcinogens. The present review summarizes current knowledge about UGT isozymes and about factors controlling glucuronide formation in cells. The roles of UGT isozymes in the metabolism and disposition of carcinogens are discussed with aromatic hydrocarbons and arylamines as examples. Regulation of UGT expression is described with phenol UGTs of the phenol/bilirubin UGT gene complex as examples. For other aspects, the reader is referred to previous reviews (Dutton, 1980; Burchell and Coughtrie, 1989; Jansen *et al.*, 1992; Mulder,

Advances in Pharmacology, Volume 27

1992). To limit the number of references, review articles are sometimes used as the sources of the original articles.

II. Factors Affecting Glucuronide Formation in Cells

UGTs catalyze the conjugation of the sugar moiety of UDP-glucuronic acid (UDP-GlcA) with a variety of nucleophilic acceptor groups of endo- and xenobiotics, including hydroxyl, carboxylic acid, thiol, amine, and hydroxylamine groups. The reaction is illustrated for a substrate bearing a hydroxyl function.

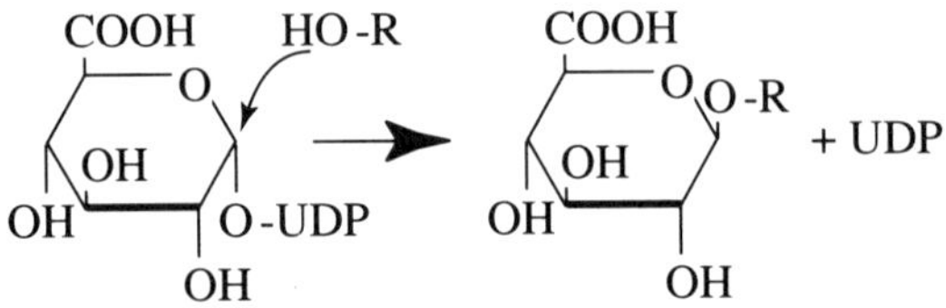

Kinetic studies are consistent with a rapid equilibrium random-order sequential mechanism (Falany *et al.*, 1987; Zakim and Dannenberg, 1992). A number of factors affect glucuronide formation in cells, such as availability of glucuronic acid, the topology of UGTs in the membrane of the endoplasmic reticulum, and the pattern of UGT isozymes.

A. Availability of UDP-Glucuronic Acid

A high rate of glucuronidation requires a sufficient supply of UDP-GlcA. Regeneration of UDP-GlcA appears to be high in the liver of fed rats, and a UDP-GlcA concentration of 0.3 μmol/g tissue wet weight is generally maintained in hepatocytes (Bock and White, 1974; Ullrich and Bock, 1984a). The concentration of UDP-GlcA is close to the K_M value of UDP-GlcA for UGTs. Hence, decreased nucleotide concentrations lead to decreased glucuronidation after treatment with various agents, such as D-galactosamine (Otani *et al.*, 1976) or ethanol (Moldéus *et al.*, 1978).

B. Membrane Factors Affecting UGT Activity

1. Membrane Topology of UGTs

Evidence derived from the gene structure of UGTs indicates that the active site of UGT is located on the luminal side of the endoplasmic reticulum (Iyanagi *et al.*, 1986; Jansen *et al.*, 1992). This orientation is supported by accumulating evidence from many laboratories, for example, by the inability of trypsin to hydrolyze or of lectins to bind UGTs in sealed microsomal vesicles (Jansen *et al.*, 1992; Yokota *et al.*, 1992). Furthermore, a lumenal orientation has recently been suggested from studies with ^{32}P-labeled azido-UDP-glucuronic acid (Drake *et al.*, 1992). The transmembrane topology indicates that UDP-glucuronic acid, synthesized in the cytoplasm, must be transported through the membrane to the active site. There is indirect evidence for carriers of UDP-glucuronic acid (Hauser *et al.*, 1988; Vanstapel and Blanckaert, 1988; Milla *et al.*, 1992).

The membrane topology of UGTs has some bearing on the phenomenon of "latency," defined as the ratio of UGT activities present in disrupted versus intact microsomes. Interestingly, latency depends upon the substrate concentration. For example, the latency factor of UGT activity toward 1-naphthol is about 26 at 0.5 m*M* and 3 at 0.002 m*M*. In contrast, the concentration of UDP-glucuronic acid does not affect latency of membrane-bound UGTs (K. W. Bock and W. Lilienblum, unpublished). The nature of the rate-limiting step remains unclear.

UDP-*N*-Acetylglucosamine has been implicated as an activator of UGTs that facilitates transport of UDP-glucuronic acid from the cytosol to the active site of UGT via the microsomal carrier of UDP-glucuronic acid (Hauser *et al.*, 1988). UGT activity in intact microsomes in the presence of UDP-*N*-acetylglucosamine may be close to that operating *in vivo* (Bock and White, 1974; Otani *et al.*, 1976; Ullrich and Bock, 1984b).

2. Interaction of UGTs with Phospholipids

When phospholipids are totally removed from purified UGT preparations, little activity remains. The type of phospholipid is critical for activity. A structure–activity study of 1-palmitoyl-*sn*-glycero-3-phosphocholines indicated that negatively charged phospholipids are inhibitory, whereas neutral or positively charged phospholipids are activators (Zakim *et al.*, 1988). Addition of phospholipids leads to conformational changes of purified UGTs (Singh *et al.*, 1982).

Interpretation of the phenomenon of latency has led to much controversy in UGT research in the past because latency can be viewed both as the result of the removal of a permeability barrier for UDP-glucuronic acid or as a constraint imposed on the enzyme by the interaction with phospholipids (Zakim and Dannenberg, 1992). Investigators have been divided into two camps: the "compartmentationalists," who believed that activation opened the compartment of restricted accessibility (e.g., the access of water-soluble glucuronic acid to the microsomal lumen), and the "conformationalists," who believed that activation removed constraints that prevented full expression of enzyme activity. So far, the controversy is not fully resolved.

C. UGT Isozymes

1. UGT Enzyme Superfamily

Many lines of evidence indicated multiplicity of UGTs in the past, for example, their differential inducibility by 3-methylcholanthrene or phenobarbital (Bock *et al.*, 1973; Lucier *et al.*, 1975; Wishart, 1978; Lilienblum *et al.*, 1982). The existence of a supergene family of isozymes has recently been established by their purification to apparent homogeneity (Bock *et al.*, 1979, 1988; Falany and Tephly, 1983) and in particular by cloning, sequencing, and expression of the cDNAs in cultured cells.

More than 26 distinct cDNAs in five mammalian species have been sequenced to date (Burchell *et al.*, 1991). A nomenclature system for UGTs similar to that developed for cytochrome P450 isozymes (Nebert and Gonzalez, 1987) has been proposed. Comparison of the deduced amino acid sequences leads to the definition of two families and a total of three subfamilies (Fig. 1). For naming each gene, the root symbol UGT was proposed, to be followed by an Arabic number denoting the family, a letter designating the subfamily, and an Arabic numeral representing the individual gene. Family 1 consists of a sole family. All of its members appear to be derived from one gene, the rat and human UGT1A or phenol/bilirubin UGT gene complex (Fig. 2; Iyanagi, 1991; Jansen *et al.*, 1992; Owens and Ritter, 1992). The four human isozymes sequenced to date share exons 2–5 and are characterized by their unique exon 1. mRNAs with different 5′-ends are formed by differential splicing. Each exon 1 is preceded by its own promotor. Four human isozymes of the gene complex have been identified: HlugP1 conjugating planar phenols (Harding *et al.*, 1988); HlugP4 conjugating planar and bulky phenols (Wooster *et al.*, 1991); and two bilirubin UGTs, Br-1 and Br-2 (Owens and Ritter, 1992). The location of exon 1 of HlugP4 in the gene complex is not yet known. Furthermore, rat A-10 (Fig. 1) has so far been isolated from Gunn rat liver as a mutated, nonfunctional protein (Iyanagi, 1991). The human phenol/bilirubin gene complex has been localized on chromosome 2 (Burchell *et al.*, 1991). Allelic variants in exons 2–5 have been shown to be responsible for multiple defects including hyperbilirubinemia in Crigler–Najjar syndromes I and II (Owens and Ritter, 1992). A similar mutation has been identified in the Gunn rat (Sato *et al.*, 1990; Iyanagi, 1991).

Individual genes of the gene complex appear to be differentially regulated, for example, rat phenol UGT (4NP UGT) and probably its human orthologue HlugP1 by 3-methylcholanthrene-type inducers (discussed above) and Br-2 by phenobarbital-type inducers (Owens and Ritter, 1992). Family 2 consists of two subfamilies (Fig. 1). Subfamily 2A consists of a

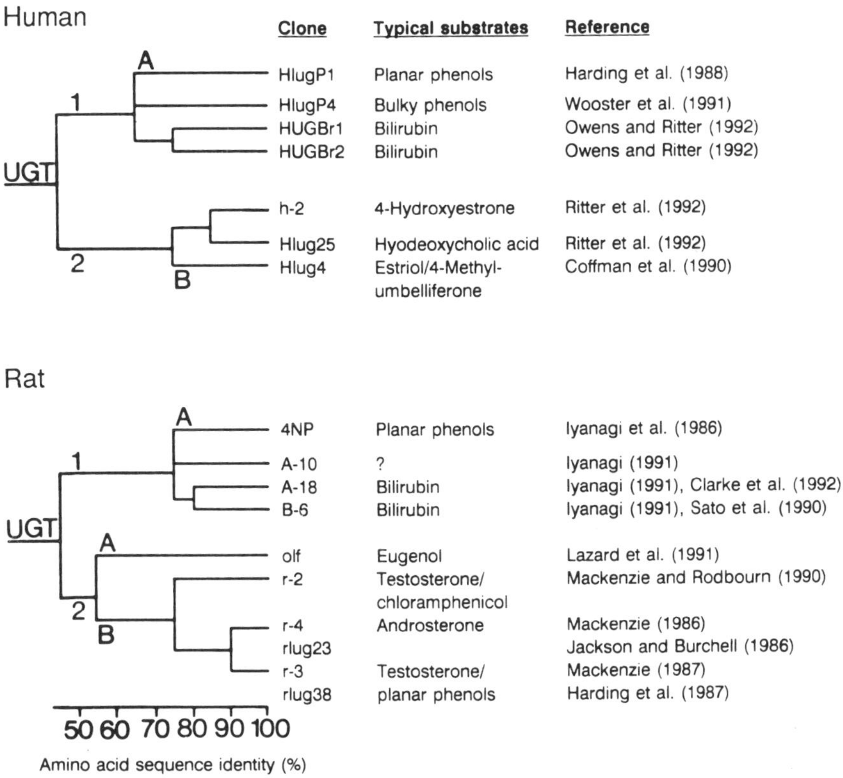

Fig. 1 Human and rat UGT families (adapted from Burchell *et al.*, 1991; Jansen *et al.*, 1992).

unique olfactory UGT present in support cells of the olfactory epithelium. This UGT2A1 is probably involved in the inactivation of odorous signals such as eugenol and borneol, but it also conjugates standard phenols such as 4-methylumbelliferone (Lazard *et al.*, 1991). Subfamily 2B is composed of multiple steroid UGTs. Overlap of substrate specificity not only includes steroids, but also xenobiotics. For example, phenobarbital-inducible testosterone/chloramphenicol UGT (UGT2B1 = r-2) is also involved in 4-methylumbelliferone and 4-hydroxybiphenyl glucuronidation (Mackenzie, 1987), and androsterone UGT (UGT2B2 = r-4) has been shown to conjugate 4-amino-biphenyl (Falany and Tephly, 1983). UGT2B3 = r-3 (previously termed 17β-hydroxytestosterone UGT) appears to conjugate standard phenols (Harding *et al.*, 1987).

Many interesting UGT isozymes have not been completely sequenced, such as rat UGT conjugating morphine (Puig and Tephly, 1986), rat UGTs

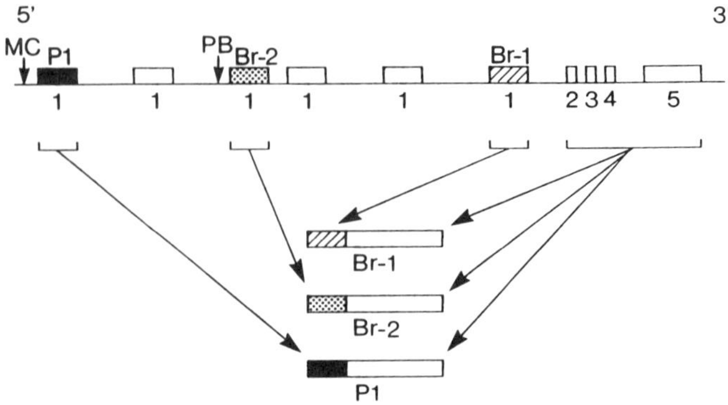

Fig. 2 Organization of the human UGT1A or phenol/bilirubin UGT gene complex on chromosome 2 (Owens and Ritter, 1992; Jansen *et al.*, 1992). The data are consistent with the existence of a large gene (>110 kb) in which at least 6 independently regulated alternative first exons share 4 common 3′-exons to form mRNAs with different 5′-ends by alternative splicing. P1, exon 1 for HlugP1; Br-2, exon 1 for bilirubin UGT2; Br-1, exon 1 for bilirubin UGT1. Each of the unique exons is preceded by its own TATA box promoter site and regulatory region. P1 is probably inducible by 3-methylcholanthrene (MC)-type inducers and Br-2 by phenobarbital (PB).

conjugating digitoxigenin-monodigitoxoside (von Meyerinck *et al.*, 1985) and 4-hydroxybiphenyl (Styczynski *et al.*, 1991). The extent to which these UGTs are involved in the metabolism of carcinogens remains to be elucidated. The pattern of UGT isozymes appears to be tissue specific. On the other hand, HlugP1 and its rat orthologue appear to be widely distributed in tissues (Bock *et al.*, 1980a, 1993). More work is needed to characterize the UGT pattern in different tissues.

2. Substrate Specificity of Phenol UGTs in Family 1A

Investigations on the substrate specificity of various UGTs purified to apparent homogeneity has provided information on their substrate specificity (Bock *et al.*, 1979; Falany and Tephly, 1983). However, due to the instability of purified UGTs (even in the presence of phospholipids), studies of substrate specificity with purified UGT isozymes are tedious. Stable transfection of UGT cDNAs offers a great opportunity in this respect, in particular in studies of low-turnover substrates (Fournel-Gigleux *et al.*, 1991; Bock *et al.*, 1992).

Knowledge about glucuronidation of carcinogens and their metabolites by individual UGT isozymes is limited. Therefore, only rat and human phenol UGT isozymes of the UGT1A gene complex will be discussed here. Simple phenols appear to be overlapping substrates for several UGTs, including 3-methylcholanthrene-inducible phenol UGT (Falany and Tephly, 1983). Phenolic polycyclic aromatic hydrocarbons (PAHs) have been suggested (on the basis of inducibility) to represent selective substrates of 3-methylcholanthrene-inducible rat phenol UGT (Lilienblum *et al.*, 1985, 1987; Bock, 1991). Particularly high induction factors were found with diglucuronide formation of 3,6-dihydroxybenzo[*a*]pyrene and 3,6-dihydroxychrysene (40- and 310-fold, respectively; Lilienblum *et al.*, 1985; Bock *et al.*, 1992).

cDNAs of UGT isozymes that have been stably transfected into cells may offer an opportunity to characterize the substrate specificity of individual isozymes. The utility of these tools has been demonstrated in studies with stably expressed rat and human recombinant phenol UGTs of family 1A (Bock *et al.*, 1992, 1993; Ebner and Burchell, 1993). Investigations with the 3-methylcholanthrene-inducible rat phenol UGT demonstrated that, in addition to small planar phenols, the enzyme conjugates phenolic metabolites of PAHs. It also catalyzes both mono- and diglucuronide formation of 3,6-dihydroxybenzo[*a*]pyrene. Studies with two human phenol UGTs revealed that HlugP1 conjugated the analgesic drug acetaminophen with higher affinity than HlugP4. In contrast, HlugP4 conjugated 4-methylumbelliferone and 6-hydroxychrysene (the latter selected as a stable and nontoxic representative of PAH phenols) with a higher affinity than HlugP1.

More work is needed with diverse endogenous and exogenous substrates to delineate the substrate specificity of UGT isozymes. Based on induction studies it has been suggested that R-naproxen (El Mouelhi and Bock, 1991), L-thyroxine (Beetstra *et al.*, 1991), and all-*trans* retinoic acid (Bank *et al.*, 1989) are substrates of 3-methylcholanthrene-inducible UGTs.

III. Regulation of UGT Isozymes

A. General Features

Similar to cytochrome P450 isozymes, rodent liver UGT isozymes are differentially regulated by prototype inducers, such as phenobarbital and 3-methylcholanthrene (Bock *et al.*, 1973, 1979, 1988; Lucier *et al.*, 1975; Wishart, 1978; Falany and Tephly, 1983; Iyanagi *et al.*, 1986), pregnenolone 16α-carbonitril (Watkins *et al.*, 1982; von Meyerinck *et al.*, 1985), clofibric acid (Lilienblum *et al.*, 1982; Sato *et al.*, 1990), and antioxidants (Bock *et al.*, 1980b; Prochaska and Talalay, 1988). All these inducers appear to switch on regulatory programs, including transcriptional activation of a number of genes.

As with other genes, it is probably the interaction of multiple cis-regulatory elements with a variety of trans-acting factors that controls the cell-specific expression of UGT isozymes (Mackenzie and Rodbourn, 1990). These factors may function with different elements at different developmental stages. In nonexpressing cells, some of these factors may be absent or modified to prevent the interactions necessary for efficient transcription. Post-transcriptional mechanisms may also have to be taken into account. Two kinds of transcriptional mechanisms will be discussed in detail, regulation of phenol UGTs by the Ah receptor (to be distinguished from UGT regulation by antioxidants) and the persistent increase of phenol UGT expression in precancerous lesions.

B. Regulation of Phenol UGTs by the Ah Receptor

Earlier genetic evidence indicated that mouse liver phenol UGT is regulated by the Ah receptor (Owens, 1977). This result in the mouse model was supported in rat liver by induction studies with Ah receptor ligands differing in induction potencies (Bock *et al.,* 1990a; Bock, 1991). Dose–response curves indicated that P4501A1 and UGT1A1 activities are induced at the same concentration of inducers that differ in potency by over 100,000-fold.

In humans, there is also some evidence for UGT isozymes responsive to 3-methylcholanthrene-type inducers. Glucuronidation of paracetamol is increased in heavy smokers (presumably exposed to 3-methylcholanthrene-type inducers; Mucklow *et al.,* 1980; Bock *et al.,* 1987). Enhanced paracetamol glucuronidation may be due to induced HlugP1, which conjugates paracetamol with higher affinity than other UGTs (Bock *et al.,* 1993; as discussed above). More work is needed on the mechanism of UGT induction by the Ah receptor. Ah receptor-controlled enzymes (termed the Ah receptor gene battery; Nebert *et al.,* 1990) may be viewed as a prototypical pleiotropic response that has been extensively studied (Poland and Knutson, 1982; Whitlock, 1990). The receptor has recently been cloned (Burbach *et al.,* 1992; Ema *et al.,* 1992). This Ah receptor response is clearly distinguishable from the response to antioxidants (Rushmore and Pickett, 1990), the latter lacking induction of cytochrome P450 isozymes (Prochaska and Talalay, 1988).

C. Persistent Alterations of Phenol UGT in Preneoplastic Liver

Similar to persistent alterations of other drug metabolizing enzymes (Farber, 1984; Pitot, 1990), 3-methylcholanthrene-inducible phenol UGT is increased in preneoplastic rat liver lesions, such as in hepatocyte foci (Fischer *et al.*, 1985) and hepatocyte nodules (Bock *et al.*, 1982; Yin *et al.*, 1982). Alterations of UGT are heterogeneous, including both UGT-positive and -negative foci (Bock *et al.*, 1989; Bock, 1991). Interestingly, UGT-positive foci predominate in rats, whereas UGT-negative foci predominate in mice. Investigation of a number of hepatocyte nodules and differentiated hepatocellular carcinomas, produced by feeding 2-acetylaminofluorene, indicated that 3-methylcholanthrene-inducible phenol UGT was selectively enhanced (Bock *et al.*, 1982, 1990b). Molecular mechanisms underlying the persistent alterations of proteins (including phenol UGT) in preneoplastic hepatocyte foci remain unknown.

IV. Roles of UGTs in Metabolism and Disposition of Carcinogens

A. Roles of UGTs in Detoxication of Carcinogens

UGTs fulfill a major role in detoxication in concert with other drug metabolizing enzymes such as the cytochrome P450 isozymes generating functional groups for conjugation. However, compounds with functional groups are also widespread in nature (for example, in plants) and are direct substrates of UGTs. One aspect that may be involved with the evolution of the described pleiotropic responses should be emphasized here, namely the protection against quinone/quinol redox cycles (Lorentzen *et al.*, 1979). Benzo[*a*]pyrene-3,6-quinol is efficiently conjugated to both mono- and diglucuronides by 3-methylcholanthrene-inducible phenol UGTs (Lilienblum *et al.*, 1985; Segura-Aquilar *et al.*, 1986; Bock *et al.*, 1992). Benzo[*a*]pyrene-3,6-quinol diglucuronide has recently been detected in rat bile after intratracheal instillation of benzo[*a*]pyrene (Bevan and Sadler, 1992). It is an intriguing question why several enzymes involved in detoxication of quinones belong to the Ah receptor gene battery (Nebert *et al.*, 1990). This battery comprises phase I genes, such as CYP1A1, CYP1A2, and NAD(P)H quinol oxidoreductase, and phase II genes, such as glutathione *S*-transferase Ya and phenol UGT. The Ah receptor gene battery may have evolved as a protection system against oxidative stress and DNA damage (Nebert *et al.*, 1990; Sies, 1991).

In addition to the formation of biologically inactive glucuronides, ''reactive'' acyl-linked glucuronides have also been demonstrated (Stogniew and Fenselau, 1982; van Breemen and Fenselau, 1986; Spahn-Langguth and Benet, 1992).

B. Glucuronides as Transport Forms of Carcinogens

Identification of 2-naphthylamine and 4-aminobiphenyl as human carcinogens led to intensive studies on the metabolism of arylamines. *N*-Hydroxy-2-naphthylamine and *N*-hydroxy-4-aminobiphenyl are converted to the corresponding *N*-glucuronides (Kadlubar *et al.*, 1977; Radomski *et al.*, 1977; Poupko *et al.*, 1979). These *N*-glucuronides represent transport forms, which are excreted via the blood into the urinary system. In the case of 2-naphthylamine, the corresponding *N*-hydroxy-*N*-glucuronide has been shown to decompose at the slightly acidic pH of urine to the hydroxylamine and to its protonated nitrenium ion. The latter readily reacts with DNA, thereby initiating bladder cancer (Fig. 3; Kadlubar *et al.*, 1981; Miller and Miller, 1981). In contrast, 1-naphthylamine is not a bladder carcinogen in experimental animals and humans (Radomski *et al.*, 1980; Purchase *et al.*, 1981). This is due to a lack of N-oxidation of 1-naphthylamine (Beland and Kadlubar, 1990), in addition to rapid glucuronidation to the *N*-glucuronide (Lilienblum and Bock, 1984; Green and Tephly, 1987; Orzechowski *et al.*, 1992).

Recently, mutagenic heterocyclic arylamines have been discovered that are formed at trace levels in foods such as meat and fish under typical household cooking practices (Sugimura, 1986). In experimental animals, these heterocyclic amines have been found to be involved in colon carcinogenicity. Their appearance in a variety of daily food may pose a significant risk for human health. As described for arylamines involved in bladder carcinogenicity, heterocyclic arylamines have to be N-oxidized in liver and are subsequently conjugated with glucuronic acid (Luks *et al.*, 1989; Wallin *et al.*, 1989; Turesky *et al.*, 1990, 1991; Alexander *et al.*, 1991). High-molecular-weight glucuronides are secreted via the bile into the intestine, and the corresponding hydroxylamines and phenolic metabolites may be liberated in the colon by bacterial β-glucuronidases (Fig. 3). The hydroxylamines are further activated by *N*-acetyltransferases, as recently discussed (Bock, 1991).

The transport function of glucuronides may be a promising tool for therapy. Recently, glucuronide pro-drugs have been shown to be activated by antibody-targeted β-glucuronidase in tumor cells (Wang *et al.*, 1992).

V. Conclusions and Future Prospects

Recent advances in the study of UGT isozymes by protein purification, molecular cloning techniques, and stable expression of the cloned isozymes established the diversity of the UGT enzyme family. Isozymes of

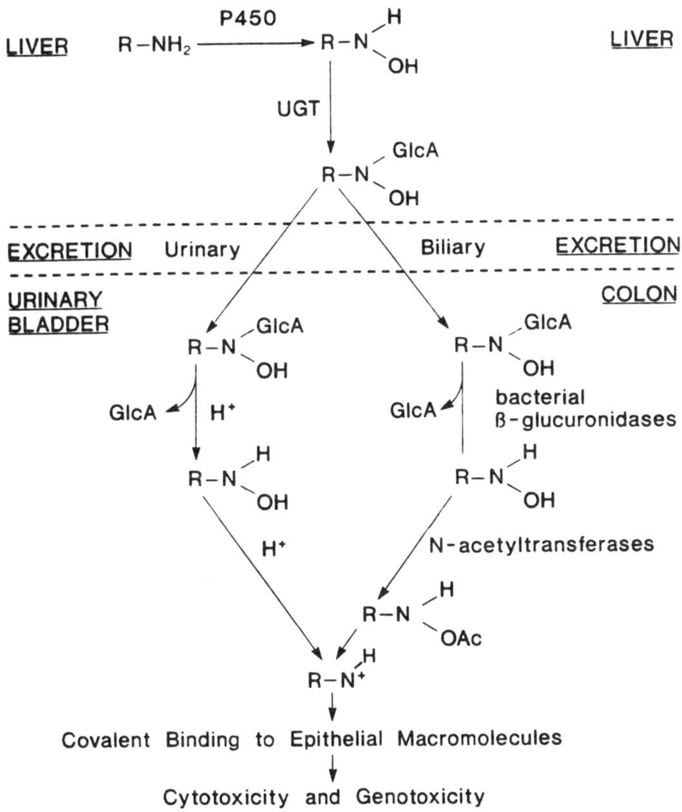

Fig. 3 Metabolism and disposition of arylamines initiating bladder and colon carcinogenesis. GlcA, glucuronic acid; Ac, acetyl.

family 1A show less than 50% amino acid sequence similarity with family 2 members. The former appear to be derived by differential splicing from one gene, the UGT1A or phenol/bilirubin UGT gene complex. Family 2 consists of various steroid UGTs. Whereas one phenol UGT of family 1A accepts a variety of bulky phenols, the other appears to be involved in glucuronidation of planar substrates, including phenolic metabolites of carcinogenic polycyclic aromatic hydrocarbons and aromatic amines and their hydroxylated congeners. Many important UGT isozymes have not been cloned, such as morphine UGT. A more comprehensive set of stably expressed UGTs will enable better delineation of the substrate specificity of UGTs, including their endogenous substrates.

Some UGT isozymes are regulated by hormones, drugs, and environmental chemicals. Therefore, knowledge needs to be improved about the mechanisms of their regulatory control and, in particular, of their participation in pleiotropic responses such as the "Ah receptor gene battery."

Although glucuronidation usually leads to biologically inactive, water-soluble compounds, glucuronide formation may also generate metastable transport forms that can be hydrolyzed enzymatically or by acid, thereby determining the target of carcinogenicity.

NOTE ADDED IN PROOF. In a recent modification of the UGT nomenclature system, UGT1 family proteins are identified by the position of their exon 1 upstream of the four conserved exons (Fig. 2). Hence, the major human bilirubin UGT (Br-1) will be designated UGT1.1h, and the phenol UGT conjugating planar phenols (P1 = UGT1A1) will be UGT1.6h. The corresponding 3-methylcholanthrene-inducible rat ortholog will be UGT1.6r. The phenobarbital-inducible rat testosterone/chloramphenicol UGT (UGT2B1) will be UGT2.01r.

References

Alexander, J., Wallin, H., Rossland, O. J., Solberg, K. E., Holme, J. A., Becher, G., Andersson, R., and Grivas, S. (1991). Formation of a glutathione conjugate and a semistable transportable glucuronide conjugate of N^2-oxidized species of 2-amino-1-methyl-6-phenylimidazo-[4,5-b]pyridine PhIP in rat liver. *Carcinogenesis (London)* **12,** 2239–2245.

Bank, P. A., Salyers, K. L., and Zile, M. H. (1989). Effect of tetrachlorodibenzo-p-dioxin (TCDD) on the glucuronidation of retinoic acid in the rat. *Biochim. Biophys. Acta* **993,** 1–6.

Beetstra, J. B., van Engelen, J. G. M., Karels, P., van der Hoek, H. J., de Jong, M., Docter, R., Krenning, E. P., Hennemann, G., Brouwer, A., and Visser, T. J. (1991). Thyroxine and 3,3′,5-triiodothyronine are glucuronidated in rat liver by different uridine diphosphate-glucuronosyltransferases. *Endocrinology (Baltimore)* **128,** 741–746.

Beland, F. A., and Kadlubar, F. F. (1990). Metabolic activation and DNA adducts of aromatic amines and nitroaromatic hydrocarbons. *In* "Handbook of Exp. Pharmacol. 94/ I" (C. S. Cooper and P. L. Grover, eds.), pp. 267–325. Springer-Verlag Berlin and New York.

Bevan, D. R., and Sadler, V. M. (1992). Quinol diglucuronides are predominant conjugated metabolites found in bile of rats following intratracheal instillation of benzo(a)pyrene. *Carcinogenesis (London)* **13,** 403–407.

Bock, K. W. (1991). Roles of UDP-glucuronosyltransferases in chemical carcinogenesis. *CRC Crit. Rev. Biochem. Mol. Biol.* **26,** 129–150.

Bock, K. W., and White, I. N. H. (1974). UDP-glucuronyltransferase in perfused rat liver and in microsomes: Influence of phenobarbital and 3-methylcholanthrene. *Eur. J. Biochem.* **46,** 451–459.

Bock, K. W., Fröhling, W., Remmer, H., and Rexer, B. (1973). Effects of phenobarbital and 3-methylcholanthrene on substrate specificity of rat liver microsomal UDP-glucuronyltransferase. *Biochim. Biophys. Acta* **327,** 46–56.

Bock, K. W., Josting, D., Lilienblum, W., and Pfeil, H. (1979). Purification of rat liver microsomal UDP-glucuronyltransferase. *Eur. J. Biochem.* **98,** 19–26.

Bock, K. W., von Clausbruch, U. C., Kaufmann, R., Lilienblum, W., Oesch, F., Pfeil, H., and Platt, K. L. (1980a). Functional heterogeneity of UDP-glucuronyltransferase in rat tissues. *Biochem. Pharmacol.* **29,** 495–500.

Bock, K. W., Kahl, R., and Lilienblum, W. (1980b). Induction of rat hepatic UDP-glucuronyltransferase by dietary ethoxyquin. *Naunyn-Schmiedeberg's Arch. Pharmacol.* **310,** 249–252.

Bock, K. W., Lilienblum, W., Pfeil, H., and Eriksson, L. C. (1982). Increased uridine diphosphate-glucuronyltransferase activity in preneoplastic liver nodules and Morris hepatomas. *Cancer Res.* **42,** 3747–3752.

Bock, K. W., Wiltfang, J., Blume, R., Ullrich, D., and Bircher, J. (1987). Paracetamol as a test drug to determine glucuronide formation in man. Effects of inducers and of smoking. *Eur. J. Clin. Pharmacol.* **31,** 677–683.

Bock, K. W., Schirmer, G., Green, M. D., and Tephly, T. R. (1988). Properties of a 3-methylcholanthrene-inducible phenol UDP-glucuronosyltransferase from rat liver. *Biochem. Pharmacol.* **37,** 1439–1443.

Bock, K. W., Kobusch, A. B., and Fischer, G. (1989). Heterogeneous alterations of UDP-glucuronosyltransferases in mouse hepatic foci. *J. Cancer Res. Clin. Oncol.* **115,** 285–289.

Bock, K. W., Lipp, H. P., and Bock-Hennig, B. S. (1990a). Induction of drug-metabolizing enzymes by xenobiotics. *Xenobiotica* **20,** 1101–1111.

Bock, K. W., Münzel, P. A., Röhrdanz, E., Schrenk, D., and Eriksson, L. C. (1990b). Persistently increased expression of a 3-methylcholanthrene-inducible phenol UDP-glucuronosyltransferase in rat hepatocyte nodules and hepatocellular carcinomas. *Cancer Res.* **50,** 3569–3573.

Bock, K. W., Gschaidmeier, H., Seidel, A., Baird, S., and Burchell, B. (1992). Mono- and diglururonide formation of chrysene and benzo(a)pyrene phenols by 3-methylcholanthrene-inducible phenol UDP-glucuronosyltransferase (UGT1A1). *Mol. Pharmacol.* **42,** 613–618.

Bock, K. W., Forster, A., Gschaidmeier, H., Brück, M., Münzel, P., Schareck, W., Fournel-Gigleux, S., and Burchell, B. (1993). Paracetamol glucuronidation by recombinant rat and human phenol UDP-glucuronosyltransferases. *Biochem. Pharmacol.* **45,** 1809–1814.

Burbach, K. M., Poland, A., and Bradfield, C. A. (1992). Cloning of the Ah-receptor cDNA reveals a distinctive ligand-activated transcription factor. *Proc. Natl. Acad. Sci. U.S.A.* **89,** 8185–8189.

Burchell, B., and Coughtrie, M. W. H. (1989). UDP-glucuronosyltransferases. *Pharmacol. Ther.* **43,** 261–289.

Burchell, B., Nebert, D. W., Nelson, D. R., Bock, K. W., Iyanagi, T., Jansen, P. L. M., Lancet, D., Mulder, G. J., Roy Chowdhury, J., Siest, G., Tephly, T. R., and Mackenzie, P. I. (1991). The UDP glucuronosyltransferase gene superfamily. Suggested nomenclature based on evolutionary divergence. *DNA Cell Biol.* **10,** 487–494.

Clarke, D. J., Keen, J. N., and Burchell, B. (1992). Isolation and characterisation of a new hepatic bilirubin UDP-glucuronosyltransferase. *FEBS Lett.* **299,** 183–186.

Coffman, B. L., Tephly, T. R., Irshaid, Y. M., Green, M. D., Smith, C., Jackson, M. R., Wooster, R., and Burchell, B. (1990). Characterization and primary sequence of a human hepatic microsomal estriol UDP-glucuronosyltransferase. *Arch. Biochem. Biophys.* **281,** 170–175.

Drake, R. R., Igari, Y., Lester, R., Elbein, A. D., and Radominska, A. (1992). Application of 5-Azido-UDP-glucose and 5-azido-UDP-glucuronic acid photoaffinity probes for the determination of the active site orientation of microsomal UDP-glucosyltransferases and UDP-glucuronosyltransferases. *J. Biol. Chem.* **267,** 11360–11365.

Dutton, G. J. (1980). "Glucuronidation of Drugs and Other Compounds," pp. 23–28. CRC Press, Boca Raton, FL.

Ebner, T., and Burchell, B. (1993). Substrate specificities of two stably expressed human liver UDP-glucuronosyltransferases of the UGT1 gene family. *Drug Metab. Dispos.* **21,** 50–55.

El Mouelhi, M., and Bock, K. W. (1991). Stereoselective (S)- and (R)-naproxen glucuronosyltransferase of rat liver. *Drug Metab. Dispos.* **19,** 304–308.

Ema, M., Sogawa, K., Watanabe, N., Chujoh, Y., Matsushita, N., Gotoh, O., Funae, Y., and Fujii-Kuriyama, Y. (1992). cDNA Cloning and structure of mouse putative Ah receptor. *Biochem. Biophys. Res. Commun.* **184,** 246–253.

Falany, C. N., and Tephly, T. R. (1983). Separation, purification and characterization of three isoenzymes of UDP-glucuronosyltransferase from rat liver microsomes. *Arch. Biochem. Biophys.* **227,** 248–258.

Falany, C. N., Green, M. D., and Tephly, T. R. (1987). The enzymatic mechanism of glucuronidation catalyzed by two purified rat liver steroid UDP-glucuronosyltransferases. *J. Biol. Chem.* **262,** 1218–1222.

Farber, E. (1984). Cellular biochemistry of the stepwise development of cancer with chemicals. *Cancer Res.* **44,** 5463–5474.

Fischer, G., Ullrich, D., and Bock, K. W. (1985). Effects of N-nitrosomorpholine and phenobarbital on UDP-glucuronyltransferase in putative preneoplastic foci of rat liver. *Carcinogenesis (London)* **6,** 605–609.

Fournel-Gigleux, S., Sutherland, L., Sabolovic, N., Burchell, B., and Siest, G. (1991). Stable expression of two human UDP-glucuronosyltransferase cDNAs in V79 cell cultures. *Mol. Pharmacol.* **39,** 177–183.

Green, M. D., and Tephly, T. R. (1987). N-Glucuronidation of carcinogenic aromatic amines catalyzed by rat hepatic microsomal preparations and purified rat liver uridine 5′-diphosphate-glucuronosyltransferases. *Cancer Res.* **47,** 2028–2031.

Harding, D., Wilson, S. M., Jackson, M. R., Burchell, B., Green, M. D., and Tephly, T. R. (1987). Nucleotide and deduced amino-acid sequence of rat liver 17β-hydroxysteroid UDP-glucuronosyltransferase. *Nucleic Acids Res.* **15,** 3936.

Harding, D., Fournel-Gigleux, S., Jackson, M. R., and Burchell, B. (1988). Cloning and substrate specificity of a human phenol UDP-glucuronosyltransferase expressed in COS-7 cells. *Proc. Natl. Acad. Sci. U.S.A.* **85,** 8381–8385.

Hauser, S. C., Ziurys, J. C., and Gollan, J. L. (1988). A membrane transporter mediates access of uridine 5′-diphosphoglucuronic acid from the cytosol into the endoplasmic reticulum of rat hepatocytes: Implications for glucuronidation reactions. *Biochim. Biophys. Acta* **967,** 149–157.

Iyanagi, T. (1991). Molecular basis of multiple UDP-glucuronosyl transferase isoenzyme deficiencies in the hyperbilirubinemic rat (Gunn rat). *J. Biol. Chem.* **266,** 24048–24052.

Iyanagi, T., Haniu, M., Sogawa, K., Fujii-Kuriyama, Y., Watanabe, S., Shively, J. E., and Anan, K. F. (1986). Cloning and characterization of cDNA encoding 3-methylcholanthrene inducible rat mRNA for UDP-glucuronosyltransferase. *J. Biol. Chem.* **261,** 15607–15614.

Jackson, M. R., and Burchell, B. (1986). The full length coding sequence of rat liver androsterone UDP-glucuronyltransferase cDNA and comparison with other members of this gene family. *Nucleic Acids Res.* **14,** 779–795.

Jansen, P. L. M., Mulder, G. J., Burchell, B., and Bock, K. W. (1992). New developments in glucuronidation research: Report of a workshop on 'Glucuronidation, its Role in Health and Disease.' *Hepatology* **15,** 532–544.

Kadlubar, F. F., Miller, J. A., and Miller, E. C. (1977). Hepatic microsomal N-glucuronidation and nucleic acid binding of N-hydroxy arylamines in relation to urinary bladder carcinogenesis. *Cancer Res.* **37,** 805–814.

Kadlubar, F. F., Unruh, L. E., Flammang, T. J., Sparks, D., Mitchum, R. K., and Mulder, G. J. (1981). Alteration of urinary levels of the carcinogen, N-hydroxy-2-naphthylamine, and its N-glucuronide in the rat by control of urinary pH, inhibition of metabolic sulfation, and changes in biliary excretion. *Chem.-Biol. Interact.* **33,** 129–147.

Lazard, D., Zupko, K., Poria, Y., Nef, P., Lazarovits, J., Horn, S., Khen, M., and Lancet,

D. (1991). Odorant signal termination by olfactory UDP glucuronosyl transferase. *Nature (London)* **349,** 790–793.

Lilienblum, W., and Bock, K. W. (1984). N-Glucuronide formation of carcinogenic aromatic amines in rat and human liver microsomes. *Biochem. Pharmacol.* **33,** 2041–2046.

Lilienblum, W., Walli, A. K., and Bock, K. W. (1982). Differential induction of rat liver microsomal UDP-glucuronosyltransferase activities by various inducing agents. *Biochem. Pharmacol.* **31,** 907–913.

Lilienblum, W., Bock-Hennig, B. S., and Bock, K. W. (1985). Protection against toxic redox cycles between benzo(a)pyrene-3,6-quinone and its quinol by 3-methylcholanthrene-inducible formation of the quinol mono- and diglucuronide. *Mol. Pharmacol.* **27,** 451–458.

Lilienblum, W., Platt, K. L., Schirmer, G., Oesch, F., and Bock, K. W. (1987). Regioselectivity of rat liver microsomal UDP-glucuronosyltransferase activities towards phenols of benzo(a)pyrene and dibenzo(a,h)anthracene. *Mol. Pharmacol.* **32,** 173–177.

Lorentzen, R. J., Lesko, S. A., McDonald, K., and Ts'o, P. O. P. (1979). Toxicity of metabolic benzo(a)pyrenediones to cultured cells and the dependence upon molecular oxygen. *Cancer Res.* **39,** 3194–3198.

Lucier, G. W., McDaniel, O. S., and Hook, G. E. R. (1975). Nature of the enhancement of hepatic uridine diphosphate glucuronyltransferase activity by 2,3,7,8-tetrachlorodibenzo-p-dioxin in rats. *Biochem. Pharmacol.* **24,** 325–334.

Luks, H. J., Spratt, T. E., Vavrek, M. T., Roland, S. F., and Weisburger, J. H. (1989). Identification of sulfate and glucuronic acid conjugates of the 5-hydroxy derivative as major metabolites of 2-amino-3-methylimidazo[4,5-f]quinoline in rats. *Cancer Res.* **49,** 4407–4411.

Mackenzie, P. I. (1986). Rat liver UDP-glucuronosyltransferase. cDNA sequence and expression of a form glucuronidating 3-hydroxyandrogens. *J. Biol. Chem.* **261,** 14112–14117.

Mackenzie, P. I. (1987). Rat liver UDP-glucuronosyltransferase. Identification of cDNAs encoding two enzymes which glucuronidate testosterone, dihydrotestosterone, and β-estradiol. *J. Biol. Chem.* **262,** 9744–9749.

Mackenzie, P. I., and Rodbourn, L. (1990). Organization of the rat UDP-glucuronosyltransferase, UDPGTr-2, gene and characterization of its promoter. *J. Biol. Chem.* **265,** 11328–11332.

Milla, M. E., Clairmont, C. A., and Hirschberg, C. B. (1992). Reconstitution into proteoliposomes and partial purification of the golgi apparatus membrane UDP-galactose, UDP-xylose, and UDP-glucuronic acid transport activities. *J. Biol. Chem.* **267,** 103–107.

Miller, E. C., and Miller, J. A. (1981). Searches for ultimate chemical carcinogens and their reactions with cellular macromolecules. *Cancer (Philadelphia)* **47,** 2327–2345.

Moldéus, R., Andersson, B., and Norling, A. (1978). Interaction of ethanol oxidation with glucuronidation in isolated hepatocytes. *Biochem. Pharmacol.* **27,** 2583–2588.

Mucklow, J. C., Fraser, H. S., Bulpitt, C. J., Kahn, C., Mould, G., and Dolery, C. T. (1980). Environmental factors affecting paracetamol metabolism in London factory and office workers. *Br. J. Clin. Pharmacol.* **10,** 67–74.

Mulder, G. J. (1992). Glucuronidation and its role in regulation of biological activity of drugs. *Annu. Rev. Pharmacol. Toxicol.* **32,** 25–49.

Nebert, D. W., and Gonzalez, F. J. (1987). P450 genes: Structure, evolution, and regulation. *Annu. Rev. Biochem.* **56,** 945–993.

Nebert, D. W., Petersen, D. D., and Fornace, A. J. (1990). Cellular responses to oxidative stress: The [Ah] gene battery as a paradigm. *Environ. Health Perspect.* **88,** 13–25.

Orzechowski, A., Schrenk, D., and Bock, K. W. (1992). Metabolism of 1- and 2-naphthylamine in isolated rat hepatocytes. *Carcinogenesis (London)* **13,** 2227–2232.

Otani, G., Abou-El-Makarem, M. M., and Bock, K. W. (1976). UDP-glucuronyltransferase in perfused rat liver and in microsomes. III. Effects of galactosamine and carbon tetrachloride on the glucuronidation of 1-naphthol and bilirubin. *Biochem. Pharmacol.* **25,** 1293–1297.

Owens, I. S. (1977). Genetic regulation of UDP-glucuronosyl-transferase induction by polycyclic aromatic compounds in mice. *J. Biol. Chem.* **252,** 2827–2833.

Owens, I. S., and Ritter, J. K. (1992). The novel bilirubin/phenol UDP-glucuronosyltransferase UGT1 gene locus: Implications for multiple nonhemolytic familial hyperbilirubinemia phenotypes. *Pharmacogenetics* **2,** 93–108.

Pitot, H. C. (1990). Altered hepatic foci: Their role in murine hepatocarcinogenesis. *Annu. Rev. Pharmacol. Toxicol.* **30,** 465–500.

Poland, A., and Knutson, J. C. (1982). 2,3,7,8-Tetrachlorodibenzo-p-dioxin and related halogenated hydrocarbons: Examination of the mechanism of toxicity. *Annu. Rev. Pharmacol. Toxicol.* **22,** 517–554.

Poupko, J. M., Hearn, W. L., and Radomski, J. L. (1979). N-Glucuronidation of N-hydroxy aromatic amines: A mechanism for their transport and bladder-specific carcinogenicity. *Toxicol. Appl. Pharmacol.* **50,** 479–484.

Prochaska, H. J., and Talalay, P. (1988). Regulatory mechanisms of monofunctional and bifunctional anticarcinogenic enzyme inducers in murine liver. *Cancer Res.* **48,** 4776–4782.

Puig, J. F., and Tephly, T. R. (1986). Isolation and purification of rat liver morphine UDP-glucuronosyltransferase. *Mol. Pharmacol.* **30,** 558–565.

Purchase, I. F. H., Kalinowski, A. E., Ishmael, J., Wilson, J., Gore, C. W., and Chart, I. S. (1981). Lifetime carcinogenicity study of 1- and 2-naphthylamine in dogs. *Br. J. Cancer* **44,** 892–901.

Radomski, J. L., Hearn, W. L., Radomski, T., Moreno, H., and Scott, W. E. (1977). Isolation of the glucuronic acid conjugate of N-hydroxy-4-aminobiphenyl from dog urine and its mutagenic activity. *Cancer Res.* **37,** 1757–1762.

Radomski, J. L., Deichmann, W. B., Altman, N. H., and Radomski, T. (1980). Failure of pure 1-naphthylamine to induce bladder tumors in dogs. *Cancer Res.* **40,** 3537–3539.

Ritter, J. K., Chen, F., Sheen, Y. Y., Lubet, R. A., and Owens, I. S. (1992). Two human liver cDNAs encode UDP-glucuronosyltransferases with 2 log differences in activity toward parallel substrates including hyodeoxycholic acid and certain estrogen derivatives. *Biochemistry* **31,** 3409–3414.

Rushmore, T. H., and Pickett, C. B. (1990). Transcriptional regulation of the rat glutathione S-transferase Ya subunit gene. *J. Biol. Chem.* **265,** 14648–14653.

Sato, H., Koiwai, O., Tanabe, K., and Kashiwamata, S. (1990). Isolation and sequencing of rat liver bilirubin UDP-glucuronosyltransferase cDNA: Possible alternate splicing of a common primary transcript. *Biochem. Biophys. Res. Commun.* **169,** 260–264.

Segura-Aquilar, J. E., Barreiro, V., and Lind, C. (1986). Dicoumarol-sensitive glucuronidation of benzo(a)pyrene metabolites in rat liver microsomes. *Arch. Biochem. Biophys.* **251,** 266–275.

Sies, H. (1991). Oxidative stress: From basic research to clinical application. *Am. J. Med.* **91,** Suppl. 3C, 31S–38S.

Singh, O. M. P., Graham, A. B., and Wood, G. C. (1982). The phospholipid dependence of UDP-glucuronosyltransferase: Conformation/reactivity studies with purified enzyme. *Biochem. Biophys. Res. Commun.* **107,** 345–349.

Spahn-Langguth, H., and Benet, L. Z. (1992). Acyl glucuronides revisited: Is the glucuronidation process a toxification as well as a detoxification mechanism? *Drug Metab. Rev.* **24,** 5–48.

Stogniew, M., and Fenselau, C. (1982). Electrophilic reactions of acyl-linked glucuronides. Formation of clofibrate mercapturate in humans. *Drug Metab. Dispos.* **10,** 609–613.

Styczynski, P., Green, M., Puig, J., Coffman, B., and Tephly, T. (1991). Purification and properties of a rat liver phenobarbital-inducible 4-hydroxybiphenyl UDP-glucuronosyltransferase. *Mol. Pharmacol.* **40,** 80–84.

Sugimura, T. (1986). Past, present, and future of mutagens in cooked foods. *Environ. Health Perspect.* **67,** 5–10.

Turesky, R. J., Bracco-Hammer, I., Markovic, J., Richli, U., Kappeler, A. M., and Welti, D. H. (1990). The contribution of N-oxidation to the metabolism of the food-borne carcinogen 2-amino-3,8-dimethylimidazo[4,5-f]quinoxaline in rat hepatocytes. *Chem. Res. Toxicol.* **3,** 524–535.

Turesky, R. J., Lang, N. P., Butler, M. A., Teitel, C. H., and Kadlubar, F. F. (1991). Metabolic activation of carcinogenic heterocyclic aromatic amines by human liver and colon. *Carcinogenesis (London)* **12,** 1839–1845.

Ullrich, D., and Bock, K. W. (1984a). Glucuronide formation of various drugs in liver microsomes and in isolated hepatocytes from phenobarbital- and 3-methylcholanthrene-treated rats. *Biochem. Pharmacol.* **33,** 97–101.

Ullrich, D., and Bock, K. W. (1984b). Inhibition of glucuronide formation by D-galactosone or D-galactosamine in isolated hepatocytes. *Biochem. Pharmacol.* **33,** 1827–1830.

van Breemen, R. B., and Fenselau, C. C. (1986). Reaction of 1-O-acyl glucuronides with 4-(p-nitrobenzyl)pyridine. *Drug. Metab. Dispos.* **14,** 197–201.

Vanstapel, F., and Blanckaert, N. (1988). Topology and regulation of bilirubin UDP-glucuronyltransferase in sealed native microsomes from rat liver. *Arch. Biochem. Biophys.* **263,** 216–225.

von Meyerinck, L., Coffman, B. L., Green, M. D., Kirkpatrick, R. B., Schmoldt, A., and Tephly, T. R. (1985). Separation, purification, and characterization of digitoxigenin-monodigitoxoside UDP-glucuronosyltransferase activity. *Drug Metab. Dispos.* **13,** 700–704.

Wallin, H., Holme, J. A., Becher, G., and Alexander, J. (1989). Metabolism of the food carcinogen 2-amino-3,8-dimethylimidazo[4,5-f]quinoxaline in isolated rat liver cells. *Carcinogenesis (London)* **10,** 1277–1283.

Wang, S.-M., Chern, J.-W., Yeh, M.-Y., Ng, J. C., Tung, E., and Roffler, S. R. (1992). Specific activation of glucuronide prodrugs by antibody-targeted enzyme conjugates for cancer therapy. *Cancer Res.* **52,** 4484–4491.

Watkins, J. B., Gregus, Z., Thompson, T. N., and Klaassen, C. D. (1982). Induction studies on the functional heterogeneity of rat liver UDP-glucuronosyltransferases. *Toxicol. Appl. Pharmacol.* **64,** 439–446.

Whitlock, J. P. (1990). Genetic and molecular aspects of 2,3,7,8-tetrachlorodibenzo-p-dioxin action. *Annu. Rev. Pharmacol. Toxicol.* **30,** 251–277.

Wishart, G. J. (1978). Demonstration of functional heterogeneity of hepatic uridine diphosphate glucuronosyltransferase activities after administration of 3-methylcholanthrene and phenobarbital to rats. *Biochem. J.* **174,** 671–672.

Wooster, R., Sutherland, L., Ebner, T., Clarke, D., Da Cruz e Silva, O., and Burchell, B. (1991). Cloning and stable expression of a new member of the human liver phenol/bilirubin: UDP-glucuronosyltransferase cDNA family. *Biochem. J.* **278,** 465–469.

Yin, Z., Sato, K., Tsuda, H., and Ito, N. (1982). Changes in activities of uridine diphosphate-glucuronyltransferase during chemical hepatocarcinogenesis. *Gann* **73,** 239–248.

Yokota, H., Yuasa, A., and Sato, R. (1992). Topological disposition of UDP-glucuronyltransferase in rat liver microsomes. *J. Biochem. (Tokyo)* **112,** 192–196.

Zakim, D., and Dannenberg, A. J. (1992). How does the microsomal membrane regulate UDP-glucuronosyltransferases? *Biochem. Pharmacol.* **43,** 1385–1393.

Zakim, D., Cantor, M., and Eibl, H. (1988). Phospholipids and UDP-glucuronosyltransferase. Structure/function relationships. *J. Biol. Chem.* **263,** 5164–5169.

Bioactivation by Glucuronide-Conjugate Formation

Parnian Zia-Amirhosseini,* Hildegard Spahn-Langguth,† and Leslie Z. Benet*

**Department of Pharmacy*
University of California, San Francisco
San Francisco, California 94143

†Department of Pharmacology
University of Frankfurt
D-6000 Frankfurt, Germany

I. Introduction

A. Types of Glucuronides

Conjugation with D-glucuronic acid is a metabolic pathway for compounds containing the following moieties: mildly acidic carbon centers, alcohols, thiols, amines, and carboxylic acids. C-glucuronide formation has been observed for a number of compounds, such as phenylbutazone, *p*-hydroxyphenylbutazone, and oxyphenylbutazone. These glucuronides are not cleaved by β-glucuronidase and are chemically stable. Conjugation of glucuronic acid with alkyl alcohols, phenols, and enolic compounds gives rise to ether, phenolic, or enolic glucuronides. Upon enzymatic reaction with uridine diphosphoglucuronic acid, chloramphenicol, trichloroethanol, and *tert*-butanol form ether glucuronides. Morphine, paracetamol, and salicylamide are drugs that form phenolic glucuronides. 4-Hydroxycoumarin forms an enolic glucuronide. Ether linkages are relatively stable. Enolic glucuronides are less stable than phenolic glucuro-

Advances in Pharmacology, Volume 27

nides (Dutton, 1980). Compounds containing sulfhydryl groups conjugate with glucuronic acid to form thiolic or carbodithiolic glucuronides (-S.GA, -C.S.S.GA). These glucuronides have stabilities similar to ether and acyl *O*-glucuronides respectively. Thiophenol, 2-mercaptobenzothiazole,and diethyldithiocarbamic acid are examples of compounds that form such glucuronides (Marsh, 1966; Caldwell, 1985). *N*-Glucuronides are unstable below netural pH. Examples of drugs that form *N*-glucuronides are dapsone (aromatic amino), sulfasoxazole (heterocyclic imino), meprobamate (ureido), and sulfathiazole (sulfonimido) (Smith and Williams, 1966). Glucuronic acid conjugates with acidic compounds to form acyl β-1-*O*-glucuronides (ester glucuronides). Many acidic compounds, such as benoxaprofen, clofibric acid, fenoprofen, flunoxaprofen, ketoprofen, naproxen, tolmetin, zomepirac (Spahn-Langguth and Benet, 1992), salicylic acid, indomethacin, and nicotinic acid (Marsh, 1966; Smith and Williams, 1966), form this type of glucuronide. These glucuronides undergo chemical and enzymatic hydrolysis, transesterification, and isomer formation (acyl migration). They have reactivities similar to those of carbodithioic *S*-glucuronides and are by far the most reactive class of glucuronides.

B. Pharmacologic Activity

A number of glucuronides have been found to have pharmacologic activity. Examples are described below.

1. Digitoxin and Digoxin Glucuronides

After cleavage of the digitoxose residues of these cardiac glycosides, a metabolic pathway for both compounds is the formation of the 16′-glucuronide. These glucuronides were shown to be active when myocardial force of contraction in the guinea pig isolated papillary muscle was measured and in human isolated ventricular heart muscle strips (Scholz and Schmitz, 1984).

2. Morphine Glucuronides

The two predominant metabolites of morphine are morphine-3-glucuronide and morphine-6-glucuronide. Morphine-6-glucuronide has been shown to be 20–40 times more potent as an analgesic than morphine (Pasternak *et al.*, 1987); the 3-glucuronide was observed to be a potent antagonist of morphine-6-glucuronide (M. T. Smith *et al.*, 1990). Like morphine, the 6-conjugate is relatively selective for μ receptors, with a higher affinity for μ_1 and μ_2 than for κ and δ receptors (Paul *et al.*, 1989; Hanna *et al.*, 1990).

3. All-*trans*-Retinoic Acid β-D-Glucuronide (Retinoyl β-D-Glucuronide)

The action of retinoic acid in acne has been attributed to decreased cohesion between epidermal cells and increased epidermal cell turnover. Importantly its glucuronide conjugate (Miller and DeLuca, 1986) has been shown to have biological activity in a growth assay (Nath and Olson, 1967) and in the rat vaginal smear assay (Stephens-Jarnagin *et al.*, 1983) similar to or even higher than that of retinoic acid. In addition, the human promyelocytic leukemic cell line HL-60 can be induced to differentiate in the presence of retinoic acid. Zile *et al.* (1987) have observed a 55–75% inhibition of cell proliferation in HL-60 by retinoyl-β-D-glucuronide (1 μM). Furthermore, the glucuronide was less cytotoxic than retinoic acid and was not converted to the aglycone during the incubation, indicating that the pharmacologic effect was not due to free retinoic acid released by hydrolysis of the glucuronide.

4. Benzo[*a*]pyrene Oxidative Metabolites and *N*-Hydroxyl-4-Aminobiphenyl

It has been suggested that glucuronides of oxidative metabolites of benzo[*a*]pyrene can form and that they appear to result in reduction of cytotoxicity (Recio and Hsie, 1984). Addition of UDPGA to an incubation medium containing metabolizing enzymes and oxidative metabolites of benzo[*a*]pyrene led to a reduction in cytotoxicity, but not mutagenicity, in a mammalian cell gene mutational assay (Recio and Hsie, 1984). The isolated glucuronic acid conjugate of *N*-hydroxy-4-aminobiphenyl was found to be an active mutagen in two of four *Salmonella typhimurium* strains (Radomski *et al.*, 1977).

II. Reactivity of Acyl Glucuronides

As mentioned earlier, ester glucuronides are reactive compounds. The ester bond undergoes chemical hydrolysis and enzymatic hydrolysis by β-glucuronidase and nonspecific esterases. As reviewed by Spahn-Langguth and Benet (1992), these reactions have been observed for acyl β-1-*O*-glucuronides of many drugs. These glucuronides also undergo an isomerization reaction whereby the aglycone migrates from position 1 on the glucuronic acid ring to positions 2, 3, and 4 (Fig. 1). Acyl glucuronides of a number of drugs have also been shown to react with nucleophilic centers of small molecules (Stogniew and Fenselau, 1982) and macromolecules to form covalent adducts *in vitro* and *in vivo*.

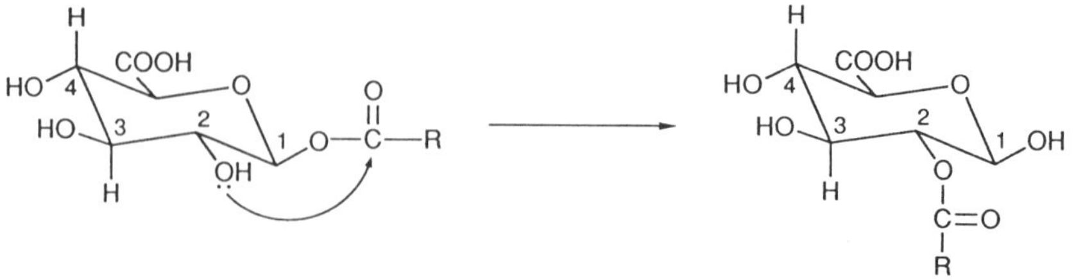

Fig. 1 Migration of the acyl group of the β-1-acyl glucuronide from C1 to C2. The acyl group can then migrate to C3 and subsequently to C4.

A. Acyl Migration and Hydrolysis

Carbohydrate chemists have thoroughly described acyl migration on sugars (Haines, 1976). The stability of acyl glucuronides for different drug molecules is quite variable and influenced by many factors. As studied in detail for zomepirac (Hasegawa *et al.*, 1982; Smith *et al.*, 1985a) and tolmetin (Munafo *et al.*, 1990), the rates of acyl migration and hydrolysis of their respective glucuronides depend upon the pH, temperature, and composition of the medium. Acyl glucuronides are most stable at pH 2–4, at low temperatures, and in presence of esterase inhibitors. The hydrolysis rate of oxaprozin glucuronide was accelerated in the presence of plasma proteins and albumin (Ruelius *et al.*, 1986). An opposite (stabilizing) effect of human serum albumin (HSA) was observed for tolmetin glucuronide, but bovine serum albumin (BSA) caused an increase in the rate of hydrolysis (Munafo *et al.*, 1990). The degradation rate of (*S*)carprofen glucuronide in the presence of fraction V and fatty acid-free HSA at 37°C and pH 7.4 was higher than in the absence of these proteins. The degradation rate of (*R*)carprofen glucuronide decreased in the presence of fatty acid-free HSA, whereas fraction V HSA had no effect on the degradation rate (Iwakawa *et al.*, 1988). Thus, the effect of protein on glucuronide stability is extremely variable. Degradation half-lives for the β-1-*O*-acyl glucuronides of various compounds in the physiologic pH range have been summarized by Spahn-Langguth and Benet (1992).

B. Reaction with Small Molecules

Clofibrate glucuronide was reported to react with the nucleophile ethanethiol to form the clofibrate thioester, whereas no thioesters were formed upon incubation of clofibrate or clofibric acid with the nucleophile. Also, a mercapturic acid conjugate of clofibrate was isolated from human urine after administration of clofibrate. Based on the ethanethiol result and the fact that initial conjugation of xenobiotics with glutathione usually leads to their excretion as mercapturic acid (*N*-acetylcysteine) conjugates in human urine, the *in vivo* finding was interpreted as evidence for a glutathione conjugate precursor (Stogniew and Fenselau, 1982).

C. Reaction with Macromolecules

Irreversible covalent binding to macromolecules (proteins) has been observed both *in vitro* and *in vivo* for acyl glucuronides of a number of xenobiotics (Spahn-Langguth and Benet, 1992). For example, a large number of nonsteroidal anti-inflammatory drugs (benoxaprofen, indomethacin, flufenamic acid, oxaprozin, zomepirac, tolmetin, carprofen, fenoprofen, diflunisal, and suprofen), the uricosuric drug probenecid, the antihyperlipoproteinemic compounds clofibric acid (van Breemen and Fenselau, 1985) and beclobrate, the diuretic agent furosemide, and the antiepileptic drug valproic acid form irreversible protein adducts. The majority of the studies have been performed with HSA.

1. *In Vitro* Studies on Covalent Protein Binding

The *in vitro* irreversible binding of acyl glucuronides was shown to be dependent on glucuronide concentration (Munafo *et al.*, 1990), pH (Smith *et al.*, 1986; Munafo *et al.*, 1990; Smith and Liu, 1993), and time (van Breemen and Fenselau, 1985; Wells *et al.*, 1987). For oxaprozin glucuronide, the highest yield of protein adduct was obtained after the glucuronide and HSA were incubated at pH 7 for approximately 1 h at 37°C. Similarly, maximum covalent binding to HSA for zomepirac glucuronide occurred after 1 h of incubation at pH 9, although the level of protein adduct decreased rapidly after this time due to its instability at this pH. High concentrations of adduct were also observed after a 6-h coincubation of zomepirac glucuronide and HSA at pH 7 and 8 at 37°C (Smith *et al.*, 1986). Covalent binding of diflunisal glucuronide was greater with fatty acid-free human serum albumin than with human and rat plasma proteins (Watt and Dickinson, 1990). The covalent binding of suprofen glucuronide to HSA increased with increasing pH at 37°C and was time-dependent (Smith and Liu, 1993).

The isomeric conjugates of the β-1-*O*-acyl glucuronides also form protein adducts. Isomers of suprofen glucuronide exhibited time-dependent binding. This binding was 38% higher than that of the β-1-glucuronide (Smith and Liu, 1993). Covalent binding of valproic acid β-1-glucuronide to proteins was minor in buffered HSA solution and human plasma; however, approximately 7% of the isomers of valproic acid glucuronide formed covalent adducts with proteins (Williams *et al.*, 1992). Thus, certain isomeric conjugates seem to be more reactive toward protein than the β-1-glucuronide.

Benet and co-workers have observed a linear correlation (r^2 = .995) between the rates of degradation of 9 acyl glucuronides and the extent of drug covalent binding (moles of drug irreversibly bound per mole of protein) to HSA. Moreover, as the number of substituents on the carbon α to the carboxylic acid of the aglycone decreased, the degradation rate constant of its glucuronide (reactivity) and the number of moles bound per mole of protein (protein covalent binding) increased (Benet *et al.*, 1993). Three of the drugs that exhibit this structure–activity relationship (beclobrate, carprofen, fenoprofen) have a chiral center, thus each producing a pair of enantiomeric glucuronides. The other three (furosemide, tolmetin, zomepirac) do not possess any chiral centers. The drugs exhibiting the greatest degree of binding, tolmetin and zomepirac, contain an unsubstituted α-carbon, whereas the compounds with an intermediate degree of binding, fenoprofen and carprofen, have a substituent on the α-carbon. The drugs with the least degree of binding, furosemide and beclobrate, are completely substituted at the α-carbon.

2. *In Vivo* Binding Studies

The irreversible plasma protein binding for beclobrate, carprofen, fenoprofen, tolmetin, and zomepirac have been studied after single-dose oral administration of the drugs at their usual therapeutic doses to five different sets of healthy volunteers. A linear relationship has been observed between the amount of drug irreversibly bound and the extent of glucuronide present measured as the area under the curve (AUC) for each drug. Furthermore, after the bound ratio (moles of drug per mole of protein) was normalized to AUC of the corresponding glucuronide metabolite, a significant linear correlation (r^2 = .873) was obtained between the normalized bound amount and the *in vitro* degradation rate constant of the glucuronide (Table I) (Benet *et al.*, 1993). Thus, it may be possible to predict the *in vivo* covalent binding of acidic drugs to plasma proteins in humans based on the degradation rate constant of the glucuronide metabolite after the amount of drug bound is normalized to the levels of glucuronide present in plasma.

Valproic acid protein adducts were measured in the plasma of epileptic patients on chronic drug therapy (Williams *et al.*, 1992). Coadministration of probenecid and zomepirac resulted in an increase in the amount of irreversibly bound zomepirac and an increase in exposure to zomepirac glucuronide as measured by AUC of the zomepirac glucuronide plasma concentration–time curve (Smith *et al.*, 1985b). Covalent binding of diflunisal, probenecid, and tolmetin have been investigated after adminis-

Table I

In Vivo Irreversibly Bound Drug, Area under the Plasma Drug Glucuronide Concentration Time Curve and *in Vitro* Acyl Glucuronide Degradation Rates[a]

Parent compound	Bound drug (mol/mol protein) × 10^4	AUC glucuronide (mol × h/liter) × 10^6	Bound/AUC 10^{-2}	k (h^{-1})
Tolmetin	2.77 ± 1.54	3.72 ± 0.95	0.75	1.78
Zomepirac	2.33 ± 0.45	6.41 ± 2.14	0.36	1.54
(*R*)Fenoprofen	1.02 ± 0.32	6.31 ± 5.65	0.16	0.71
(*S*)Fenoprofen	3.23 ± 0.85	60.4 ± 24.7	0.054	0.36
Racemic carprofen	1.92 ± 1.28	40.9 ± 7.3	0.047	0.32
(−)-Beclobric Acid	0.12 ± 0.03	8.16 ± 1.34	0.15	0.031
(+)-Beclobric Acid	0.20 ± 0.11	8.31 ± 1.63	0.24	0.027

[a] Measurement of drugs covalently bound to human serum albumin and area under the plasma concentration time curve for the glucuronide conjugates measured in five different groups of healthy volunteers following oral dosing of either 400 mg of tolmetin, 100 mg of zomepirac, 600 mg of racemic fenoprofen, 50 mg of racemic carprofen, or 100 mg of racemic beclobric acid. When covalently bound drug is normalized to area under the curve for the respective glucuronide conjugates, an excellent correlation with the *in vitro* degradation rate constant (k) is obtained with an r^2 of .873. From Benet *et al.* (1993).

tration of multiple doses of the drugs. After a 6-day regimen in healthy human volunteers of oral diflunisal with concomitant administration of oral probenecid during the last 2 days, measureable covalent binding of both drugs via their acyl glucuronide metabolite has been observed (McKinnon and Dickinson, 1989). Significant accumulation of protein adducts of tolmetin has been observed in healthy human volunteers after a 10-day multiple dosing regimen of tolmetin. The bound levels after administration of multiple doses were approximately 10 times higher than those after a single dose was given to the same subjects (Zia-Amirhosseini *et al.*, 1994). The half-lives of all three protein adducts, diflunisal (10 ± 0.9 days), probenecid (13.5 ± 0.3 days), and tolmetin (4.8 ± 0.9 days), were much longer than that of the parent drugs or their metabolites.

3. Mechanism of Binding and Determination of Binding Sites on the Protein

Two mechanisms have been proposed to describe the irreversible binding of acyl glucuronides to proteins. The first is a nucleophilic displacement mechanism whereby a nucleophilic group (NH_2, OH, SH) of the protein attacks the carbonyl group of the aglycone (Fig. 2). This leads to the regeneration of the glucuronic acid and formation of a bond (amide, ester, or thioester) between the protein and the drug. The second mechanism

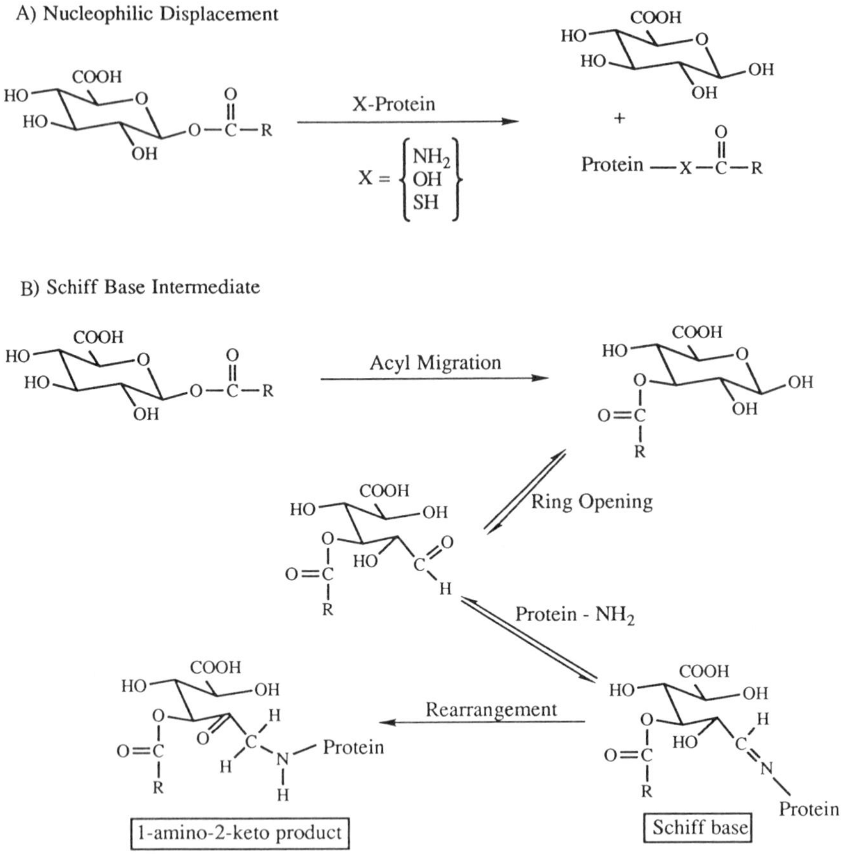

Fig. 2 Proposed mechanisms for covalent binding of acyl glucuronides to protein.

involves the formation of a Schiff base (imine) intermediate (with the amino groups on the protein), which can then undergo rearrangement to a more stable 1-amino-2-keto product (Fig. 2). In this mechanism, the glucuronic acid moiety is retained as a component of the protein adduct.

Three different approaches have been utilized to characterize the modified residues on the protein: (*a*) blockage of sites on albumin with compounds that bind to the protein reversibly at known sites (Wells *et al.*, 1987); (*b*) blockage of various nucleophiles on albumin by specific chemical reactions by conventional (Sokolovsky *et al.*, 1966; Walker, 1976; Fehske *et al.*, 1980) or newer labeling procedures (Yu and Perel, 1980; Han *et al.*, 1987); (*c*) mass spectrometric analysis of tryptic digests of albumin adducts of acyl glucuronides (Ding *et al.*, 1993a).

Different binding sites have been implicated or proven to participate in the covalent binding of albumin to the acyl glucuronide of a number of drugs. For example, reversibly bound ligands (including oxaprozin) that bind specifically to the benzodiazepine binding site on albumin were found to inhibit the irreversible binding of oxaprozin glucuronide (Ruelius *et al.*, 1986). Furthermore, irreversible blockage of sulfhydryl groups of albumin had no effect on irreversible binding of oxaprozin glucuronide, whereas chemical modification of lysine residues led to a minor decrease in the extent of binding. A significant reduction in irreversible binding and degradation of oxaprozin glucuronide, but not reversible binding, resulted from reacting the tyrosine at the benzodiazepine binding site (Tyr-411) with tetranitromethane, diisopropylfluorophosphate, and *p*-nitrophenylacetate (Wells *et al.*, 1987). Ruelius *et al.* (1986) have also implicated tyrosine as the binding moiety for oxaprozin glucuronide. Van Breemen and Fenselau (1985) hypothesized that sulfhydryls act as the primary nucleophiles; however, these same authors found that blocking the sulfhydryls on BSA could not prevent all of the binding of flufenamic glucuronide. Moreover, blockage of thiols on HSA with *p*-hydroxymercuribenzoate did not have any effect on the extent of binding of zomepirac glucuronide (Smith *et al.*, 1990).

Recently, Ding and co-workers (1993a) have investigated the binding of tolmetin glucuronide to HSA in presence of an imine trapping agent (sodium cyanoborohydride). At least six of the peptides generated by tryptic digestion of the adduct contained tolmetin. After the isolated peptides were subjected to liquid secondary-ion mass spectrometry (LSIMS) and collision-induced dissociation mass spectrometry (CIDMS), all six peptide fragments were found to contain tolmetin linked covalently via a glucuronic acid to a lysine residue. The modified residues were lysines 195, 199, 525, 137, 351, and 541. More recent data from our laboratory show that both the nucleophilic displacement reaction and the Schiff base formation pathway occur simultaneously in the absence of an imine trapping agent (Ding *et al.*, 1993b). It is possible that both mechanisms occur concurrently *in vivo*.

III. Possible Role of Acyl Glucuronides in the Toxicity of Drugs

It has been hypothesized that acyl glucuronides, due to their reactive nature, may have a role in the observed toxicities associated with administration of a number of acidic compounds. It is striking that 6 of the 25 drugs withdrawn from the U.S. and British markets between 1964 and 1983 due to severe toxicity were carboxylic acids for which acyl glucuronide formation is a metabolic route (Bakke *et al.*, 1984). Five of these, alclofenac, benoxaprofen, ibufenac, indoprofen, and zomepirac, were NSAIDs of the aryl-alkyl acid class. Recently another member of this class, suprofen, has also been withdrawn from the market (Hart *et al.*, 1987; Strom *et al.*, 1989). The drugs were withdrawn due to either severe organ toxicity (i.e., liver or kidney) or severe anaphylactic or anaphylactoid reactions. Some other acidic drugs that are still on the market but for which adverse reactions have been observed are diflunisal (Cook *et al.*, 1988), probenecid (Kickler *et al.*, 1986), aspirin (Amos *et al.*, 1971), and tolmetin (U.S. Congress and Committee on Government Operations, 1982; Bretza, 1985).

Two hypotheses have been put forth regarding the possible role of acyl glucuronides in manifestation of such toxicities. One suggests that reactivity of acyl glucuronides toward smaller nucleophiles, such as glutathione, may have a role in the observed toxicity of some acidic drugs and may also enhance the toxicity of other xenobiotics by glutathione depletion (Faed, 1984; Stogniew and Fenselau, 1982). The other hypothesis suggests that both the anaphylactic reactions and organ toxicities are immune-based reactions. Here the acyl glucuronide is hypothesized to act as a hapten that upon covalent binding to a protein becomes immunogenic and causes antibody production. When IgE antibodies are produced, an anaphylactic reaction ensues. Organ toxicity may result from covalent binding to tissue proteins and subsequent antigen–antibody complex formation (Abbas *et al.*, 1992; Spahn-Langguth and Benet, 1992). Drug-specific antibodies have been observed in aspirin hypersensitive patients (Amos *et al.*, 1971) and in patients receiving valproic acid therapy (Williams *et al.*, 1992). The low levels of valproic acid-specific antibodies are consistent with the relatively low toxicity of this drug. To our knowledge, evidence linking the acyl glucuronides to the clinical hypersensitivity states is lacking.

Acknowledgment

Preparation of this review and studies carried out in the authors' laboratories were supported in part by National Institutes of Health Grant GM 36633.

References

Abbas, A. K., Lichtman, A. H., and Pober, J. S. (1991). "Cellular and Molecular Immunology," pp. 50–52. Saunders, Philadelphia.

Amos, H. E., Wilson, D. V., Taussig, M. J., and Carlton, S. J. (1971). Hypersensitivity reactions to acetylsalicylic acid. *Clin. Exp. Immunol.* **8,** 563–572.

Bakke, O. M., Wardell, W. M., and Lasagna, L. (1984). Drug discontinuations in the United Kingdom and the United States, 1964 to 1983: Issues of safety. *Clin. Pharmacol. Ther.* **35,** 559–567.

Benet, L. Z., Spahn-Langguth, H., Iwakawa, S., Volland, C., Mizuma, T., Mayer, S., Mutschler, E., and Lin, E. T. (1993). Predictability of the covalent binding of acidic drugs in man. *Life Sci.* **53,** PL141–146.

Bretza, M. (1985). Anaphylactoid reactions to tolmetin. *West. J. Med.* **143,** 55–59.

Caldwell, J. (1985). Glucuronic acid conjugation in the context of the metabolic conjugation of xenobiotics. *In* "Advances in Glucuronide Conjugation" (S. Matern, K. W. Bock, and W. Gerok, eds.), pp. 7–20. MTP Press, Lancaster, PA.

Cook, D. J., Achong, M. R., and Murphy, F. R. (1988). Three cases of diflunisal hypersensitivity. *Can. Med. Assoc. J.* **138,** 1029–1030.

Ding, A., Ojingwa, J. C., McDonagh, A. F., Burlingame, A. L., and Benet, L. Z. (1993a). Evidence for the covalent binding of acyl glucuronides to serum albumin via an imine mechanism as revealed by tandem mass spectrometry. *Proc. Natl. Acad. Sci. U.S.A.* **90,** 3979–3801.

Ding, A., Zia-Amirhosseini, P., Wong, H., Burlingame, A. L., Benet, L. Z. (1993b). Covalent binding characteristics of tolmetin glucuronide to human serum albumin in vitro. *Pharm. Res.* **10,** S363.

Dutton, G. J. (1980). "Glucuronidation of Drugs and Other Compunds," pp. 13–16. CRC Press, Boca Raton, FL.

Faed, E. M. (1984). Properties of acyl glucuronides: Implications for studies of the pharmacokinetics and metabolism of acidic drugs. *Drug Metab. Rev.* **15,** 1213–1249.

Fehske, K. J., Muller, W. E., and Woller, U. (1980). Direct demonstration of the highly reactive tyrosine residue of human serum albumin located in fragment 299-585. *Arch. Biochem. Biophys.* **205,** 217–221.

Haines, A. H. (1976). Relative reactivities of hydroxyl groups in carbohydrates: Migration of substituents-1. Acyl migration. *Adv. Carbohydr. Chem. Biochem.* **33,** 101–107.

Han, K. K., Delacourte, A., and Hemon, B. (1987). Chemical modification of thiol group(s) in protein: Application to the study of anti-microtubular drug binding. *Comp. Biochem. Physiol. B* **88B,** 1057–1065.

Hanna, M. H., Peat, S. J., Woodham, M., Knibb, A., and Fung, C. (1990). Analgesic efficacy and CSF pharmacokinetics of intrathecal morphine-6-glucuronide: Comparison with morphine. *Br. J. Anaesth.* **64,** 547–550.

Hart, D., Ward, M., and Lifschitz, M. D. (1987). Suprofen-related nephrotoxicity—a distinct clinical syndrome. *Ann. Intern. Med.* **106,** 235–238.

Hasegawa, J., Smith, P. C., and Benet, L. Z. (1982). Apparent intramolecular acyl migration of zomepirac glucuronide. *Drug Metab. Dispos.* **10,** 469–473.

Iwakawa, S., Spahn, H., Benet, L. Z., and Lin, E. T. (1988). Carprofen glucuronides: Stereoselective degradation and interaction with human serum albumin. *Pharm. Res.* **5,** Suppl, S-214.

Kickler, T. S., Buck, S., Ness, P., Shirey, R. S., and Sholar, P. W. (1986). Probenecid induced immune hemolytic anemia. *J. Rheumatol.* **13,** 208–209.

Marsh, C. A. (1966). Chemistry of D-glucuronic acid and its glycosides. *In* "Glucuronic Acid" (G. J. Dutton, ed.), pp. 4–133. Academic Press, New York.

McKinnon, G. E., and Dickinson, R. G. (1989). Covalent binding of diflunisal and probenecid to plasma protein in humans: Persistence of the adducts in the circulation. *Res. Commun. Chem. Pathol. Pharmacol.* **66,** 339–354.

Miller, D. A., and DeLuca, H. F. (1986). Biosynthesis of retinoyl-β-glucuronide, a biologically active metabolite of all-trans-retinoic acid. *Arch. Biochem. Biophys.* **244,** 179–186.

Munafo, A., McDonagh, A. F., Smith, P. C., and Benet, L. Z. (1990). Irreversible binding of tolmetin glucuronic acid esters to albumin in vitro. *Pharm. Res.* **7,** 21–27.

Nath, K., and Olson, J. A. (1967). Natural occurrence and biological activity of vitamin A derivatives in rat bile. *J. Nutr.* **93,** 461–469.

Pasternak, G. W., Bodnar, R. J., Clark, J. A., and Inturrisi, C. E. (1987). Morphine-6-glucuronide, a potent mu agonist. *Life Sci.* **41,** 2845–2849.

Paul, D., Standifer, K. M., Inturrisi, C. E., and Pasternak, G. W. (1989). Pharmacological characterization of morphine-6 beta-glucuronide, a very potent morphine metabolite. *J. Pharmacol. Exp. Ther.* **251,** 477–483.

Radomski, J. L., Hearn, W. L., Radomski, T., Moreno, H., and Scott, W. E. (1977). Isolation of the glucuronic acid conjugate of N-hydroxy-4-aminobiphenyl from dog urine and its mutagenic activity. *Cancer Res.* **37,** 1757–1762.

Recio, L., and Hsie, A. W. (1984). Glucuronide conjugation reduces the cytotoxicity but not the mutagenicity of benzo(a)pyrene in the CHO/HGPRT assay. *Teratogenesis Carcinog. Mutagen.* **4,** 391–402.

Ruelius, H. W., Kirkman, S. K., Young, E. M., and Janssen, F. W. (1986). Reactions of oxaprozin-1-O-acyl glucuronide in solutions of human plasma and albumin. *Adv. Exp. Med. Biol.* **197,** 431–441.

Scholz, H., and Schmitz, W. (1984). Positive iontropic effects of digitoxin- and digoxin-glucuronide in human isolated ventricular heart muscle preparations. *Basic Res. Cardiol.* **79,** 134–139.

Smith, M. T., Watt, J. A., and Cramond, T. (1990). Morphine-3-glucuronide—a potent antagonist of morphine analgesia. *Life Sci.* **47,** 579–585.

Smith, P. C., and Liu, J. H. (1993). Covalent binding of suprofen acyl glucuronide to albumin in vitro. *Xenobiotica* **23,** 337–348.

Smith, P. C., Hasegawa, J., Langendijk, P. N. J., and Benet, L. Z. (1985a). Stability of acyl glucuronides in blood, plasma, and urine: Studies with zomepirac. *Drug. Metab. Dispos.* **13,** 110–112.

Smith, P. C., Langendijk, P. N. J., Bosso, J. A., and Benet, L. Z. (1985b). Effect of probenecid on the formation and elimination of acyl glucuronides: Studies with zomepirac. *Clin. Pharmacol. Ther.* **38,** 121–127.

Smith, P. C., McDonagh, A. F., and Benet, L. Z. (1986). Irreversible binding of zomepirac to plasma protein in vitro and in vivo. *J. Clin. Invest.* **77,** 934–939.

Smith, P. C., Benet, L. Z., and McDonagh, A. F. (1990). Covalent binding of zomepirac glucuronide to proteins: Evidence for a Schiff base mechanism. *Drug Metab. Dispos.* **18,** 639–644.

Smith, R. L., and Williams, R. T. (1966). Implications of the conjugation of drugs and other exogenous compounds. *In* "Glucuronic Acid" (G. J. Dutton, ed.), pp. 457–491. Academic Press, New York.

Sokolovsky, M., Riordan, J. F., and Vallee, B. L. (1966). Tetranitromethane: A reagent for the nitration of tyrosyl residues in proteins. *Biochemistry* **5,** 3582–3589.

Spahn-Langguth, H., and Benet, L. Z. (1992). Acyl glucuronides revisited: Is the glucuronidation process a toxification as well as a detoxification mechanism? *Drug Metab. Rev.* **24,** 5–48.

Stephens-Jarnagin, A., Sietsema, W. K., Miller, D. A., and DeLuca, J. F. (1983). The biological activity of rat intestine retinoic acid metabolites in the vaginal smear assay. *Arch. Biochem. Biophys.* **220,** 502–508.

Stogniew, M., and Fenselau, C. (1982). Electrophilic reactions of acyl-linked glucuronides: Formation of clofibrate mercapturate in humans. *Drug Metab. Dispos.* **10,** 609–613.

Strom, B. L., West, S. L., Sim, E., and Carson, J. L. (1989). The epidemiology of the acute flank pain syndrome from suprofen. *Clin. Pharmacol. Ther.* **46,** 693–699.

U.S. Congress and Committee on Government Operations, Intergovernment Relations and

Human Resources (1982). "FDA's Regulation of Zomax®." 98th Congress, Washington, DC.

van Breemen, R. B., and Fenselau, C. (1985). Acylation of albumin by 1-O-acyl glucuronides. *Drug Metab. Dispos.* **13,** 318–320.

Walker, J. E. (1976). Lysine residue 199 of human serum albumin is modified by acetylsalicylic acid. *FEBS Lett.* **66,** 173–175.

Watt, J. A., and Dickinson, R. G. (1990). Reactivity of diflunisal acyl glucuronide in human and rat plasma and albumin solutions. *Biochem. Pharmacol.* **39,** 1067–1075.

Wells, D. S., Janssen, F. W., and Ruelius, H. W. (1987). Interactions between oxaprozin glucuronide and human serum albumin. *Xenobiotica* **17,** 1437–1449.

Williams, A. M., Worrall, S., De Jersey, J., and Dickinson, R. G. (1992). Glucuronide-derived adduct of valproic acid and plasma protein and anti-adduct antibodies in humans. *Biochem. Pharmacol.* **43,** 745–755.

Yu, T. F., and Perel, J. (1980). Pharmacokinetic and clinical studies of carprofen in gout. *J. Clin. Pharmacol.* **20,** 347–351.

Zia-Amirhosseini, P., Ojingwa, J., Spahn-Langguth, H., McDonagh, A. F., and Benet, L. Z. (1994). Enhanced covalent binding of tolmetin to proteins in humans after multiple dosing. *Clin. Pharmacol. Ther.* (in press).

Zile, M. H., Cullum, M. E., Simpson, R. U., Barua, A. B., and Swartz, D. A. (1987). Induction of differentiation of human promyelocytic leukemia cell line HL-60 by retinoyl glucuronide, a biologically active metabolite of vitamin A. *Proc. Natl. Acad. Sci. U.S.A.* **84,** 2208–2212.

Part IV

Bioactivation and Bioconversion

N-Acetyltransferases, *O*-Acetyltransferases, and *N*,*O*-Acetyltransferases: Enzymology and Bioactivation

Patrick E. Hanna
Departments of Medicinal Chemistry and Pharmacology
University of Minnesota
Minneapolis, Minnesota 55455

I. Introduction

Acetyltransferases play a central role in the metabolic disposition, detoxication, and bioactivation of a diverse group of drugs, carcinogens, and other xenobiotics (Weber and Hein, 1985; Weber, 1987). Acetylation is a major metabolic pathway for primary aromatic amines (arylamines, $ArNH_2$), hydrazines ($RNHNH_2$), and hydrazides ($—CONHNH_2$).

It was observed in the early 1950s that acetylation was the principal mechanism for termination of the tuberculostatic action of isoniazid (**I**, Fig. 1). The latter finding prompted clinical investigations that led to the discovery of both interindividual variability in isoniazid acetylation and the relevance of this metabolic polymorphism to therapeutic outcomes. The importance of human acetylation capacity is further illustrated by the propensity of phenotypically slow acetylator patients to experience methemoglobinemia or drug-induced lupus erythematosus, or both, when treated either with arylamine drugs or with various agents that contain the hydrazine moiety (Weber and Hein, 1985). Epidemiological studies have revealed an excess of slow acetylators among bladder cancer patients and a higher incidence of colorectal carcinoma among rapid acetylators

Advances in Pharmacology, Volume 27

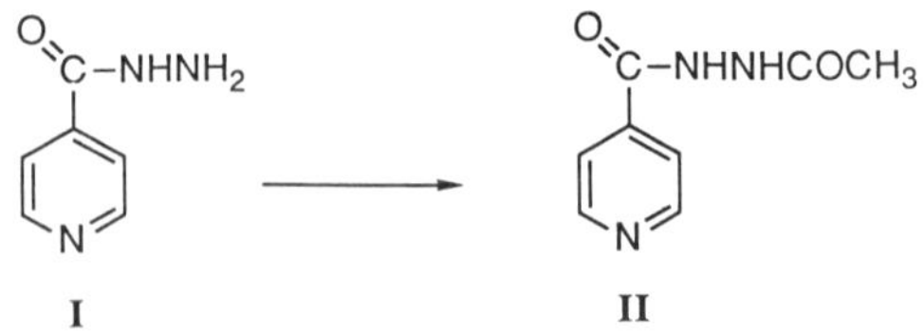

Fig. 1 Metabolic acetylation of isoniazid (**I**).

(Hein, 1988). Although the relationship of acetylator phenotype to either bladder cancer or colon cancer may be rationalized on the basis of differing capacities for arylamine metabolism, the reported correlation between acetylator phenotype and such spontaneous disorders as diabetes mellitus have not been explained. Evans (1989) has provided an extensive review of the clinical consequences of acetylation pharmacogenetics and of the associations between acetylator phenotype and spontaneous disorders.

The mutagenicity and carcinogenicity of arylamines, along with the documented environmental and dietary exposure to these agents, have stimulated numerous investigations of arylamine metabolism. Thus, the premise that formation of unreactive arylamides by N-acetylation of arylamines is primarily a detoxification process and that oxidative N-hydroxylation is a toxification reaction is supported by a substantial body of evidence (Miller and Miller, 1985; Hanna and Banks, 1985; Kadlubar *et al.*, 1992). However, the demonstration that arylamine *N*-acetyltransferases are versatile enzymes that can catalyze the conversion of both *N*-arylhydroxylamines and *N*-arylhydroxamic acids to reactive, electrophilic metabolites has broadened the scope of research on acetyltransferases, which are now included among those conjugation systems that play important roles in the bioactivation of xenobiotics (Beland and Kadlubar, 1990).

II. *N*-Acetyltransferase

N-Acetylation of hydrazines, arylamines, and related substances is catalyzed by acetyl coenzyme A:arylamine *N*-acetyltransferase (NAT) (EC 2.3.1.5). NATs are cytosolic proteins with molecular masses of approximately 33,000 Da. The enzymes utilize a "ping-pong Bi-Bi" mechanism, which requires a covalent acetyl-enzyme intermediate, as illustrated in the following equations (Weber, 1971; Steinberg *et al.*, 1971).

$$\text{AcCoA} + \text{NAT} \rightleftharpoons \text{CoA} + \text{Ac-NAT}$$
$$\text{Ac-NAT} + \text{ArNH}_2 \rightleftharpoons \text{ArNHCOCH}_3 + \text{NAT}$$

Several lines of evidence, including early results that demonstrated the sensitivity of NAT activity to sulfhydryl reagents, establish that the covalent acetyl-enzyme intermediate is a thioester. Andres *et al.* (1988) found that performic acid oxidation of a [2-^{3}H]acetyl-NAT intermediate caused the release of acetic acid, a result that is attributable to oxidation of a thioester. The latter authors also reported that amino acid analysis of rabbit liver NAT, which had been treated with an active-site-directed irreversible inhibitor, indicated the loss of a single cysteine residue. Cheon *et al.* (1992) demonstrated that 2-(bromoacetylamino)fluorene (Br-AAF) is an isozyme-selective, active-site-directed inactivator of hamster hepatic monomorphic NAT. HPLC analysis of the acid hydrolysate of the [^{14}C]Br-AAF-labeled NAT resulted in the identification and characterization of a modified cysteine residue, providing further evidence for the presence of a thiol group in the active site of NAT.

NAT activity is highest in liver, with various levels of activity being expressed in most other mammalian tissues. Most species, with the exception of the dog, exhibit N-acetylation capability (Lower and Bryan, 1973). Major advances in the elucidation of both species variability and the biological basis of interindividual variability in N-acetylation are due to the research of Weber and colleagues (Weber, 1987; Weber *et al.*, 1990). In particular, the hereditary N-acetylation polymorphisms of the rabbit, mouse, and hamster have been studied in detail (reviewed by Hein, 1988).

A. Multiple *N*-Acetyltransferases

Two cytosolic NAT isozymes are present in hamster liver (Hein *et al.*, 1985; Smith and Hanna, 1986; Ozawa *et al.*, 1990; Trinidad *et al.*, 1990), hamster bladder (Yerokun *et al.*, 1989), hamster intestine and colon (Smith and Hanna, 1986; Hein *et al.*, 1993b), and hamster peripheral blood (Hein *et al.*, 1986a). Two NAT isozymes also have been identified in human liver cytosol (Ohsako and Deguchi, 1990; Grant *et al.*, 1991); in the liver, lung, and duodenum of rabbits (Reeves *et al.*, 1991), in mouse liver (Martell *et al.*, 1992), and in rat liver (Wick and Hanna, 1990; Hein *et al.*, 1991a).

Studies with animal models, such as the rabbit, hamster, and mouse models, have shown that expression of the two NAT isozymes is under independent regulation by monomorphic and polymorphic genes (Sim *et al.*, 1992). Thus, it is the expression of the acetylator genotype-dependent (polymorphic) NAT that determines interindividual phenotypic variability in acetylation capacity. Similarly, it has been generally accepted that the two human liver NATs are monomorphic and polymorphic isozymes. The latter conclusion is consistent with the observation that some NAT substrates do not exhibit the bimodal or trimodal frequency of distribution of metabolism that is characteristic of polymorphic substrates. However, Vatsis and Weber (1993) have provided evidence that the human NAT1 gene locus, which has been considered to be monomorphic, is, in fact, polymorphic. The implications of this finding with regard to the clearance of "monomorphic" substrates by humans remain to be determined. The probable polymorphism of human NAT1 may provide an explanation for the several reports of an apparent polymorphic acetylation of *p*-aminobenzoic acid (PABA), a prototypical "monomorphic" substrate in humans, by human liver and colon cytosols (Kilbane *et al.*, 1991; Land *et al.*, 1989; Kirlin *et al.*, 1991).

The amino acid sequences of approximately 15 arylamine NATs have been deduced from genomic and cDNA clones (summarized in Watanabe *et al.*, 1992; Martell *et al.*, 1991; Abu-Zeid *et al.*, 1991). The mammalian NATs exhibit 65–85% sequence homology. The high degree of sequence conservation adjacent to Cys^{68} makes this residue likely to be the essential active site thiol. Site-directed mutagenesis of human polymorphic NAT in which cysteine was replaced by glycine at position-68 resulted in loss of all NAT activity (Dupret and Grant, 1992). The latter result is consistent with the site-directed mutagenesis results obtained with *Salmonella typhimurium* *O*-acetyltransferase (OAT) and suggests that Cys^{68} is catalytically essential (Watanabe *et al.*, 1992).

The cloned monomorphic and polymorphic NAT genes from inbred rapid and slow acetylator mouse strains have been transiently expressed in COS-1 cells (Martell *et al.*, 1992). The deduced amino acid sequences indicate that monomorphic NAT is identical in rapid and slow acetylator mouse strains, but that polymorphic NAT differs between the two strains by a single amino acid. The two polymorphic NATs also differ in heat stability and in rates of translation. Similarly, point mutations, which destabilize the polymorphic NAT proteins, appear to be responsible for the human slow acetylator phenotype, whereas deletion of the polymorphic NAT gene produces the slow acetylator phenotype in rabbits (Deguchi, 1992; Blum *et al.*, 1989).

Although both monomorphic and polymorphic NATs have been identified in human, rabbit, mouse, hamster, and rat tissues, it is important to note that there are significant species differences with regard to the substrate specificity of the isozymes. For example, sulfamethazine is a polymorphic substrate in humans and rabbits, but is acetylated monomorphically by hamsters, and is not a substrate for any of the mouse NATs (Weber and Hein, 1985; Martell *et al.*, 1992). *p*-Aminobenzoic acid has heretofore been classified as a monomorphic NAT substrate in humans (Vatsis and Weber, 1993) and rabbits, but is a polymorphic substrate for mouse, hamster, and rat liver NATs (Weber and Hein, 1985; Hein *et al.*, 1991b).

Because of the dominant role played by acetylation in the detoxification of primary arylamines, several laboratories have undertaken studies of the substrate specificities of NATs for carcinogenic arylamines. The prototypical arylamine carcinogen 2-aminofluorene (2-AF) is preferentially, but not exclusively, N-acetylated by partially purified human monomorphic NAT (NAT1) from liver cytosol. The same selectivity was observed with NAT1 expressed in mammalian and bacterial cell lines (Grant *et al.*, 1991, 1992; Minchin *et al.*, 1992a; Hein *et al.*, 1993a). Benzidine is converted to monoacetyl benzidine primarily by human NAT1, but 4-aminobiphenyl is a better substrate for the human polymorphic enzyme (NAT2) than for NAT1 (Grant *et al.*, 1992; Minchin *et al.*, 1992a). 2-AF is N-acetylated at comparable rates by mouse monomorphic NAT and the polymorphic NAT of slow acetylators, but it is a somewhat better substrate for the polymorphic NAT of rapid acetylator mice (Martell *et al.*, 1992). The rate of acetylation of 2-AF by hamster liver polymorphic NAT is reported to be several-fold greater than that exhibited by the monomorphic isozyme (Ozawa *et al.*, 1990; Trinidad *et al.*, 1990). Of potential significance for human health are the findings that the mutagenic–carcinogenic heterocyclic arylamines formed during the cooking of meat and fish are generally poor substrates for NATs (Shinohara *et al.*, 1984; Trinidad *et al.*, 1990; Minchin *et al.*, 1992a; Turesky *et al.*, 1991). Thus, in the absence of an efficient detoxification route, the heterocyclic arylamines may exhibit an increased tendency to undergo N-oxidative bioactivation *in vivo*.

Lipophilic carbocyclic arylamine carcinogens, such as 2-AF and 4-aminobiphenyl, are N-acetylated by human liver, colon, ureter, and bladder cytosols at relative rates that correspond to acetylator phenotype (Flammang *et al.*, 1987; Kirlin *et al.*, 1989, 1991; Turesky *et al.*, 1991; Pink *et al.*, 1992). Hein *et al.* (1993b,c) also reported genotype-dependent acetylation of carcinogenic arylamines by liver, colon, and bladder cytosols of hamsters, that are congenic at the polymorphic NAT gene locus. Polymorphic N-acetylation of 2-AF is exhibited by colon and liver cytosols of inbred mice (Levy *et al.*, 1993).

III. *O*-Acetyltransferase

A. Mammalian *O*-Acetyltransferases

Both arylamines (**III,** Fig. 2) and nitroaromatic compounds (**IV**) are metabolically converted in mammalian tissues to *N*-hydroxyarylamines (**V**). The latter compounds are considered to be proximate mutagens and carcinogens, which may be transformed to electrophilic reactants by either enzymatic or nonenzymatic mechanisms. The chemical (nonenzymatic)

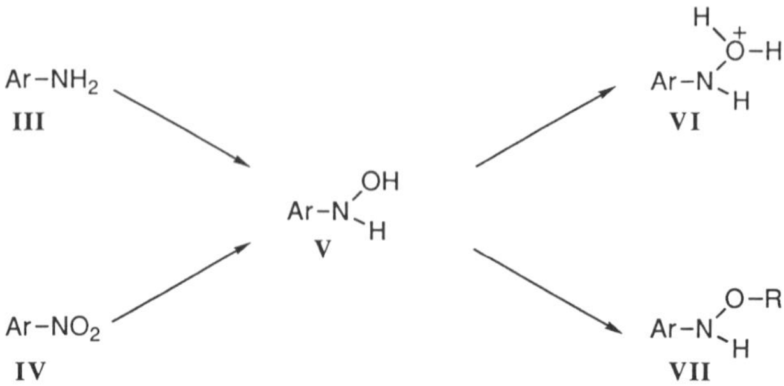

Fig. 2 Bioactivation of arylamines and nitroaromatic compounds. R = $COCH_3$, SO_3H, COC_4H_8N (prolyl), or COC_2H_6NO (seryl).

process that has been identified for activation of *N*-hydroxyarylamines (**V**) involves acid catalysis (**VI,** Fig. 2). The enzyme-dependent pathways that are capable of transforming *N*-hydroxyarylamines to electrophilic metabolites include conjugation reactions that produce acetyl, sulfonate, prolyl, and seryl esters (**VII,** Fig. 2). The bioactivation of arylamines and nitroaromatic compounds, as well as the DNA adducts formed by the reactive metabolites, has been reviewed by Beland and Kadlubar (1990).

Incubation of *N*-hydroxy-3,2′-dimethyl-4-aminobiphenyl (**VIII**) with rat liver cytosol in the presence of AcCoA and DNA resulted in the formation of DNA adducts that were identical to those obtained after administration of the carcinogenic parent compound, 3,2′-dimethyl-4-aminobiphenyl, to rats (Flammang *et al.,* 1985). The major adduct, both *in vitro* and *in vivo,* was *N*-(deoxyguanosin-8-yl)-3,2′-dimethyl-4-aminobiphenyl (**IX**). Arylamine-C^8 deoxyguanosine adducts, such as **IX,** are the most common of the adducts formed through reaction of bioactivated *N*-hydroxyarylamines with DNA (Beland and Kadlubar, 1990).

O-Acetylation of **VIII** to form a reactive *N*-acetoxyarylamine (Fig. 2, **VII,** R = $COCH_3$) was deemed to be a likely bioactivation pathway *in vivo* because **VIII** itself was very weakly reactive with DNA at neutral pH and because it was not a substrate for rat cytosolic sulfotransferase. A bioactivation pathway involving N-acetylation of **VIII** was considered less plausible because the *N*-acetyl derivative of **VIII** (i.e., the hydroxamic

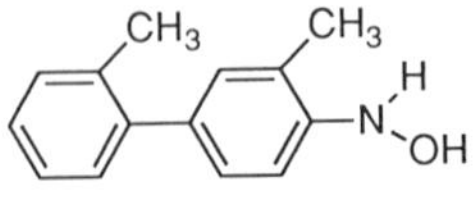

VIII

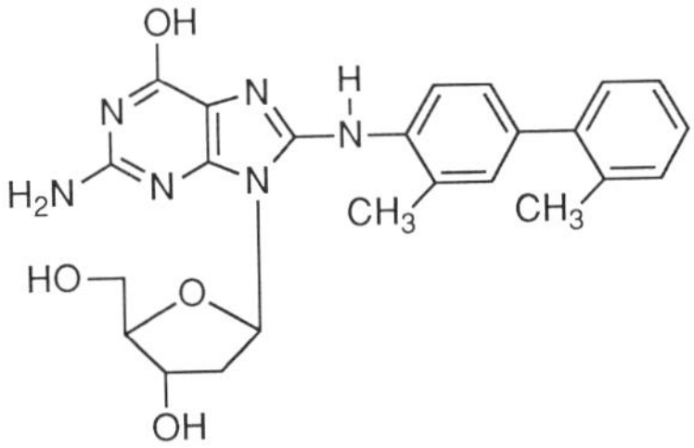

IX

acid) was not activated by enzymatic intramolecular *N,O*-acetyltransfer (see Section IV) (Flammang *et al.,* 1985). The lack of potent and selective inhibitors of cytosolic acetyltransferases complicates the effort to define their roles in the *in vivo* bioactivation of *N*-hydroxyarylamines and *N*-arylhydroxamic acids (Shinohara *et al.,* 1986a).

In addition to **VIII,** the N-hydroxylated derivatives of several other carcinogenic arylamines, including 2-AF, 4-aminobiphenyl, *N'*-acetylbenzidine, and 2-naphthylamine, are activated to DNA-binding reactants in the presence of rat liver cytosol and acetyl coenzyme A (AcCoA) (Frederick *et al.,* 1985; Flammang and Kadlubar, 1986). The hydroxylamines produced by cytosolic reduction of 1,6- and 1,8-dinitropyrene also were converted to reactive electrophiles when AcCoA was included in the incubation mixture (Djurić *et al.,* 1985). This OAT (*O*-acetyltransferase) bioactivation activity is present in rat liver, intestine, mammary gland, and kidney cytosols and in liver cytosols obtained from rabbits, hamsters, guinea pigs, mice, and humans. OAT activity is present in hamster skin cytosol (Kawakubo *et al.,* 1988), but was not found in hepatic cytosol of dogs, a species that also lacks arylamine NAT activity (Flammang and Kadlubar, 1986; Djurić *et al.,* 1985). AcCoA, *p*-nitrophenyl acetate, acetoacetyl CoA, and propionyl CoA were effective as acyl donors for rat liver OAT-catalyzed activation of *N*-hydroxyarylamines. Two *N*-arylhydroxamic acids, *N*-hydroxy-2-acetylaminofluorene (N-OH-AAF) and *N*-hydroxy-4-acetylaminobiphenyl, also functioned as acetyl donors (Flammang and Kadlubar, 1986).

As mentioned previously (Section II,A), there is substantial human exposure to a group of mutagenic–carcinogenic heterocyclic arylamines that are formed as protein and amino acid pyrolysates during the cooking of meat and fish (Sugimura, 1992; Wakabayashi *et al.,* 1992). Representative of the approximately 20 such compounds that have been characterized are 2-amino-3-methylimidazo-[4,5-*f*]quinoline (IQ,**X**), 2-amino-1-methyl-6-phenylimidazo-[4,5-*b*]pyridine (PhIP,**XI**), and 2-amino-6-methyldipyrido-[1,2-*a*:3',2'-*d*]imidazole (Glu-P-1, **XII**). Although the heterocyclic arylamines are not N-acetylated efficiently by mammalian NATs, they are converted to *N*-hydroxyarylamines by mammalian microsomal mixed function oxidases and, therefore, are candidates for further metabolic activation by OAT (Rich *et al.*, 1992; Turesky *et al.*, 1991).

Hashimoto *et al.* (1980) synthesized the N-hydroxylated derivative of Glu-P-1 (**XII**) and found that it did not form covalent adducts readily with DNA under neutral conditions. However, treatment of *N*-hydroxy-Glu-P-1 with ketene resulted in formation of the *N*-acetoxy ester, which did react with DNA. Subsequently, Yamazoe *et al.* (1982) reported that the N-hydroxylated tryptophan pyrolystate, 3-hydroxyamino-1-methyl-5*H*-pyrido[4,3-*b*]indole (*N*-hydroxy-Trp-P-2) was converted to a DNA-binding reactant upon incubation with partially purified rat liver cytosol in the presence of AcCoA. Liver, lung, kidney, intestinal, and bladder cytosols from both hamsters and rats catalyzed the AcCoA-dependent bioactivation of *N*-hydroxy-Glu-P-1 and *N*-hydroxy-Trp-P-2, with the highest activity being exhibited by hamster liver cytosol. When liver cytosols from several species were compared, AcCoA-dependent binding to tRNA was highest for hamsters and rats if either *N*-hydroxy-Glu-P-1 or *N*-hydroxy-Trp-P-2 was the substrate, but hepatic cytosol from rapid acetylator rabbits catalyzed the AcCoA-dependent activation of *N*-hydroxy-2-AF to the greatest extent. AcCoA was the most effective acetyl donor studied (Shinohara *et al.*, 1986b).

The possibility was considered that the AcCoA-dependent activation of the heterocyclic *N*-hydroxyarylamines could be the result of *N*-acetylation of the *N*-hydroxyarylamines to form *N*-arylhydroxamic acids, which then might be activated by intramolecular *N,O*-acetyl transfer (see

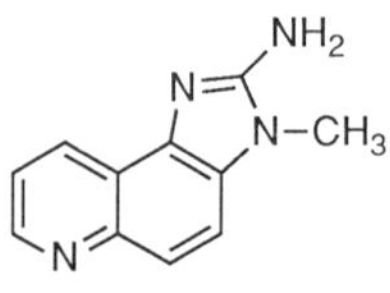

X (IQ)

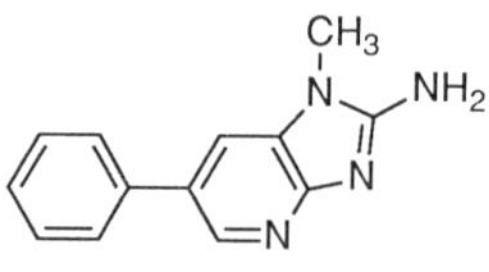

XI (PhIP)

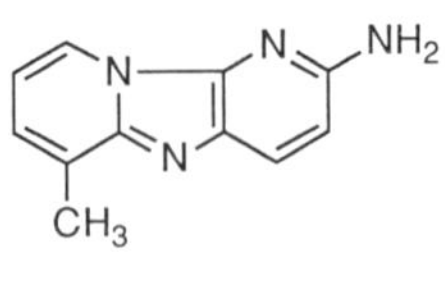

XII (Glu-P-1)

Section IV). However, analysis of liver cytosol in which *N*-hydroxy-Glu-P-1 had been incubated with AcCoA did not indicate the presence of the hydroxamic acid *N*-acetyl-*N*-hydroxy-Glu-P-1. In addition, the hydroxamic acid derivative of Glu-P-1 did not undergo bioactivation via *N,O*-acetyl transfer in the presence of liver cytosol. The enzymatic conversion of the heterocyclic *N*-hydroxyarylamines to electrophilic reactants in the presence of AcCoA is, therefore, believed to occur primarily through the direct OAT-catalyzed acetylation of the oxygen atom of the substrate (Fig. 2) (Shinohara *et al.*, 1986b).

The results of several investigations demonstrate that OAT activity is associated with NAT proteins. Homogenous preparations of hamster hepatic monomorphic NAT catalyzed both the AcCoA-dependent N-acetylation of 2-AF and the AcCoA-dependent activation of *N*-hydroxy-Glu-P-1 (Saito *et al.*, 1986; Ozawa *et al.*, 1990). Although the purified hamster liver polymorphic NAT did not catalyze the AcCoA-dependent activation of *N*-hydroxy-Glu-P-1, the polymorphic isozyme activated *N*-hydroxy-2-AF in the presence of AcCoA (Trinidad *et al.*, 1990). Additionally, expression of the cloned monomorphic enzyme from hamsters afforded a protein that catalyzes both N-acetylation of arylamines and O-acetylation of *N*-hydroxyarylamines (Abu-Zeid *et al.*, 1991; Ferguson *et al.*, 1993). Purified mouse liver NAT exhibits OAT activity (Mattano *et al.*, 1989), and human recombinant monomorphic and polymorphic isozymes exhibit both NAT and OAT activity, but differ in their substrate specificities (Minchin *et al.*, 1992a; Hein *et al.*, 1993a).

The substrate specificity of human NATs was further demonstrated by Probst *et al.* (1992), who expressed cytochrome P4501A2 (CYP1A2), the principal isozyme responsible for arylamine N-hydroxylation, in COS-1 cells along with either NAT1 or NAT2. The level of DNA adducts resulting from activation of IQ (**X**) by CYP1A2 and NAT2 was three to six times greater than that produced by CYP1A2 and NAT1.

The epidemiological evidence which indicates that human susceptibility to bladder and colorectal cancer is influenced by acetylation capacity (Hayes *et al.*, 1993; Cartwright *et al.*, 1982; Wohlleb *et al.*, 1990), the effect of acetylator genotype on the extent of protein and DNA adduct formation in animals exposed to carcinogenic arylamines (Flammang *et al.*, 1992a,b; Levy and Weber, 1989), and the greater amount of DNA damage induced by arylamines in hepatocytes of rapid acetylator rabbits than in slow acetylator rabbit hepatocytes (McQueen *et al.*, 1983) may reflect acetylation genotype-related bioactivation. Indeed, a number of studies demonstrate a polymorphic OAT-catalyzed activation of *N*-hydroxyarylamines in mammalian tissues. Genotype-dependent OAT activity was found, with *N*-hydroxy-2-AF as the substrate, in liver, intestine, kidney, and lung cytosols of inbred hamsters and in colon cytosol of Syrian hamsters congenic at the polymorphic NAT gene locus (Hein *et al.*, 1987, 1993b). However, activation of **VIII** by OAT in hamster tissue cytosols could not be correlated with genotype (Hein *et al.*, 1987). Ilett *et al.* (1991) reported that OAT activity with both **VIII** and *N*-hydroxy-2-AF as substrates is distributed similarly to polymorphic NAT in the liver and intestine of rapid acetylator rabbits.

Flammang *et al.* (1987) found that human liver cytosol from phenotypic rapid acetylators catalyzed the AcCoA-dependent activation of *N*-hydroxy-2-AF at a consistently higher rate than did liver cytosol from slow acetylators. Similarly, the OAT-catalyzed activation of *N*-hydroxy-4-aminobiphenyl and several *N*-hydroxyheterocyclic amines by human liver cytosols exhibited a polymorphic distribution, but a phenotypic dependence of OAT activity in human colon cytosols could not be demonstrated (Turesky *et al.*, 1991). Human bladder cytosols activated **VIII** to a DNA-binding reactant through an AcCoA-dependent, OAT-mediated reaction, which correlated with the polymorphic expression of arylamine NAT activity (Kirlin *et al.*, 1989). These results lend support to the contention that acetylation capacity may influence susceptibility to arylamine-induced cancer.

B. Bacterial *O*-Acetyltransferase

Numerous aromatic nitro compounds and arylamines have been evaluated for their mutagenic potency in the Ames *S. typhimurium* assay. The bacterial mutagenicity of these agents is dependent upon their conversion to *N*-hydroxyarylamines and, in most cases, subsequent esterification of the *N*-hydroxyarylamines by an AcCoA-dependent bacterial OAT (Snyderwine *et al.*, 1988; Beland and Kadlubar, 1990). The requirement for bacterial OAT-catalyzed activation of numerous mutagenic *N*-hydroxyarylamines has been established through the use of *S. typhimurium* tester strain TA98/1,8DNP$_6$, which lacks OAT activity and which is resistant to the mutagenic action of a variety of aromatic nitro compounds and arylamines (McCoy *et al.*, 1982, 1983; Orr *et al.*, 1985; Nagao *et al.*, 1983; Saito *et al.*, 1983). *S. typhimurium* strains that exhibit elevated levels of OAT activity are reported to exhibit increased ability to detect the mutagenicity of arylamines and nitroarenes (Einistö *et al.*, 1991).

Salmonella OAT exhibits both OAT and NAT activity, but did not catalyze intramolecular *N,O*-acetyl transfer when the *N*-arylhydroxamic acid *N*-hydroxy-2-acetylaminofluorene (N-OH-AAF) was evaluated as a substrate (Saito *et al.*, 1985; McCoy *et al.*, 1983; Watanabe *et al.*, 1992). A partially purified preparation of the enzyme exhibited several characteristics in common with mammalian OATs, including a preference for AcCoA as the acyl donor, inhibition by sulfhydryl blocking agents, and the ability to activate structurally diverse *N*-hydroxyarylamines to nucleic acid-binding reactants (Saito *et al.*, 1985). Both bacterial and mammalian OATs are inhibited by pentachlorophenol and 1-nitro-2-naphthol, but only the bacterial enzyme is inhibited by thiolactomycin. Neither type of OAT is inhibited by organophosphates, such as paraoxon (Saito *et al.*, 1985; Shinohara *et al.*, 1986a).

The OAT of *S. typhimurium* does not exhibit a high degree of sequence homology with the corresponding enzymes of several mammalian species, although a five-residue sequence, which includes the presumed active site cysteine residue, is highly conserved (Watanabe *et al.*, 1992). Studies by Yamazoe *et al.* (1989) of the relative capacities of rat hepatic cytosol and *S. typhimurium* TA98 cytosol to catalyze AcCoA-dependent DNA adduct formation by *N*-hydroxyarylamines revealed apparent differences in the substrate specificities. The apparent structure–activity relationships must be interpreted cautiously because of probable differences in enzyme concentration between the two cytosols and because of the likelihood that more than one rat hepatic acetyltransferase contributes to catalysis of the bioactivation reaction.

The mutagen–carcinogen PhIP (**XI**), which is found in greater quantities in cooked meats than most other heterocyclic amines, is mutagenic in the *S. typhimurium* assay after oxidation to *N*-hydroxy-PhIP. The hydroxylamine metabolite of PhIP is characterized as a "direct-acting" bacterial mutagen because O-acetylation is not required for conversion to its ultimate mutagenic form (Holme *et al.*, 1989; Buonarati and Felton, 1990). However, O-acetylation of *N*-hydroxy-PhIP has been identified as an activation process in mammalian tissues (Kaderlik *et al.*, 1993).

Differences in the substrate specificity of bacterial and human OATs may compromise the relevance of *Salmonella* mutagenicity tester strains for detection of human mutagens and carcinogens, even when human enzymes are used as the exogenous activating systems for conversion of arylamines to *N*-hydroxyarylamines. A potentially important advance in the development of bacterial test systems was reported by Grant *et al.* (1992), who expressed human monomorphic and polymorphic NATs (OATs) in *S. typhimurium*. The enzymes were functional, exhibited the expected substrate specificities, and supported the bioactivation of arylamines to mutagens.

IV. *N*-Arylhydroxamic Acid *N,O*-Acyltransferase

A. Cytosolic *N,O*-Acyltransferase

In addition to exhibiting NAT and OAT activities, cytosolic acetyltransferases may catalyze both intermolecular transfer of acyl groups from *N*-arylhydroxamic acids to arylamines and intramolecular transfer from nitrogen to oxygen within a hydroxamic acid (Fig. 3). The intermolecular arylamine acylation reaction (Fig. 3A) was described by Booth (1966) and is usually referred to as *N,N*-acetyl transfer (N,NAT activity). The enzymatic activity responsible for catalysis of the intramolecular reaction (Fig. 3B) is called *N*-arylhydroxamic acid *N,O*-acyltransferase (AHAT or N,OAT). The latter reaction results in the formation of the same type of reactive, electrophilic acyloxyarylamine that is formed in the AcCoA-dependent, OAT-catalyzed, esterification of *N*-hydroxyarylamines (Section III).

King and Phillips (1968, 1969) found that incubation of N-OH-AAF and guanosine with an enzyme fraction obtained from rat hepatic cytosol yielded *N*-(guanosin-8-yl)-2-aminofluorene, an adduct that was proposed to result from reaction of the deacetylation product, *N*-hydroxy-2-AF, with guanosine. However, the authors noted that the extent of adduct formation was greater than would have been anticipated for a reaction with *N*-hydroxy-2-AF at pH 7.4. Bartsch *et al.* (1972) provided important insight into this bioactivation process by demonstrating that incubation of an *N*-arylhydroxamic acid and an *N*-hydroxyarylamine with rat liver cytosol resulted in generation of electrophilic reactants and formation of arylamine adducts derived from both the hydroxamic acid and the hydroxylamine. Adducts were obtained only if the *N*-arylhydroxamic acid was present, but inclusion of the *N*-hydroxyarylamine in the incubation mixture was not essential. Further, the bioactivation activity was found to cochromatograph with the *N*-arylhydroxamic acid-dependent arylamine N,NAT activity reported by Booth (1966) (Fig. 3A). It was, therefore, suggested that the cytosolic bioactivation of *N*-arylhydroxamic

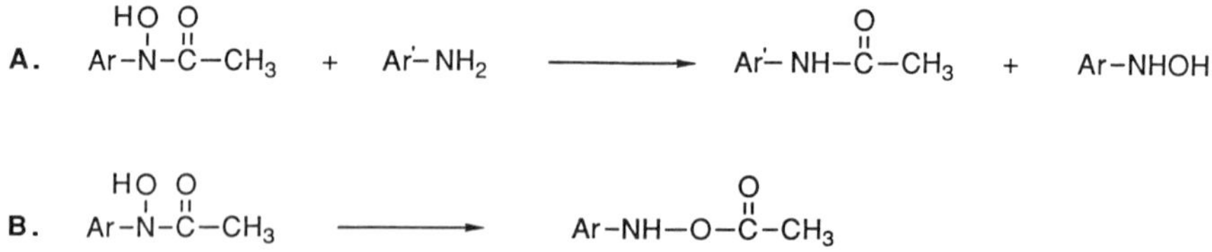

Fig. 3 (A) *N*-Arylhydroxamic acid-dependent *N,N*-acetyltransfer (N,NAT). (B) *N*-Arylhydroxamic acid *N,O*-acyltransfer (AHAT or N,OAT).

acids involved formation of reactive *N*-acetoxyarylamines (Fig. 3B). The latter proposal was supported by the finding that esterification of *N*-hydroxyarylamines with acetic anhydride produced reactants that yielded the same adducts that were formed when the corresponding *N*-arylhydroxamic acids were incubated with rat liver cytosol (Bartsch *et al.*, 1972). King (1974) proposed a mechanism involving an acylated enzyme intermediate to account for the N,OAT process (Fig. 4). The mechanism is analogous to that of the AcCoA-dependent NAT reaction in which an acetyl-enzyme intermediate is formed (Section II).

Extensive structure–activity investigations have been conducted with N,OAT substrates (reviewed by Hanna and Banks, 1985). Studies with *N*-arylhydroxamic acids having *N*-acyl groups of various electronic and steric properties revealed a consistent preference for the *N*-acetyl group in assays for N,NAT and N,OAT activities (Weeks *et al.*, 1980; Yeh and Hanna, 1982). The observation that a rat liver cytosolic N,OAT exhibited specificity for *N*-hydroxy-*N*-formyl-2-aminofluorene (Beland *et al.*, 1980) probably was due to the presence of microsomal N,OAT in the cytosolic preparation (Allaben *et al.*, 1982). Although *N*-arylhydroxamic acids with a variety of lipophilic *N*-aryl groups are capable of serving as acetyl donors for cytosolic N,NATs and N,OATs, hydroxamic acids with substituents in a position ortho to the *N*-hydroxy-*N*-acetyl group are not good substrates (Kumano *et al.*, 1986; Flammang and Kadlubar, 1986), and no evidence was found for metabolism of the acetohydroxamic acid derivative of the heterocyclic arylamine Glu-P-1 (**XII**) by hamster liver N,OAT (Shinohara *et al.*, 1986b).

N,OAT activity is present in numerous mammalian tissues, and it is now evident that cytosolic NAT, OAT, N,NAT, and N,OAT (AHAT) activities are frequently associated with single proteins. In rabbit liver, a single cytosolic protein catalyzes N,OAT and NAT activity (Glowinski *et al.*, 1980). Similarly, NAT, OAT, and N,OAT activities were copurified from rapid acetylator mouse liver cytosol (Mattano *et al.*, 1989) and hamster liver cytosol (Saito *et al.*, 1986).

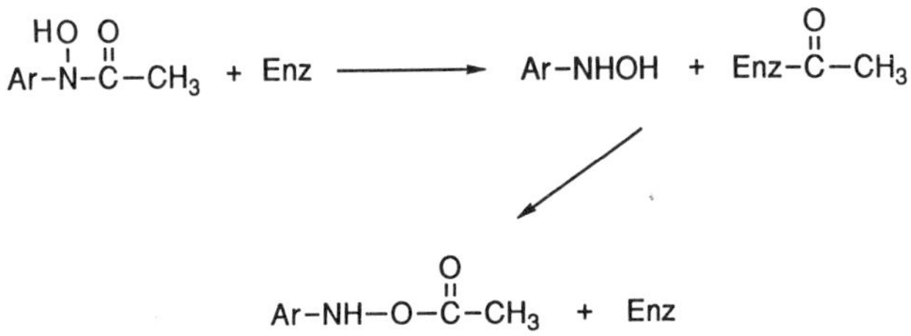

Fig. 4 Mechanism of *N*-arylhydroxamic acid *N,O*-acyltransfer.

Allaben and King (1984) purified rat liver N,OAT to homogeneity. Subsequently, Wick and Hanna (1990) used mechanism-based irreversible inactivation (suicide inactivation) and affinity chromatography to identify and resolve two forms of NAT, designated NAT I and NAT II, from rat liver cytosol. NAT I readily catalyzed the N-acetylation of PABA; NAT II catalyzed the AcCoA-dependent acetylation of PABA very poorly, but exhibited all of the AcCoA-dependent acetylation activity for procainamide (PA), all of the *N*-arylhydroxamic acid-dependent N,NAT activity, and all of the N,OAT activity. The results of extensive purification of NAT II by chromatofocusing confirmed the association of PA NAT activity, N,NAT activity, and N,OAT activity with NAT II. The results of a study of the acetylator phenotype-dependent and -independent expression of NAT isozymes in inbred rat liver indicate that the isozyme designated by Wick and Hanna (1990) as NAT I is polymorphic and that NAT II is monomorphic (Hein *et al.*, 1991a). In addition to the use of mechanism-based inactivation as a means of detecting multiple acetyltransferase isozymes in rat liver preparations, Wick *et al.* (1990) reported that 1-(fluoren-2-yl)-2-propen-1-one (VFK, **XIII**) is a useful probe for acetyltransferase multiplicity. Compound XIII, which is structurally related to the N,OAT substrate N-OH-AAF (**XIV**), is a potent, irreversible, and selective inactivator (affinity label) of the rat hepatic isozyme that exhibits N,OAT activity.

Hamster liver cytosol exhibits relatively high levels of N,OAT activity (Bartsch *et al.*, 1972; King and Olive, 1975). Treatment of a partially purified fraction of soluble protein from hamster liver with N-OH-AAF caused irreversible loss of N,OAT activity, N,NAT activity, AcCoA-dependent sulfamethazine N-acetylation activity, and, to a somewhat lesser extent, loss of AcCoA-dependent PABA NAT activity by a mechanism-based inactivation process (Banks and Hanna, 1970; Hanna *et al.*, 1982). Inclusion of the nucleophile cysteine in the incubation mixture resulted in protection of only the PABA NAT activity against inactivation by N-OH-AAF. It was proposed that hamster hepatic N,NAT, N,OAT, and AcCoA-dependent sulfamethazine NAT activities might be associated with the same protein, but that AcCoA-dependent acetylation of PABA is catalyzed by a distinct protein. A similar selective inactivation of liver acetyltransferases *in vivo* was observed when hamsters were given N-

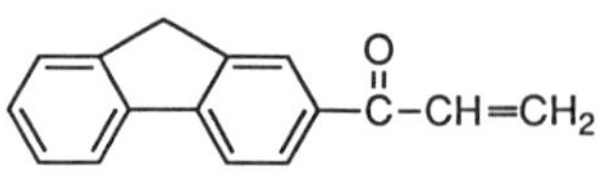

XIII (VFK)

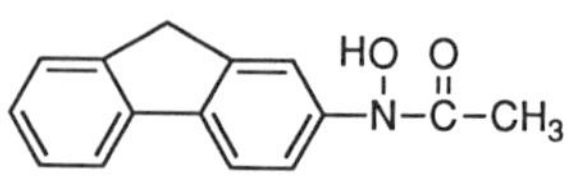

XIV (N-OH-AAF)

OH-AAF (Smith and Hanna, 1988). The association of virtually all of the N,OAT and N,NAT activity of hamster liver and intestine with an isozyme distinct from that which catalyzes the N-acetylation of PABA was confirmed by use of affinity chromatography to resolve and partially purify the activities (Smith and Hanna, 1986). The liver enzyme responsible for N,OAT and N,NAT activity corresponds to the isozyme that was determined to be genetically monomorphic by Hein *et al.* (1985, 1986b). Both the monomorphic and the polymorphic isozymes, partially purified from colon cytosol of Syrian hamsters congenic at the polymorphic NAT gene locus, appear to exhibit significant amounts of N,OAT activity (Hein *et al.*, 1993b). The latter results differ from those obtained with partially purified liver and intestinal isozymes and with the results obtained with highly purified liver preparations in which only the monomorphic isozyme exhibits substantial N,OAT and N,NAT activity (Ozawa *et al.*, 1990; Trinidad *et al.*, 1990).

King *et al.* (1975) reported the presence of *N*-arylhydroxamic acid N,OAT activity in human liver cytosol, but the activity was not detected in a later study of 35 human liver samples (Flammang *et al.*, 1987). Both Abu-Zeid *et al.* (1992) and Land *et al.* (1989) found that human liver cytosol exhibited N,OAT activity, although the latter workers concluded that 30–40% of the cytosolic activity was due to a microsomal component. Based on the extent of DNA adduct formation by *N*-acetyl-IQ in COS-1 cells in which CYP1A2 and either human monomorphic NAT or polymorphic NAT had been expressed, Probst *et al.* (1992) concluded that both NATs exhibit N,OAT activity, but that the polymorphic enzyme (NAT2) possesses approximately 10 times as much activity as the monomorphic isozyme. In contrast, Hein *et al.* (1993a) evaluated the N,OAT activities of human NATs from a bacterial expression system and found that the monomorphic isozyme (NAT1) exhibited activity with N-OH-AAF (**XIV**) as the substrate, but that the polymorphic isozyme (NAT2) had little capability for catalyzing the activation of **XIV.** Studies with a wider range of structurally varied *N*-arylhydroxamic acids should clarify the substrate specificities of the human isozymes with regard to N,OAT activity.

B. Microsomal *N,O*-Acyltransferase

A number of studies establish that microsomal carboxylesterases (esterases) (EC 3.1.1.1) exhibit N,OAT (AHAT) activity as well as AcCoA-dependent OAT and NAT activities. Thus, microsomal enzymes, as well as the previously discussed cytosolic acetyltransferases, may play significant roles in the bioactivation and detoxification of arylamines and arylamides, as well as their N-hydroxylated metabolites. Microsomal esterases contain a catalytically essential active site serine hydroxyl group and exhibit relatively broad substrate specificities (Heymann, 1980). The enzymes are inhibited by organophosphates, such as diethyl *p*-nitrophenyl phosphate (paraoxon) and bis(*p*-nitrophenyl)phosphate (BNPP). Paraoxon and BNPP are commonly used to differentiate microsomal esterase activity from cytosolic acetyltransferase activities, which are not sensitive to the organophosphates.

Rat liver and mammary gland cytosols were reported to contain two fractions of *N*-arylhydroxamic acid bioactivation activity, one of which was sensitive to inhibition by paraoxon and exhibited selectivity for activation of *N*-hydroxy-*N*-formyl-2-aminofluorene over N-OH-AAF (**XIV**). Microsomal fractions of the two tissues also exhibited specificity for the *N*-formyl hydroxamic acid (Shirai *et al.,* 1981; Allaben *et al.,* 1982). It was suggested that the cytosolic enzyme that exhibited specificity for the *N*-formyl hydroxamic acid was a solubilized microsomal component. Related experiments with guinea pig hepatic microsomes provided evidence that the conversion of *N*-arylhydroxamic acids to nucleic acid-binding reactants involved *N,O*-acyl transfer to yield *N*-acyloxyarylamines (Glowinski *et al.,* 1983; Vaught *et al.,* 1985). Such a process is mechanistically plausible and, presumably, involves formation of an acyl-enzyme intermediate as illustrated in Fig. 4 for cytosolic N,OAT-catalyzed reactions.

An AcCoA-dependent NAT activity occurs in liver microsomes of dogs, a species that does not have cytosolic NAT activity. Three microsomal enzymes, which exhibited varying degrees of NAT, N,OAT, and deacetylase (esterase) activities, were resolved, and the N,OAT and NAT activities were shown to be inhibited by paraoxon (Sone *et al.,* 1991). Microsomal preparations from both dog liver and dog bladder catalyzed the AcCoA-dependent activation of *N*-hydroxy-4-aminobiphenyl to reactants that formed nucleic acid adducts (OAT activity). 4-Acetylaminobiphenyl and *N*-hydroxy-4-acetylaminobiphenyl also served as acetyl donors for the microsomal OAT (Hatcher and Swaminathan, 1992).

Wang *et al.* (1992) demonstrated that the N-OH-AAF (**XIV**)-dependent N,OAT activity of rat liver microsomes is inducible, a characteristic of microsomal carboxylesterases. Four distinct rat liver microsomal proteins were purified to homogeneity, and each was shown to exhibit N,OAT (AHAT) activity and deacetylase (esterase) activity.

Human liver and urinary bladder microsomes exhibit AcCoA-dependent OAT-catalyzed and AcCoA-independent N,OAT-catalyzed activation of *N*-hydroxyarylamines and *N*-arylhydroxamic acids, respectively. Both activities are inhibited by paraoxon (Land *et al.,* 1989). Microsomal preparations obtained from human urothelial cell cultures also catalyze the paraoxon-sensitive activation of *N*-hydroxy-4-acetylaminobiphenyl to nucleic acid-binding reactants (Swaminathan and Reznikoff, 1992). The bioactivation capabilities of bladder-tissue esterases are of particular interest because the urinary bladder is a principal target organ for aromatic amine carcinogens in humans.

Although substantial evidence in support of microsomal esterase-catalyzed activation has appeared since Irving's report on microsomal deacetylation of N-OH-AAF (**XIV**) (Irving, 1966), a specific relationship between microsomal bioactivation and arylamine mutagenicity–carcinogenicity has not been established. Investigations of possible correlations between genetic differences in esterase activity and the biological effects of arylamines may be helpful in this regard (Hultin and Weber, 1987). Lai *et al.* (1988) demonstrated that treatment of infant male mice with the carboxylesterase inhibitor BNPP before administration of N-OH-AAF (**XIV**) inhibited the initiation of liver tumors. The latter result indicates the necessity of microsomal esterase activity for induction of carcinogenesis in this animal model, but it does not identify a specific bioactivation role (i.e., N,OAT or OAT) for the esterases. Deacetylation of N-OH-AAF (**XIV**) and *N*-hydroxyphenacetin also is required for the manifestation of mutagenicity in *S. typhimurium* test systems (Vaught *et al.*, 1981; Wirth *et al.*, 1982).

Carboxylesterase activity present in the outer membrane of cell nuclei may be an important factor in the genotoxocity of *N*-arylhydroxamic acids because of the proximity of the enzymes to nuclear DNA. Several reports describe the covalent binding of N-OH-AAF (**XIV**) to the nuclear DNA of cell-free nuclei through a process that does not require added cofactors and which is inhibited by paraoxon (Sakai *et al.*, 1978; Stout *et al.*, 1980; Meerman *et al.*, 1986). Characterization of these bioactivation enzymes remains to be accomplished.

V. *N*-Acyloxyarylamines

A. Synthesis

Although in 1972 Bartsch *et al.* demonstrated that reactive *N*-acetoxyarylamines could be generated *in situ* by treatment of *N*-hydroxyarylamines with acetic anhydride in aqueous buffer, very few publications describing either alternative synthetic methods or studies on the reactivity of the *O*-acetoxy compounds appeared during the following decade. In addition to acetic anhydride, acetylating agents used for the conversion of *N*-hydroxyarylamines to *N*-acetoxyarylamines include ketene (Hashimoto *et al.*, 1982), a mixed anhydride (Lutgerink *et al.*, 1989), acetylsalicylic acid (Minchin *et al.*, 1992b), 2-acylthiazolium salts (Ferreira *et al.*, 1991), and acetyl cyanide (Lobo *et al.*, 1987; Bosold and Boche, 1990).

Two reports indicate that reaction conditions significantly influence the relative yields of N-acetylated (i.e., the hydroxamic acid) and O-acetylated products when acetic anhydride is the acetylating agent. Frandsen *et al.* (1992) found that reaction of *N*-hydroxy-PhIP with acetic anhydride in the presence of acetic acid yielded primarily the O-acetylated product, whereas reaction with acetic anhydride in the presence of pyridine afforded exclusively the unreactive N-acetylated compound. Huggett *et al.*, (1987) reacted *N*-hydroxy-2,4-dinitrophenylamine with acetic anhydride to form *N*-acetoxy-2,4-dinitrophenylamine but inclusion of pyridine in the reaction mixture yielded the N,O-diacetylated product. Although the factors that determine the relative rates of N- and O-acetylation of *N*-hydroxyarylamines have not been elucidated in detail, aqueous or slightly acidic conditions, or both, as well as the use of an acyl cyanide, often favor O-acylation.

Meier and Boche (1990) and Boche *et al.* (1988) produced several reactive *N*-acyloxyarylamines by treatment of the corresponding *N*-arylhydroxamic acids with an organic base to catalyze an intramolecular *N,O*-acyl transfer analagous to the reaction depicted in Fig. 3B. The transformation was accomplished with *N*-acetyl-, *N*-pivaloyl-, and *N*-(α-aminoacyl)-hydroxamic acids, each of which yielded products that reacted with nucleophilic amines or 2′-deoxyguanosine, or both. Reaction with 2′-deoxyguanosine afforded the same types of *N*-(deoxyguanosin-8-yl)aminoarene adducts that are formed when *N*-acyloxyarylamines are generated enzymatically in the presence of 2′-deoxyguanosine.

B. Chemical Reactivity

It has been widely accepted that *N*-acetoxyarylamines form arylamine adducts with biological nucleophiles through reaction of positively charged arylnitrenium/carbenium ions produced by heterolytic cleavage of the N—O bond (S_N1 mechanism) (Fig. 5A) (Hanna and Banks, 1985). Alternative mechanisms include bimolecular reaction of nucleophilic groups either with the nitrogen atom (S_N2) (Fig. 6A) or with an aromatic ring carbon atom (S_N2') (Fig. 6B), homolytic cleavage of the N—O bond to form an arylamine radical (Fig. 5B), or base-catalyzed deprotonation of the *N*-acetoxy compound followed by α-elimination of the acetoxy group to yield an arylnitrene (Fig. 5C).

With regard to the widespread acceptance of the nitrenium ion process, Novak *et al.* (1989) have pointed out that arylnitrenium ions rarely undergo

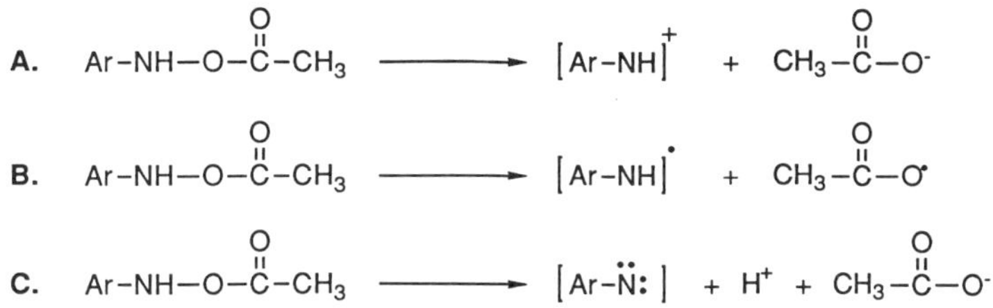

Fig. 5 Reactive intermediates that may be formed from *N*-acetoxyarylamines. (A) Arylnitrenium/carbenium ion (S_N1 mechanism). (B) Arylamine radical. (C) Arylnitrene.

nucleophilic attack at the nitrogen atom to produce N-substituted adducts, such as **IX,** which are the most common types of adduct formed in reactions between *N*-acyloxyarylamines and biological target molecules. Smith *et al.* (1985) studied the reaction of enzymatically generated *N*-acetoxy-2-aminofluorene with tRNA, polyguanylic acid, and *N*-acetylmethionine and suggested that adduct formation occurs through a bimolecular mechanism. Hashimoto *et al.* (1982) proposed that *N*-acetoxy-Glu-P-1, generated *in situ,* reacts with DNA in an S_N2-like process because the disappearance of the *N*-acetoxy compound was much faster in the presence of DNA than in its absence. Similarly, on the basis of product and kinetic studies, Ulbrich *et al.,* (1990) concluded that *N*-acetoxy-4-aminobiphenyl reacts with nucleophiles through an S_N2 mechanism. A detailed analysis of the mechanism of reaction of nucleophiles with *N*-acyloxyarylamines, including *N*-pivaloyloxy-4-aminobiphenyl, revealed an S_N2 mechanism, which occurs in competition with an S_N1 solvolysis, and led Helmick *et al.* (1991) to propose that carcinogenic *N*-acyloxyarylamines may form adducts *in vivo* through S_N2 mechanisms.

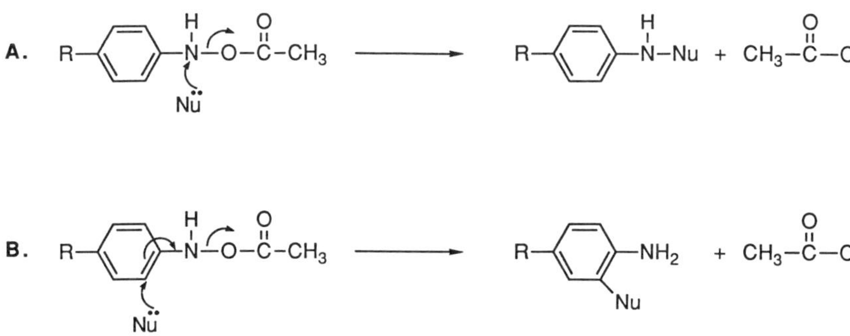

Fig. 6 Reaction of nucleophiles (Nü) with *N*-acyloxyarylamines. (A) S_N2 mechanism. (B) S_N2' mechanism.

The general acceptance of the nitrenium ion mechanism appears to be based primarily on the findings that *N*-acetoxy-*N*-arylacetamides undergo solvolytic (S_N1) reactions and on the frequent observation that acid-catalyzed reactions of *N*-hydroxyarylamines produce the same adducts with either nucleosides or DNA as do reactions of *N*-acyloxyarylamines (Scribner *et al.*, 1970; Scribner, 1977; Beland and Kadlubar, 1990). However, it was pointed out by Scribner *et al.* (1970) that the rate of loss of the *N*-acetoxy group from *N*-acetoxy-*N*-arylacetamides did not always correspond to the extent of reaction with nucleophiles, an indication that the latter reactions may not proceed exclusively through nitrenium ion mechanisms. With regard to the acid-catalyzed reactions of *N*-hydroxyarylamines, it is important to note that even if the protonated intermediate (**VI,** Fig. 2) solvolyzes to generate an arylnitrenium ion prior to reaction with biological nucleophiles, the resulting products (e.g., **IX**) can be explained by invoking either S_N2 mechanisms or the intermediacy of either free radicals or nitrenes when *N*-acyloxyarylamines are the reactants. For example, the proposed mechanism of reaction of *N*-acetoxy-2-aminofluorene with guanine nucleosides, involving formation of an N^7-adduct followed by rearrangement to the C^8-adduct, can be accommodated readily by an initial S_N1 or S_N2 reaction (Humphreys *et al.*, 1992). The involvement of S_N2 processes in adduct formation also may account in part for the enhancement of bacterial mutagenicity by electronegative substituents in 2-nitrofluorenes and for the results of molecular orbital calculations, which show that the bacterial mutagenicity of aromatic nitro compounds is not correlated with structural features that stabilize nitrenium ions (Vance *et al.*, 1987; Lopez de Compadre *et al.*, 1990).

The chemical mechanism of reaction of *N*-acyloxyarylamines with biological nucleophiles is determined principally by the electronic and steric properties of the *N*-acyloxy compounds and, to a certain extent, by the properties of the nucleophiles and the reaction medium. These concepts are illustrated by the findings that the electronic characteristics that favor reaction of *N,O*-acyltransferase-generated *N*-acetoxyarylamines with N,OAT itself are different from those that favor reaction with nucleosides and DNA and that the highly electronegative 7-acetyl-*N*-acetoxy-2-aminofluroene reacts with DNA and 2′-deoxyguanosine through a nitrene intermediate, but appears to react with *N*-acetylmethionine through a nucleophilic mechanism (Markevka *et al.*, 1984; Elfarra and Hanna, 1985; Boteju, 1990; Boteju and Hanna, 1993). The influence of structural properties on reactivity also is exemplified by the chemical behavior of 2-acetoxyamino-5-phenylpyridine, which is an unexpectedly stable crystalline solid (Lutgerink *et al.*, 1989), for which no reaction with 2′-deoxyguanosine could be detected in solution at 40°C (R. A. Iyer and P. E. Hanna, University of Minnesota, unpublished results).

The acquisition of a detailed understanding of the bioorganic reaction mechanisms involved in adduct formation between *N*-acyloxyarylamines and biological macromolecules should contribute to the development of agents that can intervene in adduct formation and thereby serve as chemopreventatives. It cannot be expected, however, that a single, unified reaction mechanism will account for the fate of all *N*-acyloxyarylamines.

Acknowledgments

The author is grateful to Mr. Ramaswamy A. Iyer for assistance with preparation of structures and figures and to Ms. Laura Wiebers for assistance with preparation of the text.

References

Abu-Zeid, M., Nagata, K., Masaaki, M., Ozawa, S., Fukuhara, M., Yamazoe, Y., and Kato, R. (1991). An arylamine acetyltransferase (AT-I) from Syrian Golden Hamster liver: Cloning, complete nucleotide sequence, and expression in mammalian cells. *Mol. Carcinog.* **4,** 81–88.

Abu-Zeid, M., Yamazoe, Y., and Kato, R. (1992). Sulfotransferase-mediated DNA binding of N-hydroxyarylamines (amide) in the liver cytosols from human and experimental animals. *Carcinogenesis (London)* **13,** 1307–1314.

Allaben, W. T., and King, C. M. (1984). The purification of rat liver arylhydroxamic acid N,O-acyltransferase. *J. Biol. Chem.* **259,** 12128–12134.

Allaben, W. T., Weeks, C. E., Weis, C. C., Burger, G. T., and King, C. M. (1982). Rat mammary gland carcinogenesis after local injection of N-hydroxy-N-acyl-2-aminofluorenes: Relationship to metabolic activation. *Carcinogenesis (London)* **3,** 233–240.

Andres, H. H., Klem, A. J., Schopfer, L. M., Harrison, J. K., and Weber, W. W. (1988). On the active site of liver acetyl-CoA arylamine N-acetyltransferase from rapid acetylator rabbits. *J. Biol. Chem.* **263,** 7521–7527.

Banks, R. B., and Hanna, P. E. (1979). Arylhydroxamic acid N,O-acyltransferase. Apparent suicide inactivation by carcinogenic N-arylhydroxamic acids. *Biochem. Biophys. Res. Commun.* **91,** 1423–1429.

Bartsch, H., Dworkin, M., Miller, J. A., and Miller, E. C. (1972). Electrophilic N-acetoxyaminoarenes derived from carcinogenic N-hydroxy-N-acetylaminoarenes by enzymatic deacetylation and transacetylation in liver. *Biochim. Biophys. Acta* **286,** 272–298.

Beland, F. A., and Kadlubar, F.F. (1990). Metabolic activation and DNA adducts of aromatic amines and nitroaromatic hydrocarbons. *In* "Handbook of Experimental Pharmacology" (C. S. Cooper and P. L. Grover, eds.), Vol. 94, Part I, pp. 267–325. Springer-Verlag, New York.

Beland, F. A., Allaben, W. T., and Evans, F. E. (1980). Acyltransferase-mediated binding of N-hydroxyarylamides to nucleic acids. *Cancer Res.* **40,** 834–840.

Blum, M., Grant, D. M., Demierre, A., and Meyer, U. A. (1989). N-Acetylation pharmacogenetics: A gene deletion causes absence of arylamine N-acetyltransferase in liver of slow acetylator rabbits. *Proc. Natl. Acad. Sci. U.S.A.* **86,** 9554–9557.

Boche, G., Bosold, F., and Schröder, S. (1988). N-Aryl-O-acylhydroxylamines: Preparation by O-acylation or N,O-transacylation and reaction with amines; model reactions for key steps connected with the carcinogenicity of aromatic amines. *Angew. Chem., Int. Ed. Engl.* **27,** 973–974.

Booth, J. (1966). Acetyltransfer in arylamine metabolism. *Biochem. J.* **100,** 745–753.

Bosold, F., and Boche, G. (1990). The ultimate carcinogen O-acetyl-N-(2-fluorenyl)hydroxylamine (N-acetoxy-2-aminofluorene), and its reaction *in vtiro* to form 2-[N-(deoxyguanosin-8-yl)amino]fluorene. *Angew. Chem., Int. Ed. Engl.* **29,** 63–64.

Boteju, L. W. (1990). Bioactivation of N-arylhydroxamic acids: Studies on the relationships between physicochemical properties and covalent binding to biological macromolecules. Ph.D. Thesis, University of Minnesota, Minneapolis.

Boteju, L. W., and Hanna, P. E. (1993). Bioactivation of N-hydroxy-2-acetylaminofluorenes by N,O-acyltransferase: Substituent effects on covalent binding to DNA. *Carcinogenesis (London)* **14,** 1651–1657.

Buonarati, M. H., and Felton, J. S. (1990). Activation of 2-amino-1-methyl-6-phenylimidazo[4,5-b]pyridine (PhIP) to mutagenic metabolites. *Carcinogenesis (London)* **11,** 1133–1138.

Cartwright, R. A., Glashen, R. W., Rogers, H. J., Ahmed, R. A., Barham-Hall, D., Higgins, E., and Kahn, M. (1982). Role of N-acetyltransferase phenotypes in bladder carcinogenesis: A pharmacogenetic epidemiological approach to bladder cancer. *Lancet* **2,** 842–846.

Cheon, H. G., Boteju, L. W., and Hanna, P. E. (1992). Affinity alkylation of hamster hepatic arylamine N-acetyltransferases: Isolation of a modified cysteine residue. *Mol. Pharmacol.* **42,** 82–93.

Deguchi, T. (1992). Physiology and molecular biology of arylamine N-acetyltransferases. *Biomed. Res.* **13,** 231–242.

Djurić, Z., Fifer, E. K., and Beland, F. A. (1985). Acetyl coenzyme A-dependent binding of carcinogenic and mutagenic dinitropyrenes to DNA. *Carcinogenesis (London)* **6,** 941–944.

Dupret, J. M., and Grant, D. M. (1992). Site-directed mutagenesis of recombinant human arylamine N-acetyltransferase expressed in *Escherichia coli*. *J. Biol. Chem.* **267,** 7381–7385.

Einistö, P., Watanabe, M., Ishidate, M., Jr., and Nohmi, T. (1991). Mutagenicity of 30 chemicals in *Salmonella typhimurium* strains possessing different nitroreductase or O-acetyltransferase activities. *Mutat. Res.* **259,** 95–102.

Elfarra, A. A., and Hanna, P. E. (1985). Substituent effects on the bioactivation of 2-(N-hydroxyacetamido)fluorenes by N-arylhydroxamic acid N,O-acyltransferase. *J. Med. Chem.* **28,** 1453–1460.

Evans, D. A. P (1989). N-acetyltransferase. *Pharmacol. Ther.* **42,** 157–234.

Ferguson, R. J., Doll, M. A. Rustan, T. D., Baumstark, B. R., and Hein, D. W. (1993). Expression of recombinant hamster arylamine N-acetyltransferase NAT1 in *E. coli*. *Toxicologist* **13,** 328.

Ferreira, L. M., Lobo, A. M., Prabhakar, S., Marcelo-Curto, M. J., Rzepa, H. S., and Yi, M. Y. (1991). 2-Acyl thiazolium salts as selective agents for the O-acylation of aromatic hydroxylamines. *J. Chem. Soc., Chem. Commun.* pp. 1127–1128.

Flammang, T. J., and Kadlubar, F. F. (1986). Acetyl coenzyme A-dependent metabolic activation of N-hydroxy-3,2′-dimethyl-4-aminobiphenyl and several carcinogenic N-hydroxyarylamines in relation to tissue and species differences, other acyl donors, and arylhydroxamic acid-dependent acyltransferase. *Carcinogenesis (London)* **7,** 919–926.

Flammang, T. J., Westra, J. G., Kadlubar, F. F., and Beland, F. A. (1985). DNA adducts formed from the probable proximate carcinogen, N-hydroxy-3,2′-dimethyl-4-

aminobiphenyl, by acid catalysis or S-acetyl coenzyme A-dependent enzymatic esterification. *Carcinogenesis (London)* **6,** 251–258.

Flammang, T. J., Yamazoe, Y., Guengerich, F. P., and Kadlubar, F. F. (1987). The S-acetyl coenzyme-A dependent metabolic activation of the carcinogen N-hydroxy-2-aminofluorene by human liver cytosol and its relationship to the aromatic amine N-acetyltransferase phenotype. *Carcinogenesis (London)* **8,** 1967–1970.

Flammang, T. J., Yerokun, T., Bryant, M. S., Couch, L. H., Kirlin, W. G., Lee, K. J., Ogolla, F., Ferguson, R. J., Talaska, G., and Hein, D. W. (1992a). Hemoglobin adduct and hepatic and urinary bladder-DNA adduct levels in rapid and slow acetylator Syrian inbred hamsters administered 2-aminofluorene. *J. Pharmacol. Exp. Ther.* **260,** 865–871.

Flammang, T. J., Couch, L. H., Levy, G. N., Weber, W. W., and Wise, C. K. (1992b). DNA adduct levels in congenic rapid and slow acetylator mouse strains following chronic administration of 4-aminobiphenyl. *Carcinogenesis (London)* **13,** 1887–1891.

Frandsen, H., Grivas, S., Andersson, R., Dragsted, L., and Larsen, J. C. (1992). Reaction of the N^2-acetoxy derivative of 2-amino-1-methyl-6-phenylimidazo[4,5-b]pyridine (PhIP) with 2′-deoxyguanosine and DNA. Synthesis and identification of N^2-(2′-deoxyguanosin-8-yl)-PhIP. *Carcinogenesis (London)* **13,** 629–635.

Frederick, C. B., Weis, C. C., Flammang, T. J., Martin, C. N., and Kadlubar, F.F. (1985). Hepatic N-oxidation, acetyl-transfer and DNA-binding of the acetylated metabolites of the carcinogen, benzidine. *Carcinogenesis (London)* **6,** 959–965.

Glowinski, I. B., Weber, W. W., Fysh, J. M., Vaught, J. B., and King, C. M. (1980). Evidence that arylhydroxamic N,O-acyltransferase and the genetically polymorphic N-acetyltransferase are properties of the same enzyme in rabbit liver. *J. Biol. Chem.* **255,** 7883–7890.

Glowinski, I. B., Savage, L., Lee, M.-S., and King, C. M. (1983). Relationship between nucleic acid adduct formation and deacylation of arylhydroxamic acids. *Carcinogenesis (London)* **4,** 67–25.

Grant, D. M., Blum, M., Beer, M., and Meyer, U. A. (1991). Monomorphic and polymorphic human arylamine N-acetyltransferases: A comparison of liver isozymes and expressed products of two cloned genes. *Mol. Pharmacol.* **39,** 184–191.

Grant, D. M., Josephy, P. D., Lord, H. L., and Morrison, L. D. (1992). *Salmonella typhimurium* strains expressing human arylamine N-acetyltransferases: Metabolism and mutagenic activation of aromatic amines. *Cancer Res.* **52,** 3961–3964.

Hanna, P. E., and Banks, R. B. (1985). Arylhydroxylamines and arylhydroxamic acids: Conjugation reactions. *In* "Bioactivation of Foreign Compounds" (M. W. Anders, ed.), pp. 375–401, Academic Press, Orlando, FL.

Hanna, P. E., Banks, R. B., and Marhevka, V. C. (1982). Suicide inactivation of hamster hepatic arylhydroxamic acid N,O-acyltransferase: A selective probe of N-acetyltransferase multiplicity. *Mol. Pharmacol.* **21,** 159–165.

Hashimoto, Y., Shudo, K., and Okamoto, T. (1980). Metabolic activation of a mutagen, 2-amino-6-methyldipyrido[1,2-a:3′,2′-d]imidazole, Identification of a 2-hydroxyamino-6-methyldipyrido[1,2-a:3′,2′-d]imidazole and its reaction with DNA. *Biochem. Biophys. Res. Commun.* **92,** 971–976.

Hashimoto, Y., Shudo, K., and Okamoto, T. (1982). Modification of DNA with potent mutacarcinogenic 2-amino-6-methyldipyrido[1,2-a:3′,2′-d]imidazole isolated from a glutamic acid pyrolysate: Structure of the modified nucleic acid base and initial chemical event caused by the mutagen. *J. Am. Chem. Soc.* **104,** 7636–7640.

Hatcher, J. F., and Swaminathan, S. (1992). Microsome-mediated transacetylation and binding of N-hydroxy-4-aminobiphenyl to nucleic acids by hepatic and bladder tissues from dog. *Carcinogenesis (London)* **13,** 1705–1711.

Hayes, R. B., Wenfang, B., Rothman, N., Broly, F., Caporaso, N., Feng, P., You, X., Yin, S., Woosley, R. L., and Meyer, U. A. (1993). N-Acetylation phenotype and genotype and risk of bladder cancer in benzidine-exposed workers. *Carcinogenesis* (*London*) **14,** 675–678.

Hein, D. W. (1988). Acetylator genotype and arylamine-induced carcinogenesis. *Biochim. Biophys. Acta* **948,** 37–66.

Hein, D. W., Kirlin, W. G., Ferguson, R. J., and Weber, W. W. (1985). Biochemical investigation of the basis for the genetic N-acetylation polymorphism in the inbred hamster. *J. Pharmacol. Exp. Ther.* **234,** 358–364.

Hein, D. W., Kirlin, W. G., Ferguson, R. J., Thompson, L. K., and Ogolla, F. (1986a). Identification and inheritance of inbred hamster N-acetyltransferase in peripheral blood. *J. Pharmacol. Exp. Ther.* **239,** 823–828.

Hein, D. W., Kirlin, W. G., Ogolla, F., Trinidad, A., Thompson, L. K., and Ferguson, R. J. (1986b). The role of acetylator genotype on hepatic and extrahepatic acetylation, deacetylation, and sulfation of 2-aminofluorene, 2-acetylaminofluorene, and N-hydroxy-2-acetylaminofluorene in the inbred hamster. *Drug Metab. Dispos.* **14,** 566–573.

Hein, D. W., Flammang, T. J., Kirlin, W. G., Trinidad, A., and Ogolla, F. (1987). Acetylator genotype-dependent metabolic activation of carcinogenic N-hydroxyarylamines by S-acetylcoenzyme A-dependent enzymes of inbred hamster tissue cytosols: Relationship to arylamine N-acetyltransferase. *Carcinogenesis* (*London*) **8,** 1767–1774.

Hein, D. W., Rustan, T. D., Bucher, K. D., Martin, W. J., and Furman, E. J. (1991a). Acetylator phenotype-dependent and -independent expression of arylamine N-acetyltransferase isozymes in rapid and slow acetylator inbred rat liver. *Drug Metab. Dispos.* **19,** 933–937.

Hein, D. W., Rustan, T. D., Bucher, K. D., Furman, E. J., and Martin, W. J. (1991b). Extrahepatic expression of the N-acetylation polymorphism toward arylamine carcinogens in tumor target organs of an inbred rat model. *J. Pharmacol. Exp. Ther.* **258,** 232–236.

Hein, D. W., Doll, M. A., Rustan, T. D., Gray, K., Feng, Y., Ferguson, R. J., and Grant, D. M. (1993a). Metabolic activation and deactivation of arylamine carcinogens by recombinant human NAT1 and polymorphic NAT2 acetyltansferases. *Carcinogenesis* **14,** 1633–1638.

Hein, D. W., Doll, M. A., Gray, K., Rustan, T. D., and Ferguson, R. J. (1993b). Metabolic activation of N-hydroxy-2-aminofluorene and N-hydroxy-2-acetylaminofluorene by monomorphic N-acetyltransferase (NAT1) and polymorphic N-acetyltransferase (NAT2) in colon cytosols of Syrian hamsters congenic at the NAT2 locus. *Cancer Res.* **53,** 509–514.

Hein, D. W., Rustan, T. D., Doll, M. A., Bucher, K. D., Ferguson, R. J., Feng, Y., Furman, E. J., and Gray, K. (1993c). Acetyltransferases and susceptibility to chemicals. *Toxicol. Lett.* **64/65,** 123–130.

Helmick, J. S., Martin, K. A., Heinrich, J. L., and Novak, M. (1991). Mechanism of the reaction of carbon and nitrogen nucleophiles with model carcinogens O-pivaloyl-N-arylhydroxylamines: Competing S_N2 substitution and S_N1 solvolysis. *J. Am. Chem. Soc.* **113,** 3459–3466.

Heymann, E. (1980). Carboxylesterases and amidases. *In* "Enzymatic Basis of Detoxification" (W. B. Jacoby, ed.), Vol. 2, pp. 291–322. Academic Press, New York.

Holme, J. A., Wallin, H., Brunborg, G., Søderlund, E. J., Hongslo, J. K., and Alexander, J. (1989). Genotoxicity of the food mutagen 2-amino-1-methyl-6-phenylimidazo[4,5-b]pyridine (PhIP): Formation of 2-hydroxyamino-PhIP, a directly acting gentoxic metabolite. *Carcinogenesis* (*London*) **10,** 1389–1396.

Huggett, A. C., Cone, J. L., Thorgeirsson, S. S., and Roller, P. P. (1987). Novel synthesis and spectral characterization of an N-acetoxyarylamine: N-Acetoxy-2,4-dinitrophenylamine. *J. Org. Chem.* **52,** 4933–4937.

Hultin, T. A., and Weber, W. W. (1987). Genetic variability in deacetylation of 2-acetylaminofluorene and N-hydroxy-2-acetylaminofluorene in inbred strains of mice. *Carcinogenesis (London)* **8,** 1939–1941.

Humphreys, W. G., Kadlubar, F. F., and Guengerich, F. P. (1992). Mechanism of C^8 alkylation of guanine residues by activated arylamines: Evidence of initial adduct formation at the N^7 position. *Proc. Natl. Acad. Sci. U.S.A* **89,** 8278–8282.

Ilett, K. F., Reeves, P. T., Minchin, R. F., Kinnear, B. F., Watson, H. F.,and Kadlubar, F. F. (1991). Distribution of acetyltransferase activities in the intestines of rapid and slow acetylator rabbits. *Carcinogenesis (London)* **12,** 1465–1469.

Irving, C. C. (1966). Enzymatic deacetylation of N-hydroxy-2-acetylaminofluorene by liver microsomes. *Cancer Res.* **26,** 1390–1396.

Kaderlik, R. K., Mulder, G. J., Casciano, D. A., Ilett, K. F., Minchin, R. F., Daugaard-Jensen, M., Teitel, C. H., and Kadlubar, F. F. (1993). Metabolic activation of the colon carcinogen PhIP in rats. *Proc. Am. Assoc. Cancer Res.* **34,** 158.

Kadlubar, F. F., Butler, M. A., Kaderlik, K. R., Chou, H.-C., and Lang, N. P. (1992). Polymorphisms for aromatic amine metabolism in humans: Relevance for human carcinogenesis. *Environ. Health Perspect.* **98,** 69–74.

Kawakubo, Y., Manabe, S., Yamazoe, Y., Nishikawa, T., and Kato, R. (1988). Properties of cutaneous acetyltransferase catalyzing N- and O-acetylation of carcinogenic arylamines and N-hydroxyarylamine. *Biochem. Pharmacol.* **37,** 265–270.

Kilbane, A. J., Petroff, T., and Weber, W. W. (1991). Kinetics of acetyl CoA: Arylamine N-acetyltransferase from rapid and slow acetylator human liver. *Drug Metab. Dispos.* **19,** 503–507.

King, C. M. (1974). Mechanism of reaction, tissue distribution and inhibition of arylhydroxamic acid acyltransferase. *Cancer Res.* **34,** 1503–1515.

King, C. M., and Olive, C. W. (1975). Comparative effects of strain, species, and sex on the acyltransferase- and sulfotransferase-catalyzed activations of N-hydroxy-N-2-fluorenylacetamide. *Cancer Res.* **35,** 906–912.

King, C. M., and Phillips, B. (1968). Enzyme-catalyzed reactions of the carcinogen N-hydroxy-2-fluorenylacetamide with nucleic acid. *Science.* **159,** 1351–1353.

King, C. M., and Phillips, B. (1969). N-Hydroxy-2-fluorenylacetamide. Reaction of the carcinogen with guanosine, ribonucleic acid, deoxyribonucleic acid, and protein following enzymatic deacetylation or esterification. *J. Biol. Chem.* **244,** 6209–6216.

King, C. M., Olive, C. W., and Cardona, R. A. (1975). Activation of carcinogenic arylhydroxamic acids by human tissues. *J. Natl. Cancer Inst. (U.S.)* **55,** 285–287.

Kirlin, W. G., Trinidad, A., Yerokun, T., Ogolla, F., Ferguson, R. J., Andrews, A. F., Brady, P. K., and Hein, D. W. (1989). Polymorphic expression of acetyl coenzyme A-dependent arylamine N-acetyltransferase and acetyl coenzyme A-dependent O-acetyltransferase-mediated activation of N-hydroxyarylamines by human bladder cytosol. *Cancer. Res.* **49,** 2448–2454.

Kirlin, W. G., Ogalla, F., Andrews, A. F., Trinidad, A., Ferguson, R. J., Yerokun, T., Mpezo, M., and Hein, D. W. (1991). Acetylator genotype-dependent expression of arylamine N-acetyltransferase in human colon cytosol from non-cancer and colorectal cancer patients. *Cancer Res.* **51,** 549–555.

Kumano, T., Yoshioka, T., and Uematsu, T. (1986). Comparative effect of chemical structure of chlorinated N-hydroxy-N-acyl-aminobiphenyl ethers and their related compounds on rat liver cytosol-catalyzed transacylation. *Drug. Metab. Dispos.* **14,** 487–493.

Lai, C.-C., Miller, E. C., Miller, J. A., and Liem, A. (1988). The essential role of microsomal deacetylase activity in the metabolic activation, DNA-(deoxyguanosin-8-yl)-2-aminofluorene adduct formation and initiation of liver tumors by N-hydroxy-2-acetylaminofluorene in the livers of infant male $B6C3F_1$ mice. *Carcinogenesis (London)* **9,** 1295–1302.

Land, S. J., Zukowski, K., Lee, M., Debiec-Rychter, M., King, C. M., and Wang, C. Y. (1989). Metabolism of aromatic amines: Relationships of N-acetylation, O-acetylation, N,O-acetyltransfer and deacetylation in human liver and urinary bladder. *Carcinogenesis (London)* **10,** 727–731.

Levy, G. N., and Weber, W. W. (1989). 2-Aminofluorene-DNA adduct formation in acetylator congenic mouse lines. *Carcinogenesis (London)* **10,** 705–709.

Levy, G. N., Martell, K. J., and Weber, W. W. (1993). Polymorphic N-acetylation of 2-aminofluorene by cell-free colon extracts from inbred mice. *Pharmacogenetics* **3,** 71–76.

Lobo, A. M., Marques, M. M., Prabhakar, S., and Rzepa, H. S. (1987). Tetrahedral intermediates formed by nitrogen and oxygen attack of aromatic hydroxylamines on acetyl cyanide. *J. Org. Chem.* **52,** 2925–2927.

Lopez de Compadre, R. L., Debnath, A. K., Shusterman, A. J., and Hansch, C. (1990). LUMO energies and hydrophobicity as determinants of mutagenicity by nitroaromatic compounds in Salmonella typhimurium. *Environ. Mol. Mutagen.* **15,** 44–45.

Lower, G. M., Jr., and Bryan, G. T. (1973). Enzymatic N-acetylation of carcinogenic aromatic amines by liver cytosol of species displaying different organ susceptibilities. *Biochem. Pharmacol.* **22,** 1581–1588.

Lutgerink, J. T., Stavenuiter, J. F. C., Zomer, G., Hamzink, M., van Dijk, P., Westra, J. G., and Kriek, E. (1989). Synthesis of heterocyclic N-acetoxyarylamines and their reactivity with DNA. *Carcinogenesis (London)* **10,** 1957–1960.

Marhevka, V. C., Ebner, N. A., Sehon, R. D., and Hanna, P. E. (1985). Mechanism-based inactivation of N-arylhydroxamic acid N,O-acyltransferase by 7-substituted-N-hydroxy-2-acetamidofluorenes. *J. Med. Chem.* **28,** 18–24.

Martell, K. J., Vatsis, K. P., and Weber, W. W. (1991). Molecular genetic basis of rapid and slow acetylation in mice. *Mol. Pharmacol.* **40,** 218–227.

Martell, K. J., Levy, G. N., and Weber, W. W. (1992). Cloned mouse N-acetyltransferases: Enzymatic properties of expressed NAT-1 and NAT-2 gene products. *Mol. Pharmacol.* **42,** 265–272.

Mattano, S. S., Land, S., King, C. M., and Weber, W. W. (1989). Purification and biochemical characterization of hepatic arylamine N-acetyltransferase from rapid and slow acetylator mice: Identity with arylhydroxamic N,O-acyltransferase and N-hydroxyarylamine O-acetyltransferase. *Mol. Pharmacol.* **35,** 599–609.

McCoy, E. C., McCoy, G. D., and Rosenkranz, H. S. (1982). Esterification of arylhydroxylamines: Evidence for a specific gene product in mutagenesis. *Biochem. Biophys. Res. Commun.* **108,** 1362–1367.

McCoy, E. C., Anders, M., and Rosenkranz, H. S. (1983). The basis of the insensitivity of *Salmonella typhimurium* strain TA98/1,8-DNP_6 to the mutagenic action of nitroarenes. *Mutat. Res.* **121,** 17–23.

McQueen, C. A., Malansky, C. J., and Williams, G. M. (1983). Role of the acetylation polymorphism in determining susceptibility of cultured rabbit hepatocytes to DNA damage by aromatic amines. *Cancer Res.* **43,** 3120–3123.

Meerman, J. N., Tijdens, R. B., and Mulder, G. J. (1986). The binding of reactive metabolites of the carcinogen N-hydroxy-2-acetylaminofluorene to DNA and protein in isolated rat liver nuclei: Effects of glutathione and methionine. *J. Biochem. Toxicol.* **1,** 1–10.

Meier, C., and Boche, G. (1990). N-(α-Aminoacyloxy)-N-arylamines: Activation of aromatic amines to ultimate carcinogens by amino acids. *Tetrahedron Lett.* **31,** 1685–1688.

Miller, E. C., and Miller, J. A. (1985). Some historical perspectives on the metabolism of xenobiotic chemicals to reactive electrophiles. *In* "Bioactivation of Foreign Compounds" (M. W. Anders, ed.), pp. 3–28. Academic Press, Orlando, FL.

Minchin, R. F., Reeves, P. T., Teitel, C. H., McManus, M. E., Mojarrabi, B., Ilett, K. F., and Kadlubar, F. F. (1992a). N- and O-acetylation of aromatic and heterocyclic amine carcinogens by human monomorphic and polymorphic acetyltransferases expressed in COS-1 cells. *Biochem. Biophys. Res. Commun.* **185,** 839–844.

Minchin, R. F., Ilett, K. F., Teitell, C. H., Reeves, P. T., and Kadlubar, F. F. (1992b). Direct O-acetylation of N-hydroxy arylamines by acetylsalicyclic to form carcinogen-DNA adducts. *Carcinogenesis (London)* **13,** 663–667.

Nagao, M., Fujita, Y., Wakabayashi, K., and Sugimura, T. (1983). Ultimate forms of mutagenic and carcinogenic heterocyclic amines produced by pyrolysis. *Biochem. Biophys. Res. Commun.* **114,** 626–631.

Novak, M., Martin, K. A., and Heinrich, J. L. (1989). S_N2 reactions of a carbon nucleophile with N-aryl-O-pivaloylhydroxylamines: A model for *in vivo* reactions of carcinogenic metabolites of aromatic amines. *J. Org. Chem.* **54,** 5430–5431.

Ohsako, S., and Deguchi, T. (1990). Cloning and expression of cDNAs for polymorphic and monomorphic arylamine N-acetyltransferases from human liver. *J. Biol. Chem.* **265,** 4630–4634.

Orr, J. C., Bryant, D. W., McCalla, D. R., and Quilliam, M. A. (1985). Dinitropyrene-resistant *Salmonella typhimurium* are deficient in an acetyl-CoA acetyltransferase. *Chem.-Biol. Interact.* **54,** 281–288.

Ozawa, S., Abu-Zeid, M., Kawakubo, Y., Toyama, S., Yamazoe, Y., and Kato, R. (1990). Monomorphic and polymorphic isozymes of arylamine N-acetyltransferases in hamster liver: Purification of the isozymes and genetic basis of N-acetylation polymorphism. *Carcinogenesis (London)* **11,** 2137–2144.

Pink, J. C., Messing, E. M., Reznikoff, C. A., Bryan, G. T., and Swaminathan, S. (1992). Correlation between N-acetyltransferase activities in uroepithelia and *in vivo* acetylator phenotype. *Drug Metab. Dispos.* **20,** 559–565.

Probst, M. R., Blum, M., Fasshauer, I., D'Orazio, D., Meyer, U. A., and Wild, D. (1992). The role of the human acetylation polymorphism in the metabolic activation of the food carcinogen 2-amino-3-methylimidazo[4,5-f]quinoline (IQ). *Carcinogenesis (London)* **13,** 1713–1717.

Reeves, P. T., Kinnear, B. F., Minchin, R. F., and Ilett, K. F. (1991). Immunological evidence for N-acetyltransferase isozymes in the rabbit. *Mol. Pharmacol.* **39,** 42–48.

Rich, K. J., Murray, B. P., Lewis, I., Rendell, N. B., Davies, D. S., Gooderham, N. J., and Boobis, A. R. (1992). N-Hydroxy-MeIQx is the major microsomal oxidation product of the dietary carcinogen MeIQx with human liver. *Carcinogenesis (London)* **13,** 2221–2226.

Saito, K., Yamazoe, Y., Kamataki, T., and Kato, R. (1983). Mechanism of activation of proximate mutagens in Ames' tester strains: The acetyl-CoA dependent enzyme in *Salmonella typhimurium* TA98 deficient in TA98/1,8DNP_6 catalyzes DNA-binding as the cause of mutagenicity. *Biochem. Biophys. Res. Commun.* **116,** 141–147.

Saito, K., Shinohara, A., Kamataki, T., and Kato, R. (1985). Metabolic activation of mutagenic N-hydroxyarylamines by O-acetyltransferase in *Salmonella typhimurium* TA98. *Arch. Biochem. Biophys.* **239,** 286–295.

Saito, K., Shinohara, A., Kamataki, T., and Kato, R. (1986). N-Hydroxyarylamine O-acetyltransferase in hamster liver: Identity with arylhydroxamic acid N,O-acetyltransferase and arylamine N-acetyltransferase. *J. Biochem. (Tokyo)* **99,** 1689–1697.

Sakai, S. R., Reinhold, C. E., Wirth, P. J., and Thorgeirsson, S. S. (1978). Mechanism of *in vitro* mutagenic activation and covalent binding of N-hydroxy-2-acetylaminofluorene in isolated liver cell nuclei from rat and mouse. *Cancer Res.* **38,** 2058–2067.

Scribner, J. D. (1977). Conversion of the carcinogen N-acetoxy-2-acetamidofluorene to 4-hydroxy-2-acetamidofluorene. *J. Am. Chem. Soc.* **99,** 7383–7384.

Scribner, J. D., Miller, J. A., and Miller, E. C. (1970). Nucleophilic substitution on carcinogenic N-acetoxy-N-arylacetamides. *Cancer Res.* **30,** 1570–1579.

Shinohara, A., Yamazoe, Y., Saito, K., Kamataki, T., and Kato, R. (1984). Species differences in the N-acetylation by liver cytosol of mutagenic heterocyclic aromatic amines in protein pyrolysates. *Carcinogenesis (London)* **5,** 683–686.

Shinohara, A., Saito, K., Yamazoe, Y., Kamataki, T., and Kato, R. (1986a). Inhibition of acetyl coenzyme A-dependent activation of N-hydroxyarylamines by phenolic compounds. Pentachlorophenol and 1-nitro-2-naphthol. *Chem.-Biol. Interact.* **60,** 275–285.

Shinohara, A., Saito, K., Yamazoe, Y., Kamataki, T., and Kato, R. (1986b). Acetyl coenzyme A dependent activation of N-hydroxy derivatives of carcinogenic arylamines: Mechanism of activation, species difference, tissue distribution, and acetyl donor specificity. *Cancer Res.* **46,** 4363–4367.

Shirai, T., Fysh, J. M., Lee, M.-S., Vaught, J. B., and King, C. M. (1981). Relationship of metabolic activation of N-hydroxy-N-acylarylamines to biological response in the liver and mammary gland of the female CD rat. *Cancer Res.* **41,** 4346–4354.

Sim, E., Hickman, D., Coroneos, E., and Kelly, S. L. (1992). Arylamine N-acetyltransferase. *Biochem. Soc. Trans.* **20,** 304–309.

Smith, B. A., Guttman, H. R., and Springfield, J. R. (1985). Interaction of nucleophiles with the enzymatically-activated carcinogen, N-hydroxy-2-acetylaminofluorene, and with the model ester, N-acetoxy-2-acetylaminofluorene. *Carcinogenesis (London)* **6,** 271–277.

Smith, T. J., and Hanna, P. E. (1986). N-Acetyltransferase multiplicity and the bioactivation of N-arylhydroxamic acids by hamster hepatic and intestinal enzymes. *Carcinogenesis (London)* **7,** 697–702.

Smith, T. J., and Hanna, P. E., (1988). Hepatic N-acetyltransferases: Selective inactivation *in vivo* by a carcinogenic N-arylhydroxamic acid. *Biochem. Pharmacol.* **37,** 427–434.

Snyderwine, E. G., Wirth, P. J., Roller, P. P., Adamson, R. H., Sato, S., and Thorgeirsson, S. S. (1988). Mutagenicity and *in vitro* covalent DNA binding of 2-hydroxyamino-3-methylimidazolo[4,5-f]quinoline. *Carcinogenesis (London)* **9,** 411–418.

Sone, T., Zukowski, K., Land, S. J., King, C. M., and Wang, C. Y. (1991). Acetylation of 2-aminofluorene derivatives by dog hepatic microsomes. *Carcinogenesis (London)* **12,** 1887–1891.

Steinberg, M. S., Cohen, S. N., and Weber, W. W. (1971). Isotope exchange studies on rabbit liver N-acetyltransferase. *Biochim. Biophys. Acta* **235,** 89–98.

Stout, D. L., Hemminki, K., and Becker, F. F. (1980). Covalent binding of 2-acetylaminofluorene and N-hydroxy-2-acetylaminofluorene to rat liver nuclear DNA and protein *in vivo* and *in vitro*. Cancer Res. **40,** 3579–3584.

Sugimura, T. (1992). Multistep carcinogenesis: A 1992 perspective. *Science* **258,** 603–607.

Swaminathan, S., and Reznikoff, C. A. (1992). Metabolism and nucleic acid binding of N-hydroxy-4-acetylaminobiphenyl and N-acetoxy-4-acetylaminobiphenyl by cultured human uroepithelial cells. *Cancer Res.* **52,** 3286–3294.

Trinidad, A., Hein, D. W., Rustan, T. D., Ferguson, R. J., Miller, L. S., Bucher, K. D., Kirlin, W. G., Ogolla, F., and Andrews, A. F. (1990). Purification of hepatic polymorphic arylamine N-acetyltransferase from homozygous rapid acetylator inbred hamster: Identity with polymorphic N-hydroxyarylamine O-acetyltransferase. *Cancer Res.* **50,** 7942–7949.

Turesky, R. J., Lang, N. P., Butler, M. A., Teitel, C. H., and Kadlubar, F. F. (1991).

Metabolic activation of carcinogenic heterocyclic aromatic amines by human liver and colon. *Carcinogenesis (London)* **12,** 1839–1845.

Ulbrich, R., Famulok, M., Basold, F., and Boche, G. (1990). S_N2 at nitrogen: The reaction of N-(4-cyanophenyl)-O-diphenylphosphinoylhydroxylamine with N-methylaniline. A model for the reactions of ultimate carcinogens of aromatic amines with (bio)nucleophiles. *Tetrahedron Lett.* **31,** 1689–1692.

Vance, W. A., Wang, Y. Y., and Okamato, H. S. (1987). Disubstituted amino-, nitroso-, and nitrofluorenes: A physicochemical basis for structure-activity relationships in *Salmonella typhimurium. Environ. Mutagen.* **9,** 123–141.

Vatsis, K. P., and Weber, W. W. (1993). Structural heterogeneity of caucasian N-acetyltransferase at the NAT1 gene locus. *Arch. Biochem. Biophys.* **301,** 71–76.

Vaught, J. B., McGarvey, P. B., Lee, M.-S., Garner, C. D., Wang, C. Y., Linsmaier-Bednar, E. M., and King, C. M. (1981). Activation of N-hydroxyphenacetin to mutagenic and nucleic acid-binding metabolites by acyltransfer, deacylation and sulfate conjugation. *Cancer Res.* **41,** 3424–3429.

Vaught, J. B., Glowinski, I. B., and King, C. M. (1985). Metabolism of N-hydroxy-2-acetylaminofluorene and N-hydroxy-phenacetin by guinea pig microsomal enzymes. *Carcinogenesis (London)* **6,** 495–499.

Wakabayashi, K., Nagao, M., Esumi, H., and Sugimura, T. (1992). Food-derived mutagens and carcinogens. *Cancer Res., Suppl.* **52,** 2092s–2098s.

Wang, C. Y., Zukowski, K., Lee, M.-S., and Sone, T. (1992). Purification and characterization of rat hepatic microsomal N,O-acyltransferases. *Carcinogenesis (London)* **13,** 2017–2020.

Watanabe, M., Sofuni, T., and Nohmi, T. (1992). Involvement of Cys^{69} residue in the catalytic mechanism of N-hydroxyarylamine O-acetyltransferase of *Salmonella typhimurium. J. Biol. Chem.* **267,** 8429–8436.

Weber, W. W. (1971). Acetylating, deacetylating and amino acid conjugating enzymes. *In* "Handbook of Experimental Pharmacology" (B. B. Brodie and J. R. Gillette, eds.), Vol. 28, Part 2, pp. 564–582. Springer-Verlag, New York.

Weber, W. W. (1987). "The Acetylator Genes and Drug Response." Oxford Univ. Press, Oxford and New York.

Weber, W. W., and Hein, D. W. (1985). N-Acetylation pharmacogenetics. *Pharmacol. Rev.* **37,** 25–79.

Weber, W. W., Levy, G. N., and Hein, D. W. (1990). Acetylation. *In* "Conjugation Reactions in Drug Metabolism: An Integrated Approach" (G. J. Mulder, ed.), pp. 163–191. Taylor & Francis, New York.

Weeks, C. E., Allaben, W. T., Tresp, N. M., Louie, S. C., Lazear, E. J., and King, C. M. (1980). Effects of structure of N-acyl-N-2-fluorenylhydroxylamines on arylhydroxamic acid acyltransferase, sulfotransferase and deacylase activities, and on mutations in *Salmonella typhimurium* TA 1538. *Cancer Res.* **40,** 1204–1211.

Wick, M. J., and Hanna, P. E. (1990). Bioactivation of N-arylhydroxamic acids by rat hepatic N-acetyltransferase. Detection of multiple enzyme forms by mechanism-based inactivation. *Biochem. Pharmacol.* **39,** 991–1003.

Wick, M. J., Yeh, H. M., and Hanna, P. E. (1990). An isozyme-selective affinity label for rat hepatic acetyltransferases. *Biochem. Pharmacol.* **40,** 1389–1398.

Wirth, P. J., Alewood, P., Calder, I., and Thorgeirsson, S. S. (1982). Mutagenicity of N-hydroxy-2-acetylaminofluorene and N-hydroxyphenacetin and their respective deacetylated metabolites in nitroreductase deficient *Salmonella* TA98FR and TA100FR. *Carcinogenesis (London)* **3,** 167–170.

Wohlleb, J. C., Hunter, C. F., Blass, B., Kadlubar, F. F., Chu, D. Z. J., and Lang, N. P. (1990). Aromatic amine acetyltransferase as a marker for colorectal cancer: Environmental and demographic associations. *Int. J. Cancer* **46,** 22–30.

Yamazoe, Y., Shimada, M., Kamataki, T., and Kato, R. (1982). Covalent binding of N-hydroxy-Trp-P-2 to DNA by cytosolic proline-dependent system. *Biochem. Biophys. Res. Commun.* **107,** 165–172.

Yamazoe, Y., Abu-Zeid, M., Gong, D., Staiano, N., and Kato, R. (1989). Enzymatic acetylation and sulfation of N-hydroxyarylamines in bacteria and rat livers. *Carcinogenesis (London)* **10,** 1675–1679.

Yeh, H. M., and Hanna, P. E. (1982). Arylhydroxamic acid bioactivation via acyl group transfer. Structural requirements for transacylating and electrophile-generating activity of N-(2-fluorenyl)hydroxamic acids and related compounds. *J. Med. Chem.* **25,** 842–846.

Yerokun, T., Kirlin, W. G., Trinidad, A., Ferguson, R. J., Ogolla, F., Andrews, A. F., Brady, P. K., and Hein, D. W. (1989). Identification and kinetic characterization of acetylator genotype-dependent and -independent arylamine carcinogen N-acetyltransferases in hamster bladder cytosol. *Drug Metab. Dispos.* **17,** 213–237.

Aminoacylases

M. W. Anders* and Wolfgang Dekant†

**Department of Pharmacology*
University of Rochester
Rochester, New York 14642

†Institut für Toxikologie
Universität Würzburg
97078 Würzburg
Germany

I. Introduction

Aminoacylases catalyze the hydrolysis of *N*-acyl-L-amino acids to give fatty acids and amino acids as products. Although the catalytic activities ascribed to aminoacylases have been known for decades, the physiological role of these enzymes is still poorly understood. It has been proposed that aminoacylases may participate in the catabolism of terminal *N*-acylpeptides or in the salvage of N-acetylated amino acids (Endo, 1980; Gade and Brown, 1981). The aminoacylases do play a role in xenobiotic detoxication and bioactivation and are important in the interorgan processing of xenobiotic-derived amino acid conjugates. Mercapturates (S-substituted *N*-acetyl-L-cysteine conjugates), which are formed by the acetylation of L-cysteine *S*-conjugates that are obtained by the hydrolysis of glutathione *S*-conjugates (see Dekant *et al.*, this volume), are cleaved by aminoacylases.

Aminoacylases participate in conjugation-dependent toxicity: for example, mercapturates may serve to deliver nephrotoxic cysteine *S*-conjugates to the kidney. In this mechanism, mercapturates delivered via the circulation to the kidney may be taken up by the probenecid-sensitive anion

Advances in Pharmacology, Volume 27

transport system located on the basolateral surface of renal proximal-tubular cells. In renal proximal-tubular cells, mercapturates may undergo hydrolysis to afford cysteine *S*-conjugates that may be bioactivated by cysteine conjugate β-lyase. Several examples of this mechanism have been described and will be discussed below (Boogaard *et al.*, 1989; Commandeur *et al.*, 1987, 1988, 1991a,b; Dekant *et al.*, 1988a; Lock and Ishmael, 1985; Pratt and Lock, 1988; Reichert and Schütz, 1986; Vamvakas *et al.*, 1987; Wolfgang *et al.*, 1989; Zhang and Stevens, 1989).

Deacetylation of mercapturates is not necessarily required for the expression of toxicity: the intact mercapturates *S*-(3-oxopropyl)-*N*-acetyl-L-cysteine and *S*-(3-chloro-2-propenyl)-*N*-acetyl-L-cysteine apparently undergo flavoprotein monooxygenase-catalyzed bioactivation reactions that afford cytotoxic products (Hashmi *et al.*, 1992; Park *et al.*, 1992).

Although aminoacylase activity is found in many tissues, this chapter will focus largely on renal aminoacylases, which are the only aminoacylases that have been linked to xenobiotic bioactivation reactions. Aryl acylamidases (EC3.5.1.13), which catalyze the conversion of anilides to anilines and fatty acid residues, will not be discussed. A brief review of aminoacylases has appeared (Anders and Dekant, 1993).

II. Enzymology of the Aminoacylases

Aminoacylases catalyze the conversion of *N*-acyl α-amino acids to amino acids and fatty acid residues (Fig. 1). Several aminoacylases have been identified: acylase I (EC 3.5.1.14, *N*-acylamino acid amidohydrolase); aspartoacylase or acylase II (EC 3.5.1.15, *N*-acyl-L-aspartate amidohydrolase); acyllysine deacylase (EC 3.1.5.17, N^{ε}-acyl-L-lysine amidohydrolase); and acylase III, which catalyzes preferentially the N-deacetylation of *N*-acyl aromatic amino acids.

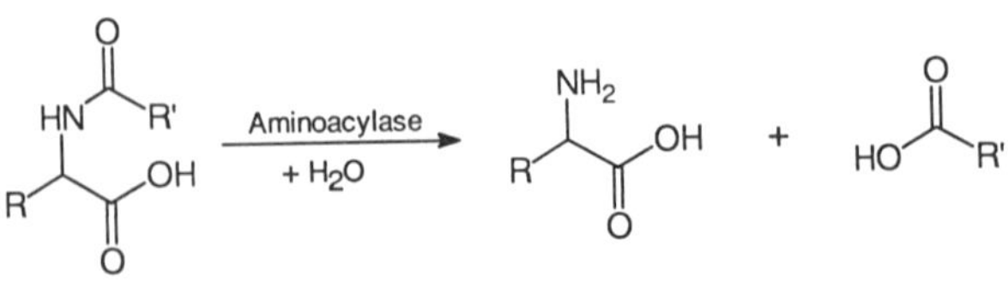

Fig. 1 Aminoacylase-catalyzed deacylation of *N*-acyl α-amino acids. For typical substrates, R = alkyl, aromatic, or aralkyl groups and R′ = methyl or chloromethyl groups.

A. Acylase I

Acylase I catalyzes the hydrolysis of a range of *N*-acyl α-amino acids (Fig. 2). Early studies provided several generalizations about the susceptibility of *N*-acyl amino acids to hydrolysis by acylase I: (*a*) Rates of hydrolysis vary with the nature of the amino acid; (*b*) when the amino acid moiety is held constant, the highest rates of hydrolysis are observed with *N*-chloroacetyl amino acids; (*c*) with straight-chain amino acids, the rates of hydrolysis increase from glycine to valine and then decrease as the chain-length is increased beyond valine; (*d*) branched-chain amino acids are hydrolyzed slower than straight-chain amino acids; (*e*) *N*-acyl-D-amino acids are resistant to hydrolysis; and (*f*) the rates of hydrolysis of allostereoisomers of threonine and isoleucine are different (Birnbaum *et al.*, 1952). The enzyme shows a preference for neutral, aliphatic *N*-acyl α-amino acids, such as *N*-acyl alanine, 2-aminobutanoic acid, valine, leucine, norleucine, 2-aminoheptanoic acid, and methionine (Birnbaum *et al.*, 1952; Fones and Les, 1953). *N*-Acetyl-L-histidine is hydrolyzed *in vitro* by mouse kidney tissue extracts, and, in mice given *N*-acetyl-L-histidine, both the acyl amino acid and histidine accumulate in the kidney (Endo, 1980). In a series of *N*-chloroacetyl amino acids, *N*-chloroacetyl-DL-valine showed the highest rate of cleavage, and the rates decreased as the length of the side chain was increased (Fu and Birnbaum, 1953). The electronegativity of the acyl group has a prominent effect on aminoacylase-catalyzed deacetylation reactions: the rates of hydrolysis of *N*-haloacetyl-DL-alanines followed the order trifluoroacetyl > fluoroacetyl > chloroacetyl > acetyl, and *N*-trichloroacetyl-DL-alanine was not hydrolyzed (Fones and Lee, 1953). In addition, some dipeptides, such as glycyl-L-alanine, glycyl-L-norvaline, L-alanyl-L-

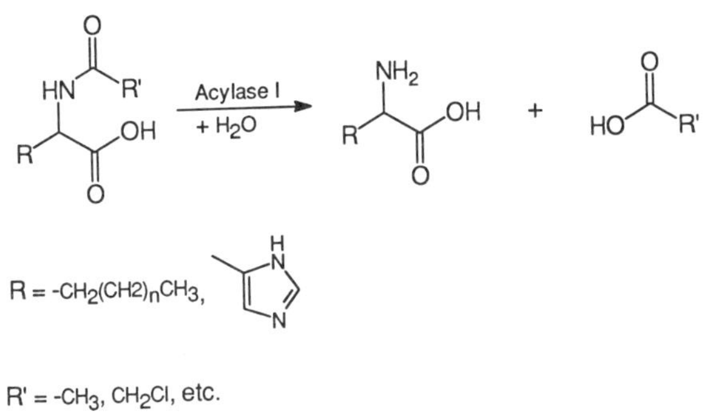

Fig. 2 Acylase I-catalyzed hydrolysis of *N*-acyl α-amino acids. The structures of preferred substrates are shown.

norvaline, and L-norvalyl-L-norvaline, are hydrolyzed (Moravcsik *et al.*, 1977). *N*-[3-(2-Furyl)acryloyl]-, *N*-(thienyl-acryloyl)-, and *N*-(cinnamoyl)amino acids are useful substrates for studying acylase I because hydrolysis releases products that absorb intensely at 310 to 340 nm, thus affording a facile means to measure acylase I activity (Löffler *et al.*, 1988). A report of the action of acylase I on over 50 unnatural or rarely occurring *N*-acyl amino acids has recently appeared (Chenault *et al.*, 1989).

Bray and James (1960) observed the deacetylation of mercapturates *in vivo* in rabbits, rats, and guinea pigs and by rabbit, rat, and guinea pig kidney and liver tissue extracts. *N*-Acetyl-*S*-benzyl-L-cysteine, *N*-acetyl-*S*-(2-chloro-4-nitrophenyl)-L-cysteine, and *N*-acetyl-*S*-butyl-L-cysteine were all hydrolyzed by kidney tissue extracts, and hydrolytic rates were higher in the guinea pig than in the rat or rabbit. Rates of hydrolysis of the mercapturates were generally lower in liver tissue extracts than in kidney tissue extracts. The available data do not, however, indicate unequivocally that the deacetylation observed in kidney tissue extracts can be attributed to acylase I.

Mercapturates of nephrotoxic and cytotoxic cysteine *S*-conjugates are deacetylated. Rat cytosol catalyzes the deacetylation of *N*-acetyl-*S*-(pentachlorobutadienyl)-L-cysteine, *N*-acetyl-*S*-(trichlorovinyl)-L-cysteine, and *N*-acetyl-*S*-(1,2-dichlorovinyl)-L-cysteine; increases the covalent binding of radioactivity from [*butadienyl*-^{14}C]*N*-acetyl-*S*-(pentachlorobutadienyl)-L-cysteine to rat kidney proteins; and increases the mutagenicity of *N*-acetyl-*S*-(pentachlorobutadienyl)-L-cysteine, *N*-acetyl-*S*-(trichlorovinyl)-L-cysteine, and *N*-acetyl-*S*-(1,2-dichlorovinyl)-L-cysteine (Pratt and Lock, 1988; Vamvakas *et al.*, 1987). In rat renal proximal–tubular cells, *N*-acetyl-*S*-(1,1,2,2-tetrafluoroethyl)-L-cysteine and *N*-acetyl-*S*-(2-chloro-1,1,2-trifluoroethyl)-L-cysteine are deacetylated more rapidly than *N*-acetyl-*S*-(2,2-dichloro-1,1-difluoroethyl)-L-cysteine and *N*-acetyl-*S*-(2,2-dibromo-1,1-difluoroethyl)-L-cysteine (Boogaard *et al.*, 1989). *In vivo* [*acetyl*-$^{2}H_3$]*N*-acetyl-L-cysteine *S*-conjugates are converted to unlabeled mercapturates, indicating significant deacetylation and reacetylation(Commandeur *et al.*, 1991b). Although these studies demonstrate the *in vitro* and *in vivo* deacetylation of mercapturates, data showing the involvement of acylase I in these reactions are lacking. The aminoacylase-dependent bioactivation of xenobiotics is discussed in detail below.

Acylase I has been well characterized. The enzyme was purified 300-fold from pig kidney (Bruns and Schulze, 1962); with *N*-acetyl-DL-methionine as the substrate, its pH optimum was 7 and its K_m was 2 m*M*. The enzyme was inhibited by the organomercurial mersalyl and, weakly, by iodine and iodoacetate. Paper electrophoresis showed a single band with aminoacylase activity. The functional enzyme is a homodimer with an apparent molecular weight of 86,000 (Kördel and Schneider, 1976).

Renal acylase I is a zinc-containing enzyme (Kördel and Schneider, 1977). The enzyme is inactivated by dialysis in the presence of 1,10-phenanthroline, and activity is restored by addition of zinc ions. The enzyme contains one zinc ion per subunit (Kördel and Schneider, 1977), which can be exchanged for Co^{2+}, Mn^{2+}, Ni^{2+}, or Cd^{2+} (Löffler *et al.*, 1986). Acylase I contains seven cysteine residues per subunit (Heese and Röhm, 1989); three sulfhydryl groups are readily modified by 4-hydroxymercuribenzoate, but only one of these sulfhydryl groups is apparently directly involved in acylase activity. Subcellular fractionation studies show that acylase I is a soluble enzyme (Greenhough and Turner, 1991), rather than a glycosyl-phosphatidylinositol-anchored ectoenzyme, as previously reported (Heese *et al.*, 1988).

cDNAs encoding the complete amino acid sequence of acylase I have recently been reported (Jakob *et al.*, 1992; Mitta *et al.*, 1992). These data show that porcine kidney acylase I consists of two identical subunits with a M_r of 45,260 and a single chain of 406 amino acids. A porcine liver acylase I was also cloned and was identical with the porcine kidney acylase I, and Northern blot analysis indicated that acylase I is more highly expressed in the kidney than in the liver (Mitta *et al.*, 1992). An enzyme from bovine liver whose molecular weight and substrate selectivities are similar to renal acylase I has been purified and characterized (Gade and Brown, 1981).

Human aminoacylase I has been assigned to distal chromosome 3p21.1 (Miller *et al.*, 1990). This region is deleted in several neoplasms, including small-cell lung cancer and renal cell carcinoma, and it has been suggested that the acylase I and small-cell lung cancer tumor suppressor genes are closely linked (Miller *et al.*, 1989; Zhar *et al.*, 1987).

Although acylase I activity is high in the kidney of many species, activity is present in other organs, including liver, adrenal medulla, pancreas, spleen, thymus, brain, lung, seminal vesicles, and heart and skeletal muscle (Albert and Szewczuk, 1972; Miller and Kao, 1989). Immunocytochemical studies show that acylase I is localized in the distal tubule of pig kidney (Löffler *et al.*, 1982).

Several inhibitors of acylase I are available. *N*-Hydroxy-α-amino acids, such as *N*-hydroxy-2-aminobutanoate and *N*-hydroxy-2-aminoheptanoate, are competitive inhibitors of acylase I with K_i values in the range of 1 to 10 μM (Löffler and Schneider, 1987). Several Boc-substituted amino acids also inhibit acylase I; the *K* values for *N*-(*tert*-butyloxycarbonyl)-L-valine, *N*-(*tert*-butyloxycarbonyl)-L-leucine, *N*-(*tert*-butyloxycarbonyl)-L-isoleucine, and *N*-(*tert*-butyloxycarbonyl)-L-methionine range from 1.2 to 25 m*M* (Heese *et al.*, 1990). *N*-(*p*-Toluenesulfonyl)-L-phenylalanine is a competitive aminoacylase inhibitor with a K_i of less than 1 m*M* (Heese *et al.*, 1990). Butylmalonate is a slow-binding inhibitor of pig kidney acylase I with a K_i value of about 100 μM (Röhm, 1989); the analog benzylmalonate is a competitive inhibitor with a K_i of 6.2 mM. These inhibitors have been used in *in vitro* experiments with purified pig kidney acylase I, but it is not known whether inhibitory concentrations of these compounds could be achieved *in vivo*. The development of acylase I inhibitors that are effective both in cellular preparations and *in vivo* would allow the exploration of the role of the enzyme in detoxication and bioactivation.

B. Aspartoacylase

Early studies on the selectivity of aminoacylases showed that porcine kidney fractions cleaved many *N*-acyl amino acids, but that *N*-acetyl- or *N*-chloroacetyl-DL-aspartate was not cleaved (Birnbaum, 1955; Birnbaum *et al.*, 1952). Additional studies indicated that *N*-acetylated aspartates were hydrolyzed by an enzyme other than acylase I. A protein fraction was isolated that readily hydrolyzed *N*-acetyl- and *N*-chloroacetyl-DL-aspartate (Fig. 3), but *N*-chloroacetyl-DL-glutamate, *N*-chloroacetyl-DL-alanine, *N*-chloroacetyl-DL-leucine, *N*-chloroacetyl-DL-serine, and *N*-acetyl-DL-methionine were hydrolyzed at very low rates; this enzyme was termed acylase II.

No xenobiotic-derived amino acid conjugates have apparently been shown to serve as substrates for aspartoacylase. Aspartoacylase has been purified from bovine brain (Kaul *et al.*, 1991). The enzyme is a 58-kDa monomeric protein whose activity is increased by divalent cations, nonionic detergents, and dithiothreitol.

The function of aspartoacylase is not known, but it may play a role in brain biology. *N*-Acetyl-L-aspartate is present in human brain (Tallan *et al.*, 1956). Interestingly, a deficiency of aspartoacylase activity and *N*-acetyl-L-aspartic aciduria are hallmarks of Canavan disease, an autosomal recessive disorder characterized by spongy degeneration of the brain (Matalon *et al.*, 1988).

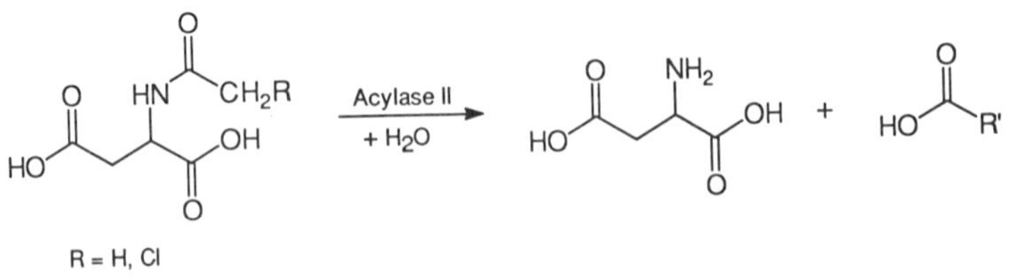

Fig. 3 Aspartoacylase-catalyzed hydrolysis of *N*-acyl-L-aspartates.

C. Acyllysine Deacylase

The observation that N^{ε}-acetyl-L-lysine, but not N^{α}-acetyl-L-lysine, was almost as effective as L-lysine in promoting weight gain in rats lead to the proposal that N^{ε}-acetyl-L-lysine may be enzymatically cleaved to afford L-lysine. Subsequent studies showed that an enzyme present in rat kidney and liver hydrolyzed N^{ε}-acetyl-L-lysine to give L-lysine as a product (Fig. 4) (Paik *et al.*, 1957). Apparently the enzyme has not been investigated further.

Protein-bound N^{ε}-trifluoroacetyl-L-lysine has been identified in rat liver as a covalently bound metabolite of halothane and 2,2-dichloro-1,1,1-trifluoroethane (HCFC-123) (Harris *et al.*, 1991). Recent studies show that N^{ε}-trifluoroacetyl-L-lysine is hydrolyzed to give L-lysine as a product (M. W. Anders, S. Kennedy, and H. Heyn, unpublished observations). Hence acyllysine deacylase may catalyze the hydrolysis of xenobiotic-modified lysine residues.

D. Acylase III

A comparison of the rates of hydrolysis of a range of *N*-acyl amino acids by purified acylase I and by kidney homogenates indicated the presence of an aminoacylase that catalyzed the hydrolysis of *N*-acyl-L-aromatic amino acids (Greenstein and Winitz, 1961); this enzyme was termed acylase III.

Endo (1978) partially purified an aminoacylase from rat kidney that preferentially cleaved *N*-acyl aromatic acids, including *N*-acetyl-L-tryptophan, *N*-acetyl-L-tyrosine, and *N*-acetyl-L-phenylalanine (Fig. 5). The purified enzyme has a M_r of about 55,000, is weakly inhibited by *p*-chloromercuribenzoate, and has a pH optimum of about 8. With *N*-acetyl-L-tryptophan as the substrate, activity in liver was about 10% of that in the kidney. In addition, when *N*-acetyl-L-tryptophan was given to mice, L-tryptophan accumulated in the kidney, indicating significant *in vivo* deacetylation (Endo, 1980). Developmental studies in mice indicated that the activities of both acylases I and III are low at birth and reach a plateau at about 5 weeks of age.

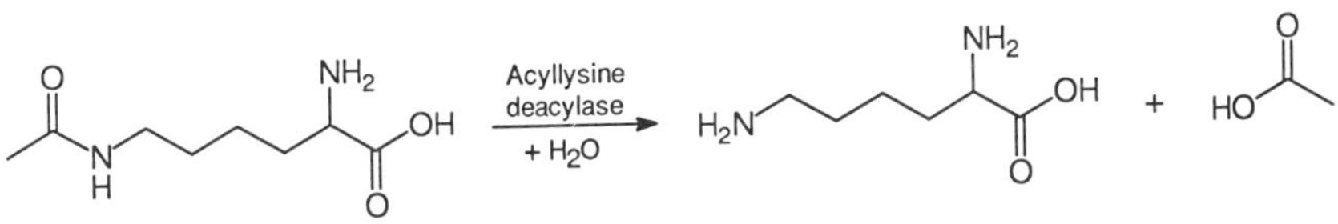

Fig. 4 Acyllysine deacylase-catalyzed hydrolysis of N^{ε}-acetyl-L-lysine.

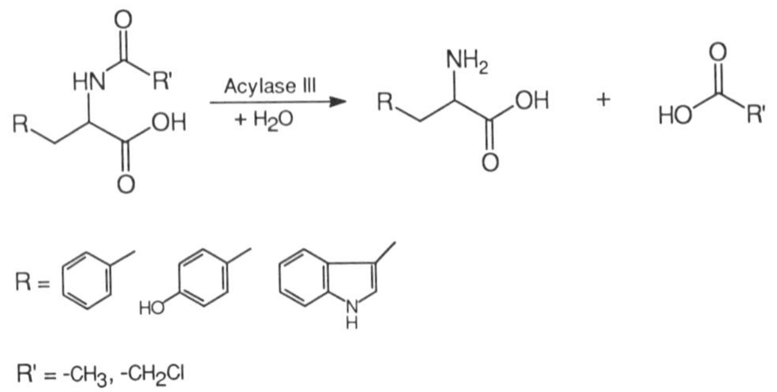

Fig. 5 Acylase III-catalyzed hydrolysis of *N*-acyl aromatic amino acids. The structures of preferred substrates are shown.

E. Liver Aminoacylases

The liver enzymes catalyzing the deacetylation of mercapturates have been characterized and a new aminoacylase has been identified (Suzuki and Tateishi, 1981). The rat liver aminoacylase has a M_r of about 35,000 by sodium dodecyl sulfate–gel electrophoresis and a M_r of about 145,000 by size-exclusion chromatography, indicating that the enzyme is likely a homotetramer. The substrate selectivities of the rat liver aminoacylase and rat kidney acylase III were similar, and *N*-acetyl-*S*-aryl- and *N*-acetyl-*S*-aralkyl-L-amino acids were the preferred substrates (Fig. 6). *N*-acetyl-L-aspartate and N^{ε}-acetyl-L-lysine were not substrates. Mercury (II), *N*-ethylmaleimide, and *p*-chloromercuribenzoate inhibited the liver aminoacylase.

These studies show that an array of aminoacylases with overlapping substrate selectivities catalyze the deacetylation of *N*-acyl α-amino acids. Activities are usually higher in kidney than in other tissues. Although

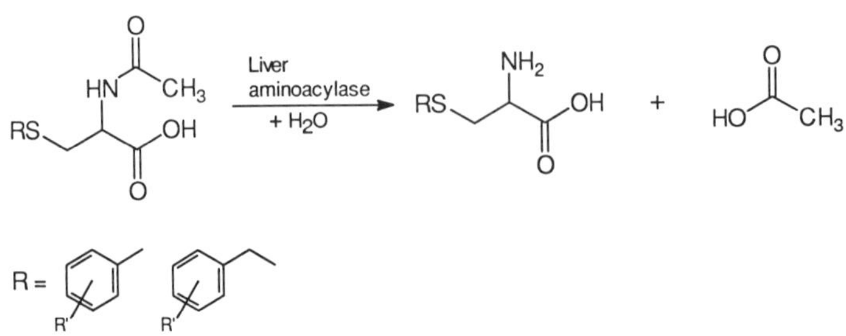

Fig. 6 Liver aminoacylase-catalyzed deacetylation of mercapturates. The structures of preferred substrates are shown.

acylase I has been thoroughly characterized and studied, the relationship of kidney acylase I to kidney and liver acylase III has not been clarified. The enzymes have similar, but not identical, substrate selectivities. The enzymes responsible for the deacetylation of xenobiotic-derived mercapturates have not been fully identified. Kidney acylase III catalyzes the deacetylation of a range of mercapturates (Suzuki and Tateishi, 1981) and may account for the deacetylation of mercapturates observed earlier (Bray and James, 1960). The role of acylase I in the deacetylation of mercapturates has apparently not been investigated. Studies designed to clarify the enzymology of xenobiotic-derived mercapturates are needed to understand fully the enzymatic processing of mercapturates.

III. Xenobiotic Bioactivation by Aminoacylases

Aminoacylases play an important role in the bioactivation of polyhalogenated alkenes to nephrotoxic intermediates. These alkenes are biotransformed to glutathione conjugates in the liver and excreted in bile (see Dekant *et al.*, this volume). The glutathione *S*-conjugates are thought to be hydrolyzed in the bile duct and in the gut to cysteine *S*-conjugates and, after reabsorbtion, *N*-acetylated in the liver by *N*-acetyl transferases. The mercapturates thus formed are then transported to the kidney and accumulated there by the renal organic anion transporter. The enzyme catalyzing the formation of ultimate toxic metabolites in this pathway, cysteine conjugate β-lyase, is a pyridoxal phosphate-dependent enzyme, which requires substrates with a free amino group. Hence mercapturates formed from halogenated alkenes require deacetylation before β-lyase-catalyzed formation of reactive intermediates can occur in the kidney. Several studies indicate an important role for aminoacylases in the activation of polyhalogenated alkenes and their *S*-conjugates.

A. Toxicity of Haloalkene-Derived Mercapturates

When *N*-acetyl-*S*-(pentachlorobutadienyl)-L-cysteine is given to rats, renal proximal-tubular damage is observed (Lock and Ishmael, 1985; Nash *et al.*, 1984). The lesions were identical to those seen after giving the nephrotoxin hexachlorobutadiene, which is metabolized to *N*-acetyl-*S*-(pentachlorobutadienyl)-L-cysteine (Reichert and Schütz, 1986) and occurred at lower doses. Moreover, *N*-acetyl-*S*-(1-chloro-1,2,2-trifluoroethyl)-L-cysteine, *N*-acetyl-*S*-(1,1-dibromo-2,2-difluoroethyl)-L-cysteine, *N*-acetyl-*S*-(1,1-dichloro-2,2-difluoroethyl)-L-cysteine, *N*-acetyl-*S*-(1,1,2,2-tetrafluoroethyl)-L-cysteine, and *N*-acetyl-*S*-(1,2-

dichlorovinyl)-L-cysteine are nephrotoxic in rats *in vivo* and in renal cells *in vitro*. Aminooxyacetic acid, a β-lyase inhibitor, decreased the toxicity of these haloalkene-derived mercapturates, indicating β-lyase-dependent bioactivation after aminoacylase-catalyzed deacetylation (Boogaard *et al.*, 1989; Commandeur *et al.*, 1988). No studies investigating the role of aminoacylase inhibitors in the toxicity of haloalkene-derived mercapturates have apparently been published.

B. The Role of Aminoacylases in the Mutagenicity of Haloalkene-Derived Mercapturates

Many haloalkene-derived mercapturates are mutagenic in *Salmonella typhimurium*. Their mutagenicity is inhibited by aminooxacetic acid and is also dependent on aminoacylase-catalyzed biotransformation (Vamvakas *et al.*, 1987). To elucidate the reactions involved in the bacterial mutagenicity of haloalkene-derived mercapturates, the biotransformation of [1,2,3,4-*butadienyl*-^{14}C]*N*-acetyl-*S*-(pentachlorobutadienyl)-L-cysteine, *N*-acetyl-*S*-(1,2,2-trichlorovinyl)-L-cysteine, and *N*-acetyl-*S*-(1,2-dichlorovinyl)-L-cysteine by subcellular fractions from male rat liver and kidney homogenates and the role of biotransformation in the mutagenicity of the mercapturates were studied. [1,2,3,4-*butadienyl*-^{14}C]*N*-Acetyl-*S*-(pentachlorobutadienyl)-L-cysteine was extensively metabolized by cytosolic, but not by microsomal, enzymes from rat liver and kidney. The major metabolite identified by GC/MS was [1,2,3,4-*butadienyl*-^{14}]*S*-(pentachlorobutadienyl)-L-cysteine; the amount produced was higher in kidney cytosol than in liver cytosol. Biotransformation of [1,2,3,4-*butadienyl*-^{14}C]*N*-acetyl-*S*-(pentachlorobutadienyl)-L-cysteine by kidney and liver cytosol resulted in covalent binding of radioactivity to proteins, and binding of radioactivity was reduced in the presence of the cysteine conjugate β-lyase inhibitor aminooxyacetic acid. *N*-Acetyl-*S*-(1,2,2-trichlorovinyl)-L-cysteine and *N*-acetyl-*S*-(1,2-dichlorovinyl)-L-cysteine were also biotransformed by cytosolic enzymes to the corresponding cysteine conjugates *S*-(1,2,2-trichlorovinyl)-L-cysteine and *S*-(1,2-dichlorovinyl)-L-cysteine. The three mercapturates were potent mutagens in *S. typhimurium* only after addition of rat kidney cytosol: in the absence of rat kidney cytosol, *N*-acetyl-*S*-(1,2,2-trichlorovinyl)-L-cysteine and *N*-acetyl-*S*-(1,2-dichlorovinyl)-L-cysteine were weak mutagens, and *N*-acetyl-*S*-(pentachlorobutadienyl)-L-cysteine was not mutagenic under these conditions. In contrast to *N*-acetyl-*S*-(pentachlorobutadienyl)-L-cysteine, both *N*-acetyl-*S*-(1,2,2-trichlorovinyl)-L-cysteine and *N*-acetyl-*S*-(1,2-dichlorovinyl)-L-cysteine were biotransformed to pyruvate by 100,000× *g* supernatants from bacterial (*S. typhimurium* TA100) homogenates. Pyruvate is the expected endproduct of the biotransformation of these mercapturates by aminoacylases followed by cysteine conjugate β-lyase-dependent bioactivation (Dekant *et al.*, 1986). Pyruvate is formed in concentrations equimolar to those of the reactive intermediate. These experiments indicate that haloalkene-derived mercapturates are deacetylated to the corresponding cysteine conjugates by cytosolic [*N*-acetyl-*S*-(pentachlorobutadienyl)-L-cysteine, *N*-acetyl-*S*-(1,2,2-trichlorovinyl)-L-cysteine, and *N*-acetyl-*S*-(1,2-dichlorovinyl)-L-cysteine] and bacterial enzymes [*N*-acetyl-*S*-(1,2,2-trichlorovinyl)-L-cysteine and *N*-acetyl-*S*-(1,2-dichlorovinyl)-L-cysteine] and then cleaved to reactive and mutagenic intermediates by mammalian and bacterial β-lyases. High cysteine conjugate β-lyase activity is present in *S. typhimurium* (Dekant *et al.*, 1986).

C. The Role of Aminoacylases in Haloalkene-Derived Mercapturate Toxicity

Comparative studies on the role of aminoacylases in the biotransformation and nephrotoxicity of mercapturates were performed with mercapturates derived from 1,1-difluoro-2,2-dihaloethenes, including *N*-acetyl-*S*-(1,1,2,2-tetrafluoroethyl)-L-cysteine, *N*-acetyl-*S*-(1-chloro-1,2,2-trifluoroethyl)-L-cysteine, *N*-acetyl-*S*-(1,1-dichloro-2,2-difluoroethyl)-L-cysteine, and *N*-acetyl-*S*-(1,1-dibromo-2,2-difluoroethyl)-L-cysteine (Boogaard *et al.*, 1989; Commandeur *et al.*, 1988, 1989, 1991b). All four mercapturates were nephrotoxic in rats *in vivo* and were cytotoxic in renal proximal-tubular cells in suspension, as indicated by a reduced uptake of α-methylglucose. The threshold doses for the induction of nephrotoxicity with the four mercapturates varied and were strongly dependent on halogen substitution. In studies with *N*-($^2H_3$-acetyl)-labeled mercapturates, 20 to 30% of the doses of the less toxic compounds [*acetyl*-$^2H_3$]*N*-acetyl-*S*-(1,1-dichloro-2,2-difluoroethyl)-L-cysteine and [*acetyl*-$^2H_3$]*N*-acetyl-*S*-(1,1-dibromo-2,2-difluroethyl)-L-cysteine was recovered in urine. Also, significant amounts of *N*-acetyl-*S*-(1,1-dichloro-2,2-difluoroethyl)-L-cysteine and *N*-acetyl-*S*-(1,1-dibromo-2,2-difluoroethyl)-L-cysteine were excreted after giving the deuterium-labeled compounds, indicating that aminoacylase-dependent deacetylation followed by efficient N-acetylation of the formed cysteine *S*-conjugates had occurred. Only small amounts of the more toxic *N*-acetyl-*S*-(1,1,2,2-tetrafluoroethyl)-L-cysteine and *N*-acetyl-*S*-(1-chloro-1,2,2-trifluoroethyl)-L-cysteine were recovered in urine, indicating extensive metabolism by cysteine conjugate β-lyase and other pathways.

When the biotransformation of mercapturates was investigated in renal and hepatic subcellular fractions, the rates of deacetylation of *N*-

acetyl-*S*-(1,1-dichloro-2,2-difluoroethyl)-L-cysteine and *N*-acetyl-*S*-(1,1-dibromo-2,2-difluoroethyl)-L-cysteine, the less toxic mercapturates, were much lower than the rates of deacetylation observed with *N*-acetyl-*S*-(1,1,2,2-tetrafluoroethyl)-L-cysteine and *N*-acetyl-S-(2-chloro-1,1,2-trifluoroethyl)-L-cysteine, which are more toxic. This less efficient deacetylation may contribute to the relatively high excretion rates of unchanged mercapturates after *N*-acetyl-*S*-(1,1-dichloro-2,2-difluoroethyl)-L-cysteine and *N*-acetyl-*S*-(1,1-dibromo-2,2-difluoroethyl)-L-cysteine are given. The same rank order of cytotoxicity and deacetylation was observed with these mercapturates in renal cells. The much higher activities of aminoacylase activity observed in renal fractions, together with the ability of the kidney to concentrate these mercapturates, may help explain the kidney-selective toxicity of these compounds. These results demonstrate the important role of aminoacylases in the renal toxicity of haloalkene-derived mercapturates and indicate a major influence of substrate structure on the extent of deacylation of mercapturates.

N-Acetyl-*S*-(pentachlorobutadienyl)-L-cysteine is presumed to be the major toxic metabolite of hexachlorobutadiene (Dekant *et al.*, 1988b; Reichert and Schütz, 1986). As with hexachlorobutadiene, *N*-acetyl-*S*-(pentachlorobutadienyl)-L-cysteine is more nephrotoxic to female rats than to male rats (Pratt and Lock, 1988). The deacetylation of [1,2,3,4-*butadienyl*-^{14}C]*N*-acetyl-*S*-(pentachlorobutadienyl)-L-cysteine to [1,2,3,4-*butadienyl*-^{14}C]*S*-(pentachlorobutadienyl)-L-cysteine and the binding of radiolabeled metabolites to protein do not account for the observed sex differences in hexachlorobutadiene-induced nephrotoxicity in rats. *N*-Acetyl-*S*-(pentachlorobutadienyl)-L-cysteine is rapidly biotransformed by renal cytosol, but the rate of biotransformation is similar with renal cytosol from both male and female rats, and the major metabolite is *S*-(pentachlorobutadienyl)-L-cysteine. *N*-Acetylation of *S*-(pentachlorobutadienyl)-L-cysteine to give *N*-acetyl-*S*-(pentachlorobutadienyl)-L-cysteine was not detected in the presence of cytosol obtained from the kidney of either male or female rats. *N*-Acetylation of *S*-(pentachlorobutadienyl)-L-cysteine is, however, observed in renal microsomes in presence of acetyl-CoA (Birner *et al.*, 1992). Covalent binding of radioactivity from [1,2,3,4-*butadienyl*-^{14}C]*N*-acetyl-*S*-(pentachlorobutadienyl)-L-cysteine to cytosolic proteins was detected after a 5-min incubation, and the extent of binding was similar in cytosol from both male and female rats. Covalent binding was prevented by aminooxyacetic acid, an inhibitor of cysteine conjugate β-lyase, providing further support for the pathway outlined in Fig. 7.

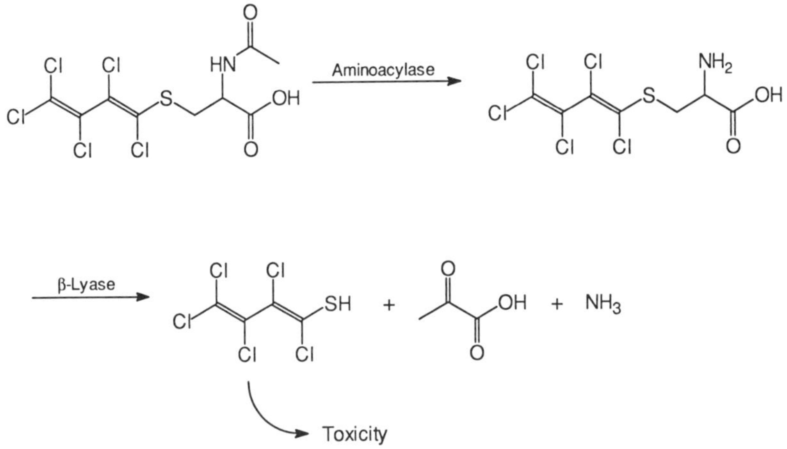

Fig. 7 Aminoacylase- and cysteine conjugate β-lyase-dependent bioactivation of *N*-acetyl cysteine *S*-conjugates, with *N*-acetyl-*S*-(pentachlorobutadienyl)-L-cysteine as an example.

D. Contribution of Aminoacylases to the Toxicity of Hydroquinone-Derived Mercapturates

Aminoacylase-dependent reactions may modulate the toxicity of hydroquinone-derived mercapturates. 2-Bromo-3-(*N*-acetyl-L-cystein-*S*-yl)hydroquinone, 2-bromo-5-(*N*-acetyl-L-cystein-*S*-yl)hydroquinone, and 2-bromo-6-(*N*-acetyl-L-cystein-*S*-yl)hydroquinone are less nephrotoxic than the corresponding cysteine *S*-conjugates (Monks *et al.*, 1991). Deacetylation of the mercapturates to the corresponding cysteine S-conjugates may contribute to the nephrotoxicity of these thioethers, because substitution of hydroquinones with cysteine lowers their redox potentials and thus renders the hydroquinone moieties more prone to oxidation to the electrophilic quinones (Monks *et al.*, 1988, 1991). A role for aminoacylase and the balance between deacetylation and reacetylation are important in the nephrotoxicity of 2,3,5-(triglutathion-*S*-yl)hydroquinone in different species. The extent of the nephrotoxicity of this *S*-conjugate does not correlate with the renal activities of γ-glutamyltransferase in Fischer R344-rats, mice, and guinea pigs. Guinea pigs are highly susceptible to the nephrotoxicity of 2,3,5-(triglutathion-*S*-yl)hydroquinone, and also exhibit the highest renal cytosolic acylase activities and the lowest microsomal *N*-acetyltransferase activities, when compared with the rat and with B6C3F1 and BALB/c mice. The high acylase activities in guinea pigs may

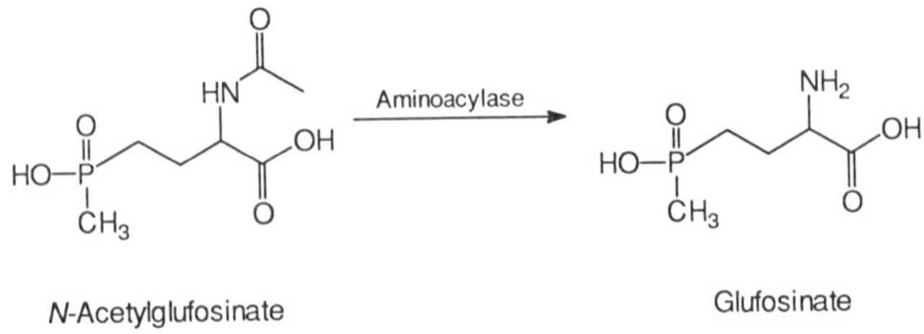

Fig. 8 Aminoacylase-dependent bioactivation of *N*-acetylglufosinate to give the glutamine-synthetase inhibitor glufosinate.

explain the sensitivity of the guina pig to 2,3,5-(triglutathion-*S*-yl)hydroquinone-induced nephrotoxicity (Kleiner *et al.*, 1993).

E. Aminoacylase-Dependent Activation of the Herbicide *N*-Acetylglufosinate

Glutamine synthetase catalyzes the conversion of glutamate to glutamine and plays a central role in nitrogen metabolism. Glutamine synthetase is a ubiquitous enzyme and is present in both prokaryotes and eukaryotes. The enzyme is inhibited by L-methione-(*S*)-sulfoximine and by low concentrations of phosphinothricin, which is marketed as the herbicide glufosinate. *N*-Acetylglufosinate is a prodrug of glufosinate and serves as another example for the role of aminoacylase in bioactivation (Fig. 8). Deacetylation of *N*-acetylglufosinate to the active herbicide glufosinate must occur before herbicidal action is expressed (De Block *et al.*, 1987; Logusch *et al.*, 1988).

Acknowledgments

Research in the authors' laboratories was supported by NIEHS Grants ES03127 and E05407 to M.W.A. by Deutsche Forschungsgemeinschaft SFB 172 to W.D., and by NATO Grant 901032 to M.W.A. and W. D. The authors thank Ms. Sandra Morgan for her assistance in preparing the manuscript.

References

Albert, Z., and Szewczuk, A. (1972). Histochemical and biochemical studies on Co^{++}-activated acylase in rodent tissues. *Acta Histochem.* **42,** 68–76.

Anders, M. W., and Dekant, W. (1993). Renal xenobiotic metabolism: Role in bioactivation of nephrotoxic xenobiotics. *In* "Renal Disposition and Nephrotoxicity of Xenobiotics" (M. W. Anders, W. Dekant, D. Henschler, H. Oberleithner, and S. Silbernagl, eds.), pp. 155–183. Academic Press, San Diego, CA.

Birnbaum, S. M. (1955). Amino acid acylases I and II from hog kidney. *In* "Methods in

Enzymology" (S. Colowick and N. Kaplan, eds.), Vol. 2, pp. 115–119. Academic Press, New York.

Birnbaum, S. M., Levintow, L., Kingsley, R. B., and Greenstein, J. P. (1952). Specificity of amino acid acylases. *J. Biol. Chem.* **194,** 455–470.

Birner, G., Werner, M., and Dekant, W. (1992). The role of renal N-acetyltransferase in the metabolism of haloalkenes. *Toxicologist* **12,** 343.

Boogaard, P. J., Commandeur, J. N. M., Mulder, G. J., Vermeulen, N. P. E., and Nagelkerke, J. F. (1989). Toxicity of the cysteine-S-conjugates and mercapturic acids of four structurally related difluoroethylenes in isolated proximal tubular cells from rat kidney: Uptake of the conjugates and activation to toxic metabolites. *Biochem. Pharmacol.* **38,** 3731–3741.

Bray, H. G., and James, S. P. (1960). The formation of mercapturic acids. 4. Deacetylation of mercapturic acids by the rabbit, rat and guinea pig. *Biochem. J.* **74,** 394–397.

Bruns, F. H., and Schulze, C. (1962). Acylase I. Reindarstellung, physikalisch-chemische Eigenschaften and Identifizieren mit "Hippurikase." *Biochem. Z.* **336,** 162–181.

Chenault, H. K., Dahmer, J., and Whitesides, G. M. (1989). Kinetic resolution of unnatural and rarely occurring amino acids: Enantioselective hydrolysis of N-acyl amino acids catalyzed by acylase I. *J. Am. Chem. Soc.* **111,** 6354–6364.

Commandeur, J. N. M., Oostendorp, R. A. J., Schoofs, P. R., Xu, B., and Vermeulen, N. P. E. (1987). Nephrotoxicity and hepatotoxicity of 1,1-dichloro-2,2-difluoroethylene in the rat. *Biochem. Pharmacol.* **36,** 4229–4237.

Commandeur, J. N. M., Brakenhoff, J. P. G., De Kanter, F. J. J., and Vermeulen, N. P. E. (1988). Nephrotoxicity of mercapturic acids of three structurally related 2,2-difluoroethylenes in the rat. *Biochem. Pharmacol.* **37,** 4495–4504.

Commandeur, J. N. M., De Kanter, F. J. J., and Vermeulen, N. P. E. (1989). Bioactivation of the cysteine-S-conjugate and mercapturic acid of tetrafluoroethylene to acylating reactive intermediates in the rat: Dependence of activation and deactivation activities on acetyl coenzyme A availability. *Mol. Pharmacol.* **36,** 654–663.

Commandeur, J. N. M., Boogaard, P. J., Mulder, G. J., and Vermeulen, N. P. E. (1991a). Mutagenicity and cytotoxicity of two regioisomeric mercapturic acids and cysteine S-conjugates of trichloroethylene. *Arch. Toxicol.* **65,** 373–380.

Commandeur, J. N. M., Stijntjes, G. J., Wijngaard, J., and Vermeulen, N. P. E. (1991b). Metabolism of L-cysteine S-conjugates and N-(trideuteroacetyl)-L-cysteine S-conjugates of four fluoroethylenes in the rat. Role of balance of deacetylation and acetylation in relation to the nephrotoxicity of mercapturic acids. *Biochem. Pharmacol.* **42,** 31–38.

De Block, M., Botterman, J., Vandewiele, M., Dockx, J., Thoen, C., Gosselé, V., Movva, N. R., Thompson, C., Van Montagu, M., and Leemans, J. (1987). Engineering herbicide in plants by expression of a detoxifying enzyme. *EMBO J.* **6,** 2513–2518.

Dekant, W., Vamvakas, S., Berthold, K., Schmidt, S., Wild, D., and Henschler, D. (1986). Bacterial β-lyase mediated cleavage and mutagenicity of cysteine conjugates derived from the nephrocarcinogenic alkenes trichloroethylene, tetrachloroethylene and hexachlorobutadiene. *Chem.-Biol. Interact.* **60,** 31–45.

Dekant, W., Berthold, K., Vamvakas, S., and Henschler, D. (1988a). Thioacylating agents as ultimate intermediates in the β-lyase catalyzed metabolism of S-(pentachlorobutadienyl)-L-cysteine. *Chem.-Biol. Interact.* **67,** 139–148.

Dekant, W., Schrenk, D., Vamvakas, S., and Henschler, D. (1988b). Metabolism of hexachloro-1,3-butadiene in mice: *In vivo* and *in vitro* evidence for activation by glutathione conjugation. *Xenobiotica* **18,** 803–816.

Endo, Y. (1978). N-Acyl-L-aromatic amino acid deacylase in animal tissues. *Biochim. Biophys. Acta* **523,** 207–214.

Endo, Y. (1980). *In vivo* deacetylation of N-acetyl amino acids by kidney acylases in mice and rats. A possible role of acylase system in mammalian kidneys. *Biochim. Biophys. Acta* **628,** 13–18.
Fones, W. S., and Lee, M. (1953). Hydrolysis of N-acyl derivatives of alanine and phenylalanine by acylase I and carboxypeptidase. *J. Biol. Chem.* **201,** 847–856.
Fu, S.-C. J., and Birnbaum, S. M. (1953). The hydrolytic action of acylase I on N-acylamino acids. *J. Am. Chem. Soc.* **75,** 918–920.
Gade, W., and Brown, J. L. (1981). Purification, characterization and possible function of α-N-acylamino acid hydrolase from bovine liver. *Biochim. Biophys. Acta* **662,** 86–93.
Greenhough, K. J., and Turner, A. J. (1991). Aminoacylase I is not a glycolipid-anchored ectoenzyme in pig kidney. *Biochim. Biophys. Acta* **1076,** 364–368.
Greenstein, J. P., and Winitz, M. (1961). "Chemistry of the Amino Acids," Vol. 2, pp. 1753–1767. Wiley, New York.
Harris, J. W., Pohl, L. R., Martin, J. L., and Anders, M. W. (1991). Tissue acylation by the chlorofluorocarbon substitute 2,2-dichloro-1,1,1-trifluoroethane (HCFC-123). *Proc. Natl. Acad. Sci. U.S.A.* **88,** 1407–1410.
Hashmi, M., Vamvakas, S., and Anders, M. W. (1992). Bioactivation mechanism of S-(3-oxopropyl)-N-acetyl-L-cysteine, the mercapturic acid of acrolein. *Chem. Res. Toxicol.* **5,** 360–365.
Heese, D., and Röhm, K.-H. (1989). Reactivities of sulfhydryl groups in native and metal-free aminoacylase I. *Biol. Chem. Hoppe-Seyler* **370,** 607–612.
Heese, D., Löffler, H.-G., and Röhm, K.-H. (1988). Further characterization of porcine kidney aminoacylase I reveals close similarity to 'renal dipeptidase.' *Biol. Chem. Hoppe-Seyler* **369,** 559–566.
Heese, D., Berger, S., and Röhm, K.-H. (1990). Nuclear magnetic relaxation studies of the role of the metal ion in Mn^{2+}-substituted aminoacylase I. *Eur. J. Biochem.* **188,** 175–180.
Jakob, M., Miller, Y. E., and Röhm, K.-H. (1992). Cloning and sequence analyses of cDNAs encoding aminoacylase I from porcine kidney. *Biol. Chem. Hoppe-Seyler* **373,** 1227–1231.
Kaul, R., Casanova, J., Johnson, A. B., Tang, P., and Matalon, R. (1991). Purification, characterization, and localization of aspartoacylase from bovine brain. *J. Neurochem.* **56,** 129–135.
Kleiner, H. E., Monks, T. J., and Lau, S. S. (1993). N-Acetylation/N-deacetylation as a determinant of 2,3,5-(triglutathion-S-yl)-hydroquinone mediated nephrotoxicity. *Toxicologist* **13,** 241.
Kördel, W., and Schneider, F. (1976). Chemical investigations on pig kidney aminoacylase. *Biochim. Biophys. Acta* **445,** 446–457.
Kördel, W., and Schneider, F. (1977). Renal aminoacylase, a zinc enzyme. *Z. Naturforsch. C* **32C,** 342–344.
Lock, E. A., and Ishmael, J. (1985). Effect of the organic acid transport inhibitor probenecid on renal cortical uptake and proximal tubular toxicity of hexachloro-1,3-butadiene and its conjugates. *Toxicol. Appl. Pharmacol.* **81,** 32–42.
Löffler, H.-G., and Schneider, F. (1987). Aminoacylase from hog kidney: Essential carboxyl groups, inhibition by α-N-OH-acids, lack of esterase activity. *Biol. Chem. Hoppe-Seyler* **368,** 1072–1073.
Löffler, H. G., Schneider, F., Aumüller, G., and Unsicker, K. (1982). Immunzyto-chemische Untersuchungen zur Lokalisation der Aminoacylase I (EC 3.5.1.14) in der Schweineniere. *Acta Histochem., Suppl.* **25,** 57–60.
Löffler, H. G., Röhm, K. H., and Schneider, F. (1986). Studies on the metal ion dependence, kinetics and SH (S-S) groups of acylamino acid amidohydrolase. *In* "Zinc Enzymes" (I. Bertini, C. Luchinat, W. Maret, and M. Zeppezauer, eds.), Vol. 1, pp. 281–288. Birkhäuser, Boston.

Löffler, H. G., Kroll, M., and Schneider, F. (1988). 3-(2-Furyl)acryloyl-, theinyl-acryloyl- and cinnamoyl-amino acids: Novel substrates for aminoacylase I (EC 3.5.1.14). *Biol. Chem. Hoppe-Seyler* **369,** 866.

Logusch, E. W., Walker, D. M., McDonald, J. F., and Franz, J. E. (1988). Substrate variability as a factor in enzyme inhibitor design: Inhibition of ovine brain glutamine synthetase by α- and γ-substituted phosphinothricins. *Biochemistry* **28,** 3043–3051.

Matalon, R., Michals, K., Sebasta, D., Deanching, M., Gashkoff, P., and Casanova, J. (1988). Aspartoacylase deficiency and N-acetylaspartic aciduria in patients with Canavan disease. *Am. J. Med. Genet.* **29,** 463–471.

Miller, Y. E., and Kao, B. (1989). Monoclonal antibody based immunoassay for human aminoacylase-1. *J. Immunoassay* **10,** 129–152.

Miller, Y. E., Minna, J. D., and Gazdar, A. F. (1989). Lack of expression of aminoacylase-1 in small cell lung cancer. Evidence for inactivation of genes encoded by chromosome 3p. *J. Clin. Invest.* **83,** 2120–2124.

Miller, Y. E., Drabkin, H., Jones, C., and Fisher, J. H. (1990). Human aminoacylase-1: Cloning, regional assignment to distal chromosome 3p1.1, and identification of a cross-hydridizing sequence on chromosome 18. *Genomics* **8,** 149–154.

Mitta, M., Ohnogi, H., Yamamoto, A., Kato, I., Sakiyama, F., and Tsunasawa, S. (1992). The primary structure of porcine aminoacylase 1 deduced from cDNA sequence. *J. Biochem.* (*Tokyo*) **112,** 737–742.

Monks, T. J., Highet, R. J., and Lau, S. S. (1988). 2-Bromo-(diglutathion-S-yl)hydroquinone nephrotoxicity: Physiological, biochemical, and electrochemical determinants. *Mol. Pharmacol.* **34,** 492–500.

Monks, T. J., Jones, T. W., Hill, B. A., and Lau, S. S. (1991). Nephrotoxicity of 2-bromo-(cystein-S-yl)hydroquinone and 2-bromo-(N-acetyl-L-cystein-S-yl)hydroquinone thioethers. *Toxicol. Appl. Pharmacol.* **111,** 279–298.

Moravcsik, E., Telegdi, J., Tüdös, H., Kömives, K., and Ötvös, L. (1977). Substrate specificity of acylase-I-catalyzed dipeptide hydrolysis. *Acta Biochim. Biophys. Acad. Sci. Hung.* **12,** 399–402.

Nash, J. A., King, L. J., Lock, E. A., and Green, T. (1984). The metabolism and disposition of hexachloro-1 : 3-butadiene in the rat and its relevance to nephrotoxicity. *Toxicol. Appl. Pharmacol.* **73,** 124–137.

Paik, W. K., Bloch-Frankenthal, L., Birnbaum, S. M., Winitz, M., and Greenstein, J. P. (1957). ε-Lysine acylase. *Arch. Biochem. Biophys.* **69,** 56–66.

Park, S. B., Osterloh, J. D., Vamvakas, S., Hashmi, M., Anders, M. W., and Cashman, J. R. (1992). Flavin-containing monooxygenase-dependent stereoselective S-oxygenation and cytotoxicity of cysteine S-conjugates and mercapturates. *Chem. Res. Toxicol.* **5,** 193–201.

Pratt, I. S., and Lock, E. A. (1988). Deacetylation and further metabolism of the mercapturic acid of hexachloro-1,3-butadiene by rat kidney cytosol *in vitro*. *Arch. Toxicol.* **62,** 341–345.

Reichert, D., and Schütz, S. (1986). Mercapturic acid formation is an activation and intermediary step in the metabolism of hexachlorobutadiene. *Biochem. Pharmacol.* **35,** 1271–1275.

Röhm, K.-H. (1989). Butylmalonate is a transition state analogue for aminoacylase I. *FEBS Lett.* **250,** 191–194.

Suzuki, S., and Tateishi, M. (1981). Purification and characterization of a rat liver enzyme catalyzing N-deacetylation of mercapturic acid conjugates. *Drug Metab. Dispos.* **9,** 573–577.

Tallan, H. H., Moore, S., and Stein, W. H. (1956). N-Acetyl-L-aspartic acid in brain. *J. Biol. Chem.* **219,** 257–264.

Vamvakas, S., Dekant, W., Berthold, K., Schmidt, S., Wild, D., and Henschler, D. (1987).

Enzymatic transformation of mercapturic acids derived from halogenated alkenes to reactive and mutagenic intermediates. *Biochem. Pharmacol.* **36,** 2741–2748.

Wolfgang, G. H. I., Gandolfi, A. J., Stevens, J. L., and Brendel, K. (1989). N-Acetyl S-(1,2-dichlorovinyl)-L-cysteine produces a similar toxicity to S-(1,2-dichlorovinyl)-L-cysteine in rabbit renal slices: Differential transport and metabolism. *Toxicol. Appl. Pharmacol.* **101,** 205–219.

Zhang, G., and Stevens, J. L. (1989). Transport and activation of S-(1,2-dichlorovinyl)-L-cysteine and N-acetyl-S-(1,2-dichlorovinyl)-L-cysteine in rat kidney proximal tubules. *Toxicol. Appl. Pharmacol.* **100,** 51–61.

Zhar, B., Brauch, H., Talmadge, C., and Linehan, M. (1987). Loss of alleles of loci on the short arm of chromosome 3 in renal cell carcinoma. *Nature (London)* **327,** 721–724.

Bioactivation by S-Adenosylation, S-Methylation, or N-Methylation

Jerald L. Hoffman
Department of Biochemistry
School of Medicine
University of Louisville
Louisville, Kentucky 40292

I. Introduction: Synthesis and Utilization of AdoMet[1]

This chapter will review the activation of xenobiotic methionine analogs by S-adenosylation and the activation of xenobiotic sulfur- and nitrogen-containing molecules by methylation. These topics are related by their involvement of AdoMet or its analogs. Research on AdoMet synthesis and utilization is now in its fifth decade, since this molecule was first described forty years ago by Cantoni (1953). AdoMet is second only to ATP in the number of different group-transfer reactions in which it acts as a group donor. AdoMet is best known as a methyl donor, and among the types of molecules that act as methyl acceptors in a variety of species, including higher animals, are DNA, RNA, proteins, lipids, amino acids, catecholamines, various other nitrogen compounds, and sulfur and selenium compounds. AdoMet is also a group donor during polyamine biosynthesis. It is first decarboxylated to form dcAdoMet, which donates one aminopropyl group to putrescine to form spermidine. Spermidine then accepts a second aminopropyl group from dcAdoMet to form spermine.

[1] Abbreviations used: AdoMet, S-adenosylmethionine; dcAdoMet, decarboxylated S-adenosylmethionine; AdoHcy, S-adenosylhomocysteine; DMSO, dimethyl sulfoxide; DEAE, diethylaminoethyl; m^5C, 5-methylcytosine; MAO, monoamine oxidase. Other abbreviations are given in the figures.

Advances in Pharmacology, Volume 27

In addition to the above functions as a direct group donor, AdoMet is an essential metabolite in the flow of sulfur from methionine through AdoHcy, homocysteine, cystathionine, and cysteine to glutathione. The quantitative importance of this pathway in maintaining glutathione levels has been demonstrated in several studies. Beatty and Reed (1980) showed that in cultured hepatocytes the sulfur of ^{35}S-labeled L-methionine was incorporated in substantial amounts into glutathione, and this incorporation was blocked by the γ-cystathionase inhibitor propargylglycine. In hepatocytes depleted of glutathione by diethyl maleate, methionine supported the repletion of glutathione. The importance of methionine in maintaining glutathione levels during toxic stress was emphasized by the work of Miners *et al.* (1984). They found that mice treated with paracetamol had greatly decreased levels of glutathione in the liver. Supplementation with methionine maintained higher glutathione levels and increased the urinary excretion of glutathione-derived conjugates of paracetamol metabolites.

The enzymatic biosynthesis of AdoMet is of interest due to the importance of the product and also because of the unique nature of the reaction (Fig. 1). AdoMet synthesis is one of only two known adenosylation reactions utilizing ATP, the other being the biosynthesis of adenosylcobalamin. AdoMet is synthesized from ATP and methionine in a reaction catalyzed by ATP:L-methionine *S*-adenosyltransferase (EC 2.5.1.6), commonly known as AdoMet synthetase. The triphosphate group of ATP is released as inorganic tripolyphosphate, which remains enzyme-bound. AdoMet synthetase subsequently hydrolyzes the tripolyphosphate to pyrophosphate and phosphate, with the phosphate originating from the γ-phosphate group of ATP. This hydrolysis (and the hydrolysis of pyrophosphate to two molecules of phosphate by ubiquitous pyrophosphatases) provides the free energy for the synthesis of the high-energy sulfonium compound AdoMet.

Despite the importance of AdoMet synthetase, it was 24 years between the discovery of AdoMet and the first purification to homogeneity of the enzyme by Chiang and Cantoni (1977). These authors worked with yeast and obtained two forms of the enzyme, but the yields were low. Markham *et al.* (1980) solved the yield problem by using a strain of recombinant *Escherichia coli* containing a plasmid carrying extra copies of the gene for the bacterial enzyme. This strain gave yields of AdoMet synthetase on the order of 1 mg/g wet cells.

The first homogeneous purification of a mammalian (rat liver) AdoMet synthetase was reported in preliminary form by Hoffman and Kunz (1980). During this work it was realized that the conflicting reports in the literature regarding the physical and kinetic properties of mammalian AdoMet syn-

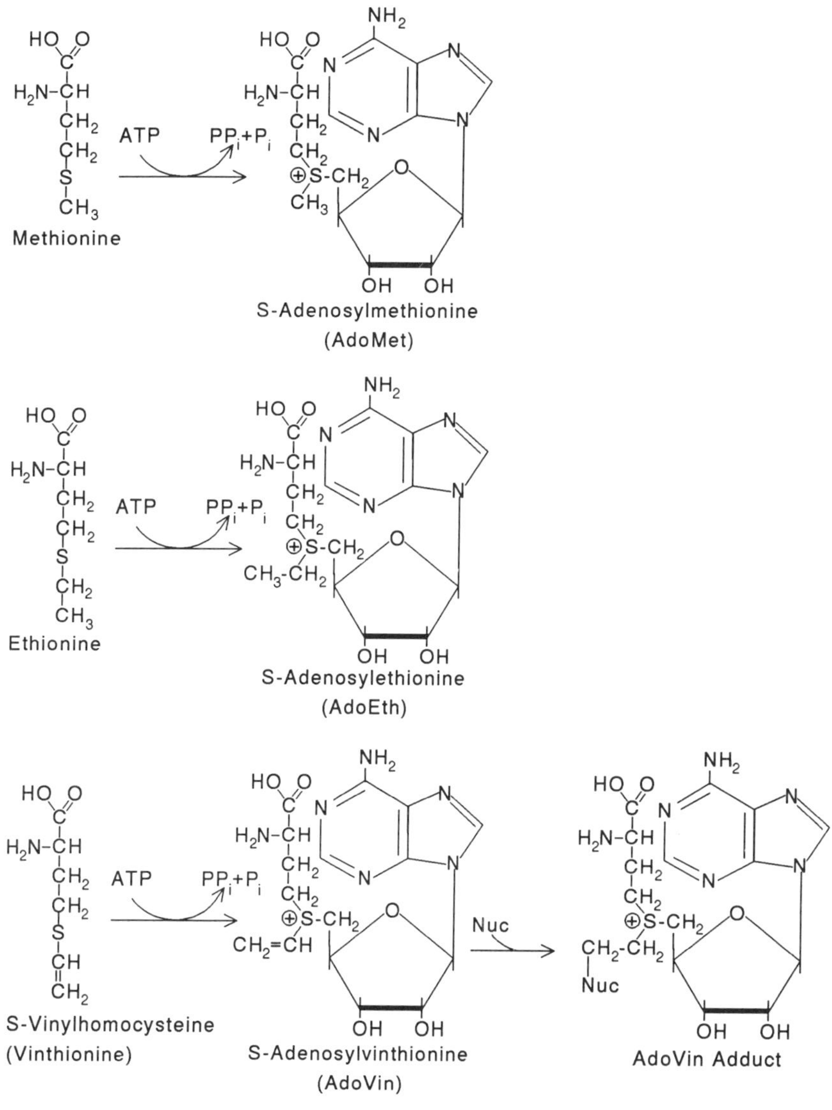

Fig. 1 S-Adenosylation of methionine and its xenobiotic analogs as catalyzed by AdoMet synthetase. In this and the remaining figures, abbreviations used in the text are shown either alone or in parentheses accompanying the full name of the compound. "Nuc" indicates a nucleophilic group reacting with AdoVin.

thetase were due to the presence of tissue-specific forms of the enzyme. Liau *et al.* (1977) had previously suggested this based on the kinetic complexity observed with crude preparations of AdoMet synthetase from rat liver. Hoffman and Kunz (1977) used a combination of chromatographic fractionation and enzyme kinetics with and without various effectors to characterize the different forms of the enzyme. One unusual effector that proved to be quite useful was DMSO. This organic solvent at 10% (v/v) in assay mixtures activated AdoMet synthetase in crude extracts from rat liver but not from other tissues. When rat liver extracts were chromatographed on columns of either DEAE-cellulose or Sephadex G-150, the single broad peaks of activity in each chromatogram were asymmetrically activated by DMSO, indicating the presence of closely eluting forms of the enzyme. Kunz *et al.* (1980) found that hydrophobic chromatography on phenyl–Sepharose gave baseline separation of two peaks of AdoMet synthetase activity from rat liver with the later-eluting form being strongly activated by DMSO.

Further refinement of the hydrophobic and gel filtration chromatography procedures by Sullivan and Hoffman (1983) enabled them to distinguish three forms of AdoMet synthetase from rat liver (I, II, and III), of which only form II was found in rat kidney. AdoMet synthetase III was purified to homogeneity and was shown to consist of two apparently identical subunits of M_r 47,000. This form was highly activated by DMSO and was also unusual in being activated by its product, AdoMet, whereas forms I (M_r 208,000) and II (M_r 120,000) were virtually unaffected by DMSO and strongly inhibited by AdoMet. Using similar procedures, Okada *et al.* (1981) found two forms of AdoMet synthetase from rat liver, which they designated α (corresponding to I) and β (corresponding to III), and one form from kidney, which they designated γ (corresponding to II). The kinetic properties and molecular weights of these forms determined by this group agreed well with previous observations.

The DNA-derived amino acid sequences are known for a number of AdoMet synthetases. The first such sequence was determined by Markham *et al.* (1984) from the *E. coli MetK* gene that codes for AdoMet synthetase. Sequences are known for both of the genes from yeast that code for AdoMet synthetase, *SAM1* (Thomas and Surdin-Kerjan, 1987) and *SAM2* (Thomas *et al.*, 1988). Finally, Tsukuda's laboratory has sequenced cDNAs from the rat coding for the liver-specific (Horikawa *et al.*, 1989) and kidney-specific (Horikawa *et al.*, 1990) forms. In the liver, both forms I (α) and III (β) are apparently products of the same gene, with form I being a tetramer and form III a dimer of identical subunits. This is consistent with the finding by Cabrero and Alemany (1988) that AdoMet synthetase I can be converted to form III by high concentrations of LiBr. The physiological factors that regulate interconversion of forms I and III are not known. Horikawa *et al.* (1990) have compared the above five AdoMet synthetase sequences and made the following observations. The rat kidney enzyme sequence of 395 amino acids was 85% similar to the liver sequence of 397 amino acids, whereas the nucleotide sequences in the coding regions of the two cDNAs were 73% identical. Degrees of identity of other amino acid sequences of AdoMet synthetases with that for the rat kidney enzyme were *SAM1,* 67% (382 amino acids); *SAM2,* 68% (384 amino acids); and *MetK,* 52% (384 amino acids). *SAM1* and *SAM2* have 83% identity between their DNA base sequences and 92% similarity between the amino acid sequences of the encoded proteins (Thomas *et al.,* 1988). All five AdoMet synthetases contain a typical ATP-binding sequence, which is Gly–Ala–Gly–Asp–Gln–Gly from 131 to 136 and Lys at 159 in the kidney sequence. These seven residues are at similar positions in the other sequences and are identical, except for the conservative replacement of Lys by Arg in the *E. coli* enzyme.

The existence of AdoMet synthetase isoforms in human liver similar to I and III in rat liver was established by Cabrero *et al.* (1988), in the course of studies on cirrhosis. They found that gel filtration of extracts from functionally normal human liver gave two peaks of AdoMet synthetase activity, a high-M_r, DMSO-insensitive form and a low-M_r, DMSO-activated form. Similar analysis of extracts from cirrhotic liver showed a specific reduction of about sixfold in the high-M_r form with little change in the low-M_r form. This caused the total AdoMet synthetase activity, as assayed under physiological concentrations of ATP and methionine, to be reduced in cirrhotic patients to 37% of that in cholecystectomy patients with otherwise normal liver function. This explained the long-standing observation that methionine clearance from the plasma is slowed by liver damage (Kinsell *et al.,* 1947).

There are also several reports of genetic deficiencies of human hepatic AdoMet synthetase, of which the case described by Gahl *et al.* (1987) is typical. A 31-year-old man was examined due to a complaint of a persistent odd odor to his breath with no other symptoms. His plasma methionine was elevated about 30 times normal (716 μM vs 15–40 μM). His hepatic AdoMet synthetase activity was 28% of normal, was not stimulated by 10% DMSO, and was not increased at higher methionine concentration. This indicates that this person has a mutation in the gene for hepatic AdoMet synthetase preventing the form-III dimer from aggregating into a form-I tetramer. Since form I is more active than form III at physiological concentrations of methionine (Kunz *et al.,* 1980), this amino acid accumulates as a result of form-I deficiency and is degraded by other pathways to methanethiol and other malodorous sulfur compounds, which are exhaled.

II. Bioactivation of Xenobiotic Methionine Analogs by S-Adenosylation Catalyzed by AdoMet Synthetase

A. Ethionine Induction of Fatty Liver

Figure 1 shows those methionine analogs known to be toxic or carcinogenic by virtue of their S-adenosylation by AdoMet synthetase. The most widely studied of these methionine analogs is ethionine, which has two major biological effects in animals in that it causes fatty liver and is carcinogenic. The metabolic sequelae occurring in female rats after ethionine treatment leading to development of fatty liver are well understood and have been reviewed by Farber (1967). The following general discussion of ethionine-induced fatty liver is based on this review with references from the primary literature given only for key papers. L-Ethionine or compounds that can be metabolically converted to L-ethionine, such as D-ethionine and *N*-acetyl-L-ethionine, are capable of causing this pathological condition. Ethionine can be given either in the diet or by intraperitoneal injection. The latter method is used most commonly due to the better control achieved of the timing and particularly the dose, since ethionine in the diet causes a variable loss of appetite.

In female rats, a minimum single dose of ethionine of 100 mg/kg body weight causes the concentration of liver triglycerides to increase approximately 20-fold by 24 h after treatment. The peak of lipid concentration occurs between 24 and 48 h and then declines to normal values by 72 to 120 h. No liver lipids other than triglycerides show consistent changes, including phospholipids, total cholesterol, or cholesterol esters. The increase in liver lipids is mirrored by a decrease in plasma concentrations of all lipids and lipoproteins, except for free fatty acids. The increase in free fatty acids in plasma is apparently a secondary effect due to the induction of hypoglycemia by ethionine, since blood glucose concentrations decline to 50–60 mg% by 4 h after ethionine treatment. If this hypoglycemia is prevented by administration of glucose or sucrose to ethionine-treated rats, the occurrence of fatty liver is prevented. In the liver of ethionine-treated rats compared with controls there is no deficiency of uptake of radioactive fatty acids ([1-^{14}C]palmitate) nor their incorporation into triglycerides. However, in normal rats the radioactivity in triglycerides falls off rapidly starting 5 min after label injection, whereas in ethionine-treated rats it continues to rise for over 30 min and then declines slowly. This is accompanied by a large decline in the appearance of labeled triglycerides in the plasma of ethionine-treated rats in comparison to the controls. These data indicate that ethionine causes a defect in the transport of triglycerides from the liver to the plasma.

There are four potential metabolic fates of ethionine: sulfoxidation and N-acetylation; transamination and oxidation to CO_2; incorporation into protein; and S-adenosylation. The latter appears to be of primary importance in the induction of fatty liver. The key to understanding the biochemical pathogenesis of ethionine-induced fatty liver came (1962) when Shull reported that ethionine caused a rapid decline of liver ATP to about 20% of control within 2 h. This was associated with a decrease in RNA and protein syntheses and a breakup of polyribosomes into monomeric ribosomes. This decline in ATP and the effects on RNA and protein syntheses were prevented by administration of ATP precursors such as adenine, inosine, or 5-amino-4-imidazole carboxamide. The ATP concentrations were restored if one of these precursors was given within 24 h after ethionine treatment. Prevention or early reversal of the decline in liver ATP concentrations also prevents the appearance of fatty liver. Shull *et al.* (1966) went on to show that these effects of ethionine on ATP concentrations were due to the formation of AdoEth catalyzed by AdoMet synthetase, as shown in Fig. 1. AdoEth is used by methyltransferases at a much lower rate than is AdoMet, and so it accumulates in the liver, which has high levels of AdoMet synthetase activity. The resulting decline in ATP concentrations causes a temporary stimulation of *de novo* synthesis of adenine (as ATP) starting from glycine and ribose-5-phosphate. However, since this biosynthetic pathway requires energy obtained from the phosphate–phosphate bonds of ATP, it rapidly becomes unable to replenish the declining pool of this nucleotide. As might be expected, the natural substrate for AdoMet synthetase, methionine, when given at about one-fourth the dose of ethionine prevents the development of fatty liver.

In summary the primary biochemical event responsible for ethionine-induced fatty liver is the AdoMet synthetase-catalyzed formation of AdoEth at a rate that exceeds the capacity of the liver to maintain adequate ATP concentrations by *de novo* biosynthesis. The accompanying decreases in RNA and protein synthesis lead to insufficient amounts of apolipoproteins for export of triglycerides from the liver to the plasma, resulting in accumulation of this class of lipid in the liver and its deficiency in plasma.

B. Ethionine Carcinogenesis

Although the induction of fatty liver is a consequence of acute ethionine treatment, the second major biological effect of this amino acid, carcinogenesis, results from chronic administration. Popper *et al.* (1953) described the development of tumor-like nodules in the livers of rats fed a diet containing 0.5% ethionine. This concentration of ethionine is fairly toxic, which may have prevented rats from surviving long enough to develop full-blown cancer. When Farber (1956) reduced the dietary concentration of ethionine to 0.25%, the rats survived longer and developed outright hepatomas, some of which were invasive and metastatic. The sequence of cellular and tissue changes during ethionine feeding is as follows: the appearance of oval cells at 1–3 weeks; bile duct proliferation within 2 months; nodular hyperplasia after 6–12 weeks; and, finally, development of hepatomas after 5–6 months (Farber, 1963). A large number of studies have confirmed the sensitivity of rats of various strains and both sexes to ethionine carcinogenesis. Mice also develop hepatomas when fed ethionine, but Syrian golden hamsters are resistant (Farber, 1963).

As in the case of ethionine induction of fatty liver, the carcinogenicity of ethionine appears to require its conversion to AdoEth. Classical metabolically activated carcinogens are thought to react with DNA to initiate carcinogenesis, and ethionine administration does lead to ethylation of nucleic acids (reviewed by Farber, 1963). However, a careful quantitative study showed that, when rats were given [*ethyl*-1-^{14}C]ethionine, liver DNA contained only one ethyl group per 2×10^7 bases. This is several orders of magnitude less than the amount of DNA adduct formation found with other carcinogens, suggesting that ethylation of DNA is not the biochemical event necessary for ethionine carcinogenesis.

A more likely mechanism for the carcinogenic activity of ethionine is that the accumulation of AdoEth and the suppression of AdoMet lead to hypomethylation of oncogenes in DNA triggering their inappropriate overexpression. General aspects of DNA methylation and its effects on gene expression have been summarized by Cedar (1988). About 3.5% of the cytosine residues in mammalian DNA are postreplicatively converted to m^{5}C. Nearly all of these m^{5}C residues occur in the sequence m^{5}CpG and so will be paired with the identical sequence on the opposite strand. After DNA replication, DNA methyltransferase uses AdoMet to methylate the C residues in the hemimethylated double-stranded sites generated by DNA polymerase. In general there is an inverse relationship between methylation of genes and their expression. Several examples are known of naturally programmed site-specific demethylation associated with increased gene expression, including the α-actin gene in myoblasts, the IgG K gene in B cells, and many developmentally regulated genes in liver. This inverse relationship has also been observed in experiments with cultured cells into which genes in various states of methylation are stably transfected or when the DNA methyltransferase inhibitor 5-azacytidine is used to block DNA methylation. The mechanism of repression of gene transcription by m^{5}C has been clarified by the recent work by Boyes and Bird (1991). They have found that a specific m^{5}C-binding protein (MeCP-1) exists, and if the 5′-flanking region near the promoter of a gene contains m^{5}C residues, binding of MeCP-1 inhibits transcription of the gene.

In reviewing the role of DNA methylation in cancer, Jones and Buckley (1990) pointed out that many alkylating carcinogens inhibit methylation of DNA, leading to activation of genes, malignant transformation of cells, and an increase in the rate of tumor diversification. What is most striking in relating DNA methylation to cancer is that prolonged intake by rats of methyl-deficient diets (low in methionine and choline) with no added carcinogens results in the development of liver tumors. A possible rationale for this was established by the first report linking methyl-deficient diets to hypomethylation and increased expression of oncogenes in rat liver. Wainfan *et al.* (1990) found that feeding methyl-deficient diets to rats gave large increases in liver mRNAs encoded by c-*myc* and c-*fos*, with a lesser increase in c-Ha-*ras* mRNA. These increases in mRNA levels were accompanied by hypomethylation of the respective oncogenes. Wainfan and Poirier (1992) have reviewed several additional studies showing that such methyl-deficient diets (low in methionine and choline) cause depletion of AdoMet pools, hypomethylation of DNA, and increases in expression of oncogenes in liver.

Ethionine also affects DNA methylation and oncogene expression. Cox and Irving (1977) found that treatment of partially hepatectomized rats with ethionine caused hypomethylation of DNA in the regenerating liver. Shivapurkar *et al.* (1984) fed ethionine to rats and also observed hypomethylation of liver DNA. Yaswen *et al.* (1985) fed a choline-deficient diet containing 0.1% ethionine to rats and examined the effects on expression of six oncogenes. The levels of mRNAs for c-Ki-*ras*, c-Ha-*ras*, and c-*myc* all increased by 2 weeks on the diet (the earliest time point examined). The increase in c-Ha-*ras* mRNA was transient, whereas the other two oncogene mRNAs remained elevated throughout the 35 weeks on the ethionine diet and were also elevated in a primary tumor. None of these changes were found in rats fed the choline-deficient diet lacking ethionine. The mRNA for c-*src* was detectable but unchanged during the period of ethionine feeding, whereas mRNAs for c-*abl* and c-*mos* were not detected in preneoplastic nor neoplastic liver.

Two questions remain to be investigated regarding the relationships among ethionine treatment, DNA hypomethylation, and oncogene expression. The first is whether the generalized hypomethylation of DNA caused by ethionine treatment extends to specific oncogenes as well. The second is whether feeding ethionine in a methyl-sufficient diet will cause hypomethylation and overexpression of specific oncogenes. However, these observations on ethionine treatment combined with the above discussion of the relationship between DNA methylation and cancer suggest the following biochemical events in ethionine carcinogenesis. Chronic feeding of ethionine causes the persistence of AdoEth and the suppression of AdoMet synthesis in the liver. The combined effects of high concentrations of the inhibitor AdoEth and low concentrations of the substrate AdoMet for DNA methyltransferase result in hypomethylation of DNA in general, including various oncogenes. This leads to transcriptional activation of selected oncogenes and progression to liver cancer.

C. Mutagenicity and Carcinogenicity of Vinthionine

Vinthionine was first synthesized in the Millers' laboratory for investigation as a possible proximate carcinogen of ethionine. Although no metabolic relationship between these two methionine analogs has been demonstrated, vinthionine proved to have xenobiotic properties apparently resulting from its S-adenosylation. In their first report, Leopold *et al.* (1979) compared the mutagenicity of vinthionine and ethionine in the Ames test. Vinthionine and *N*-acetylvinthionine methyl ester were both shown to be direct mutagens for *Salmonella typhimurium* T100, whereas ethionine was inactive. Addition of rat liver microsomes and cytosol (13,000 $\times$ g supernatant) had no effect of the mutagenicity of vinthionine or ethionine but increased the mutagenicity of *N*-acetylvinthionine methyl ester. This latter effect was ascribed to conversion of the blocked amino acid to vinthionine by amidases and esterases. The mutagenicity of vinthionine at 60 nmol/plate was inhibited by methionine, with 50% inhibition at 230 nmol/plate and 100% inhibition at 1000 nmol/plate. Ethionine at 230 nmol/plate had little effect and was toxic at or above 500 nmol/plate. The importance of the amino acid portion of vinthionine was emphasized by the finding that several non-amino acid analogs of vinthionine, including ethyl vinyl sulfide and sulfone, were nonmutagenic. Neither ethionine nor vinthionine caused lung adenomas upon chronic treatment of mice. However, this study was inconclusive, since ethionine is quite liver-specific in its carcinogenicity.

In an expanded study comparing ethionine and vinthionine, Leopold *et al.* (1982) found the latter to be a liver carcinogen. Both methionine analogs, when fed at 0.1% of a choline-supplemented diet, caused equivalent high rates of occurrence of liver tumors in male rats. Tumors appeared sooner in ethionine-fed rats, and a methyl-deficient diet facilitated ethionine carcinogenesis more than vinthionine. Both male and female CD-1 mice developed hepatomas when fed either vinthionine or ethionine at 0.1%, with females being more susceptible to both amino acids. A single injected dose of vinthionine lowered rat liver ATP concentrations, but not to the same extent as an equimolar dose of ethionine. These authors were unable to demonstrate the presence of AdoVin in livers of vinthionine-treated rats or the biosynthesis of AdoVin *in vitro* from ATP and vinthionine with a yeast extract active with methionine and ethionine.

They did report a personal communication from Sullivan and Hoffman describing the use of more sensitive methods that enabled the identification of AdoVin in liver extracts from vinthionine-treated mice and the biosynthesis of AdoVin by mouse liver extracts containing AdoMet synthetase. In a later report, Sullivan and Hoffman (1982) showed that vinthionine was a strong competitive inhibitor vs methionine of rat AdoMet synthetases I and II but activated form III at subsaturating concentrations of methionine. Leopold *et al.* (1982) also synthesized [*vinyl*-^{14}C]vinthionine and injected it into rats to test for *in vivo* binding of radioactivity to liver protein, DNA, and RNA. All three macromolecules became highly labeled, and the amount of binding to DNA was comparable to that found for strongly alkylating carcinogens.

On the basis of these observations, Leopold *et al.* (1982) hypothesized two possible routes of activation and vinyl adduct formation by vinthionine. The first of these is illustrated in Fig. 1, where the vinyl group of AdoVin is a reactive electrophile due to the presence of the vinyl sulfonium group. In this case the entire AdoVin structure would become incorporated as an AdoEth residue linked to nucleophiles by C-2 of its ethyl group. The second possible route for vinyl adduct formation involves the γ-cleavage of vinthionine releasing vinyl mercaptan, which could form 2-mercaptoethyl adducts with nucleophiles. It should be possible to determine which of these two mechanisms predominates by testing for the presence of the *S*-adenosyl residue in macromolecules from vinthionine-treated animals which, if found, would indicate the prevalence of the former route involving AdoVin. Consideration of this mechanism indicates that AdoVin has potential as an affinity-labeling reagent for AdoMet-binding proteins such as AdoMet synthetase, AdoMet decarboxylase, or the various methyltransferases if these enzymes have nucleophilic groups in or near their active sites.

Alks *et al.* (1992) have reported the synthesis of another methionine analog with the potential to form a vinyl sulfonium compound by S-adenosylation. This analog is (*Z*)-L-2-amino-4-methylthio-3-butenoic acid [$CH_3SCH{=}CHCH(NH_2)COOH$, abbreviated AMTB]. Synthesis of AMTB was prompted by the previous work of Sufrin *et al.* (1982) showing that the oxygen-containing analog (*Z*)-L-2-amino-4-methoxy-3-butenoic acid was a potent inhibitor of all isoforms of AdoMet synthetase from rat liver and Novikoff hepatoma. If AMTB is a substrate for S-adenosylation by AdoMet synthetase, the resulting *S*-adenosyl-AMTB would be reactive due to its vinyl sulfonium content. *S*-Adenosyl-AMTB should also be a suicide inhibitor for AdoMet decarboxylase similar to the effects of other 3,4-unsaturated amino acids on their respective decarboxylases. Such amino acids covalently inactivate their respective decarboxylases after the initial formation of a Schiff's base between the amino group of the analog and a carbonyl group on either a pyridoxal-phosphate cofactor or a pyruvate group on the decarboxylase (Jung and Koch-Weser, 1981).

D. Liver Specificity of Methionine Analogs

The requirement for S-adenosylation by AdoMet synthetase to potentiate the xenobiotic effects of ethionine and vinthionine draws attention to the role of the various isoforms of this enzyme in the liver specificity of these amino acids. Abe *et al.* (1980) tested ethionine as a substrate for the liver-specific forms, AdoMet synthetases I and III, from the rat and the effects of ethionine administration on the relative amounts of these two forms. In normal rat liver, 50 μM ethionine was a good substrate for both forms of the enzyme and, as with methionine, form III was highly activated by 10% DMSO when assayed at this concentration of ethionine. The finding that livers of ethionine-treated rats accumulate high concentrations at AdoEth may be due to the lack of *S*-adenosyl product inhibition of AdoMet synthetase III. This was found by Sullivan and Hoffman (1983) for the natural product AdoMet, which actually activated this isoform. Assuming that AdoEth acts similarly, its activation of AdoMet synthetase III could be a major factor in the high rate of AdoEth synthesis in the liver, which leads to ATP depletion and the persistence of AdoEth at levels inhibitory to methyltransferases. A similar argument can be presented for the liver-specific S-adenosylation of vinthionine.

Abe *et al.* (1980) also found that daily administration of ethionine to rats at a dose of 250 mg/kg body weight for 9 days caused a transient increase at 2 days in liver AdoMet synthetase I followed by a decrease to slightly less than normal activity by 9 days. AdoMet synthetase III showed only a decline to nearly undetectable activity by the end of the 9 days of ethionine treatment. These changes were not due to alterations in ATP since this was prevented by giving adenine daily at a dose of 120 mg/kg body weight. This regimen of ethionine is similar to that used to induce fatty liver, and it has not been investigated whether chronic feeding of lower daily amounts of ethionine, as used in carcinogenesis studies, has the same effects. However, Matsumoto *et al.* (1984) found that feeding a diet of 3% methionine to rats for 6 days increased liver AdoMet synthetase I activity by 4- to 5-fold, but form III activity increased only 1.5-fold. The activity of kidney AdoMet synthetase II did not change. These data suggest that methionine and its analog ethionine, or their *S*-adenosyl derivatives, regulate the level of activity of AdoMet synthetases I and III in the liver. The question whether this regulation is transcriptional, post-transcriptional, translational, or post-translational awaits further investigation.

III. Xenobiotic Bioactivation by N-Methylation of Azaheterocyclic Compounds

A. Biosynthesis of Paraquat

Several N-methylated azaheterocyclic compounds are known to be toxic, which raises the question whether bioactivation can occur by N-methylation of their desmethyl derivatives. One such N-methylated compound that has been studied for some time is paraquat, which is toxic due to its participation in one-electron redox cycling that generates reactive oxygen species. The structure of paraquat is shown in Fig. 2 along with a proposed pathway for conversion of 4,4′-bispyridyl to paraquat by sequential N-methylation of the two pyridyl residues from AdoMet. Two groups independently showed that cytosolic *N*-methyltransferases could catalyze one of these methylation steps *in vitro*. Godin *et al.* (1986) used dialyzed cytosols from several rabbit tissues as a source of enzyme along with [*methyl*-^{3}H]AdoMet. The *methyl*-^{3}H labeled products were identified by cochromatography of radioactivity on cation-exchange HPLC with authentic standards. Cytosols from lung, kidney, and liver monomethylated 3,3′- and 4,4′-bispyridyl, but 2,2′-bispyridyl was not a substrate. The methyltransferase activity in these three tissues declined in the order given, and brain cytosol was inactive with all substrates tested. Ansher

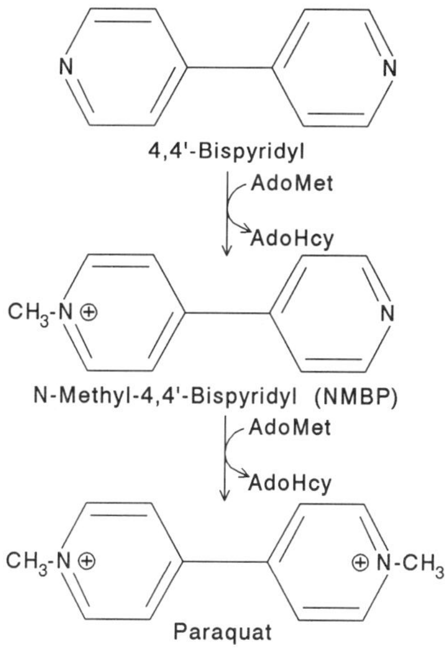

Fig. 2 Sequential N-methylation of 4,4′-bispyridyl to form paraquat.

et al. (1986) carried out a similar study using two homogeneous *N*-methyltransferases (designated A and B) previously isolated from rabbit liver (Ansher and Jakoby, 1986). They also found that these rabbit enzymes formed only the monomethyl derivative of 4,4′-bispyridyl, NMBP. Form A was most active with a K_m for 4,4′-bispyridyl of 0.3 m*M* and a V_{max} of 161 pmol/min/mg protein compared with the values for form B of 4.6 m*M* and 52 pmol/min/mg protein. The two groups that carried out these studies later joined forces for a more thorough evaluation of the substrate specificities of the *N*-methyltransferases A and B from rabbit liver (Crooks *et al.*, 1988). Among other findings they verified the previous conclusion that both enzymes could monomethylate 3,3′- and 4,4′-bispyridyl.

Godin and Crooks (1989) tested for the *in vivo* methylation of 4,4′-bispyridyl after its injection in rabbits and guinea pigs by preparative HPLC fractionation of urine followed by analysis of appropriate fractions. Their findings verified that *N*-methyltransferases in the rabbit are capable of monomethylation of 4,4′-bispyridyl, since only the monomethyl derivative, NMBP, was found in the urine at about 1% of the dose. However, in urine from treated guinea pigs, the dimethyl derivative, paraquat, was found at about 3% of the dose along with NMBP at about 1% of the dose, thus providing the first demonstration of toxic activation by N-methylation. These *in vitro* and *in vivo* results showed that the desmethyl derivative of paraquat 4,4′-bispyridyl can be methylated to either the mono- or the dimethyl derivative in a species-specific fashion. Oral administration of 4,4′-bispyridyl to rats causes pulmonary hemorhage similar to that caused by paraquat, indicating probable toxic activation by N-methylation in this species (Croce and Kimbrough, 1982).

B. N-Methylation of Environmental or Endogenous Toxins in Parkinson's Disease

1. N-Methylated Azaheterocyclic Compounds and Parkinsonian Syndrome

Although most cases of Parkinson's disease remain idiopathic, the last 10 years of research on this disease have been marked by considerable progress in understanding the potential etiological roles of environmental or endogenous toxins. Research along these lines was triggered by an initial single case report (Davis *et al.*, 1979) and later an additional report of four cases (Langston *et al.*, 1983) describing the development of Parkinsonism in humans after self-administration of a "designer drug" contaminated with MPTP (see Fig. 3 for structures). The clinical signs in these

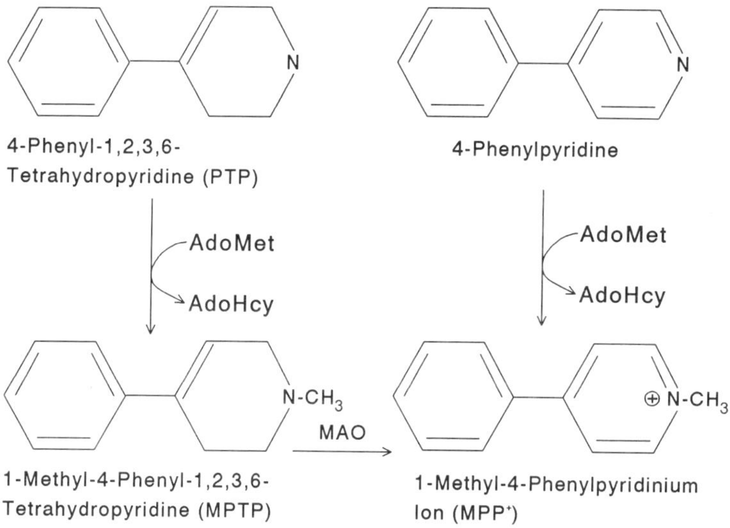

Fig. 3 Metabolic conversion of PTP and 4-phenylpyridine to the Parkinsonian neurotoxin MPP^+.

cases of akinesia, rigidity, resting tremor, fixed posture, etc., were identical to those found in patients with idiopathic Parkinson's disease. Biochemical findings included lower than normal homovanillic acid in cerebrospinal fluid. A neuropathological study was conducted following the death of the first patient due to a drug overdose 2 years after the diagnosis of chemically induced Parkinsonism. Destruction of dopaminergic neurons in the substantia nigra was seen comparable to that in the idiopathic disease. The clinical status of patients with the chemically induced syndrome was generally improved by administration of the dopamine precursor L-DOPA or dopamine agonists.

These findings stimulated the development of animal models and the conduct of biochemical toxicological studies that have produced our current understanding of the role of MPTP in Parkinsonism (Heikkila *et al.*, 1989; Maret *et al.*, 1990; Sayre, 1989; Singer and Ramsey, 1990). It has been shown that MPTP is oxidized by monoamine oxidase in the brain to MPP^+, which is selectively taken up into the dopaminergic cells of the nigrostriatum via the dopamine reuptake transporter. Once in the cells' pathway, MPP^+ becomes concentrated in mitochondria, as driven by the transmembrane potential, where it causes cell death either by inhibiting respiration at complex I (NADH dehydrogenase) or by generating reactive oxygen species. As above with paraquat, the toxicity of an N-methylated azaheterocyclic compound, MPP^+, suggests that the desmethyl derivative might be activated by a methyltransferase.

2. N-Methylation of MPP^+ Precursors

Two obvious choices for desmethyl precursors of MPP^+ are PTP, which could be converted to MPP^+ by methylation and oxidation, and 4-phenylpyridine, which would require only methylation (Fig. 3). Ansher *et al.* (1986) found that both of these compounds were substrates for *N*-methyltransferases in crude cytosols from brains of humans, monkeys, rabbits, mice, and rats as well as for the homogeneous *N*-methyltransferases A and B from rabbit liver. The K_m values for PTP and 4-phenylpyridine ranged from 0.2 to 0.8 m*M* for the various enzyme preparations. In comparing V_{max} values between the pure enzymes, *N*-methyltransferase A was about 10-fold more active than B with both substrates. A later study by Crooks *et al.* (1988) confirmed that 3- and 4-phenylpyridine were substrates for these methyltransferases. Godin *et al.* (1986) tested the ability of cytosolic preparations from rabbit lung, liver, kidney, and brain to methylate 2-, 3-, or 4-phenylpyridine. The 2-phenylpyridine was inactive as a substrate with enzyme extracts from all four tissues. Both 3- and 4-phenylpyridine were substrates for the enzyme extracts from lung, liver, and kidney but not for that from brain. One possible reason that this study, unlike that of Ansher *et al.* (1986), did not detect an *N*-methyltransferase from brain capable of methylating phenylpyridines is that the previous work used ultrafiltration to concentrate the brain extract, thereby increasing the sensitivity of the assay.

Godin and Crooks (1989) tested for *in vivo* methylation of 4-phenylpyridine and found that rabbits and guinea pigs treated with this compound did not excrete MPP^+ in their urine, although as noted above in this same study both of these species excreted *N*-methylpyridinium derivatives upon treatment with 4,4′-bispyridyl. This lack of urinary excretion of MPP^+ after 4-phenylpyridine treatment does not necessarily mean that this compound is not methylated *in vivo,* since tissue retention or additional metabolic alterations might occur. In a speculative note on environmental toxins and Parkinson's disease, Snyder and D'Amato (1985) pointed out that 4-phenylpyridine as well as many other pyridine derivatives are found in foods and in the environment. They also indicated that 4-phenylpyridine alters dopamine-related biochemical parameters in neural cell cultures and mice in a manner similar to that of MPTP. Despite the inability to show methylation *in vivo,* these studies on 4-phenylpyridine have stimulated further searches for environmental or endogenous protoxins capable of being converted by N-methylation to toxins causing Parkinsonism. The results of this research have led to the discovery of two classes of such protoxins, β-carbolines and tetrahydroisoquinolines, and the findings on each are summarized separately in the following two sections.

3. N-Methylation of β-Carbolines

β-Carbolines (BC, see Fig. 4 for structures and metabolic reactions) are formed by oxidation of THBCs initially produced from condensation of aldehydes with tryptophan or tryptamine and occur widely in nature (Bloom *et al.*, 1983). Shortly after MPP^+ was discovered, Collins and Neafsey (1985) noted its structural resemblance to 2-MeBCs and suggested that BCs were good candidates for Parkinsonian protoxins. Hoppel *et al.* (1987) found that, among many pyridine derivatives tested, the BCs harmine and norharmine were potent inhibitors of mitochondrial oxidation

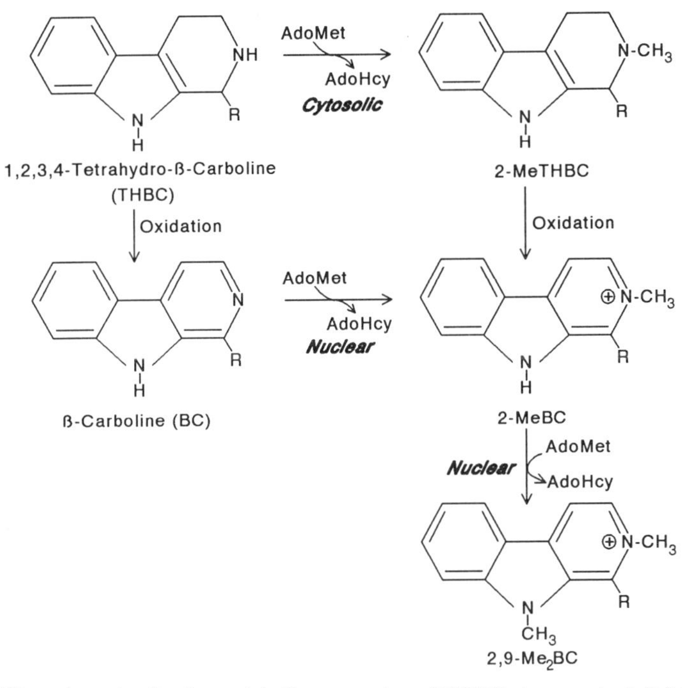

Fig. 4 Alternate routes for the metabolic conversion of THBCs to neurotoxic 2,9-Me_2BCs. The subcellular locations in brain of the *N*-methyltransferases involved in each methylation reaction are indicated in italics. Two specific BCs whose methylation is discussed in the text are harman (R = —CH_3) and norharman (R = H). Other BCs are known with substituents on the phenyl ring.

of NADH. Albores *et al.* (1990) reported on an expanded study of mitochondrial respiratory inhibition by BCs. They found that various 2-MeBCs and their 3,4-dihydro derivatives were strong inhibitors of mitochondrial respiration. However, these compounds lacked one feature of MPP^+ in that neutral anhydronium forms generated by deprotonation of the indole nitrogen (atom 9), rather than purely cationic forms, were shown to be involved in mitochondrial entry. The best mimics of MPP^+ in regard to their mitochondrial sequestration and respiratory inhibition proved to be 2,9-Me_2BCs such as 2,9-bismethylnorharman. Arora *et al.* (1990) compared several MPP^+ analogs, including 2-methylharmine, to MPP^+ with regard to their dopaminergic toxicity after intranigral infusion in rats, ability to induce dopamine release from striatal synaptosomes, and *in vitro* inhibition of mitochondrial respiration. The BC derivative 2-methylharmine was active in all three assays.

These studies prompted a search for enzymes capable of N-methylating BCs or their 1,2,3,4-tetrahydro derivatives on the 2- or 9-nitrogens, or both. Collins *et al.* (1992) found that 2,9-Me_2BCs inhibited mitochondrial respiration and had toxic effects on neural cell cultures with potencies similar to those of MPP^+. They also showed that unfractionated whole-brain extracts from guinea pigs and rats could sequentially methylate norharman to form 2,9-bismethylnorharman. In an expanded study, Matsubara *et al.* (1992b) examined the subcellular distribution and other details regarding BC-*N*-methyltransferase activity in rat brain extracts. In contrast to the cytosolic *N*-methyltransferase activity described in Section III,B,2, the enzyme (or enzymes) acting on these BCs was found primarily in nuclear fractions. There was an obligatory order of addition of the two methyl groups during the formation of 2,9-Me_2BCs, with the 2-nitrogen methylated first followed by the 9-nitrogen. This methylation of the 9-nitrogen appears to be the first demonstration of methylation of an indole nitrogen. The authors suggested that this obligatory order is due to the initial formation of the quaternary nitrogen in 2-MeBC, which enhances the nucleophilicity of the indole 9-nitrogen during its methyltransferase-catalyzed SN_2 attack on the methyl group of AdoMet. The authors also suggested that a single methyltransferase was responsible for both 2- and 9-N-methylation of BCs, since both activities had similar subcellular and regional brain distributions as well as similar kinetic constants for substrates.

As noted at the beginning of this section, BCs may be synthesized by oxidation of THBCs. THBC derivatives have been found in many biological sources, including rat tissues (Barker *et al.*, 1981; Fukushima *et al.*, 1992), as well as in foods and in human urine and milk (Adachi *et al.*, 1991). Endogenous formation of THBCs was demonstrated by Susilo and

Rommelspacher (1987), who found 1-methyl-THBC-1-carboxylic acid in rat brains after intracerebral injection of tryptamine and pyruvate. In the conversion of THBCs to 2,9-Me_2BCs, the oxidation and methylation reactions could conceivably occur in any order. This was explored by Matsubara *et al.* (1992a) in a parallel study to their BC methylation work examining the ability of brain extracts to methylate THBCs. They found that brain extracts from both rats and guinea pigs could methylate THBC on the 2-nitrogen but not the 9-nitrogen. Although this was similar to their findings with BCs, these authors also found two important differences in comparing methylation of THBC and BCs. First, the *N*-methyltransferase active on the 2-nitrogen of THBC was cytosolic rather than nuclear as with BCs, suggesting that this activity may be the same as the cytosolic enzymes described in Section III,B,2. Second, the 2-MeTHBC formed was not a substrate for methylation on the 9-nitrogen even in whole-brain extracts that contained the nuclear *N*-methyltransferase capable of methylating this atom in 2-MeBCs. The reason given for this is that 2-MeTHBC contains a tertiary amine group rather than a quaternary *N*-methylpyridinium nitrogen as found in 2-MeBCs. Because of this, the 2-methylation of THBC does not increase the nucleophilicity of the 9-nitrogen as does the 2-methylation of BCs. Accordingly Matsubara *et al.* (1992a) have indicated that the conversion of THBCs to 2,9-Me_2BCs may follow the alternate routes shown in Fig. 4, where the 2-methylation and oxidation of THBCs may occur in either order to yield 2-MeBCs, but 9-methylation will take place only after the biosynthesis of the 2-MeBCs.

4. N-Methylation of Isoquinolines

Isoquinoline derivatives constitute the second class of candidate compounds for the role of protoxins or toxins capable of causing idiopathic Parkinson's disease (see Fig. 5 for structures and metabolic transformations). Similar to the BCs discussed above, isoquinolines can be formed by condensation of an aldehyde with an arylethylamine to form a tetrahydro derivative, which is then oxidized. In the case of the isoquinolines, the aryl group is a phenyl derivative rather than an indole, as in the BCs. Methylation of isoquinolines produces *N*-methylisoquinolinium compounds structurally similar to MPP^+ in containing a *N*-methylpyridinium group, suggesting that they may likewise cause Parkinsonism.

The simplest *N*-methylisoquinolinium compound, $NMIQ^+$, has been shown to be neurotoxic to dopaminergic neurons (Naoi *et al.*, 1989a). This raises the question whether TIQ can be found endogenously and converted to $NMIQ^+$. The presence of TIQ in brains of humans with Parkinsonism was established by Niwa *et al.* (1989) by gas chromatogra-

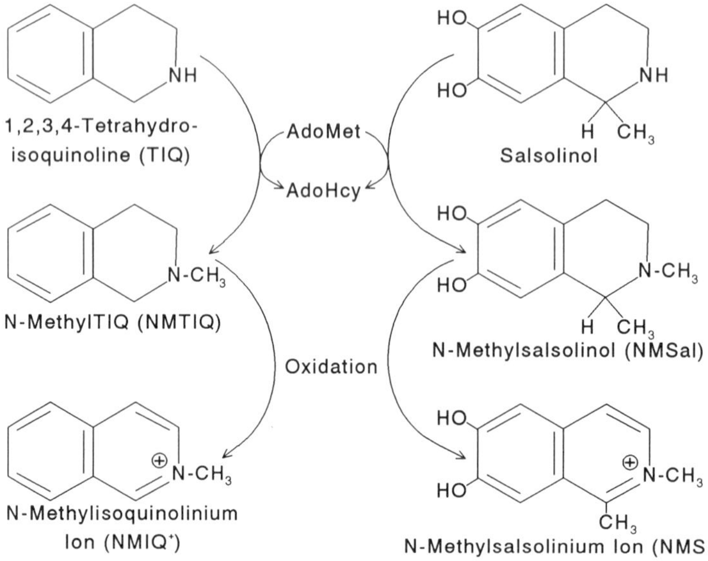

Fig. 5 Metabolic conversion of TIQ and salsolinol to the neurotoxic compound $MNIQ^+$ and the potential neurotoxin $NMSI^+$. The proposed oxidation of NMSal to $NMSI^+$ has not been demonstrated.

phy–mass spectrometry. Nagatsu and Yoshida (1988) showed that TIQ caused Parkinsonism in monkeys, although it was not as potent as MPTP. The methylation and oxidation steps shown in Fig. 5 converting TIQ to $NMIQ^+$ have also been demonstrated. Naoi *et al.* (1989b) found that human brain extracts were capable of N-methylating TIQ *in vitro*. In addition NMTIQ was demonstrated in the brains of monkeys after systemic administration of TIQ (Niwa *et al.*, 1990), indicating that TIQ is N-methylated *in vivo*. NMTIQ was shown to be oxidized *in vitro* by monoamine oxidase to form $NMIQ^+$ (Naoi *et al.*, 1989c). Finally, Hirata *et al.* (1990) have demonstrated $NMIQ^+$ uptake by the dopamine transporter. Although these results are suggestive about the potential for isoquinoline derivatives to be involved in Parkinsonism, several problems remain, such as the inability to show accumulation of TIQ in human neurons. Therefore the search has turned to other isoquinolines more likely to have increased specificity for dopaminergic neurons.

Dopamine itself can be converted to isoquinoline derivatives, and one of these, salsolinol, has been found in human brains (Sjöquist *et al.*, 1982) as well as cerebrospinal fluid and urine (Sjöquist *et al.*, 1981). The amount of salsolinol in human urine is increased after treatment with L-DOPA (Dostert *et al.*, 1989). The carbon atom bearing the methyl group at position

1 of salsolinol is chiral, and synthesis of the *R*-enantiomer predominates in the human brain. Dostert *et al.* (1990) found that (*R*)-salsolinol was synthesized in humans by decarboxylation of the product formed by condensation of dopamine with pyruvate. It has not been possible to induce Parkinsonism by systemic administration of (*R*)-salsolinol, probably because this compound cannot pass the blood–brain barrier (Origitano *et al.*, 1981). Despite this, Niwa *et al.* (1991) have reported the presence of NMSal in brains of normal and Parkinsonian humans, suggesting its endogenous synthesis.

This prompted an elegant study by Maruyama *et al.* (1992), who used *in vivo* microdialysis to establish the occurrence of N-methylation of (*R*)-salsolinol in various regions of rat brain. This was done by perfusing with (*R*)-salsolinol and then using HPLC with multichannel electrochemical detection to analyze the dialysate for NMSal. Methylating activity was found in the highest amount in the substantia nigra, with lesser amounts in the striatum, hypothalamus, and hippocampus. These authors indicated that this regional dopaminergic specificity of NMSal biosynthesis and the synthesis of (*R*)-salsolinol from dopamine make this latter compound an attractive candidate for the neurotoxin causing idiopathic Parkinson's disease. They also found that perfusion of (*R*)-salsolinol caused decreases in dopamine and serotonin and increases in 3,4-dihydroxyphenylacetic acid, homovanillic acid, and 5-hydroxyindole acetic acid in the dialysate of the striatum. These changes paralleled those reported in a similar study using MPP^+ (Ozaki *et al.*, 1987) and are indicative of the dopamine release and inhibition of type A monoamine oxidase typical of dopaminergic-specific toxicity.

Although much remains to be done to establish more firmly the role of N-methylation of endogenous protoxins in the etiology of idiopathic Parkinson's disease in humans, the work reviewed above is reflective of the current excitement and activity in this field. It is anticipated that many more such interesting results will be forthcoming in the very near future.

IV. Xenobiotic Bioactivation by S-Methylation

There are three enzymes known to use AdoMet to methylate the sulfur atom in low-molecular-weight compounds. Two of these enzymes methylate sulfhydryl groups to form methyl thioether products, and both of these have been reviewed with regard to their biochemistry (Weinshilboum, 1989a) and pharmacogenetics (Weinshilboum, 1989b). They differ in that thiopurine methyltransferase (EC 2.1.1.67) is cytoplasmic and methylates the sulfur atom in aromatic and heterocyclic thiols typified by

6-mercaptopurine, whereas thiol methyltransferase (EC 2.1.1.9) is microsomal and methylates the sulfur atom in aliphatic thiols typified by 2-mercaptoethanol. Both of these are generally considered to be detoxifying enzymes. Thiol methyltransferase methylates the sulfur in drugs such as D-penicillamine and captopril, which greatly reduces their biological activity. Thiopurine methyltransferase methylates the sulfur in drugs such as 6-mercaptopurine, and individuals with genetically determined, low activity of this enzyme are at higher risk from thiopurine toxicity.

The third mammalian enzyme that methylates sulfur atoms is *S*-adenosylmethionine:thioether *S*-methyltransferase, commonly called thioether methyltransferase. The discovery and purification of this enzyme were results of a search for the enzyme catalyzing the formation of trimethyl selenonium ion, a urinary excretion product of selenium metabolism. Mozier *et al.* (1988) devised an assay for the enzyme by measuring the formation of radioactive trimethyl selenonium ion from [*methyl*-^{3}H]AdoMet and dimethyl selenide. They showed that the enzyme was cytosolic, and that mouse lung had the highest specific activity among several tissues tested. The enzyme was purified from mouse lung and shown to be a monomer with an M_r of 28,000. Although originally assayed as a selenoether methyltransferase, the pure enzyme was also shown to methylate the sulfur atom in a variety of thioethers and so was named thioether methyltransferase. The enzyme has broad substrate specificity, since it was found to be capable of methylating atom "X" in compounds of the type R-X-R′, where X could be S, Se, or Te. After variety of thioethers were tested, it was concluded that thioether methyltransferase would not methylate compounds containing a carboxyl group nor ones having the sulfur attached directly to an aromatic ring or conjugated double bond system. In a preliminary report, thioether methyltransferase activity and immunoreactivity with a rabbit antiserum against the mouse lung enzyme were found in tissues from rabbits, dogs, pigs, guinea pigs, hamsters, rats, and chimpanzees (Warner *et al.*, 1989). Human tissues have not been tested for thioether methyltransferase activity or immunoreactivity, but the enzyme must be present since humans excrete its product, trimethyl selenonium ion. Mozier and Hoffman (1990) showed that mice injected with [*methyl*-^{3}H]methionine (to label the AdoMet pool) along with various thioethers contained the corresponding [*methyl*-^{3}H]sulfonium ions in their liver, lungs, and urine, but not their kidneys. This tissue distribution paralleled the tissue distribution of thioether methyltransferase activity determined previously and indicated that thioether methyltransferase was active *in vivo* in methylating thioethers prior to urinary excretion of the water-soluble methyl sulfonium ion.

The ability of animals to biosynthesize sulfonium ions creates the possibility that this might be a route to toxic bioactivation. There is a profusion of references in the chemical literature regarding the reactivity of sulfonium ions exemplified by the need to use two volumes to review this field (Stirling, 1981). The sulfonium ions MEVS, CEEMS, and DAMS are products of thioether methyltransferase (Mozier *et al.*, 1988; Mozier and Hoffman, 1990) and have the potential for further reactivity, as shown in Fig. 6.

The reactivity of vinyl sulfonium ions is due to stabilization by the positively charged sulfonium group of neighboring carbonions or carbanions at C-1 in the transition state during nucleophilic addition to C-2 of the vinyl group, as shown in Fig. 6 (Stirling, 1977). A likely protein target for covalent inactivation by vinyl sulfonium compounds is thioether methyltransferase itself, since MEVS is a product of this enzyme and thus has an affinity for binding. Warner (1992) investigated this by measuring the suicide inactivation of pure thioether methyltransferase from mouse lung upon reaction with AdoMet and ethyl vinyl sulfide. He found that such inactivation occurred dependent on both time and the concentration of ethyl vinyl sulfide. Covalent linkage of MEVS to thioether methyltransfer-

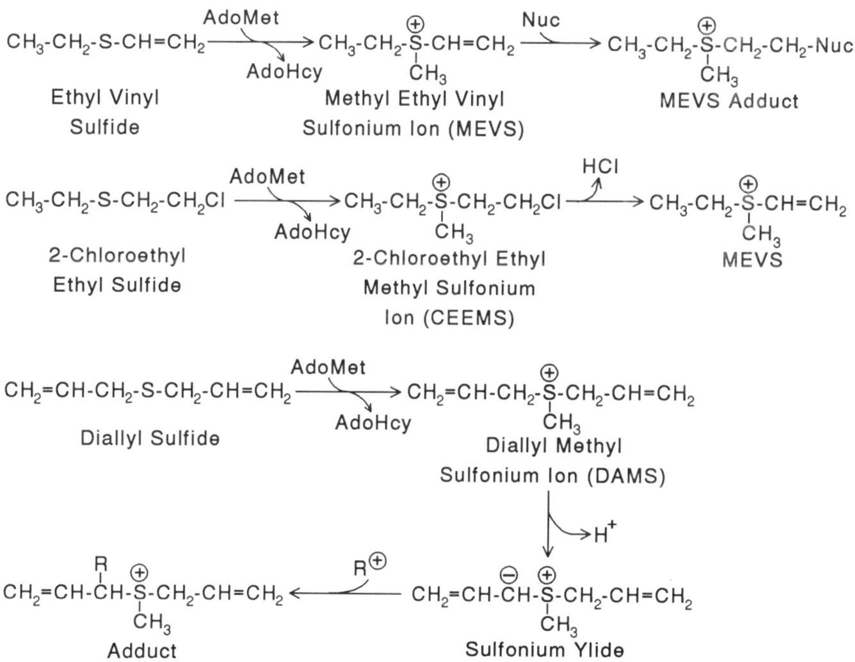

Fig. 6 The biosynthesis and potential reactivities of methyl sulfonium ions produced by thioether methyltransferase. "Nuc" indicates a nucleophilic group reacting with MEVS. "R^+" in the diallyl sulfide pathway indicates a carbonium ion or other electrophilic group.

ase was inferred from experiments in which the enzyme was incubated with [*methyl*-^{3}H]AdoMet and ethyl vinyl sulfide followed by analysis of the reaction mixture by gel filtration under denaturing conditions. Radioactivity coeluted with the protein only if the reaction mixture contained ethyl vinyl sulfide and not in its absence either with or without dimethyl sulfide. Attempts at overnight digestion of the labeled protein with trypsin prior to purification and sequencing of the labeled peptide led to virtually complete loss of label from the peptide products. A further search for digestion conditions consistent with retention of the label has not been completed. Despite this it has been shown that the vinyl sulfonium ion MEVS will react with at least one protein, thioether methyltransferase. It remains to be established whether treatment of animals with vinyl sulfur compounds such as MEVS or its precursor, ethyl vinyl sulfide, will lead to more widespread covalent damage to other cellular macromolecules.

Sulfur mustards (2-chloroethyl thioethers) are substrates for thioether methyltransferase both *in vitro* and *in vivo*, and when this was first found it was suggested that this would result in detoxification of these compounds since the methyl sulfonium product could not form the reactive episulfonium ion (Mozier and Hoffman, 1990). However, as shown in Fig. 6, 2-chloroethyl groups in sulfonium ions such as CEEMS can be converted to vinyl groups by elimination of HCl. The resulting vinyl sulfonium ion would be reactive as described above. This sequence of reactions remains to be demonstrated in a biological system, but might possibly contribute to the systemic toxic effects of sulfur mustards.

Finally, sulfonium ion products of thioether methyltransferase that contain allyl groups may also be reactive as shown for DAMS in Fig. 6. The hydrogen on C-1 of the allyl residue is acidic due to the resonance and sulfonium charge stabilization of the ylide resulting from deprotonation at this carbon. The ready release of this proton accelerates addition reactions of electrophiles to C-1 of the allyl group. Such addition reactions have not been searched for in biological systems.

Xenobiotic bioactivation by S-methylation is a relatively new and unexplored field. However, the established biosynthesis of sulfonium ions by thioether methyltransferase combined with the known chemical reactivities of sulfonium ions bearing various groups suggests that this field is worthy of further study.

Acknowledgment

Research in the author's laboratory on thioether methylation was supported by Research Grant ES-4887 from the National Institute of Environmental Health Sciences, U.S. Public Health Service.

References

Abe, T., Yamano, H., Teraoka, H., and Tsukada, K. (1980). Changes in the activity of S-adenosylmethionine synthetase isozymes from rat liver on ethionine administration. *FEBS Lett.* **121,** 29–32.

Adachi, J., Mizui, Y., Naito, T., Ogawa, Y., Uetani, Y., and Ninomiya, I. (1991). Identification of tetrahydro-β-carboline-3-carboxylic acid in foodstuffs, human urine, and human milk. *J. Nutr.* **121,** 646–652.

Albores, R., Neafsey, E. J., Drucker, G., Fields, J. Z., and Collins, M. A. (1990). Mitochondrial respiratory inhibition by N-methylated beta-carboline derivatives structurally resembling N-methyl-4-phenylpyridine. *Proc. Natl. Acad. Sci. U.S.A.* **87,** 9368–9372.

Alks, V., Keith, D. D., and Sufrin, J. R. (1992). Synthesis of the constrained L-methionine analog, (Z)-L-2-amino-4-methylthio-3-butenoic acid. *Synthesis* **7,** 623–625.

Ansher, S., and Jakoby, W. B. (1986). Amine N-methyltransferases from rabbit liver. *J. Biol. Chem.* **261,** 3996–4001.

Ansher, S., Cadet, J., Jakoby, W. B., and Baker, J. K. (1986). Role of N-methyltransferases in the neurotoxicity associated with the metabolites of 1-methyl-4-phenyl-1,2,3,6-tetrahydropyridine (MPTP) and other 4-substituted pyridines present in the environment. *Biochem. Pharmacol.* **35,** 3359–3363.

Arora, P. K., Riachi, N., Fiedler, G. C., Singh, M. P., Abdallah, F., Harik, S. I., and Sayre, L. M. (1990). Structure-neurotoxicity trends of analogues of 1-methyl-4-phenylpyridinium ion (MPP^+), the cytotoxic metabolite of the dopaminergic neurotoxin MPTP. *Life Sci.* **46,** 379–390.

Barker, S. A., Harrison, R. E., Monti, J. A., Brown, G. B., and Christian, S. T. (1981). Identification and quantification of 1,2,3,4-tetrahydro-β-carboline, 2-methyl-1,2,3,4-tetrahydro-β-carboline, and 6-methoxy-1,2,3,4-tetrahydro-β-carboline as in vivo constituents of rat brain and adrenal gland. *Biochem. Pharmacol.* **30,** 9–17.

Beatty, P. W., and Reed, D. J. (1980). Involvement of the cystathionine pathway in the biosynthesis of glutathione by isolated rat hepatocytes. *Arch. Biochem. Biophys.* **204,** 80–87.

Bloom, F., Barchas, J., Sandler, M., and Usdin, E., eds. (1983). "Beta-Carbolines and Tetrahydroquinolines." Alan R. Liss, New York.

Boyes, J., and Bird, A. (1991). DNA methylation inhibits transcription indirectly via a methyl-CpG binding protein. *Cell (Cambridge, Mass.)* **64,** 1123–1134.

Cabrero, C., and Alemany, S. (1988). Conversion of rat liver S-adenosyl-L-methionine synthetase from high-M_r to low-M_r form by LiBr. *Biochim. Biophys. Acta* **952,** 277–281.

Cabrero, C., Duce, A. M., Ortiz, P., Alemany, S., and Mato, J. M. (1988). Specific loss of the high-molecular-weight form of S-adenosyl-L-methionine synthetase in human liver cirrhosis. *Hepatology* **8,** 1530–1534.

Cantoni, G. L. (1953). S-Adenosylmethionine: A new intermediate formed enzymatically from L-methionine and adenosinetriphosphate. *J. Biol. Chem.* **204,** 402–416.

Cedar, H. (1988). DNA methylation and gene activity. *Cell (Cambridge, Mass.)* **53,** 3–4.

Chiang, P. K., and Cantoni, G. L. (1977). Activation of methionine for transmethylation: Purification of the S-adenosylmethionine synthetase of yeast and its separation into two forms. *J. Biol. Chem.* **252,** 4506–4513.

Collins, M. A., and Neafsey, E. J. (1985). Beta-carboline analogs of N-methyl-4-phenyl-1,2,3,6-tetrahydropyridine (MPTP); Endogenous factors underlying idiopathic parkinsonism? *Neurosci. Lett.* **55,** 179–184.

Collins, M. A., Neafsey, E. J., Matsubara, K., Cobuzzi, R., Jr., and Rollema, H. (1992).

Indole N-methylated β-carbolinium ions as potential brain-bioactivated neurotoxins. *Brain Res.* **570,** 154–160.
Cox, R., and Irving, C. C. (1977). Inhibition of DNA methylation by S-adenosylethionine with production of methyl-deficient DNA in regenerating rat liver. *Cancer Res.* **37,** 222–225.
Croce, D. F., and Kimbrough, R. D. (1982). Acute and subacute toxicity in Sherman strain rats exposed to 4,4′- and 2,2′-dipyridyl. *J. Toxicol. Environ. Health* **10,** 363–372.
Crooks, P. A., Godin, C. S., Damani, L., Ansher, S., and Jakoby, W. B. (1988). Formation of quaternary amines by N-methylation of azaheterocycles with homogeneous amine N-methyltransferases. *Biochem. Pharmacol.* **37,** 1673–1677.
Davis, C. G., Williams, A. C., Markey, S. P., Ebert, M. H., Caine, E. D., Reichert, C. M., and Kopin, I. J. (1979). Chronic parkinsonism secondary to intravenous injection of meperidine analogs. *Psychiatry Res.* **1,** 249–254.
Dostert, P., Strolin Benedetti, M., Dordain, G., and Vernay, D. (1989). Enantiomeric composition of urinary salsolinol in Parkinsonian patients after Madopar. *J. Neural Transm.* **1** (P-D Sect.), 269–278.
Dostert, P., Strolin Benedetti, M., Bellotti, V., Allevi, C., and Dordain, G. (1990). Biosynthesis of salsolinol, a tetrahydroisoquinoline alkaloid, in healthy subjects. *J. Neural Transm.* **81,** 215–233.
Farber, E. (1956). Similarities in the sequence of early histological changes induced in the liver of the rat by ethionine, 2-acetylaminofluorene, and 3′-methyl-4-dimethylaminoazobenzene. *Cancer Res.* **16,** 142–148.
Farber, E. (1963). Ethionine carcinogenesis. *Adv. Cancer Res.* **7,** 383–474.
Farber, E. (1967). Ethionine fatty liver. *Adv. Lipid Res.* **5,** 119–183.
Fukushima, S., Matsubara, K., Akane, A., and Shiono, H. (1992). 1-Methyltetrahydro-β-carboline-3-carboxylic acid is present in rat brain and is not increased after acute ethanol injection with cyanamide treatment. *Alcohol* **9,** 31–35.
Gahl, W. A., Finkelstein, J. D., Mullen, K. D., Bernardini, I., Martin, J. J., Backlund, P., Ishak, K. G., Hoofnagle, J. H., and Mudd, S. H. (1987). Hepatic methionine adenosyltransferase deficiency in a 31-year-old man. *Am. J. Hum. Genet.* **40,** 39–49.
Godin, C. S., and Crooks, P. A. (1989). N-Methylation as a toxication route for xenobiotics. II. *In vivo* formation of N,N′-dimethyl-4,4′-bipyridyl ion (paraquat) from 4,4′-bipyridyl in the guinea pig. *Drug Metab. Dispos.* **17,** 180–185.
Godin, C. S., Crooks, P. A., and Damani, L. A. (1986). N-Methylation of phenylpyridines and bispyridyls as a potential toxication route: Tissue distribution of azaheterocycle N-methyltransferase activity in the rabbit. *Toxicol. Lett.* **34,** 217–222.
Heikkila, R. E., Sieber, B. A., Manzino, L., and Sonsalla, P. (1989). Some features of the nigrostriatal dopaminergic neurotoxin 1-methyl-4-phenyl-1,2,3,6-tetrahydropyridine (MPTP) in the mouse. *Mol. Cell. Neuropathol.* **10,** 171–185.
Hirata, Y., Minami, M., Naoi, M., and Nagatsu, T. (1990). Studies on the uptake of N-methylisoquinolinium ion into rat striatal slices using high-performance liquid chromatography with fluorimetric detection. *J. Chromatogr.* **503,** 189–195.
Hoffman, J. L., and Kunz, G. L. (1977). Differential activation of rat liver methionine adenosyltransferase isozymes by dimethylsulfoxide. *Biochem. Biophys. Res. Commun.* **77,** 1231–1236.
Hoffman, J. L., and Kunz, G. L. (1980). Purification and properties of the liver-specific, high K_m, positively cooperative isozyme of rat methionine adenosyltransferase. *Fed. Proc. Fed. Am. Soc. Exp. Biol.* **39,** Abst. 448.
Hoppel, C. L., Greenblatt, D., Kowk, H., Arora, P., Singh, M. P., and Sayre, L. M. (1987). Inhibition of mitochondrial respiration by analogs of 4-phenylpyridine and 1-methyl-4-

phenyl pyridinium cation (MPP^+), the neurotoxic metabolite of MPTP. *Biochem. Biophys. Res. Commun.* **148,** 684–693.

Horikawa, S., Ishikawa, M., Ozasa, H., and Tsukada, K. (1989). Isolation of a cDNA encoding the rat liver S-adenosylmethionine synthetase. *Eur. J. Biochem.* **184,** 497–501.

Horikawa, S., Sasuga, J., Shimizu, K., Ozasa, H., and Tsukada, K. (1990). Molecular cloning and nucleotide sequence of cDNA encoding the rat kidney S-adenosylmethionine synthetase. *J. Biol. Chem.* **265,** 13683–13686.

Jones, P. A., and Buckley, J. D. (1990). The role of DNA methylation in cancer. *Adv. Cancer Res.* **54,** 1–23.

Jung, M. J., and Koch-Weser, J. (1981). Enzyme-activated irreversible inhibitors of α-amino acid decarboxylases. *In* "Molecular Basis of Drug Action" (T. P. Singer and R. N. Ondarza, eds.), pp. 135–150. Elsevier/North-Holland, New York.

Kinsell, L. W., Harper, H. A., and Marton, H. C. (1947). Rate of disappearance from plasma of intravenously administered methionine in patients with liver damage. *Science* **106,** 589–590.

Kunz, G. L., Hoffman, J. L., Chia, C.-S., and Stremel, B. (1980). Separation of rat liver methionine adenosyltransferase isozymes by hydrophobic chromatography. *Arch. Biochem. Biophys.* **202,** 565–572.

Langston, J. W., Ballard, P., Tetrud, J. W., and Irwin, I. (1983). Chronic parkinsonism in humans due to a product of meperidine-analog synthesis. *Science* **219,** 979–980.

Leopold, W. R., Miller, J. A., and Miller, E. C. (1979). S-Vinyl homocysteine, an analog of ethionine that is highly mutagenic for *S. typhimurium*. *Biochem. Biophys. Res. Commun.* **88,** 395–401.

Leopold, W. R., Miller, J. A., and Miller, E. C. (1982). Comparison of some carcinogenic, mutagenic, and biochemical properties of S-vinylhomocysteine and ethionine. *Cancer Res.* **42,** 4364–4374.

Liau, M. C., Lin, G. W., and Hurlbert, R. B. (1977). Partial purification and characterization of tumor and liver S-adenosylmethionine synthetases. *Cancer Res.* **37,** 427–435.

Maret, G., Testa, B., Jenner, P., El Tayar, N., and Carrupt, P.-A. (1990). The MPTP Story: MAO activates tetrahydropyridine derivatives to toxins causing parkinsonism. *Drug Metab. Rev.* **22,** 291–332.

Markham, G. D., Hafner, E. W., Tabor, C. W., and Tabor, H. (1980). S-Adenosylmethionine synthetase from *Escherichia coli*. *J. Biol. Chem.* **255,** 9082–9092.

Markham, G. D., De Parisis, J., and Gatmaitan, J. (1984). The Sequence of MetK, the structural gene for S-adenosylmethionine synthetase in *Escherichia coli*. *J. Biol. Chem.* **259,** 14505–14507.

Maruyama, W., Nakahara, D., Ota, M., Takahashi, T., Takahashi, A., Nagatsu, T., and Naoi, M. (1992). N-Methylation of dopamine-derived 6,7-dihydroxy-1,2,3,4-tetrahydroisoquinoline, (R)-salsolinol, in rat brains: *In vivo* microdialysis study. *J. Neurochem.* **59,** 395–400.

Matsubara, K., Collins, M. A., and Neafsey, E. J. (1992a). Mono-N-methylation of 1,2,3,4-tetrahydro-β-carbolines in brain cytosol: Absence of indole methylation. *J. Neurochem.* **59,** 505–510.

Matsubara, K., Neafsey, E. J., and Collins, M. A. (1992b). Novel S-adenosylmethionine-dependent indole N-methylation of β-carbolines in brain particulate fractions. *J. Neurochem.* **59,** 511–518.

Matsumoto, C., Suma, Y., and Tsukada, K. (1984). Changes in the activities of S-adenosylmethionine synthetase isozymes from rat liver with dietary methionine. *J. Biochem. (Tokyo)* **95,** 285–290.

Miners, J. O., Drew, R., and Birkett, D. J. (1984). Mechanism of paracetamol protective agents in mice *in vivo*. *Biochem. Pharmacol.* **33,** 2995–3000.

Mozier, N. M., and Hoffman, J. L. (1990). Biosynthesis and urinary excretion of methyl sulfonium derivatives of the sulfur mustard analog, 2-chloroethyl ethyl sulfide, and other thioethers. *FASEB J.* **4,** 3329–3333.

Mozier, N. M., McConnell, K. P., and Hoffman, J. L. (1988). S-Adenosyl-L-methionine:thioether S-methyltransferase, a new enzyme in sulfur and selenium metabolism. *J. Biol. Chem.* **263,** 4527–4531.

Nagatsu, T., and Yoshida, M. (1988). An endogenous substrate of brain, tetrahydroisoquinoline, produces parkinsonism in primates with decreased dopamine, tyrosine hydroxylase, and biopterin in the nigrostriatal regions. *Neurosci. Lett.* **87,** 178–182.

Naoi, M., Takahashi, T., Parvez, H., Kabeya, R., Taguchi, E., Yamaguchi, H., Hirata, Y., Minami, M., and Nagatsu, T. (1989a). N-methylisoquinolinium ion as an inhibitor of tyrosine hydroxylase, aromatic L-amino acid decarboxylase, and monoamine oxidase. *Neurochem. Int.* **15,** 315–320.

Naoi, M., Matsuura, S., Takahashi, T., and Nagatsu, T. (1989b). An N-methyltransferase in human brain catalyzes N-methylation of 1,2,3,4-tetrahydroisoquinoline into N-methyl-1,2,3,4-tetrahydroisoquinoline, a precursor of a dopaminergic neurotoxin, N-methylisoquinolinium ion. *Biochem. Biophys. Res. Commun.* **161,** 1213–1219.

Naoi, M., Matsuura, S., Parvez, H., Takahashi, T., Hirata, Y., Minami, M., and Nagatsu, T. (1989c). Oxidation of N-methyl-1,2,3,4-tetrahydroisoquinoline into N-methylisoquinolinium ion by monoamine oxidase. *J. Neurochem.* **52,** 653–655.

Niwa, T., Takeda, N., Sasaoka, T., Kaneda, N., Hashizume, Y., Yoshizumi, H., Tatematsu, A., and Nagatsu, T. (1989). Detection of tetrahydroisoquinoline in parkinsonism as an endogenous amine by use of gas chromatography-mass spectrometry. *J. Chromatogr.* **491,** 397–403.

Niwa, T., Matsuura, S., Yoshida, M., Kawachi, M., Naoi, M., and Nagatsu, T. (1990). Endogenous synthesis of N-methyl-1,2,3,4-tetrahydroisoquinoline, a precursor of N-methylisoquinolinium ion, in the brains of primates with parkinsonism after systemic administration of 1,2,3,4-tetrahydroisoquinoline. *J. Chromatogr.* **533,** 145–151.

Niwa, T., Takeda, T., Yoshizumi, H., Tatematsu, A., Yoshida, M., Dostert, P., Naoi, M., and Nagatsu, T. (1991). Presence of 2-methyl-6,7-dihydroxy-1,2,3,4-tetrahydroisoquinoline and 1,2-dimethyl-6,7-dihydroxy-1,2,3,4-tetrahydroisoquinoline, novel endogenous amines, in parkinsonian and normal human brains. *Biochem. Biophys. Res. Commun.* **177,** 603–609.

Okada, G., Teraoka, H., and Tsukada, K. (1981). Multiple species of S-adenosylmethionine synthetase. Partial purification and characterization. *Biochemistry* **20,** 934–940.

Origitano, T., Hannigan, J., and Collins, M. (1981). Rat brain salsolinol and blood-brain barrier. *Brain Res.* **224,** 446–451.

Ozaki, N., Nakahara, D., Kaneda, N., Kiuchi, K., Okada, T., Kasahara, Y., and Nagatsu, T. (1987). Acute effects of 1-methyl-4-phenylpyridinium ion (MPP^+) on dopamine and serotonin metabolism in rat striatum as assayed *in vivo* by a microdialysis technique. *J. Neural Transm.* **70,** 241–250.

Popper, H., de la Huerga, J., and Yesnick, C. (1953). Hepatic tumors due to prolonged ethionine feeding. *Science* **118,** 80–82.

Sayre, L. M. (1989). Biochemical mechanism of action of the dopaminergic neurotoxin 1-methyl-4-phenyl-1,2,3,6-tetrahydropyridine (MPTP). *Toxicol. Lett.* **48,** 121–149.

Shivapurkar, N., Wilson, M. J., and Poirier, L. A. (1984). Hypomethylation of DNA in ethionine-fed rats. *Carcinogenesis (London)* **5,** 989–992.

Shull, K. H. (1962). Hepatic phosphorylase and adenosine triphosphate levels in ethionine-treated rats. *J. Biol. Chem.* **237,** PC1734–PC1735.

Shull, K. H., McConomy, J. M., Vogt, M., Castillo, A., and Farber, E. (1966). On the mechanism of induction of hepatic adenosine triphosphate deficiency by ethionine. *J. Biol. Chem.* **241,** 5060–5070.

Singer, T. P., and Ramsey, R. R. (1990). Mechanism of the neurotoxicity of MPTP. *FEBS Lett.* **274,** 1–8.

Sjöequist, B., Borg, S., and Kvande, H. (1981). Salsolinol and methylated salsolinol in urine and cerebrospinal fluid from healthy volunteers. *Subst. Alcohol. Actions/Misuse* **2,** 73–77.

Sjöequist, B., Eriksson, A., and Winblad, B. (1982). Salsolinol and catecholamines in human brain and their relation to alcoholism. *Prog. Clin. Biol. Res.* **90,** 57–67.

Snyder, S. H., and D'Amato, R. J. (1985). Predicting Parkinson's disease. *Nature* (*London*) **317,** 198–199.

Stirling, C. J. M. (1977). Sulfonium salts. *In* "Organic Chemistry of Sulfur" (S. Oae, ed.), pp. 473–525. Plenum, New York and London.

Stirling, C. J. M., ed. (1981). "The Chemistry of the Sulphonium Group," Parts 1 and 2. Wiley, Chichester.

Sufrin, J. R., Lombardini, J. B., and Keith, D. D. (1982). L-2-Amino-4-methoxy-cis-but-3-enoic acid, a potent inhibitor of the enzymatic synthesis of S-adenosylmethionine. *Biochem. Biophys. Res. Commun.* **106,** 251–255.

Sullivan, D. M., and Hoffman, J. L. (1982). Differential properties and regulation of methionine adenosyltransferase isoforms from the rat. *In* "Biochemistry of S-Adenosylmethionine and Related Compounds" (E. Usdin, R. T. Borchardt, and C. R. Creveling, eds.), pp. 683–686. Macmillan, London and Basingstoke.

Sullivan, D. M., and Hoffman, J. L. (1983). Fractionation and kinetic properties of rat liver and kidney methionine adenosyltransferase isozymes. *Biochemistry* **22,** 1636–1641.

Susilo, R., and Rommelspacher, H. (1987). Formation of a β-carboline (1,2,3,4-tetrahydro-1-methyl-β-carboline-1-carboxylate) following intracerebroventricular injection of tryptamine and pyruvic acid. *Arch. Pharmacol.* **335,** 70–76.

Thomas, D., and Surdin-Kerjan, Y. (1987). SAM1, the structural gene for one of the S-adenosylmethionine synthetases in *Saccharomyce cerevisiae*. Sequence and expression. *J. Biol. Chem.* **262,** 16704–16709.

Thomas, D., Rothstein, R., Rosenberg, N., and Surdin-Kerjan, Y. (1988). SAM2 encodes the second methionine S-adenosyl transferase in *Saccharomyces cerevisiae*: Physiology and regulation of both enzymes. *Mol. Cell. Biol.* **8,** 5132–5139.

Wainfan, E., and Poirier, L. A. (1992). Methyl groups in carcinogenesis: Effects on DNA methylation and gene expression. *Cancer Res., Suppl.* **52,** 2071s–2077s.

Wainfan, E., Dizik, M., Sheikhnejad, G., and Christman, J. K. (1990). Early changes in nucleic acid methylation and gene expression in livers of rats fed cancer-promoting, methyl-deficient diets. *FASEB J.* **4,** Abstr. 1043.

Warner, D. R. (1992). Cloning and active site labeling of thioether methyltransferase. Ph.D. Dissertation, University of Louisville, Louisville, KY.

Warner, D. R., Mozier, N. M., and Hoffman, J. L. (1989). Species distribution of S-adenosylmethionine:thioether S-methyltransferase. *J. Cell Biol.* **107,** Abstr. 3640.

Weinshilboum, R. (1989a). Thiol S-methyltransferases. I. Biochemistry. *In* "Sulfur-Containing Drugs and Related Organic Compounds" (L. A. Damani, ed.), Vol. 2, Part A, pp. 121–142. Ellis Horwood, Chichester.

Weinshilboum, R. (1989b). Thiol S-methyltransferases. II. Pharmacogenetics. *In* "Sulfur-Containing Drugs and Related Organic Compounds" (L. A. Damani, ed.), Vol. 2, Part A, pp. 144–157. Ellis Horwood, Chichester.

Yaswen, P., Goyette, M., Shank, P. R., and Fausto, N. (1985). Expression of c-Ki-*ras*, c-Ha-*ras*, and c-*myc* in specific cell types during hepatocarcinogenesis. *Mol. Cell. Biol.* **5,** 780–786.

Bioconversion of Prodrugs by Conjugate-Processing Enzymes

Spyridon Vamvakas* and M. W. Anders†

**Institut für Toxikologie*
Universität Würzburg
97078 Würzburg
Germany

†*Department of Pharmacology*
University of Rochester
Rochester, New York 14642

I. Introduction

A major goal in drug research is the development of tissue-selective drugs that enhance therapeutic effectiveness and minimize side effects. For successful targeting, the properties of the drug must meet specific requirements without altering the pharmacodynamic profile of the drug in the target tissue. Drug targeting has evolved into an important subdiscipline of experimental pharmacy and pharmacology, but clinical experience with targeted therapy is limited.

Two general approaches have been used to improve delivery to selected tissues:

1. In the carrier approach, the drug is covalently bound to macromolecules, i.e., monoclonal antibodies, or encapsuled in nanoparticles or liposomes. The pharmacokinetic fate of such drugs is dictated by the properties of the carrier. This approach may be advantageous if the drug-carrier moiety is preferentially taken up by target tissues compared with nontarget tissues, where toxicity may occur, and if the active drug is liberated and retained in the target tissue. The carrier approach for drug delivery has

been reviewed (Eichler, 1991; Gregoriadis, 1989; Meijer *et al.*, 1992; Speiser, 1991).

2. In the prodrug approach, a chemical derivative of the drug is synthesized that has more favorable pharmacokinetic properties or is preferentially bioconverted to the active drug in the target tissue, or both. Prodrugs are pharmacologically inactive forms of drugs that undergo spontaneous or enzyme-catalyzed conversion to pharmacologically active forms. One focus of prodrug development is to improve delivery of drugs to their sites of action and thereby improve efficacy and reduce toxicity.

Prodrugs should fulfill several requirements: the prodrug should not undergo presystemic metabolic cleavage or systemic metabolic alteration of the pharmacodynamically active moiety during the distribution phase; the prodrug should be preferentially taken up and retained in the target tissue; the prodrug should be cleaved to the pharmacodynamically active drug in the target tissue; and uptake and metabolism of the prodrug to the active drug in other tissues should be limited. Several volumes and review articles attest to the level of interest in the development of effective prodrugs (Bundgaard, 1985; Juliano, 1980; Roche, 1987).

The present chapter will deal with the prodrug approach and will be largely limited to prodrugs that are bioconverted by conjugate-processing enzymes. Drug conjugates intended to increase lipophilicity and absorption or to prolong the biological half life in plasma will not be included.

II. Bioconversion of Prodrugs by Conjugate-Processing Enzymes

A. Aminoacylases

1. *N*-Acetyl-L-cysteine

N-Acetyl-L-cysteine is used in the clinical management of acetaminophen intoxication. Its antidotal efficacy could depend on a direct reaction with electrophilic metabolites of acetaminophen or on provision of L-cysteine as a precursor for glutathione biosynthesis. These possibilities were tested by comparing the efficacy of *N*-acetyl-L-cysteine and *N*-acetyl-D-cysteine in experimental acetaminophen intoxication in mice (Corcoran and Wong, 1986). *N*-Acetyl-L-cysteine, but not the D-enantiomer, protected mice against acetaminophen-induced hepatotoxicity, indicating that *N*-acetyl-L-cysteine is a prodrug of L-cysteine, which supports glutathione biosynthesis. Similarly, *N*-acetyl-L-cysteine is more effective than the D-enantiomer as an antidote against acrylonitrile toxicity (Nerland *et al.*, 1989).

N-Acetyl-L-cysteine is biotransformed to L-cysteine in several tissues, including liver, lung, and intestine (Chasseaud, 1974; Cotgreave *et al.* 1987; Sheffner *et al.*, 1966; Sjödin *et al.*, 1989) (Fig. 1), but the enzymes catalyzing the deacetylation of *N*-acetyl-L-cysteine have not been characterized. Neither purified rat hepatic aminoacylase nor rat kidney acylase III catalyzes the deacetylation of *N*-acetyl-L-cysteine (Suzuki and Tateishi, 1981). Also, bovine liver α-*N*-acylamino acid hydrolase fails to hydrolyze *N*-acetylcysteine (Gade and Brown, 1981).

2. *N*-Acetyl-L-cysteine *S*-Conjugates

Several mercapturates of nephrotoxic cysteine *S*-conjugates are cytotoxic and nephrotoxic. These mercapturates apparently undergo an aminoacylase-catalyzed deacetylation to yield cysteine *S*-conjugates, which are bioactivated by cysteine conjugate β-lyase. This bioactivation pathway is reviewed elsewhere in this volume (see Dekant *et al.*).

B. Cysteine Conjugate β-Lyase

Cysteine conjugate β-lyase activity is high in kidney cytosol and mitochondria (for reviews, see Cooper and Anders, 1990; Cooper, this volume). The ability of the kidney to transport and accumulate L-cysteine *S*-conjugates and the high renal activity of β-lyase account for the organ-selective toxicity of a range of L-cysteine *S*-conjugates (see Dekant *et al.*, this volume). Hence the same factors that account for the organ-selective toxicity of L-cysteine *S*-conjugates may be exploited to target therapeutic agents to the kidney. β-Lyase catalyzes the β-elimination reactions of L-cysteine *S*-conjugates with pyruvate, ammonia, and thiols as products, and it is possible, in principle, to target thiol-based therapeutic agents to the kidney.

Renal cell carcinomas account for 2 to 3% of all malignant neoplasms in adults, and the incidence has increased over the last 60 years (Dayal and Wilkinson, 1989; Jensen *et al.*, 1990). Chemotherapeutic agents effective in the therapy of renal cell carcinomas are not available, and surgical

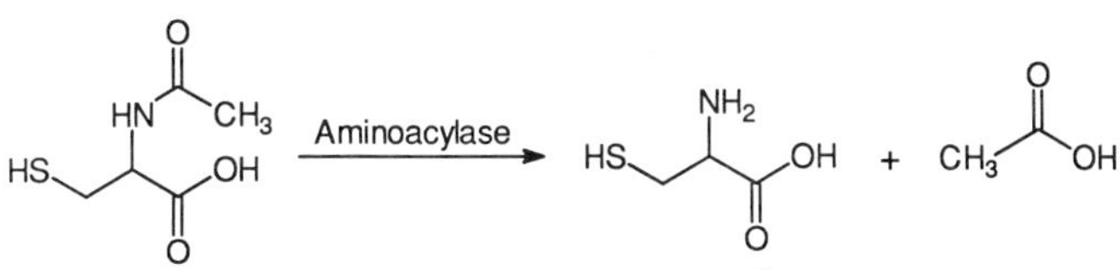

1 2

Fig. 1 Aminoacylase-catalyzed bioconversion of *N*-acetyl-L-cysteine **1** to L-cysteine **2**.

intervention is the mainstay of the treatment of renal cell carcinoma. Compared with other types of cancer, renal carcinoma cells display low proliferative rates, and only a small percentage of the cell population is in the S-phase of the cell cycle at a given time. Hence, for most chemotherapeutic agents to be effective, they must be continuously present in the kidney. This requirement cannot be met by increasing the dose or the dose frequency, or both, because this clinical approach would be expected to cause severe side effects and toxicity to nontarget tissues. Selective targeting of chemotherapeutic agents to the kidney is an attractive goal to overcome these problems.

The elucidation of the multistep bioactivation pathway responsible for the nephrotoxicity and nephrocarcinogenicity of halogenated alkenes gave rise to the synthesis of L-cysteine-based prodrugs. This development is based on the observation that *S*-alkyl-, *S*-alkenyl-, and *S*-aryl-substituted L-cysteines are substrates for the β-lyase. Hwang and Elfarra (1989, 1991) prepared *S*-(6-purinyl)-L-cysteine as a potential β-lyase substrate, the expectation being that *S*-(6-purinyl)-L-cysteine would yield the antitumor and immunosuppressant drug 6-mercaptopurine upon β-lyase-catalyzed cleavage (Fig. 2). After giving *S*-(6-purinyl)-L-cysteine to rats, kidney concentrations of 6-mercaptopurine and its metabolites 6-methylmercaptopurine and 6-thiouric acid were nearly 90-fold higher than in plasma and 2.3-fold higher than in liver. Probenecid and aminooxyacetic acid, inhibitors of the renal organic acid transport system and of β-lyase, respectively, blocked the accumulation of metabolites in the kidney and supported a role for β-lyase in the metabolism of *S*-(6-purinyl)-L-cysteine. Furthermore, the rate of biotransformation of *S*-(6-purinyl)-L-cysteine with renal mitochondrial β-lyase was nearly 3-fold higher than with the corresponding liver enzyme (Hwang and Elfarra, 1991).

The acute nephrotoxicity of *S*-(6-purinyl)-L-cysteine in rats was evaluated by measuring blood urea nitrogen concentrations, urine volume, and plasma glucose concentrations 24 h after 1200 μmol/kg was given; this

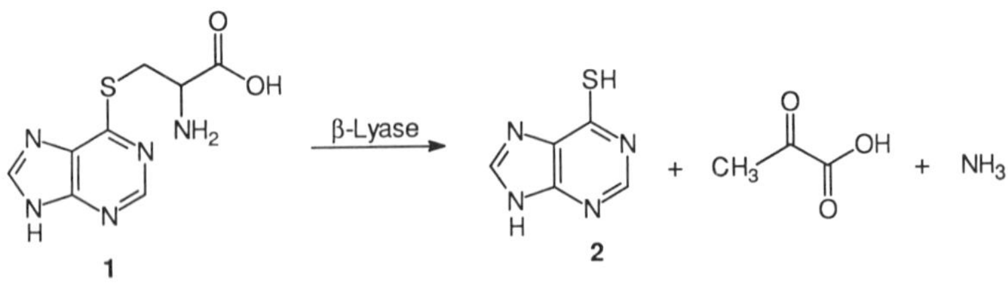

Fig. 2 Cysteine conjugate β-lyase-catalyzed bioconversion of *S*-(purinyl)-L-cysteine **1** to 6-mercaptopurine **2**.

treatment did not induce toxic alterations in the above parameters. Studies to assess fully the acute, subchronic, and chronic toxicity of *S*-(6-purinyl)-L-cysteine are required, as are studies to compare the efficacy of *S*-(6-purinyl)-L-cysteine and 6-mercaptupurine. Nevertheless, *S*-(6-purinyl)-L-cysteine is an important first step in the development of prodrugs that undergo β-lyase-catalyzed bioconversion.

C. γ-Glutamyltransferase and Dipeptidases

1. Prodrugs

Prodrugs and protoxins that contain a L-γ-glutamyl or *N*-acyl-L-γ-glutamyl moiety have received considerable attention. *N*-Acyl-L-γ-glutamyl compounds require hydrolysis to L-γ-glutamyl compounds, which may be hydrolyzed by γ-glutamyltransferase (EC 2.3.2.2). γ-Glutamyltransferase is highly selective for γ-glutamyl groups and variably selective for substituents at the α-carboxyl group. Magnan *et al.* (1982) found considerable variability in the rates of hydrolysis of a range of *N*-L-γ-glutamyl derivatives; for example, *N*-L-γ-glutamyl-2-aminobutyric acid was readily cleaved, but *N*-L-γ-glutamyl-4-aminobutyric acid (γ-glutamyl-GABA) was a poor substrate. The enzyme is selectively distributed with high activities in certain epithelial tissues, such as renal proximal tubular cells (Hinchman and Ballatori, 1990). Hence several attempts have been made to develop drugs targeted to the kidney. Moreover, the observation that some tumors contain high γ-glutamyltransferase activity has led to attempts to exploit this activity to target drugs to tumor cells.

a. *N*-(γ-Glutamyl)-L-3,4-dihydroxyphenylalanine (γ-Glutamyl-dopa) and *N*-(γ-Glutamyl)-3,4-dihydroxyphenethylamine (γ-Glutamyl-dopamide) The effects of low doses of dopamine, which include increases in renal blood flow, glomerular filtration, and urinary sodium excretion in man, are mediated by renal dopamine receptors. In contrast, high doses of dopamine activate α- and β-adrenergic receptors, causing an increase in peripheral resistance and blood pressure (Goldberg, 1972). Hence a derivative of dopamine that is selectively accumulated in the kidney may have advantages over dopamine itself. Thus attempts have been made to target dopamine to the kidney by development of γ-glutamyl derivatives of dopamine and dopamine precursors, including γ-glutamyl-dopamide (Minard *et al.*, 1980) and γ-glutamyl-dopa (Wilk *et al.*, 1978).

γ-Glutamyl-dopa (Fig. 3, **1**) was developed with the expectation that it would be sequentially converted by γ-glutamyltransferase and L-dopa decarboxylase to dopamine (Fig. 3, **4**). The renal concentration of dopamine after giving γ-glutamyl-dopa to rats and mice was almost 5 times

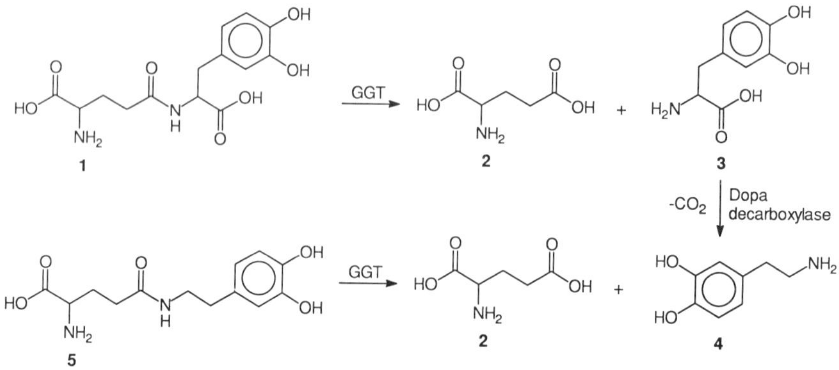

Fig. 3 γ-Glutamyltransferase-catalyzed bioconversion of γ-glutamyl-dopa **1** to L-glutamate **2** and 3,4-dihydroxyphenylalanine **3** and of γ-glutamyl-dopamide **5** to L-glutamate **2** and dopamine **3**. 3,4-Dihydroxyphenylalanine **3** is converted to dopamine **4** by dopa decarboxylase. GGT, γ-glutamyltransferase.

higher than that seen after giving an equivalent dose of L-dopa. Moreover, γ-glutamyl-dopa increased renal blood flow, glomerular filtration, and urinary sodium excretion in both experimental animals and man, whereas equimolar doses of L-dopa failed to produce these effects. An additional advantage of the prodrug is that the dose of γ-glutamyl-dopa required to induce a pressor effect by activation of adrenergic receptors is approximately 20 times higher than the renal vasodilating and natriuretic dose. Renal tubular receptors for dopamine may be upregulated in essential hypertension in man, and prolonged infusion of γ-glutamyl-dopa induces a greater fall in blood pressure in hypertensives than in normotensives (Lee, 1990). Unfortunately, when γ-glutamyl-dopa is given orally, it is largely converted to 3,4-dihydroxyphenylacetic acid by the successive actions of γ-glutamyltransferase, dopa decarboxylase, and monoamine oxidase in the gut and liver. Its mean oral bioavailability is less than 2%, indicating that oral γ-glutamyl-dopa does not have therapeutic potential as an oral diuretic and antihypertensive agent.

γ-Glutamyl-dopamide (Fig. 3, **5**) has also been studied as a prodrug of dopamine (Fig. 3, **4**) (Minard *et al.*, 1980). γ-Glutamyl-dopamide was hydrolyzed by γ-glutamyltransferase, as shown by stimulation by glycylglycine and inhibition by serine-borate, to afford dopamine as a product. Rats given γ-glutamyl-dopamide intraperitoneally showed higher renal dopamine concentrations than rats given dopamine itself or L-DOPA. Intragastric administration of γ-glutamyl-dopamide increased renal dopamine concentrations, but higher renal dopamine concentrations were observed in rats given L-DOPA or γ-glutamyl-dopamide *n*-octyl ester. γ-

Glutamyl-dopamide given intravenously to dogs produced less cardiac stimulation than dopamine, indicating a kidney-selective release of dopamine from γ-glutamyl-dopamide.

b. γ-Glutamyl-sulfamethoxazole The high renal activities of aminoacylase and γ-glutamyltransferase have been exploited in an attempt to target sulfamethoxazole to the kidney (Orlowski *et al.*, 1980). γ-Glutamyl and *N*-acyl-γ-glutamyl derivatives of sulfamethazole have been prepared and tested in mice. Sulfamethoxazole was released enzymatically from γ-glutamyl-sulfamethoxazole (Fig. 4) and from *N*-acetyl-, *N*-butyryl-, *N*-chloroacetyl-, and *N*-glycyl-L-γ-glutamyl-sulfamethoxazole in kidney homogenates. Similarly, sulfamethoxazole was formed when L-γ-, D-γ-, or L-α-glutamyl-sulfamethoxazole was incubated with kidney homogenates. Some preferential accumulation of sulfamethoxazole in the kidney was observed when the prodrugs were given to mice, but relatively high concentrations of sulfamethoxazole were found in other tissues. Kidney-selective accumulation of sulfamethoxazole was observed, however, when *N*-acyl derivatives were given, indicating that selectivity is improved when two renal enzymes are involved in the bioconversion of the prodrugs.

c. *p*-[Bis-(2-chloroethyl)amino]-L-γ-glutamylanilide (γ-Glutamyl-*p*-phenylenediamine Mustard) Attempts have been made to target alkylating agents to tumor cells with high γ-glutamyltransferase activities. γ-Glutamyl-*p*-phenylenediamine mustard is a substrate for γ-glutamyltransferase (Fig. 5) and is more cytotoxic to JB1 cells, which express high γ-glutamyltransferase activities, than to BL8L cells, which have lower activities (Manson *et al.*, 1981; Smith *et al.*, 1984).

d. L-γ-Glutaminyl-4-hydroxybenzene Several phenolic amines have antimelanoma activity, which is apparently related to the presence of tyrosinase in melanocytes that oxidizes the phenolic amines to cytotoxic

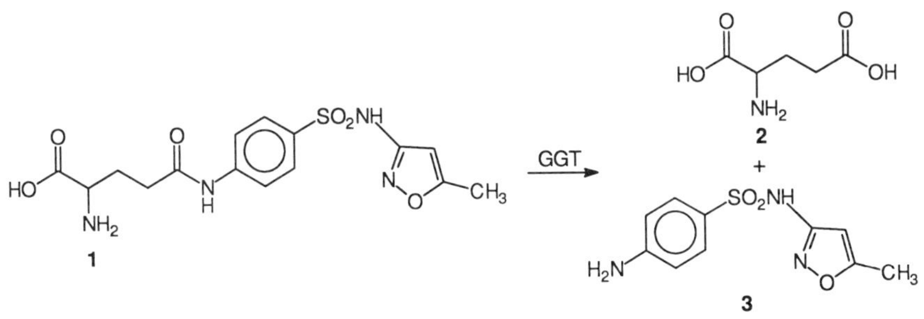

Fig. 4 γ-Glutamyltransferase-catalyzed bioconversion of γ-glutamyl-sulfamethoxazole **1** to L-glutamate **2** and sulfamethoxazole **3**. GGT, γ-glutamyltransferase.

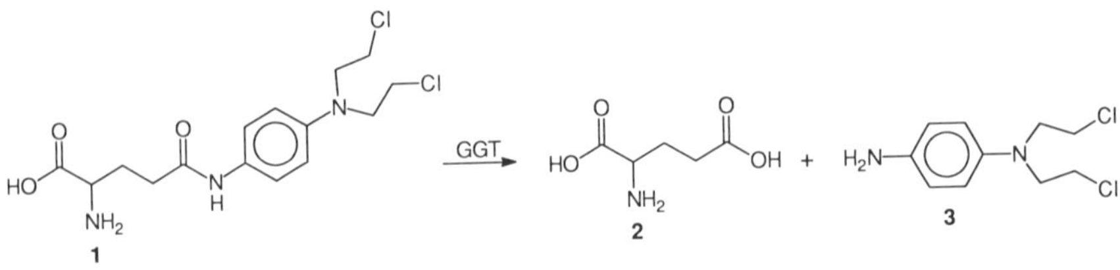

Fig. 5 γ-Glutamyltransferase-catalyzed bioconversion of γ-glutamyl-*p*-phenylenediamine mustard **1** to L-glutamate **2** and *p*-phenylenediamine mustard **3**. GGT, γ-glutamyltransferase.

intermediates. L-γ-Glutaminyl-4-hydroxybenzene occurs in the mushroom *Agaricus bisporus* and is active against animal models of melanoma (Vogel *et al.*, 1977, 1979). The mechanism of action includes γ-glutamyltransferase-catalyzed hydrolysis and tyrosinase-catalyzed oxidation of aminophenols to cytotoxic products (Boekelheide *et al.*, 1980; Prezioso *et al.*, 1993) (Fig. 6). The cytotoxicity of the analog L-γ-glutaminyl-4-hydroxy-3-iodobenzene in melanoma cells correlates with γ-glutamyltransferase activities (Prezioso *et al.*, 1993).

2. Protoxins

Several γ-glutamyl-based protoxins, which occur as natural products or as synthetic materials, are known and will be discussed briefly. γ-Glutamyltransferase- and dipeptidase-catalyzed processing of glutathione conjugates affords *S*-conjugates that may undergo bioactivation by cysteine conjugate β-lyase; this bioactivation mechanism is discussed elsewhere in this volume (see Dekant *et al.*) and will not be reviewed here.

γ-Glutamyl amino acids are metabolized in the kidney and certain other epithelial tissues (Orlowski and Wilk, 1976). Accordingly, γ-glutamyl derivatives of nephrotoxic cysteine *S*-conjugates would be expected to be

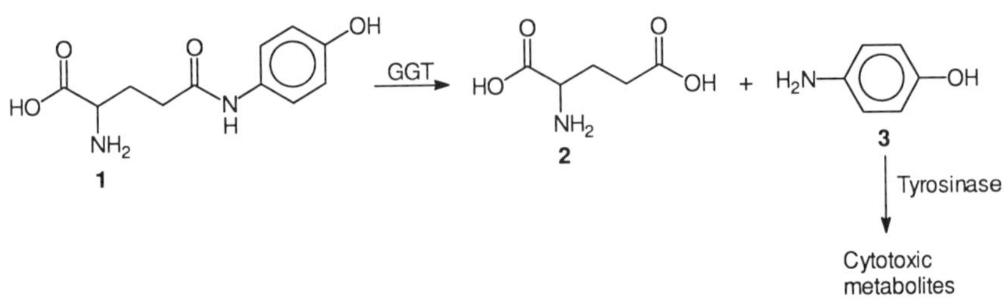

Fig. 6 γ-Glutamyltransferase-catalyzed bioconversion of L-γ-glutaminyl-4-hydroxybenzene **1** to L-glutamate **2** and 4-aminophenol **3**, which is converted to cytotoxic metabolites by tyrosinase. GGT, γ-glutamyltransferase.

delivered to the kidney where the cysteine *S*-conjugate should be released and available for bioactivation by cysteine conjugate *β*-lyase.

Ridgewell *et al.* (1992) prepared *N*-(L-γ-glutamyl)-*S*-(1,2-dichlorovinyl)-L-cysteine (Fig. 7, **1**) and compared its toxicity with that of *S*-(1,2-dichlorovinyl)-L-cysteine (Fig. 7, **3**) in the pentobarbital-anesthetized dog. Both compounds were nephrotoxic and produced similar patterns of renal damage. No evidence of extrarenal toxicity was reported, indicating efficient targeting of *N*-(L-γ-glutamyl)-*S*-(1,2-dichlorovinyl)-L-cysteine (Fig. 7, **1**) to the kidney. The toxicity of *S*-(1,2-dichlorovinyl)-L-cysteine is associated with its cysteine conjugate *β*-lyase-dependent bioactivation to dichloroethenethiol (Fig. 7, **4**) (see Dekant *et al.,* this volume).

Hypoglycin A [L-(methylenecyclopropyl)alanine] (Fig. 8, **3**) and hypoglycin B [L-γ-glutamyl-L-(methylenecyclopropyl)alanine] (Fig. 8, **1**) are found in the unripe fruit of the Jamaican ackee tree *Blighia sapida* (Hassall and John, 1959; Hassall and Reyle, 1955). Hypoglycins are toxic to humans and caused an estimated 5000 deaths between 1886 and 1950 (Sherratt, 1986). Presumably hypoglycin B is toxic because it undergoes γ-glutamyltransferase-catalyzed conversion to hypoglycin A (Fig. 8). Hypoglycin A is biotransformed to methylenecyclopropylacetic acid (Fig. 8, **4**), which is a suicide inhibitor of fatty acid *β*-oxidation. The hallmarks of hypoglycin intoxication are severe hypoglycemia, vomiting, depletion of liver glycogen concentrations, fatty liver, elevated plasma fatty acid concentrations, dicarboxylic aciduria, and organic acidemia, which are associated with hypoglycin-induced alterations in fatty acid metabolism (Sherratt, 1986; Stewart and Hanley, 1969).

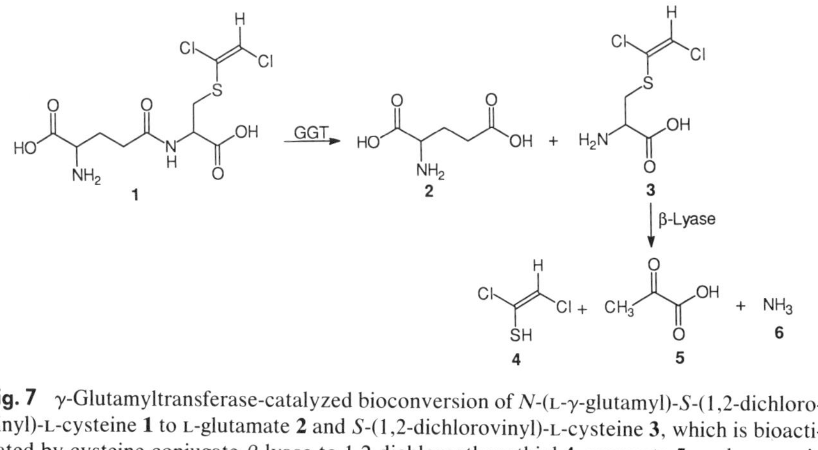

Fig. 7 γ-Glutamyltransferase-catalyzed bioconversion of *N*-(L-γ-glutamyl)-*S*-(1,2-dichlorovinyl)-L-cysteine **1** to L-glutamate **2** and *S*-(1,2-dichlorovinyl)-L-cysteine **3**, which is bioactivated by cysteine conjugate *β*-lyase to 1,2-dichloroethenethiol **4**, pyruvate **5**, and ammonia **6**. GGT, γ-glutamyltransferase.

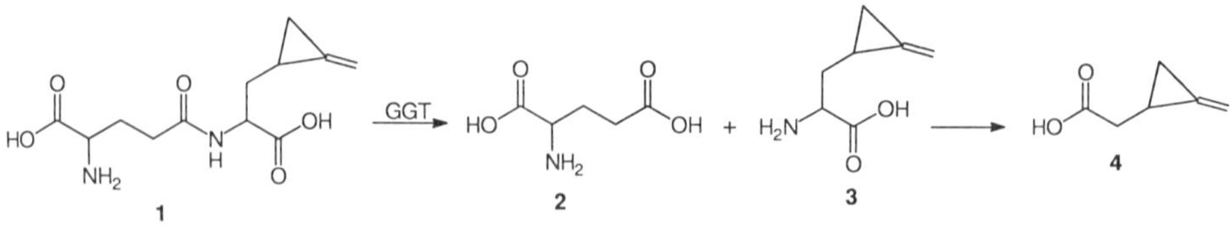

Fig. 8 γ-Glutamyltransferase-catalyzed bioconversion of hypoglycin B (L-γ-glutamyl-L-(methylenecyclopropyl)alanine, **1**) to L-glutamate **2** and hypoglycin A (L-(methylenecyclopropyl)alanine, **3**). **3** is converted to methylenecyclopropylacetic acid **4**, which is a suicide inhibitor of fatty acid β-oxidation. GGT, γ-glutamyltransferase.

D. Amino- and Carboxypeptidases

Aminopeptidases catalyze the cleavage of peptides containing a free amino terminal group. The enzymes, like γ-glutamyltransferase, will accept a range of peptides and peptide-based compounds as substrates and have been used in an attempt to develop prodrugs.

1. Aminopeptidases

Aminopeptidases catalyze the hydrolysis of peptides containing a free amino group. *N,N*-Bis(2-chloroethyl)-*N'*-glycyl-*p*-phenylenediamine (Fig. 9, **1**) and related *N*-acyl derivatives were prepared as prodrugs of the nitrogen mustard *N,N*-bis(2-chloroethyl)-*p*-phenylenediamine (Fig. 9, **3**) (Dalton and Hebborn, 1965). *N,N*-Bis(2-chloroethyl)-*N'*-glycyl-*p*-phenylenediamine was efficiently hydrolyzed (Fig. 9), particularly by kidney homogenates, whereas hydrolysis was much slower with the *N*-acetyl, *N*-dichloroacetyl, and *N*-methylcarbamate analogs. For the compounds studied, there was a linear relationship between the rate of hydrolysis and the LD_{50} in rats. The therapeutic index (LD_{50}/ED_{90}), however, was similar for *N,N*-bis(2-chloroethyl)-*N'*-glycyl-*p*-phenylenediamine (Fig. 9, **1**) and *N,N*-bis(2-chloroethyl)-*p*-phenylenediamine (Fig. 9, **3**), but was improved in the *N*-acetyl derivative.

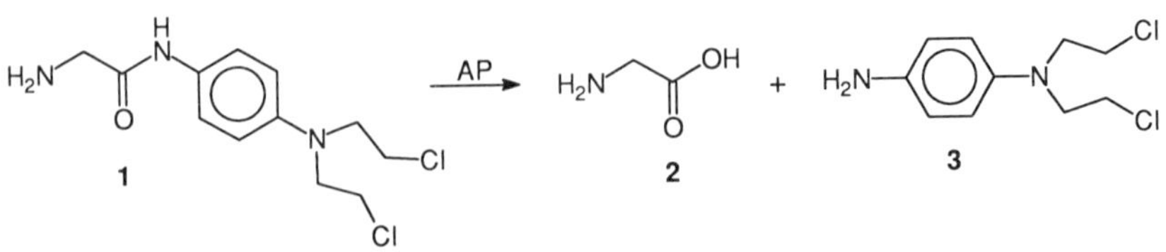

Fig. 9 Aminopeptidase-catalyzed bioconversion of *N,N*-bis(2-chloroethyl)-*N'*-glycyl-*p*-phenylenediamine **1** to glycine **2** and *N,N*-bis(2-chloroethyl)-*p*-phenylenediamine **3**.

2. Carboxypeptidases

Carboxypeptidases catalyze the hydrolysis of peptides containing a free carboxylic acid group and can be used to bioconvert peptide-based prodrugs. Springer *et al.* (1990) prepared prodrugs of bifunctional alkylating agents in which the activating effect of the carboxyl group was blocked by an amide bond to the α-amino group of L-glutamic acid. The activation of the prodrugs by carboxypeptidase G2 was studied. The cytotoxicity of 4-[bis[2-(mesyloxy)ethyl]amino]benzoyl-L-glutamic acid, 4-[(2-chloroethyl)[2-(mesyloxy)ethyl]amino]benzoyl-L-glutamic acid, and 4-[bis(2-chloroethyl)amino]benzoyl-L-glutamic acid was increased in two human cell lines after incubation with carboxypeptidase G2 (Fig. 10).

E. Oxoprolinase

Because of the cytoprotective role of glutathione, there is much interest in manipulating cellular glutathione concentrations to test hypotheses about the role of glutathione in xenobiotic-induced cell damage. Moreover, restoration or elevation of cellular glutathione concentrations is important in the therapy of intoxications, particularly in acetaminophen intoxication. Reviews about cysteine and glutathione delivery systems have appeared (Meister, 1985, 1988; Meister *et al.*, 1986).

1. L-2-Oxothiazolidine-4-carboxylate and Related Compounds

Because L-cysteine is required for glutathione biosynthesis, considerable attention has been given to prodrugs that release L-cysteine. As was discussed above, *N*-acetyl-L-cysteine is used to support the synthesis of glutathione in the clinical management of acetaminophen intoxication. 5-Oxoprolinase, an enzyme in the γ-glutamyl cycle (Meister and Anderson,

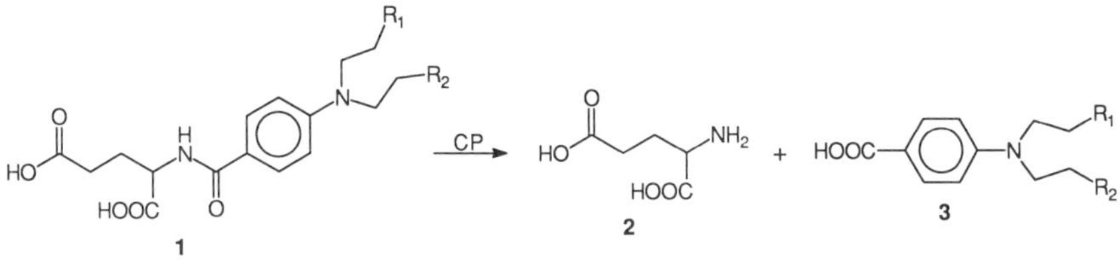

Fig. 10 Carboxypeptidase-catalyzed bioconversion of 4-[bis[2-(mesyloxy)ethyl]amino]-benzoyl-L-glutamic acid (**1**, $R_1 = R_2 = CH_3SO_2O$—), 4-[(2-chloroethyl)[2-(mesyloxy)ethyl]amino]benzoyl-L-glutamic acid (**1**, $R_1 = ClCH_2CH_2$—, $R_2 = CH_3SO_2O$—), and 4-[bis(2-chloroethyl)amino]benzoyl-L-glutamic acid (**1**, $R_1 = R_2 = ClCH_2CH_2$—) to L-glutamate **2** and the corresponding 4-carboxyphenyl mustards **3**.

1983; Meister *et al.*, 1986), catalyzes the enzymatic conversion of 5-oxo-L-proline to L-glutamate, which is coupled to the hydrolysis of ATP to ADP (Van Der Werf *et al.*, 1971). L-2-Oxothiazolidine-4-carboxylate, an analog of 5-oxo-L-proline in which sulfur is substituted for the 4-methylene carbon, and ATP are also substrates for the enzyme and yield L-cysteine and ADP as products (Williamson and Meister, 1981, 1982) (Fig. 11, **1** → **2**). Moreover, administration of L-2-oxothiazolidine-4-carboxylate to fasted mice increases hepatic glutathione concentrations to about twice those of controls (Williamson and Meister, 1981) and protects against acetaminophen-induced lethality (Hazelton *et al.*, 1986; Williamson *et al.*, 1982). L-2-Oxothiazolidine-4-carboxylate administration is reported to increase brain cysteine, but not glutathione, concentrations (Anderson and Meister, 1989); however, others have reported an increase in brain glutathione concentrations after giving L-2-oxothiazolidine-4-carboxylate (Mesina *et al.*, 1989).

Whereas L-2-oxothiazolidine-4-carboxylate requires the action of 5-oxoprolinase for conversion to L-cysteine, 2-alkyl-substituted analogs liberate L-cysteine at physiological temperature and pH without enzymatic intervention (Nagasawa *et al.*, 1982) (Fig. 11, **3** → **2**; R = CH_3—, C_3H_7—, C_6H_5—, C_5H_4N—). 2(*R,S*)-Methylthiazolidine-4(*R*)-carboxylate is as effective as *N*-acetyl-L-cysteine and 2-oxothiazolidine-4(*R*)-carboxylate in protecting against acetaminophen-induced hepatotoxicity and lethality, but is less toxic than 2-oxothiazolidine-4(*R*)-carboxylate. These studies were extended to include thiazolidine-4-carboxylates with a range of substituents at the 2-position (Nagasawa *et al.*, 1984). 2(*R,S*)-Methyl,

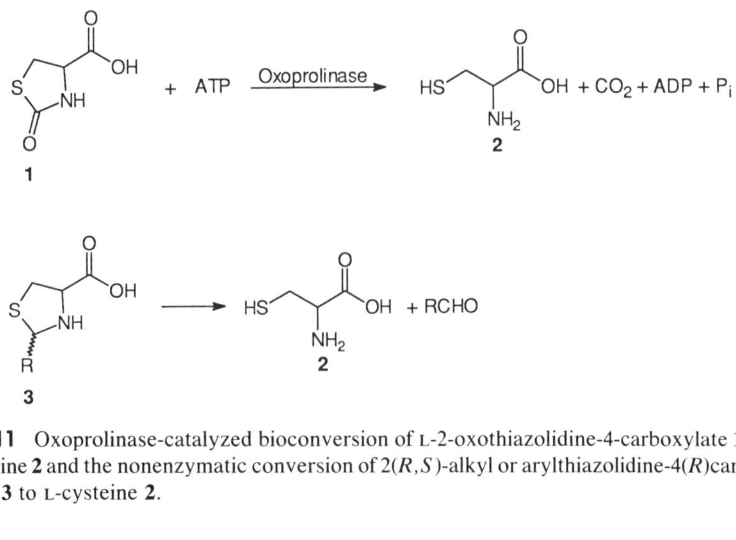

Fig. 11 Oxoprolinase-catalyzed bioconversion of L-2-oxothiazolidine-4-carboxylate **1** to L-cysteine **2** and the nonenzymatic conversion of 2(*R,S*)-alkyl or arylthiazolidine-4(*R*)carboxylates **3** to L-cysteine **2**.

2(*R,S*)-*n*-propyl-, and 2(*R,S*)-*n*-pentylthiazolidine-4(*R*)-carboxylates were almost equipotent in protecting against acetaminophen-induced liver damage, whereas the 2(*R,S*)-ethyl-, 2(*R,S*)-phenyl-, and 2-(*R,S*)-(4-pyridyl)thiazolidine-4-carboxylates were less effective, and 2(*R,S*)-methylthiazolidine-4(*S*)-carboxylate was ineffective. A series of 2-(polyhydroxyalkyl)- and 2-(polyacetoxyalkyl)thiazolidine-4(*R*)-carboxylates have also been prepared as prodrugs for L-cysteine (Roberts *et al.*, 1987). 2(*R,S*)-D-Ribo-(1′,2′,3′,4′-tetrahydroxybutyl)thiazolidine-4(*R*)-carboxylate (RibCys), which was prepared from D-ribose and L-cysteine, was the most effective analog in elevating glutathione concentrations in isolated hepatocytes and in protecting against acetaminophen-induced hepatotoxicity. The 2-(polyacetoxyalkyl)thiazolidine-4-(*R*)-carboxylates studied proved to be toxic. RibCys also elevated tissue glutathione concentrations in mice (Roberts and Francetic, 1991a,b).

F. *β*-Glucuronidase

The observation that some tumors contain high *β*-glucuronidase activities has led to attempts to exploit this activity in developing glucuronide-based prodrugs. Connors and Whisson (1966) observed a good correlation between the therapeutic efficacy of aniline mustard and tumor-cell *β*-glucuronidase activity. They proposed that aniline mustard ((*N,N*-di-2-chloroethyl)aniline) was metabolized in the liver to the highly cytotoxic *p*-hydroxyaniline mustard, which was, in turn, metabolized to the glucuronide. In tumor cells, the glucuronide may be hydrolyzed to afford the cytotoxic *p*-hydroxyaniline mustard (Fig. 12). Hence this is an example of a biosynthetic prodrug, whose selectivity resides in the high *β*-glucuronidase activities expressed in tumor cells. The results of clinical treatment of patients with aniline mustard have been disappointing, but some evidence indicating a partial correlation between glucuronidase activity in tumor tissue and tumor regression and that clinical relapse was associated with loss of *β*-glucuronidase activity has been obtained (Kyle *et al.*, 1973; Young *et al.*, 1976).

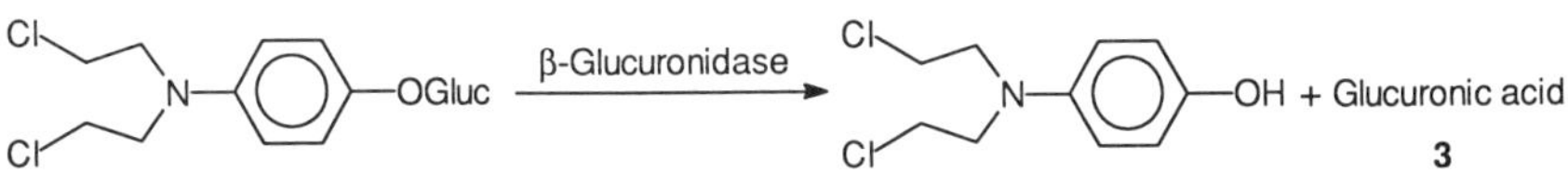

1 2 3

Fig. 12 *β*-Glucuronidase-catalyzed bioconversion of *p*-hydroxyaniline mustard glucuronide **1** to *p*-hydroxyaniline mustard **2** and glucuronic acid **3**.

This problem has recently been reinvestigated, and a glucuronide prodrug/enzyme–monoclonal antibody system has been developed to target *p*-hydroxyaniline mustard to tumor cells (Roffler *et al.*, 1991). β-Glucuronidase was conjugated to a monoclonal antibody that binds to COLO 205 human colon carcinoma cells but does not bind to WISH cells, and *p*-hydroxyaniline mustard glucuronide was prepared by synthesis. Although *p*-hydroxyaniline mustard was highly cytotoxic to both COLO 205 and WISH cells, *p*-hydroxyaniline mustard glucuronide was 30 to 50 times less toxic than *p*-hydroxyaniline mustard; incubation of COLO 205 or WISH cells with *p*-hydroxyaniline mustard glucuronide and β-glucuronidase gave cytotoxicity comparable with that obtained with *p*-hydroxyaniline mustard (Fig. 12). These results were extended to HepG2 human hepatoma and AS-30D rat hepatoma cells (Wang *et al.*, 1992). Again, the cytotoxicity of *p*-hydroxyaniline mustard was similar to the cytotoxicity of *p*-hydroxyaniline mustard glucuronide in the presence of β-glucuronidase, and *p*-hydroxyaniline mustard glucuronide was about 1000 and 150 times less cytotoxic to AS-30D and HepG2, respectively, than *p*-hydroxyaniline mustard. Incubation of a conjugate of β-glucuronidase to a monoclonal antibody against a cell-surface protein present in AS-30D cells reduced the IC_{50} of *p*-hydroxyaniline mustard glucuronide about 1000-fold. These interesting results indicate that targeting of conjugate-processing enzymes to tumor cells along with administration of conjugates of chemotherapeutic agents may allow development of prodrugs for cancer chemotherapy.

G. Esterases

1. Glutathione Esters

As discussed above, cysteine prodrugs, such as *N*-acetyl-L-cysteine, L-2-oxothiazolidine-4-carboxylate, and 2(*R*,*S*)-alkylthiazolidine-4(*R*)-carboxylates, have been developed to restore or increase cellular glutathione concentrations. Although such drugs have demonstrated utility, the increase in cellular glutathione concentrations may be limited by feedback inhibition of γ-glutamylcysteine synthetase by glutathione (Richman and Meister, 1975). Because intact glutathione is apparently not taken up by cells, attempts were made to develop a derivative of glutathione that would be taken up by cells and release glutathione.

Administration of L-γ-glutamyl-L-cysteinylglycyl monomethyl (glutathione monomethyl ester) or monoethyl (glutathione monoethyl ester) ester led to increases in liver and kidney glutathione concentrations in fasted mice and in mice given buthionine sulfoximine, indicating that glutathione esters are converted to glutathione and that synthesis of glutathione was not involved in the observed enhancement of liver and kidney glutathione concentrations (Puri and Meister, 1983) (Fig. 13). In addition, glutathione monomethyl ester increased hepatic glutathione concentrations in acetaminophen-treated mice and protected against acetaminophen-induced lethality. Studies demonstrating the uptake of glutathione esters into erythrocytes, lymphoid cells, and fibroblasts followed by conversion to glutathione have appeared (Anderson *et al.*, 1985; Wellner *et al.*, 1984). In addition to liver and kidney, glutathione esters increase glutathione concentrations in spleen, pancreas, and heart, but not in brain (Anderson *et al.*, 1985). Glutathione isopropyl ester protects newborn rats and mice against buthionine sulfoximine-induced cataracts and lens epithelial cell damage (Mårtensson *et al.*, 1989) and against acetaminophen- or allyl alcohol-induced liver damage (Uhlig and Wendel, 1990).

2. Carbamates

Thiols, such as glutathione, react reversibly with isocyanates and isothiocyanates. The thiocarbamate and dithiocarbamate conjugates thus formed can be considered reversible prodrugs; the metabolic and toxicological properties of thiocarbamates and dithiocarbamates are reviewed elsewhere in this volume (see Baillie and Kassahum).

Esters of polar drugs have been prepared to increase lipophilicity and absorption. This strategy is, however, difficult to exploit, because esterases are widely distributed and relatively nonselective. Carboxylesterases are mostly membrane bound and are abundant in the endoplasmic reticulum of the liver. Acetylcholinesterase is also predominantly membrane bound, whereas arylesterases and nonspecific cholinesterases are soluble enzymes, and activities are high in the plasma. The physiological substrates are known only for a few esterases, including acetylcholinesterase and certain lipases.

The targeting of ester conjugates to nonspecific cholinesterases has recently gained interest. Nonspecific cholinesterase is the key enzyme involved in the design of the bronchodilator bambuterol, which is the bis-*N*,*N*-dimethylcarbamate derivative of the bronchodilator terbutaline (Svensson and Tunek, 1988).

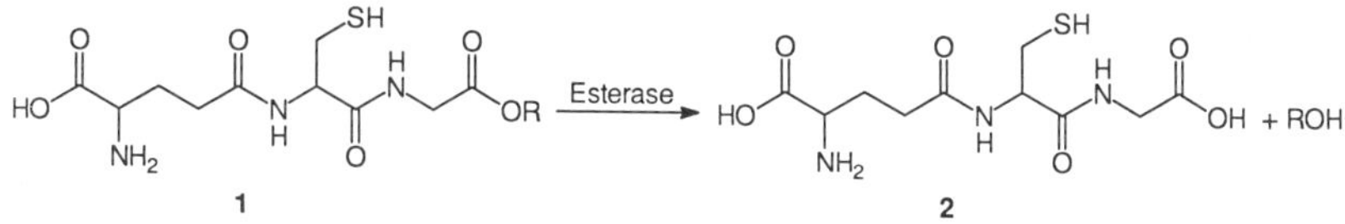

Fig. 13 Esterase-catalyzed bioconversion of glutathione monoester **1** to glutathione **2**.

Terbutaline (Fig. 14, **2**) is an effective bronchodilator, provided that it is absorbed and reaches the lung before biotransformation occurs in the gut wall and in the liver. Terbutaline has major pharmacokinetic limitations: due to its high hydrophilicity, gastrointestinal absorption after oral administration ranges from 25 to 80% of the dose. In addition, up to 70% of the absorbed terbutaline undergoes first-pass conjugation with sulfate, resulting in bioavailability between 7 and 26%. Carbamylation of the phenolic hydroxyl groups of the resorcinol moiety, which undergo sulfate-conjugate formation, should increase lipophilicity and absorption and also reduce the first-pass metabolic inactivation of terbutaline. Hence terbutaline was converted to its dimethylcarbamate derivative (Fig. 14, **1**). This strategy was designed to exploit the known reversible inhibition of nonspecific cholinesterases by carbamates. Moreover, carbamates are generally poor substrates for the nonspecific carboxylesterases, which are concentrated in the intestine and liver. The overall result should be a retardation of the formation of the active drug due to the slow enzymatic hydrolysis of the carbamate. Furthermore, inhibition of nonspecific cholinesterases by bambuterol is dose-dependent; consequently, increasing doses of bambuterol should lead to less terbutaline formation.

Bambuterol showed both prolongation of the pharmacologic effect and increased site selectivity compared with terbutaline. In addition to hydrolysis, bambuterol is also oxidized at the carbamate methyl groups by hepatic cytochromes P450-dependent monooxygenases. Because the terbutaline moiety of bambuterol is not attacked by hepatic monooxygenases, the hydroxylated metabolites formed in this pathway are still prodrugs that may be cleaved to give terbutaline. Bambuterol is concentrated sevenfold in quinea pig lung tissue, whereas lung uptake of the less lipophilic terbutaline is significantly lower (Svensson, 1991).

Results obtained in clinical studies confirmed the advantageous properties of bambuterol. After a single daily dose of bambuterol is given, the bronchodilating effect is sustained for the entire 24-h dose interval. Compared with patients given terbutaline, plasma terbutaline concentrations are lower after oral administration of bambuterol. Because side effects,

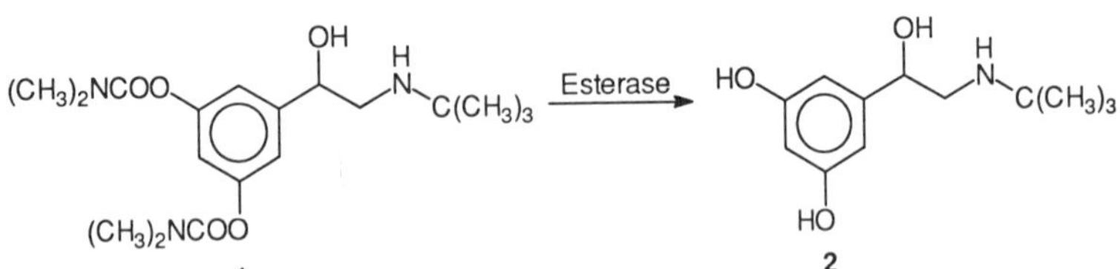

1 2

Fig. 14 Esterase-catalyzed bioconversion of bambuterol **1** to terbutaline **2**.

such as tremor and palpitations, are proportional to the plasma concentrations of the β-mimetic agent, undesirable effects should be less frequent with bambuterol; this remains to be confirmed in long-term studies with sufficient numbers of patients. In summary, conversion of terbutaline to bambuterol facilitates the delivery of terbutaline to the lung, so that plasma levels and side effects of the β-mimetic agent are kept low.

III. Future Possibilities for Development of Conjugate-Based Prodrugs

The data reviewed above indicate that xenobiotic conjugates may be exploited to deliver selectively therapeutic agents to their targets. A particularly attractive approach is to use antibodies to direct xenobiotic conjugate processing enzymes to target cells followed by administration of conjugate-based prodrugs, as was described above (see Section II,F).

Because of the importance of glutathione as a cytoprotective agent and its role in the development of resistance to cancer chemotherapeutic drugs, much attention will likely be given to studies designed to manipulate cellular glutathione concentrations. 2-Substituted thiazolidine-4-carboxylates, *N*-acyl-L-cysteines, and glutathione esters have already been developed (see Sections II,A; II,E; and II,G), and additional work may be expected in this area. A particularly challenging area is the delivery of glutathione to the brain; attempts to modulate brain glutathione concentrations have met with limited success thus far, and more work is needed in this area.

Cysteine conjugate β-lyase-dependent bioactivation is a key step in the kidney-selective toxicity of cysteine *S*-conjugates, and this pathway has been studied as a possible mechanism to deliver drugs to the kidney (see Section II,B). Homocysteine *S*-conjugates are also substrates for the β-lyase and undergo transamination reactions followed by retro-Michael reactions (Lash *et al.*, 1990). In contrast to cysteine *S*-conjugates, homocysteine *S*-conjugates lower glutathione concentrations in isolated rat kidney cells (Lash *et al.*, 1986). Thus it may be possible to exploit this action to both deliver a chemotherapeutic agent and increase the susceptibility of kidney cells to its action.

It is clear, however, that conjugate-based prodrugs offer promise and merit research in the future. The availability of site-selective drugs will afford improved efficacy and reduced nontarget toxicity of pharmacological agents, thereby overcoming major limiting factors in the clinical use of drugs.

Acknowledgments

Research in the authors' laboratories was supported by NIEHS Grants ES03127 and E05407 to MWA, by Deutsche Forschungsgemeinschaft SFB 172 to SV and by NATO Grant 901032 to MWA. The authors thank Ms. Sandra Morgan for her assistance in preparing the manuscript.

References

Anderson, M. E., and Meister, A. (1989). Marked increase of cysteine levels in many regions of the brain after administration of 2-oxothiazolidine-4-carboxylate. *FASEB J.* **3,** 1632–1636.

Anderson, M. E., Powrie, F., Puri, R. N., and Meister, A. (1985). Glutathione monoethyl ester: Preparation, uptake by tissues, and conversion to glutathione. *Arch. Biochem. Biophys.* **239,** 538–548.

Boekelheide, K., Graham, D. G., Mize, P. D., and Koo, E. H. (1980). Melanocytotoxicity and the mechanism of activation of γ-L-glutaminyl-4-hydroxybenzene. *J. Invest. Dermatol.* **75,** 322–327.

Bundgaard, H., ed. (1985). "Design of Prodrugs." Elsevier, Amsterdam.

Chasseaud, L. F. (1974). Reaction with electrophiles after enzyme-catalysed deacetylation of N-acetylcysteine. *Biochem. Pharmacol.* **23,** 1133–1134.

Connors, T. A., and Whisson, M. E. (1966). Cure of mice bearing advanced plasma cell tumours with aniline mustard: The relationship between glucuronidase activity and tumour sensitivity. *Nature (London)* **210,** 866–867.

Cooper, A. J. L., and Anders, M. W. (1990). Glutamine transaminase K and cysteine conjugate β-lyase. *Ann. N.Y. Acad. Sci.* **585,** 118–127.

Corcoran, G. B., and Wong, B. K. (1986). Role of glutathione in prevention of acetaminophen-induced hepatotoxicity by N-acetyl-L-cysteine *in vivo:* Studies with N-acetyl-D-cysteine in mice. *J. Pharmacol. Exp. Ther.* **238,** 54–61.

Cotgreave, I. A., Berggren, M., Jones, T. W., Dawson, J., and Moldéus, P. (1987). Gastrointestinal metabolism of N-acetylcysteine in the rat, including an assay for sulfite in biological systems. *Biopharm. Drug Dispos.* **8,** 377–386.

Dalton, C., and Hebborn, P. (1965). Sequential enzymic activation as a means of increasing therapeutic efficiency of nitrogen mustards. *Biochem. Pharmacol.* **14,** 1567–1576.

Dayal, H. H., and Wilkinson, G. S. (1989). Epidemiology of renal cell cancer. *Semin. Urol.* **7,** 139–143.

Eichler, H. G. (1991). Clinical experience of targeted therapy. *Biotherapy* **3,** 77–85.

Gade, W., and Brown, J. L. (1981). Purification, characterization and possible function of α-N-acylamino acid hydrolase from bovine liver. *Biochim. Biophys. Acta* **662,** 86–93.

Goldberg, L. I. (1972). Cardiovascular and renal actions of dopamine. Potential clinical applications. *Pharmacol. Rev.* **24,** 1–29.

Gregoriadis, G. (1989). Targeting of drugs: Implications in medicine. *In* "Drug Carrier Systems" (F. H. D. Roerdink and A. M. Kroon, eds.), pp. 1–31. Wiley, New York.

Hassall, C. H., and John, D. I. (1959). The constitution of hypoglycin B. *Tetrahedron Lett.*, pp. 7–8.

Hassall, C. H., and Reyle, K. (1955). Hypoglycin A and B, two biologically active polypeptides from *Blighia sapida*. *Biochem. J.* **60,** 334–339.

Hazelton, G. A., Hjelle, J. J., and Klaassen, C. D. (1986). Effects of cysteine pro-drugs on acetaminophen-induced hepatotoxicity. *J. Pharmacol. Exp. Ther.* **237,** 341–349.

Hinchman, C. A., and Ballatori, N. (1990). Glutathione-degrading capacities of liver and kidney in different species. *Biochem. Pharmacol.* **40,** 1131–1135.

Hwang, I. Y., and Elfarra, A. A. (1989). Cysteine S-conjugates may act as kidney-selective prodrugs: Formation of 6-mercaptopurine by the renal metabolism of S-(6-purinyl)-L-cysteine. *J. Pharmacol. Exp. Ther.* **251,** 448–454.

Hwang, I. Y., and Elfarra, A. A. (1991). Kidney-selective prodrugs of 6-mercaptopurine: Biochemical basis of the kidney selectivity of S-(6-purinyl)-L-cysteine and metabolism of new analogs in rats. *J. Pharmacol. Exp. Ther.* **258,** 171–177.

Jensen, O. M., Esteve, J., Moller, H., and Renard, H. (1990). Cancer in the European community and its member states. *Eur. J. Cancer* **26,** 1167–1256.

Juliano, R. L., ed. (1980). "Drug Delivery Systems: Characteristics and Biochemical Applications." Oxford Univ. Press, New York.

Kyle, R. A., Costa, G., Cooper, M. R., Ogawa, M., Silver, R. T., Glidewell, O., and Holland, J. F. (1973). Evaluation of aniline mustard in patients with multiple myeloma. *Cancer Res.* **33,** 956–960.

Lash, L. H., Elfarra, A. A., and Anders, M. W. (1986). S-(1,2-Dichlorovinyl)-L-homocysteine-induced cytotoxicity in isolated rat kidney cells. *Arch. Biochem. Biophys.* **251,** 432–439.

Lash, L. H., Elfarra, A. A., Nemeth, D. R., and Anders, M. W. (1990). Bioactivation mechanism of cytotoxic homocysteine S-conjugates. *Arch. Biochem. Biophys.* **276,** 322–330.

Lee, M. R. (1990). Five years' experience with γ-L-glutamyl-L-dopa: A relatively renally specific dopaminergic prodrug in man. *J. Auton. Pharmacol.* **10,** Suppl. 1, s103–s108.

Magnan, S. D. J., Shirota, F. N., and Nagasawa, H. T. (1982). Drug latentiation by γ-glutamyl transpeptidase. *J. Med. Chem.* **25,** 1018–1021.

Manson, M. M., Legg, R. F., Watson, J. V., Green, J. A., and Neal, G. E. (1981). An examination of the relative resistances to aflatoxin B_1 and susceptibilities to γ-glutamyl p-phenylene diamine mustard of γ-glutamyl transferase negative and positive cell lines. *Carcinogenesis (London)* **2,** 661–670.

Mårtensson, J., Steinherz, R., Jain, A., and Meister, A. (1989). Glutathione ester prevents buthionine sulfoximine-induced cataracts and lens epithelial damage. *Proc. Natl. Acad. Sci. U.S.A.* **86,** 8727–8731.

Meijer, D. K. F., Jansen, R. W., and Molema, G. (1992). Drug targeting systems for antiviral agents: Options and limitations. *Antiviral Res.* **18,** 215–258.

Meister, A. (1985). The fall and rise of cellular glutathione levels: Enzyme-based approaches. *Curr. Top. Cell. Regul.* **26,** 383–394.

Meister, A. (1988). Novel drugs that affect glutathione metabolism. *In* "Mechanisms of Drug Resistance in Neoplastic Cells" (P. V. Woolley, III and K. D. Tew, eds.), pp. 99–126. Academic Press, San Diego, CA.

Meister, A., and Anderson, M. E. (1983). Glutathione. *Annu. Rev. Biochem.* **52,** 711–760.

Meister, A., Anderson, M. E., and Hwang, O. (1986). Intracellular cysteine and glutathione delivery systems. *J. Am. Coll. Nutr.* **5,** 137–151.

Mesina, J. E., Page, R. H., Hetzel, F. W., and Chopp, M. (1989). Administration of L-2-oxothiazolidine-4-carboxylate increases glutathione levels in rat brain. *Brain Res.* **478,** 181–183.

Minard, F. N., Grant, D. S., Cain, J. C., Jones, P. H., and Kyncl, J. (1980). Metabolism of g-glutamyl dopamide and its carboxylic acid esters. *Biochem. Pharmacol.* **29,** 69–75.

Nagasawa, H. T., Goon, D. J. W., Zera, R. T., and Yuzon, D. L. (1982). Prodrugs of L-cysteine as liver-protective agents. 2(RS)-Methylthiazolidine-4(R)-carboxylic acid, a latent cysteine. *J. Med. Chem.* **25,** 489–491.

Nagasawa, H. T., Goon, D. J. W., Muldoon, W. P., and Zera, R. T. (1984). 2-Substituted thiazolidine-4(R)-carboxylic acids as prodrugs of L-cysteine. Protection of mice against acetaminophen hepatotoxicity. *J. Med. Chem.* **27,** 591–596.

Nerland, D. E., Benz, F. W., and Babiuk, C. (1989). Effects of cysteine isomers and derivatives on acute acrylonitrile toxicity. *Drug Metab. Rev.* **20,** 233–246.

Orlowski, M., and Wilk, S. (1976). Metabolism of γ-glutamyl amino acids and peptides in mouse liver and kidney *in vivo*. *Eur. J. Biochem.* **71,** 549–555.

Orlowski, M., Mizoguchi, H., and Wilk, S. (1980). N-Acyl-γ-glutamyl derivatives of sulfamethoxazole as models of kidney-selective prodrugs. *J. Pharmacol. Exp. Ther.* **212,** 167–172.

Prezioso, J. A., Damodaran, K. M., Wang, N., and Bloomer, W. D. (1993). Mechanism(s) regulating inhibition of thymidylase synthase and growth by γ-L-glutaminyl-4-hydroxy-3-iodobenzene, a novel melanin precursor, in melanogenic melanoma cells. *Biochem. Pharmacol.* **45,** 473–481.

Puri, R. N., and Meister, A. (1983). Transport of glutathione, as γ-glutamylcysteinylglycyl ester, into liver and kidney. *Proc. Natl. Acad. Sci. U.S.A.* **80,** 5258–5260.

Richman, P. G., and Meister, A. (1975). Regulation of γ-glutamyl-cysteine synthetase by nonallosteric feedback inhibition by glutathione. *J. Biol. Chem.* **250,** 1422–1426.

Ridgewell, R. E., Krejci, M. E., and Koechel, D. A. (1992). Acute effects of a γ-glutamylated derivative of S-(1,2-dichlorovinyl)-L-cysteine on renal function and ultrastructure in pentobarbital-anesthetized dogs: Site-specific toxicity involving S_1 and S_2 cells of the proximal tubule. *Am. J. Vet. Res.* **53,** 840–846.

Roberts, J. C., and Francetic, D. J. (1991a). Mechanisms of chemoprotection by RibCys, a thiazolidine prodrug of L-cysteine. *Med. Chem. Res.* **1,** 213–219.

Roberts, J. C., and Francetic, D. J. (1991b). Time course for the elevation of glutathione in numerous organs of L1210-bearing CDF1 mice given the L-cysteine prodrug, RibCys. *Toxicol. Lett.* **59,** 245–251.

Roberts, J. C., Nagasawa, H. T., Zera, R. T., Fricke, R. F., and Goon, D. J. W. (1987). Prodrugs of L-cysteine as protective agents against acetaminophen-induced hepatotoxicity. 2-(Polyhydroxyalkyl)- and 2-(polyacetoxyalkyl)thiazolidine-4(R)-carboxylic acids. *J. Med. Chem.* **30,** 1891–1896.

Roche, E. B., ed. (1987). "Bioreversible Carriers in Drug Design: Theory and Application." Pergamon, New York.

Roffler, S. R., Wang, S.-M., Chern, J.-W., Yeh, M.-Y., and Tung, E. (1991). Anti-neoplastic glucuronide prodrug treatment of human tumor cells targeted with a monoclonal antibody-enzyme conjugate. *Biochem. Pharmacol.* **42,** 2062–2065.

Sheffner, A. L., Medler, E. M., Bailey, K. R., Gallo, D. G., Mueller, A. J., and Sarett, H. P. (1966). Metabolic studies with acetylcysteine. *Biochem. Pharmacol.* **15,** 1523–1535.

Sherratt, H. S. A. (1986). Hypoglycin, the famous toxin of the unripe Jamaican ackee fruit. *Trends Pharmacol. Sci.* **7,** 186–191.

Sjödin, K., Nilsson, E., Hallberg, A., and Tunek, A. (1989). Metabolism of N-acetyl-L-cysteine. Some structural requirements for the deacetylation and consequences for the oral bioavailability. *Biochem. Pharmacol.* **38,** 3981–3985.

Smith, G. D., Chakravarty, P. K., Connnors, T. A., and Peters, T. J. (1984). Synthesis and preliminary characterization of a novel substrate for γ-glutamyl transferase. A potential anti-hepatoma drug. *Biochem. Pharmacol.* **33,** 527–529.

Speiser, P. P. (1991). Nanoparticles and liposomes. A state of the art. *Methods Find. Exp. Clin. Pharmacol.* **13,** 337–342.

Springer, C. J., Antoniw, P., Bagshawe, K. D., Searle, F., Bisset, G. M. F., and Jarman, M. (1990). Novel prodrugs which are activated to cytotoxic alkylating agents by carboxypeptidase G2. *J. Med. Chem.* **33,** 677–681.

Stewart, G. A., and Hanley, T. (1969). Other oral hypoglycaemic agents. *In* "Oral Hypoglycaemic Agents" (G. D. Campbell, ed.), pp. 348–407. Academic Press, London.

Suzuki, S., and Tateishi, M. (1981). Purification and characterization of a rat liver enzyme catalyzing N-deacetylation of mercapturic acid conjugates. *Drug Metab. Dispos.* **9,** 573—577.

Svensson, L. Å. Mechanism of action of bambuterol: A β-agonist prodrug with sustained lung affinity. *Agents Actions* **34,** Suppl., 71–78.

Svenson, L. Å., and Tunek, A. (1988). The design and bioactivation of presystemically stable prodrugs. *Drug Metab. Rev.* **19,** 165–194.

Uhlig, S., and Wendel, A. (1990). Glutathione enhancement in various mouse organs and protection by glutathione *iso*propyl ester against liver injury. *Biochem. Pharmacol.* **39,** 1877–1881.

Van Der Werf, P., Orlowski, M., and Meister, A. (1971). Enzymatic conversion of 5-oxo-L-proline (L-pyrrolidone carboxylate) to L-glutamate coupled with cleavage of adenosine triphosphate to adenosine diphosphate, a reaction in the γ-glutamyl cycle. *Proc. Natl. Acad. Sci. U.S.A.* **68,** 2982–2985.

Vogel, F. S., Kemper, L. A. K., Jeffs, P. W., Cass, M. W., and Graham, D. G. (1977). γ-L-Glutaminyl-4-hydroxybenzene, an inducer of crytobiosis in *Agaricus bisporus* and a source of specific metabolic inhibitors for melanogenic cells. *Cancer Res.* **37,** 1133–1136.

Vogel, F. S., Kemper, L. A. K., Boekelheide, K., Graham, D. G., and Jeffs, P. W. (1979). Intracellular activation of γ-L-glutaminyl-4-hydroxybenzene by tyrosinase, a mechanism for selective cytotoxicity against melanocarcinoma. *Cancer Res.* **39,** 1490–1493.

Wang, S.-M., Chern, J.-W., Yeh, M.-Y., Ng, J. C., Tung, E., and Roffler, S. R. (1992). Specific activation of glucuronide prodrugs by antibody-targeted enzyme conjugates for cancer chemotherapy. *Cancer Res.* **52,** 4484–4491.

Wellner, V. P., Anderson, M. E., Puri, R. N., Jensen, G. L., and Meister, A. (1984). Radioprotection by glutathione ester: Transport of glutathione ester into human lymphoid cells and fibroblasts. *Proc. Natl. Acad. Sci. U.S.A.* **81,** 4732–4735.

Wilk, S., Mizoguchi, H., and Orlowski, M. (1978). γ-Glutamyl dopa: A kidney-specific dopamine precursor. *J. Pharmacol. Exp. Ther.* **206,** 227–232.

Williamson, J. M., and Meister, A. (1981). Stimulation of hepatic glutathione formation by administration of L-2-oxothiazolidine-4-carboxylate, a 5-oxo-L-prolinase substrate. *Proc. Natl. Acad. Sci. U.S.A.* **78,** 936–939.

Williamson, J. M., and Meister, A. (1982). New substrates of 5-oxo-L-prolinase. *J. Biol. Chem.* **257,** 12039–12042.

Williamson, J. M., Boettcher, B., and Meister, A. (1982). Intracellular cysteine delivery system that protects against toxicity by promoting glutathione synthesis. *Proc. Natl. Acad. Sci. U.S.A.* **79,** 6246–6249.

Young, C. W., Yagoda, A., Bittar, E. S., Smith, S. W., Grabstald, H., and Whitmore, W. (1976). Therapeutic trial of aniline mustard in patients with advanced cancer. Comparison of therapeutic response with cytochemical assessment of tumor cell β-glucuronidase activity. *Cancer (Philadelphia)* **38,** 1887–1985.

Index

Contents of Previous Volumes

Volume 21

Volume 22

Volume 23

Volume 24

Volume 25

Volume 26

ISBN 0-12-032927-1